Model Organisms for Microbial Pathogenesis, Biofilm Formation and Antimicrobial Drug Discovery

Busi Siddhardha • Madhu Dyavaiah •
Asad Syed
Editors

Model Organisms for Microbial Pathogenesis, Biofilm Formation and Antimicrobial Drug Discovery

Springer

Editors
Busi Siddhardha
Department of Microbiology
School of Life Sciences
Pondicherry University
Puducherry, India

Madhu Dyavaiah
Department of Biochemistry and Molecular Biology, School of Life Sciences
Pondicherry University
Puducherry, India

Asad Syed
Botany and Microbiology Department
King Saud University
Riyadh, Saudi Arabia

ISBN 978-981-15-1694-8 ISBN 978-981-15-1695-5 (eBook)
https://doi.org/10.1007/978-981-15-1695-5

This Springer imprint is published by the registered company Springer Nature Singapore Pte Ltd.
The registered company address is: 152 Beach Road, #21-01/04 Gateway East, Singapore 189721, Singapore

Preface

Currently, one of the greatest challenges faced by medical field is infectious diseases. A large number of infectious agents and associated diseases came into the picture which led to major changes in the clinical practice, microbiological services, public health activities, and biomedical research. Also, we are running out of antimicrobial agents effective against these infectious diseases. One of the major reasons behind this problem is due to the emerging antibiotic-resistant pathogens. These antibiotic-resistant strains are attracting much of the attention of research as we have reached to a point where barely any treatment choices for infections are left. According to the current rate of emergence and spread of antibiotic resistance, it is expected that the annual loss of life would reach around 10 million deaths with an estimated economic loss of 100 trillion dollar by 2050. If we want to effectively mitigate these resistance strains, a multifaceted approach is recommended with a justifiable and sustainable use of antimicrobials. These approaches foster the innovation of new therapeutic methods and diagnostic tools which impedes the spread of diseases. Still, there are several questions to be answered regarding the diagnosis, treatment, and use of novel antimicrobial agents. In addition, we are having too little knowledge of the pathophysiology of major infectious diseases and the strategies to mitigate them effectively.

Much of what we know about infections, their pathogenicity, and treatment comes from the studies using model organisms. Model systems are directed toward the development of more sophisticated tools for scientists and clinicians to diagnose and treat the disease. Model organisms are important to study microbial pathogenesis, biofilm formation, and antimicrobial drug discovery. Model organisms are inimitable tools of biomedicine and clinical research which helped our scientists to stockpile a massive amount of knowledge on microbial pathogenesis. They are frequently employed in the research of human diseases and their prevention owing to their similarity to humans in terms of anatomy, physiology, and genetics. Hence they are often known as a proxy for better understanding of the biology of human pathogens. Also, animal models are often desirable for experimental disease research because of their boundless supply and ease of manipulation. Many drugs, treatments, and anti-virulence strategies may be developed through the help of model organisms which is difficult to complete with in vitro strategies alone.

The effectiveness of biological models in providing insights into disease mechanisms, pathogenesis, diagnostics, and treatment is unquestionable. Definitely, the list of groundbreaking visions into human diseases and highly efficient drugs developed is based on model organisms. Even though basic medical research relies on these biological models, clinicians more closely engaged in applied biomedical research are inclined toward the models that are similar to humans both physiologically and genetically. In all circumstances, biological models acquired an undisputed supreme position as a model system for studying human diseases. From the evolutionary perspectives, a wide range of model organisms starting from invertebrates to mammalian model systems are being considered to study the pathogenesis profile and provide strategic platforms to combat such infections. The use of model organisms in the development of novel drug candidates also provides preliminary yet advantageous characteristics before the developed drug candidates are being considered for clinical investigations followed by introduction of such drug moieties into the market for public use.

This book not only discusses in detail about the advantages, successes, and promises of the model systems as the most prominent and widely used for microbial pathogenesis but also enlightens antimicrobial drug discovery. Rather, this book is evidence that the research community does not put all of its model systems in one basket; instead it considers the opportunities that lie in the great biodiversity of different model systems for each infectious disease. Information contained in this book may provide new perceptions in pathogenesis and new diagnostic approaches and perhaps new therapies may be derived. In light of this, book will provide comprehensive knowledge of different model systems used so far for evaluating disease biology and to develop new therapeutic methods which will benefit future scientists to focus more on the aspects related to the emerging pathogens.

Puducherry, India — Busi Siddhardha
Puducherry, India — Madhu Dyavaiah
Riyadh, Saudi Arabia — Asad Syed

Contents

Editors and Contributors

About the Editors

Busi Siddhardha is Assistant Professor in the Department of Microbiology, Pondicherry University, Puducherry, India. He has more than 12 years of research experience in the field of microbiology, antimicrobial drug discovery, and nanobiotechnology. He has worked for the last 10 years on antimicrobial compounds and antimicrobial drug discovery. He obtained his PhD from Biology Division, CSIR-IICT, Hyderabad, India. He completed two research projects on antiquorum sensing and antibiofilm activity of natural products funded by the Government of India. He is extensively working in the field of nanotechnology, especially applications of nanotechnology in drug discovery, drug delivery, sustained release, and photodynamic therapy. Currently, his group is working on bacterial quorum sensing, biofilms, antimicrobial photodynamic therapy, and nanobiotechnology. He is editorial board member of several reputed journals. He has also published more than 60 research articles in peer-reviewed international journals and authored or coauthored numerous book chapters. He is a member of many national and international scientific societies. He has more than 7 years of teaching and research experience at the university level.

Madhu Dyavaiah is an Assistant Professor in the Department of Biochemistry and Molecular Biology, Pondicherry University, Puducherry. He has earlier served as Research Scientist in the College of Nanoscale Science and Engineering, Albany, USA (2010–2012), postdoctoral associate at Gen'NY'sis Center for Excellence in Cancer Genomic Rensselaer, USA (2006–2010), and postdoctoral fellow at Wadsworth Center, USA (2003–2006) and IISc Bangalore, India (2002–2003). His research interest includes DNA damage response, tRNA modification and translation regulation, and aging biology. He has worked in the area of nanotechnology in the USA, and he has research experience in different model systems including plants, *S. cerevisiae*, cell lines, and mice. Currently, he is working on *S. cerevisiae* and mice model to study the effect of natural compounds on age-related diseases. He has been conferred with various prestigious awards. He has served as referee for a number of international journals. He has more than 15 years of research and 6 years of teaching experience in genomics, proteomics, molecular biology, clinical biochemistry biology, and drug discovery. He has also published more than 20 research articles in the peer-reviewed international journal and authored numerous book chapters. He is a member of many scientific societies and organizations.

Asad Syed is Assistant Professor at Botany and Microbiology Department, College of Science, King Saud University, Riyadh, Saudi Arabia. He has been awarded Doctorate from the University of Pune, whereas he has completed his research work from CSIR-National Chemical Laboratory, Pune, India, where he served as Research Associate and Senior Research Fellow in the Biochemical Sciences Division. His research interest is in fabrication of nanomaterials, by utilizing fungi, bacteria, plant extracts, protein, etc. Another area of his expertise is secondary metabolite from endophytic fungi. He is serving as reviewer for international peer journals that include materials science and microbiology journals. He is teaching basic microbiology and laboratory skills to bachelor's students and involved in mentoring master's and PhD students at Botany and Microbiology Department, KSU, Riyadh. His has published his findings in international peer-reviewed materials science and microbiology journals.

Contributors

Dharnappa Sannejal Akhila Nitte University Centre for Science Education and Research, NITTE (Deemed to be University), Mangaluru, Karnataka, India

Kotian Akshatha Nitte University Centre for Science Education and Research, NITTE (Deemed to be University), Mangaluru, Karnataka, India

Fawaz Alasmari Department of Pharmacology and Toxicology, College of Pharmacy, King Saud University, Riyadh, Saudi Arabia

Mohammed S. Alqahtani Department of Pharmaceutics, College of Pharmacy, King Saud University, Riyadh, Saudi Arabia

V. T. Anju Department of Biochemistry and Molecular Biology, School of Life Sciences, Pondicherry University, Puducherry, India

Sivaraman Balaji Department of Biotechnology, Bharathiar University, Coimbatore, India

Sayak Bhattacharya School of Environmental Studies, Jadavpur University, Kolkata, India

Suparna Chakraborty Division of Clinical Medicine, National Institute of Cholera and Enteric Diseases, Kolkata, India

Indranil Chattopadhyay Department of Life Sciences, Central University of Tamil Nadu, Thiruvarur, Tamil Nadu, India

Santasabuj Das Division of Clinical Medicine, National Institute of Cholera and Enteric Diseases, Kolkata, India

Saprativ P. Das Department of Biotechnology, Brainware University, Kolkata, West Bengal, India

Shivika Datta Department of Zoology, Doaba College, Jalandhar, Punjab, India

Vijaya Kumar Deekshit Nitte University Centre for Science Education and Research, NITTE (Deemed to be University), Mangaluru, Karnataka, India

V. A. Iswarya Deepthi Department of Biotechnology, Andhra University, Visakhapatnam, India

Daljeet Singh Dhanjal Department of Biotechnology, Lovely Professional University, Phagwara, Punjab, India

Madhu Dyavaiah Department of Biochemistry and Molecular Biology, School of Life Sciences, Pondicherry University, Puducherry, India

Poulomi Ghosh Department of Biotechnology, Brainware University, Kolkata, West Bengal, India

Nishant Gopalan V.V.P. Engineering College, Rajkot, Gujarat, India

Vipin Gopinath Malabar Cancer Centre, Kannur, Kerala, India

Parijat Hazarika Department of Microbiology, Assam University, Silchar, India

V. Lakshmi Kalpana Department of Biotechnology, College of Science and Technology, Andhra University, Visakhapatnam, India

K. Meganathan Kannan Division of Blood and Vascular Biology, Department of Life Sciences, School of Life Sciences, Central University of Tamil Nadu, Tiruvarur, India

Mojibur Khan Molecular Biology and Microbial Biotechnology Laboratory, Division of Life Sciences, Institute of Advanced Study in Science and Technology (IASST), Guwahati, Assam, India

Anushka Khasnobish Laboratory of Animal Applied Microbiology, Graduate School of Environmental and Life Science, Okayama University, Okayama, Japan

Arun Kumar Molecular Biology and Microbial Biotechnology Laboratory, Division of Life Sciences, Institute of Advanced Study in Science and Technology (IASST), Guwahati, Assam, India

K. Vinodhini Krishna Kumar Division of Blood and Vascular Biology, Department of Life Sciences, School of Life Sciences, Central University of Tamil Nadu, Tiruvarur, India

Sanjay Kumar Regional Advanced Water Testing Laboratory, Phase-II, S.A.S. Nagar, Punjab, India

Punjab Biotechnology Incubator (PBTI), Phase-V, S.A.S. Nagar, Punjab, India

Vijay Kumar Regional Ayurveda Research Institute for Drug Development, Gwalior, Madhya Pradesh, India

Ranjith Kumavath Department of Genomic Science, School of Biological Sciences, Central University of Kerala, Kasaragod, Kerala, India

Ramatchandirane Mahesh Department of Biochemistry and Molecular Biology, Pondicherry University, Puducherry, India

Sarangam Majumdar Dipartimento di Ingegneria Scienze Informatiche e Matematica, Università degli Studi di L'Aquila, L'Aquila, Italy

Ayesha Mateen Department of Clinical Laboratory Sciences, College of Applied Medical Sciences, King Saud University, Riyadh, Saudi Arabia

Himani Meena Department of Microbiology, School of Life Sciences, Pondicherry University, Puducherry, India

Hidetoshi Morita Laboratory of Animal Applied Microbiology, Graduate School of Environmental and Life Science, Okayama University, Okayama, Japan

Joydeep Mukherjee School of Environmental Studies, Jadavpur University, Kolkata, India

K. Madhavan Nampoothiri MPTD, CSIR-NIIST, Thiruvananthapuram, Kerala, India

Sukla Pal Department of Physics, Centre for Quantum Science, Dodd-Walls Centre for Photonic and Quantum Technologies, University of Otago, Dunedin, New Zealand

Rajesh Pamanji Molecular Reproduction, Development and Genetics, Indian Institute of Science, Bangalore, India

Paramanantham Parasuraman Department of Microbiology, School of Life Sciences, Pondicherry University, Puducherry, India

Subhaswaraj Pattnaik Department of Microbiology, School of Life Sciences, Pondicherry University, Puducherry, India

Angela E. Peter Department of Biotechnology, College of Science and Technology, Andhra University, Visakhapatnam, India

Sudhakar Pola Department of Biotechnology, Andhra University, Visakhapatnam, India

J. Prarthana SDM PG College, Ujire, Karnataka, India

Ashwath Priyanka Nitte University Centre for Science Education and Research, NITTE (Deemed to be University), Mangaluru, Karnataka, India

Jobina Rajkumari Department of Microbiology, School of Life Sciences, Pondicherry University, Puducherry, India

Akella Vijayaramya Department of Biotechnology, Andhra University, Visakhapatnam, India

B. Ganga Rao College of Pharmaceutical Science, Andhra University, Visakhapatnam, India

T. Eswara Rao Department of Genomic Science, School of Biological Sciences, Central University of Kerala, Kasaragod, Kerala, India

Tingirikari Jagan Mohan Rao Department of Biotechnology, NIT Andhra Pradesh, Tadepalligudem, India

Subhasree Saha Infectious Disease Biology, Institute of Life Sciences, Bhubaneswar, India

Department of Biotechnology, School of Earth Biological and Environmental Sciences, Central University of South Bihar, Gaya, Bihar, India

Pavani Sanapala Department of Biotechnology, Andhra University, Visakhapatnam, India

B. V. Sandeep Department of Biotechnology, College of Science and Technology, Andhra University, Visakhapatnam, India

Preeti C. Sangave School of Pharmacy and Technology Management, SVKM's NMIMS, MPTP, Shirpur, India

Muthu Vijaya Sarathy Department of Microbial Biotechnology, Bharathiar University, Coimbatore, India

Youssef Sari Department of Pharmacology and Experimental Therapeutics, College of Pharmacy and Pharmaceutical Sciences, University of Toledo, Toledo, OH, USA

Indu Sharma Department of Microbiology, Assam University, Silchar, India

Priyanka Sharma Biocon Bristol Myers Squibb Syngene International Pvt Ltd, Bangalore, India

Busi Siddhardha Department of Microbiology, School of Life Sciences, Pondicherry University, Puducherry, India

Daljeet Singh Department of Biotechnology, Lovely Professional University, Jalandhar, Punjab, India

Durg Vijai Singh Infectious Disease Biology, Institute of Life Sciences, Bhubaneswar, India

Department of Biotechnology, School of Earth Biological and Environmental Sciences, Central University of South Bihar, Gaya, Bihar, India

Joginder Singh Department of Biotechnology, Lovely Professional University, Phagwara, Punjab, India

Satyender Singh Regional Advanced Water Testing Laboratory, Mohali, Punjab, India

Shyam Babu Singh Department of Ayurveda, Regional Ayurveda Research Institute for Drug Development, Gwalior, Madhya Pradesh, India

Simranjeet Singh Regional Advanced Water Testing Laboratory, Phase-II, S.A.S. Nagar, Punjab, India

Punjab Biotechnology Incubator (PBTI), Phase-V, S.A.S. Nagar, Punjab, India

Department of Biotechnology, Lovely Professional University, Phagwara, Punjab, India

Brijesh Singh Sisodia Department of Biochemistry, Regional Ayurveda Research Institute for Drug Development, Gwalior, MP, India

Kitlangki Suchiang Department of Biochemistry and Molecular Biology, Pondicherry University, Puducherry, India

Pola Sudhakar Department of Biotechnology, College of Science and Technology, Andhra University, Visakhapatnam, India

Asad Syed Department of Botany and Microbiology, College of Sciences, King Saud University, Riyadh, Saudi Arabia

Rabbani Syed Department of Pharmaceutics, College of Pharmacy, King Saud University, Riyadh, Saudi Arabia

Department of Clinical Laboratory Sciences, College of Applied Medical Sciences, King Saud University, Riyadh, Saudi Arabia

Sagolsem Yaiphathoi Department of Microbiology, Assam University, Silchar, India

Microbial Infections and Virulence Factors

1

Sayak Bhattacharya and Joydeep Mukherjee

Contents

Abstract

Infection is defined as invasion and colonization processes of pathogenic or harmful microorganisms on host cells. A diverse array of microorganisms such as bacteria, virus, fungi, and protozoa attack or infect host cell. They deploy enormous strategies to target or manipulate the host cell or to escape from host immunity. Virulence factors have been secreted as a either cell associated or secreted out of cells upon infection. Infection also depends on health of host cells. Pathogens secrete different types of toxins or enzymes or some molecules to destroy the immunity of the host cells. These are encoded by either chromo-

S. Bhattacharya · J. Mukherjee (✉)
School of Environmental Studies, Jadavpur University, Kolkata, India
e-mail: joydeep.mukherjee@jadavpuruniversity.in

B. Siddhardha et al. (eds.), *Model Organisms for Microbial Pathogenesis, Biofilm Formation and Antimicrobial Drug Discovery*,
https://doi.org/10.1007/978-981-15-1695-5_1

some or plasmid of host cells. Even drug resistant property is one of the tactics to destroy the antibiotics. Sometimes they modify themselves or mimic like host molecules upon infection and hence immune cells do not identify them. Nowadays these peculiar behaviors of microbial pathogen draw attention to many scientists worldwide. They not only affect humans, but they also infect other organisms that are economically important for human welfare. Some deadly diseases such as swine flu or dengue or other fatal diseases are really concern for human society. Every year these deadly diseases affect more than half of the people have been infected by these deadly diseases. In this section, we will discuss and shed the light on human pathogens and their disease characteristic. We will even discuss the virulence factors that aid the pathogen to progress the disease.

Keywords
Infection · Pathogen · Immune compromised · Virulence factor

1.1 Introduction

Infection is defined as multiplication or colonization of microorganisms on the host system or invading into the host cells. Despite the presence of substantial number of bacteria in the human body, their occurrence in healthy persons is usually limited to some body parts like the skin, vagina, the mucosae of nasal and buccal cavities and, most significantly, the gastrointestinal part (Schommer and Gallo 2013; Costello et al. 2009; Lemon et al. 2010). However, opportunistic pathogens colonize on the host cells for prolonged period and they cause diseases occasionally. Among the microorganisms, most deadly microorganism is virus which has the ability to hijack the host system to replicate themselves. It is very difficult to diagnose as well as to eradicate the viral diseases completely. Fungi deploy different cell associated virulence factors or secrete virulence factors to modify the host cell completely. Vector borne microbial infection is another global problem. Here we will summarize some human microbial diseases and their virulence factors.

1.1.1 Colonization of Host Surfaces by Pathogenic Microorganisms

Roughly 300–400 m^2 of surface area are represented by respiratory, digestive, and urogenital mucosae (200 times more than that of the skin) and hence these sites make maximum contact with bacteria. These sites consist of three layers: an epithelium, a loose connective tissue layered lamina propria, and the thin layered smooth muscles. These mucosal surfaces constitute frontline barriers and hence limit the invasion by both pathogenic and commensal bacteria. These sites are not suitable for colonization of the pathogens. Ciliary movement has been observed of the epithelial cells that aid to clear the pathogens from the host system. Various molecular strategies have been deployed by pathogenic bacteria for adherence to these epithelia and to proliferate at their surface by circumventing the high defense barriers.

1.1.2 Initiation and Maintenance of Pathogens in Intracellular Lifestyle

Bacterial pathogens have diverse advantages by maintaining intracellular lifestyle, i.e., they are not attacked by humoral and complement-mediated immune system; a broad range of nutrients are accessible to them and they avoid shear stress-induced clearance. However, these intracellular bacteria are targeted by different mechanisms employed by the host cells. Thus, intracellular pathogens have evolved different strategies to successfully establish and maintain an intracellular infection for a prolonged period.

1.1.3 Crossing of Host Barriers by Pathogens

Diverse types of sentinel cells like dendritic cells (DCs) and M cells continuously sense the existence of pathogenic bacteria rich mucosal environment. Though there is coordination between innate and adaptive immune response to utmost the colonization of pathogens in the host, sometimes pathogens use as entry portals. Specialized cells such as M cells are found in the intestinal epithelium and other epithelia in humans. M cells transport antigens from the lumen to cells of the immune system. Thus, they initiate an immune response or tolerance. Their function is different from that of their adjacent epithelial cells. Antigens of the mucosal environment are recognized by DCs and play a middle role in the adaptive immunity. Mucosal tissues are enriched with these cells but sometimes these cells may migrate to mesenteric lymph nodes, where they interconnect with lymphocytes.

1.2 Microbial Infections

1.2.1 Bacterial Infections

Staphylococcus aureus is a classic opportunistic pathogen which causes skin and soft tissue infection (SSTI) worldwide. Various infections are caused by them, ranging from self-limiting skin infections to extreme life-threatening pneumonia, bacteremia, and endocarditis (Moran et al. 2006; David and Daum 2010; Talan et al. 2011). There are several microorganisms which cause skin infection. Life-threatening condition with high mortality happens when bacteria infect the central nervous system (CNS). Additionally, it may cause permanent neurological deficits in survivors (Geyer et al. 2019). Sometimes, some diseases are caused by two different genera of microorganisms like cystic fibrosis (CF). CF lung is first colonized by *S. aureus*, but in the adulthood stage, *Pseudomonas aeruginosa* acts as the second colonizer, the most common isolated bacterium causing chronic lung infections (LiPuma 2010). More specifically, methicillin-resistant *S. aureus* (MRSA) is an infamous cause of healthcare-associated and community-associated disease along with *Acinetobacter baumannii*, with an estimated prevalence of about 30%. They colonize in chronic obstructive lung disease (COPD) affected hospitalized patients (Furuno et al. 2008; David and Daum 2010).

Bacterial meningitis, another deadly disease in sub-Saharan Africa, is caused by both *Neisseria meningitides* and *Streptococcus pneumoniae*. *S. pneumoniae,* the second most common pathogen, causes higher morbidity and mortality than *N. meningitides* in the same region (Ramakrishnan et al. 2009; Gessner et al. 2010; Mihret et al. 2016). *Haemophilus influenzae* also contributes to 2% of meningitides (Mihret et al. 2016). Brucellosis, a zoonotic disease, in humans the disease is identified by recurrent undulant fever, endocarditis, debilitating arthritis, and meningitis (Corbel 1997; Godfroid et al. 2011). It causes considerable economic loss and becomes a major public health burden (Pappas et al. 2005; Pappas 2010). Keratitis, a common form of corneal blindness across the globe, is caused by both bacteria and fungi. *P. aeruginosa,* an opportunistic human pathogen, causes chronic pulmonary infections in cystic fibrosis (CF) patients (Winstanley and Fothergill 2009). *P. aeruginosa* also affects immune-compromised individuals having human immunodeficiency virus (HIV) or undergoing cancer chemotherapy and those with burn wounds (Lau et al. 2004).

Clostridia, strict anaerobic bacteria, have been isolated from necrotizing infections in humans (Zhao-Fleming et al. 2017). In recent times, several bacterial species especially Clostridia like *Clostridium butyricum, C. perfringens*, and *C. neonatale* have been associated with necrotizing enterocolitis (NEC) outbreaks (Hosny et al. 2017; Roze et al. 2017). Life-threatening diseases in humans and animals are caused by pathogens, carried by well-known house fly. More than 100 pathogens including bacteria, viruses, fungi, and parasites (protozoans and metazoans) have been connected with the insect (Tsagaan et al. 2015; Nassiri et al. 2015). *S. pneumoniae* (Pneumococcus) is a gram-positive, α-hemolytic, and facultative anaerobic organism which inhabits the nasopharynx. Upper or lower meningitis, respiratory infections, and septicemia are caused by *S. pneumoniae*. *S. pneumoniae* operates major diseases with a considerably high mortality (Manco et al. 2006). *Yersinia pestis* causes another deadly disease, plague. It was the causative agent of epidemics in Europe during the first and second pandemics, including the Black Death, infamous for their widespread mortality and lasting social and economic impact (Bramanti et al. 2019). A life-threatening disease called Rocky Mountain spotted fever is caused by *Rickettsia rickettsii,* an obligately intracellular bacterium. Ticks spread this disease to human beings (Dantas-Torres 2007). Gas gangrene or clostridial myonecrosis is caused by *C. perfringens* type A. This disease causes rapid spread of tissue necrosis combined with a lack of leukocyte infiltration at the site of infection (Rood 1998). A chronic infectious disease called leprosy or

Hansen's disease is caused by *Mycobacterium leprae*. However, it was discovered that *M. lepromatosis* caused diffuse lepromatous leprosy (DLL) in human (Han et al. 2008; Scollard 2016). A potentially deadly disease tularemia in mammals including humans is caused by the intracellular pathogen *Francisella tularensis* (Tärnvik 1989). Psittacosis, an animal vector-borne disease, is caused by *Chlamydia psittaci.* Transmission of this fatal disease can happen to humans coming in close association with a variety of birds, most frequently Psittacidae (parrots, lories, parakeets, and cockatoos) or Columbiformes (pigeons). Exposure to equine placental material as a risk factor for transmission of Psittacosis has also been discovered recently (Polkinghorne and Greub 2017). Lyme borreliosis (LB), a tick-borne disease, is also endemic and causes serious public health problems in USA and Europe (Rosenberg 2018; Sprong et al. 2018). LB is caused by the genus Borrelia, a spirochete. Persistent infections both in the vertebrate host and tick vector are established by these pathogens (Rego et al. 2019).

1.2.2 Fungal Infections

Not only bacteria but also fungi are similarly efficient to produce several diseases in human system. Filamentous fungi such as *Aspergillus fumigatus* and the *Scedosporium apiospermum* species complex (16–58% and 9–10% respectively), the most common pathogens, are isolated from cystic fibrosis (CF) patients (Noni et al. 2017; Reece et al. 2017). *Scedosporium boydii*, *S. apiospermum*, and *S. aurantiacum* are frequently involved with CF patients whereas *S. ellipsoidium* and *S. minutisporum* are periodically involved with CF patients, suffering from CF lung infections with a geographically variable prevalence (Blyth et al. 2010; Zouhair et al. 2013; Sedlacek et al. 2015). Human lung mycetoma, caused by uncommon CF pathogen *S. Angustum,* is a previously reported disease (Kravitz et al. 2011). *Fusarium* spp. causes from mild superficial infections to invasive systemic infections. Localized diseases such as onychomycosis and keratitis are also caused by *Fusarium* spp. in normal hosts. Invasive fusariosis affected the lungs, sinuses, and visceral organs and necrotic skin lesions and positive blood cultures are developed in 60% of individuals with immune-compromised conditions (Nelson et al. 1994; Nucci et al. 2003; Lionakis and Kontoyiannis 2004). Not only healthy individual, sometimes immuno-compromised individual may also be affected. *Aspergillus*, most ubiquitous in nature, infects immune-compromised host and causes invasive aspergillosis which is limited to lungs (Steinbach et al. 2012; Camargo and Husain 2014; Husain et al. 2017).

1.2.3 Viral Infections

Viruses are the most crucial human pathogens among all others microorganisms. Recently, Zika virus infection has been highlighted. Upon Zika virus infection, flu-like symptoms appear; mild infection with joint pain, low to moderate fever, headache, fatigue, and rash (Posen et al. 2016; Kindhauser et al. 2016; Shuaib et al.

2016). Another deadly virus is human cytomegalovirus (HCMV) which causes a chronic infection with lifelong latency in humans. HCMV, an opportunistic pathogen, causes infection in immunosuppressed individuals causing birth defects (Mocarski et al. 2007). Japanese encephalitis (JE) causes uncontrolled inflammatory disease to the central nervous system. Neurotropic flavivirus, JE virus (JEV) cause Japanese encephalitis (JE) (Misra and Kalita 2010). Transmission of JEV occurs by mosquito vector. JEV transmits in a zoonotic cycle where pig acts as an amplifier and water bird acts as reservoir hosts (Solomon 2004). Dengue is another example of mosquito-borne disease and a global public health problem. Approximately 390 millions are affected annually due to dengue infections (Bhatt et al. 2013) (Fig. 1.1). Dengue virus DENV infections occur from asymptomatic cases to life-threatening hypovolemic shock (WHO 2009). Some viral diseases resolve within 2–6 weeks like hepatitis in humans. However, the hepatitis can become chronic in pregnant women and in immune-compromised individuals (Purcell and Emerson 2008; Aggarwal and Jameel 2011; Pérez-Gracia et al. 2017). Hepatitis E virus is transmitted with the consumption of meat, liver, sausages, and offal products derived from domestic pig, deer, and wild boar (Tei et al. 2004; Colson et al. 2010, 2012). The bite of infected mosquitoes can sometimes be dangerous as it transmits yellow fever virus (YFV), a member of the Flavivirus genus, to susceptible hosts like humans or non-human primates. Main availability of YFB is in endemic parts of Africa and South America, including Brazil (Barrett and Monath 2003). Rift valley fever is caused by Rift Valley fever virus (RVFV) in ruminants as well in humans (Bird et al. 2009). The large enveloped DNA virus poxvirus family infects a wide variety of hosts. Poxvirus outbreaks are caused by zoonotic infections of cowpox virus, monkeypox 70 virus, and recently discovered poxvirus species (Di Giulio and Eckburg 2004; Vora et al. 2015). Tick-borne encephalitis is the most occured viral disease in the Central Europe. It is one of the most famous arthropod borne diseases. The TBE virus (TBEV) belongs to the genus Flavivirus (family Flaviviridae). It consists of three different subtypes: (a) the European subtype, transmitted mainly by *Ixodes ricinus*, (b) the Siberian subtype, and (c) the Far Eastern subtype, which are mainly transmitted by *I. persulcatus* (Süss 2011). Tick and their vertebrate hosts spread the TBE virus, within particular geographical area (Dobler et al. 2011).

1.3 Virulence Factors Associated with Microbial Infections

Virulence factors are produced by pathogenic microorganisms to promote diseases (Table 1.1). These factors are either protein or carbohydrate or lipid in nature. Virulence factors are either secreted or cell mediated or cytosolic in nature and they can interfere the immune system of host cells. The cytosolic factors aid the bacterium to undertake quick adaptive—physiological, metabolic, and morphological

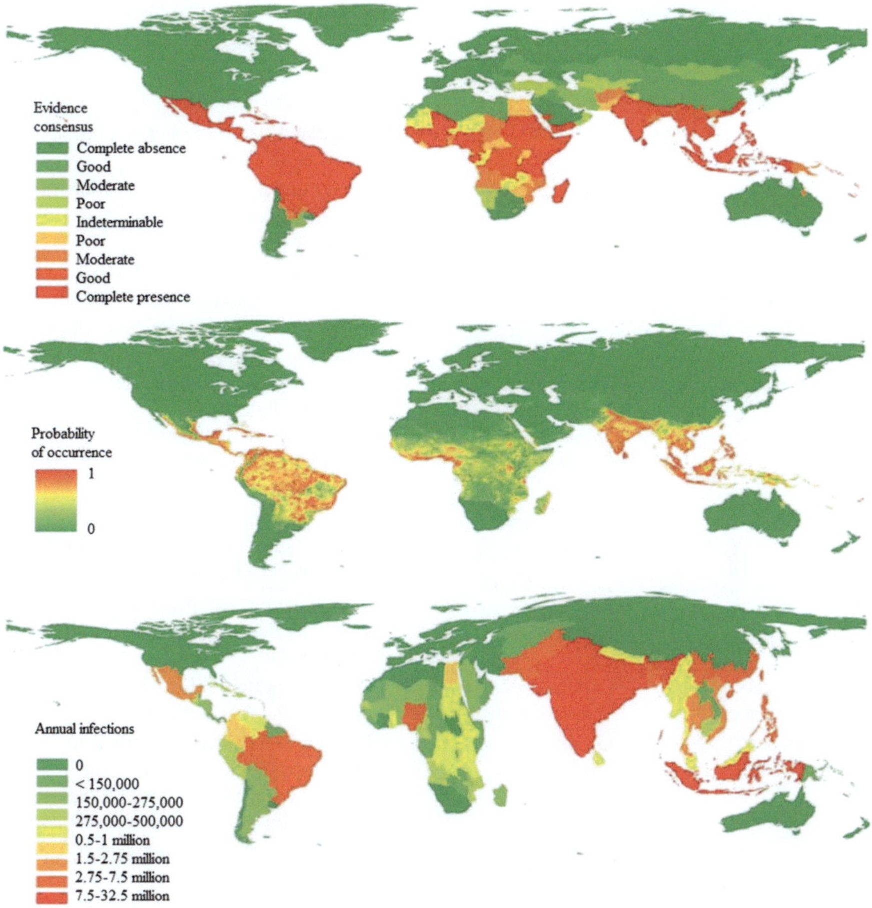

Fig. 1.1 Global evidence consensus, risk, and burden of dengue in 2010 (Bhatt et al. 2013). (Reprinted with permission [License number: 4595801418447])

shifts for the survival within host cells. The bacterium adheres and evades the host cell by the membrane-associated virulence factors. The secretory factors help the bacterium to avoid the innate and adaptive immune response mounted within the host. The host cells are killed by the synergistic effect of secretory virulence factors of extracellular pathogens. Sometimes, they intoxicate food items by secreting toxins into them and upon injecting this contaminated food item, human will be diseased. Virulence property of an organism enables to infect the host cell and causes a disease.

Table 1.1 Microbial virulence factors involved in bacterial pathogen dissemination through mammalian host

Virulence factors	Microorganism involved	Mechanism of action
Coagulase	*S. aureus*	Coagulates the fibrinogen in plasma (Shaw et al. 1951)
Collagenase	*Clostridium* spp.	Breaks down the collagen to allow the pathogen to spread (Wiinsch and Heidrich 1963)
Deoxyribonuclease (along with calcium and magnesium)	*Staphylococci*, *C. perfringens*	Lowers viscosity of exudates and allowing the pathogen more mobility (Erickson and Deibel 1973; Stern and Warrack 1964)
Hyaluronidase	Groups A, B, C, and G *Streptococci*, *Staphylococci*, *Clostridia*	Hydrolyzes hyaluronic acid and renders the intercellular spaces amenable to passage by the pathogen (Wessels and Bronze 1994; Canard et al. 1994; Hart et al. 2009)
Leukocidins	*Staphylococci*, *Streptococci*	Forms pore on leukocytes and causes degranulation of lysosomes within leukocytes (Gillet et al. 2002)
Porins	*Salmonella typhimurium*	Inhibits leukocytes phagocytosis by activating the adenylate cyclase system (Tufano et al. 1984)
Pyrogenic exotoxin B (cysteine protease)	Group A *Streptococci*, *Streptococcus pyogenes*	Degrades proteins (Kuo et al. 1998)
Streptokinase (fibrinolysin)	Groups A, C, and G *Streptococci*, *Staphylococci*	Plasmin is activated by binding to plasminogen and thus allows the pathogen to move from the clotted area (Cederholm-Williams et al. 1979)
Lecithinase or phospholipase	*Clostridium* spp.	Destroys lecithin and allows pathogen to spread (Hayward 1943)
Protein A	*S. aureus*	Binds with IgG and thus prevents complement from interacting with bound IgG (Forsgren and Sjöquist 1966)
Hemolysin	*Escherichia coli*, *C. perfringens*	Lyses erythrocytes and makes iron available for microbial growth (Mitsui et al. 1973; Mackman and Holland 1984)
Elastase and alkaline protease	*P. aeruginosa*	Cleaves laminin associated with basement membranes (Morihara and Homma 1985)

1.3.1 Extracellular Virulence Factors

Many pathogenic microorganisms secrete extracellular proteases and cell wall degrading enzymes, toxins that perform as important virulence factors. They are either encoded by plasmid or chromosome. Moreover, conditions like pH, bile, bicarbonate, mucus, alkalinity, and high osmolarity alter the expression of virulence factors in the gastrointestinal tract due to modulation of virulence factors of some enteropathogenic pathogenic strains by different host factors (Hofmann 1999; Begley et al. 2005; Chiang 2013). Two main virulence factors like the cytotoxin-associated gene A and the vacuolating cytotoxin A are present in gram-negative *Helicobacter pylori* (Fig. 1.2).

An increased risk of developing gastric cancer is caused by these virulence factors. Thus, they regulate their virulence property to the most conducive niche of infection. *E. coli* (STEC) causes food-borne disease worldwide by producing Shiga toxin. Pathogenic STEC or enterohemorrhagic *E. coli* (EHEC) causes intestinal disorders including watery or bloody diarrhea. These disorders may ultimately develop to life-threatening diseases such as thrombotic thrombocytopenic purpura or hemolytic uremic syndrome (HUS). Keystone pathogen *Porphyromonas*

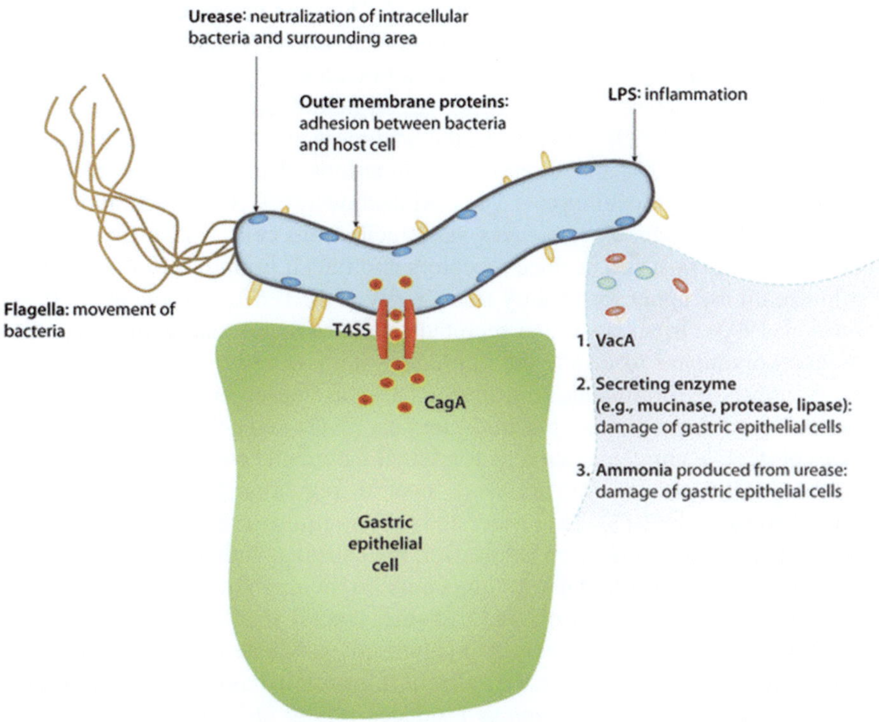

Fig. 1.2 Pathogenic virulence factors of *H. pylori* and their functions. *CagA* cytotoxin-associated gene A, *T4SS* type IV secretion system, *VacA* vacuolating cytotoxin, *LPS* lipopolysaccharide (Kim 2016). (Reprinted with permission [License number: 4593700407377])

gingivalis causes periodontal dysbiosis (Hajishengallis 2014). It processes several virulence factors that act as immunogenic molecules. *P. gingivalis* produces most important protease, termed as Gingipains (cysteine proteases) (Guo et al. 2010). Heme acquisition is one of the main functions of the gingipains protease (Smalley and Olczak 2017). Elastase, rhamnolipids, alginate, and lipopolysaccharide (LPS) of *Pseudomonas* facilitate *acute Pseudomonas* infection into a chronic infection (Bjarnsholt et al. 2010; Girard and Bloemberg 2008).

Clathrin-coated vesicle trafficking is modulated by *Chlamydial trachomatis* CT229. It controls the trafficking of both transferrin and the mannose-6-phosphate receptor for proper development. CT229 regulates several host vesicular trafficking pathways, essential for chlamydial infection (Faris et al. 2019). Adherence is another crucial virulence factor for most of the pathogens. Adhesin-deficient *S. aureus* was eliminated rapidly and it suggested adherence played crucial function for maintaining colonization process (Mulcahy et al. 2012; Weidenmaier et al. 2004). *C. trachomatis* and *L. monocytogenes* secrete the second messenger cyclic dimeric (c-di)-AMP that binds straight to stimulator of interferon (IFN) genes (STING) (Woodward et al. 2010; Barker et al. 2013), whereas STING is activated by several other pathogens in a cyclic GMP–AMP synthase (cGAS)-dependent manner (Zhang et al. 2014; Watson et al. 2015). However, *Brucella abortus* functions differently. STING directly detects bacterial cyclic dinucleotides (CDNs) and hence response of type I IFN is triggered. This leads to the upregulation of several IFN-related genes, including guanylate-binding proteins (GBPs) (Costa Franco et al. 2018).

Majority of *S. aureus* isolates secrete alpha-hemolysin (Hla), a toxin which forms pore (Li et al. 2009). Hla hijacks the host molecule ADAM10, a disintegrin and metalloprotease 10, and disrupts cell junctions. Thus it infects invariably (Kennedy et al. 2010; Inoshima et al. 2012; Malachowa et al. 2013). Neuraminidase, produced by *S. Pneumoniae,* cleaves sialic acid from cell surface glycans and mucin. It exposes host cell surface receptors and thus it promotes *S. Pneumoniae* to colonize on the upper respiratory tract (Kelly et al. 1967; O'Toole et al. 1971; Paton et al. 1993). Toxin is another secreted virulence factor that facilitates pathogenic microorganisms to invade the immune system. *Bacillus anthracis,* a gram-positive endospore forming android shaped bacteria, causes Anthrax. *B. anthracis* carries two extrachromosomal plasmids, namely pXO1 and pXO2. Plasmid pXO1 encodes protective antigen, edema factor, lethal factor, and anthrax toxin activator A (AtxA). All these toxins act as a central regulator for toxin synthesis (Hammerstrom et al. 2011). These secreted toxins require a short duration of time to establish a systemic infection. Moreover, the capsule synthesizing proteins are encoded by plasmid pXO2. Disruption occurs on the membrane integrity of host cells due to pore forming anthrolysin O, another crucial virulence factor (Shannon et al. 2003). *S. dysgalactiae* secretes hyaluronidase whose molecular weight is approximately 55 kDa. However, the value of this protein has not been determined (Sting et al. 1990). *S. zooepidemicus* produces a lytic enzyme called zoocin A (zooA). The cell walls of some closely related streptococcal species are specifically targeted by zooA. The sequence of zooA was determined by cloning it (Simmonds et al. 1997). Group B *Streptococcus* strains produce 12 kDa molecular

Table 1.2 Major differences between exotoxin and endotoxin

Exotoxin	Endotoxin
Gram-positive and gram-negative microorganisms	Integral part of cell wall of gram-negative microorganisms
Protein (polypeptide)	Lipopolysaccharide (LPS)
Diffusible, secreted by living cells	Nondiffusible and released on cell lysis
Highly antigenic	Poorly antigenic
Very high toxicity	Low toxicity
Unstable at temperature ˃60 °C and toxicity destroyed rapidly	Stable at ˃60 °C or more than that for several hours without losing toxicity
It can be converted into toxoid	No effect, cannot be converted

weight novel pyrogenic toxin which causes streptococcal toxic shock-like syndrome (TSLS) (Schlievert et al. 1993). *C. perfringens* type A produces the most toxic extracellular enzyme alpha-toxin which is strictly required for its pathogenicity (Awad et al. 1995; Ellemor et al. 1999). Alpha-toxin hydrolyzes both major constituents of eukaryotic cell membranes phosphatidylcholine and sphingomyelin due to its phospholipase activity (Rood 1998; Titball and Rood 2000). Enterotoxin, produced by *S. aureus*, causes several diseases and food-borne diseases are one of them. Staphylococcal food-borne diseases are the most occurred (Archer and Young 1988; Bean et al. 1990; Bunning et al. 1997). Botulism toxin is a neurotoxin and causes severe food-borne illness. Another neurotoxin is tetanus toxin which consists of a heavy chain and light chain. Light chain translocates to the cytosol by binding neuroselectively followed by internalization and intraneuronal sorting. Light chain cleaved at a single site of Synaptobrevin and SNAP-25 with unique selectivity and thus synaptic transmission is catalytically inhibited (Poulain et al. 1988; Bittner et al. 1989; Mochida et al. 1990; Kurazono et al. 1992; Niemann et al. 1994). Neurotoxin complex of *C. botulinum* type A consists of a core neurotoxin protein, several toxin-associated hemagglutinin (HA) proteins, and a non-toxin non-hemagglutinin (NTNH) protein. The continuous advancement in the knowledge of the botulinum toxin's molecular mechanism has aided to proceed parallel in their clinical use (Cordivari et al. 2004; Grumelli et al. 2005).

All of these above discussed toxins are exotoxins. Gram negative bacteria contain another type of toxin and this has been called endotoxin as it is cell associated and not secreted out of the cells (Table 1.2). Lipopolysaccharide (LPS) acts as an endotoxin secreted from gram-negative microorganisms. LPS is made of three parts: core polysaccharide, lipid A, and O antigen. Among these, O antigen is the most variable part and as a result, antibody could not recognize O antigen. Nowadays, LPS is well-known as a crucial factor responsible for toxic indication of severe gram-negative infections and generalized inflammation (Alving 1993). Both types of toxins have many differences (Table 1.2).

1.3.2 Cell Associated Virulence Factors

Capsule is a classic example among cell associated virulence factors. The capsule of *B. anthracis* consists of poly-γ-d-glutamic acid. It blocks the phagocytosis *of B. anthracis* during infection and hence it is weakly immunogenic in nature (Makino et al. 2002). Host cell phagocytic receptors and/or specific pattern recognition receptors (PRRs) recognize various cell surface ligands of mycobacteria like HSP70, phosphatidylinositol mannoside (PIM), 19 kDa lipoarabinomannan (LAM), and lipoprotein (Dorhoi et al. 2011). However, uptake by some of the receptors is advantageous for the pathogen's survival. Another encapsulated bacteria is *Streptococcus* which causes some serious invasive infections, including septicemia, pneumonia, and meningitis. They contain capsular material which contains sialic acid and it acts as a virulence factor (Jacques et al. 1990).

1.4 Conclusion

Microbial infection is one of the serious problems worldwide and it needs to be focused. More than half of the human death occurs annually due to microbial infections. Poor hygiene, overuse of antibiotics, and lack of education are the main reasons for microbial infection. However, very few pathogens have been discovered till date. Microbes deploy several novel strategies to attack host cells and sometimes it is very hard or almost impossible task to treat these microbial diseases. Microbes cause infections by secreting various virulence factors that manipulate the host cell system. It can be either secreted or cell associated. Sometimes, they cause disease indirectly like in case of food-borne illness where microbes secrete toxins into the food and the humans are infected with those toxin-contaminated foods. Further studies on the pathogenic microorganisms aid to discover more novel pathways and virulence factors that harm human host cells.

References

Aggarwal R, Jameel S (2011) Hepatitis E. Hepatology 54:2218–2226

Alving CR (1993) Lipopolysaccharide, lipid A, and liposomes containing lipid A as immunologic adjuvants. Immunobiology 187:430–446

Archer DL, Young FE (1988) Contemporary issues: diseases with a food vector. Clin Microbiol Rev 1:377–398

Awad MM, Bryant AE, Stevens DL, Rood JI (1995) Virulence studies on chromosomal α-toxin and θ-toxin mutants constructed by allelic exchange provide genetic evidence for the essential role of α-toxin in *Clostridium perfringens*-mediated gas gangrene. Mol Microbiol 15:191–202

Barker JR, Koestler BJ, Carpenter VK, Burdette DL, Waters CM, Vance RE, Valdivia RH (2013) STING-dependent recognition of cyclic di-AMP mediates type I interferon responses during *Chlamydia trachomatis* infection. MBio 4:e00018–e00013

Barrett AD, Monath TP (2003) Epidemiology and ecology of yellow fever virus. Adv Virus Res 61:291–315

Bean NH, Griffin PM, Goulding JS, Ivey CB (1990) Foodborne disease outbreaks, 5-year summary, 1983–1987. MMWR CDC Surveill Summ 39:15–57
Begley M, Gahan CG, Hill C (2005) The interaction between bacteria and bile. FEMS Microbiol Rev 29:625–651
Bhatt S, Gething PW, Brady OJ, Messina JP, Farlow AW, Catherine L, Moyes CL, Drake JM, Brownstein JS, Hoen AG, Sankoh O, Myers MF, George DB, Jaenisch T, Wint GRW, Simmons CP, Scott TW, Farrar JJ, Hay SI (2013) The global distribution and burden of dengue. Nature 496:504–507
Bird BH, Ksiazek TG, Nichol ST, MacLachlan NJ (2009) Rift Valley fever virus. J Am Vet Med Assoc 234:883–893
Bittner MA, DasGupta BR, Holz RW (1989) Isolated light chains of botulinum neurotoxins inhibit exocytosis. J Biol Chem 264:10354–10360
Bjarnsholt T, Jensen PØ, Jakobsen TH, Phipps R, Nielsen AK, Rybtke MT, Nielsen TT, Givskov M, Høiby N, Ciofu O, Scandinavian Cystic Fibrosis Study Consortium (2010) Quorum sensing and virulence of *Pseudomonas aeruginosa* during lung infection of cystic fibrosis patients. PLoS One 5:e10115
Blyth CC, Middleton PG, Harun A, Sorrell TC, Meyer W, Chen SC (2010) Clinical associations and prevalence of *Scedosporium* spp. in Australian cystic fibrosis patients: identification of novel risk factors? Med Mycol 48:S37–S44
Bramanti B, Dean KR, Walløe L, Chr Stenseth N (2019) The third plague pandemic in Europe. Proc Biol Sci 286(1901):20182429
Bunning VK, Lindsay JA, Archer DL (1997) Chronic health effects of microbial foodborne disease. World Health Stat Q 50:51–56
Camargo JF, Husain S (2014) Immune correlates of protection in human invasive aspergillosis. Clin Infect Dis 59:569–577
Canard B, Garnier T, Saint-Joanis B, Cole ST (1994) Molecular genetic analysis of the nagH gene encoding a hyaluronidase of *Clostridium perfringens*. Mol Gen Genet 243:215–224
Cederholm-Williams SA, Decock F, Lijnen HR, Collen O (1979) Kinetics of the reactions between streptokinase, plasmin, and lX2-antiplasmin. Eur J Biochem 100:125–132
Chiang JYL (2013) Bile acid metabolism and signaling. Compr Physiol 3:1191–1212
Colson P, Borentain P, Queyriaux B, Kaba M, Moal V, Gallian P, Heyries L, Raoult D, Gerolami R (2010) Pig liver sausage as a source of hepatitis E virus transmission to humans. J Infect Dis 202:825–834
Colson P, Romanet P, Moal V, Borentain P, Purgus R, Benezech A, Motte A, Gérolami R (2012) Autochthonous infections with hepatitis E virus genotype 4, France. Emerg Infect Dis 18:1361–1364
Corbel MJ (1997) Brucellosis: an overview. Emerg Infect Dis 3:213–221
Cordivari C, Misra VP, Catania S, Lees AJ (2004) New therapeutic indications for botulinum toxins. Mov Disord 8:S157–S161
Costa Franco MM, Marim F, Guimaraes ES, Assis NRG, Cerqueira DM, Alves-Silva J, Harms J, Splitter G, Smith J, Kanneganti TD, de Queiroz NMGP, Gutman D, Barber GN, Oliveira SC (2018) *Brucella abortus* triggers a cGAS-independent STING pathway to induce host protection that involves guanylate-binding proteins and inflammasome activation. J Immunol 200:607–622
Costello EK, Lauber CL, Hamady M, Fierer N, Gordon JI, Knight R (2009) Bacterial community variation in human body habitats across space and time. Science 326:1694–1697
Dantas-Torres F (2007) Rocky Mountain spotted fever. Lancet Infect Dis 7:724–732
David MZ, Daum RS (2010) Community-associated methicillin-resistant *Staphylococcus aureus*: epidemiology and clinical consequences of an emerging epidemic. Clin Microbiol Rev 23(3):616–687
Di Giulio DB, Eckburg PB (2004) Human monkeypox: an emerging zoonosis. Lancet Infect Dis 4:15–25
Dobler G, Hufert F, Pfeffer M, Essbauer S (2011) Tick-borne encephalitis: from microfocus to human disease. Progress in parasitology. Parasitol Res Monogr 2:323–331

Dorhoi A, Reece ST, Kaufmann SH (2011) For better or for worse: the immune response against *Mycobacterium tuberculosis* balances pathology and protection. Immunol Rev 240:235–251

Ellemor D, Baird R, Awad M, Boyd R, Rood J, Emmins J (1999) Use of genetically manipulated strains of *Clostridium perfringens* reveals that both alpha-toxin and theta-toxin are required for vascular leukostasis to occur in experimental gas gangrene. Infect Immun 67:4902–4907

Erickson A, Deibel RH (1973) Turbidimetric assay of staphylococcal nuclease. Appl Microbiol 25:337–341

Faris R, Merling M, Andersen SE, Dooley CA, Hackstadt T, Weber MM (2019) *Chlamydia trachomatis* CT229 subverts Rab GTPase-dependent CCV trafficking pathways to promote chlamydial infection. Cell Rep 26:3380–3390

Forsgren A, Sjöquist J (1966) "Protein A" from *S. aureus*: I. Pseudo-immune reaction with human γ-globulin. J Immunol 97:822–827

Furuno JP, Hebden JN, Standiford HC, Perencevich EN, Miller RR, Moore AC, Strauss SM, Harris AD (2008) Prevalence of methicillin-resistant *Staphylococcus aureus* and *Acinetobacter baumannii* in a long-term acute care facility. Am J Infect Control 36:468–471

Gessner BD, Mueller JE, Yaro S (2010) African meningitis belt pneumococcal disease epidemiology indicates a need for an effective serotype 1 containing vaccine, including for older children and adults. BMC Infect Dis 10:22–31

Geyer S, Jacobs M, Hsu NJ (2019) Immunity against bacterial infection of the central nervous system: an astrocyte perspective. Front Mol Neurosci 12:57–65

Gillet Y, Issartel B, Vanhems P, Fournet J-C, Lina G, Bes M, Vandenesch F, Piémont Y, Brousse N, Floret D, Etienne J (2002) Association between *Staphylococcus aureus* strains carrying gene for Panton-Valentine leukocidin and highly lethal necrotising pneumonia in young immunocompetent patients. Lancet 359:753–759

Girard G, Bloemberg G (2008) Central role of quorum sensing in regulating the production of pathogenicity factors in *Pseudomonas aeruginosa*. Future Microbiol 3:97–106

Godfroid J, Scholz HC, Barbier T, Nicolas C, Wattiau P, Fretin D, Whatmore AM, Cloeckaert A, Blasco JM, Moriyon I, Saegerman C, Muma JB, Al Dahouk S, Neubauer H, Letesson JJ (2011) Brucellosis at the animal/ecosystem/human interface at the beginning of the 21st century. Prev Vet Med 102:118–131

Grumelli C, Verderio C, Pozzi D, Rossetto O, Montecucco C, Matteoli M (2005) Internalization and mechanism of action of clostridial toxins in neurons. Neurotoxicology 26:761–767

Guo Y, Nguyen KA, Potempa J (2010) Dichotomy of gingipains action as virulence factors: from cleaving substrates with the precision of a surgeons knife to a meat chopper-like brutal degradation of proteins. Periodontol 2000 54:15–44

Hajishengallis G (2014) Immunomicrobial pathogenesis of periodontitis: keystones, pathobionts, and host response. Trends Immunol 35:3–11

Hammerstrom TG, Roh JH, Nikonowicz EP, Koehler TM (2011) *Bacillus anthracis* virulence regulator AtxA: oligomeric state, function and CO(2) signalling. Mol Microbiol 82:634–647

Han XY, Seo YH, Sizer KC, Schoberle T, May GS, Spencer JS, Li W, Nair RG (2008) A new *Mycobacterium* species causing diffuse lepromatous leprosy. Am J Clin Pathol 130:856–864

Hart ME, Hart MJ, Roop AJ (2009) Genotypic and phenotypic assessment of hyaluronidase among type strains of a select group of Staphylococcal species. Int J Microbiol 2009:61437

Hayward NJ (1943) The rapid identification of *Cl. welchii* by Nagler tests in plate cultures. J Pathol Bacteriol 55:285–293

Hofmann AF (1999) Bile acids: the good, the bad, and the ugly. News Physiol Sci 14:24–29

Hosny M, Cassir N, La Scola B (2017) Updating on gut microbiota and its relationship with the occurrence of necrotizing enterocolitis. Hum Microb J J4:14–19

Husain S, Silveira FP, Azie N, Franks B, Horn D (2017) Epidemiological features of invasive mold infections among solid organ transplant recipients: PATH Alliance® registry analysis. Med Mycol 55:269–277

Inoshima N, Wang Y, Bubeck Wardenburg J (2012) Genetic requirement for ADAM10 in severe *Staphylococcus aureus* skin infection. J Invest Dermatol 132:1513–1516

Jacques M, Gottschalk M, Foiry B, Higgins R (1990) Ultrastructural study of surface components of *Streptococcus suis*. J Bacteriol 172:2833–2838
Kelly RT, Farmer S, Greiff D (1967) Neuraminidase activities of clinical isolates of *Diplococcus pneumoniae*. J Bacteriol 94:272–273
Kennedy AD, Bubeck Wardenburg J, Gardner DJ, Long D, Whitney AR, Braughton KR, Schneewind O, DeLeo FR (2010) Targeting of alpha-hemolysin by active or passive immunization decreases severity of USA300 skin infection in a mouse model. J Infect Dis 202:1050–1058
Kim JM (2016) H. pylori virulence factors: toxins (CagA, VacA, DupA, OipA, IceA). In: Helicobacter pylori. Springer, Singapore, pp 77–88
Kindhauser MK, Allen T, Frank V, Santhana RS, Dye C (2016) Zika: the origin and spread of a mosquito-borne virus. Bull World Health Organ 94:675–686C
Kravitz JN, Steed LL, Judson MA (2011) Intracavitary voriconazole for the treatment of hemoptysis complicating *Pseudallescheria angusta* pulmonary mycetomas in fibrocystic sarcoidosis. Med Mycol 249:198–201
Kurazono H, Mochida S, Binz T, Eisel U, Quanz M, Grebenstein O, Poulain B, Tauc L, Niemann H (1992) Minimal essential domains specifying toxicity of the light chains of tetanus toxin and botulinum neurotoxin type A. J Biol Chem 267:14721–14729
Kuo CF, Wu JJ, Lin KY, Tsai PJ, Lee SC, Jin YT, Lei HY, Lin YS (1998) Role of Streptococcal pyrogenic exotoxin B in the mouse model of group A streptococcal infection. Infect Immun 66:3931–3935
Lau G, Hassett D, Ran H, Kong F (2004) The role of pyocyanin in *Pseudomonas aeruginosa* infection. Trends Mol Med 10:599–606
Lemon KP, Klepac-eraj V, Schiffer HK, Brodie EL, Lynch SV, Kolter R (2010) Comparative analyses of the bacterial microbiota of the human nostril and oropharynx. MBio 1:e00129–e00110
Li M, Diep BA, Villaruz AE, Braughton KR, Jiang X, DeLeo FR, Chambers HF, Lu Y, Otto M (2009) Evolution of virulence in epidemic community-associated methicillin-resistant *Staphylococcus aureus*. Proc Natl Acad Sci U S A 106:5883–5888
Lionakis MS, Kontoyiannis DP (2004) Fusarium infections in critically ill patients. Semin Respir Crit Care Med 25(2):159–169
LiPuma JJ (2010) The changing microbial epidemiology in cystic fibrosis. Clin Microbiol Rev 23:299–323
Mackman N, Holland IB (1984) Functional characterization of a cloned haemolysin determinant from *E. coli* of human origin encoding information for the secretion of a 107 K polypeptide. Mol Gen Genet 196:129–134
Makino S, Wataraj M, Cheun HI, Shirahata T, Uchida I (2002) Effect of the lower molecular capsule released from the cell surface of *Bacillus anthracis* on the pathogenesis of anthrax. J Infect Dis 186:227–233
Malachowa N, Kobayashi SD, Braughton KR, DeLeo FR (2013) Mouse model of *Staphylococcus aureus* skin infection. Methods Mol Biol 1031:109–116
Manco S, Hernon F, Yesilkaya H, Paton JC, Andrew PW, Kadioglu A (2006) Pneumococcal neuraminidases A and B both have essential roles during infection of the respiratory tract and sepsis. Infect Immunol 74:4014–4020
Mihret W, Lema T, Merid Y, Kassu A, Abebe W, Moges B, Tenna A, Woldegebriel F, Yidnekachew M, Mekonnen W, Ahmed A, Yamuah L, Silamsaw M, Petros B, Oksnes J, Rosenqvist E, Ayele S, Aseffa A, Caugant DA, Norheim G (2016) Surveillance of bacterial meningitis, Ethiopia, 2012–2013. Emerg Infect Dis 22:75–78
Misra UK, Kalita J (2010) Overview: Japanese encephalitis. Prog Neurobiol 91:108–120
Mitsui K, Mitsui N, Hase J (1973) *Clostridium perfringens* exotoxins II. Purification and some properties of θ-toxin. Jpn J Exp Med 43:377–391
Mocarski ES, Shenk T, Pass RF (2007) Cytomegaloviruses. In: Knipe DM, Howley PM, Griffin DE, Lamb RA, Martin MA et al (eds) Fields virology, 5th edn. Lippincott, Williams & Wilkins, Philadelphia, pp 2701–2772

Mochida S, Poulain B, Eisel U, Binz T, Kurazono H, Niemann H, Tauc L (1990) Exogenous mRNA encoding tetanus or botulinum neurotoxins expressed in Aplysia neurons. Proc Natl Acad Sci U S A 87:7844–7848

Moran GJ, Krishnadasan A, Gorwitz RJ, Fosheim GE, McDougal LK, Carey RB, Talan DA (2006) Methicillin-resistant *S. aureus* infections among patients in the emergency department. N Engl J Med 355(7):666–674

Morihara K, Homma JY (1985) Pseudomonas proteases. In: Holder IA (ed) Bacterial enzymes and virulence. CRC Press, Boca Raton, FL, pp 41–79

Mulcahy ME, Geoghegan JA, Monk IR, O'Keeffe KM, Walsh EJ, Foster TJ, McLoughlin RM (2012) Nasal colonisation by *Staphylococcus aureus* depends upon clumping factor B binding to the squamous epithelial cell envelope protein loricrin. PLoS Pathog 8:e1003092

Nassiri H, Zarrin M, Veys-Behbahani R, Faramarzi S, Nasiri A (2015) Isolation and identification of pathogenic filamentous fungi and yeasts from adult house fly (Diptera: Muscidae) captured from the hospital environments in Ahvaz city, Southwestern Iran. J Med Entomol 52:1351–1356

Nelson PE, Dignani MC, Anaissie EJ (1994) Taxonomy, biology, and clinical aspects of *Fusarium* species. Clin Microbiol Rev 7:479–504

Niemann H, Blasi J, Jahn R (1994) Clostridial neurotoxins: new tools for dissecting exocytosis. Trends Cell Biol 4:179–185

Noni M, Katelari A, Kapi A, Stathi A, Dimopoulos G, Doudounakis SE (2017) *Scedosporium apiospermum* complex in cystic fibrosis; should we treat? Mycoses 60:594–599

Nucci M, Anaissie EJ, Queiroz-Telles F, Martins CA, Trabasso P, Solza C, Mangini C, Simões BP, Colombo AL, Vaz J, Levy CE, Costa S, Moreira VA, Oliveira JS, Paraguay N, Duboc G, Voltarelli JC, Maiolino A, Pasquini R, Souza CA (2003) Outcome predictors of 84 patients with hematologic malignancies and Fusarium infection. Cancer 98:315–319

O'Toole RD, Goode L, Howe C (1971) Neuraminidase activity in bacterial meningitis. J Clin Investig 50:979–985

Pappas G (2010) The changing Brucella ecology: novel reservoirs, new threats. Int J Antimicrob Agents 36(Suppl. 1):S8–S11

Pappas G, Akritidis N, Bosilkovski M, Tsianos E (2005) Brucellosis. N Engl J Med 352:2325–2336

Paton JC, Andrew PW, Boulnois GJ, Mitchell TJ (1993) Molecular analysis of the pathogenicity of *Streptococcus pneumoniae*: the role of pneumococcal proteins. Annu Rev Microbiol 47:89–115

Pérez-Gracia MT, Suay-García B, Mateos-Lindemann ML (2017) Hepatitis E and pregnancy: current state. Rev Med Virol 27:e1929

Polkinghorne A, Greub G (2017) A new equine and zoonotic threat emerges from an old avian pathogen, *Chlamydia psittaci*. Clin Microbiol Infect 23:693–694

Posen HJ, Keystone JS, Gubbay JB, Morris SK (2016) Epidemiology of Zika virus. 1947–2007. BMJ Glob Health 1:e000087

Poulain B, Tauc L, Maisey EA, Wadsworth JDF, Mohan PM, Dolly JO (1988) Neurotransmitter release is blocked intracellularly by botulinum neurotoxin, and this requires uptake of both toxin polypeptides by a process mediated by the larger chain. Proc Natl Acad Sci U S A 85:4090–4094

Purcell RH, Emerson SU (2008) Hepatitis E: an emerging awareness of an old disease. J Hepatol 48:494–503

Ramakrishnan M, Ulland AJ, Steinhardt LC, Moïsi JC, Were F, Levine OS (2009) Sequelae due to bacterial meningitis among African children: a systematic literature review. BMC Med 7:47

Reece E, Segurado R, Jackson A, McClean S, Renwick J, Greally P (2017) Co-colonisation with *Aspergillus fumigatus* and *Pseudomonas aeruginosa* is associated with poorer health in cystic fibrosis patients: an Irish registry analysis. BMC Pulm Med 17:70

Rego ROM, Trentelman JJA, Anguita J, Nijhof AM, Sprong H, Klempa B, Hajdusek O, Tomás-Cortázar J, Azagi T, Strnad M, Knorr S, Sima R, Jalovecka M, Havlíková SF, Ličková M, Sláviková M, Kopacek P, Grubhofer L, Hovius JW (2019) Counter attacking the tick bite: towards a rational design of anti-tick vaccines targeting pathogen transmission. Parasit Vectors 12(1):229

Rood JI (1998) Virulence genes of *Clostridium perfringens*. Annu Rev Microbiol 52:333–360

Rosenberg R (2018) Vital signs: trends in reported vectorborne disease cases—United States and territories, 2004–2016. MMWR Morb Mortal Wkly Rep 67:496–501

Rozé JC, Ancel PY, Lepage P, Martin-Marchand L, Al Nabhani Z, Delannoy J, Picaud JC, Lapillonne A, Aires J, Durox M, Darmaun D, Neu J, Butel MJ, Chollat C, Nutrition EPIPAGE 2 study group; EPIFLORE Study Group (2017) Nutritional strategies and gut microbiota composition as risk factors for necrotizing enterocolitis in very-preterm infants. Am J Clin Nutr 36:821–830

Schlievert PM, Gocke JE, Deringer JR (1993) Group B streptococcal toxic shock-like syndrome: report of a case and purification of an associated pyrogenic toxin. Clin Infect Dis 17:26–31

Schommer NN, Gallo RL (2013) Structure and function of the human skin microbiome. Trends Microbiol 21:660–668

Scollard DM (2016) Infection with *Mycobacterium lepromatosis*. Am J Trop Med Hyg 95:500–501

Sedlacek L, Graf B, Schwarz C, Albert F, Peter S, Würstl B, Wagner S, Klotz M, Becker A, Haase G, Laniado G, Kahl B, Suerbaum S, Seibold M, Tintelnot K (2015) Prevalence of *Scedosporium* species and *Lomentospora prolificans* in patients with cystic fibrosis in a multicenter trial by use of a selective medium. J Cyst Fibros 14:237–241

Shannon JG, Ross CL, Koehler TM, Rest RF (2003) Characterization of anthrolysin O, the *Bacillus anthracis* cholesterol-dependent cytolysin. Infect Immun 71:3183–3189

Shaw C, Stitt JM, Cowan ST (1951) Staphylococci and their classification. J Gen Microbiol 5:1010

Shuaib W, Stanazai H, Abazid AG, Mattar AA (2016) Re-emergence of Zika virus: a review on pathogenesis, clinical manifestations, diagnosis, treatment, and prevention. Am J Med 129:879.e7–879.e12

Simmonds RS, Simpson WJ, Tagg JR (1997) Cloning and sequence analysis of zooA, a *Streptococcus zooepidemicus* gene encoding a bacteriocin-like inhibitory substance having a domain structure similar to that of lysostaphin. Gene 189:255–261

Smalley JW, Olczak T (2017) Heme acquisition mechanisms of *Porphyromonas gingivalis*—strategies used in a polymicrobial community in a heme-limited host environment. Mol Oral Microbiol 32:1–23

Solomon T (2004) Flavivirus encephalitis. N Engl J Med 351:370–378

Sprong H, Azagi T, Hoornstra D, Nijhof AM, Knorr S, Baarsma ME, Hovius JW (2018) Control of Lyme borreliosis and other Ixodes ricinus-borne diseases. Parasit Vectors 11:145

Steinbach WJ, Marr KA, Anaissie EJ, Azie N, Quan SP, Meier-Kriesche HU, Apewokin S, Horn DL (2012) Clinical epidemiology of 960 patients with invasive aspergillosis from the PATH Alliance registry. J Infect 65:453–464

Stern M, Warrack GH (1964) The types of *Clostridium perfringens*. J Pathol Bacteriol 88:279–282

Sting R, Schaufuss P, Blobel H (1990) Isolation and characterization of hyaluronidases from *Streptococcus dysgalactiae*, *S. zooepidemicus* and *S. equi*. Zentralbl Bakteriol 272:276–282

Süss J (2011) Tick-borne encephalitis 2010 epidemiology, risk areas, and virus strains in Europe and Asia—an overview. Ticks Tick Borne Dis 2:2–15

Talan DA, Krishnadasan A, Gorwitz RJ, Fosheim GE, Limbago B, Albrecht V, Moran GJ, EMERGEncy ID Net Study Group (2011) Comparison of *Staphylococcus aureus* from skin and soft-tissue infections in US emergency department patients, 2004 and 2008. Clin Infect Dis 53:144–149

Tärnvik A (1989) Nature of protective immunity to *Francisella tularensis*. Rev Infect Dis 11:440–451

Tei S, Kitajima N, Ohara S, Inoue Y, Miki M, Yamatani T, Yamabe H, Mishiro S, Kinoshita Y (2004) Consumption of uncooked deer meat as a risk factor for hepatitis E virus infection: an age- and sex-matched case-control study. J Med Virol 74:67–70

Titball R, Rood J (2000) Bacterial phospholipases. In: Aktories K, Just I (eds) Bacterial protein toxins. Springer, Berlin, pp 529–556

Tsagaan A, Kanuka I, Okado K (2015) Study of pathogenic bacteria detected in fly samples using universal primer-multiplex PCR. Mong J Agric Sci 15:27–32

Tufano MA, Berlingieri MT, Sommese L, Galdiero F (1984) Immune response in mice and effect on cells by outer membrane porins from *Salmonella typhimurium*. Microbiologica 7:353–366

Vora NM, Li Y, Geleishvili M, Emerson GL, Khmaladze E, Maghlakelidze G, Navdarashvili A, Zakhashvili K, Kokhreidze M, Endeladze M, Mokverashvili G, Satheshkumar PS, Gallardo-Romero N, Goldsmith CS, Metcalfe MG, Damon I, Maes EF, Reynolds MG, Morgan J, Carroll DS (2015) Human infection with a zoonotic orthopoxvirus in the country of Georgia. N Engl J Med 372:1223–1230

Watson RO, Bell SL, MacDuff DA, Kimmey JM, Diner EJ, Olivas J, Vance RE, Stallings CL, Virgin HW, Cox JS (2015) The cytosolic sensor cGAS detects *Mycobacterium tuberculosis* DNA to induce type I interferons and activate autophagy. Cell Host Microbe 17:811–819

Wessels MR, Bronze MS (1994) Critical role of group A streptococcal capsule in pharyngeal colonization and infection in mice. Proc Natl Acad Sci U S A 91:12238–12242

Weidenmaier C, Kokai-Kun JF, Kristian SA, Chanturiya T, Kalbacher H, Gross M, Nicholson G, Neumeister B, Mond JJ, Peschel A (2004) Role of teichoic acids in *Staphylococcus aureus* nasal colonization, a major risk factor in nosocomial infections. Nat Med 10:243–245

Winstanley C, Fothergill J (2009) The role of quorum sensing in chronic cystic fibrosis *Pseudomonas aeruginosa* infections. FEMS Microbiol Lett 290:1–9

Woodward JJ, Iavarone AT, Portnoy DA (2010) c-di-AMP secreted by intracellular *Listeria monocytogenes* activates a host type I interferon response. Science 328:1703–1705

World Health Organization (2009) Dengue guidelines for diagnosis, treatment, prevention and control. WHO, Geneva, pp 1–160

Wiinsch E, Heidrich HG (1963) Zur quantitaiven Bestimung der Kollagenase. Hoppe Seylers Z Physiol Chem 333:149–151

Zhang Y, Yeruva L, Marinov A, Prantner D, Wyrick PB, Lupashin V, Nagarajan UM (2014) The DNA sensor, cyclic GMP-AMP synthase, is essential for induction of IFN-b during *Chlamydia trachomatis* infection. J Immunol 193:2394–2404

Zhao-Fleming H, Dissanaike S, Rumbaugh K (2017) Are anaerobes a major, underappreciated cause of necrotizing infections? Anaerobe 45:65–70

Zouhair R, Rougeron A, Razafimandimby B, Kobi A, Bouchara JP, Giraud S (2013) Distribution of the different species of the *Pseudallescheria boydii/Scedosporium apiospermum* complex in French patients with cystic fibrosis. Med Mycol 51:603–613

Microbial Pathogenesis: Virus Pathogen–Host Interactions

2

Vinodhini Krishnakumar and Meganathan Kannan

Contents

Abstract

Any microorganism which is capable of bringing illness in a host organism is termed as pathogen. Pathogens manipulate the cellular mechanisms of host organisms via pathogen–host interactions (PHIs) in order to take advantage of the capabilities of host cells, leading to infections. This chapter is restricted to human viral pathogens, though plant and animal pathogens are also extensive in nature. When a pathogenic microorganism (especially virus) infects the human host, a fight develops between the host's innate and adaptive immune systems and the pathogen's assorted virulence mechanisms. The battle results in establishing how well the host survives and recovers. Failures in vaccine production and inadequacy in antiviral therapeutics are the recent challenges what virology researchers facing now. For years, the discipline of virology has been excessively

V. Krishnakumar (✉) · M. Kannan
Division of Blood and Vascular Biology, Department of Life Sciences, School of Life Sciences, Central University of Tamil Nadu, Tiruvarur, India

B. Siddhardha et al. (eds.), *Model Organisms for Microbial Pathogenesis, Biofilm Formation and Antimicrobial Drug Discovery*,
https://doi.org/10.1007/978-981-15-1695-5_2

concentrated on pathogens than host response. We are aware that host response is the determinant factor for the eventual development of the pathological outcome of the infection. Viral pandemic and epidemic infections like acquired immunodeficiency syndrome (AIDS), swine flu, dengue, and other chronic and acute infections are posing a huge threat. We are in a need to develop biological approaches with sophisticated computational strategies by understanding the components, interactions, and dynamics of a biological system in a comprehensive, quantitative, and integrative fashion.

Keywords
Pathogen–host interaction · Virus · Pathogens · Infections

2.1 Introduction

Existing and emerging infections and infectious diseases are a major distress to human health. It has been an unsolved serious issue for the past two decades (Southwood and Ranganathan 2016). Pathogens influence host cellular mechanisms by means of pathogen–host interactions (PHIs) to benefit the facilities of the host cells, leading to infections. Hence PHI plays a crucial role in understanding the mechanism and developing new and effective measures. Rejecting the traditional approach of considering the host and pathogen separately, a system-level approach, considering the PHI system as a whole is very important to explain the mechanisms of infection. The importance of host–pathogen interactions lies in the more knowledge we gain about pathogens and how they are interacting with their personal host systems; the easier it is to hit upon new ways of defeating them. Because the conflict against infectious disease is a classic and demanding war, which has not yet been overcome since decades. The real challenge what the researchers have now is to come up with the efficient methods which can analyze the host–pathogen interactions to give answers on our prevailing questions such as how pathogens invade, escape from the immune system, multiply, and how do they establish infections in their hosts. This collective information will give a permanent solution for developing new therapeutics, treatment strategies, methods of diagnosis, and vaccines to prevent and arbitrate infection. Considerable developments have been made in recent years with the help of genomics and proteomics technology. This strongly suggests that host–pathogen interactions are center piece of the developmental research (Urban et al. 2014). The currently known pathogens are viruses, bacteria, fungi, protozoa, and parasites. Among these infectious diseases caused by human pathogenic viruses remains a huge threat to global health. The mechanisms of how these pathogenic viruses cause human disease differ widely. Recent trends of research currently focused on mouse models to understand the mechanism of pathogenesis (Krishnakumar et al. 2019). Current chapter mainly focuses on three main concepts. The first section deals with the fundamentals of host–virus interaction and some terminology of pathogenesis, and the contention around how it is defined. The second section focuses on the consequences of the host system in response to the virus interaction. The last section deals with the factors affecting the virus–host interaction.

2.2 Historical Views of Host–Microbe Interactions

Host–microbe interaction has been in use for nearly a century. Near the beginning of this field, microbes were considered to be a principal invader that presides over the host–pathogen interaction, resulting in disease. As the technology grows, new information about the characteristics of microbes and their hosts ends up in the understanding that the host–pathogen interaction does not always result in disease development. This recognition, in turn, showed the way to explain the terms to (1) microbes surviving within the host causing obviously no disease and (2) why some microbes only cause disease in certain hosts (Stebbing and Gazzard 2003). Commensalism, carrier state, and opportunist are the terms which describe the behavior of microbes and conditions that were associated with disease rather than to define a more general host–microbe relationship. Casadevall and Pirofski (1999) reviewed the concepts of virulence and pathogenicity and described how these terms have changed over the period of time. According to the concept, the host damage was the most predominant and appropriate result of the host–pathogen interaction, they redefined the terms pathogen, pathogenicity, and virulence (Stebbing and Gazzard 2003). In order to understand the interaction between virus and host, we must first need to understand the eventual outcome of host–virus interactions (Fig. 2.1), namely infection, commensalism, colonization, persistence, and disease. *Infection* can be defined as invasion of disease causing virus in to the host cell; *Commensalism* is a long-term biological interaction between a host and a pathogen in which pathogenic virus gain benefits while the host neither benefited nor harmed; *Colonization* implies the presence of detectable concentration of pathogenic organism; yet the virus is causing symptoms or not in the susceptible host; *Persistence*: pathogenic viruses establish a persistent infection by actively curtailing the host's antiviral immune response, e.g., chronic infections are a type of persistent infection. The following pages cover mainly the eventual process that underlies between the human host and pathogenic virus interaction. Host–virus interactions are the battle between the viruses and the susceptible host (Tennant et al. 2018); viruses fight with the host system in order to ensure their replication, and the host cell elicits its defense in order to restrict the infection by an intruding agent. The expected outcome is not always convinced and positive; it purely depends on the balance of actions and end results between host resistance and the viruses escape mechanisms. This battle run after some of the principal trends in its eventual process (Tennant et al. 2018).

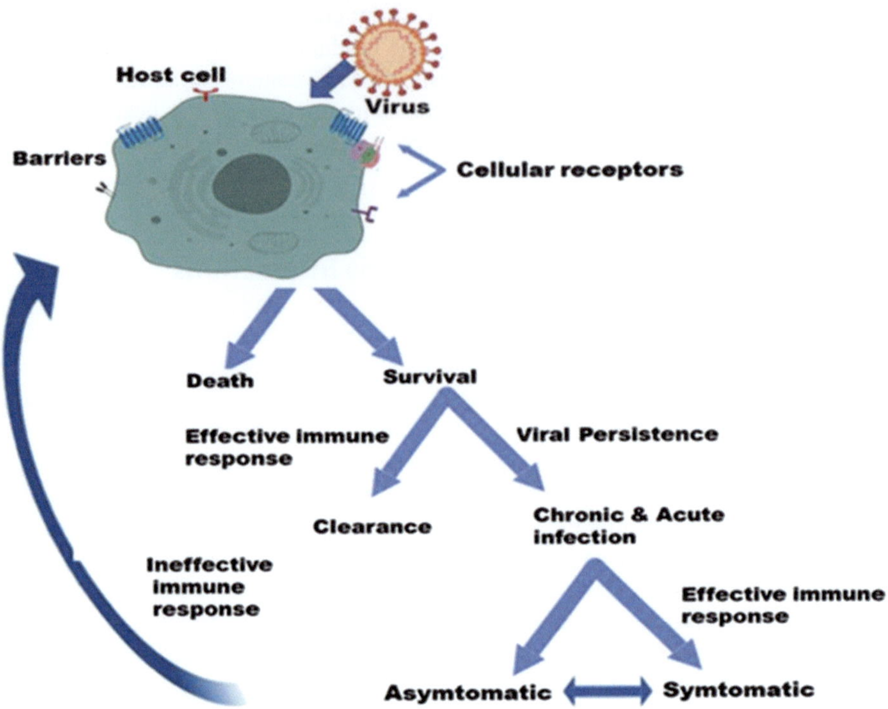

Fig. 2.1 Representative illustration of virus–host interaction. It is an eventual process that depends on individuals' immunity to develop a disease or it can be a normal commensal

2.2.1 Fundamental Properties of Viruses

Virus is a small biological parasite, which has self-directed infectious particles that differ greatly from other microorganisms (bacteria, fungi, protozoans, etc.) in a number of characteristics, i.e., they have no cellular structure, contains only nucleic acids, i.e., deoxyribonucleic acid (DNA) and ribonucleic acid (RNA) and proteins. They are devoid of intrinsic metabolic systems and depend upon the living hosts for cellular metabolic process (Kayser et al. 2004). Virus genomes do not encode all the proteins and RNAs, such as transfer RNA, ribosomal RNA, and ribosomal proteins, needed for replication. Even it lacks the genes necessary for its own energy production. In order to accomplish all its needs, it infects a susceptible host cell, where it can insert its genetic material into the host. Literally a virus can direct the cell machinery or take over the host functions. An infected host cell produces more viral proteins and genetic materials instead of its usual. A few viruses may stay dormant inside host cells for extended periods, causing no observable change in their host cells (a stage known as the lysogenic phase). But when a dormant virus is stimulated, it enters the lytic phase: new viruses are produced, self-assembled, and burst out of the host cell, killing the cell and going on to infect other cells (Lodish et al. 2000; Suttle 2007; Breitbart 2012). A wealth of subsequent research has enabled us for the better understanding of the virus structure and function (Lodish et al. 2000; Koonin et al. 2015). The continuous co-existence of viruses with their hosts can be viewed as a molecular race between the virus and host eradication mechanisms. It also shapes the immune system and in turn viruses have manipulated host immune control mechanisms to facilitate their propagation. The following four basic principles are considered most important for host–pathogen interactions (Stebbing and Gazzard 2003).

2.2.1.1 Principal Rules of Virus–Host Interactions

1. Viruses can be broadly divided into two categories depending on the type of genetic material (DNA or RNA) they contain.
2. All viruses can exist in two forms (extracellular/intracellular).
3. Production of antigenic proteins can be with sequence homology or without homology.
4. Virus and their subtypes can have different effects on the same host.

Figure 2.1 best describes the process of virus–host interaction and the consequences.

2.2.2 Host Defense Mechanisms

The host cell brings out its defense by the following four important aspects (Roitt 1974).

Mutated Virus Receptor Where the host cannot recognize the pathogen, hence the host resistance can be expressed by altering the virus replication. On the other hand, mutations in the virus permit the receptor recognition and enable the entry of virus, and infection.

Host Innate Immunity Virus replication can be effectively prohibited by host innate immunity. Viruses usually neutralize host innate immunity by escaping or exploiting the fragility in the innate immunity of host system.

Adaptive Immunity: Acquired Immunity Is Used to Differentiate Self-Antigens from Other By means of which host cells can collect information on the virus and impede or reduce its replication in a highly selective and effective manner. Viruses, on the other hand, can even avoid and manage its recognition.

Programmed Cell Death (PCD) Infected cells will undergo prompt self-destruction in order to avoid infection of other cells. Of note, the same system that induces PCD can induce dormancy (stasis) in an organism.

In this chapter, we will inspect the (1) interactions between a virus and its host (significant for the establishment and maintenance of infection), (2) the array of events in host antiviral mechanisms, and (3) the virus evasion mechanisms, which impose equal importance in understanding host–pathogen interactions. Eukaryotic viruses exhibit more challenges in terms of the destruction and damage they cause than that of prokaryotic viruses. Hence this chapter committed to virus and host interactions that lead to disease does not cover the multitude of phenotypic manifestations of virus diseases.

2.2.2.1 Defensive Host and Unwearied Pathogens

Resistance and recovery to viral infections is mostly relied on how the interactions are happening between the virus and the host. The resistance raised by the host may directly act on the infectious virus or indirectly altering the cellular mechanism or assassinating the infected cell. The infected host can handle the viral entry by means of two forms: (1) specific host defense (SHD) and (2) non-specific host defense (NSHD). SHD works in post infection and in recovery of immunity to consequent challenges, whereas NSHD works on early infection to encounter and prevent viral infection (Murray et al. 2015).

The principal determinants of virus–host interactions are (1) barriers to infection, (2) host defense mechanisms, (3) virus-induced immunopathology, and (4) immunosuppression. In order to infect the host, the virus has to overcome number of barriers which are inherently present in the organism. They represent the first line of defense which functions to prevent or limit infection, e.g., skin and membrane receptors. The skin acts as a formidable barrier to most viruses and only after this barrier is breached viruses will be able to infect the host. Lack of membrane receptors: Viruses gain entry into host cells by first binding to specific receptors on cells' list of infection barriers and host defenses against viral infections and its targets for each of these defenses. Defensive responses are tabulated in Tables 2.1 and 2.2 (Murray et al. 2015). Along with the above stated defenses the host also exerts other types of defenses like induced immunopathology. It is the immunity to the infection that is the direct basis of tissue injury. For example, children infected with cytomegalovirus (CMV) results in arthritis and glomerular nephritis. On the other hand, immunosuppression where the conditions in which virus are able to suppress immune responses and thereby overcome or minimize host defenses (e.g., human immunodeficiency virus (HIV) (Fig. 2.2).

Table 2.1 Barriers to infection

Inherent barriers	Skin, lack of membrane receptors, mucus, ciliated epithelium, low pH	Roitt, I.M., 1989. *Essential immunology* (No. sixth edition)
Induced barriers	Fever, low pH	Murray, Medical Microbiology, fifth Ed, Mims, medical microbiology, Zinsser, microbiology, 20th Ed Roitt, immunology fifth Ed.
	Non-specific humoral components (interferon (IFN), complements, and cytokines)	
	Specific humoral components (IgG, IgM, and IgA antibodies)	
	Non-specific cellular components (macrophages and NK cells)	
	Specific cellular components (T-cells)	

Table 2.2 Host defenses against viral infections (Baron et al. 1994)

Host defense	Effectors	Target of effectors
Early non-specific responses	Fever	Virus replication
	Phagocytosis	Virus
	Inflammation	Virus replication
	NK cell activity	Virus-infected cell
	Interferon	Virus replication, immunomodulation
	Cytotoxic T-lymphocytes	Virus-infected cell
	Activated macrophages	Virus, virus-infected cell
Immune responses mediated by cells	Lymphokines	Virus-infected cells, immunomodulation
	ADCC	Virus-infected cell
	Antibody	Virus, virus-infected cell
Humoral immune responses	Antibody + complement	Virus, virus-infected cell

2.2.3 Different Types of Virus–Host Interactions

There are five different patterns of infections, through which the interactions of the DNA can be determined: (1) transient, (2) acute, (3) latent, (4) persistent asymptomatic, and (5) persistent symptomatic (Fig. 2.3).

2.2.3.1 Transient Infections

Transient infection effects within the host cell last only for a short period. Transfection of viral genetic materials may be very transient, such that the cell communicates the foreign genetic information only temporarily, without replication of the nucleic acid. Transient infections were otherwise known as "transitory infections," although "transitory detection." Examples of transient viral infection are human papilloma virus infections (Alizon et al. 2017).

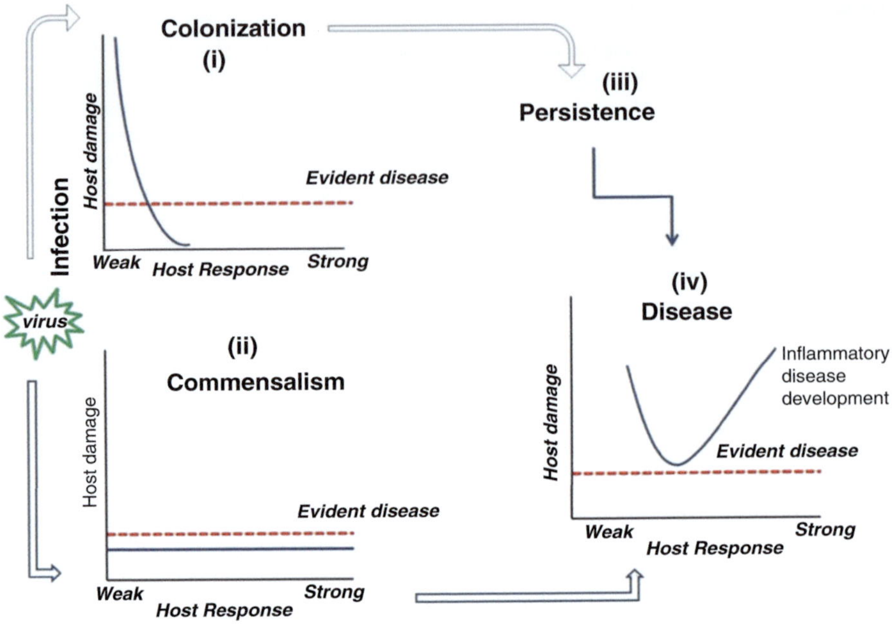

Fig. 2.2 An orderly event of a viral infection. (i) Describes the detectable concentration of virus. (ii) A long-term process of interaction between the virus and host. (iii) Long-term persistence of virus within the host by escaping from the host immune response. (iv) End result of disease development by defeating the individual immunity

2.2.3.2 Acute Infections

Acute viral infections are characterized by microbe living inside a host for a very limited period of time, typically less than 6 months , and rapid inception of disease, brief period of symptoms, finally resolves within days. It is usually accompanied by production and elimination of infectious virions and infections by the host immune system (Mason et al. 2018). Examples of acute pathogenic viral infections are influenza, rhinovirus, and Ebola hemorrhagic fever. For Ebola, the course of disease is extraordinarily severe. Mostly acute infections are in apparent infections and cause very little or no symptoms. During this phase, replication of viruses will occur adequately to induce antiviral antibodies; these infections are vital for the spread of infection to new hosts, for the reason that they are not easily detected (Martelli and Russo 1985).

2.2.3.3 Latent Infections

In latent infections, viruses can able to hide themselves from the immune responses generated against them and to set up a state where they are not visible to the immune system. Diseases are not obviously produced, but the viruses will not be eradicated. This balanced state between host and pathogenic virus is achieved in different ways.

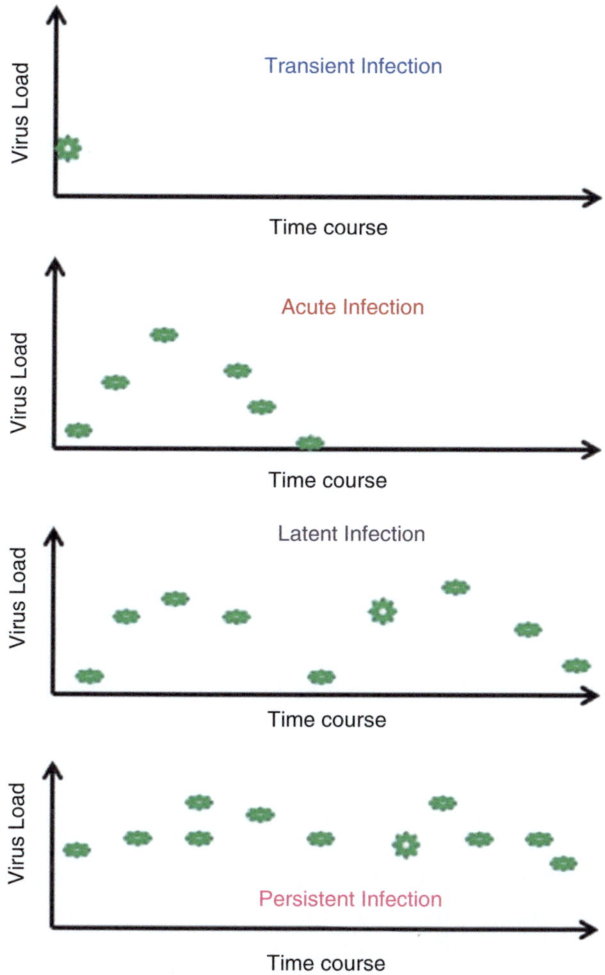

Fig. 2.3 Pictorial representation of different types of viral infection and its corresponding viral loads

The virus may exist in a truly latent non-infectious occult form, possibly as an integrated genome or an episomal agent, or as an infectious and continuously replicating agent, termed as a persistent viral infection. Such viruses can escape from B lymphocytes, T cells, CD4+ T cells, neurons, germinal epithelial cells, and others (Stevceva 2015). They can reactivate at times when the immunity starts to cooperate. The adverse effects of these dormant viruses on the immune systems are not well established, some may contribute to malignant transformation, some may (cytomegalovirus) involve in driving the T cells to terminal differentiation and exhaustion. In addition, other latent viruses such as Bornavirus are thought to be involved in the pathogenesis of human psychiatric diseases such as bipolar disorder and depression (Lafon 2009). An example of latent infection includes chronic congenital rubella (CCR), Epstein–Barr virus (EBV), hepatitis B, CMV, and HIV.

2.2.3.4 Persistent Infections

Most viral pathogens cause acute, self-limiting infections whereby the virus replicates fast and spreads to other organism prior to immune clearance or may lead to the death of the host. Inversely, some viruses are able to cause persistent infections all the way through adoption of complicated relationships with their hosts and exploiting a wide variety of cellular mechanisms for their own advantage. Persistence can occur through non-productive infection, e.g., herpesvirus latency (Knipe and Cliffe 2008; Kane and Golovkina 2010), proviral integration into the host genome, e.g., retroviruses (Bushman et al. 2005; Goff 2007), and/or continuous viral replication, e.g., flaviviruses, arenaviruses, and polyomaviruses (Imperiale and Major 2007; Buchmeier 2007). At the same time as each virus has chosen different pathways to facilitate persistence, there are some common theme in the establishment of permanent infection. Such as (1) assortment of cell subsets which are suitable for long-term maintenance of the viral genome, (2) pattern of viral gene expression, (3) viral cellular apoptotic pathways and its subversion, and (4) ignoring the clearance by the immune system. These observed commonalities among the complexity of host–virus interactions provide a new insight for researchers to treat these infections (Kane and Golovkina 2010). Figure 2.2 represents the different types of viral infections with respect to the viral load and time course of infection. Viruses widely use the host intracellular mechanism for replication, expression of viral genes, and initiating of infection. As an outcome of the infection process, they interact deeply with the host during their biological cycle. Along with the host biological activity, pathogenic viruses also initiate some sequential events to establish the infection.

2.2.4 Virus–Host Interactions

2.2.4.1 Virus–Host Interactions: Attachment and Entry

Virion targets their hosts either by physical or biological forces by exploiting fundamental cellular processes to achieve entry, principally defined by the interactions between virus particles and their receptors at the cell surface. Figure 2.1 best describes the virus–host interactions. These interactions are highly tissue tropism, that is, the types of cells in which the virus is able to replicate (Rawls et al. 1981). Virus–host interactions result in conformational changes in cell receptors that favor the penetration and membrane association. The process of infection by the virus and its spread from the infected cell to another are the two different complicated processes that mostly prefer to have a different signaling cascade to achieve so. Depending on the host it infects, the pathogenic virus preferred to choose different types of interactions. Direct introduction of virions into the host cell; in case of plant cells by means of creating wounds (Sanchez and Lagunoff 2015). Mycoviruses achieve cell entry by cytoplasmic exchange during cell division, mating and anastomosis, and by spores (Roossinck 2010). Some pathogenic DNA viruses like Adeno-, Papilloma-, Polyoma-, Pox-, and Herpesvirus families which do not deliver their virions directly prefer to have uncoating, i.e., release of viral genomes from its capsid or capsid-envelope coats. Once it successfully delivers its genetic material into the host, then it can step into the next experimental process known as genome replication and virion assembly which occurs in specific intracellular sections referred to as virus factories. Virus factories can be imagined as "organelles" synthesized newly from recruited cellular and viral components such as cell membranes, cytoskeletal elements, and mitochondria. These important structural and energy components are providing benefit of protection to the virus from the host's defence and cellular degradation machinary (Rawls et al. 1981; Sanchez and Lagunoff 2015).

2.2.4.2 Virus–Host Interactions: Reprogramming the Metabolism

Viruses are non-living entities and as such do not inherently have their own metabolism. It relies on the host cell mechanism for the production of viral components such as nucleic acid, proteins, and other necessary membranes (Martinez-Martin 2017). Recent reports have evidenced that viruses significantly modify cellular metabolism upon entry into a host cell. Mostly viruses are either enveloped (have both capsid proteins and a lipid/protein membrane) or non-enveloped viruses (have only capsid proteins). Possibly for several reasons, viruses are evolved to stimulate metabolic pathways like replication/reproduction. In order to attain a successive replication it requires nucleotides, amino acids, and energy products produced by extremely coordinated metabolic pathways. Though viruses do not program a metabolic network within their genome, it has the ability to significantly alter their host's metabolism by generating a new framework (Purdy 2019; Munger et al. 2008; Vastag et al. 2011; Rabinowitz et al. 2011; Delgado et al. 2012; Fontaine et al. 2014, 2015). Principally virus replication occurs in the host whose metabolism is lenient to infection. By binding the virions to the host receptors, clathrin-mediated endocytosis or micropinocytosis is triggered to facilitate the internalization of virions and subsequent escape from the endosomal vacuole into the cytosol (Birungi et al. 2010). Here, the viral genome is delivered and transported to other cellular compartments, where viral replication occurs: mostly DNA viruses and a few RNA viruses (e.g., influenza virus) enter the nucleus, whereas most RNA viruses stay in the cytosol (Dogra et al. 2019; Mercer et al. 2010; Schelhaas 2010). Once the viral genome and proteins are synthesized, they assembled to form new virus particles (virions), and subsequently a complex release process from the host cell is initiated (Heaton and Randall 2010; Eisenreich et al. 2019; Khan et al. 2015). Formation of new viruses solely depends on the metabolic pathways of the host cell to provide the necessary metabolites, i.e., nucleotides, amino acids, and fatty acids (FAs)/lipids.

2.2.4.2.1 General Trends of Metabolic Changes

Some of the general fashions of metabolic changes are discussed here, but they are not common or the collective changes observed among all types of viruses; each virus has its meticulous molecular needs, hence uses varieties of ways to meet up its metabolic requirements. Table 2.2 gives the collective information on the name of the viruses and the type of metabolic changes it causes on the host cell. Researchers started to tunnel out the floor on virus-induced metabolic responses in the early 1950s. Series of publications have reported the virus-induced metabolic changes of different viruses, namely poliovirus (Eagle and Habel 1956; Levy and Baron 1956), avian retrovirus (Figard and Levine 1966), Rous sarcoma virus (Venuta and Rubin 1973; Steck et al. 1968), lentivirus, and HIV-1 and HIV-2 (Hollenbaugh et al. 2016). Increase in the levels of glucose and glycolytically generated adenosine triphosphates (ATPs) are required for polio virus replication; two- to tenfold increase in the flow of carbons from glucose and acetate to lipids for avian retrovirus and Rous sarcoma virus; higher rate of glucose introduction and glycolysis are needed for Rous sarcoma virus; and recent report evidenced that the HIV-1 and HIV-2 alter the activity of glycolysis, tricarboxylic acid cycle (TCA), and nucleotide metabolism.

1. *Glycolysis*: Metabolic changes concerned with glycolysis are in the lead of virus infection, the glycolytic pathways are induced, resulting in improved glucose intake, breakdown of glucose (glycolysis), and lactic acid production (Sanchez and Lagunoff 2015). Induced glycolysis can go along with the process of apoptosis, but this is not universal, definitely changes with the virus and host cell type. Augmented glycolysis results not only in better generation of ATP but also improves the generation of metabolites that feed into other metabolic pathways like fatty acid biosynthesis which is mandatory for virion assembly (Fermin and Tennant 2018).
2. *Fatty Acid Synthesis*: Many viruses induce and require fatty acids biosynthesis at some point of their life cycle in infected cells by redirecting the metabolic pathway. Studies on carbon flux of human cytomegalovirus (HCMV) infected cells during lytic replication provoke the change of glucose-derived carbon into fatty acid synthesis in primary human foreskin fibroblast (HFF) cells (Munger et al. 2008). Significant increase in the conversion rate of carbon in to its metabolites citrate and malonyl CoA indicates the utilization. Hence glucose is fed as an energy source for fatty acid synthesis as well as the TCA cycle in HCMV infected cells.
3. *Glutaminolysis*: Poliovirus and Vaccinia virus replication is proficient in glutamine metabolism than glucose metabolism (Sanchez et al. 2017; Sanchez and Lagunoff 2015).

2.2.5 Virus–Host Interactions: Reprogramming the Changes in Structure and Function of Cell Organelles

Cytopathogenic virus–host interactions result not only in metabolic changes but also affect structure and function of some cell organelles (Islinger et al. 2015). The reported examples of these cytopathogenic effects are (1) rounding up of the infected cells, (2) defect in contact inhibition, (3) formation of giant cells (syncytia) by forming a fusion with neighboring cells, (4) inclusion bodies formation within the nucleus or cytoplasm, which indicates either altered host cell structures or gathering of viral components (Beltran et al. 2017). Viruses will either deform, rearrange endoplasmic reticulum (ER) outer membrane during the processes of virus entry, replication, or release. Viruses cleverly seize the cell intramembrane proteases involved in regulated intramembrane proteolysis (RIP) to facilitate their proliferation (Inoue and Tsai 2013). The reported evidence states that for several viral infections, e.g., flavivirus, a member of the family Flaviviridae (Weissenhorn et al. 2013), consist of more than 70 viruses which are epidemiologically very important viruses including dengue virus (DENV), West Nile virus (WNV) and others; replication, virion assembly and budding process of cytoskeleton protein filaments and its rearrangements plays a major role (Foo and Chee 2015). In recent research the unclear mechanism responsible for dengue severity encourages researchers to study cytoskeleton disorganization (Kanlaya et al. 2009; Wang et al. 2010). The significance and the extent of cytoskeletal reformation differ with different virus families (Foo and Chee 2015). Almost all DNA viruses reproduce themselves in the nucleus of their host cells. Access into the nucleus is attained by several viral proteins (reviewed by Fulcher and Jans 2011), which helps them for replication and formation of capsids and other subparticles. For most of the DNA viruses nucleus is the site of replication. As an initial step they deliver their genome into the nucleus of host cell. Nuclear transportation of molecules is highly specific and strongly regulated process that takes place with the help of nuclear pore complex (NPC). Each type of viruses developed its own strategy to enter into the nucleus due to the diversity. For example, baculoviruses aim to hit NPC with their DNA-containing capsid and subsequently enter the nucleus undamaged, whereas the hepatitis B virus capsid crosses the NPC but disassociates at the nuclear side of the NPC. Though herpes simplex virus (HSV) and adenovirus both target NPC, they have their mechanism for delivering their genome into the nucleus. To the extreme, other DNA viruses, such as parvoviruses and human papillomaviruses, access the entry into the nucleus through an NPC-independent mechanism (Fay and Panté 2015). These viruses exploit the processes occurring in this organelle, like the use of enzyme polymerases and other replication factors. Conversely, viruses other than DNA viruses also make use of the nucleus to fulfill their replication needs and can provoke morphological alteration, thus affecting its normal functioning as like other organelles (Islinger et al. 2015).

2.2.6 Future Directions of Virus–Host–Pathogen Interactions

In recent research, bioinformatics plays a major role in helping the researchers out and reducing their burden in gathering information. In the field of host–pathogen interactions, different databases are available which focus on a virus–virus, virus–host, and host–host interaction, e.g., VirHostNet 2.0 (Guirimand et al. 2014; Navratil et al. 2008) network database. While other virus databases include VirusMentha (Calderone et al. 2013), ViralZone (Hulo et al. 2010), and ViTa (Hsu et al. 2006). Important databases for particular life threatening viruses also exist, such as the HIV-1, Interaction Databases and for human immunodeficiency virus (HIV Database Tools LANL) (http://www.ncbi.nlm.nih.gov/genome/viruses/

retroviruses/hiv-1/interactions/), for hepatitis C virus (HCV Database Tools LANL) (https://www.hiv.lanl.gov/content/index/), for hemorrhagic fever virus or Ebola (HFV Database Tools LANL) (https://hcv.lanl.gov/content/index/), and for human papilloma virus hpvPDB (Kumar et al. 2013). Also we are provided with other databases that specifically focus on specific virus interactions, e.g., Influenza Research Database (IRD) focused on influenza viruses and hosts. Much progress has been attained in recent years in the study of host–pathogen interactions, but still there are a number of imperative issues that must be dealt with. In broad terms, lack of thorough understanding of what is host–pathogen interaction. There are still uncertainities are exists, how the virulence mechanisms are differing with each other across the pathogens belonging to the same family. As same strains may target different hosts, there occur the significant differences in between in terms of how they attack, reproduce, and harm their hosts (Converse and Cox 2005).

2.3 Conclusion

To understand the infectious disease, treatment and prevention; host–pathogen interactions are very important. Though the chapter briefly elaborates the key points on how virus–host interacts with each other, still other problems occur every day with the evolution of the pathogens and the way how they interact with the host will continue to play a large role in expanding our knowledge in this field. Traditional alignment methods, structural analysis, and machine learning methods are all essential to improve our understanding of host–pathogen interactions, and how such interactions can be utilized to treat disease. The futuristic development of new drugs, vaccines, and other therapeutics measures highly depends on the information gained from investigating host–pathogen interactions.

References

Alizon S, Murall C, Bravo I (2017) Why human papillomavirus acute infections matter. Viruses 9:293

Baron EJ, Chang RS, Howard DH et al (1994) A short course. J Med Microbiol 1057(QR 46):M45

Beltran PM, Cook KC, Cristea IM (2017) Exploring and exploiting proteome organization during viral infection. J Virol 91:00268–00217

Birungi G, Chen SM, Loy BP et al (2010) Metabolomics approach for investigation of effects of dengue virus infection using the EA. hy926 cell line. J Proteome Res 9:6523–6534

Breitbart M (2012) Marine viruses- truth or dare. Annu Rev Mar Sci 4:425–448

Buchmeier MJ (2007) Arenaviridae the viruses and their replication. Field Virol 2:1792–1827

Bushman F, Lewinski M, Ciuffi A (2005) Genome-wide analysis of retroviral DNA integration. Nat Rev Microbiol 3:848

Calderone A, Castagnoli L, Cesareni G (2013) Mentha: A resource for browsing integrated protein-interaction networks. Nature Methods 10(8):690–691

Casadevall A, Pirofski LA (1999) Host-pathogen interactions-redefining the basic concepts of virulence and pathogenicity. Nat Immunol 67:3703–3713

Converse SE, Cox JS (2005) A protein secretion pathway critical for Mycobacterium tuberculosis virulence is conserved and functional in *Mycobacterium smegmatis*. J Bacteriol 187:1238–1245
Delgado T, Sanchez EL, Camarda R et al (2012) Global metabolic profiling of infection by an oncogenic virus: KSHV induces and requires lipogenesis for survival of latent infection. PLoS Pathog 8:1002866
Dogra P, Butner JD, Chuang YL et al (2019) Mathematical modeling in cancer nanomedicine-a review. Biomed Microdevices 21:40
Eagle H, Habel K (1956) The nutritional requirements for the propagation of poliomyelitis virus by the HeLa cell. J Exp Med 104:271
Eisenreich W, Rudel T, Heesemann J, Goebel W (2019) How viral and intracellular bacterial pathogens reprogram the metabolism of host cells to allow their intracellular replication. Front Cell Infect Microbiol 9:42
Fay N, Panté N (2015) Nuclear entry of DNA viruses. Front Microbiol 6:467
Fermin G, Tennant P (2018) Host–Virus interactions-battles between viruses and their hosts. Nucleic Acids 12:245
Figard PH, Levine AS (1966) Incorporation of labeled precursors into lipids of tumors induced by Rous sarcoma virus. Biochimica et Biophysica Acta (BBA)-Lipids and Lipid Metabolism 125(3):428–434
Fontaine KA, Camarda R, Lagunoff M (2014) Vaccinia virus requires glutamine but not glucose for efficient replication. J Virol 88:4366–4374
Fontaine KA, Sanchez EL, Camarda R et al (2015) Dengue virus induces and requires glycolysis for optimal replication. J Virol 89:2358–2366
Foo KY, Chee HY (2015) Interaction between flavivirus and cytoskeleton during virus replication. Biomed Res Int 2015:427814
Goff SP (2007) Retroviridae-the retroviruses and their replication. J Virol 2:1999–2069
Guirimand T, Delmotte S, Navratil V (2014) VirHostNet 2.0- surfing on the web of virus/host molecular interactions data. Nucleic Acids Res 43:583–587
Heaton NS, Randall G (2010) Dengue virus-induced autophagy regulates lipid metabolism. Cell Host Microbe 8:422–432
Hollenbaugh JA, Montero C, Schinazi RF et al (2016) Metabolic profiling during HIV-1 and HIV-2 infection of primary human monocyte-derived macrophages. Virology 491:106–114
Hsu PW, Lin LZ, Hsu SD (2006) ViTa- prediction of host micro RNAs targets on viruses. Nucleic Acids Res 35:381–385
Hulo C, De Castro E, Masson P et al (2010) ViralZone: a knowledge resource to understand virus diversity. Nucleic Acids Res 39:576–582
Imperiale MJ, Major EO (2007) Polyomaviruses. J Virol 5:2263–2298
Inoue T, Tsai B (2013) How viruses use the endoplasmic reticulum for entry, replication, and assembly. Cold Spring Harb Perspect Biol 5:013250
Islinger M, Godinho LF, Costello J et al (2015) The different facets of organelle interplay—an overview of organelle interactions. Front Cell Dev Biol 3:56
Kane M, Golovkina T (2010) Common threads in persistent viral infections. J Virol 84:4116–4123
Kanlaya R, Pattanakitsakul SN, Sinchaikul S et al (2009) Alterations in actin cytoskeletal assembly and junctional protein complexes in human endothelial cells induced by dengue virus infection and mimicry of leukocyte transendothelial migration. J Proteome Res 8:2551–2562
Kayser FH, Bienz KA, Eckert J, Zinkernagel RM (2004) Medical microbiology, 1st edn. Georg Thieme Verlag, Stuttgart., ISBN-13: 978-1588902450, p 724
Khan M, Syed GH, Kim SJ, Siddiqui A (2015) Mitochondrial dynamics and viral infections-a close nexus. Biochim Biophys Acta 1853:2822–2833
Knipe DM, Cliffe A (2008) Chromatin control of herpes simplex virus lytic and latent infection. Nat Rev Microbiol 6:211
Koonin EV, Dolja VV, Krupovic M (2015) Origins and evolution of viruses of eukaryotes-the ultimate modularity. J Virol 479:2–5

Krishnakumar V, Durairajan SS, Alagarasu K et al (2019) Recent updates on mouse models for human immunodeficiency, influenza, and dengue viral infections. Viruses 11:252
Kumar S, Lingaraja Jena SD et al (2013) hpvPDB- an online proteome reserve for human papillomavirus. Genomics Inform 11:289
Lafon M (2009) Latent viral infections of the nervous system-role of the host immune response. Rev Neurol 165:1039–1044
Levy HB, Baron S (1956) Some metabolic effects of poliomyelitis virus on tissue culture. Nature 178:1230
Lodish H, Berk A, Zipursky SL et al (2000) Molecular cell biology, 4th edn. W. H. Freeman, New York
Martelli GP, Russo M (1985) Virus-host relationships. In: The plant viruses. Springer, Boston, pp 163–205
Martinez-Martin N (2017) Technologies for proteome-wide discovery of extracellular host-pathogen interactions. J Immunol Res 2017:2197615
Mason S, Devincenzo J, Toovey S et al (2018) Comparison of antiviral resistance across acute and chronic viral infections. Antivir Res 158:103–112
Mercer J, Schelhaas M, Helenius A (2010) Virus entry by endocytosis. Annu Rev Biochem 79:803–833
Munger J, Bennett BD, Parikh A et al (2008) Systems-level metabolic flux profiling identifies fatty acid synthesis as a target for antiviral therapy. Nat Biotechnol 26:1179
Murray PR, Rosenthal KS, Pfaller MA (2015) Med Microbiol. Mosby, Philadelphia
Navratil V, de Chassey B, Meyniel L et al (2008) VirHostNet-a knowledge base for the management and the analysis of proteome-wide virus–host interaction networks. Nucleic Acids Res 37:661–668
Purdy JG (2019) Pathways to understanding virus-host metabolism interactions. Curr Clin Microbiol Rep 6:34–43
Rabinowitz JD, Purdy JG, Vastag L et al (2011) Metabolomics in drug target discovery. Cold Spring Harb Symp Quant Biol. 76:235–246
Rawls WE, Chan MA, Gee SR (1981) Mechanisms of persistence in arenavirus infections: a brief review. Can J Microbiol 27:568–574
Roitt IM (1974) Essential immunology, 2nd edn. Blackwell Scientific Publications, Oxford
Roossinck MJ (2010) Lifestyles of plant viruses. Philos Trans R Soc Lond Ser B Biol Sci 365:1899–1905
Sanchez EL, Lagunoff M (2015) Viral activation of cellular metabolism. Virology 479:609–618
Sanchez EL, Pulliam TH, Dimaio TA et al (2017) Glycolysis, glutaminolysis, and fatty acid synthesis are required for distinct stages of Kaposi's sarcoma-associated herpesvirus lytic replication. J Virol 91:02237–02216
Schelhaas M (2010) Come in and take your coat off–how host cells provide endocytosis for virus entry. Cell Microbiol 12:1378–1388
Southwood D, Ranganathan S (2016) Host-pathogen interactions. In: Encyclopedia of Bioinformatics and Computational Biology, vol 3. Elsevier, Amsterdam, pp 103–112
Stebbing J, Gazzard B (2003) Virus host interactions. Obstet Gynecol 5:103–106
Steck TL, Kaufman S, Bader JP (1968) Glycolysis in chick embryo cell cultures transformed by Rous sarcoma virus. Cancer Res 28:1611–1619
Stevceva L (2015) Latent viral infections in humans. In: Vaccines for latent viral infections. Bentham Science, Sharjah, p 3
Suttle CA (2007) Marine viruses—major players in the global ecosystem. Nat Rev Microbiol 5:801
Tennant P, Fermin G, Foster JE (2018) Viruses- molecular biology, host interactions, and applications to biotechnology. Academic Press, New York
Urban M, Pant R, Raghunath A et al (2014) The pathogen-host interactions database (PHI-base)-additions and future developments. J Nucleic Acids Res 43:D645–D655
Vastag L, Koyuncu E, Grady SL et al (2011) Divergent effects of human cytomegalovirus and herpes simplex virus-1 on cellular metabolism. PLoS Pathog 7:1002124

Venuta S, Rubin H (1973) Sugar transport in normal and Rous sarcoma virus-transformed chick-embryo fibroblasts. Proc Natl Acad Sci 70:653–657

Wang JL, Zhang JL, Chen W et al (2010) Roles of small GTPase Rac1 in the regulation of actin cytoskeleton during dengue virus infection. PLoS Negl Trop Dis 4:809

Weissenhorn W, Poudevigne E, Effantin G et al (2013) How to get out-ssRNA enveloped viruses and membrane fission. J Virol 3:159–167

A Physical Insight of Biofilms

3

Sarangam Majumdar and Sukla Pal

Contents

Abstract

Microbial communities are the assemblies of microorganisms which live collectively on a surface being encapsulated in a matrix of extracellular polymeric substances (EPS). This particular form of the lifestyle which is collective in nature and shares the common living space is known as biofilm—the most familiar form of bacterial growth. In the last few decades, a considerable number of researches have been dedicated towards the search for mechanisms behind the biofilm growth. In this chapter, we provide a description of the collective and social microbial behavior of biofilm with underlying mechanism for better understanding of the fundamental mechanisms behind the biofilm formation. Our aim is to describe the collective behavior of the microbial community with specific emphasis on the physical process of the biofilm development on the basis of statistical theory and hydrodynamics. The physical process of biofilm

S. Majumdar (✉)
Dipartimento di Ingegneria Scienze Informatiche e Matematica, Università degli Studi di L'Aquila, L'Aquila, Italy

S. Pal
Department of Physics, Centre for Quantum Science, Dodd-Walls Centre for Photonic and Quantum Technologies, University of Otago, Dunedin, New Zealand

B. Siddhardha et al. (eds.), *Model Organisms for Microbial Pathogenesis, Biofilm Formation and Antimicrobial Drug Discovery*,
https://doi.org/10.1007/978-981-15-1695-5_3

formation often encounters the phenomena of pattern formation and is influenced by the external parameters such as environmental stress conditions and nutrient limitation. Finally, we accumulate some recent significant observations on biofilms.

Keywords
Bacteria · Biofilms · Environmental stress · Pattern formation

3.1 Introduction

Bacteria are the oldest living microorganisms in the entire world, which cause different microbial infection diseases in the body such as infections of gastrointestinal tract, eye, dental implants, urogenital tract, lung tissue, and many more (Majumdar and Roy 2018b). Bacteria as a single cell is not harmful, but when behaves collectively, it becomes pathogenic to humans. We find many pathogenic bacteria (i.e., *Pseudomonas aeruginosa*, *Staphylococcus aureus*, *Escherichia coli*) in the different parts of the human body. The bacterial coordinated collective behavior (or bacterial communication process) is known as quorum sensing. It is a density-dependent complex biochemical phenomenon and is mediated by autoinducers (small chemical diffusible signaling molecules). Bacteria emit and receive autoinducers in order to mediate gene expression which governs the communication process. When the concentration of the autoinducers reaches (at high bacterial cell density) a threshold level, a qualitative change in the bacterial communication process occurs. The single cell bacterial behavior at low cell density switches to multicellular network when the cell density reaches the threshold level (see Fig. 3.1b). It activates the genes expression (Gray et al. 1994; Fuqua et al. 1996; Shapiro 1998; Williams et al. 2007; Majumdar and Mondal 2016; Majumdar and Pal 2016; Majumdar and Pal 2017b; Majumdar et al. 2017; Majumdar and Roy 2018a). Various types of quorum sensing circuits such as ComD/ComE, AgrC/AgrA, ComP/ComA, LuxI/LuxR, TraI/TraR, LasI/LasR–RhlI/RhlR, ExpI/ExpR–CarI/CarR, AhyI/AhyR, CepI/CepR, EsaI/EsaR, EagI/EagR, YenI/YenR, and YtbI/YtbR system have been observed in gram-positive and gram-negative bacteria (Miller and Bassler 2001). Besides intraspecies communication, bacteria can communicate with other bacterium using autoinducers-2. The production of autoinducers-2 is controlled by *lux*S gene (interspecies communication) (Majumdar et al. 2012; Majumdar and Pal 2017a). This cell-to-cell communication mechanism also regulates several other biological phenomena, which includes biofilm formation, antibiotic production, virulence, motility, sporulation, symbiosis, competence, and conjugation (Miller and Bassler 2001; Majumdar and Pal 2018).

"Biofilms are aggregates of microbial cells, which are embedded in extracellular polymeric substances (EPS) that are adherent to each other and a surface" (Vert et al. 2012). This bacterial lifestyle is completely different from free living bacteria

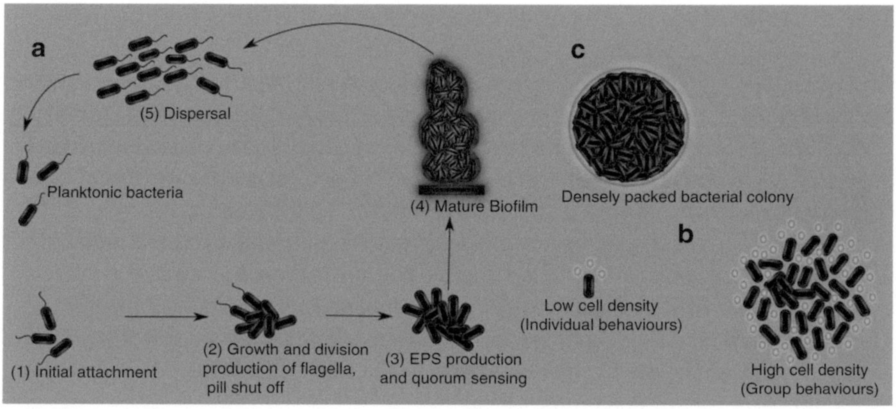

Fig. 3.1 (**a**) Schematic visualization of bacterial lifestyle and a multistage process of bacterial biofilm formation, (**b**) In cell to-cell communication process, a single bacterial cell secretes autoinducers, but we can't see any quorum sensing at low cell density. On the other hand, bacteria emit autoinducers, which are received by surrounding bacterial cells at high cell density (cell-to-cell communication occurs), (**c**) illustration of densely packed bacterial cells, which can be considered as coarse-grained system (Adapted from Majumdar and Roy 2019)

while having the emergent properties as well (see Fig. 3.1a). Fundamental emergent properties of the biofilm include social interaction, novel structures, patterns, and physical interaction (biofilm formation) (Flemming et al. 2016). Biofilms are allied with many human diseases, contamination of several medical devices, plant infections, animal infections, wastewater treatment, food technology, and many more. Biofilm consists of high cell number density (10^8–10^{11} cells per gram wet weight) and undergoes differentiation.

3.2 Single Particle Tracking Method

One can use microfluidic devices (Hohne et al. 2009), standard rheometers (Pavlovsky et al. 2013), atomic force microscopy (Aggarwal and Hozalski 2010), and combination of all to understand the physical properties of the biofilms. It is very difficult to understand several details of three-dimensional biofilm architecture in a different biofilm stage. We can study properties of reconstituted EPS (Cheong et al. 2009), motion of non-flagellated and flagellated bacteria (Rogers et al. 2008), and the effects of environment (Galy et al. 2012) using the single particle tracking. Charge interaction plays a significant role in mediating mobility inside the biofilms. We were observed that *E. coli* biofilms show height dependent charge density over time by microrheological concept and single particle tracking. We find some novel insight of the interconnecting micron scale channels (related to nutrient transfer) by the statistical analysis of bead trajectory. Moreover, this method provides a significant evidence of the biofilm structure (i.e., permeability) and its properties over time (Birjiniuk et al. 2014).

3.3 Relevance of Physical Properties in Biofilms Formation

Bacteria live inside the biofilm, which is considered as a bacterial response to external stress (osmolarity, shear, pH, starvation). Inside the biofilms, bacteria are densely packed and resistant to the hostile environment (i.e., antibiotics). Biofilm formation begins with collective cells and inert EPS matrix (as a response of external stress) (Branda et al. 2005; Monds and O'Toole 2009). This bacterial lifestyle is very stressful and heterogeneous because of nutrient limitation and limited metabolites diffusion in the matrix. In the initial stage of biofilm formation, bacteria conversation takes place, which is mediated by autoinducers (quorum sensing mechanism). This cell-to-cell communication process induces cell aggregation as well as biofilm formation. We study the chemoattraction (neglecting cell-to-cell communication) by the following set of equations:

$$\frac{\partial \rho_b}{\partial t} = -D_b \nabla^2 \rho_b + k \nabla \rho_f \tag{3.1}$$

$$\frac{\partial \rho_f}{\partial t} = -D_f \nabla^2 \rho_f - C_{\rho b \rho f} \tag{3.2}$$

where ρ_b is the local bacterial density, ρ_f is the local food density, k represents chemoattractive sensitivity, C represents bacterial consumption rate of food, D_b and D_f represent the effective diffusion coefficient of bacteria at a coarse-grained level (see Fig. 3.1c) and nutrients, respectively. The behavior of the solutions of the set of Eqs. (3.1) and (3.2) is thus highly influenced by the external environmental conditions. We notice that bacteria diffuse over time in the absence of chemotaxis (or consumption) and follow the food gradient (in presence of food). Thus, it is failed to describe the cell aggregation (Lambert et al. 2014).

Now, we consider the cell-to-cell communication process and c_t represents signaling field (quorum sensing) for chemotaxis-based bacterial movement within each micro-habitat patch (MHP) and follow the equation below:

$$T = -D.\nabla \psi_t + \chi.\psi_t.\nabla c_t \tag{3.3}$$

$$\partial_t \psi_t = \mathcal{G} + \nabla T \tag{3.4}$$

where ψ_t represents bacterial density, $\mathcal{G}$ is the local growth, and $\nabla \mathcal{T}$ is the chemotactic spatial coupling. The first term of the above Eq. (3.3) describes the balance between dispersive forces and other term shows chemotaxis-based aggregation (depend on density), at local scale (Lambert et al. 2014).

Later, Keller–Segel equations (for similar situation) are described as follows:

$$\frac{\partial \rho}{\partial t} = D_b \nabla^2 \rho - \nabla.[k \rho \nabla c] + \alpha \rho \tag{3.5}$$

$$\frac{\partial c}{\partial t} = D_c \nabla^2 c + \beta f \rho \tag{3.6}$$

$$\frac{\partial f}{\partial t} = D_f \nabla^2 f - \gamma \rho \tag{3.7}$$

where bacteria consume food at a rate γ, signaling molecule is produced by the food at a rate β, net growth rate is α, f is the concentration of the food, and c is the concentration of the chemoattractant molecule. We find instability in the bacterial population by perturbative analysis of the Keller–Segel equations. This happens at the early stage of the biofilm formation (Lambert et al. 2014).

The Keller–Segel equations give an instability at the very beginning of the biofilm formation and trigger bacterial population to crowd into small volume. We can investigate the whole developmental process from the single species of bacteria to biofilm inside micro-habitat (see detail in Lambert et al. 2014). Moreover, we find the viscoelastic properties of non-adherent biofilms, competition dynamics, and horizontal gene transfer between isolated biofilm populations (Lambert et al. 2014).

3.4 Branching Patterns

Branching patterns are common in nature, which is fascinating for both a mathematician and a biologist. We observe branching patterns in different cases, which includes circulatory system, lungs in vertebrates (Metzger et al. 2008), networks of slime molds (Tero et al. 2010), the branched colonies of corals (Helmuth et al. 1997), and roots of trees (de Smet and Jürgens 2007). Swarming is considered as population dispersal process. Bacterial colonies migrate over surface (agar plate) due to swarming. Bacterium (i.e., *P. aeruginosa*) swarming colonies can form a branching pattern (see Fig. 3.2). The dispersal ability is growing and hyperswarming has been noticed, which destroy the patterns. This experimental evidence reconstructs by computer simulation of SIMSWARM and ecological model of colony dispersal (Deng et al. 2014).

Now, we describe the mathematical formalism of branching patterns. Let us assume $L(d)$ be a spatial kernel which represents the colonization of patch at a distance d to the focal point. Using integro-difference technique, we can find out the dynamics of colonization of focal patch, N as

$$N_{t+1} = \int \int_{-\infty}^{+\infty} L\left(d_{x,y}\right) N_t\left(x,y\right) dxdy \tag{3.8}$$

The niche is located in coordinate (x, y) and $d_{x,\,y}$ represents the distance between focal niche. Shape and variance are separated by the below exponential equation for the kernel.

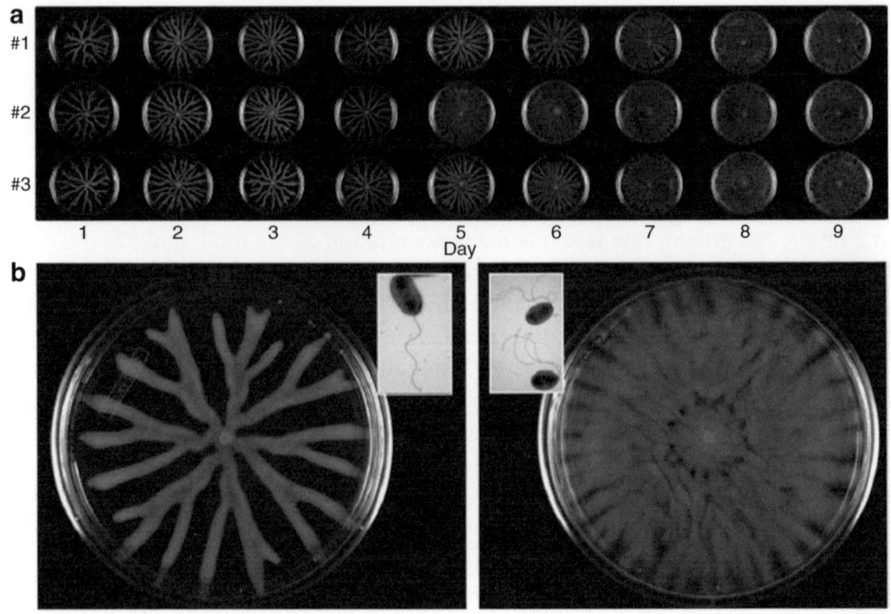

Fig. 3.2 (**a**) The experimental evidence of *P. aeruginosa* swarming colonies. It demonstrates branching patterns and how it is disrupted by hyperswarmers. (**b**) *P. aeruginosa* (wild-type) have a single flagellum (left panel). Lab experiment explores a highly dispersive hyperswarmer mutants (with multiple flagella) and the bacterial colonies without branching patterns (right panel) (Adapted from Deng et al. 2014)

$$L_{+}(d) \propto 2^{-\left(\frac{d}{d_1}\right)^{h_1}} \tag{3.9}$$

where h_1 sets the shape. The d_1 is the distance where the colonization rate $L_{+}(d)$ is half of its maximal. Let us consider the second process (counteracts dispersal) as follows:

$$L_{-}(d) \propto 2^{-\left(\frac{d}{d_2}\right)^{h_2}} \tag{3.10}$$

This process implies a negative effect (i.e., crowding, repulsion). The resulting spatial kernel can be represented by

$$f(d) = b \times 2^{-\left(\frac{d}{d_1}\right)^{h_1}} - 2^{-\left(\frac{d}{d_2}\right)^{h2}} \tag{3.11}$$

where b scales the strength of the positive process relative to the negative. One can apply cellular automata model with this spatial kernel (SIMSWARM). Then the simulation generates a range of colony shapes as shown in Fig. 3.3a. If the shape factor is high, the colony becomes round. On the other hand, branched colony appears at low shape factor. In particular, the colony branching is possible (with

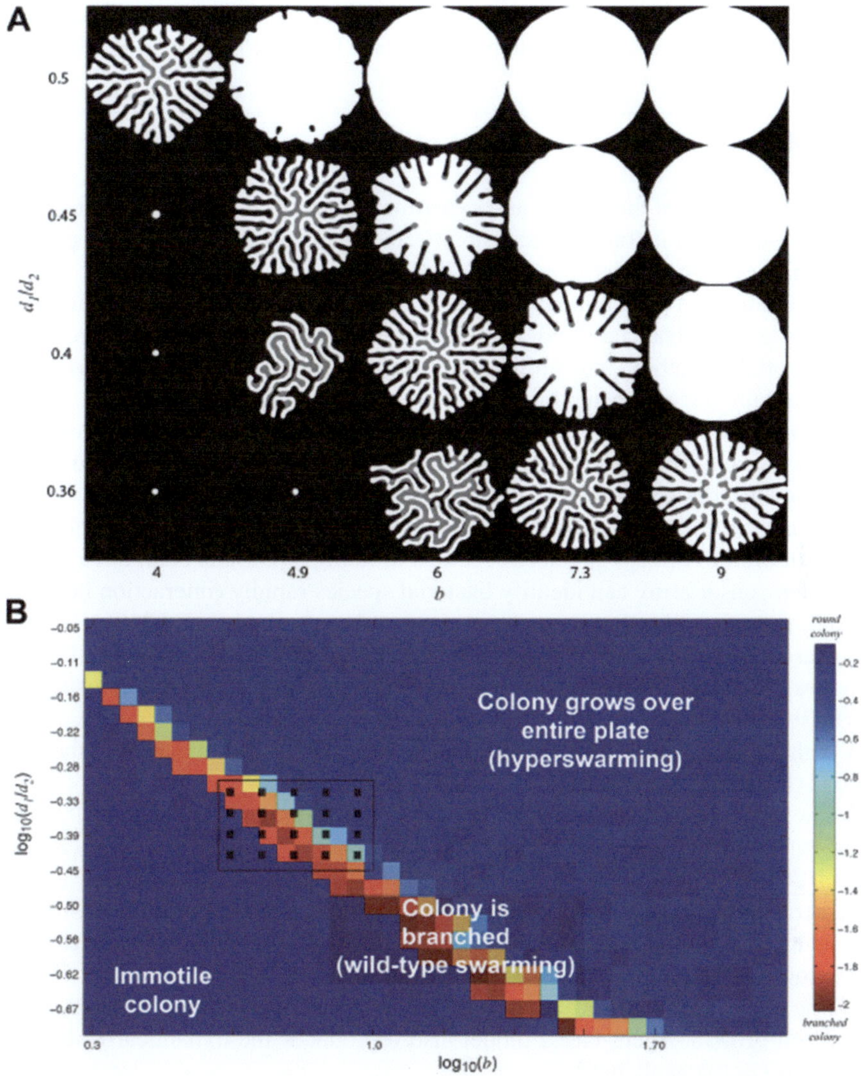

Fig. 3.3 Bacterial colony morphology induces by spatial kernel. (**A**) The colony shapes aregenerated by SIMSWARM. Parameter b (strength of the positive process) and $\frac{d_1}{d_2}$ (scale ofpositive interaction) varies across row and column respectively. The white color shows the bacteria occupied niches (currently) and grey color shows niches that were at some point occupied by bacteria. (**B**) All stimulation is conducted with $h_1 = h_2 = \infty$. Simulations shows that for each value of $\frac{d_1}{d_2}$ branching occurs in narrow windows of b (Adapted from Deng et al. 2014)

$h_1 = h_2 = \infty$) in a very narrow region of parameter space. One can find this parameter region by plotting circularity of colony shapes (see Fig. 3.2b) (Deng et al. 2014).

3.5 Recent Observations on Biofilms

- The combination of theoretical and experimental results describes that the effective cell–cell interaction potential captures the emergent architecture and growth dynamics of biofilm (Hartmann et al. 2019).
- The printable and programmable *B. subtilis* biofilms are made, which represents a completely advanced kind of living functional material. They are self-regenerating, tunable, and multifunctional. This living functional material has application in medicine and biotechnology (Huang et al. 2019).
- *P. aeruginosa* biofilm formation on the cornea is responsible for keratitis. *P. aeruginosa* biofilm infection (corneal) is limited by the neutrophilic recruitment (Thanabalasuriar et al. 2019).
- In vitro model can be useful for fundamental studies on electrochemical treatment of biofilm infections (electroceutical dressings) (Dusane et al. 2019).
- Polymer sensor array can identify bacterial species rapidly (interaction between biofilm matrix and polymer sensor elements) (Ngernpimai et al. 2019).
- Bacterial evolution and ecology are shaping by the bacteriophages. Phages induce biofilms. Biofilm formation is considered as the bacterial defense strategy (Hansen et al. 2019).
- A high intensity focused ultrasound can be used to distort the matrix of the biofilms (Bharatula et al. 2019).
- A new technological advancement (lattice light sheet microscopy) is used to find out details of biofilms development and gives crucial information on it (Zhang et al. 2019).
- Bacterial biofilms are active matter systems, which have large-scale ordered structures. Biofilms have mechanical instabilities, intrinsic length scales, and topological defects (Yaman et al. 2019).
- Exotic mechanical state is observed by the interplay between cell verticalization and cell growth. Agent-based model also reproduces the dynamical feature of biofilms (Beroz et al. 2018).
- Distant biofilm communities couple through ion-channel mediated electrical signaling mechanism. A synchronization is observed in the growth dynamics of the couple biofilms. Moreover, nutrient competition between biofilms is resolved by time-sharing behavior (Liu et al. 2017).

References

Aggarwal S, Hozalski RM (2010) Determination of biofilm mechanical properties from tensile tests performed using a micro-cantilever method. Biofouling 26(4):479–486

Beroz F, Yan J, Meir Y, Sabass B, Stone HA, Bassler BL, Wingreen NS (2018) Verticalization of bacterial biofilms. Nat Phys 14(9):954

Bharatula LD, Rice S, Marsili E, Kwan JJ (2019) Sounding out bacteria: microstructural effects of therapeutic ultrasound on bacterial biofilms. J Acoust Soc Am 145(3):1894

Birjiniuk A, Billings N, Nance E, Hanes J, Ribbeck K, Doyle PS (2014) Single particle tracking reveals spatial and dynamic organization of the Escherichia coli biofilm matrix. New J Phys 16(8):085014

Branda SS, Vik Å, Friedman L, Kolter R (2005) Biofilms: the matrix revisited. Trends Microbiol 13(1):20–26

Cheong FC, Duarte S, Lee SH, Grier DG (2009) Holographic microrheology of polysaccharides from *Streptococcus mutans* biofilms. Rheol Acta 48(1):109–115

de Smet I, Jürgens G (2007) Patterning the axis in plants–auxin in control. Curr Opin Genet Dev 17(4):337–343

Deng P, de Vargas Roditi L, van Ditmarsch D, Xavier JB (2014) The ecological basis of morphogenesis: branching patterns in swarming colonies of bacteria. New J Phys 16(1):015006

Dusane DH, Lochab V, Jones T, Peters CW, Sindeldecker D, Das A, Roy S, Sen CK, Subramaniam VV, Wozniak DJ, Prakash S, Stoodly P (2019) Electroceutical treatment of *Pseudomonas aeruginosa* biofilms. Sci Rep 9(1):2008

Flemming HC, Wingender J, Szewzyk U, Steinberg P, Rice SA, Kjelleberg S (2016) Biofilms: an emergent form of bacterial life. Nat Rev Microbiol 14(9):563

Fuqua C, Winans SC, Greenberg EP (1996) Census and consensus in bacterial ecosystems: the LuxRLuxI family of quorum-sensing transcriptional regulators. Ann Rev Microbiol 50:727–751

Galy O, Latour-Lambert P, Zrelli K, Ghigo JM, Beloin C, Henry N (2012) Mapping of bacterial biofilm local mechanics by magnetic microparticle actuation. Biophys J 103(6):1400–1408

Gray KM, Passador L, Iglewski BH, Greenberg EP (1994) Interchangeability and specificity of components from the quorum-sensing regulatory systems of *Vibrio fischeri* and *Pseudomonas aeruginosa*. J Bacteriol 176:3076–3080

Hansen MF, Svenningsen SL, Røder HL, Middelboe M, Burmølle M (2019) Big impact of the tiny: bacteriophage–Bacteria interactions in biofilms. Trends Microbiol. https://doi.org/10.1016/j.tim.2019.04.006

Hartmann R, Singh PK, Pearce P, Mok R, Song B, Díaz-Pascual F, Dunken J, Drescher K (2019) Emergence of three-dimensional order and structure in growing biofilms. Nat Phys 15(3):251

Helmuth BS, Sebens KP, Daniel TL (1997) Morphological variation in coral aggregations: branch spacing and mass flux to coral tissues. J Exp Mar Biol Ecol 209(1–2):233–259

Hohne DN, Younger JG, Solomon MJ (2009) Flexible microfluidic device for mechanical property characterization of soft viscoelastic solids such as bacterial biofilms. Langmuir 25(13):7743–7751

Huang J, Liu S, Zhang C, Wang X, Pu J, Ba F, Xue S, Ye H, Zhao T, Li K, Wang Y, Zhang J, Wang L, Fan C, Lu TK, Zhong C (2019) Programmable and printable Bacillus subtilis biofilms as engineered living materials. Nat Chem Biol 15(1):34

Lambert G, Bergman A, Zhang Q, Bortz D, Austin R (2014) Physics of biofilms: the initial stages of biofilm formation and dynamics. New J Phys 16(4):045005

Liu J, Martinez-Corral R, Prindle A, Dong-yeon DL, Larkin J, Gabalda-Sagarra M, Garcia-Ojalvo J, Suel GM (2017) Coupling between distant biofilms and emergence of nutrient time-sharing. Science 356(6338):638–642

Majumdar S, Mondal S (2016) Conversation game: talking bacteria. J Cell Commun Signal 10(4):331–335

Majumdar S, Pal S (2016) Quorum sensing: a quantum perspective. J Cell Commun Signal 10(3):173–175

Majumdar S, Pal S (2017a) Cross-species communication in bacterial world. J Cell Commun Signal 11(2):187–190

Majumdar S, Pal S (2017b) Bacterial intelligence: imitation games, time-sharing and long - range quantum coherence. J Cell Commun Signal 11(3):281–284

Majumdar S, Pal S (2018) Information transmission in microbial and fungal communication: from classical to quantum. J Cell Commun Signal 12(2):491–502

Majumdar S, Roy S (2018a) Relevance of quantum mechanics in bacterial communication. NeuroQuantol 16(3):1–6

Majumdar S, Roy S (2018b) Mathematical model of quorum sensing and biofilm. In: Implication of quorum sensing system in biofilm formation and virulence. Springer, Singapore, pp 351–368

Majumdar S, Roy S (2019) Bacterial intelligence. Acta Sci Neurol 2(4):7–9

Majumdar S, Datta S, Roy S (2012) Mathematical modelling of quorum sensing and bioluminescence in bacteria. Int J Adv Appl Sci 1(3):139–146

Majumdar S, Roy S Llinas R (2017) Bacterial conversations and pattern formation. bioRxiv 098053. https://doi.org/10.1101/098053

Metzger RJ, Klein OD, Martin GR, Krasnow MA (2008) The branching programme of mouse lung development. Nature 453(7196):745

Miller MB, Bassler BL (2001) Quorum sensing in bacteria. Ann Rev Microbiol 55(1):165–199

Monds RD, O'Toole GA (2009) The developmental model of microbial biofilms: ten years of a paradigm up for review. Trends Microbiol 17(2):73–87

Ngernpimai S, Geng Y, Makabenta JM, Landis RF, Keshri P, Gupta A, Li CH, Chompoosor A, Rotello VM (2019) Rapid identification of biofilms using a robust multichannel polymer sensor array. ACS Appl Mater Interface 11(12):11202–11208

Pavlovsky L, Younger JG, Solomon MJ (2013) In situ rheology of *Staphylococcus epidermidis* bacterial biofilms. Soft Matter 9(1):122–131

Rogers SS, van der Walle C, Waigh TA (2008) Microrheology of bacterial biofilms in vitro: *Staphylococcus aureus* and *Pseudomonas aeruginosa*. Langmuir 24(23):13549–13555

Shapiro JA (1998) Thinking about bacterial populations as multicellular organisms. Ann Rev Microbiol 52:81–104

Tero A, Takagi S, Saigusa T, Ito K, Bebber DP, Fricker MD, Yumiki K, Kobayashi R, Nakagaki T (2010) Rules for biologically inspired adaptive network design. Science 327(5964):439–442

Thanabalasuriar A, Scott BNV, Peiscler M, Willson ME, Zeng Z, Warrener P, Keller AE, Surewaard BGJ, Dozier EA, Korhonen JT, Cheng L, Gadjeva M, Stover CK, DiGiandomenico A, Kubes P (2019) Neutrophil extracellular traps confine *Pseudomonas aeruginosa* ocular biofilms and restrict brain invasion. Cell Host Microbe 25(4):526–536

Vert M, Doi Y, Hellwich KH, Hess M, Hodge P, Kubisa P, Rinaudo M, Schue F (2012) Terminology for biorelated polymers and applications (IUPAC recommendations 2012). Pure Appl Chem 84(2):377–410

Williams P, Winzer K, Chan WC, Camara M (2007) Look who's talking communication and quorum sensing in the bacterial world. Philos Trans R Soc Lond Ser B Biol Sci 362(1483):1119–1134

Yaman YI, Demir E, Vetter R, Kocabas A (2019) Emergence of active nematics in chaining bacterial biofilms. Nat Commun 10(1):2285

Zhang M, Zhang J, Wang J, Achimovich AM, Aziz AA, Corbitt J, Acton ST, Gahlmann A (2019) 3D imaging of single cells in bacterial biofilms using lattice light-sheet microscopy. Biophys J 116(3):25a

A Review on Microbial Pathogenesis and Host Response

4

Himani Meena, Asad Syed, and Busi Siddhardha

Contents

Abstract

Although the prominent convalesce in understanding of microbial pathogenicity and host immune response has been acknowledged still many obstacles has to face for proper solution for the pathogenicity problem. In the present chapter, we have tried to conclude every possible mechanism behind pathogenicity as well as the host immune responses. The relationship between microbe and host is subdivided into mutualism, commensalism, and parasitism. Microbial interaction with host cell is regulated by different routes such as adhesion, internalization,

H. Meena · B. Siddhardha (✉)
Department of Microbiology, School of Life Sciences, Pondicherry University, Puducherry, India

A. Syed
Botany and Microbiology Department, College of Science, King Saud University, Riyadh, Saudi Arabia

B. Siddhardha et al. (eds.), *Model Organisms for Microbial Pathogenesis, Biofilm Formation and Antimicrobial Drug Discovery*,
https://doi.org/10.1007/978-981-15-1695-5_4

colonization, and secretion of toxic molecules and enables bacterial cell development and growth within the host tissue. Host immune system is composed of innate and adaptive immunity. Innate immunity system involves macrophages, neutrophils, and monocytes whereas adaptive immune system provides a detailed mechanism of defense through B lymphocytes, dendritic cells, complement system, and T lymphocytes ($CD4^+$ and $CD8^+$) cells. Therefore, understanding of microbial pathogenesis and host response will enable the scientific community to come up with strategies to fight with dangerous pathogens without any reluctance towards new drug design and development.

Keywords
Microbial infection · Pathogenicity · Cellular immunity · T lymphocytes · Disease

4.1 Introduction

Microbes are the most abundant survivors on the earth with a variety of tactics to subsidize their existence on non-living and living objects. Microbial interaction with the host cell is important for establishment of the infection that lead to infectious disease and eventually death of the affected patients. World Health Organization has declared microbial infection as the second most cause of the deaths worldwide. Microbial infection can be divided into acute and persistent forms; acute infection can be cured within days of infection whereas persistent infection may have a long term effect on host system (Thakur et al. 2019). Outbreak of the pre-occurred disease with the time and antibiotic resistance towards existing drugs has increased the stress in the scientific community in urge to excavate the underlying mechanism of microbial pathogenesis. Exploring microbial pathogenesis is essential to understand the crucial course of microbial adhesion, invasion, and damage to host system. Deciphering the interaction between microbial pathogenic determinants and the factors associated with host defense system has put light on different aspects of treatment for the disease and recovery from the damage caused by the previous medication process (Solanki et al. 2018). This chapter highlights the relationship between host–pathogen, routes of infection and progression, host immune response and host mediated pathogenesis.

4.1.1 Host–Microbe Relationship

Skin is the outer most layer of the host defense system which provides early exposure to microbes and helps immune system to block the infection at initial level of defense. There are three kinds of symbiosis that occurs between pathogen and host cell depending on their constructive arrangement for survival in the host system.

4.1.1.1 Mutualism

Mutualism refers to the relationship between two individual where the host provides nutrition and habitat for the endurance and microbe supplies energy and nutrient for host development and growth. Gut microbiota shares specific relationship with the host body system. The environment and host metabolic routine has an impact on the microbiome survival in the host; a slight variation in the environment or host metabolism can cause severe changes in microbial behavior that may led to different aspects of illness (Visconti et al. 2019; Wu and Wang 2019). Gut microbiota has numerous bacterial species due to its nutrition-rich environment and can be divided into two groups, dominant (*Bacteroidetes*, *Actinobacteria*, *Firmicutes*) and facultative species (*Enterococci*, *Lactobacilli*, *Enterobacteriaceae,* and *Streptococci*). *Bifidobacteria* is a gram-positive, anaerobic bacteria with different beneficial traits such as gut immunostimulation, immunomodulation, reduced lactose intolerance, and cholesterol level in serum (Candela et al. 2008). *Bifidobacterium longum* is an intrinsic bacterium that dominates the gut intestinal tract (GIT) microbiota. Enzymes secreted by *B. longum*, i.e., glycosyl hydrolases and phosphotransferase are linked with carbohydrate utilization and offers high metabolic rate for food consumption and maintain nutritional balance throughout the food cycle (O'Callaghan and van Sinderen 2016). Another well-known microbe, *Lactobacillus* is a part of complex gut microbiome ecosystem but only 4% of the world population harbors *Lactobacillus reuteri*. Apart from playing important role in the digestive process, *L. reuteri* also prevent chances of diarrhea, prevent IgE-associated eczema and sensitization, and boost production of CD4 lymphocytes (Walter et al. 2018).

4.1.1.2 Commensalism

Commensalism represents the relationship between microbe and host where harmless parasite resides in host body without affecting host health. Commensal microbe can be present on skin (epidermis layer) and mucus layer of respiratory and gastrointestinal tract, for example, *Streptococcus pyogenes* and *Lactobacilli*. *S. pyogenes* is a commonly found bacteria on skin flora and digestive tract that maintain acidic pH level and inhibit other pathogenic strains growth. Under stress condition, *S. pyogenes* move from the normal flora to the epiglottis region and cause swelling and inflammation that lead to sore throat problem (Eloe-Fadrosh and Rasko 2013). Commensal *Lactobacilli* species (*L. iners*, *L. crispatus*, and *L. jensenii*) dwell in the vaginal system, maintain low pH condition, and release toxic compounds for extrusion of harmful microbes to prevent women from urinary tract infection (*Pseudomonas aeruginosa*, *Escherichia coli*, etc.), sexually transmitted disease (STD), and bacterial vaginosis (Antikainen et al. 2007; Eloe-Fadrosh and Rasko 2013).

4.1.1.3 Parasitism

Human provides microbe a favorable environment, protein- and nutrient-rich conditions for proper growth and development. Parasitism can be defined as co-evolution of microbe and host where microbe is totally dependent on host body for its survival. Microbial parasites, *Leishmania*, *Schistosoma*, *Toxoplasma,* and

Trypanosoma, are some of the rare disease causing agents which exist in human system and consume food and energy through host metabolic pathways (Tedla et al. 2019). Parasites do not harm the host system as it is required for their own survival. Two main protozoan species, *Schistosoma mansoni* (Philippsen and DeMarco 2019) and *Leishmania major* are able to develop cutaneous leishmaniasis and schistosomiasis. Establishment of the infection is mediated by microbial penetration of dermis layer through secretion of enzyme, i.e., protein disulfide isomerases (PDI) which dissolve extracellular matrix (ECM) in the skin. Matrycryptin endostatin released through PDI and ECM interaction mediates binding of *Leishmania* promastigotes to host cell and promotes their growth (Miele et al. 2019; Philippsen and DeMarco 2019).

4.2 Routes of Infection and Progression

Microbial infection can be transmitted through diverse routes based on their establishment of the disease in the host. Microbial adhesion on outer surface, invasion, colonization, and toxin secretion are some of the well-known routes based on the available literatures (Fig. 4.1).

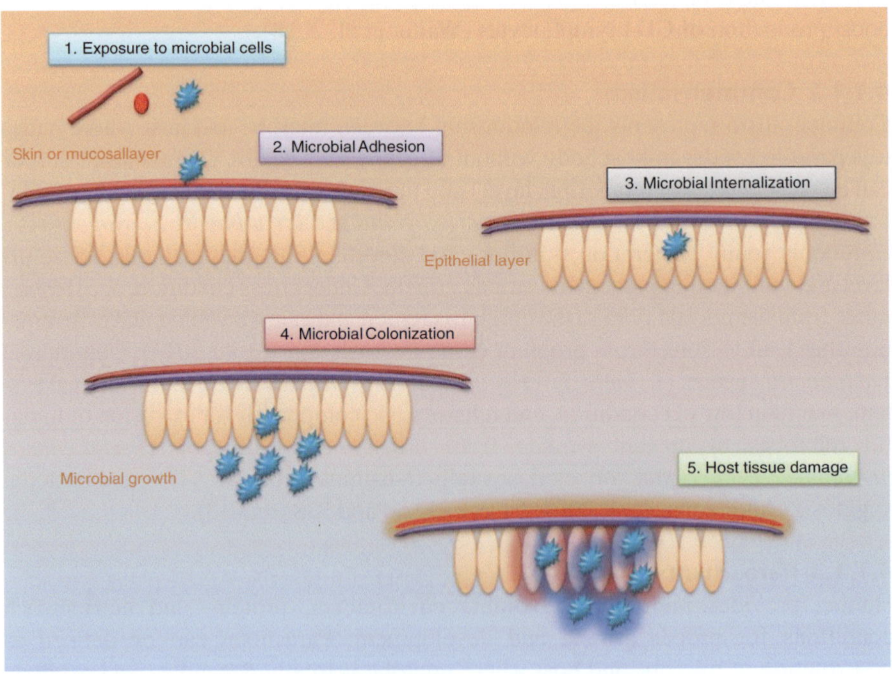

Fig. 4.1 Graphical representation of microbial infection to the host cells via different routes of pathogenesis

4.2.1 Adhesion

Microbial adhesion on the host cell membrane is the most traditional mode of establishing an infection in the host. Microbial surface protein recognizes cell surface protein that initiates microbial infection, understands different signaling pathways involved in pathogenesis, helps to achieve persistent infection, and enhances microbial invasion (Boyle and Finlay 2003). Bacterial adherence is mediated by lectin projectors that specifically bind to glycan molecules present on the host cell surface. Bacterial lectins can modify with changing glycome dynamics and environmental conditions. Lectins produced from different microorganisms are responsible for different assignment, for example, Lectin A or B from *P. aeruginosa* help to establish biofilm or botulinum from *Clostridium botulinum* which act as bacteriocins (Moonens and Remaut 2017). Surface anchored molecules are reported for their adherence towards host tissues such as in *Campylobacter flaA*, *ciaB*, *cadF*, and *pldA* and are identified to be involved in adherence and invasion of the bacteria in epithelial cells. Flagellin encoded by *flaA* gene participates in the cell adherence and fibronectin-associated surface protein is encoded by *cadF* gene. Another outer membrane phospholipase synthesizer is reported in *Campylobacter* for cell invasion (Ganan et al. 2010). Seven cell adhesion genes *ompA*, *inv*, *sip*, *aut*, *hly*, *fliC*, and *cpa* were distinguished among *Cronobacter sakazakii* and *C. malonaticus* species based on their role in adhesion and invasion process (Holy et al. 2019). Enteropathogenic *E. coli* (EPEC) one of the causative agent of diarrhea expressed *pilS* gene during adherence to intestinal epithelium and microvilli effacement and is responsible for histopathological attaching and effacing" (A/E) lesion (Garcia et al. 2019). Another example of cell adhesion protein is extracellular adherence protein (Eap) (Palankar et al. 2018) from *Staphylococcus aureus*.

4.2.2 Internalization

Bacterial invasion or internalization facilitates microbial entry to host cells and allows to multiply which results in bacterial pathogenesis. Based on the genetic and physiological features, gram-negative and gram-positive bacteria possess different structural arrangement on the cell surface. Adhesion and internalization in gram-negative bacteria is regulated through membrane channel which provides translocation of cellular molecule to host cell. Bacterial invasion system can be categorized in two specific types based on the intrusion mechanism used by the pathogen, i.e., trigger and zipper mechanism. Type III secretion system (T3SS) allows bacterial cells to invade host cell using trigger mechanism. In the process, bacterial cell accumulates their effectors into cytoplasm, releases effector protein to host cell, rearranges the host cell membrane, and allows invasion without encountering the outer membrane defense system (for example, type III secretory systems of *P. aeruginosa* and *Salmonella*) (Bohn et al. 2019, Kochut and Dersch 2012). Additional bacterial species which follows this invasion technique to infiltrate the host cells are *Salmonella*, *Shigella*, *Yersinia*, *E. coli*, and opportunistic pathogen *Citrobacter*

(Hu and Wai 2017). Human α-defensin protein is responsible for intestinal homeostasis and innate immunity. Then, it channelizes the entry of *Shigella* for adhesion, internalization, and hence enhances the pathogenicity (Xu et al. 2018). *Yersinia pestis*, *Y. enterocolitica*, and *Y. pseudotuberculosis* mediate effector translocation via cell surface appendages such as invasin (Inv), *Yersinia* adhesion A (YadA), and attachment and invasion locus (Ail). This combination of adhesion and invasion protein confers host cell interaction through leukocytes directing towards myeloid cells. On the other hand, gram-positive bacteria possess invasion molecules which are attached at C-terminus with LPXTG motif to interact with host cell receptors reservoir. Zipper mechanism involves ligand–protein binding approach. The zipper mechanism could be categorized into three different phases, first phase: interaction between bacterial cell-anchored ligand with specific cell receptor on host cell surface, second phase: activation of signaling pathways, third phase: involvement of membrane bound vacuole that help bacterial cell to be internalized by the host cell (Kochut and Dersch 2012). Gram-positive bacteria, *Listeria monocytogenes,* causative agent for listeriosis, exploit zipper mechanism involving two surface invasion factors, InlA and InlB and pore forming enzyme listeriolysin O which help bacterial cells to infect epithelial cell or blood–cerebrospinal fluid barrier and allow to proliferate in the host cytoplasm by promoting the phagolysis (Phelps et al. 2018; Grundler et al. 2013). Another bacterial surface molecule with C-terminus anchored PfbA (plasmin and fibronectin-binding protein A) via LPKTG motif from *Streptococcus pneumoniae* identifies the host cell receptors such as fibronectin, plasminogen, and serum albumin and modulates infection process (Beulin et al. 2014).

4.2.3 Colonization

Microbial colonization is defined as establishment of microorganisms on host tissue and they eventually overcome the host defense system for their proliferation and growth. Bacterial colonization can be categorized in three phases, adherence to the host cell surface, invading the host cell and suppress the host immune system for their proper reproduction or multiplication (Ribet and Cossart 2015). Failure in cell surface recognition patterns and cellular signaling pathways in host during adhesion and invasion process might have negative impact on bacterial metabolic rate ultimately leading to lack of production of virulence factors and debilitated colonization (Stones and Krachler 2016). Plaque formation on tooth enamel is one of the best examples for bacterial colonization. Bacterium, *Porphyromonas gingivalis* regulates microbial succession in the dental health where it allows growth of other bacterial species such as *S. oralis*, *S. gordonii*, *Actinomyces oris*, *Veillonella* sp., *Fusobacterium nucleatum,* and *Aggregatibacter actinomycetemcomitans* and forms biofilm on enamel (Periasamy and Kolenbrander 2009). *Bifidobacterium* is the primary bacterial species transmitted through mother's placenta to infant during delivery. It is mostly present in the epithelial layer of skin, urinary tract, or gastrointestinal tract. Several factors are available that favors colonization of *B. longum* such as availability for carbohydrate consumption via glycosyl hydrolases and phosphotransferase, adaptation towards bile salt concentration, and abundance of adhesins and pili for bacterial entrapment on mucus layer (Gonzalez-Rodriguez et al. 2013; Grimm et al. 2014; Zhang et al. 2019a). Gut microbiome forms microbial network which maintains specific niche for microbial succession and provides nutritional support (Cui et al. 2019; Suskind et al. 2019). Gram-negative bacteria, *E. coli* colonizes epithelial layer of skin, respiratory tract, and soft tissue via penetration. *E. coli* is a causative agent for bacterial infection in urinary tract, meningitis, and bacteremia (Alfaro-Viquez et al. 2019). *Enterococcus faecalis* and *E. faecium* occupies human gastrointestinal tract and can have transitional state changes between beneficial microorganism to pathogenic strain based on the environment conditions (Banla et al. 2019). *P. aeruginosa* causes infection and injury in the epithelial layer of cornea, further followed by microbial keratitis. Microbial keratitis in another opprobrium of biofilm and colonization on the epithelial cells (Wu et al. 2019). Gram-positive bacteria, *S. aureus* is a well-known colonizer that resides within epithelial layer of nasal cavity via zipper mechanism. Bacterial cell surface adhesins, iron-regulated surface determinant A and clumping factor B (ClfB) interacts with loricrin, cytokeratin 10 (K10), involucrin, filaggrin, and small proline-rich proteins to establish an intense infection (Mulcahy and McLoughlin 2016).

4.2.4 Microbial Secretion System

P. aeruginosa produces a variety of toxins during infection but redox-active phenazine compounds (pyocyanin and pyoverdine), cell-lytic enzymes (protease, elastase, chitinase), exotoxin A, and exoenzymes are main virulence factors secreted during critical infection phase. Type 3 secretory system is a well described system known as needle complex in *P. aeruginosa* to passage effectors protein to the host cell through needle-shaped pipeline (Fig. 4.2). First, formation of the basal compartment on the bacterial surface, followed by accumulation of effector proteins in cytoplasm and formation of needle-shaped bridge outward by initiating configuration changes on the host cell surface to form a pore. In the end, energy driven translocation of the effector proteins from the bacterial cell to host cell cytoplasm (Lombardi et al. 2019). T3SS system is regulated by an operon known as exsCEBA operon whose expression is controlled by ExsA transcriptional regulator to facilitate the translocation process. ExsD acts as an anti-activator and ExsC serves as anti-antiactivator synchronized under the influence of calcium concentration in the cellular medium. Among four effector proteins, ExoU is known for phospholipase activity and possess cytotoxic effect which causes immunosuppression and activates inflammatory cascades. ExoY harbors an adenylate cyclase activity and it maintains cyclic adenosine monophosphate (cAMP) level in the cell. Bifunctional effector proteins, ExoS and ExoT, are responsible for rounding of cell and have detrimental effect on wound healing process (Dela Ahator and Zhang 2019; Sawa 2014). Pathogenic bacterium, *Vibrio cholera* spread through contaminated water consumption and produce cholera toxin, a virulence factor determined to have great impact on pathogenicity. Pili-mediated bacterial cell adhesion to the microvilli of small intestinal mucus layer leads to the release of cholera toxin from *V. cholera* at

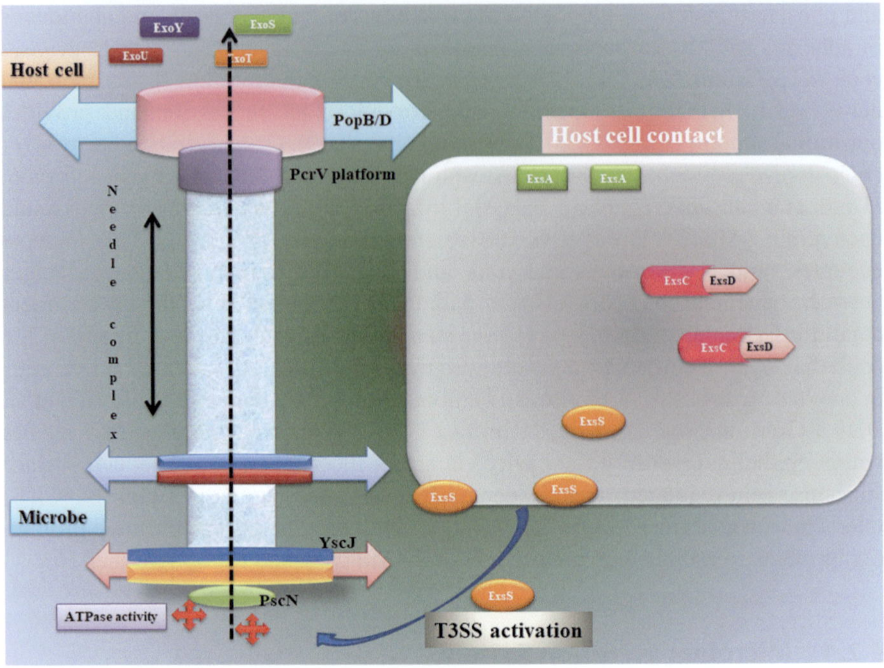

Fig. 4.2 Type III secretion system (T3SS) mediated pathogenicity in gram-negative bacteria, *Pseudomonas aeruginosa* that requires activation via exsACDE operon. During interaction with host cell, ExsC binds with ExsD to form complex and release bound ExsE for activation of T3SS

host body temperature. Cholera toxin is mucolytic by nature that disrupts the mucus layer integrity and controls bacterial penetration to the host cell. Cholera toxin is a complex structure composed of single A subunit and five B subunits. Both subunits are assigned with specific functions such as A subunit possess adenylate cyclase activity while B subunit acts as binding factor. Binding subunits binds to host surface receptor of mucus layer, called as GM1 ganglioside which forms complex to initiate phagocytosis. The cAMP concentration reaches to a threshold level and mediates release of internal Na^+, H_2O, and K^+ through increased permeability of chloride channel. Accumulation of Na^+, H_2O, and K^+ in the cytoplasm intensifies dehydration level in the infected host and causes severe diarrhea (Tirumale and Tessy 2018).

4.3 Microbial Pathogenesis and Host Immune Response

4.3.1 Innate Immune Response

The skin layer is a first line of defense as it covers the entire body as a physical barrier against environmental factors and regulates secretion of various immune response under stress conditions. It is composed of three layers, epidermis, dermis, and hypodermis, which maintain tissue homeostasis and provoke innate defense system during infection. The outermost layer of the epidermis region is made up of stratum corneum responsible for keratinization. Keratinocytes are capable of converting themselves into corneocytes by replacing outer membrane with cornified envelope under special conditions. Corneocytes retrieve water from the external environment and maintain membrane integrity by providing moisture. Corneocytes prevents bacterial colonization by supplanting themselves via desquamation process within 2–4 weeks (Egawa and Kabashima 2018). Keratinocytes are present at epidermis layer that secretes immunostimulators like cytokines, chemokines, and Toll-like receptors and are able to recruit innate immune cells which provide temporary tolerance towards pathogens (Guttman-Yassky et al. 2019) (Fig. 4.3). Mucosal layer is primary defense line for internal organs, gives protection against pathogenic agents. It possesses typical immune cells, i.e., dendritic cell and macrophages which trigger immune responses. Immune cells identify microbial cells using pattern recognition receptors (PPRs) which determine presence of microbes via microorganism-associated molecular patterns (MAMPs). PPRs can be categorized into three different classes: surface cell-anchored molecules (CD14, CD209, CLEC4E, TLR4) and intracellular-associated (NAIP, IFIH1, DAI) and soluble (PTX4, SAA1) based on their structural motifs (Gonzalez et al. 2018). Inflammation response is an organized mechanism where chemical factors released from different immune cells activate inflammatory response system that triggers pain sensors,

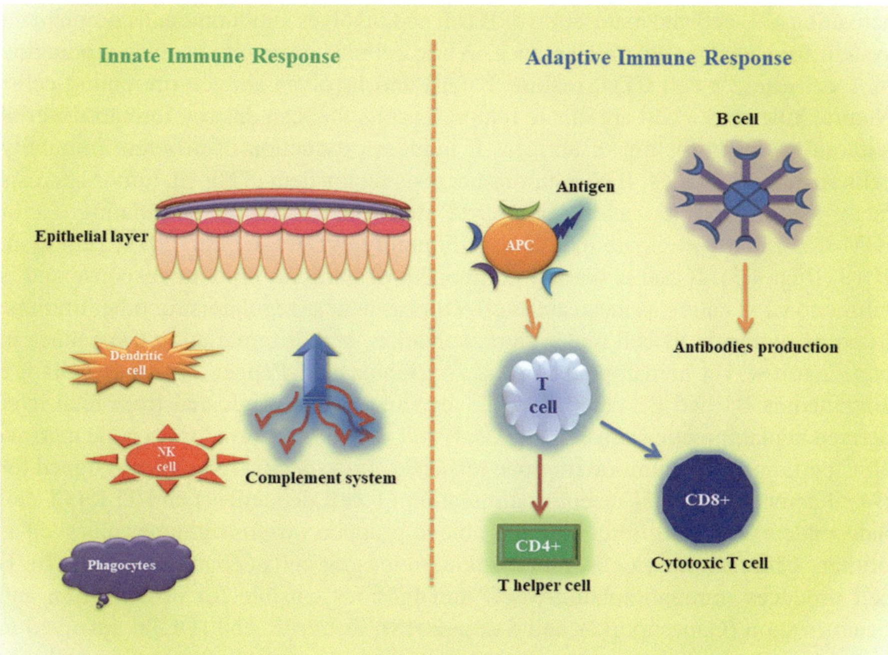

Fig. 4.3 Illustration of host immune system comprised of innate immune response and adaptive immune response

expands the blood vessels, and recruits phagocytic cells for proper immune response. Cytokines and chemokines enable the activation of macrophages and neutrophils that further magnify the immune system by involving adaptive immune response (leukocytes, lymphocytes). Innate immune response involves monocytes, macrophages, neutrophils, and natural killer cells that mediate overall production of immune response molecule for protection. Toll-like receptors (TLRs, i.e., TLR3, TLR8, TLR9) and RIG-like receptors (RLRs, i.e., RIG1) are able to identify the Zika virus and activate innate immune response against the virus. Zika virus attachment on host cell leads to high expression of interferon production (IFN-α, IFN-β, IFN-γ), cytokines (IL-1β, IL-2, IL-4, IL-6, IL-9, IL-10, IL-13, IL-17, and TNF-α,), and chemokines (CXCL-10, CXCL-12, CCL-2, and CCL-3) which have stimulatory effect on STAT1 and STAT2 phosphorylation. STAT1 and STAT2 proteins are involved in JAK-STAT pathway where signaling pathway is initiated by phosphorylation of heterodimer, STAT1 and STAT2 combined with IFN-regulatory factor 9 (IRF9) which activates transcription for anti-viral immune response (da Silva et al. 2019).

4.3.2 Adaptive or Specific Immune Responses

Adaptive immunity is an acquired immunity mechanism and can be divided into two sub-categories, humoral and cellular immunity. Humoral immunity deals with activation of B cell and maturation of B cell and involves antibodies and complement system for specific antigen clearance. While cellular immunity enables production of T cell (helper cell CD4, mature T cell) and involves antigen-presenting cells. Natural killer (NK) cell are innate lymphoid cells that can destroy foreign material without any perturbation of antigen. It induces production of different immunity cells such as cytokines, IFN-γ, tumor necrosis factor-beta (TNF-β), tumor necrosis factor-alpha (TNF-α), and granulocyte macrophage-colony stimulating factor (GM-CSF) that participate in adaptive immunity (Veerman et al. 2019; Han et al. 2019) (Fig. 4.3). B cell is the key component of humoral immune response and is bifunctional in nature as it can act as effectors as well as regulators for other immune response elements. B cell maturation, activation, and differentiation takes place in bone marrow via hematopoiesis process (Sebina and Pepper 2018). B cells are classified as B1 and B2 cells; B1 cells are innate cells originated from fetal liver derived hematopoietic stem cell (HSC). B cell which is nurtured from bone marrow HSC participates in humoral immune response. Two specific routes are assigned for B cell activation: B cell receptor stimulation (T cell dependent) and TLRs (T cell independent). B cell differentiation is able to produce various immune cells such as primary cells, memory cells, and cytokine releasing cells (Zhang et al. 2019b). B cell produces immunoglobulins, IgG and IgM responsible for opsonization and neutralization (Ganeshpurkar and Saluja 2018). Dendritic cell (DC) is involved in both innate and adaptive immune system and acts as antigen-presenting cells for T cells. After infection, DC incorporates the whole antigen or specific peptides on the cell surface and represents it to antigen specific T cell for further immune response for microbial clearance (Heath et al. 2019). DC complex with major histocompatibility complex (MHC)-peptide stimulates production of $CD4^+$ (helper cell) and $CD8^+$ (cytotoxic cell) T cell production. Activation of cytotoxic T cell may require interaction with mature DC cell associated with $CD4^+$ cell to produce $CD8^+$ memory cells. Production of T follicular helper lineage (Tfh cells) takes place when B lymphocytes are presented with surface associated MHC II complex (Christoffersson and von Herrath 2019). Moreover, Tfh cells are linked with B cell maturation and differentiation in bone marrow with concomitant production of interleukin 4 (IL-4), interleukin 21 (IL-21), and membrane-anchored CD40 ligand (Or-Guil et al. 2018; Park et al. 2018).

4.4 Conclusion

In the book chapter, authors have tried to put lights on the interaction between microbial pathogenesis and host immune response. The microbial interaction with immune response elements has raised many questions towards scientific community due to their well-organized system. Science has many scientific solutions to overcome antibiotic resistance through interrupting microbial pathogenicity routes but still some hidden mechanisms are unknown which have to be discovered. Various unknown mechanism are present contributing to the microbial pathogenicity and host response towards the infection, hence, new approaches has to approve for prevention and control.

References

Alfaro-Viquez E, Esquivel-Alvarado D, Madrigal-Carballo S, Krueger CG, Reed JD (2019) Proanthocyanidin-chitosan composite nanoparticles prevent bacterial invasion and colonization of gut epithelial cells by extra-intestinal pathogenic *Escherichia coli*. Int J Biol Macromol 135:630–636

Antikainen J, Kuparinen V, Lähteenmäki K, Korhonen TK (2007) Enolases from gram-positive bacterial pathogens and commensal *Lactobacilli* share functional similarity in virulence-associated traits. FEMS Immunol Med Microbiol 51(3):526–534

Banla LI, Salzman NH, Kristich CJ (2019) Colonization of the mammalian intestinal tract by *Enterococci*. Curr Opin Microbiol 47:26–31

Beulin DS, Yamaguchi M, Kawabata S, Ponnuraj K (2014) Crystal structure of PfbA, a surface adhesin of *Streptococcus pneumoniae*, provides hints into its interaction with fibronectin. Int J Biol Macromol 64:168–173

Bohn E, Sonnabend M, Klein K, Autenrieth IB (2019) Bacterial adhesion and host cell factors leading to effector protein injection by type III secretion system. Int J Med Microbiol 309(5):344–350

Boyle EC, Finlay BB (2003) Bacterial pathogenesis: exploiting cellular adherence. Curr Opin Cell Biol 15(5):633–639

Candela M, Biagi E, Turroni S, Vitali B, Brigidi P (2008) Mechanisms involved in the intestinal interaction between host and bifidobacteria. Microb Ecol Health Dis 20(4):189–192

Christoffersson G, von Herrath M (2019) Regulatory immune mechanisms beyond regulatory T cells. Trends Immunol 40(6):482–491

Cui QQ, Yang J, Niu YN, Qiang CX, Li ZR, Xu KY, Li RX, Shi DY, Wei HL, Zhao XZ, Wang XM, Sun SJ, Zhao JH (2019) Epidemiological investigation of *Clostridioides difficile* colonization in Chinese community infants. Anaerobe 56:116–123

da Silva MHM, Moises RNC, Alves BEB, Pereira HWB, de Paiva AAP, Morais IC, Nascimento YM, Monteiro JD, de Souto JT, Nascimento MSL, de Araujo JMG, da Guedes PMM, Fernandes JV (2019) Innate immune response in patients with acute Zika virus infection. Med Microbiol Immunol 208:703

Dela Ahator S, Zhang L (2019) Small is mighty-chemical communication systems in *Pseudomonas aeruginosa*. Annu Rev Microbiol 73:559

Egawa G, Kabashima K (2018) Barrier dysfunction in the skin allergy. Allergol Int 67(1):3–11

Eloe-Fadrosh EA, Rasko DA (2013) The human microbiome: from symbiosis to pathogenesis. Annu Rev Med 64:145–163

Ganan M, Campos G, Munoz R, Carrascosa AV, de Pascual-Teresa S, Martinez-Rodriguez AJ (2010) Effect of growth phase on the adherence to and invasion of Caco-2 epithelial cells by *Campylobacter*. Int J Food Microbiol 140(1):14–18

Ganeshpurkar A, Saluja AK (2018) Protective effect of catechin on humoral and cell mediated immunity in rat model. Int Immunopharmacol 54:261–266

Garcia BG, Castro FS, Vieira MAM, Girao DM, Uenishi LT, Cergole-Novella MC, Dos Santos LF, Piazza RMF, Hernandes RT, Gomes TAT (2019) Distribution of the *pilS* gene in *Escherichia coli* pathovars, its transfer ability and influence in the typical enteropathogenic *E. coli* adherence phenotype. Int J Med Microbiol 309(1):66–72

Gonzalez OA, Kirakodu S, Novak MJ, Stromberg AJ, Orraca L, Gonzalez-Martinez J, Burgos A, Ebersole JL (2018) Comparative analysis of microbial sensing molecules in mucosal tissues with aging. Immunobiol 223(3):279–287

Gonzalez-Rodriguez I, Ruiz L, Gueimonde M, Margolles A, Sánchez B (2013) Factors involved in the colonization and survival of Bifidobacteria in the gastrointestinal tract. FEMS Microbiol Lett 340(1):1–10

Grimm V, Westermann C, Riedel CU (2014) Bifidobacteria-host interactions—an update on colonisation factors. Biomed Res Int 2014:960826

Grundler T, Quednau N, Stump C, Orian-Rousseau V, Ishikawa H, Wolburg H, Schroten H, Tenenbaum T, Schwerk C (2013) The surface proteins InlA and InlB are interdependently required for polar basolateral invasion by *Listeria monocytogenes* in a human model of the blood-cerebrospinal fluid barrier. Microbes Infect 15(4):291–301

Guttman-Yassky E, Zhou L, Krueger JG (2019) The skin as an immune organ: tolerance versus effector responses and applications to food allergy and hypersensitivity reactions. J Allergy Clin Immunol 144:362–374

Han C, Jiang Y, Wang Z, Wang H (2019) Natural killer cells involved in tumour immune escape of hepatocellular carcinoma. Int Immunopharmacol 73:10–16

Heath WR, Kato Y, Steiner TM, Caminschi I (2019) Antigen presentation by dendritic cells for B cell activation. Curr Opin Immunol 58:44–52

Holy O, Cruz-Cordova A, Xicohtencatl-Cortes J, Hochel I, Parra-Flores J, Petrzelova J, Facevicova K, Forsythe S, Alsonosi A (2019) Occurrence of virulence factors in *Cronobacter sakazakii* and *Cronobacter malonaticus* originated from clinical samples. Microb Pathog 127:250–256

Hu L, Wai TT (2017) Comparing invasive effects of five foodborne bacterial pathogens in human embryonic intestine 407 cells and human ileocecum HCT-8 cells. Asian Pac J Trop Biomed 7(10):937–944

Kochut A, Dersch P (2012) Bacterial invasion factors: tools for crossing biological barriers and drug delivery? Eur J Pharm Biopharm 84(2):242–250

Lombardi C, Tolchard J, Bouillot S, Signor L, Gebus C, Liebl D, Fenel D, Teulon JM, Brock J, Habenstein B, Pellequer JL, Faudry E, Loquet A, Attree I, Dessen A, Job V (2019) Structural and functional characterization of the type three secretion system (T3SS) needle of *Pseudomonas aeruginosa*. Front Microbiol 10:573

Miele AE, Badaoui S, Maugliani L, Salza R, Boumis G, Chichiarelli S, Duclos B, Ricard-Blum S (2019) A comparative analysis of secreted protein disulfide isomerases from the tropical co-endemic parasites *Schistosoma mansoni* and *Leishmania major*. Sci Rep 9(1):9568

Moonens K, Remaut H (2017) Evolution and structural dynamics of bacterial glycan binding adhesins. Curr Opin Struct Biol 44:48–58

Mulcahy ME, McLoughlin RM (2016) Host-bacterial crosstalk determines *Staphylococcus aureus* nasal colonization. Trends Microbiol 24(11):872–886

O'Callaghan A, van Sinderen D (2016) Bifidobacteria and their role as members of the human gut microbiota. Front Microbiol 7:925

Or-Guil M, Boyd SD (2018) Systems immunology of human humoral immunity. Curr Opin Sys Biol 12:70–77

Palankar R, Binsker U, Haracska B, Wesche J, Greinacher A, Hammerschmidt S (2018) Interaction between the *Staphylococcus aureus* extracellular adherence protein Eap and its subdomains with platelets. Int J Med Microbiol 308(6):683–691

Park JS, Kim JH (2018) Role of non-classical T cells in skin immunity. Mol Immunol 103:286–292

Periasamy S, Kolenbrander PE (2009) Mutualistic biofilm communities develop with Porphyromonas gingivalis and initial, early, and late colonizers of enamel. J Bacteriol 191(22):6804–6811

Phelps CC, Vadia S, Arnett E, Tan Y, Zhang X, Pathak-Sharma S, Gavrilin MA, Seveau S (2018) Relative roles of listeriolysin O, InlA, and InlB in *Listeria monocytogenes* uptake by host cells. Infect Immun 86(10):e00555–e00518

Philippsen GS, DeMarco R (2019) Impact of transposable elements in the architecture of genes of the human parasite *Schistosoma mansoni*. Mol Biochem Parasitol 228:27–31

Ribet D, Cossart P (2015) How bacterial pathogens colonize their hosts and invade deeper tissues. Microbes Infect 17(3):173–183

Sawa T (2014) The molecular mechanism of acute lung injury caused by *Pseudomonas aeruginosa*: from bacterial pathogenesis to host response. J Intensive Care 2(1):10

Sebina I, Pepper M (2018) Humoral immune responses to infection: common mechanisms and unique strategies to combat pathogen immune evasion tactics. Curr Opin Immunol 51:46–54

Solanki V, Tiwari M, Tiwari V (2018) Host-bacteria interaction and adhesin study for development of therapeutics. Int J Biol Macromol 112:54–64

Stones DH, Krachler AM (2016) Against the tide: the role of bacterial adhesion in host colonization. Biochem Soc Trans 44(6):1571–1580

Suskind DL, Lee D, Solan P, Wahbeh G, Hayden H, Brittnacher M, Nuding M, Miller S (2019) Dietary therapy for clostridium difficile colonization: a case series. Anaerobe 57:1–3

Tedla MG, Every AL, Scheerlinck JPY (2019) Investigating immune responses to parasites using transgenesis. Parasites Vectors 12:303

Thakur A, Mikkelsen H, Jungersen G (2019) Intracellular pathogens: host immunity and microbial persistence strategies. J Immunol Res 2019:1356540

Tirumale S, Tessy AT (2018) Pathogenesis of cholera: recent prospectives in rapid detection and prevention of cholera. In: Bacterial pathogenesis and antibacterial control, pp 129–144

Veerman RE, Gucluler Akpinar G, Eldh M, Gabrielsson S (2019) Immune cell-derived extracellular vesicles – functions and therapeutic applications. Trends Mol Med 25(5):382–394

Visconti A, Le Roy CI, Rosa F, Rossi N, Martin TC, Mohney RP, Weizhong L, de Rinaldis E, Bell JT, Venter JC, Nelson KE, Spector TD, Falchi M (2019) Interplay between the human gut microbiome and host metabolism. bioRxiv 561787

Walter J, Britton RA, Roos S (2018) Host-microbial symbiosis in the vertebrate gastrointestinal tract and the *Lactobacillus reuteri* paradigm. Proc Natl Acad Sci U S A 108(1):4645–4652

Wu YT, Truong TN, Tam C, Mendoza MN, Zhu L, Evans DJ, Fleiszig SMJ (2019) Impact of topical corticosteroid pretreatment on susceptibility of the injured murine cornea to *Pseudomonas aeruginosa* colonization and infection. Exp Eye Res 179:1–7

Wu ZA, Wang HX (2019) A systematic review of the interaction between gut microbiota and host health from a symbiotic perspective. SN Comprehensive Clin Med 1:224–235

Xu D, Liao C, Zhang B, Tolbert WD, He W, Dai Z, Zhang W, Yuan W, Pazgier M, Liu J, Yu J, Sansonetti PJ, Bevins CL, Shao Y, Lu W (2018) Human enteric α-Defensin 5 promotes *Shigella* infection by enhancing bacterial adhesion and invasion. Immunity 48(6):1233–1244.e6

Zhang H, Wang L, Chu Y (2019a) Reactive oxygen species: the signal regulator of B cell. Free Radic Biol Med 142:16

Zhang C, Yu Z, Zhao J, Zhang H, Zhai Q, Chen W (2019b) Colonization and probiotic function of *Bifidobacterium longum*. J Funct Foods 53:157–165

Role of Quorum Sensing in Microbial Infections and Biofilm Formation

5

T. Eswara Rao and Ranjith Kumavath

Contents

Abstract

The infections and pathogenesis are continuously evolving processes to avoid the effects of hazardous situations and antibiotic susceptibility. In this process, the microbes could percept and conscientiousness towards its circumstances by their cellular and physiological mechanisms. All the bacterial populations can maintain phenotypes and genotype monitors in their competitive environment. To maintain all these community behaviors, the bacterial population utilize various natural signaling pathways under specific microbial language. The bacterial population was well regulated their extracellular or intercellular cooperative communication mechanisms known as quorum sensing. The quorum-sensing (QS)

T. Eswara Rao · R. Kumavath (✉)
Department of Genomic Science, School of Biological Sciences, Central University of Kerala, Kasaragod, Kerala, India
e-mail: RNKumavath@cukerala.edu.in

B. Siddhardha et al. (eds.), *Model Organisms for Microbial Pathogenesis, Biofilm Formation and Antimicrobial Drug Discovery*,
https://doi.org/10.1007/978-981-15-1695-5_5

mechanism responds through small diffusible signal molecules; these signaling molecules were synthesized and secreted into intercellular or extracellular micro-environment at different phage of bacterial growth. However, we focus on this chapter microbial signal communications and also QS mechanisms in biofilm formation.

Keywords
Quorum sensing · Biofilm · Bacterial Infection · Signal molecules

5.1 Introduction

Microbial infections habitually resulted from the complex interplay of multiple virulence factors and various social behaviors. The commitment of microbial population in social activities, including the organization of the group, perform via quorum-sensing signals and biofilm formation (Bodelón et al. 2016). Moreover, these signaling pathways encompass to fascinating at different social lives. The microbial population also uses several signal mechanisms in their lifetime as eukaryotic organisms (Nadell et al. 2008). The chemical signaling by various growth phases and competence of their surveillance in the presence of antibiotics, starvation, and harsh conditions (Jayaraman and Wood 2008). Both quorum sensing and biofilm coordinately concerned with disease-causing and pathogenicity of bacterial species. Most of the microorganisms become infectious pathogens by communicating through quorum sensing and biofilm formation (Claesson 2010; Brackman et al. 2016). Intra and Intercellular interaction in real-world situations on the source of molecules secreted to its external environment.

The diffusible molecules based communication to be widespread among microbes, QS molecules used as cooperation and communication to future action an evolutionary perspective. In addition, the QS-regulated genes have extensive scope of physiological processes such as cell differentiation, stress tolerance, antibiotic production, virulence, or biofilm formation, all of which are crucial for survival and pathogenicity of individual bacterial cells (Diggle et al. 2007; Bodelón et al. 2016). Microbial biofilms are the gathering of individual cells to form complex self-organized communities of microbial cells where they can survive over unfavorable environment conditions (Ramadhan and Hegedus 2005).

5.2 The Bacterial Chatter's

All living systems from prokaryotic to eukaryotic use a variety of communication methods to convey the message from one to one or one to the group. But the understanding of microbial communication it took some time to find by microbiologists was teaming up with chemists and engineers to find ways to undermine the microbes by interfering with the signals they used to communicate (Marx 2014; Visick and Fuqua 2005). In the late 1970s, they termed quorum sensing is a system in certain traits of bacterial species which enables to the signaling channels through as chemical messages (Chen and Chen 2000; Bassler and Losick 2006). The production of active signal molecules during the growth of the bacterial population it maintains homeostasis in between interspecies communication and its individual population development (Gallio et al. 2002). The bacterial communication systems is a simplest mechanism to detect the accumulation of a signaling molecule known as Autoinducer (AI) (Gallio et al. 2002; Bassler and Losick 2006). The autoinducers in gram-positive bacteria are processed oligo-peptides and in gram-negative bacteria acylated homoserine lactones act as AIs for cell-cell communication (Sureshchandra 2010).

5.3 Quorum Sensing: The Bacterial Language

Microbial populations have diverse habitats during host bacterial relationships in both pathogenic and symbiotic. To address this communication they broadly adopt and reveal the physiological importance (Xavier and Bassler 2005). Cell-to-cell signaling mechanism using chemical signaling molecules called autoinducers (AIs) bacteria can speak to each other. AIs produced in the environment in high population density (Fig. 5.1), those communities used as the information to examine the cell numbers and alter the gene expression collectively. The first effort to identify and understand this bacterial language processing system has begun three decades past in two marine bacterial species *Vibrio harveyi* and *V. fischeri* (Bassler and Losick 2006). Quorum sensing is working on the standard when a bacterium releases autoinducers (AIs) based on its growth environment. The autoinducers concentration reaches a certain intensity when adequate bacteria are present, they regulate

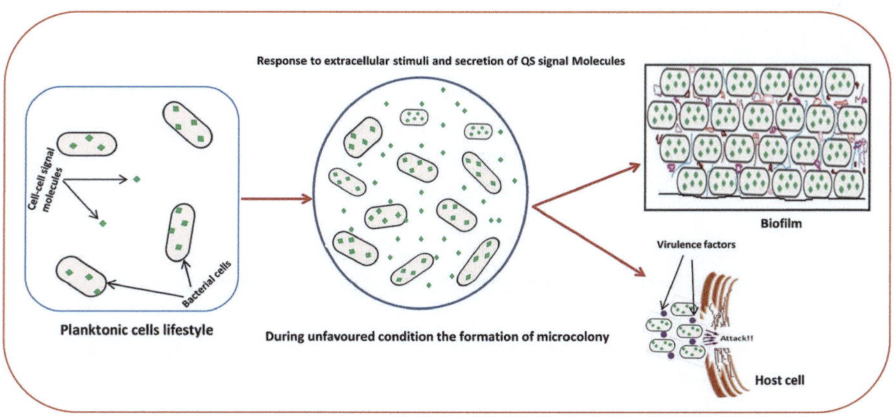

Fig. 5.1 A schematic lifestyle of the bacterial population allowing biofilm formation by activation and secretion of density polymers

their further living mechanism to activate or suppress a specific gene target (Miller and Bassler 2001). By the secretion of AIs, some bacterial species are able to respond and communicate between intra and interspecies bacterial communities.

An initial study of quorum sensing in *Vibrio* bacterial strains revealed the involvement of increase of secreted autoinducer concentration in light production, by stimulating operon luxCDABE various genes listed in Table 5.1, homologous of various bacterial species. It varies along with species to species to some extent by maintaining the homology; autoinducer also varies between these species. *Pseudomonas aeruginosa* (LasI/LasR), *V. anguillarum* (Van I/Van R), etc. are following this strategy as mentioned before. In gram-positive bacteria species, auto-inducing peptides (AIPs) serve as signal molecules (Umesha and Shivakumar 2013) in their quorum-sensing system. QS in certain gram-positive bacteria is regulated by an influx of AIPs into the cell cytoplasm where they contact with numerous transcription regulating factors and expression of genes cascade (Miller and Bassler 2001).

5.4 Biofilm Formation and Bacterial Infections Control by Quorum Sensing

The most important responsibility in controlling different class of microbial actions, such as pathogenesis, virulence molecules synthesis, and formation of a strong biofilm by a system quorum sensing that impacts on human health, industrial production (Choudhary and Schmidt-Dannert 2010).

The formation of biofilms is an orchestrated process and controlled specific genes expression by quorum-sensing mechanism (McDougald et al. 2012). Moreover, it also controls the turn of phrase in virulence genes expression in several micro-molecules during human infections (Jayaraman and Wood 2008). Many previous well-studied QS molecules were essential determinant to QS (Table 5.2). Besides, understanding the different types of micro-chemical signal molecules resulting in the identification of quorum sensing-related processes, including biofilm formation and QS-biofilm community-related infections.

However, the mass interaction like interkingdom will provides a consequence benefits in host bacteria colony formation, defense against immunity, adaptation at changing microenvironment, and the biofilm formation. The coordination of the bacterial infection and quorum sensing between microbes and their hosts is still an unrevealed mechanism (Antunes et al. 2010).

5.5 Applications of Quorum Sensing in Biotechnology

The microbial QS has wide range applications in biotechnology and it is considered to be regulators of the signaling networks in the multi-species/multi-kingdom communities. In this context, the signaling networks between species and kingdoms interact in several ways on host responses to bacterial species (Atkinson and Williams 2009). The bacterial QS signaling biosensors

Table 5.1 Various bacterial species quorum-sensing signal regulating operon homologs to the quorum-sensing gene pairs of the luxR-luxI family

Bacterium	Homologs	Major AHL	Phenotype regulation	Reference cited
Burkholderia vietnamiensis	*BviI/BviR*	N-decanoyl-homoserine lactone (DHL)	Transcription regulator	Malott and Sokol (2007)
Chromobacterium violaceum	*CviR/CviI*	C6-HSL	Exoenzymes and antibiotics	Stauff and Bassler (2011)
Enterobacter agglomerans	*EagR/EagI*	3-Oxo-C6-HSL	Endophytic life	Shrout and Parsek (2006)
Pseudomonas fluorescent	*MupI/ MupR*	N-acyl homoserine lactone	Proteolytic activity	Lerat and Moran (2004)
Erwinia carotovora Pectobacterium carotovorum	*EcbI/EcbR*	N-(3-Oxohexanoyl)-L-homoserinelactone	To switch on virulence	Sjöblom (2006)
Erwinia chrysanthemi	*ExpR/ ExpIEchR/ EchI*	3-Oxo-C6-HSL	Synthesis of pectinases during cell growth	Nasser et al. (1998)
Obesumbacterium proteus	*OprR/ OprI*	3-Oxo-C6-HSL	To control biofilm formation	Swift et al. (1999)
Escherichia coli.	*Csrb/Csrc*	N-acyl Homoserinelactone	Epithelial cell invasion	Suzuki et al. (2006)
Hafnia alvei	*HalI/HalR*	N-butyryl-l-homoserine lactone	Biofilm formation	Hou et al. (2017)
Nitrosomonas europaea	*Hdts*	3-Oxo-C6-HSL	Emergence in lag phase survival	Burton et al. (2005)
Pantoea stewartii	*EsaR/EsaI*	3-Oxo-C6-HSL	Production of exopolysaccharides	Minogue et al. (2002)
Pseudomonas aeruginosa	*LasR/LasI*	3-Oxo-C12-HSL	Biofilm formation, cell-cell spacing	Seed et al. (1995)
Vibrio anguillarum	*VanR/VanI*	3-Oxo-C10-HSL	Conjunction	Williams et al. (2000)
Pseudomonas fluorescens	*AfmI/ AfmR*	N-Acyl homoserine lactone	Proteolytic activity and swarming motility	Lerat and Moran (2004)
Pseudomonas syringae pv. tabaci	*PsyR/PsyI*	AHL	Regulators of epiphytic fitness	Swift et al. (1999)
Pseudomonas syringae	*AhlI*	N-Acyl homoserine lactone	Motility and exopolysaccharide production	Dulla and Lindow (2008)
Pseudomonas syringae	*PsmI/ PsmR*	N-Acyl homoserine lactone	Production of Violacein	Quiñones et al. (2004)
Rhizobium leguminosarum	*RhiR*	3-Hydroxy-7-Cis-C14-HSL	Survival, Bacteriocin and nodulation	Gray et al. (1996)
Serratia liquefaciens	*SwrR/SwrI*	C4-HSL	Proteases production and swarming motility	Rasmussen et al. (2000)

(continued)

Table 5.1 (continued)

Bacterium	Homologs	Major AHL	Phenotype regulation	Reference cited
Vibrio cholerae	*Csra/?*	Luxo	Biofilm formation	Lenz et al. (2005)
Pseudomonas aeruginosa	*RhlR/RhlI (Vsmr/ Vsmi)*	C4-HSL	Type 4Pili, Lectins, Rhamnolipid and exoenzymes	Chapon-Hervé et al. (1997)
Yersinia enterocolitica	*YenR/YenI*	C6-HSL	Swarming motility	Tsai and Winans (2011)
Pseudomonas aureofaciens	*PhzR/PhzI*	C6-HSL	Phenazine antibiotic	Khan et al. (2005)
Yersinia pseudotuberculosis	*YtbR/YtbI*	C8-HSL	Swimming motility	Atkinson et al. (2008)
Escherichia coli	*Lsrk/Lsrr*	Phospho-AI-2	Biofilms were regulated	Gao et al. (2016)
Aeromonas hydrophila	*AhyR/AhyI*	C4-HSL	Biofilm formation	Swift et al. (1999)
Erwinia carotovora Subsp. carotovora	*Carr/Expr Expi/ (Cari)*	3-Oxo-C6-HSL	Antibiotics	Andersson et al. (2000)
Aeromonas hydrophila	*AhyR/AhyI*	C4-HSL	Opportunistic	Kirkeet al. (2004)
Rhodobacter sphaeroides	*Cerr/Ceri*	7-Cis-C14-HSL	Quorum-sensing community	Puskas et al. (1997)
Aeromonas salmonicida	*Asar/Asai*	C4-HSL	Production of extracellular proteases	Swift et al. (1997)
Erwinia stewartii Pantoea stewartii	*Esai/Esar*	N-acylhomoserine lactone autoinducer.	Activation of RNA polymerase	von Bodman and Farrand (1995)
Aeromonas salmonicida	*Asai/Asar /Ahyr*	C4-HSL	Protease production	Lerat and Moran (2004)
Pseudomonas syringae	*PsyI/PsyR*	N-Acyl homoserine lactone synthase	Sigma factor regulation	Chatterjee et al. (2007)
Agrobacterium tumefaciens	*TraR/TraI*	3-Oxo-C8-HSL	Coalesce	Fuqua and Winans (1996)
Burkholderia cepacia	*CepR/ CepI*	C8-HSL	Protease, Siderophore	Subramoni and Sokol (2012)
Yersinia pseudotuberculosis	*YpsR/YpsI*	3-Oxo-C6-HSL	Biofilm and cells motility,	Atkinson et al.(1999)
Vibrio fischeri	*LuxR/LuxI*	3-Oxo-C6-HSL	Bioluminescence	Lerat and Moran (2004)

molecules are a novel, inexpensive, and portable filter paper-based strip biosensor that developed a fast and convenient method for the detection of clinical samples for pathogenicity and biofilm, which could be employed for first-level screening of a variety of infections (Umesha and Shivakumar 2013). Components of bacterial QS systems constitute an important in the construction of QS-based microbial biosensors were good progress in biological applications to create novel biological systems to recognize pathogenic microbes present in the environment and diseased host organisms. One such biological device produces fluorescence at critical AHL density. The engineered QS systems could be very useful in industrial production biomolecules, tissues engineering, and also extremely designing mixed-species fermentation methods. An exciting application of QS is to create engineered bacteria capable of invading cancer cells. It is possible to envision the creation of novel anti-cancer therapeutics by the addition of cancer-destructing modules to these microbial biosensors (Choudhary and Schmidt-Dannert 2010).QS has also been reported to be of use in bio-ethanol and biodiesel production. Generation of bioelectricity by microbial fuel cells and H2 and biomethane by microbial electrolysis cells requires strong biofilms. Certain toxins, cytokines, tumor antigens can be exploited as a novel and effective treatment of cancer. Industrial products microbial QSS produces various extracellular products such as enzymes, rhamnolipids, isobutanol, and 1, 3-propanediol, 2,3-butanediol with commercial applications. Engineered LuxI/LuxR system fused with antigen proteins has been used to produce vaccines (Kalia et al. 2018).

5.6 Biofilm Matrix Composition and Architecture

The biofilm is defined as a microbiologically derived static characterized by the cells that are irreversibly attached to a substratum or interface or each other and embedded in a matrix of extracellular polymeric substances (EPS) that they have produced (Baldassarri et al. 2001). Bacterial biofilms are structures consisting of single species or diverse communities of different bacteria. Biofilm formation in microbial populations gives an account for their emerging antibiotic resistance over and above stress adaptability. When bacteria adhere to surfaces, it aggregates to each other and form biofilms on a wide range of spaces. It is found that antibiotic resistance was more frequent in biofilm forming strains than the non biofilm forming strains. This is concluded as a consequence of incipient colonization by a single bacterial species which then alter the microenvironment, in such a way that enables new species of bacteria to populate (Schwartz et al. 2016). Formation of biofilm includes processes like reversible attachment of planktonic bacteria to surfaces, irreversible attachment to surfaces, formation of the external matrix, biofilms acquire a three-dimensional structure, biofilm detachment. It is reported that environmental changes are responsible for the transition from planktonic growth to this altered bacterial lifestyle (Kostakioti et al. 2013). Biofilm varies with the altered expression of surface molecules, virulence factors, and metabolic

Table 5.2 Quorum-sensing signal molecule structures and their chemical compositions

S.No.	Quorum-sensing signal molecule	Molecule chemical composition	Molecule structure	Reference
1.	*N*-Butyryl-L-homoserine lactone	BHL, C4-HSL, $C_8H_{13}NO_3$Mr 171	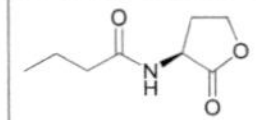	Reimmann et al. (1997)
2.	*N*-(3-Oxobutyryl)-L-homoserine lactone	OBHL, 3-oxo-C4-HSL, $C_8H_{11}NO_4$Mr 185	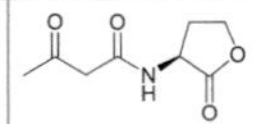	Schwarzer et al. (2015)
3.	*N*-Pentanoyl-L-homoserine lactone-d_9	PHL-d_9, C5-HSL-d_9, $C_9H_6D_9NO_3$Mr 194	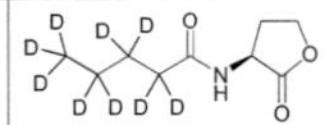	Ruparell (2012)
4.	*N*-Octanoyl-L-homoserine lactone	OHL, C8-HSL, $C_{12}H_{21}NO_3$Mr 227	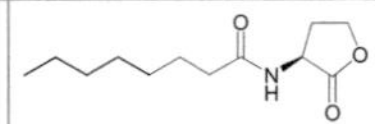	Kuo et al. (1996)
5.	*N*-Pentanoyl-L-homoserine lactone	PHL, C5-HSL, $C_9H_{15}NO_3$Mr 185	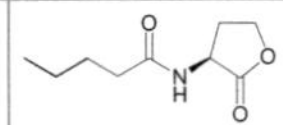	Ortori et al. (2011)
6.	*N*-(3-Oxohexanoyl)-L-homoserine lactone	OHHL, 3-oxo-C6-HSL, $C_{10}H_{15}NO_4$Mr 213	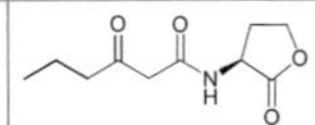	Bainton et al. (1992)
7.	*N*-(3-Hydroxyhexanoyl)-L-homoserine lactone	HHHL, 3-OH-C6-HSL, $C_{10}H_{17}NO_4$Mr 215	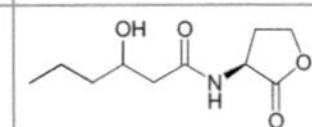	Milton et al. (2001)
8.	*N*-Hexanoyl-L-homoserine lactone	HHL, C6-HSL, $C_{10}H_{17}NO_3$Mr 199	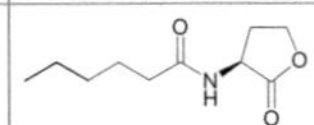	Winson et al. (1995)
9.	*N*-(3-Hydroxybutyryl)-L-homoserine lactone	HBHL, 3-OH-C4-HSL, $C_8H_{13}NO_4$Mr 187	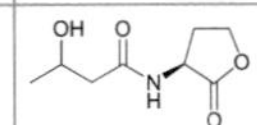	Bainton et al. (1992)
10.	*N*-(3-Oxohexanoyl)-D-homoserine lactone	D-OHHL, 3-oxo-D-C6-HSL, $C_{10}H_{15}NO_4$Mr 213	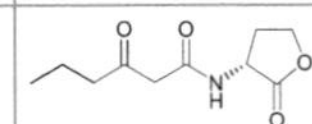	Bainton et al. (1992)
11.	*N*-(3-Oxodecanoyl)-L-homoserine lactone	ODHL, 3-oxo-C10-HSL, $C_{14}H_{23}NO_4$Mr 269	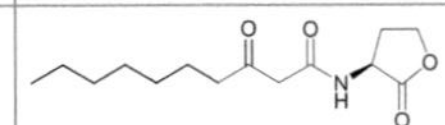	Milton et al. (1997)

12.	*N*-(3-Oxoheptanoyl)-L-homoserine lactone	OHpHL, 3-oxo-C7-HSL, $C_{11}H_{17}NO_4$Mr 227	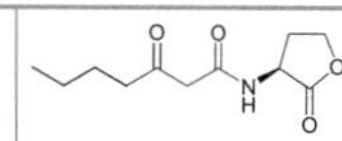	Chhabra et al. (1993)
13.	*N*-Decanoyl-L-homoserine lactone	DHL, C10-HSL, $C_{14}H_{25}NO_3$Mr 255	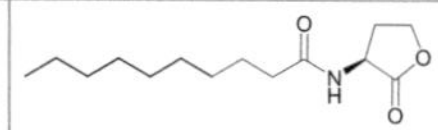	Shaw et al. (1997)
14.	*N*-(3-Oxododecanoyl)-L-homoserine lactone	OdDHL, 3-oxo-C12-HSL, $C_{16}H_{27}NO_4$Mr 297	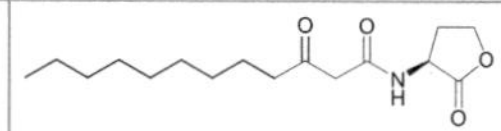	Telford et al. (1998)
15.	*N*-(3-Oxooctanoyl)-L-homoserine lactone	OOHL, 3-oxo-C8-HSL, $C_{12}H_{19}NO_4$Mr 241	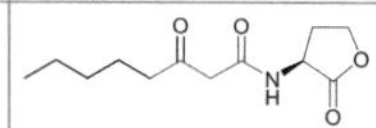	Shaw et al. (1997)
16.	*N*-Dodecanoyl-L-homoserine lactone	dDHL, C12-HSL, $C_{16}H_{29}NO_3$Mr 283	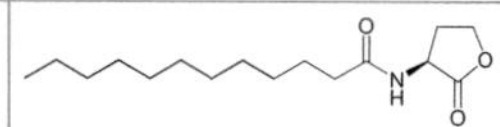	Winson et al. (1995)
17.	*N*-(3-Hydroxyoctanoyl)-L-homoserine lactone	HOHL, 3-OH-C8-HSL, $C_{12}H_{21}NO_4$Mr 243	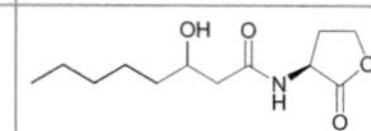	Ulrich et al. (2004)
18.	*N*-(3-Hydroxydecanoyl)-L-homoserine lactone	HDHL, 3-OH-C10-HSL, $C_{14}H_{25}NO_4$Mr 271	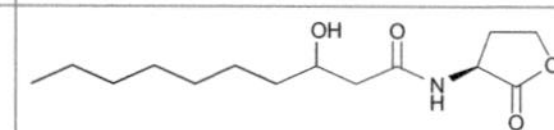	Morohoshi et al. (2008)
19.	*N-(*3-Hydroxytridecanoyl)-L-homoserine lactone	HtriDHL, 3-OH-C13-HSL, $C_{17}H_{31}NO_4$Mr 313	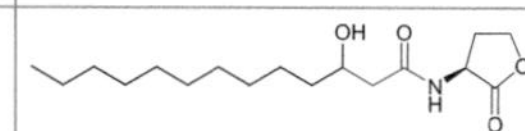	Ruysbergh et al. (2016)
20.	*N*-(3-Oxododecanoyl)-L-homocysteine thiolactone	OdDTL, 3-oxo-C12-HTL, $C_{16}H_{27}NO_3S$Mr 313	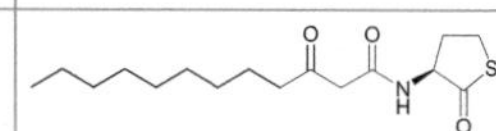	Lawrence et al. (1999)
21.	N-(3-Hydroxydodecanoyl)-L-homoserine lactone	HdDHL, 3-OH-C12-HSL, $C_{16}H_{29}NO_4$Mr 299	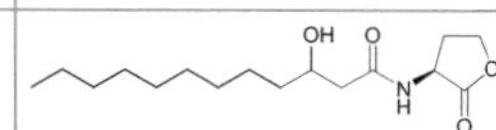	Chhabra et al. (2003)
22.	*N*-Tridecanoyl-L-homoserine lactone	triDHL, C13-HSL, $C_{17}H_{31}NO_3$Mr 297	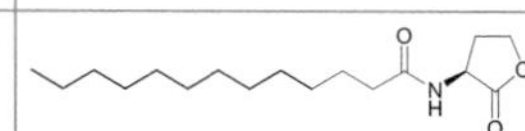	Nomura and Miyazaki (2009)

(continued)

Table 5.2 (continued)

S.No.	Quorum-sensing signal molecule	Molecule chemical composition	Molecule structure	Reference
23	N-Icosanoyl-L-homoserine lactone	IHL, C20-HSL, C24H43NO3Mr395	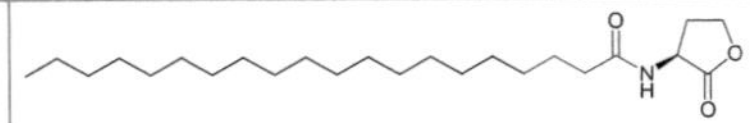	Nomura and Miyazaki (2009)
24.	N-(3-Oxooctadecanoyl)-L-homoserine lactone	OoDHL, 3-oxo-C18-HSL, $C_{22}H_{39}NO_4$Mr 381	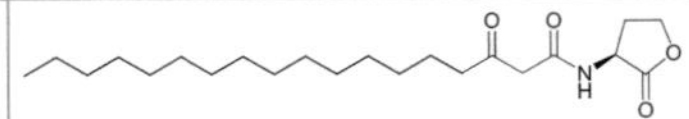	Telford et al. (1998)
25.	N-Tetradecanoyl-L-homoserine lactone	tDHL, C14-HSL, $C_{18}H_{33}NO_3$Mr311	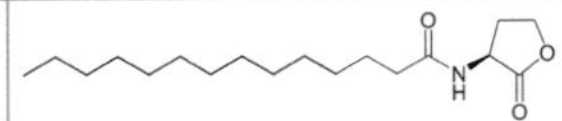	McClean et al. (1997)
26.	N-(3-Oxotetradecanoyl)-L-homoserine lactone	OtDHL, 3-oxo-C14-HSL, $C_{18}H_{31}NO_4$Mr 325	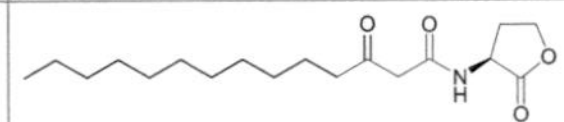	Laue et al. (2000)
27.	N-Hexadecanoyl-L-homoserine lactone	hDHL, C16-HSL, $C_{20}H_{35}NO_3$Mr339	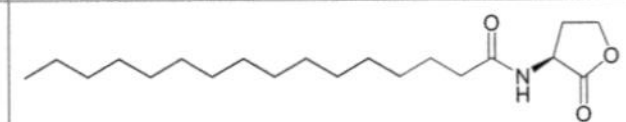	Morohoshi et al. (2008)
28.	N-decanoyl cyclopentylamine	C10-CPA, $C_{15}H_{29}NO$Mr 239	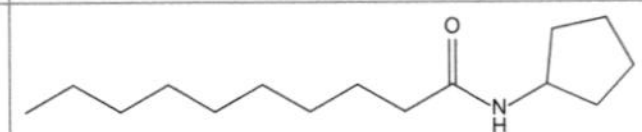	Morohoshi et al. (2007)
29.	N-(3-Oxohexadecanoyl)-L-homoserine lactone	OhDHL, 3-oxo-C16-HSL, $C_{20}H_{35}NO_4$Mr 353	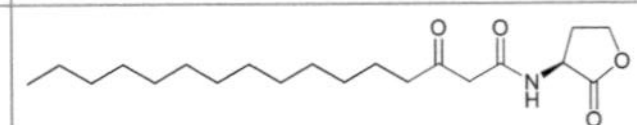	Bartels et al. (2007)
30.	N-Octadecanoyl-L-homoserine lactone	oDHL, C18-HSL, $C_{22}H_{39}NO_3$Mr367	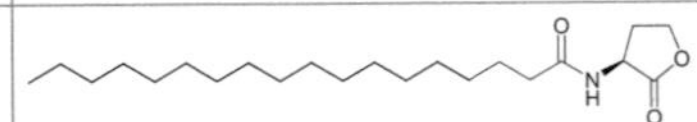	Marketon et al. (2003)
31.	N-(3-Oxoicosanoyl)-L-homoserine lactone	OIHL, 3-oxo-C20-HSL, $C_{24}H_{43}NO_4$Mr 409	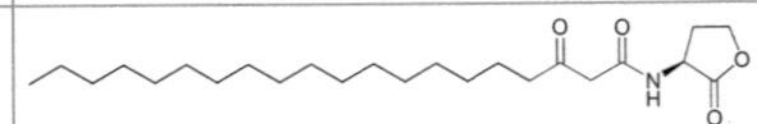	Telford et al. (1998)

status, allowing the bacteria to acquire properties that enable their survival in unfavorable conditions. It can tolerate a higher level of antibiotic stress as compared to the planktonic bacterial community. The adaptability of bacteria in biofilm towards abiotic and biotic stress conditions makes them difficult t to eliminate from living hosts and they are having a higher degree of resistance to phagocytosis (Pillai et al. 2004).

5.6.1 Biofilm Matrix Composition

Understanding of the biofilm organization and chemical composition can reveal vulnerabilities and probably a new target for disease control also leads to prospective treatment. The interaction between microcolonies and surfaces was strength by the formation of an extracellular matrix, which provides that facilitate cell sticking together and provides mechanical support. Typically 5–35% of the biofilm would be constitute extracellular polysaccharides and matrix proteins that control biofilm morphogenesis and eventually maintain the mature biofilm (Dietrich et al. 2013; Janissen et al. 2015). The nutrients and minerals are trapped from the environment through the hunting organism; however, the extracellular substances such as polysaccharides (1–2%) DNA (<1%), RNA (<1%); ions, lipoteichoic acids, lipopolysaccharides, and 97% of water (Whitchurch et al. 2002; Schuster and Markx 2013; Janissen et al. 2015; Jamal et al. 2018).

5.6.2 Biofilm Architecture

In the early days the formation of biofilms considered to be moderately homogeneous structures microcolonies evenly embedded with a gel matrix of EPS. However, the biofilms are relatively flat and smear like elaborated morphological structures were identified. By developing advanced imaging techniques like confocal Raman spectroscopy, look over the probe microscopy, epifluorescence microscopy, and confocal laser scanning microscopy (CLSM), which allows the 3-dimensional (3D) structure of biofilm images. Besides, biofilms have been seen to be more complex and compact structures from patchy clumps with detailed morphological shapes such as pillars or mushroom and with water channels in between for the exchange of materials with the surroundings and within the biofilms (Schuster and Markx 2013). Experimental comments reveal those mature biofilms may display rich surface patterns including radial ridges, branches, concentric rings, labyrinthine networks, and their combinations with distinct surface topographies. Further, the genotype characteristic, the selection of a particular pattern also depends on the physiological conditions and environmental cues of the biofilm (Zhang et al. 2017). In this biofilm architecture, the combinations of living and non-living particles are very common in individual clusters of microcolonies. The non-living particle is an essential building block of biofilms, Furthermore, the heterogeneity of biofilms Biomass were both chemical and microbial cells (Hödl et al. 2014).

The structure of the extracellular polymeric substance (EPS) matrix could compose by extracellular polysaccharides, DNA, lipopolysaccharides, and proteins, water channels. The EPS increases the binding of water and leads to a decrease the chance of dehydration of the bacterial cells in biofilms. The water channels allows nutrients, metabolites, and genetic material exchange with biofilms and also allow the water, air, and nutrients to get to all parts of the biofilm (Rabin et al. 2015). Moreover, the increase in the number of biofilms and the formation of the novel architecture were observed during spaceflight. The biofilms formed during spaceflight exhibited a column-and-canopy structure that has not been observed on Earth (Kim et al. 2013). *P. aeruginosa*, for example, during biofilm development, exhibits mushroom-like structures; microbial biofilm development is observed like filamentous streams, hemispherical mound-shaped microcolonies pellicles, flocs, or granules, wrinkled colony morphogenesis which forms an isotropic pattern on the various surfaces (Lembre et al. 2012). The biofilm architecture involves the following five stages.

- Stage 1: Initial contact/attachment to the surface: Reversible attachment of bacterial cells to a surface,
- Stage 2: Microcolony formation: irreversible attachment mediated by the formation of exopolymeric material,
- Stage 3: Maturation and formation of the architecture of the biofilm: formation of microcolonies and the beginning of biofilm maturation,
- Stage 4: Formation of a mature biofilm: with a 3-dimensional structure containing cells packed in clusters with channels between the clusters that allow transport of water and nutrients and waste removal, and
- Stage 5: Finally detachment/dispersion of the biofilm: detachment and dispersion of cells from the biofilm and initiation of new biofilm formation; dispersed cells are more similar to planktonic (that is, free swimming) cells than to mature biofilm cells.

5.6.3 Biofilm Role in Human Infections

The biofilm formation is an important mechanism for the growth of bacterial cells to survive in the host cell environment. This bacterial associated chronic infection about 80% was estimated with biofilm, these include both non-device and device-associated infections. The bacterial lifestyle was as complex that could adapt and protect from host defense, antibiotic susceptibility and environmental stresses (Flemming et al. 2016). A matured biofilms enclosed with the hydrated matrix of exopolymeric substances, polysaccharides, polypeptides, and nucleic acid fragments, the bacterial cells can stand for antibiotics at concentrations up to 1000 times higher than those required to kill non-biofilm forming cells (Hashem et al. 2017). The microcolonies were slime-enclosed with a matrix of various polymeric substances that protect from hostile environments and resist from hazardous antibiotics, thus the morphological structure is easy to be implanted on medical devices and prostheses (Douglas 2003; Hall-Stoodley et al. 2004). The medical apparatus associated infections data have been estimated for several devices, such as: 4% for pacemakers and defibrillators or mechanical heart valves; 10% for ventricular shunts; 2% for breast implants; 2% for joint prostheses; 36% native valve endocarditis (NVE); and 40% for ventricular-assisted devices (Burmolle et al. 2010; O'Toole et al. 2000; Bjarnsholt 2013). Bacterial cells embedded with dense matrix, which is more merciful of host defenses and antimicrobials when compared with single independent cells, and thus obstruct the treatment (Bjarnsholt 2013).

5.7 Applications of Biofilm in Biotechnology

Biofilms are natural forms of cell immobilization in which microorganisms attach to solid supports. Biofilms are naturally attached to a surface with communities of microorganisms. These changes are reflected in the new phenotypic characteristics developed by biofilm. On the other hand, biofilms can also be useful in human applications (Zavahir and Seneviratne 2007). The plastic composite-supporters (PCS) chips in biofilm reactors containing rings and disks were enhanced the production of ethanol and its final concentration (Demirci et al. 1997). Biofilms are used for the production of vinegar and biofilms are also potentially useful in biocatalysis. Biofilms are already extensively used in wastewater treatment and play roles in the production of biofuels, such as methane production by methanogenesis (Schuster and Markx 2013). The attached-microbial growth reactors were used in the biological nitrification/denitrification process from industrial wastewater (Chen and Chen 2000).

5.8 Conclusion

Biofilm and quorum sensing are complex and diverse. Better understanding of quorum-sensing regulatory circuits in controlling biofilm formation and infection will further facilitate the mechanistic insights and development against biofilm-related infections. In adding up together, different aspects of the human appliance also there for medical applications like diagnostics. Microbial biofilms have a strong bearing on acute and chronic bacterial infections and nations economic loss. Since biofilm formation for many organisms is QS-mediated, therapeutic strategies targeting QS systems are attracting attention to disease-causing pathogens.

Acknowledgments The authors thank to DST-SERB-EMEQ-051/2014, Govt. of India for financial assistance. ER thanks to SERB for JRF fellowship and the research facilities supported by CU Kerala.

References

Andersson RA, Eriksson AR, Heikinheimo R, Mäe A, Pirhonen M, Kõiv V, Hyytiainen H, Tuikkala A, Palva ET (2000) Quorum sensing in the plant pathogen *Erwinia carotovora* subsp. *carotovora*: the role of expREcc. Mol Plant-Microbe Interact 13(4):384–393

Antunes LCM, Ferreira RB, Buckner MM, Finlay BB (2010) Quorum sensing in bacterial virulence. Microbiology 156(8):2271–2282

Atkinson S, Williams P (2009) Quorum sensing and social networking in the microbial world. J RSoc Interface 6(40):959–978

Atkinson S, Throup JP, Stewart GS, Williams P (1999) A hierarchical quorum-sensing system in *Yersinia pseudotuberculosis* is involved in the regulation of motility and clumping. Mol Microbiol 33(6):1267–1277

Atkinson S, Chang CY, Patrick HL, Buckley CM, Wang Y, Sockett RE, Camara M, Williams P (2008) Functional interplay between the *Yersinia pseudotuberculosis* YpsRI and YtbRI quorum sensing systems modulates swimming motility by controlling expression of flhDC and fliA. Mol Microbiol 69(1):137–151

Bainton NJ, Stead P, Chhabra SR, Bycroft BW, Salmond GPC, Stewart GS, Williams P (1992) N-(3-oxohexanoyl)-L-homoserine lactone regulates carbapenem antibiotic production in *Erwinia carotovora*. Biochem J 288(3):997–1004

Baldassarri L, Cecchini R, Bertuccini L, Ammendolia MG, Iosi F, Arciola CR, Montanaro L, di Rosa R, Gherardi G, Dicuonzo G, Orefici G, Creti R (2001) *Enterococcus* spp. produces slime and survives in rat peritoneal macrophages. Med Microbiol Immunol 190(3):113–120

Bartels FW, McIntosh M, Fuhrmann A, Metzendorf C, Plattner P, Sewald N, Anselmetti D, Ros R, Becker A (2007) Effector-stimulated single molecule protein-DNA interactions of a quorum-sensing system in *Sinorhizobium meliloti*. Biophys J 92(12):4391–4400

Bassler BL, Losick R (2006) Bacterially speaking. Cell 125(2):237–246

Bjarnsholt T (2013) The role of bacterial biofilms in chronic infections. APMIS 121:1–58

Bodelón G, Montes-García V, López-Puente V, Hill EH, Hamon C, Sanz-Ortiz MN et al (2016) Detection and imaging of quorum sensing in Pseudomonas aeruginosa biofilm communities by surface-enhanced resonance Raman scattering. Nat Mater 15(11):1203

Brackman G, Breyne K, De Rycke R et al (2016) The quorum sensing inhibitor hamamelitannin increases antibiotic susceptibility of *Staphylococcus aureus* biofilms by affecting peptidoglycan biosynthesis and eDNA release. Sci Rep 6:20321

Burmolle M, Thomsen TR, Fazli M et al (2010) Biofilms in chronic infections–a matter of opportunity–monospecies biofilms in multispecies infections. FEMS Immunol Med Microbiol 59(3):324–336

Burton EO, Read HW, Pellitteri MC, Hickey WJ (2005) Identification of acyl-homoserine lactone signal molecules produced by *Nitrosomonas europaea* strain Schmidt. Appl Environ Microbiol 71(8):4906–4909

Chapon-Hervé V, Akrim M, Latifi A, Williams P, Lazdunski A, Bally M (1997) Regulation of the xcp secretion pathway by multiple quorum-sensing modulons in *Pseudomonas aeruginosa*. Mol Microbiol 24(6):1169–1178

Chatterjee A, Cui Y, Hasegawa H, Chatterjee AK (2007) PsrA, the pseudomonas sigma regulator, controls regulators of epiphytic fitness, quorum-sensing signals, and plant interactions in *Pseudomonas syringae pv.* tomato strain DC3000. Appl Environ Microbiol 73(11):3684–3694

Chen CY, Chen SD (2000) Biofilm characteristics in biological denitrification biofilm reactors. Water Sci Technol 41(4–5):147–154

Chhabra SR, Stead P, Bainton NJ, Salmond GP, Stewart GS, Williams P, Bycroft BW (1993) Autoregulation of carbapenem biosynthesis in *Erwinia carotovora* by analogues of N-(3-oxohexanoyl)-L-homoserine lactone. J Antibiot 46(3):441–454

Chhabra SR, Harty C, Hooi DS et al (2003) Synthetic analogues of the bacterial signal (quorum sensing) molecule N-(3-oxododecanoyl)-L-homoserine lactone as immune modulators. J Med Chem 46(1):97–104

Choudhary S, Schmidt-Dannert C (2010) Applications of quorum sensing in biotechnology. Appl Microbiol Biotechnol 86(5):1267–1279

ClaessonC (2010) Staphylococci and Enterococci: studies on activity of antimicrobial agents and detection of genes involved in biofilm formation. Doctoral dissertation, Linköping University Electronic Press

Demirci A, Pometto AL III, Ho KG (1997) Ethanol production by *Saccharomyces cerevisiae* in biofilm reactors. J Ind Microbiol Biotechnol 19(4):299–304

Dietrich LE, Okegbe C, Price-Whelan A, Sakhtah H, Hunter RC, Newman DK (2013) Bacterial community morphogenesis is intimately linked to the intracellular redox state. J Bacteriol 195(7):1371–1380

Diggle SP, Gardner A, West SA, Griffin AS (2007) Evolutionary theory of bacterial quorum sensing: when is a signal not a signal? Philos Trans RSoc B BiolSci 362(1483):1241–1249

Douglas L (2003) Candida biofilms and their role in infection. Trends Microbiol 11(1):30–36

Dulla G, Lindow SE (2008) Quorum size of *Pseudomonas syringae* is small and dictated by water availability on the leaf surface. Proc Nat AcadSci 105(8):3082–3087

Flemming HC, Wingender J, Szewzyk U, Steinberg P, Rice SA, Kjelleberg S (2016) Biofilms: an emergent form of bacterial life. Nat Rev Microbiol 14(9):563

Fuqua C, Winans SC (1996) Conserved cis-acting promoter elements are required for density-dependent transcription of *Agrobacterium tumefaciens* conjugal transfer genes. J Bacteriol 178(2):435–440

Gallio M, Sturgill G, Rather P, Kylsten P (2002) A conserved mechanism for extracellular signaling in eukaryotes and prokaryotes. Proc Nat Acad Sci 99(19):12208–12213

Gao M, Song H, Zheng H et al (2016) Culture of low density *E. coli* cells in alginate–chitosan microcapsules facilitates stress resistance by up-regulating luxS/AI-2 system. Carbohydr Polym 141:160–165

Gray KM, Pearson JP, Downie JA, Boboye BE, Greenberg EP (1996) Cell-to-cell signaling in the symbiotic nitrogen-fixing bacterium *Rhizobium leguminosarum*: autoinduction of a stationary phase and rhizosphere-expressed genes. J Bacteriol 178(2):372–376

Hall-Stoodley L, Costerton JW, Stoodley P (2004) Bacterial biofilms: from the natural environment to infectious diseases. NatRev Microbiol 2(2):95

Hashem YA, Amin HM, Essam TM, Yassin AS, Aziz RK (2017) Biofilm formation in *Enterococci*: genotype-phenotype correlations and inhibition by vancomycin. Sci Rep 7(1):5733

Hödl I, Mari L, Bertuzzo E, Suweis S, Besemer K, Rinaldo A, Battin TJ (2014) Biophysical controls on cluster dynamics and architectural differentiation of microbial biofilms in contrasting flow environments. Environ Microbiol 16(3):802–812

Hou HM, Zhu YL, Wang JY, Jiang F, Qu WY, Zhang GL, Hao HS (2017) Characteristics of N-acylhomoserine lactones produced by *Hafnia alvei* H4 isolated from spoiled instant sea cucumber. Sensors 17(4):772

Jamal M, Ahmad W, Andleeb S et al (2018) Bacterial biofilm and associated infections. J Chin Med Assoc 81(1):7–11

Janissen R, Murillo DM, Niza B, Sahoo PK, Nobrega MM, Cesar CL et al (2015) Spatiotemporal distribution of different extracellular polymeric substances and filamentation mediate *Xylella fastidiosa* adhesion and biofilm formation. Sci Rep 5:9856

Jayaraman A, Wood TK (2008) Bacterial quorum sensing: signals, circuits, and implications for biofilms and disease. Annu Rev Biomed Eng 10:145–167

Kalia VC, Patel SK, Kang YC, Lee JK (2018) Quorum sensing inhibitors as antipathogens: biotechnological applications. Biotechnol Adv 37(1):68–90

Khan SR, Mavrodi DV, Jog GJ, Suga H, Thomashow LS, Farrand SK (2005) Activation of the phz operon of *Pseudomonas fluorescens* 2-79 requires the LuxR homolog PhzR, N-(3-OH-hexanoyl)-L-homoserine lactone produced by the LuxI homolog PhzI, and a cis-acting phz box. J Bacteriol 187(18):6517–6527

Kim W, Tengra FK, Young Z, Shong J, Marchand N, Chan HK et al (2013) Spaceflight promotes biofilm formation by *Pseudomonas aeruginosa*. PLoS One 8(4):e62437

Kirke DF, Swift S, Lynch MJ, Williams P (2004) The *Aeromonas hydrophila* LuxR homologue AhyR regulates the N-acyl homoserine lactone synthase, AhyI positively and negatively in a growth phase-dependent manner. FEMS Microbiol Lett 241(1):109–117

Kostakioti M, Hadjifrangiskou M, Hultgren SJ (2013) Bacterial biofilms: development, dispersal, and therapeutic strategies in the dawn of the postantibiotic era. Cold Spring Harb Perspect Med 3(4):a010306

Kuo A, Callahan SM, Dunlap PV (1996) Modulation of luminescence operon expression by N-octanoyl-L-homoserine lactone in ainS mutants of *Vibrio fischeri*. J Bacteriol 178(4):971–976

Laue BE, Jiang Y, Chhabra SR et al (2000) The biocontrol strain *Pseudomonas fluorescens* F113 produces the Rhizobium small bacteriocin, N-(3-hydroxy-7-cis-tetradecenoyl) homoserine lactone, via HdtS, a putative novel N-acylhomoserine lactone synthase. Microbiology 146(10): 2469–2480

Lawrence RN, Dunn WR, Bycroft B et al (1999) The *Pseudomonas aeruginosa* quorum-sensing signal molecule, N-(3-oxododecanoyl)-l-homoserine lactone, inhibits porcine arterial smooth muscle contraction. Br J Pharmacol 128(4):845–848

LembreP, LorentzC, diMartinoP (2012) Exopolysaccharides of the biofilm matrix: a complex biophysical world. In: The complex world of polysaccharides. pp 371–392

Lenz DH, Miller MB, Zhu J, Kulkarni RV, Bassler BL (2005) CsrA and three redundant small RNAs regulate quorum sensing in *Vibrio cholerae*. Mol Microbiol 58(4):1186–1202

Lerat E, Moran NA (2004) Erratum: The evolutionary history of quorum-sensing systems in bacteria (Molecular Biology and Evolution (2004) 21 (903–913)). Mol Biol Evol 21(8):1612

Malott RJ, Sokol PA (2007) Expression of the bviIR and cepIR quorum-sensing systems of *Burkholderia vietnamiensis*. J Bacteriol 189(8):3006–3016

Marketon MM, Glenn SA, Eberhard A, González JE (2003) Quorum sensing controls exopolysaccharide production in *Sinorhizobium meliloti*. J Bacteriol 185(1):325–331

Marx V (2014) Cell communication: stop the microbial chatter. Nature 511:493–497

McClean KH, Winson MK, Fish L et al (1997) Quorum sensing and *Chromobacterium violaceum*: exploitation of violacein production and inhibition for the detection of N-acylhomoserine lactones. Microbiology 143(12):3703–3711

McDougald D, Rice SA, Barraud N, Steinberg PD, Kjelleberg S (2012) Should we stay or should we go: mechanisms and ecological consequences for biofilm dispersal. Nat Rev Microbiol 10(1):39

Miller MB, Bassler BL (2001) Quorum sensing in bacteria. Ann Rev Microbiol 55(1):165–199

Milton DL, Hardman A, Camara M, Chhabra SR, Bycroft BW, Stewart GS, Williams P (1997) Quorum sensing in *Vibrio anguillarum*: characterization of the vanI/vanR locus and identification of the autoinducer N-(3-oxodecanoyl)-L-homoserine lactone. J Bacteriol 179(9): 3004–3012

Milton DL, Chalker VJ, Kirke D, Hardman A, Cámara M, Williams P (2001) The LuxM homologue VanM from *Vibrio anguillarum* directs the synthesis of N-(3-Hydroxyhexanoyl) homoserine lactone and N-Hexanoyl homoserine lactone. J Bacteriol 183(12):3537–3547

Minogue TD, Trebra MWV, Bernhard F, Bodman SBV (2002) The autoregulatory role of EsaR, a quorum-sensing regulator in *Pantoea stewartii ssp. stewartii*: evidence for a repressor function. Mol Microbiol 44(6):1625–1635

Morohoshi T, Shiono T, Takidouchi K, Kato M, Kato N, Kato J, Ikeda T (2007) Inhibition of quorum sensing in *Serratia marcescens* AS-1 by synthetic analogs of N-acylhomoserine lactone. Appl Environ Microbiol 73(20):6339–6344

Morohoshi T, Kato M, Fukamachi K, Kato N, Ikeda T (2008) N-acylhomoserine lactone regulates violacein production in *Chromobacterium violaceum* type strain ATCC 12472. FEMS Microbiol Lett 279(1):124–130

Nadell CD, Xavier JB, Levin SA, Foster KR (2008) The evolution of quorum sensing in bacterial biofilms. PLoS Biol 6(1):e14

Nasser W, Bouillant ML, Salmond G, Reverchon S (1998) Characterization of the *Erwinia chrysanthemi* expI–expR locus directing the synthesis of two N-acyl-homoserine lactone signal molecules. Mol Microbiol 29(6):1391–1405

NomuraN, MiyazakiH (2009) US Patent No. 7537906. US Patent and Trademark Office, Washington

Ortori CA, Dubern JF, Chhabra SR, Cámara M, Hardie K, Williams P, Barrett DA (2011) Simultaneous quantitative profiling of N-acyl-L-homoserine lactone and 2-alkyl-4 (1H)-quinolone families of quorum-sensing signaling molecules using LC-MS/MS. Anal Bioanal Chem 399(2):839–850

O'Toole G, Kaplan HB, Kolter R (2000) Biofilm formation as microbial development. Ann Rev Microbiol 54(1):49–79

Pillai SK, Sakoulas G, Eliopoulos GM, Moellering RC Jr, Murray BE, Inouye RT (2004) Effects of glucose on fsr-mediated biofilm formation in *Enterococcus faecalis*. J Infect Dis 190(5):967–970

Puskas A, Greenberg EP, Kaplan S, Schaefer AL (1997) A quorum-sensing system in the free-living photosynthetic bacterium *Rhodobacter sphaeroides*. J Bacteriol 179(23):7530–7537

Quiñones B, Pujol CJ, Lindow SE (2004) Regulation of AHL production and its contribution to epiphytic fitness in *Pseudomonas syringae*. Mol Plant Microbe Interact 17(5):521–531

Rabin N, Zheng Y, Opoku-Temeng C, Du Y, Bonsu E, Sintim HO (2015) Biofilm formation mechanisms and targets for developing antibiofilm agents. Future Med Chem 7(4):493–512

Ramadhan AA, Hegedus E (2005) Biofilm formation and esp gene carriage in enterococci. J Clin Pathol 58(7):685–686

Rasmussen TB, Manefield M, Andersen JB, Eberl L, Anthoni U, Christophersen C et al (2000) How Delisea pulchra furanones affect quorum sensing and swarming motility in *Serratia liquefaciens* MG1. Microbiology 146(12):3237–3244

Reimmann C, Beyeler M, Latifi A, Winteler H, Foglino M, Lazdunski A, Haas D (1997) The global activator GacA of *Pseudomonas aeruginosa* PAO positively controls the production of the autoinducer N-butyryl-homoserine lactone and the formation of the virulence factors pyocyanin, cyanide, and lipase. Mol Microbiol 24(2):309–319

RuparellA (2012) Interplay between quorum sensing and metabolism in *Pseudomonas aeruginosa*. Doctoral dissertation, University of Nottingham

Ruysbergh E, Stevens CV, de Kimpe N, Mangelinckx S (2016) Synthesis and analysis of stable isotope-labelled N-acyl homoserine lactones. RSC Adv 6(77):73717–73730

Schuster JJ, Markx GH (2013) Biofilm architecture. In: Productive biofilms. Springer, Cham, pp 77–96

Schwartz K, Ganesan M, Payne DE, Solomon MJ, Boles BR (2016) Extracellular DNA facilitates the formation of functional amyloids in *Staphylococcus aureus* biofilms. Mol Microbiol 99(1):123–134

Schwarzer C, Fu Z, Morita T, Whitt AG, Neely AM, Li C, Machen TE (2015) Paraoxonase 2 serves a proapopotic function in mouse and human cells in response to the *Pseudomonas aeruginosa* quorum-sensing molecule N-(3-oxododecanoyl)-homoserine lactone. J Biol Chem 290(11):7247–7258

Seed PC, Passador L, Iglewski BH (1995) Activation of the *Pseudomonas aeruginosa* lasI gene by LasR and the *Pseudomonas* autoinducer PAI: an autoinduction regulatory hierarchy. J Bacteriol 177(3):654–659

Shaw PD, Ping G, Daly SL, Cha C, Cronan JE, Rinehart KL, Farrand SK (1997) Detecting and characterizing N-acyl-homoserine lactone signal molecules by thin-layer chromatography. Proc Nat Acad Sci 94(12):6036–6041

Shrout JD, Parsek MR (2006) Quorum sensing: coordinating group behavior through intercellular signals. In: Molecular paradigms of infectious disease. Springer, Boston, pp 404–437

Sjöblom S, Brader G, Koch G, Palva ET (2006) Cooperation of two distinct ExpR regulators controls quorum sensing specificity and virulence in the plant pathogen *Erwinia carotovora*. Mol Microbiol 60(6):1474-1489

Stauff DL, Bassler BL (2011) Quorum sensing in *Chromobacterium violaceum*: DNA recognition and gene regulation by the CviR receptor. J Bacteriol 193(15):3871–3878

Subramoni S, Sokol PA (2012) Quorum sensing systems influence *Burkholderia cenocepacia* virulence. Future Microbiol 7(12):1373–1387

Sureshchandra B (2010) Quorum sensing-cell to cell communication in bacteria. J Endod 22(2):97–101

Suzuki K, Babitzke P, Kushner SR, Romeo T (2006) Identification of a novel regulatory protein (CsrD) that targets the global regulatory RNAs CsrB and CsrC for degradation by RNase E. Genes Dev 20(18):2605–2617

Swift S, Karlyshev AV, Fish L, Durant EL, Winson MK, Chhabra SR, Williams P, Macintyre S, Stewart GS (1997) Quorum sensing in *Aeromonas hydrophila* and *Aeromonas salmonicida*: identification of the LuxRI homologs AhyRI and AsaRI and their cognate N-acylhomoserine lactone signal molecules. J Bacteriol 179(17):5271–5281

Swift S, Lynch MJ, Fish L, Kirke DF, Tomás JM, Stewart GS, Williams P (1999) Quorum sensing-dependent regulation and blockade of exoprotease production in *Aeromonas hydrophila*. Infect Immun 67(10):5192–5199

Telford G, Wheeler D, Williams P, Tomkins PT, Appleby P, Sewell H, Stewart GS, Bycroft BW, Pritchard DI (1998) The *Pseudomonas aeruginosa* quorum-sensing signal molecule n-(3-oxododecanoyl)-L-homoserine lactone has immunomodulatory activity. Infect Immun 66(1):36–42

Tsai CS, Winans SC (2011) The quorum-hindered transcription factor YenR of *Yersinia enterocolitica* inhibits pheromone production and promotes motility via a small non-coding RNA. Mol Microbiol 80(2):556–571

Ulrich RL, Hines HB, Parthasarathy N, Jeddeloh JA (2004) Mutational analysis and biochemical characterization of the *Burkholderia thailandensis* DW503 quorum-sensing network. J Bacteriol 186(13):4350–4360

Umesha S, Shivakumar J (2013) Bacterial quorum sensing and its application in biotechnology. Int J Pharm Bio Sci 4:850–861

Visick KL, Fuqua C (2005) Decoding microbial chatter: cell-cell communication in bacteria. J Bacteriol 187(16):5507–5519

von Bodman SB, Farrand SK (1995) Capsular polysaccharide biosynthesis and pathogenicity in *Erwinia stewartii* require induction by an N-acylhomoserine lactone autoinducer. J Bacteriol 177(17):5000–5008

Whitchurch CB, Tolker-Nielsen T, Ragas PC, Mattick JS (2002) Extracellular DNA required for bacterial biofilm formation. Science 295(5559):1487

Williams P, Camara M, Hardman A, Swift S, Milton D, Hope VJ, Winzer K, Middleton B, Pritchard DI, Bycroft BW (2000) Quorum sensing and the population-dependent control of virulence. Philos Trans R Soc Lond Ser B Biol Sci 355(1397):667–680

Winson MK, Camara M, Latifi A, Foglino M, Chhabra SR, Daykin M, Bally M, Chapon V, Salmond GP, Bycroft BW (1995) Multiple N-acyl-L-homoserine lactone signal molecules regulate production of virulence determinants and secondary metabolites in *Pseudomonas aeruginosa*. Proc Nat Acad Sci 92(20):9427–9431

Xavier KB, Bassler BL (2005) Interference with AI-2-mediated bacterial cell–cell communication. Nature 437(7059):750

Zavahir JS, Seneviratne G (2007) Potential of developed microbial biofilms in generating bioactive compounds. Res J Microbiol 2(4):397–401

Zhang C, Li B, Tang JY, Wang XL, Qin Z, Feng XQ (2017) Experimental and theoretical studies on the morphogenesis of bacterial biofilms. Soft Matter 13(40):7389–7397

Microbial Pathogenesis and Antimicrobial Drug Resistance

6

Indranil Chattopadhyay

Contents

Abstract

Antimicrobial drug resistance has become a serious threat and it caused the death of 700,000 individuals in 2016. Gram-negative bacteria such as Acinetobacter, Pseudomonas, Enterobacter spp. *Enterococcus faecium*, *Staphylococcus aureus*, and *Klebsiella pneumoniae* are insensitive to antibiotics. *E. faecium*, *S. aureus*, *K. pneumoniae*, *Acinetobacter baumannii*, *Pseudomonas aeruginosa*, and *Enterobacter spp.* are called as "ESKAPE" group of pathogens which have multidrug resistance property. Multidrug-resistant (MDR) bacteria are involved in

I. Chattopadhyay (✉)
Department of Life Sciences, Central University of Tamil Nadu, Thiruvarur, Tamil Nadu, India
e-mail: indranil@cutn.ac.in

B. Siddhardha et al. (eds.), *Model Organisms for Microbial Pathogenesis, Biofilm Formation and Antimicrobial Drug Discovery*,
https://doi.org/10.1007/978-981-15-1695-5_6

increased mortality and produce economic burden in the world. Bacteria escape the toxic action of antibiotics by increasing the efflux of antibiotics, inactivation of antibiotics through chemical modifications and modification of antibiotic targets. Bacteria use a genetic mechanism to avoid antibiotic effect through mutations in the gene(s) which are associated with antibiotic action and acquisition of resistance genes through horizontal gene transfer (HGT). The antibiotic resistance property in bacteria occurred through horizontal gene transfer such as transformation, transduction, and conjugation of plasmids or transposons, and mutations in the existing genes. Efflux pumps contribute antibiotic resistance at three levels such as intrinsic, acquired, and phenotypic. The human microbiome is considered as a reservoir of antibiotic resistance genes. Development of antibiotic resistance should be considered as an adaptive response in Darwinian's principles of evolution. Therefore, understanding the molecular mechanisms and evolution of multidrug-resistant bacteria are need to be studied.

Keywords
Antibiotics · Multiple drug resistance · Gram-positive and gram-negative bacteria · Efflux pumps · Antibiotic resistance genes

6.1 Introduction

Antimicrobial drug resistance is a phenomenon by which bacteria, parasites, and viruses modify themselves to bypass the action of antibiotics, antiviral, and antiparasitic drugs. Due to antimicrobial drug resistance, medical treatments such as surgery, organ transplants, chemotherapy, and diabetes management became a serious threat (WHO. Antimicrobial resistance. www.who.int/mediacentre/factsheets/fs194/en/. Accessed February 26, 2018). 700,000 deaths due to antimicrobial resistance were reported in 2016 and this number may increase to ten million annual deaths by 2050 (Bello and Dingle 2018). Antimicrobial resistance (AMR) has become a major global concern observed in gram-negative bacteria such as *Acinetobacter*, *Pseudomonas*, and *Enterobacter spp. Enterococcus faecium*, *Staphylococcus aureus*, and *Klebsiella pneumoniae* (Pham et al. 2019). *Acinetobacter baumannii*, *Campylobacter jejuni*, *Clostridium difficile*, *Enterobacter spp. Enterococcus faecium*, *E. faecalis*, *Escherichia coli*, *Haemophilus influenzae*, *K. pneumoniae*, *Pseudomonas aeruginosa*, *Salmonella spp. S. aureus*, *S. epidermidis*, and *Streptococcus pneumoniae* are commonly found in hospitals and develop resistance to variety of antibiotics. *Mycobacterium tuberculosis* is also extremely drug-resistant (Davies and Davies 2010). *E. faecium*, *S. aureus*, *K. pneumoniae*, *A. baumannii*, *P. aeruginosa*, and *Enterobacter* spp. are called as "ESKAPE" group of pathogens which have multidrug resistance property (Rice

Table 6.1 List of key antibiotics used in treatment of infections due to gram-negative bacteria

Name of antibiotics	Class of antibiotics	Applied in clinical condition
Amoxycillin, ampicillin	Penicillin	Respiratory tractinfections
Imipenem, meropenem, doripenem ertapenem	Carbapenems	Infections of gram-negative bacteria
Doxycycline, minocycline	Tetracyclines	Minor infections in respiratory tract
Gentamicin	Aminoglycosides	Endocarditis
Norfloxacin, ciprofloxacin, Moxifloxacin	Quinolones	Infections in urinary tract
Polymyxin B, colistin	Polymyxins	Treatment of multi-resistant gram-negative infections
Chloramphenicol	Phenicols	Bacterial meningitis
Erythromycin	Macrolides	Treatment of minor infections Due to gram-positive bacteria

2008). These bacteria developed multidrug resistance property through inactivation of drug, modification of drug target site, and enhancement of efflux of drug (Santajit and Indrawattana 2016). Biofilm formation by these bacteria inhibits the action of antibiotics against these bacteria (Lewis 2007). Antibiotics (Table 6.1) inhibit the growth of gram-negative bacteria by crossing the cell envelope. Bacteria escape the toxic action of antibiotics by increasing the efflux of antibiotics, inactivation of antibiotics through chemical modifications and modification of antibiotic targets (Kohanski et al. 2010). The outer membrane (OM) of gram-negative bacteria (GNB) prevents the movement of amphipathic drugs whereas an inner membrane of GNB inhibits the transport of hydrophilic drugs (Masi et al. 2017). 6-Deoxynybomycin amine is effective against multidrug-resistant gram-negative bacteria such as *E. coli*, *K. pneumoniae*, and *A. baumannii* (Richter et al. 2017). Antimicrobial peptide, Lassomycin from *Lentsea kentuckyensis* inhibits ATP-dependent protease complex ClpP1P2C1 protease in *Mycobacterium tuberculosis*. Teixobactin which is the product of *Eleftheria terrae* prevents the biosynthesis of cell wall in gram-positive bacteria (Pham et al. 2019). *P. aeruginosa* developed antibiotic resistance during long antibiotic treatment of cystic fibrosis patients (Horrevorts et al. 1990). Twenty-eight genomic islands encoding antibiotic resistance property have been reported in *A. baumannii* (Barbe et al. 2004). More than 20,000 potential resistance genes are discovered in available bacterial genome sequences (Liu and Pop 2009). Extended-spectrum-β-lactamase (ESBL)-producing Enterobacteriaceae, vancomycin-resistant enterococci (VRE), carbapenem-resistant Enterobacteriaceae (CRE), and *Clostridium difficile* are considered as a serious threat within hospitals (Pamer 2016). Discovery of antibiotics and its resistance property against bacteria are thrust area of research in academic and pharmaceutical world. In this chapter, antibacterial drug resistance mechanisms in relevant bacterial pathogens are emphasized.

6.2 Demographic Distribution of Antimicrobial Drug Resistance Strains

The incidence of AMR has been highly reported in Asia, Africa, and the Middle East. In India, China, and Pakistan, people consume large number of antibiotics. In Africa, high prevalence of chloramphenicol, trimethoprim-sulfamethoxazole, and tetracycline resistant gram-negative bacteria was reported. These strains are susceptible to third-generation cephalosporins and fluoroquinolones. Travel across international borders enhances the exposure of antimicrobial resistance strains. Multidrug-resistant (MDR) *S. pneumoniae* strain spreads from South Africa to Europe whereas drug-resistant gonorrhea spreads from Asia to the Pacific and North America. New Delhi Metallo-beta-lactamase 1 (blaNDM-1) gene dramatically migrated from India and Pakistan to Europe. The colistin-resistant (*mcr-1*) gene was first reported from people in China in 2016 and later on it migrated to the USA. Migration of population is a major source of burden of AMR in Europe and other developed countries. Over burden of population in refugee camps induces transmission of AMR. Carriers of extended-spectrum β-lactamase enzyme (ESBL-E) genes are reported in 1.1 billion population in Southeast Asia, 280 million in Western Pacific region, and 110 million in Africa. Patients hospitalized in Asia, sub-Saharan Africa, and Latin America were more prone to get infected with methicillin-resistant *S. aureus* (MRSA). It has been reported about infection of carbapenem-resistant *A. baumannii* among soldiers in Iraq during gulf war (Semret and Haraoui 2019). Quinolones and cephalosporins resistant strains have been reported in Pakistan in 2016 (Wain et al. 2015). Global Gonococcal Antimicrobial Surveillance Program in participating countries revealed that 97% gonococcal isolates are resistant to ciprofloxacin (Unemo and Shafer 2014). Vancomycin-resistant Enterococci (VRE) are reported in the Europe (van den Bogaard et al. 1997). Quinolone-resistant *Salmonella enterica* are reported in Denmark and Taiwan (Molbak et al. 2002). Hospital wastewater and wastewater treatment plants which are contaminated by antibiotic contribute to the development of bacterial resistance (Johnning et al. 2013) (Table 6.2). Infections of carbapenem-resistant bacterial strains such as *E. coli*, *K. pneumoniae*, *P. aeruginosa* and *A. baumannii* were reported in blood of 40 patients at Tata Medical Center, Kolkata, India. Out of 40 patients, 21 patients had hematologic malignancy and 19 patients had solid tumor (Exner et al. 2017). *E. coli*, *K. pneumoniae*, *K. oxytoca*, and *P. mirabilis* which are positive for ESBL are reported in south-eastern Asian countries such as Thailand, Singapore, Malaysia, Vietnam, Indonesia, Philippines, Laos, Cambodia, Myanmar, and Brunei. The percentage of carbapenem-resistant (CR) *A. baumannii* and *P. aeruginosa* was 76% and 23%, respectively (Suwantarat and Carroll 2016). The pathogens responsible for nosocomial infections such as *E. coli*, *E. faecalis*, *E. faecium*, *Enterococci*, *S. aureus*, and *C. difficile* resistant to antibiotics such as cephalosporins, penicillins, and carbapenems were reported in Germany in 2011 with frequency of 18, 13.2, 13.1, and 8.1%, respectively (Behnke et al. 2013).

Table 6.2 Sources of antibiotic resistance bacteria

Antibiotic resistance bacterial species	Common types of antibiotic resistance	Source	Types of infection
Streptococcus pneumonia	Penicillin, macrolides, cephalosporins, tetracyclines	Childcare facilities	Pneumonia
Streptococcus pyogenes	Macrolides, tetracyclines	Schools	Pharyngitis
Staphylococcus aureus	Methicillin, cephalosporins, macrolides	Hospitals, soldiers	Pneumonia, sepsis
Enterococcus spp.	Ampicillin, vancomycin, aminoglycosides	Hospitals	Urinary tract
Neisseria gonorrhoeae	Penicillin, cephalosporins, quinolones	Sex workers	Inflammatory disease
Salmonella spp.	Cephalosporins, quinolones, tetracyclines	Poultry	Diarrhea
Campylobacter jejuni	Erythromycin, quinolones	Poultry	Gastroenteritis
Escherichia coli	Trimethoprim, sulphonamides, quinolones	Childcare facilities	Urinary tract, diarrhea

6.3 Molecular Mechanism of Antibiotic Resistance

Bacteria use genetic mechanism to avoid antibiotic effect through mutations in gene(s) which are associated with antibiotic action and acquisition of resistance genes through horizontal gene transfer (HGT). The antibiotic resistance property in bacteria occurred through horizontal gene transfer such as transformation, transduction, and conjugation of plasmids or transposons, and mutations in the existing genes (Giedraitienė et al. 2011). The antimicrobial resistance occurred through the molecular mechanisms such as modifications of the antibiotic target site that reduces the affinity for the antibiotics, reduced uptake of antibiotic, activation of efflux pumps and alterations of metabolic pathways (Munita and Arias 2016). Conjugation of mobile genetic elements (MGEs) such as plasmids and transposons occurs at high frequency in the gastrointestinal tract of humans in patients undergoing antibiotic treatment in the hospital (Manson et al. 2010). Other antimicrobial resistance genes are integrons which are site-specific recombination systems. Integrons are responsible for the incorporation of new genes into bacterial chromosomes (Thomas and Nielsen 2005). Fluoroquinolone (FQ) resistance developed in the bacteria through mutations in DNA gyrase and topoisomerase IV and overexpression of efflux pumps. Enzymes which are involved in chemical alterations of antibiotics developed antibiotic resistance in both gram-negative and gram-positive bacteria. They catalyzed acetylation (aminoglycosides, chloramphenicol, and streptogramins), phosphorylation (aminoglycosides, chloramphenicol), and adenylation (aminoglycosides, lincosamides) (Munita and Arias 2016). Aminoglycoside modifying enzymes (AMEs) are involved in covalent modification in hydroxyl or amino groups of the aminoglycoside molecule. Aminoglycoside acetyltransferases are reported in *Providencia stuartii*, *E. faecium*, and *S. marcescens* (Ramirez and Tolmasky 2010). Phosphotransferase (APH) is responsible to altered kanamycin and streptomycin in gram-positive and gram-negative bacteria. Acetyltransferase

(ACC) in *Enterobacteriaceae*, *Pseudomonas*, and *Acinetobacter* effects aminoglycosides including amikacin and gentamicin. High-level gentamicin and vancomycin resistance are detected in enterococci and methicillin-resistant in *S. aureus* (Ramirez and Tolmasky 2010; Hollenbeck and Rice 2012). Chloramphenicol acetyltransferases (CATs) are involved in chemical modification of chloramphenicol which inhibits protein synthesis by binding peptidyl-transfer center of the 50S ribosomal subunit. High-level resistance type A cat gene and low-level resistance type B cat gene are reported in both gram-positive and gram-negative bacteria (Schwarz et al. 2004).

Efflux of antibiotics into bacterial cell before it reaches into the target site is due to overexpression of transmembrane multidrug efflux pumps (Alcalde-Rico et al. 2016). In *P. aeruginosa*, mutation in porins results in the development of resistance against antibiotics such as carbapenems by reducing the permeability of the cell wall. In *E. coli*, transmembrane proton gradient efflux pumps expel multiple antibiotics. In gram-positive bacteria, the outer membrane inhibits transport of hydrophobic drugs (Shaikh et al. 2015). In gram-positive and gram-negative bacteria, tetracycline efflux pumps (*TetA*) protein is normally not expressed but TetR repressor protein is expressed. Tetracycline binds and inactivates TetR that drives activation of TetA which induces the efflux of antibiotic. Penicillin-resistant gene (MecA) in *Streptococcus pneumoniae*; vancomycin-resistant gene (*vanA*) in *S. aureus*; *Enterococcus* and sulfonamide-resistant gene in *S. pneumoniae*, *S. pyogenes*, *Neisseria* spp., and *E. coli* (Pelgrift and Friedman 2013) were reported. Genes encoding aminoglycoside-modifying enzymes in plasmids or transposons altered the binding affinity for 30S ribosomal subunit by modifying OH or NH_2 groups on aminoglycosides. Mutations of β-lactamase genes (Table 6.3) enhance the resistance property of bacteria against β-lactam antibiotics such as cephalosporins (Shaikh et al. 2015). *S. aureus* and *Neisseria meningitidis* developed resistance against sulfonamide by secreting para-aminobenzoic acid which binds with the active site of bacterial dihydropteroate synthetase (Yun et al. 2012). Streptomycetes produce a variety of β-lactamases (Ogawara et al. 1999). *S. pneumoniae* acquired resistance property by alterations of penicillin-binding proteins (PBPs) which reduced the binding affinity of drug (Furuya and Lowy 2006). Healthcare-associated methicillin-resistant *S. aureus* MRSA (HA-MRSA) and community-associated methicillin resistant *S. aureus* (CA-MRSA) carried mobile chromosomal element which is known as staphylococcal chromosomal cassette (SCC). SCC

Table 6.3 Key antibiotic resistance mechanism involved in gram-positive pathogens

Gene(s) involved	Antibiotic	Resistance mechanisms in pathogens
blaZ	β-Lactams (penicillins)	Penicillins are hydrolyzed by plasmid-encoded β-lactamase
mecA	β-Lactams	Low affinity of PBP2a
23S rRNA genes, L3/L4 ribosomal proteins	Linezolid	Mutations in 23S rRNA genes
mprF, dlt, *vraRS*, *yycFG*, *pgsA*, *cls*	Daptomycin	Repulsion of the antibiotic due to increased positive charge of cell envelope

carries the methicillin-resistance gene (mec). SCCmec element from groups I–III are reported in HA-MRSA whereas SCCmec type IV and SCCmec type V are reported in CA-MRSA. Among them, SCCmec type IV is more mobile. CA-MRSA is more genetically diverse as compared to HA-MRSA (Daum et al. 2002; Hiramatsu et al. 2002).

6.4 Role of Bacterial Outer Membrane in Influx and Efflux of Antibiotic

Gram-negative bacteria acquired antibiotic resistance due to cell envelope which consists of an outer membrane (OM) and inner membrane (IM). Periplasmic space separates OM and IM (Zgurskaya et al. 2015). The most abundant OM proteins such as OmpF and OmpC are reported in *E. coli* (Baslé et al. 2006). The permeability of the OM in gram-negative bacteria becomes lower to antibiotics. Multidrug efflux pumps belonging to the ABC (ATP-binding cassette family), MF (the major facilitator superfamily), SMR (small multidrug resistance family), MATE (multidrug and toxic compound extrusion family), PACE and RND (resistance-nodulation-cell division) super families are reported in bacterial genomes. Reduced numbers of porins in the OM and modular tripartite efflux pumps (ABC, MFS, or RND) in IM provide resistance against antibiotics (Nikaido and Pagès 2012). The ABC exporters have two categories such as homodimeric and heterodimeric. The heterodimeric ABC exporters do not support ATP hydrolysis. The ABC exporter MacB is responsible for the development of antibiotic resistance in *E. coli* and other gram-negative bacteria. The MATE transporters are classified into NorM, DinF (DNA-damage inducible protein F), and eukaryotic subfamilies on the basis of their amino acid sequence homology. MATE transporters efflux of polyaromatic and cationic drugs through transmembrane H+ and/or Na+ gradients. RND family such as AcrB, MexB, MtrD, and CmeB are reported in *E. coli*, *P. aeruginosa*, *N. gonorrhoeae*, and *C. jejuni*, respectively. *A. chlorhexidine* efflux protein (AceI) of PACE family developed resistance against chlorhexidine, acriflavine, proflavine, and benzalkonium. Two-component systems (TCSs) such as CpxAR in Enterobacteriaceae, AdeSR in *A. baumannii*, and AmgRS in *P. aeruginosa* control the expression of multidrug resistance efflux pumps. *A. baumannii* acquired mutations in adeSR and overexpression of the RND pump AdeAB. AdeSR is responsible for the expression of the tripartite pump system such as ade-ABC and biofilm formation. TCSs also regulate the expression of efflux pumps such as BaeSR and AdeAB and AdeIJK RND in *A. baumannii*. In *E. coli*, five out of 15 TCSs regulates the expression of drug efflux pump genes. Mutations in tetracycline repressor protein (TetR) family of transcriptional repressors (emrR, acrR, and mtrR) enhance the overexpression of efflux pumps (Du et al. 2018). Efflux pumps such as mefA and mefE are responsible to efflux erythromycin and are mainly reported in *S. pyogenes* and *S. pneumonia*. Transposon Tn1207 is located in MefA (Ross et al. 1990). RND pumps developed resistance to tetracyclines, chloramphenicol, some β-lactams, novobiocin, fusidic acid, and fluoroquinolones (Munita and Arias 2016).

6.5 Role of Colistin Resistance Gene in Microbial Pathogenesis

Colistin of *Paenibacillus polymyxa* is an antibiotic. Colistin is widely used in animal husbandry. Colistin-mediated resistance gene (*mcr-1* gene) was found in *S. enterica* and *E. coli*. Colistin resistance genes in plasmids are horizontally transmitted across the bacteria. Positively charged diaminobutyric acid (Dab) residues of colistin primarily interact with the negatively charged phosphate groups of lipid A of lipopolysaccharide (LPS) which is present in outer-membrane (OM) of gram-negative bacteria. Modification of LPS such as addition of 4-amino-4-deoxy-L-arabinose (L-Ara4N) and/or phosphoethanolamine (PEtn) to lipid A moiety of LPS induces colistin resistance through reducing the negative charge of LPS. Colistin resistance *mcr* genes encode a phosphoethanolamine transferase that induces the binding of PEtn to lipid A. This reduces binding of colistin to LPS. Eight types of mcr genes (mcr-1 to -8) have been reported in different geographical regions such as mcr-2 gene in *E. coli* from pigs and calves of Belgium, mcr-3 in *E. coli* from pigs of China, mcr-4 in *S. enterica serovar Typhimurium* from pigs of Italy, mcr-5 in *S. paratyphi* B dTa+ from poultry of Germany, mcr-6 in Moraxella spp. from pigs of Great Britain, mcr-7 in *Klebsiella pneumoniae* from chickens of China, and mcr-8 in K. pneumonia from pigs and humans of China (Lima et al. 2019). Animals are the primary source of non-typhoidal Salmonella (NTS). 93.8 million illness and 155,000 deaths annually in world due to NTS gastroenteritis have been reported (Majowicz et al. 2010). In *S. enterica*, colistin resistance was developed due to PmrA/PmrB and PhoP/PhoQ two-component regulatory systems that induced the biosynthesis of L-Ara4N and PEtn. Constitutive expression of PmrA/PmrB and PhoP/PhoQ enhanced binding of L-Ara4N and PEtn, respectively, to lipid A (Olaitan et al. 2014).

6.6 Molecular Mechanisms of AMR in Gram-Negative Bacteria

Several multidrug resistance mechanisms are involved in Enterobacteriaceae family of gram-negative bacteria such as enzymatic degradation, modification of target site, overexpression of efflux pumps, and reduction of cell permeability. β-Lactamase enzymes hydrolyze the β-lactam ring of penicillin and cephalosporins to make them inactive. AmpC β-lactamases, extended-spectrum-β-lactamases (ESBLs), and carbapenemase are responsible for the degradation of cephalosporins. Development of fluoroquinolone resistance acquired as a result of modification of enzymes such as DNA gyrase and DNA topoisomerase IV. Antimicrobials are preferred to bind with porins which are present at the outer membrane of gram-negative bacteria. Loss of porins reduces the permeability of cell wall and prevents the entry of antibiotics into the cell. Resistance-nodulation-division (RND) efflux systems ArcAB-TolC of *E. coli* and MexAB-OprM of *P. aeruginosa* are effective against cephalosporins, fluoroquinolones, penicillin, and chloramphenicol. Trimethoprim resistance is developed due to alteration of target site of trimethoprim (Mukerji et al. 2017; Verraes et al. 2013; Li and Nikaido 2009). Fluoroquinolones are used during urinary tract, respiratory tract, and gastrointestinal infections. Fluoroquinolone resistance is acquired due to mutation in DNA gyrase (*gyrA*, *gyrB*) and DNA topoisomerase IV genes (*parC*, *parE*) (Cavaco et al. 2008). Four types of β-lactamases such as class A serine β-lactamases (ESBLs, penicillinases), class B metallo-β-lactamases, class C AmpC-type-β-lactamases, and class D OXA β-lactamases are reported. Among them, ESBL cephalosporinases (CTX-M type enzymes) are common β-lactamases and these developed resistance against penicillin and cephalosporins (Bush and Jacoby 2010). Different types of carbapenemase enzymes such as class D enzymes (OXA family) of carbapenemases in Enterobacter and *P. aeruginosa*, metallo-β-lactamases (IMP, VIM family) in *Klebsiella* and *Enterobacter* spp., and non-metallo-carbapenemases (SME, IMI/NMC) in *Serratia* and *Enterobacter* spp. are reported (Carattoli 2009). The KPC enzymes in *K. pneumoniae* and *E. cloacae* are responsible for the development of resistance against all β-lactams such as cephalosporins, monobactams, and carbapenems (Livermore and Woodford 2006). Class C β-lactamases developed resistance to all penicillins and cephalosporins. AmpC (cephalosporinase) is most clinically relevant class C enzyme and this has been reported in *E. cloacae*, *E. aerogenes*, *C. freundii*, *S. marcescens*, *Providencia* sp. *Morganella morganii*, and *P. aeruginosa* (Jacoby 2009). Class D β-lactamases hydrolyze oxacillin and many OXA variants such as OXA-11 from *P. aeruginosa*, OXA-23 from *A. baumannii*, and OXA-48 *K. pneumoniae* are able to degrade third-generation cephalosporins (Evans and Amyes 2014).

6.7 Molecular Mechanisms of AMR in Gram-Positive Bacteria

MDR gram-positive organisms such as methicillin-resistant *S. aureus* (MRSA), vancomycin-resistant *E. faecium* (VRE), and drug-resistant *Streptococcus pneumoniae* are serious public threats. (Centers for Disease Control and Prevention. Antibiotic resistance threats in the United States. Available at: http://www.cdc.gov/drugresistance/threat-report-2013/index.html. Accessed 9 March 2015). Staphylococcal-type Aβ-lactamase is responsible for hydrolysis of cefazolin. Methicillin resistance depends on gaining of staphylococcal chromosomal cassette mec (SCCmec) which contains mecA. Transpeptidase PBP2a which is encoded by mecA showed low affinity for all β-lactams. Ceftaroline and ceftobiprole resistance results due to substitution mutations such as Y446N and E447K in penicillin-binding site of transpeptidase domain of PBP2a (DeLeo and Chambers 2009; Kelley et al. 2015). Pneumococcal β-lactam resistance results from alterations in native PBPs through recombination with exogenous pbp genes. The resistance property is enhanced in the presence of point mutations in pbp genes (Munita et al. 2015). Mutations in genes encode murein (*murM*), GlcNAc deacetylase (*pdgA*), and glycosyltransferase (*cpoA*) and developed β-lactam resistance in *Pneumococci* (Hakenbeck et al. 2012). Oxazolidinone resistance results due to mutations in the 23S ribosomal RNA (rRNA) genes, genes encoding L3/L4 ribosomal proteins, and methylation of the 23S rRNA gene (Munita et al. 2015). Daptomycin (DAP) is effective against gram-positive bacteria. DAP resistance in *S. aureus* results due to electrostatic repulsion of the DAP calcium complex from cell surface (Bayer et al. 2013). DAP resistance in *S. aureus* results due to increase in the content of positive charge phospholipid lysyl-phosphatidylglycerol in cell envelop through upregulation of transmembrane protein MprF (Tran et al. 2013).

6.8 Role of Efflux Pumps in Gram-Positive and Gram-Negative Bacteria

Efflux pumps (Tables 6.4 and 6.5) contribute antibiotic resistance at three levels such as intrinsic, acquired, and phenotypic (Fajardo et al. 2008). Acquired resistance can be achieved by horizontal gene transfer and mutations that over expressed chromosomally encoded efflux pumps. Bacteria developed intrinsic resistome towards antibiotics through constitutively lower expression of efflux pumps (Olivares et al. 2013). Phenotypic resistance is defined as inheritable overexpression of an efflux pump in a specific growing condition. In gram-negative bacteria, tripartite complexes are made up of inner-membrane efflux pump, an outer-membrane protein, and a membrane fusion protein. Multi AnTimicrobial Extrusion (MATE) transporters are divided into three subfamilies such as DinF, NorM, and the eukaryotic subfamilies (Lu 2015). The major facilitator superfamily (MFS) consists of importers and exporters (Law et al. 2008). MFS is most predominant in gram-positive bacteria. The *E. coli* EmrAB-TolC efflux pump is one of the examples of MFS. The Resistance Nodulation and cell Division (RND) superfamily of efflux pumps is commonly reported in gram-negative bacteria. These are composed of three different proteins such as active efflux pump, an outer-membrane protein, and a fusion protein. The RND efflux pumps which contributed to intrinsic resistance are AcrAB-TolC in Enterobacteriaceae, MexAB-OprM and MexXY in *P. aeruginosa*, and SmeDEF in *Stenotrophomonas maltophilia* (Hernando-Amado et al. 2016).

S. aureus acquired MFS efflux pumps QacA and QacB which reduced susceptibility of *S. aureus* towards antibiotics commonly used in nosocomial infections. (Wassenaar et al. 2015). MFS efflux pumps such as NorA, NorB, NorC, and NorD in *S. aureus* are responsible for the development of resistance against norfloxacin, ciprofloxacin, tetraphenylphosphonium, and cetrimide (Costa et al. 2013). NorD is responsible to develop resistance against norfloxacin, moxifloxacin, delafloxacin, levofloxacin nalidixic acid, ciprofloxacin, tetracycline, polymyxin B, trimethoprim, daptomycin, and triclosan. MdeA which belongs to MFS efflux pump is responsible to efflux norfloxacin and tetraphenylphosphonium (Yamada et al. 2006). Plasmid-borne resistance pumps such as QacG, QacH, QacJ, and Smr are observed in *S. haemolyticus* (Correa et al. 2008). Two MFS efflux pumps such as MdrL and Lde are reported in *Listeria monocytogenes*. Lde provided resistance against fluoroquinolones, whereas overexpression of MdrL provided resistance

Table 6.4 Clinically important efflux pumps present in gram-positive bacteria

Family of efflux pump	Efflux Pump	Organism	Location	Efflux of antibiotics by pump
MFS	QacA	*S. aureus*	Plasmid	Benzalkonium chloride, cetrimide, propamidine, isethionate, diaminodiphenylamine dihydrochloride, pentamidine, chlorhexidine, acriflavine
MFS	QacB	*S. aureus*	Plasmid	Chlorhexidine, benzalkonium chloride, tetraphenylphosphonium, acriflavine
MFS	NorA	*S. aureus*	Chromosome	Norfloxacin, enoxacin, ofloxacin, ciprofloxacin, pentamidine, cetrimide, benzalkonium chloride, tetraphenylphosphonium, bromide, acriflavine
MFS	NorC	*S. aureus*	Chromosome	Norfloxacin, ciprofloxacin, sparfloxacin, gemifloxacin,
MFS	NorD	*S. aureus*	Chromosome	Polymyxin B, nalidixic acid, trimethoprim, daptomycin, tetracycline, norfloxacin, daptomycin
MFS	MefA	*Enterococcus*	Chromosome	Erythromycin
MFS	Tap	*Mycobacterium*	Chromosome	Aminoglycosides, tetracycline, rifampicin, clofazimine, acriflavine
MFS	JefA	*Mycobacterium*	Chromosome	Isoniazid, ethambutol, streptomycin
MATE	MepA	*S. aureus*	Chromosome	Ciprofloxacin, norfloxacin, moxifloxacin, sparfloxacin, tigecycline, pentamidine, cetrimide, benzalkonim chloride, dequalinium tetraphenyl-phosphonium, chlorhexidine, acriflavine
ABC	MsrB & MsrA	*Enterococcus*	Chromosome	Erythromycin
MFS	PmrA	*Streptococcus pneumoniae*	Chromosome	Fluoroquinolones
ABC	PatAB	*Streptococcus pneumoniae*	Chromosome	Fluoroquinolones
MATE	PdrM	*Streptococcus pneumoniae*	Chromosome	Chloramphenicol Erythromycin

against benzalkonium (Godreuil et al. 2003; Romanova et al. 2006). Mutations in gyrA and overexpression of efflux pumps are responsible to develop ofloxacin resistance in *M. tuberculosis* (Sun et al. 2014). MFS efflux pump such as JefA contributed resistance against ethambutol and isoniazid. MFS efflux pump such as Tap is responsible for the development of resistance property against aminoglycosides and tetracyclines (Gupta et al. 2006; Ramon-Garcia et al. 2009). Tap has been detected in *Mycobacterium abscessus*, *M. chelonae*, *M. fortuitum*, *M. mageritense*, *M. peregrinum*, *M. alvei*, and *M. porcinum* (De Groote and Huitt 2006).

Table 6.5 Clinically important RND family efflux pumps present in gram-negative bacteria

Family of efflux pump	Efflux pump	Organism	Efflux of antibiotics by pump
RND	AdeABC	*Acinetobacter baumannii*	Aminoglycosides, cephalosporins, fluoroquinolones, tetracyclines-tigecycline, macrolides, chloramphenicol, and trimethoprim
RND	AdeFGH	*Acinetobacter baumannii*	Fluoroquinolones, tetracyclines-tigecycline, chloramphenicol, lincosamides, sulfonamides and trimethoprim
RND	AdeIJK	*Acinetobacter baumannii*	β-lactams, cephalosporins, fluoroquinolones, tetracyclines-tigecycline, macrolides, lincosamides, novobiocin, rifampicin, cotrimoxazole, trimethoprim, chloramphenicol and fusidic acid
RND	MtrCDE	*Neisseria gonorrhoeae*	Aminoglycosides; penicillin (β- lactams); azithromycin (macrolides); ceftriaxone
RND	MexAB–OprM	*Pseudomonas aeruginosa*	Aminoglycosides; amphenicols; β- lactams (except imipenem); fluoroquinolones; macrolides; novobiocin; sulfonamides; tetracyclines; thiolactomycin; tigecycline; trimethoprim
RND	AcrAB–TolC	*Enterobacteriaceae*	β-lactams; chloramphenicol; erythromycin; fluoroquinolones; novobiocin; tetracycline; linezolid

AcrAB-TolC contributed antibiotic resistance property of *K. pneumoniae*, *Salmonella*, and *Enterobacter* (Hernando-Amado et al. 2016). MexAB-OprM and MexXY-OprM which belonged to RND family are responsible for the development of intrinsic antibiotic resistance property in *P. aeruginosa* (Morita et al. 2001). Clinically relevant RND efflux systems such as SmeABC, SmeDEF, SmeJK, SmeVWX, and SmeYZ are reported in *S. maltophilia* (Hernando-Amado et al. 2016). SmeJK is involved in the development of resistance against aminoglycosides, tetracyclines, and fluoroquinolones whereas SmeYZ is responsible for the development of intrinsic aminoglycosides resistance (Sanchez 2015). MacABCsm which belongs to ABC efflux pump is responsible for the development of intrinsic resistance to polymyxins, macrolides, and aminoglycosides in *S. maltophilia* (Lin et al. 2014).

6.9 Role of Protection, Modification, and Enzymatic Alteration of Target Site in AMR Development

Tetracycline resistance determinants Tet(M) in Streptococcus spp. and Tet(O) in *C. jejuni* are examples of the target protection mechanism. TetO and TetM bind with the ribosome and replace tetracycline from its binding site through GTP-dependent mechanisms. TetO competes with tetracycline for the binding site at ribosome and allows to continue protein synthesis (Li et al. 2013). Plasmid-mediated fluoroquinolone resistance gene Qnr is reported in a clinical isolate of K. pneumonia. It competes for the binding site of the DNA gyrase and topoisomerase IV (Aldred et al. 2014). Modifications of target site occur through mutations in the genes encoding the target site, enzymatic alterations of antibiotic binding site such as methyl group addition and replacement of target site (Munita and Arias 2016). Rifamycin inhibits the activity of DNA-dependent RNA polymerase which is composed of α2ββ'σ subunits. Rifamycin binds with the β subunit of the RNA polymerase (rpoB). Amino acid substitutions mutations in the rpoB gene prevent the antibiotic to bind with the rpoB which develops rifampin resistance (Floss and Yu 2005). Genetic alterations in DNA gyrase and topoisomerase IV developed antibiotic resistance. Resistance to oxazolidinones such as linezolid and tedizolid developed due to mutations in genes encoding the domain V of the 23S rRNA in the 50S ribosomal subunit and/or substitutions in the ribosomal proteins L3 (rplC) and L4 (rplD) (Mendes et al. 2014). Resistance to erythromycin is due to mono- or demethylation of an adenine residue at A2058 of the domain V of the 23rRNA of the 50S ribosomal subunit (Weisblum 1995). Resistance to methicillin in *S. aureus* occurs due to gaining of staphylococcal chromosomal cassette mec (SCCmec) (Chambers and Deleo 2009). Vancomycin resistance in enterococci results due to acquiring of van gene clusters which alter D-Ala to D-lactate and D-Ala to D-serine in peptidoglycan for developing high and low resistance, respectively (Arthur 2010).

6.10 Multidrug Resistance *Mycobacterium tuberculosis* in Development of MDR Tuberculosis

As per WHO report, 457,000 multidrug-resistant TB (MDR-TB) cases were reported in 2017; out of which 8.5% cases were considered as extensively drug-resistant TB (XDR-TB) (*Global Tuberculosis Report Geneva. Geneva, Switzerland: World Health Organisation, 2018*). Mycobacterial cell wall is composed of peptidoglycan (PG), mycolic acid (MA), and arabinogalactan (AG) (Maitra et al. 2019). Drugs for MDR treatment have been classified into following group as World Health Organization (WHO) treatment guidelines: fluoroquinolones (FQ) such as levofloxacin, moxifloxacin, and gatifloxacin; amikacin (AMK), capreomycin (CAP), kanamycin (KAN), and streptomycin (STR); ethionamide (ETH), prothionamide (PTH), cycloserine (CS), terizidone, linezolid (LZD), clofazimine (CFZ); and pyrazinamide (PZA), ethambutol (EMB), high-dose isoniazid (INH), bedaquiline (BDQ), delamanid, (DLM), para-aminosalicylic acid (PAS), imipenem, cilastatin, meropenem, amoxicillin-clavulanate, and thioacetazone (Miotto et al. 2018). Rifampicin (RIF) resistant MTB strains have mutations in the codons 450, 445, and 435 of β-subunit of RNA polymerase (rpoB) (Jamieson et al. 2014). PZA resistance results from mutations in pncA. EMB resistance results due to missense mutations at codons 306, 406, and 497 of embCAB operon. Enhancement of EMB resistance results due to missense mutation in Rv3806c (ubiA) V188A, A237V, R240C, and A249G. (Miotto et al. 2018). FQ resistance in MTB is caused by mutations at

Table 6.6 Key MDR genes involved in MDR TB

Resistance-related genes	Function of gene	Name of the drugs
rpoB	RNA polymerase subunit B	Rifampicin
rpsL	Ribosomal protein S12	Streptomycin
gyrA and *gyrB*	DNA gyrase subunit A DNA gyrase subunit B	Quinolones
embB	Arabinosyl transferase	Ethambutol
Rrs	16S rRNA	Kanamycin/amikacin
ahpC	Alkyl hyperperoxide reductase	Isoniazid

codons 90, 91, and 94 of gyrA (Lu et al. 2014) (Table 6.6). KAN and AMK resistance are caused by mutations at nucleotide positions 1401 and 1402 of rrs gene (Georghiou et al. 2012). MDR-TB patients developed resistance against isoniazid and rifampicin whereas extensively drug-resistant (XDR) TB patients developed resistance against kanamycin, amikacin, or capreomycin. *M. tuberculosis* developed resistance due to chromosomal mutations. The rate of resistance mutations was estimated at 10^{-8} and 10^{-9} mutations/bacterium/cell division for isoniazid and rifampicin, respectively (Müller et al. 2013).

6.11 Antimicrobial Resistance Genes in Human Microbiome

Antimicrobial resistance genes (ARGs) are distributed through horizontal gene transfer (HGT), conjugation, phage transduction, or transformation. Human microbiome is considered as a reservoir of ARGs. Human microbiome is the source of about 3.3 million non-redundant genes. Human gut has 10^{14} microbial cells which represent 400 different bacterial phylotypes (Brinkac et al. 2017). Genes resistant to tetracycline (e.g., *tet*(*M*), *tet*(*O*), *tet*(*Q*), and *tet*(*W*)), amoxicillin, and erythromycin are predominant in oral microbiome. Streptococci are the principal carriers of AMR (tet genes) in the oral cavity of children. Veillonella showed resistance towards ampicillin and penicillin (Seville et al. 2009). Aminoglycoside and β-lactam antibiotics (BLr), tetracycline (Tcr), and methicillin (mecA) resistance genes are present in fecal samples of newborns (Gosalbes et al. 2016). Infants acquired AMR bacteria from mother. Infants and mothers showed presence of sulphonamides, spectinomycin, streptomycin, and trimethoprim resistance integrase genes (*int1*) and Tcr genes (Ravi et al. 2015). Plasmid-mediated quinolone resistance qnrA and extended-spectrum-β-lactamase resistance blaCTX-M are derived from non-pathogenic bacteria (Poirel et al. 2005).

6.12 Conclusion

Bacterial pathogens develop resistance to all antibiotics through mutation, transcriptomic alteration, and acquisition of resistance genes. Therefore, understanding the molecular mechanisms and evolution of multidrug-resistant bacteria need to be studied. Multidrug-resistant (MDR) bacteria are involved in increased mortality and produce economic burden in world. It has now become greatest threats of the twenty-first century in public health. Multidrug resistance of the infected bacterial pathogens is common in clinical settings. Development of antibiotic resistance should be considered as an adaptive response in Darwinian's principles of evolution. Research has to be focused on the development of antibiotics. It is better to understand the antibiotic resistance mechanisms in bacteria to design novel antibiotic to encounter this global threat. To solve this AMR issue, research and development are to be enhanced. It is essential to develop new antibiotics and understand the response of microbes to new antibiotics.

References

Alcalde-Rico M, Hernando-Amado S, Blanco P, Martínez JL (2016) Multidrug efflux pumps at the crossroad between antibiotic resistance and bacterial virulence. Front Microbiol 7:1483

Aldred KJ, Kerns RJ, Osheroff N (2014) Mechanism of quinolone action and resistance. Biochemist 53(10):1565–1574

Arthur M (2010) Antibiotics: vancomycin sensing. Nat Chem Biol 6(5):313–315

Barbe V, Vallenet D, Fonknechten N, Kreimeyer A, Oztas S, Labarre L, Cruveiller S, Robert C, Duprat S, Wincker P, Ornston LN, Weissenbach J, Marliere P, Cohen GN, Medigue C (2004) Unique features revealed by the genome sequence of *Acinetobacter* sp. ADP1, a versatile and naturally transformation competent bacterium. Nucleic Acids Res 32:5766–5779

Baslé A, Rummel G, Storici P, Rosenbusch JP, Schirmer T (2006) Crystal structure of osmoporin OmpC from *E. coli* at 2.0 Å. J Mol Biol 362:933–942

Bayer AS, Schneider T, Sahl HG (2013) Mechanisms of daptomycin resistance in *Staphylococcus aureus*: role of the cell membrane and cell wall. Ann N Y Acad Sci 1277:139–158

Behnke M, Hansen S, Leistner R, Diaz LA, Gropmann A, Sohr D, Gastmeier P, Piening B (2013) Nosocomial infection and antibiotic use: a second national prevalence study in Germany. Dtsch Arztebl Int 110(38):627–633

Bello A, Dingle TC (2018) What's that resistance mechanism? Understanding genetic determinants of gram-negative bacterial resistance. Clin Microbiol Newslett 40:165–174

Brinkac L, Voorhies A, Gomez A, Nelson KE (2017) The threat of antimicrobial resistance on the human microbiome. Microb Ecol 74(4):1001–1008

Bush K, Jacoby GA (2010) Updated functional classification of beta-lactamases. Antimicrob Agents Chemother 54:969–997

Carattoli A (2009) Resistance plasmid families in Enterobacteriaceae. Antimicrob Agents Chemother 53:2227–2238

Cavaco LM, Frimodt-Moller N, Hasman H, Guardabassi L, Nielsen L, Aarestrup FM (2008) Prevalence of quinolone resistance mechanisms and associations to minimum inhibitory concentrations in quinolone-resistant *Escherichia coli* isolated from humans and swine in Denmark. Microb Drug Resist 14:163–169

Chambers HF, Deleo FR (2009) Waves of resistance: *Staphylococcus aureus* in the antibiotic era. Nat Rev Microbiol 7(9):629–641

Correa JE, de Paulis A, Predari S, Sordelli DO, Jeric PE (2008) First report of qacG, qacH and qacJ genes in *Staphylococcus haemolyticus* human clinical isolates. J Antimicrob Chemother 62:956–960

Costa SS, Viveiros M, Amaral L, Couto I (2013) Multidrug efflux pumps in *Staphylococcus aureus*: an update. Open Microbiol J 7:59–71

Daum RS, Ito T, Hiramatsu K, Hussain F, Mongkolrattanothai K, Jamklang M, Boyle-Vavra S (2002) A novel methicillin-resistance cassette in community-acquired methicillin-resistant *Staphylococcus aureus* isolates of diverse genetic backgrounds. J Infect Dis 186:1344–1347

Davies J, Davies D (2010) Origins and evolution of antibiotic resistance. Microbiol Mol Biol Rev 74(3):417–433

de Groote MA, Huitt G (2006) Infections due to rapidly growing mycobacteria. Clin Infect Dis 42:1756–1763

DeLeo FR, Chambers HF (2009) Re-emergence of antibiotic-resistant Staphylococcus aureus in the genomics era. J Clin Invest 119:2464–2474

Du D, Wang-Kan X, Neuberger A, van Veen HW, Pos KM, Piddock LJV, Luisi BF (2018) Multidrug efflux pumps: structure, function and regulation. Nat Rev Microbiol 16(9): 523–539

Evans BA, Amyes SG (2014) OXA β-lactamases. Clin Microbiol Rev 27(2):241–263

Exner M, Bhattacharya S, Christiansen B, Gebel J, Goroncy-Bermes P, Hartemann P, Heeg P, Ilschner C, Kramer A, Larson E, Merkens W, Mielke M, Oltmanns P, Ross B, Rotter M, Schmithausen RM, Sonntag HG, Trautmann M (2017) Antibiotic resistance: what is so special about multidrug-resistant gram-negative bacteria? GMS Hyg Infect Control 12:Doc05

Fajardo A, Martinez-Martin N, Mercadillo M, Galan JC, Ghysels B, Matthijs S, Cornelis P, Wiehlmann L, Tummler B, Baquero F, Martinez JL (2008) The neglected intrinsic resistome of bacterial pathogens. PLoS One 3:e1619

Floss HG, Yu TW (2005) Rifamycin-mode of action, resistance, and biosynthesis. Chem Rev 105(2):621–632

Furuya EY, Lowy FD (2006) Antimicrobial-resistant bacteria in the community setting. Nat Rev Microbiol 4(1):36–45

Georghiou SB, Magana M, Garfein RS, Catanzaro DG, Catanzaro A, Rodwell TC (2012) Evaluation of genetic mutations associated with Mycobacterium tuberculosis resistance to amikacin, kanamycin and capreomycin: a systematic review. PLoS One 7:e33275

Giedraitienė A, Vitkauskienė A, Naginienė R, Pavilonis A (2011) Antibiotic resistance mechanisms of clinically important bacteria. Medicina (Kaunas) 47(3):137–146

Godreuil S, Galimand M, Gerbaud G, Jacquet C, Courvalin P (2003) Efflux pump Lde is associated with fluoroquinolone resistance in *Listeria monocytogenes*. Antimicrob Agents Chemother 47:704–708

Gosalbes MJ, Vallès Y, Jiménez-Hernández N et al (2016) High frequencies of antibiotic resistance genes in infants' meconium and early fecal samples. J Dev Orig Health Dis 7:35–44

Gupta AK, Chauhan DS, Srivastava K, Das R, Batra S, Mittal M, Goswami P, Singhal N, Sharma VD, Venkatesan K, Hasnain SE, Katoch VM (2006) Estimation of efflux mediated multi-drug resistance and its correlation with expression levels of two major efflux pumps in mycobacteria. J Commun Dis 38:246–254

Hakenbeck R, Bruckner R, Denapaite D, Maurer P (2012) Molecular mechanisms of beta-lactam resistance in *Streptococcus pneumoniae*. Future Microbiol 7:395–410

Hernando-Amado S, Blanco P, Alcalde-Rico M, Corona F, Reales-Calderón JA, Sánchez MB, Martínez JL (2016) Multidrug efflux pumps as main players in intrinsic and acquired resistance to antimicrobials. Drug Resist Updat 28:13–27

Hiramatsu K, Katayama Y, Yuzawa H, Ito T (2002) Molecular genetics of methicillin-resistant *Staphylococcus aureus*. Int J Med Microbiol 292:67–74

Hollenbeck BL, Rice LB (2012) Intrinsic and acquired resistance mechanisms in enterococcus. Virulence 3(5):421–433

Horrevorts AM, Borst J, Puyk RJ, De Ridder R, Dzoljicdanilovic G, Degener JE, Kerrebijn KF, Michel MF (1990) Ecology of *Pseudomonas aeruginosa* in patients with cystic fibrosis. J Med Microbiol 31:119–124

Jacoby GA (2009) AmpC beta-lactamases. Clin Microbiol Rev 22(1):161–182

Jamieson FB, Guthrie JL, Neemuchwala A, Lastovetska O, Melano RG, Mehaffy C (2014) Profiling of rpoB mutations and MICs for rifampin and rifabutin in *Mycobacterium tuberculosis*. J Clin Microbiol 52:2157–2162

Johnning A, Moore ERB, Svensson-Stadler L et al (2013) Acquired genetic mechanisms of a multiresistant bacterium isolated from a treatment plant receiving wastewater from antibiotic production. Appl Environ Microbiol 79:7256–7263

Kelley WL, Jousselin A, Barras C, Lelong E, Renzoni A (2015) Missense mutations in PBP2A affecting ceftaroline susceptibility detected in epidemic hospital-acquired methicillin-resistant Staphylococcus aureus clonotypes ST228 and ST247 in western Switzerland archived since 1998. Antimicrob Agents Chemother 59:1922–1930

Kohanski MA, Dwyer DJ, Collins JJ (2010) How antibiotics kill bacteria: from targets to networks. Nat Rev Microbiol 8:423435

Law CJ, Maloney PC, Wang DN (2008) Ins and outs of major facilitator superfamily antiporters. Annu Rev Microbiol 62:289–305

Lewis K (2007) Persister cells, dormancy and infectious disease. Nat Rev Microbiol 5(1):48–56

Li XZ, Nikaido H (2009) Efflux-mediated drug resistance in bacteria. Drugs 69:1555–1623

Li W, Atkinson GC, Thakor NS, Allas U, Lu CC, Chan KY, Tenson T, Schulten K, Wilson KS, Hauryliuk V, Frank J (2013) Mechanism of tetracycline resistance by ribosomal protection protein Tet(O). Nat Commun 4:1477

Lima T, Domingues S, da Silva GJ (2019) Plasmid-mediated Colistin resistance in *Salmonella enterica*: a review. Microorganisms 7(2):E55

Lin YT, Huang YW, Liou RS, Chang YC, Yang TC (2014) MacABCsm, an ABC-type tripartite efflux pump of *Stenotrophomonas maltophilia* involved in drug resistance, oxidative and envelope stress tolerances and biofilm formation. J Antimicrob Chemother 69:3221–3226

Liu B, Pop M (2009) ARDB—antibiotic resistance genes database. Nucleic Acids Res 37:D443–D447

Livermore DM, Woodford N (2006) The beta-lactamase threat in Enterobacteriaceae, *Pseudomonas* and *Acinetobacter*. Trends Microbiol 14:413–420

Lu M (2015) Structures of multidrug and toxic compound extrusion transporters and their mechanistic implications. Channels (Austin) 10(2):88–100

Lu J, Liu M, Wang Y, Pang Y, Zhao Z (2014) Mechanisms of fluoroquinolone monoresistance in *Mycobacterium tuberculosis*. FEMS Microbiol Lett 353:40–48

Maitra A, Munshi T, Healy J, Martin LT, Vollmer W, Keep NH, Bhakta S (2019) Cell wall peptidoglycan in *Mycobacterium tuberculosis*: an Achilles' heel for the TB-causing pathogen. FEMS Microbiol Rev 43:548

Majowicz SE, Musto J, Scallan E, Angulo FJ, Kirk M, O'Brien SJ, Jones TF, Fazil A, Hoekstra RM (2010) The global burden of nontyphoidal *Salmonella gastroenteritis*. Clin Infect Dis 50:882–889

Manson JM, Hancock LE, Gilmore MS (2010) Mechanism of chromosomal transfer of *Enterococcus faecalis* pathogenicity island, capsule, antimicrobial resistance, and other traits. Proc Natl Acad Sci U S A 107(27):12269–12274

Masi M, Réfregiers M, Pos KM, Pagès JM (2017) Mechanisms of envelope permeability and antibiotic influx and efflux in gram-negative bacteria. Nat Microbiol 2:17001

Mendes RE, Deshpande LM, Jones RN (2014) Linezolid update: stable in vitro activity following more than a decade of clinical use and summary of associated resistance mechanisms. Drug Resist Updat 17(1–2):1–12

Miotto P, Zhang Y, Cirillo DM, Yam WC (2018) Drug resistance mechanisms and drug susceptibility testing for tuberculosis. Respirology 23(12):1098–1113

Molbak K, Gerner-Smidt P, Wegener HC (2002) Increasing quinolone resistance in *Salmonella enterica* serotype Enteritidis. Emerg Infect Dis 8:514–515

Morita Y, Kimura N, Mima T, Mizushima T, Tsuchiya T (2001) Roles of MexXY- and MexAB-multidrug efflux pumps in intrinsic multidrug resistance of *Pseudomonas aeruginosa* PAO1. J Gen Appl Microbiol 47:27–32

Mukerji S, O'Dea M, Barton M, Kirkwood R, Lee T, Abraham S (2017) Development and transmission of antimicrobial resistance among gram-negative bacteria in animals and their public health impact. Essays Biochem 61(1):23–35

Müller B, Borrell S, Rose G, Gagneux S (2013) The heterogeneous evolution of multidrug-resistant *Mycobacterium tuberculosis*. Trends Genet 29(3):160–169

Munita JM, Arias CA (2016) Mechanisms of antibiotic resistance. Microbiol Spectr 4(2). https://doi.org/10.1128/microbiolspec.VMBF-0016-2015

Munita JM, Bayer AS, Arias CA (2015) Evolving resistance among gram-positive pathogens. Clin Infect Dis 61(2):S48–S57
Nikaido H, Pagès JM (2012) Broad-specificity efflux pumps and their role in multidrug resistance of gram-negative bacteria. FEMS Microbiol Rev 36:340–363
Ogawara H, Kawamura N, Kudo T, Suzuki KI, Nakase T (1999) Distribution of β-lactamases in actinomycetes. Antimicrob Agents Chemother 43:3014–3017
Olaitan AO, Morand S, Rolain JM (2014) Mechanisms of polymyxin resistance: acquired and intrinsic resistance in bacteria. Front Microbiol 5:643
Olivares J, Bernardini A, Garcia-Leon G, Corona F, Sanchez B, Martinez JL (2013) The intrinsic resistome of bacterial pathogens. Front Microbiol 4:103
Pamer EG (2016) Resurrecting the intestinal microbiota to combat antibiotic-resistant pathogens. Science 352(6285):535–538
Pelgrift RY, Friedman AJ (2013) Nanotechnology as a therapeutic tool to combat microbial resistance. Adv Drug Deliv Rev 65(13–14):1803–1815
Pham TN, Loupias P, Dassonville-Klimpt A, Sonnet P (2019) Drug delivery systems designed to overcome antimicrobial resistance. Med Res Rev 39:2343. https://doi.org/10.1002/med.21588
Poirel L, Rodriguez-Martinez JM, Mammeri H et al (2005) Origin of plasmid-mediated quinolone resistance determinant QnrA. Antimicrob Agents Chemother 49:3523–3525
Ramirez MS, Tolmasky ME (2010) Aminoglycoside modifying enzymes. Drug Resist Updat 13(6):151–171
Ramon-Garcia S, Martin C, Thompson CJ, Ainsa JA (2009) Role of the *Mycobacterium tuberculosis* P55 efflux pump in intrinsic drug resistance, oxidative stress responses, and growth. Antimicrob Agents Chemother 53:3675–3682
Ravi A, Avershina E, Foley SL et al (2015) The commensal infant gut meta-mobilome as a potential reservoir for persistent multidrug resistance integrons. Sci Rep 5:15317
Rice LB (2008) Federal funding for the study of antimicrobial resistance in nosocomial pathogens: no ESKAPE. J Infect Dis 197:1079–1081
Richter MF, Drown BS, Riley AP et al (2017) Predictive compound accumulation rules yield a broad-spectrum antibiotic. Nature 545:299–304
Romanova NA, Wolffs PF, Brovko LY, Griffiths MW (2006) Role of efflux pumps in adaptation and resistance of *Listeria monocytogenes* to benzalkonium chloride. Appl Environ Microbiol 72:3498–3503
Ross JI, Eady EA, Cove JH, Cunliffe WJ, Baumberg S, Wootton JC (1990) Inducible erythromycin resistance in *Staphylococci* is encoded by a member of the ATP-binding transport super-gene family. Mol Microbiol 4:1207–1214
Sanchez MB (2015) Antibiotic resistance in the opportunistic pathogen *Stenotrophomonas maltophilia*. Front Microbiol 6:658
Santajit S, Indrawattana N (2016) Mechanisms of antimicrobial resistance in ESKAPE pathogens. Biomed Res Int 2016:2475067
Schwarz S, Kehrenberg C, Doublet B, Cloeckaert A (2004) Molecular basis of bacterial resistance to chloramphenicol and florfenicol. FEMS Microbiol Rev 28(5):519–542
Semret M, Haraoui LP (2019) Antimicrobial resistance in the tropics. Infect Dis Clin N Am 33(1):231–245
Seville LA, Patterson AJ, Scott KP et al (2009) Distribution of tetracycline and erythromycin resistance genes among human oral and fecal metagenomic DNA. Microb Drug Resist 15:159–166
Shaikh S, Fatima J, Shakil S, Rizvi SM, Kamal MA (2015) Antibiotic resistance and extended spectrum beta-lactamases: types, epidemiology and treatment. Saudi J Biol Sci 22(1):90–101
Sun Z, Xu Y, Sun Y, Liu Y, Zhang X, Huang H, Li C (2014) Ofloxacin resistance in *Mycobacterium tuberculosis* is associated with efflux pump activity independent of resistance pattern and genotype. Microb Drug Resist 20:525–532
Suwantarat N, Carroll KC (2016) Epidemiology and molecular characterization of multidrug-resistant gram-negative bacteria in Southeast Asia. Antimicrob Resist Infect Control 5:15
Thomas CM, Nielsen KM (2005) Mechanisms of, and barriers to, horizontal gene transfer between bacteria. Nat Rev Microbiol 3(9):711–721

Tran TT, Panesso D, Mishra NN et al (2013) Daptomycin-resistant *Enterococcus faecalis* diverts the antibiotic molecule from the division septum and remodels cell membrane phospholipids. MBio 4:13

Unemo M, Shafer WM (2014) Antimicrobial resistance in *Neisseria gonorrhoeae* in the 21st century: past, evolution, and future. Clin Microbiol Rev 27(3):587–613

van den Bogaard AE, Jensen LB, Stobberingh EE (1997) Vancomycin-resistant *Enterococci* in turkeys and farmers. New Engl J Med 337:1558–1559

Verraes C, van Boxstael S, van Meervenne E, van Coillie E, Butaye P, Catry B et al (2013) Antimicrobial resistance in the food chain: a review. Int J Environ Res Public Health 10:2643–2669

Wain J, Hendriksen RS, Mikoleit ML et al (2015) Typhoid fever. Lancet 385(9973):1136–1145

Wassenaar TM, Ussery D, Nielsen LN, Ingmer H (2015) Review and phylogenetic analysis of qac genes that reduce susceptibility to quaternary ammonium compounds in *Staphylococcus* species. Eur J Microbiol Immunol 5:44–61

Weisblum B (1995) Erythromycin resistance by ribosome modification. Antimicrob Agents Chemother 39(3):577–585

Yamada Y, Shiota S, Mizushima T, Kuroda T, Tsuchiya T (2006) Functional gene cloning and characterization of MdeA, a multidrug efflux pump from *Staphylococcus aureus*. Biol Pharm Bull 29:801–804

Yun MK, Wu Y, Li Z et al (2012) Catalysis and sulfa drug resistance in dihydropteroate synthase: crystal structures reveal the catalytic mechanism of DHPS and the structural basis of sulfa drug action and resistance. Science 335(6072):1110–1114

Zgurskaya HI, López CA, Gnanakaran S (2015) Permeability barrier of gram-negative cell envelopes and approaches to bypass it. ACS Infect Dis 1:512–522

Pathogenesis and Antibiotic Resistance of *Staphylococcus aureus*

7

Sanjay Kumar, Simranjeet Singh, Vijay Kumar, Shivika Datta, Daljeet Singh Dhanjal, Priyanka Sharma, and Joginder Singh

Contents

Sanjay Kumar, Simranjeet Singh and Vijay Kumar contributed equally with all other contributors.

S. Kumar
Regional Advanced Water Testing Laboratory, Phase-II, S.A.S. Nagar, Punjab, India

Punjab Biotechnology Incubator (PBTI), Phase-V, S.A.S. Nagar, Punjab, India

S. Singh
Regional Advanced Water Testing Laboratory, Phase-II, S.A.S. Nagar, Punjab, India

Punjab Biotechnology Incubator (PBTI), Phase-V, S.A.S. Nagar, Punjab, India

Department of Biotechnology, Lovely Professional University, Phagwara, Punjab, India

V. Kumar
Regional Ayurveda Research Institute for Drug Development, Gwalior, Madhya Pradesh, India

S. Datta
Department of Zoology, Doaba College, Jalandhar, Punjab, India

D. S. Dhanjal · J. Singh (✉)
Department of Biotechnology, Lovely Professional University, Phagwara, Punjab, India
e-mail: joginder.15005@lpu.co.in

P. Sharma
Biocon Bristol Myers Squibb Syngene International Pvt Ltd, Bangalore, India

B. Siddhardha et al. (eds.), *Model Organisms for Microbial Pathogenesis, Biofilm Formation and Antimicrobial Drug Discovery*,
https://doi.org/10.1007/978-981-15-1695-5_7

Abstract

Staphylococcus aureus is a potent gram-positive bacterium. It is recognized as one of the major causative agents for the community- and hospital-acquired diseases. Therefore, it has long been considered as a concern for public health. It causes many types of human infections, mainly soft tissue, bone, skin, respiratory, joint, gastrointestinal, and endovascular disorders. *S. aureus* can adapt different environments inside the host to modulate virulence using complex regulatory networks to sense diverse signals. This bacterium has the capacity to cross all barriers of the host immune and defense system. Hence it also possesses a strong spectrum of virulence factors. It is the major cause behind biofilm-related infections of indwelling medical devices, which is key responsible for huge healthcare cost every year in the developed countries. *S. aureus* has various virulence factors that are implicated in their pathogenesis. It can produce various toxins such as super-antigens that result in causative agents in disease entities such as toxic-shock syndrome, staphylococcal scarlet fever, etc. and has developed acquired resistance to most of the used antibiotics. Methicillin-resistant *S. aureus* (MRSA) infections have reached epidemic levels in many parts of the world. MRSA causes severe healthcare-associated infections. It also induces various health related issues in dairy animals suffering from mastitis. This chapter describes the pathogenesis and antibiotic resistance of *S. aureus*. It also covers the recent advancement in the structural basis antibiotic multi-resistance acquisition and possible novel strategies for therapeutic intervention.

Keywords

Antimicrobial resistance · Pathogenesis · *Staphylococcus aureus* · Antibiotic therapeutics

7.1 Introduction

Staphylococcus aureus is a potent gram-positive bacteria of Micrococcaceae family. It contains cell wall with peptidoglycan which contains *N*-acetylglucosamine (NAG) and *N*-acetylmuramic (NAM) acid subunits (Leonard and Markey 2008). It has the surface proteins which contain virulence factors. It can produce numerous toxins such as cytotoxins and toxins which aid it in causing broad range of diseases in humans (Murray 2005). It is the major causative agent for both community- as well as hospital-acquired diseases (Murdoch et al. 2009). It causes wide range of diseases starting from soft-tissue infections to mild skin to bacteremia or pneumonia, chronic osteomyelitis, and life-threatening endocarditis, which are related to significant mortality and morbidity (Roberts and Chambers 2005; Mitchell and Howden 2005). Therefore, it is regarded as a major concern of public health.

The resistance of *S. aureus* against various antibodies is the serious issue for physicians in curing diseases caused by resistant strains. The extensive antibiotics usage in animal feeds has developed antibiotic-resistant species of *S. aureus* (Ortega et al. 2010). *S. aureus* started to develop antibiotic resistance from 1942 to till date. First methicillin- and penicillin-resistant strains of *S. aureus* were recognized in 1942 and 1961, respectively (McKee and Houck 1943; Jevons 1961). In addition to that in late 1980s, vancomycin- and quinolones-resistant strains of *S. aureus* were discovered (Lowy 2003).

Treatment of the staphylococci infection is very difficult for physician without knowing about the resistant strain of bacterium species. Specific types of antibiotic suitable for specific strains such as penicillin G are suitable for penicillin-sensitive *S. aureus* strain, whereas methicillin is suitable for those strains which have developed resistance against penicillin (Loomba et al. 2010). Tetracycline is the targeted drug for treating the community-acquired methicillin-resistant strains of *S. aureus* (CA-MRSA). Clindamycin is another choice for CA-MRSA. It has good oral bioavailability and tissue penetrating properties (Loewen et al. 2017).

In this chapter we have covered general feature and pathogenicity of *S. aureus*. We have also discussed antibiotic resistance mechanism of *S. aureus* as well as therapeutic approach for Staphylococci infection.

7.2 Features of *S. aureus*

S. aureus is a member of class Bacilli and Micrococcaceae family. They can easily be distinguished from other species of Staphylococcus genus by appearance of gold pigmented colonies and positive results for deoxyribonuclease coagulase, and sugar fermentation tests (Wilkinson 1997).

7.2.1 Cell Wall

Cell wall of *S. aureus* species consists of 50% peptidoglycan. Polysaccharide subunits such as *N*-acetylmuramic (NAM) and *N*-acetylglucosamine (NAG) acid are present in peptidoglycan. NAM and NAG are attached by 1,4-β linkages. It forms cross-linkage of peptidoglycan chains by pentaglycine bridge and tetrapeptide bonds attached to *N*-acetylmuramic acid (Kim et al. 2015). Peptidoglycan is linked to ribitol teichoic acids which are major component of cell wall in *S. aureus*. In the cytoplasmic membrane, lipoteichoic acid is bonded to glycolipid terminus end. Endotoxin-like activity is also associated with peptidoglycan. As, it triggers the release of cytokines which causes the aggregation of platelets, activation of complement system, and macrophages (Lee 1996).

7.2.2 Capsules

Microcapsules have been produced by most of the staphylococcal species. Till date only 11 different types of microcapsular serotypes (Polysaccharide based) are reported and discovered. Types 8 and 5 are mainly responsible for human infection and Type 5 is isolated from most of the MRSA (O'Riordan and Lee 2004).

7.2.3 Surface Proteins

S. aureus has the ability to express wide range of virulence factors found on bacterial surface which are known as surface proteins. These proteins are bonded to peptidoglycan by covalent bonds also stated as cell wall-anchored proteins (CWA) due to their attachment with the host. These surface-bound proteins are responsible for the commensal and pathogenic nature of this bacterium (Lacey et al. 2016). Among the staphylococcal species, *S. aureus* expresses 24 CWA proteins, whereas other species like *S. lugdunensis* and *S. epidermidis* express a lesser number of CWA proteins on their surface (Speziale et al. 2014). The expression of cell wall-anchored protein depends on the condition of growth media. For example, most of the cell wall-anchored proteins are expressed in iron deficient conditions; whereas some of them are found to be expressed on cell in the exponential or stationary phase (Foster et al. 2014).

CWA proteins have been classified into four groups based upon their structure–function analysis. The most significant class of CWA proteins is microbial surface component recognizing adhesive matrix molecule (MSCRAMM) (Arora et al. 2016). This class is directly associated with Immunoglobulin G-like folded domains. Other classes of CWA proteins are G5–E repeat family, three-helical bundle proteins, and near iron transporter motif proteins (NEAT) (Foster et al. 2014).

7.2.4 Genetic Materials

The genome of Staphylococcal contains round chromosome with a total of 2800 bp, plasmids, transposons, and prophases. The resistant genes are positioned on the chromosome elements, as well as on chromosomes (Novick 1990). The extra chromosomal elements are the channels for the transfer of genes between other gram-positive bacterial species and staphylococcal strains (Schaberg and Zervos 1986).

7.2.5 Toxins

Numerous toxins have been produced by staphylococci species and are differentiated on the basis of their action mechanism. The first category is cytotoxins having molecular weight of 33-kD protein-alpha toxin. They cause proinflammatory changes and pore-formation in mammalian cells (Otto 2014). The second class of toxins is pyrogenic-toxin super-antigens. They are bound with class II proteins of major histocompatibility complex (MHC), cytokine release, and extensive T-cell proliferation. Other classes of toxins are enterotoxins and exfoliative toxins. The enterotoxins are causative agents for food poisoning and toxin sock syndrome (TSC). TSC is analogous to structures of enterotoxins B and enterotoxins C. Skin erythema and separation is facilitated by the exfoliative toxins such as epidermolytic toxins A as well as epidermolytic toxins B (Pinchuk et al. 2010).

7.2.6 Enzymes and Other Components

Several enzymes have been produced by staphylococci species including hyaluronidase lipase and protease. These enzymes destroy tissue of the host species. These enzymes can also felicitate the spread of infections to adjacent tissues (Kong et al. 2016). Enzymes play an important role to gain resistance from antibiotics. For example, β-lactamase inactivates penicillin. Coagulase converts fibrinogen to fibrin and is a prothrombin activator and contributes to various bacterial virulence (Kobayashi et al. 2015).

7.3 Pathogenesis of *Staphylococcus aureus*

S. aureus contains extraordinary virulence factors, which can support it to survive in harsh conditions in human body. It regulates the expression of various virulence factors, which leads to severe infection in most of the parts of healthy person (Beceiro et al. 2013).

7.3.1 Colonization

S. aureus causes the diseases by colonization on the host surface. *S. aureus* can colonize and cause severe infection when it comes in contact with an open wound. It can easily colonize on skin or mucosal surface of humans and leads to severe infection. The mucosal surface such as nasal chamber, axillae, groin, throat line, vaginal wall, and gastrointestinal tract are the common habitats for *S. aureus* (Balasubramanian et al. 2017). Nasal chamber is the most important for colonization. Nearly about 20% of individuals are nasally colonized with the strains of the *S. aureus* whereas 30% of individuals are colonized transiently with different strains of *S. aureus* (Brown et al. 2014). Defense system and resident microorganisms in human nasal chamber are protected from colonization of *S. aureus*. Once *S. aureus* comes in contact with nasal epithelial, it can invade and adhere to epithelial cells using surface-bound proteins like MSCRAMM molecules (Johannessen et al. 2012). Apart from MSCRAMM other bacterial components are also imperative for the attachment and adhesion to epithelial cells of nasal chamber. The constituents of host immune system such as IgA, lactoferrin, lysozyme, and antimicrobial peptides protected host cell from bacterial colonization (Cole et al. 2001). It has been reported that the colonization is more frequent in patients with diabetes and HIV as well as in younger children (Wertheim et al. 2005).

7.3.2 Pathogenesis

S. aureus is an organism which can survive either inside or outside the host cell. Generally, infection occurs after contact of *S. aureus* with exposed wound. The virulence genes get upregulated on the start of exposure of *S. aureus* with host tissues (mucosal or skin surface) (Brescó et al. 2017). On the site of infection, the host microbes and epithelial cells of the mucosal tissue or skin reacts to tissue injury or either toxin produced by the bacteria activates the immune system. Inflammatory response has been stimulated strongly after *S. aureus* infection which leads to activation of first line of defense mechanism resulting in the transfer of macrophages and neutrophils to the infection site. These cells engulf the foreign pathogens, which are recognized due to antibodies on it by complement system in the host serum and aid in disposing it off (Fournier and Philpott 2005).

It has numerous surface proteins stated as MSCRAMMs. These surface proteins participate in establishment of infection by adhering to the surface of host tissues. They bind with different components such as fibrinogen, collagen, and fibronectin, etc. which allow it to adhere to the mucosal tissue or skin of the host (Josse et al. 2017). MSCRAMMs play a very important role in initiation of infection on different parts of host such as prosthetic device infections, endovascular infections, and joint and bone infections (Ghasemian et al. 2015). Infections are specific for the different types of strains of *S. aureus* because of diversity in components of MSCRAMMs in strains of *S. aureus* (Bien et al. 2011).

Once *S. aureus* habitats in skin or mucosal tissue, it starts multiplying and increase its population in numerous ways. It can form biofilm (slime) on skin or mucosal tissue, countering and avoiding the host defense systems. It can also survive and adhere to both endothelial and epithelial cells (Archer et al. 2011). Apart from biofilm it forms small-colony variants (SCVs) which are held responsible for recurrent and persistent infection. These SCVs have the ability to protect themselves in host tissues without damaging the host cell; therefore, they remain protected from defense systems of host and antibiotics (Proctor et al. 2014). Sometimes, in the later stage, SCVs transform themselves to highly virulent wild-type phenotype, which causes recurrent and persistent infection (Kahl et al. 2016).

S. aureus can avoid defense system of host by secreting the proteins which can inhibit the neutrophil chemotaxis. It secrets toxins which can damage host cell and kill the leukocytes. It causes phagocytosis by avoiding opsonization in the cells (Teng et al. 2017). *S. aureus* can also develop resistance to lysozyme which is an important part of an innate immunity to develop immunity against microbial infection and contains muramidase that cuts the glycosidic linkage between NAM and NAG of peptidoglycan cell wall. The C6 hydroxyl of muramic acid is modified by membrane-bound *O*-acetyltransferase of S. *aureus* which prevent cleavage between NAM and NAG resulted resistance against lysozyme (Fournier and Philpott 2005).

7.4 Pathogenesis of HA-MRSA

S. aureus developed resistance against methicillin because of *mecA* gene present in it. This gene encodes penicillin-binding proteins (PBP2A) which have least affinity for β-lactam antibiotics. Hospital-acquired methicillin resistance strains of *S. aureus* (HA-MRSA) contain SCC (staphylococcal cassette chromosome) *mec*, which encoded multiple antibiotic resistance genes (Gordon and Lowy 2008).

Sometimes HA-MRSA causes diseases without showing any symptoms in individuals. HA-MRSA strains live only in those surroundings where competition of HA-MRSA strains is limited by antibiotics (Eady and Cove 2003). Laurent et al. reported that strains exposed to mHA-MRSA have extended generation time in comparison with methicillin-sensitive *S. aureus* (MSSA) (Laurent et al. 2001). Voyichand and his colleagues reported HA-MRSA strains are more susceptible to neutrophils and are less pathogenic than MSSA when tested on mice (Voyich et al. 2005).

7.5 Pathogenesis of CA-MRSA

MRSA infections were investigated only in immune-suppressed patients or in the individuals that have already been exposed to various other diseases in the late 1990s. In 1997, MRSA pneumonia has been responsible for death of four children. Researchers have noticed that the causative agent for MRSA pneumonia was a new strain of MRSA (Centers for Disease Control and Prevention 1999). These strains were different from HA-MRSA, and have a SCC*mec* cassette Type IV which encodes for genes of Panton–Valentine leukocidin (PVL) (Vandenesch et al. 2003).

7.6 Development of Antibiotic Resistance in *S. aureus*

Bacterial species continuously developed an alternative path for survival and increase their growth in antimicrobial environment. Antibiotic resistance of bacterial species is the biggest problem for physicians and society in these days (Fair and Tor 2014). There is an enhancement in the resistivity of bacterial species against number of antibiotics due to the usage of antibiotics in animal fodder and feed as growth enhancer (Deurenberg and Stobberingh 2008).

Most of the gram-positive bacteria like Staphylococci, Enterococci, and Pneumococci are taken as priority for concern. These bacterial species are responsible for various life-threatening infections (Santajit and Indrawattana 2016). Among these bacterial species, *S. aureus* is the pathogenic bacteria of universal concern due to its virulence, having the ability to cause different type of infections and amend themselves according to environmental conditions (Tong et al. 2015).

S. aureus has consciously developed resistance against antibiotic since 1941 to till date (Fig. 7.1). In 1984, first *S. aureus*-resistant strain against penicillin was isolated and in 1961, first methicillin-resistant strain of *S. aureus* was reported (Lowy 2003). The strains of *S. aureus* have developed various mechanisms as shown in Table 7.1 to develop resistance against antibiotics.

7.6.1 Penicillin Resistance

The penicillin was first introduced in early 1940s. It was used as a drug for patients infected with Staphylococcus. In 1942, Staphylococcus was recognized resistant

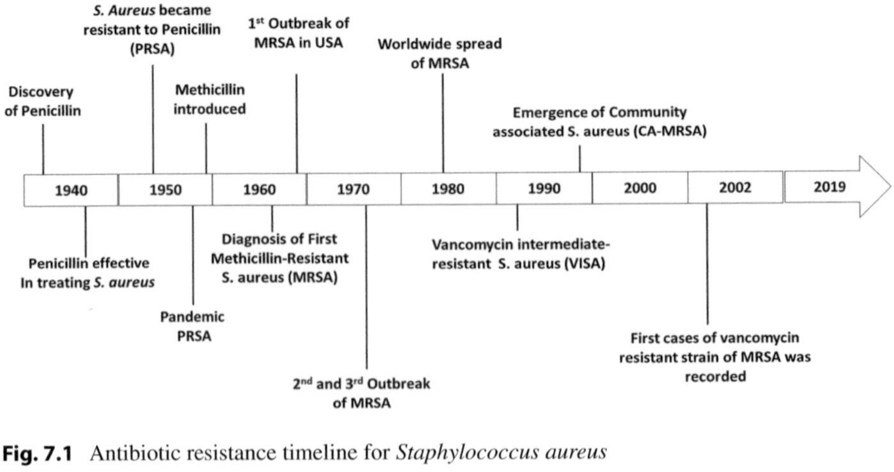

Fig. 7.1 Antibiotic resistance timeline for *Staphylococcus aureus*

Table 7.1 Mechanism of resistance of *Staphylococcus aureus* against antibiotics

Resistance gene	Antibiotic	Gene product	Mechanism of resistance	References
blaZ	Penicillin	β-Lactamase	Hydrolysis of β-lactam ring via β-lactamase	Kong et al. (2010)
mecA	Methicillin (β-lactams)	PBP2a	Change in PBP2a structure	Gordon and Lowy (2008)
aacA-aphD	Gentamicin (aminoglycosides)	Acetylation and phosphorylation enzyme	Modification of aminoglycosides by the action of acetylation and phosphorylation enzyme	Rouch et al. (1987)
aphA	Gentamicin (aminoglycosides)	Phosphorylation enzyme	Modification of aminoglycosides by the action of phosphorylation enzyme	Derbise et al. (1996)
dfrB	Trimethoprim/ sulfamethoxazole	Dihydrofolate reductase (DHFR)	Reduce affinity for DHFR	Coelho et al. (2017)
sulA	Trimethoprim/ sulfamethoxazole	Dihydropteroate synthase	Excessive synthesis of p-aminobenzoic because of Dihydropteroate synthase	Coelho et al. (2017)
gyrA	Quinolones	GyrA component of gyrase	Reduce affinity for quinolones	Aldred et al. (2014)
gyrB	Quinolones	GyrB component of gyrase	Reduce affinity for quinolones due to alteration in QRDR	Aldred et al. (2014)
parC	Quinolones	ParC component of topoisomerase (IV)	Reduce affinity for quinolones due to alteration in QRDR	Aldred et al. (2014)
ermA	Macrolides, Lincosamides, and Streptogramins antibiotics	Erythromycin ribosomal methylase	Methylation of 23 s rRNA to reduce binding affinity	Leclercq and Courvalin (1991)
ermB	Macrolides, Lincosamides, and Streptogramins antibiotics	Erythromycin ribosomal methylase	Reduces binding affinity after methylation of 23 s rRNA	Leclercq and Courvalin (1991)
ermC	Macrolides, Lincosamides, and Streptogramins antibiotics	Erythromycin ribosomal methylase	Reduces binding affinity after methylation of 23 s rRNA	Leclercq and Courvalin (1991)
VatA	Quinupristin/ dalfopristin	Acetyltransferases	Modification of dalfopristin enzyme	Dowzicky et al. (2000)
vatB	Quinupristin/ dalfopristin	Acetyltransferases	Modification of dalfopristin enzyme	Dowzicky et al. (2000)

against penicillin (Rammelkamp and Maxon 1942). It was recognized in hospitals first followed by community. In 1950s and 1960s, there was a dramatic rise in the number of penicillin-resistant staphylococci. In late 1960, 80% of staphylococcal strains were found to develop resistance against penicillin (Livermore 2000).

These strains produce an enzyme such as penicillinase which deactivated the penicillin. Bondi and Dietz comprehended the function of penicillinase enzyme in *S. aureus* (Wilson and Cockcroft 1952). Now most of the staphylococcal species produce penicillinase which is located on large plasmid where other antimicrobial resistance genes were located such as erythromycin and gentamicin (Munita and Arias 2016).

The blaZ gene is responsible to get resistance of staphylococci species against penicillin. It encodes extracellular enzyme β-lactamase which is synthesized by staphylococci on exposure to β-lactamen-rich antibiotics. The enzyme hydrolyzed the penicillin by acting on β-lactam ring which transforms it to penicilloic acid (Fig. 7.2). Penicilloic acid does not have any antibacterial characteristic. The blaZ gene is usually suppressed by blaI and antirepressor blaR1, the two adjacent regulatory genes. For the production of β-lactamase, it requires successive cleavage of two regulatory proteins blaI and blaR1for its activation (Kong et al. 2010).

7.6.2 Methicillin Resistance (MRSA)

The antibiotic, methicillin, was introduced in 1961. It was the first antibiotic which can be used for penicillin-resistant staphylococcal infection. However, *S. aureus* resistant to methicillin (MRSA) has been reported very early after introduction of antibiotic methicillin (Barber 1961). The *S. aureus* strains have touched epidemic proportions and have emerged as a global concern. Methicillin-resistant strains of *S. aureus* are also responsible for most of the staphylococcal disease. It has been widely spread in many counters at hospital as well as community level (Harkins et al. 2017). In the USA, community-associated methicillin-resistant *S. aureus* (CA-MRSA) strains are a causative agent for both soft-tissue and skin infections (David and Daum 2010). These strains have been reported worldwide mostly in developed as well developing countries such as Australia, Canada, South America, Asia, Norway, Netherlands, Denmark, and Finland. It was also reported in many Asian countries as well as African countries (Lee et al. 2018).

The gene accountable for methicillin resistance is *mecA* in strains of *S. aureus*. The *mecA* gene is fragment of staphylococcal cassette chromosome (*SCCmec*) (Katayama et al. 2000). Hiramatsu et al. characterized four sizes of SCCmec elements (21–67 kb) (Hiramatsu et al. 2001). The *mecA* gene synthesizes PBP2a which is membrane-bound protein (78-kDa), which enhances the reaction of transpeptidation. This reaction is responsible for the cross-linkage of chains of peptidoglycan unit (Ghuysen 1994). Also, different reports revealed that PBP2a has low affinity against all β-lactam antibiotics. That is why most of the *S. aureus* strains can survive in antibiotics enriched in β-lactam even in high concentrations. Once staphylococci

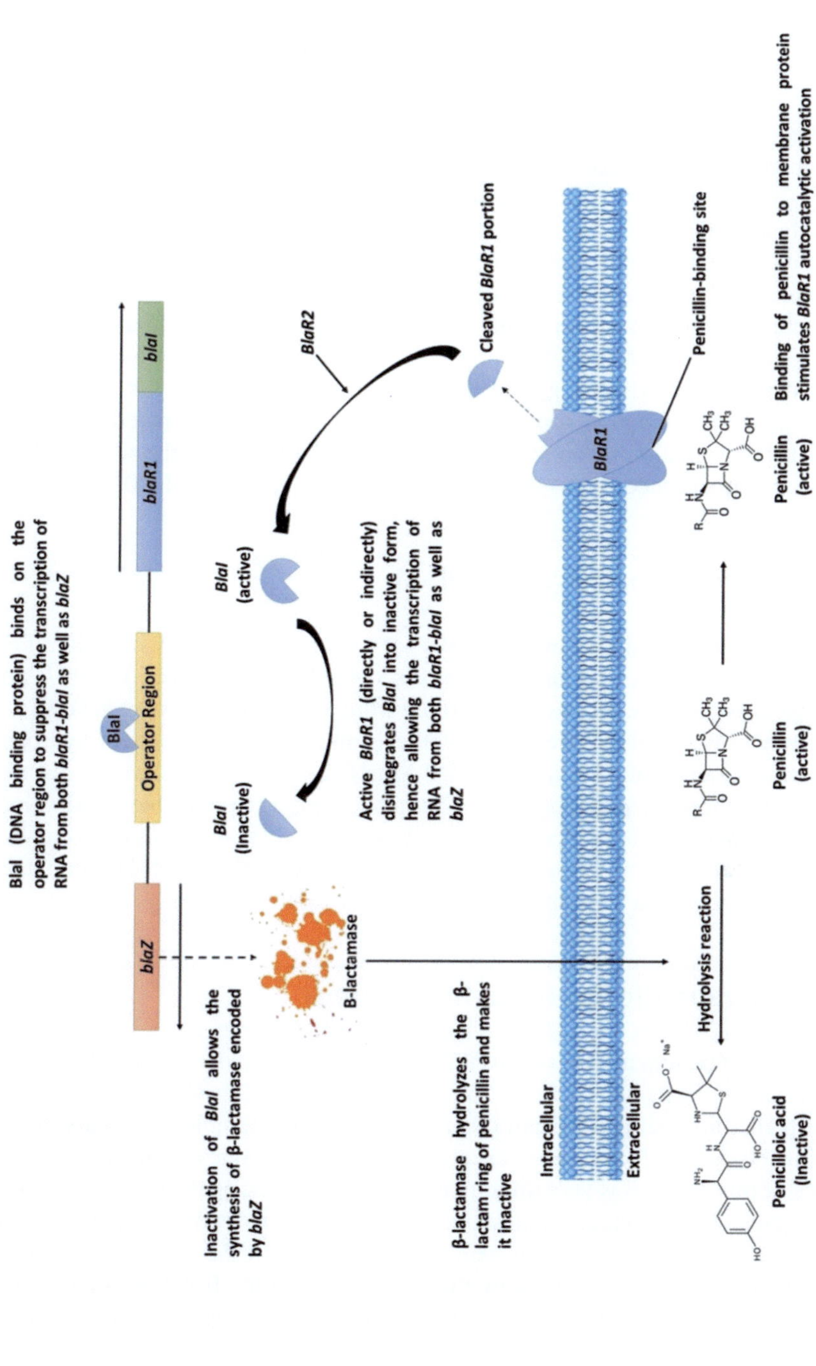

Fig. 7.2 Mechanism for penicillin resistance of *Staphylococcus aureus*

develop resistance against methicillin, they also develop resistance to all other antibiotics of β-lactam (Bæk et al. 2014).

7.6.3 Quinoline Resistance

Quinolone antibiotic, fluoroquinolones were introduced in 1980s for treating infections caused by gram-negative bacterial strains. Therefore, they have been used for treating various gram-positive bacterial infection, i.e., pneumococci and staphylococci infection (Cruciani and Bassetti 1994). Staphylococcus resistant strains are also found resistant against quinolone antibiotics. Later on, fluoroquinolones were used to reduce the staphylococci infections (Gade and Qazi 2013).

The chromosomal mutation on the target of DNA gyrase is accountable for quinolone antibiotic resistance of staphylococci. The quinolone antibiotics act on topoisomerase IV and DNA gyrase. DNA gyrase is responsible for relieving of DNA supercoiling whereas topoisomerase IV separates concatenated strands of DNA. Chromosomal mutation reduces quinolone affinity on both the sites resulting in developing resistance against quinolone antibiotics (Aldred et al. 2014).

7.6.4 Vancomycin Resistance (VISA)

Vancomycin was introduced in late 1980s for treatment of MRSA infection. Few years later resistance to vancomycin was discovered in enterococci. In 1997, first vancomycin-resistant *S. aureus* was reported from Japan (Hiramatsu et al. 1997).

The vancomycin inhibits cell wall synthesis and forms non-covalent hydrogen bonds with residues of D-Ala with UDP-MurNAc-pentapeptides which disturbs late-stage synthesis of peptidoglycan. The gene, *vanA* is responsible for development of resistance against vancomycin. *vanA* gene coded for synthesis of an enzyme which hydrolyzes the peptidoglycan precursors of dipeptide D-Ala-D-Ala, which inhibits binding of vancomycin (McGuinness et al. 2017).

7.7 Therapeutic Approach for *S. aureus*

Prior to start of treatment of *S. aureus* infection, it must be known about which types of strain are responsible for infection. Development of resistance in *S. aureus* against penicillin, methicillin, and other broad-spectrum antibiotics occurs during evolutionary periods because of excessive use of antibiotics (Liu et al. 2011). Because of this region it is very difficult to choose antibiotic against *S. aureus* infection every day. If the isolated strain shows sensitivity against penicillin (<10% of the strains), the treatment of choice is penicillin G, whereas if the strain is resistant to penicillin but sensitive to methicillin, the choice is oxacillin. If patient shows allergic responses to penicillin and strain is *S. aureus* sensitive to methicillin, a valid option is the use of cephalosporins, unless the allergy is anaphylactic, in which case the use of vancomycin is indicated (Tibavizco et al. 2007).

Fridkin et al. (2005) and Naimi et al. (2003) reported most of the isolates maintain susceptibility to vancomycin clindamycin, rifampin, linezolid, gentamicin, trimethoprim–sulfamethoxazole (TMP-SMX), and tetracycline, but their patterns differ from region to region (Naimi et al. 2003).

Fridkin et al. reported more than 80% of CA-MRSA is vulnerable to tetracycline (Fridkin et al. 2005). The long-acting tetracyclines such as minocycline and doxycycline possess greater anti-staphylococcal properties and also have excellent oral bioavailability than tetracycline (Chopra and Roberts 2001).

Clindamycin is another antibiotic which showed antibacterial activity against the isolated strains of various CA-MRSA due to its ability to inhibit toxin chemicals. Clindamycin is available for both intravenous and oral administration. It has good tissue penetration and excellent oral bioavailability. It also works against β hemolytic streptococci. Therefore, it used for empiric therapy for soft tissue and skin infections also (Leong et al. 2018).

7.8 Conclusion

S. aureus is the bacterium which may cause broad spectrum of infection from soft skin infection to life-threatening diseases like chronic osteomyelitis, endocarditis, and pneumonia. Dealing with staphylococci infection is the biggest challenge for physicians because of continuous development of resistance against marketed antibiotics. *S. aureus* has developed resistance against penicillin in 1942, followed by developing resistance against methicillin in 1961. In 1997, it has developed resistance against vancomycin. Hence, now specific antibiotics are used for treating *S. aureus* infection. The antibiotic, penicillin G is used for only penicillin-sensitive *S. aureus* strain. Similarly methicillin is used for penicillin-resistant and methicillin-sensitive strains. Tetracycline can be used for MRSA infection. The long-acting tetracyclines such as minocycline and doxycycline are more suitable to use for MRSA infection. Clindamycin is another antibiotic which has toxin inhibiting properties. It also showed activity against many CA-MRSA.

References

Aldred KJ, Kerns RJ, Osheroff N (2014) Mechanism of quinolone action and resistance. Biochemist 53(10):1565–1574

Archer NK, Mazaitis MJ, Costerton JW, Leid JG, Powers ME, Shirtliff ME (2011) *Staphylococcus aureus* biofilms: properties, regulation, and roles in human disease. Virulence 2(5):445–459

Arora S, Uhlemann AC, Lowy FD, Hook M (2016) A novel MSCRAMM subfamily in coagulase negative Staphylococcal species. Front Microbiol 7:540

Bæk KT, Gründling A, Mogensen RG, Thøgersen L, Petersen A, Paulander W, Frees D (2014) β-Lactam resistance in methicillin-resistant *Staphylococcus aureus* USA300 is increased by inactivation of the ClpXP protease. Antimicrob Agents Chemother 58(8):4593–4603

Balasubramanian D, Harper L, Shopsin B, Torres VJ (2017) *Staphylococcus aureus* pathogenesis in diverse host environments. Pathog Dis 75(1):ftx005

Barber M (1961) Methicillin-resistant Staphylococci. J Clin Pathol 14(4):385–393

Beceiro A, Tomás M, Bou G (2013) Antimicrobial resistance and virulence: a successful or deleterious association in the bacterial world? Clin Microbiol Rev 26(2):185–230

Bien J, Sokolova O, Bozko P (2011) Characterization of virulence factors of *Staphylococcus aureus*: novel function of known virulence factors that are implicated in activation of airway epithelial proinflammatory response. J Pathog 2011:601905

Brescó MS, Harris LG, Thompson K, Stanic B, Morgenstern M, O'Mahony L, Richards RG, Moriarty TF (2017) Pathogenic mechanisms and host interactions in *Staphylococcus epidermidis* device-related infection. Front Microbiol 8:1401

Brown AF, Leech JM, Rogers TR, McLoughlin RM (2014) *Staphylococcus aureus* colonization: modulation of host immune response and impact on human vaccine design. Front Immunol 4:507

Centers for Disease Control and Prevention (CDC) (1999) Four pediatric deaths from community-acquired methicillin-resistant *Staphylococcus aureus*—Minnesota and North Dakota, 1997-1999. MMWR Morb Mortal Wkly Rep 48(32):707–710

Chopra I, Roberts M (2001) Tetracycline antibiotics: mode of action, applications, molecular biology, and epidemiology of bacterial resistance. Microbiol Mol Biol Rev 65(2):232–260

Coelho C, de Lencastre H, Aires-de-Sousa M (2017) Frequent occurrence of trimethoprim-sulfamethoxazole hetero-resistant *Staphylococcus aureus* isolates in different African countries. Eur J Clin Microbiol Infect Dis 36(7):1243–1252

Cole AM, Tahk S, Oren A, Yoshioka D, Kim YH, Park A, Ganz T (2001) Determinants of *Staphylococcus aureus* nasal carriage. Clin Diagn Lab Immunol 8(6):1064–1069

Cruciani M, Bassetti D (1994) The fluoroquinolones as treatment for infections caused by gram-positive bacteria. J Antimicrob Chemother 33(3):403–417

David MZ, Daum RS (2010) Community-associated methicillin-resistant *Staphylococcus aureus*: epidemiology and clinical consequences of an emerging epidemic. Clin Microbiol Rev 23(3):616–687

Derbise A, Dyke KG, El Solh N (1996) Characterization of a *Staphylococcus* aureus Transposon, Tn5405, located within Tn5404and carrying the aminoglycoside resistance genes, aphA-3andaadE. Plasmid 35(3):174–188

Deurenberg RH, Stobberingh EE (2008) The evolution of *Staphylococcus aureus*. Infect Genet Evol 8(6):747–763

Dowzicky M, Talbot GH, Feger C, Prokocimer P, Etienne J, Leclercq R (2000) Characterization of isolates associated with emerging resistance to quinupristin/dalfopristin (Synercid®) during a worldwide clinical program. Diagn Microbiol Infect Dis 37(1):57–62

Eady EA, Cove JH (2003) Staphylococcal resistance revisited: community-acquired methicillin resistant *Staphylococcus aureus*-an emerging problem for the management of skin and soft tissue infections. Curr Opin Infect Dis 16(2):103–124

Fair RJ, Tor Y (2014) Antibiotics and bacterial resistance in the 21st century. Perspect Medicin Chem 6:25–64

Foster TJ, Geoghegan JA, Ganesh VK, Höök M (2014) Adhesion, invasion and evasion: the many functions of the surface proteins of *Staphylococcus aureus*. Nat Rev Microbiol 12(1):49–62

Fournier B, Philpott DJ (2005) Recognition of *Staphylococcus aureus* by the innate immune system. Clin Microbiol Rev 18(3):521–540

Fridkin SK, Hageman JC, Morrison M, Sanza LT, Como-Sabetti K, Jernigan JA, Harriman K, Harrison LH, Lynfield R, Farley MM (2005) Methicillin-resistant *Staphylococcus aureus* disease in three communities. N Engl J Med 352(14):1436–1444

Gade ND, Qazi MS (2013) Fluoroquinolone therapy in *Staphylococcus aureus* infections: where do we stand? J Lab Physicians 5(2):109–112

Ghasemian A, Najar Peerayeh S, Bakhshi B, Mirzaee M (2015) The microbial surface components recognizing adhesive matrix molecules (MSCRAMMs) genes among clinical isolates of *Staphylococcus aureus* from hospitalized children. Iran J Pathol 10(4):258–264

Ghuysen JM (1994) Molecular structures of penicillin-binding proteins and β-lactamases. Trends Microbiol 2(10):372–380

Gordon RJ, Lowy FD (2008) Pathogenesis of methicillin-resistant *Staphylococcus aureus* infection. Clin Infect Dis 46(5):S350–S359

Harkins C, Pichon B, Doumith M, Parkhill J, Westh H, Tomasz A, de Lencastre H, Bentley S, Kearns A, Holden M (2017) Methicillin-resistant *Staphylococcus aureus* emerged long before the introduction of methicillin into clinical practice. Genome Biol 18(1):1–30

Hiramatsu K, Hanaki H, Ino T, Yabuta K, Oguri T, Tenover F (1997) Methicillin-resistant *Staphylococcus aureus* clinical strain with reduced vancomycin susceptibility. J Antimicrob Chemother 40(1):135–136

Hiramatsu K, Cui L, Kuroda M, Ito T (2001) The emergence and evolution of methicillin-resistant *Staphylococcus aureus*. Trends Microbiol 9(10):486–493

Jevons MP (1961) "Celbenin"-resistant Staphylococci. Br Med J 1(5219):124–125

Johannessen M, Sollid JE, Hanssen AM (2012) Host-and microbe determinants that may influence the success of S. aureus colonization. Front Cell Infect Microbiol 2:56

Josse J, Laurent F, Diot A (2017) Staphylococcal adhesion and host cell invasion: fibronectin-binding and other mechanisms. Front Microbiol 8:2433

Kahl BC, Becker K, Löffler B (2016) Clinical significance and pathogenesis of Staphylococcal small colony variants in persistent infections. Clin Microbiol Rev 29(2):401–427

Katayama Y, Ito T, Hiramatsu K (2000) A new class of genetic element, staphylococcus cassette chromosome mec, encodes methicillin resistance in *Staphylococcus aureus*. Antimicrob Agents Chemother 44(6):1549–1555

Kim SJ, Chang J, Singh M (2015) Peptidoglycan architecture of gram-positive bacteria by solid-state NMR. Biochim Biophys Acta Biomembr 1848(1):350–362

Kobayashi SD, Malachowa N, DeLeo FR (2015) Pathogenesis of *Staphylococcus aureus* abscesses. Am J Pathol 185(6):1518–1527

Kong KF, Schneper L, Mathee K (2010) Beta-lactam antibiotics: from antibiosis to resistance and bacteriology. APMIS 118(1):1–36

Kong C, Neoh HM, Nathan S (2016) Targeting *Staphylococcus aureus* toxins: a potential form of anti-virulence therapy. Toxins 8(3):72

Lacey K, Geoghegan J, McLoughlin R (2016) The role of *Staphylococcus aureus* virulence factors in skin infection and their potential as vaccine antigens. Pathogens 5(1):22

Laurent F, Lelièvre H, Cornu M, Vandenesch F, Carret G, Etienne J, Flandrois J-P (2001) Fitness and competitive growth advantage of new gentamicin-susceptible MRSA clones spreading in French hospitals. J Antimicrob Chemother 47(3):277–283

Leclercq R, Courvalin P (1991) Bacterial resistance to macrolide, lincosamide, and streptogramin antibiotics by target modification. Antimicrob Agents Chemother 35(7):1267

Lee JC (1996) The prospects for developing a vaccine against *Staphylococcus aureus*. Trends Microbiol 4(4):162–166

Lee AS, de Lencastre H, Garau J, Kluytmans J, Malhotra-Kumar S, Peschel A, Harbarth S (2018) Methicillin-resistant *Staphylococcus aureus*. Nat Rev Disease 4:18033

Leonard FC, Markey BK (2008) Methicillin-resistant *Staphylococcus aureus* in animals: a review. Vet J 175(1):27–36

Leong HN, Kurup A, Tan MY, Kwa AL, Liau KH, Wilcox MH (2018) Management of complicated skin and soft tissue infections with a special focus on the role of newer antibiotics. Infect Drug Resist 11:1959–1974

Liu C, Bayer A, Cosgrove SE, Daum RS, Fridkin SK, Gorwitz RJ, Kaplan SL, Karchmer AW, Levine DP, Murray BE, Rybak MJ (2011) Clinical practice guidelines by the Infectious Diseases Society of America for the treatment of methicillin-resistant *Staphylococcus aureus* infections in adults and children. Clin Infect Dis 52(3):e18–e55

Livermore DM (2000) Antibiotic resistance in Staphylococci. Int J Antimicrob Agents 16:3–10

Loewen K, Schreiber Y, Kirlew M, Bocking N, Kelly L (2017) Community-associated methicillin-resistant *Staphylococcus aureus* infection: literature review and clinical update. Can Fam Physician 63(7):512–520

Loomba PS, Taneja J, Mishra B (2010) Methicillin and vancomycin resistant *S. aureus* in hospitalized patients. J Glob Infect Dis 2(3):275

Lowy FD (2003) Antimicrobial resistance: the example of *Staphylococcus aureus*. J Clin Invest 111(9):1265–1273

McGuinness WA, Malachowa N, DeLeo FR (2017) Vancomycin resistance in *Staphylococcus aureus*. Yale J Biol Med 90(2):269–281

McKee CM, Houck CL (1943) Induced resistance to penicillin of cultures of Staphylococci, Pneumococci and Streptococci. Proc Soc Exp Biol Med 53(1):33–34

Mitchell D, Howden B (2005) Diagnosis and management of *Staphylococcus aureus* bacteraemia. Intern Med J 35:S17–S24

Munita JM, Arias CA (2016) Mechanisms of antibiotic resistance. Microbiol Spectr 4(2). VMBF-0016-2015

Murdoch DR, Corey GR, Hoen B, Miró JM, Fowler VG, Bayer AS, Karchmer AW, Olaison L, Pappas PA, Moreillon P (2009) Clinical presentation, etiology, and outcome of infective endocarditis in the 21st century: the international collaboration on endocarditis–prospective cohort study. Arch Intern Med 169(5):463–473

Murray R (2005) *Staphylococcus aureus* infective endocarditis: diagnosis and management guidelines. Intern Med J 35:S25–S44

Naimi TS, LeDell KH, Como-Sabetti K, Borchardt SM, Boxrud DJ, Etienne J, Johnson SK, Vandenesch F, Fridkin S, O'Boyle C (2003) Comparison of community-and health care–associated methicillin-resistant *Staphylococcus aureus* infection. JAMA 290(22):2976–2984

Novick RP (1990) The staphylococcus as a molecular genetic system. In: Molecular biology of the Staphylococci, vol 204. VCH, New York, pp 1–37, 587–636

O'Riordan K, Lee JC (2004) *Staphylococcus aureus* capsular polysaccharides. Clin Microbiol Rev 17(1):218–234

Ortega E, Abriouel H, Lucas R, Gálvez A (2010) Multiple roles of *Staphylococcus aureus* enterotoxins: pathogenicity, superantigenic activity, and correlation to antibiotic resistance. Toxins 2(8):2117–2131

Otto M (2014) *Staphylococcus aureus* toxins. Curr Opin Microbiol 17:32–37

Pinchuk IV, Beswick EJ, Reyes VE (2010) Staphylococcal enterotoxins. Toxins 2(8):2177–2197

Proctor RA, Kriegeskorte A, Kahl BC, Becker K, Löffler B, Peters G (2014) *Staphylococcus aureus* small Colony variants (SCVs): a road map for the metabolic pathways involved in persistent infections. Front Cell Infect Microbiol 4:99

Rammelkamp CH, Maxon T (1942) Resistance of *Staphylococcus aureus* to the action of penicillin. Proc Soc Exp Biol Med 51(3):386–389

Roberts S, Chambers S (2005) Diagnosis and management of *Staphylococcus aureus* infections of the skin and soft tissue. Intern Med J 35:S97–S105

Rouch DA, Byrne ME, Kong YC, Skurray RA (1987) The aacA-aphD gentamicin and kanamycin resistance determinant of Tn4001 from *Staphylococcus aureus*: expression and nucleotide sequence analysis. J Gen Microbiol 133(11):3039–3052

Santajit S, Indrawattana N (2016) Mechanisms of antimicrobial resistance in ESKAPE pathogens. Biomed Res Int 2016:2475067

Schaberg DR, Zervos MJ (1986) Intergeneric and interspecies gene exchange in gram-positive cocci. Antimicrob Agents Chemother 30(6):817–822

Speziale P, Pietrocola G, Foster TJ, Geoghegan JA (2014) Protein-based biofilm matrices in Staphylococci. Front Cell Infect Microbiol 4:171

Teng TS, Ji AL, Ji XY, Li YZ (2017) Neutrophils and immunity: from bactericidal action to being conquered. J Immunol Res 2017:9671604

Tibavizco D, Rodríguez JY, Silva E, Cuervo SI, Cortés JA (2007) Therapeutic approach to *Staphylococcus aureus* bacteremia. Biomedica 27(2):294–307

Tong SY, Davis JS, Eichenberger E, Holland TL, Fowler VG (2015) *Staphylococcus aureus* infections: epidemiology, pathophysiology, clinical manifestations, and management. Clin Microbiol Rev 28(3):603–661

Vandenesch F, Naimi T, Enright MC, Lina G, Nimmo GR, Heffernan H, Liassine N, Bes M, Greenland T, Reverdy M-E (2003) Community-acquired methicillin-resistant *Staphylococcus aureus* carrying Panton-valentine leukocidin genes: worldwide emergence. Emerging Infect Dis 9(8):978–984

Voyich JM, Braughton KR, Sturdevant DE, Whitney AR, Saïd-Salim B, Porcella SF, Long RD, Dorward DW, Gardner DJ, Kreiswirth BN (2005) Insights into mechanisms used by *Staphylococcus aureus* to avoid destruction by human neutrophils. J Immunol 175(6):3907–3919

Wertheim HF, Melles DC, Vos MC, van Leeuwen W, van Belkum A, Verbrugh HA, Nouwen JL (2005) The role of nasal carriage in *Staphylococcus aureus* infections. Lancet Infect Dis 5(12):751–762

Wilkinson BJ (1997) Biology. In: Crossley KB, Archer GL (eds) The Staphylococci in human diseases. Churchill Livingstone, New York, pp. 1–37

Wilson R, Cockcroft WH (1952) Penicillin resistant staphylococcal infection. Can Med Assoc J 66(6):548–551

Pathogenesis, Virulence Factors, and Antibiotic Resistance of Group B Streptococcus

8

Subhaswaraj Pattnaik, Asad Syed, and Busi Siddhardha

Contents

Abstract

Streptococcus is an important genus in the gram-positive coccus, belonging to family Streptococcaceae. The members of the genus *Streptococcus* are biomedically relevant owing to their widespread pathogenic profile causing severe healthcare issues such as pharyngitis, pneumoniae, neonatal meningitis, sepsis, endocarditis, bacteremia, and urinary tract infections (UTIs). The diversified species of genus *Streptococcus* are basically categorized based on the inherent hemolytic properties,

S. Pattnaik · B. Siddhardha (✉)
Department of Microbiology, School of Life Sciences, Pondicherry University, Puducherry, India

A. Syed
Botany and Microbiology Department, College of Science, King Saud University, Riyadh, Saudi Arabia

B. Siddhardha et al. (eds.), *Model Organisms for Microbial Pathogenesis, Biofilm Formation and Antimicrobial Drug Discovery*,
https://doi.org/10.1007/978-981-15-1695-5_8

i.e., either facilitating the oxidation of iron in hemoglobin within the red blood cells (Alpha-hemolytic) or complete rupturing of red blood cells (beta-hemolytic). The beta-hemolytic group of *Streptococci* are further classified into several serotypes using Lancefield grouping system which is based on the presence or absence of specific carbohydrate moieties on the bacterial cell wall. Among the different serotypes, Lancefield group A (Group A Streptococcus) and Lancefield group B (Group B Streptococcus) are critically important in the medical settings based on their ability to cause life-threatening diseases in the immunocompromised individuals. Among the different species of genus *Streptococcus*, the most clinically relevant species are *Streptococcus pneumoniae* and *S. viridians* (belonging to Alpha-hemolytic group of *Streptococcus*). Apart from Alpha-hemolytic Streptococcus, Lancefield groups A and B (also known as "group A strep" and "group B strep") are also considered to be highly relevant in clinical and biomedical setup. Group B Streptococcus is an opportunistic pathogenic bacteria causing severe neonatal sepsis, meningitis, bacteremia, urinary tract infections, endometritis, maternal bacteremia, and other associated diseases. The disease severity and chronic infection profile of GBS could be attributed to the presence of specific virulence determinants such as pore-forming toxins and capsular polysaccharides. The epidemiological profile of GBS gained considerable attention owing to its ability to exhibit resistance against conventional antibiotic treatment by forming recalcitrant biofilms. In the fight against GBS infection, specific antibiotics like penicillin and vancomycin, high throughput therapeutic strategy like intrapartum antibiotic prophylaxis (IAP), and public awareness programmes are considered to be effective in controlling the bacterial infections. In addition, novel drug molecules from natural sources could also be utilized as prolific arsenal against GBS infections. Despite the development in the therapeutic strategies to control GBS infections, the mortality and morbidity caused by GBS infections remain an uphill challenge for the scientific community. In this context, it is imperative to quest for novel strategies in preventing the GBS infections and that could be implemented through public awareness programmes and prenatal screening workshops apart from conventional therapeutic approach.

Keywords
Group B Streptococcus · Pathogenesis · Antibiotic resistance · Biofilm · Maternal immunization

8.1 Introduction

Since the discovery of antibiotics to treat microbial infections, an array of antibiotics belonging to different classes have been discovered and are effectively used as an arsenal against microbial infections and related health consequences. Though there is a marked development in the field of finding novel antibiotics form last half-decade;

the evolutionary changes in the microbial spectrum also emerge as lethal challenge to the available antibiotics in the form of resistance. The incidence of resistance to conventional antibiotics becomes more prominent from last few decades, thereby microbial infections remain a significant public threat both at sociological and economical levels (Rajagopal 2009). The advent of the resistance to antibiotics has also put a significant burden to the developing countries as well as under-developed countries. Among the pathogenic bacteria causing severe public health issues and showed resistance towards the conventional antibiotics, the emergence of Streptococcal infections also gained considerable attentions due to their detrimental effects on human health such as meningitis, cellulitis, pneumonia, pericarditis, pharyngitis, and urinary tract infections (UTIs). A majority of the *Streptococcus* sp. are found to be commensal microorganisms owing to their ability to inhabit oral cavity and nasopharynx of human beings without affecting their physiological functions. However, the Streptococcal sp. also have the inherent ability to cause various chronic diseases and lethal infections in the form of superficial or systemic infection and in most cases without symptomatic infections and hence they are generally considered as opportunistic pathogens (Nobbs et al. 2015).

8.2 Grouping and Classification of Genus *Streptococcus*

The genus Streptococcus belongs to the Phylum Firmicutes and can be classified on the basis of their differential hemolytic activity, antigenic composition, growth characteristics, genetic features as well as biochemical characteristics. On the basis of hemolytic properties, the genus Streptococcus was differentiated into alpha-hemolytic, beta-hemolytic, and gamma-hemolytic types. According to the antigenic properties shown, the beta-hemolytic streptococci were grouped into different classes such as A-H and K-V. Among these groups, five groups such as A, B, C, D, and G have gained considerable attention based on their inherent potential to cause infections and disease conditions. Group B streptococci (GBS) are further sub-categorized into type I a/c, and III based on the presence of specific surface proteins as antigenic markers.

8.2.1 Group A Streptococcus (GAS)

Among the pathogenic *Streptococcus* sp., Group A Streptococcus (GAS) are known for their ability to induce severe infections such as pharyngeal and skin infections, necrotizing fasciitis, and toxic shock syndrome in humans. One of the interesting features of GAS colonization and subsequent infections is the involvement of highly complex regulatory network. The unique characteristic of this regulatory network is its differential regulation mechanisms under different environmental circumstances promoting diversity in the routes and manifestations of infection pathways in the hosts (Shelburne et al. 2008).

8.2.2 Group B Streptococcus (GBS)

Group B Streptococcus (GBS) is a beta-hemolytic group of catalase negative facultative anaerobe. The GBS has the inherent ability to colonize the oropharynx, gastrointestinal, and genitourinary tract in 10–30% of humans without any sorts of symptoms. The GBS is known for its ability to cause severe invasive neonatal infections such as neonatal bacteremia, sepsis, pneumonia, and meningitis. Apart from neonatal infections, GBS also tend to colonize adult human beings especially pregnant women with severe invasive infections such as endometritis, urinary tract infections, and occasionally, maternal bacteremia. The unique ability of GBS to cause severe infections could also be attributed due to their efficacy in infecting non-pregnant individuals in the form of sepsis and meningitis leading to increased mortality and morbidity (Rosa-Fraile and Spellerberg 2017; Skolnik et al. 2017). As per recent trends, GBS remains one of the most common causal agents for inducing severe neonatal sepsis and meningitis in the immunocompromised patients (Zimmermann et al. 2017). The lethality and severity of GBS infections could be observed from their ability to cause serious diseases not only neonates but also in pregnant women as well as non-pregnant immunocompromised individuals. In majority of cases, the GBS infections occur in the genital tract or placenta in pregnant women and lead to severe miscarriages and stillbirth issues. In addition, the immunocompromised individuals with diabetes, cancer, cirrhosis, HIV infection, and age factors could be instrumental in the onset of GBS infections (Chen et al. 2013).

8.2.3 Group B Streptococcus: An Overview

Streptococcus agalactiae (Group B Streptococcus, GBS) is an opportunistic pathogen colonizing the gastrointestinal and genitourinary tracts of health individual causing severe asymptomatic health issues. Apart from that, GBS also accounts for the leading cause of invasive neonatal infections (Lopez et al. 2018). The GBS has the inherent property of colonizing the genitourinary tract of 15–35% pregnant women who can transmit the pathogen to their neonate during childbirth and contribute to early onset disease (EOD) leading to severe sepsis and meningitis (Medugu et al. 2017). In addition to neonatal meningitis and sepsis, GBS is also associated with severe life-threatening syndromes such as necrotizing fasciitis and toxic shock syndrome (Lupo et al. 2014). The GBS colonization of the lower genital tract or vagina or rectum in pregnant women leads to asymptomatic bacteriuria, severe urinary tract infections, chorioamnionitis, postpartum endometritis, and bacteremia (Cho et al. 2019). Group B streptococcal infection in elderly individuals is strongly linked to congestive heart failure, neurologic illness, urinary tract infection, pneumonia, and soft tissue infection as the most common manifestations of infection.

Apart from human infections, GBS (*S. agalactiae*) also has the ability to infect other mammalian hosts as well as other vertebrates such as reptiles, amphibians, and fish in particular Tilapia (*Oreochromis niloticus*). The GBS infections in fish cause severe streptococcosis. The fish streptococcosis can be characterized by septicemia, exophthalmia, and meningoencephalitis and put significant impact on the development of pisciculture sectors worldwide (Chideroli et al. 2017; Zhang et al. 2018). The diversified clinical manifestations by GBS in the immunocompromised individuals could be attributed to the presence of highly specific regulatory network which has the potential to induce efficient adaptability of bacteria under different environmental conditions. The highly complex regulatory network also allows the GBS to express specific virulence determinants thereby promoting the colonization and invasion of epithelial barriers of the host cells. These virulence phenotypes also allow GBS to develop resistance to severe stress conditions by bypassing the host immune mechanisms thereby contributing to the pathogenesis of infection (Otaguiri et al. 2013).

8.3 Pathophysiology of GBS

8.3.1 Epidemiological Characteristics of GBS Infections

The epidemiological severity of GBS infections in the neonatal period and up to 90 days post-birth has provided a greatest impact on human health. The GBS infections in neonates could be categorized into two types such as early onset (EO) and late onset (LO) disease. Among these, early onset GBS disease (infection presenting in the first 6 days of life) accounts for approximately 60–70% of all GBS disease. GBS serotypes Ia, II, III, and V are responsible for most EO disease and invasive infections (Lopez et al. 2018). The pre-requisite factors for the early onset GBS disease are maternal carriage of GBS in the gastrointestinal and/or genital tracts. In majority of cases, early onset GBS disease has the inherent property of causing the disease severity through pneumonia and sepsis. The early onset disease is vertically acquired from the mother through ascending infections from lower genital tract (Kwatra et al. 2016). Despite the high throughput development in the intensive supportive care, diagnostic approaches and therapeutic advances, these early onset GBS infections remain associated with high mortality and morbidity owing to its asymptomatic properties (Melin 2011).

The process of neonatal infection (early onset disease) by GBS is highly complex and assisted by both host and bacterial virulence factors. The first stage in the pathogenesis of GBS-EOD includes the establishment of colonization in the vaginal mucosa of the pregnant woman by successful adherence to the vaginal epithelial cells and increased tolerance to host immune system. The second stage involves the invasion across host epithelial barriers and progression of bacterial colonization into the amniotic cavity followed by rapid proliferation in the skin or mucous membranes of the fetus or to enter the fetal lung through aspiration of infected amniotic fluid. The third important step is the ability of GBS to replicate within the alveoli of the neonate, adhere to pulmonary epithelium, and avoid clearance by pulmonary macrophages after birth. The invasion into the pulmonary epithelial and endothelial cells by GBS significantly allows the GBS progression into the blood stream causing severe septicemia followed by meningitis and osteomyelitis. The disease progression efficacy shows the inherent potential of GBS to bypass the host's natural immune defenses to adhere, to invade and to progress through several cell barriers (Melin 2011; Cools and Melin 2017) (Fig. 8.1). Meanwhile, the late onset GBS disease (onset between days 7 and 89 of life) is primarily caused predominantly by serotype III and is acquired perinatally, nosocomially or from community sources. In majority of cases, late onset GBS disease (50%) has the inherent property of causing the disease severity through septicemia and meningitis (Doare and Heath 2013; Joubrel et al. 2015; Kobayashi et al. 2016).

During the infection process, GBS utilizes prolific strategies to colonize the vagina of pregnant women. These strategies include the binding to the host surfaces and subverting the host immune defense mechanisms. These strategies also significantly promote the dissemination and colonization efficacy of GBS under severe stress conditions not only in pregnant women but also in the new born babies. Apart from colonization strategies, GBS strains also have the inherent potential of causing disease severity and chronic infections owing to the evolutionary acquisition of specified virulence factors. These factors not only enhance the dissemination and disease progression cycle but also allow the GBS to evade the host immune tolerance. At the cellular level, GBS regulates the expression of specific virulence determinants through highly complex, species-specific signal transduction regulatory network system. This regulatory system has the potential to sense host environments followed by adequate response to enhance the survival in the vulnerable

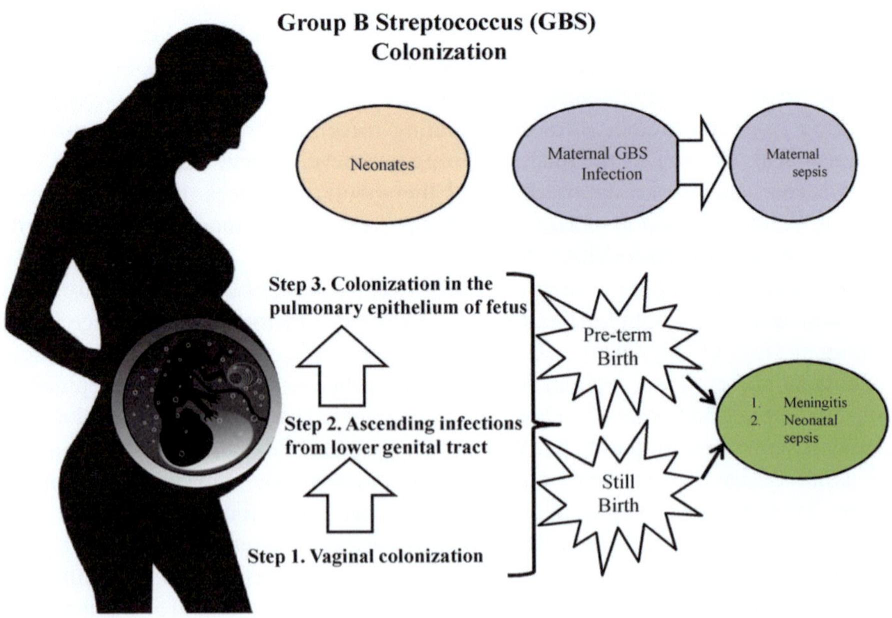

Fig. 8.1 Schematic representation of maternal GBS colonization into the neonates through ascending infections followed by proliferation in the pulmonary epithelium leading to severe meningitis and neonatal sepsis in neonates. Maternal GBS infection also causes maternal sepsis

environmental setup in the host environments outside the lower genital tract for efficient progression of infection process (Armistead et al. 2019).

8.3.2 Virulence Factors Associated with GBS Infections

In pathogenic microorganisms, one of the important factors that plays crucial role in maintaining the pathophysiology is the ability of pathogenic microorganisms to produce a myriad of virulence factors. The production and secretion of virulence factors have enabled the pathogenic microorganisms to invoke healthcare related issues and ability to cause disease. Like other pathogenic counterparts, GBS produces an array of virulence determinants that critically emphasize the pathogenicity of GBS. The important virulence determinants encoded by GBS are pore-forming toxins and the sialic acid-rich capsular polysaccharide (CPS). Apart from pore-forming toxins and CPS, other important virulence determinants in GBS infections are adhesion factors (that facilitate the binding to cells or extracellular matrix), evasion factors (that modulate the neutrophil recruitment and prevent complement binding), and other virulence factors that showed resistance to antimicrobial peptides and other conventional therapeutics (Chen et al. 2013).

The role of pore-forming toxins in GBS is to facilitate the host invasion followed by their survival and systemic dissemination. Two highly characterized pore-forming toxins such as β-hemolysin/cytolysin (β-H/C) and Christie Atkins Munch Peterson (CAMP) factor were reported in GBS. The pore-forming toxin, β-H/C not only promotes GBS invasion of host cell barriers such as the epithelial and endothelial cells of the lung and the blood–brain barrier (BBB) but also promotes immune evasion by inducing host inflammatory responses and enhanced resistance to reactive oxygen species (ROS). The expression of pore-forming toxin, β-H/C is generally regulated by the hyperpigmentation phenomenon which enables GBS to be hyperhemolytic thereby promoting invasive properties and severe pathogenesis (Lupo et al. 2014). Meanwhile, the other pore-forming toxin, i.e., CAMP factor plays a crucial role in increasing the bacterial pathogenesis by inducing septicemia and increased lethality by promoting cell lysis through targeting the susceptible target membranes (Rajagopal 2009).

Apart from pore-forming toxins, sialic acid-rich capsular polysaccharides (CPS) are important determinants of bacterial pathogenicity in GBS infections. The GBS CPSs comprise of various arrangements of monosaccharide building blocks containing glucose, galactose, and/or N-acetylglucosamine. In addition, the presence of sialic acid residue on the branching terminus of each repeating monosaccharide unit characteristically determines unique features of GBS in evading host immune system. The GBS CPSs have the inherent property to evade host recognition, complement factor decomposition, and phagocytosis owing to the presence of sialic acid and remain one of the important constituents in the GBS pathogenesis. The presence of sialic acid residue not only inhibit the complement cascade by accelerated dissociation of C3 convertase but also significantly modulate the phagocytosis and oxidative burst from neutrophils and monocytes, thereby enhancing bacterial survival (Chen et al. 2013). In GBS, ten capsular serotypes such as Ia, Ib, and II–IX have been reported. During the host infection, GBS utilizes typical CPS for adherence and invasion into the host tissue (Baker 2013; Vornhagen et al. 2017). Among the identified serotypes, type III strains are particularly important in imparting pathogenicity owing to its inherent ability in invading the brain microvascular endothelial cells as compared to its other GBS serotype counterparts (Baker 2013).

The most prevalent virulence factors found in the GBS isolates are *cyl*(E) (encoding cytolysin–hemolysin), followed by *scp*(B) (encoding an invasion with C5a peptidase activity), *rib* (Alp family surface protein Rib resistant to protease effect), and *bca* (encoding beta subunit of the C protein) with a frequency of 90.5, 75.6, 62.1, and 43.4%, respectively. These virulence genes are exclusively found in the GBS isolates isolated from human (Emaneini et al. 2016a). In addition, the virulence factors *pep*(B) (encoding oligopeptidase protein) and *bib*(A) (encoding bacterial immunogenic adhesin) are found to be observed only in neonatal isolates. The virulence factors which are observed to be prevalent in neonatal isolates are *bca* (encoding beta subunit of the C protein) and *scp*(B) (encoding an invasin). The virulence factor, i.e., *fbs*(B) (encoding fibrinogen-binding protein mediating invasivity) has been observed frequently in colonizing and pathogenic isolates as compared to neonatal isolates (Carvalho-Castro et al. 2017; Lopez et al. 2018).

In GBS related infections, the secretion of GBS hyaluronidase significantly promotes the immune evasion process by degrading the pro-inflammatory hyaluronan fragments into hyaluronan disaccharides which in turn block toll-like receptor, TLR2/4 signaling pathway (Kolar et al. 2015). Two-component systems (TCSs) also found to be crucial in inducing pathogenicity in GBS infections by virtue of their ability to detect the environmental changes and other stress conditions.

8.3.3 Antibiotic Resistance Profile of GBS

The pathogenic bacteria have the inherent ability to exhibit chronic pathogenicity by producing an array of virulence determinants as well as recalcitrant biofilms. Biofilms are defined as sessile microbial communities in which the pathogenic bacteria are embedded within a self-produced extracellular polymer matrix. The polymeric matrix is considered to generate a safe microenvironment for the embedded microbial communities from stress conditions such as extreme pH, antimicrobial therapeutics, host immunity and thereby promoting persistent colonization and infection (Boonyayatra et al. 2016; Nie et al. 2018). One of the important members of Group B Streptococcus is *S. agalactiae* which is known for its contagious pathogenicity owing to its ability to form highly resistant biofilm architecture. The inherent ability of *S. agalactiae* to form biofilms can be correlated with enhanced pathogenicity and tolerance to the conventional antibiotics. The microcolonies residing within the biofilm architecture enabled the residing *Streptococcal* microcolonies to modulate the pathophysiological responses on treatment with conventional therapeutics thereby blocking the entry of therapeutic drugs into the bacterial system, thereby minimizing the efficacy of the administered drugs (Ebrahimi et al. 2013).

No doubt the presence of capsular polysaccharides in GBS enhances the bacterial pathogenicity and associated health ailments. In addition to capsular polysaccharides, GBS also contains elaborate surface-anchored pili which enable GBS during colonization in the host cells by promoting bacterial adhesion, invasion into host cells, mitigating host immune responses, and biofilm formation (Xia et al. 2015; Khodaei et al. 2018; Perichon et al. 2019). The epidemiological prospects of surface adhesins and pili enable GBS adaptation to stress conditions and increases host specificity. In GBS, three pilus islands such as (PI)-1, PI-2a, and PI-2b were identified. Each PI encodes for three structural proteins such as a backbone protein (BP), two ancillary proteins (AP), and two pilus-specific class C sortase enzymes. The ancillary proteins allow the initiation of bacterial adherence to various host tissues. Meanwhile, the backbone proteins strictly facilitate the invasion efficacy and paracellular translocation of host cells. The role of class C sortase enzymes is to recognize LPXTG amino acid motifs on structural proteins and facilitate covalent attachment. The pili island, PI-2a is found to be crucial for biofilm formation and thus enable GBS for antibiotic resistance. Meanwhile, PI-2b promotes the intracellular survival of GBS in macrophages (Springman et al. 2014).

As per recent trends, *S. agalactiae* (GBS) exhibited immense resistance profile against aminoglycosides group of antibiotics including sulphazotrim (sulfamethoxazole with trimethoprim), tetracycline, ampicillin as well as fluoroquinolones group of antibiotics (Chideroli et al. 2017). The resistance pattern shown by GBS against conventional antibiotics, especially tetracycline could be attributed to the presence of specific virulence genes such as *scpB*, *hlyB*, and *bca* which are significantly correlated with the presence of tetracycline resistant *tetM* gene (Rato et al. 2013; Emaneini et al. 2016b). Apart from tetracycline resistance, the GBS isolated from neonates from clinical setup shown to exhibit multidrug resistance (MDR) to other traditional antibiotics such as erythromycin and clindamycin (Wang et al. 2015).

S. agalactiae adhesion to host cells constitutes an important step in colonization and frequently involves components of the extracellular matrix (ECM) such as fibronectin, fibrinogen, collagen, and laminin. The cell-wall-anchored (CWA) proteins often bind with ECM and cellular receptors and initiate the progression of chronic infections (Chuzeville et al. 2015). During GBS infections, genes encoding putative surface proteins and in particular an antigen I/II have been identified on integrative and conjugative elements (ICEs) found in *S. agalactiae*. The presence of this antigen as putative surface protein enables GBS in cell-cell aggregation and biofilm formation.

8.3.4 Therapeutic Strategies to Control GBS Infections

The prevalence of GBS infections in the neonates and the critical role of breast milk in imparting severity in the GBS infections suggested the scientific community to develop maternal immunization programme to combat GBS infections. From therapeutic perspectives, maternal immunization programme which includes the development of maternal vaccines remains instrumental in the fight against GBS infections. Till date, three generations of maternal vaccines such as native polysaccharide vaccines (first generation maternal vaccines), glycoconjugate vaccines (second generation maternal vaccines), and vaccine design with high throughput technological applications (third generation vaccines) have been developed as an arsenal against GBS infections (Chen et al. 2013). The maternal vaccination programme in the pregnant women not only prevent invasive GBS related diseases in the neonates but also significantly reduce the recto-vaginal colonization in the pregnant women, who themselves are at increased risk for developing invasive GBS disease (Madhi and Dangor 2017).

As it is evident from earlier studies that maternal colonization is one of the primary risk factors in the onset of GBS disease and associated health risks, the development of intrapartum antibiotic prophylaxis (IAP) could be considered as promising in reducing the risks associated with GBS infections. The emergence of IAP strategy has been recommended for both GBS colonized women as well as women with premature rupture of membranes, prolonged membrane rupture, fever, and preterm birth owing to its ability in reducing as much as 80% of GBS early onset disease in USA. Apart from prophylaxis strategy, prenatal screening in pregnant women is considered to be influential in reducing the incidence of GBS neonatal disease in the developed countries. However, due to limiting factors such as resources limitations and under-developed infrastructure in the developing and low-income countries, the prenatal screening settings and IAP strategy have not been implemented in these countries (Sadaka et al. 2018). Based on the promising aspects shown by prophylaxis strategy, it is important to develop this strategy in developing countries as well as low-income countries in the fight against GBS related health issues (Medugu et al. 2017).

Owing to the susceptibility shown by GBS towards penicillin, its prophylactic use could be instrumental in the fight against GBS related infections by significantly reducing the incidence of early onset diseases in neonatal individuals (Boswihi et al. 2012). The synergistic activity of gentamicin with penicillin could be explored as an alternative therapeutic approach in minimizing the severity of GBS infections (Ruppen et al. 2017). In addition, gentamicin in combination with benzylpenicillin and/or rifampicin also proved to be effective in combating GBS associated biofilm dynamics (Moreno et al. 2017). However, vancomycin could also be used as a therapeutic strategy for initial treatment of GBS infection in particular cases where the individuals showed allergic reactions to the penicillin treatment.

Apart from conventional therapeutics, the prophylactic use of naturally derived compounds (plant-derived and microbial-derived compounds) could also considered to be promising arsenal against GBS associated infections. In this context, plant-derived eugenol and mycosynthesized silver nanoparticles (AgNPs) showed synergistic activity in controlling the GBS infections. In addition, the use of eugenol in the treatment against GBS infection has a promising aspect in disrupting the biofilm forming ability of GBS and thus increases the sensitivity of GBS towards conventional antibiotics. The therapeutic efficacy of eugenol and other natural-derived drug candidates could be utilized as a promising alternative to advanced IAP-based therapeutics (Biasi-Garbin et al. 2015). Additionally, breast milk and colostrum contain antimicrobial and immunomodulatory components which have the inherent property to impair translocation of infectious pathogens including GBS. These substances not only compensate directly for deficiency in the neonatal immune system but also enhance the survival of defense agents such as secretory IgA (SIgA), lactoferrin, lysozyme, IFN-γ. In addition, these substances also prevent inflammation or enhance specific-antibody production, such as PAF-acetylhydrolase, antioxidants, interleukins, transforming growth factor (TGF), secretory leukocyte protease inhibitors (SLPIs) and defensin1 (Doare and Kampmann 2014). Recently, human milk oligosaccharides (HMOs) are also considered as alternative approach in controlling the pathogenic profile of GBS by significantly attenuating the biofilm formation and disruption of biofilm architecture of GBS thereby increasing the susceptibility of GBS to conventional therapeutics for the complete eradication of GBS infections (Ackerman et al. 2017).

8.4 Current Trends and Future Perspectives

Owing to the severity of GBS infections in neonates, it is important to take necessary measures to counteract the GBS related infections. Beyond therapeutics and development of GBS vaccines, it is highly important to generate public awareness by educating the public about the severity of GBS infections and by conducting frequent prenatal screening to promote prevention of neonatal GBS infections which could provide new avenues in the prevention of neonatal GBS infections (Burns and Plumb 2013). With the advancement in high throughput technologies and molecular techniques, it is imperative to utilize the arsenal of molecular typing of GBS isolates in order to understand the variability and epidemiology of GBS and thereby provide novel avenues to the scientific community for the development of effective control and eradication programmes in the fight against GBS infections (Reyes et al. 2017). As the capsular polysaccharides played key role as virulence factors and are important targets for the development of vaccine strategies and drug development in the process of fighting against GBS infections (Campisi et al. 2016; Jiang et al. 2016).

8.5 Conclusion

The beta-hemolytic group of gram-positive *S. agalactiae* (Group B Streptococcus, GBS) is often encountered as a part of the normal flora and considered as commensal microorganism colonizing the gastro-intestinal and the genital tract of healthy women. However, GBS could reach the new born through the birth canal and cause sepsis and/or meningitis and hence observed to be opportunistic pathogen. Apart from neonatal sepsis and meningitis, GBS is also associated with severe chronic infections such as UTIs, bacteremia, endocarditis, and other related health issues. The severity of chronic infections caused by GBS could be attributed to the diversity of virulence determinants such as pore-forming toxins and capsular polysaccharides. In addition to capsular polysaccharides, other surface associated factors like pili islands could be instrumental in inducing bacterial resistance phenomenon by its inherent ability to form persistent biofilms. As per recent trends, IAP therapeutics and conventional antimicrobial therapy are observed to be instrumental in controlling GBS infections. However, associated resistance could limit their widespread applications. In this context, it is imperative to develop effective strategies to counteract GBS infections related health issues. One of the primary strategies to prevent GBS infections is to generate public awareness through prenatal screening which could be advantageous in the fight against GBS infections.

References

Ackerman DL, Doster RS, Weitkamp JH, Aronoff DM, Gaddy JA, Townsend SD (2017) Human milk oligosaccharides exhibit antimicrobial and antibiofilm properties against Group B *Streptococcus*. ACS Infect Dis 3:595–605

Armistead B, Oler E, Adams Waldorf K, Rajagopal L (2019) The double life of Group B Streptococcus: asymptomatic colonizer and potent pathogen. J Mol Biol 431(16):2914–2931

Baker CJ (2013) The spectrum of perinatal Group B streptococcal disease. Vaccine 31S:D3–D6

Biasi-Garbin RP, Otaguiri ES, Morey AT, da Silva MF, Morguette AEB, Lancheros CAC, Kian D, Perugini MRE, Nakazato G, Duran N, Nakamura CV, Yamauchi LM, Yamada-Ogatta SF (2015) Effect of eugenol against *Streptococcus agalactiae* and synergistic interaction with biologically produced silver nanoparticles. Evid Based Complement Alternat Med 2015:861497

Boonyayatra S, Pata P, Nakharuthai P, Chaisri W (2016) Antimicrobial resistance of biofilm-forming *Streptococcus agalactiae* isolated from bovine mastitis. J Veter Sci Technol 7:5

Boswihi SS, Udo EE, Al-Sweih N (2012) Serotypes and antibiotic resistance in Group B Streptococcus isolated from patients at the Maternity Hospital, Kuwait. J Med Microbiol 61:126–131

Burns G, Plumb J (2013) GBS public awareness, advocacy, and prevention—What's working, what's not and why we need a maternal GBS vaccine. Vaccine 31S:D58–D65

Campisi E, Rosini R, Ji W, Guidotti S, Rojas-Lopez M, Geng G, Deng Q, Zhong H, Wang W, Liu H, Nan C, Margarit I, Rinaudo CD (2016) Genomic analysis reveals multi-drug resistance clusters in Group B Streptococcus CC17 hypervirulent isolates causing neonatal invasive disease in Southern mainland China. Front Microbiol 7:1265

Carvalho-Castro GA, Silva JR, Paiva LV, Custodio DAC, Moreira RO, Mian GF, Prado IA, Chalfun-Junior A, Costa GM (2017) Molecular epidemiology of *Streptococcus agalactiae* isolated from mastitis in Brazilian dairy herds. Braz J Microbiol 48:551–559

Chen VL, Avci FY, Kasper DL (2013) A maternal vaccine against group B *Streptococcus*: past, present, and future. Vaccine 31S:D13–D19

Chideroli RT, Amoroso N, Mainardi RM, Suphoronski SA, de Padua SB, Alfieri AF, Alfiieri AA, Mosela M, Moralez ATP, de Oliveira AG, Zanolo R, Di Santis GW, Pereira UP (2017) Emergence of a new multidrug-resistant and highly virulent serotype of *Streptococcus agalactiae* in fish farms from Brazil. Aquaculture 479:45–51

Cho CY, Tang YH, Chen YH, Wang SY, Yang YH, Wang TH, Yeh CC, Wu KG, Jeng MJ (2019) Group B Streptococcal infection in neonates and colonization in pregnant women: an epidemiological retrospective analysis. J Microbiol Immunol Infect 52:265–272

Chuzeville S, Dramsi S, Madec JY, Haenni M, Payot S (2015) Antigen I/II encoded by integrative and conjugative elements of *Streptococcus agalactiae* and role in biofilm formation. Microb Pathog 88:1–9

Cools P, Melin P (2017) Group B *Streptococcus* and perinatal mortality. Res Microbiol 168:793–801

Doare KL, Heath PT (2013) An overview of global GBS epidemiology. Vaccine 31S:D7–D12

Doare KL, Kampmann B (2014) Breast milk and Group B Streptococcal infection: vector of transmission or vehicle for protection? Vaccine 32:3128–3132

Ebrahimi A, Moatamedi A, Lotfalian S, Mirshokraei P (2013) Biofilm formation, hemolysin production and antimicrobial susceptibilities of *Streptococcus agalactiae* isolated from the mastitis milk of dairy cows in Shahrekord district, Iran. Veter Res Forum 4(4):269–272

Emaneini M, Khoramian B, Jabalameli F, Abani S, Dabiri H, Beigverdi R (2016a) Comparison of virulence factors and capsular types of *Streptococcus agalactiae* isolated from human and bovine infections. Microb Pathog 91:1–4

Emaneini M, Jabalameli F, Mirsalehian A, Ghasemi A, Beigverdi R (2016b) Characterization of virulence factors, antimicrobial resistance pattern and clonal complexes of group B Streptococci isolated from neonates. Microb Pathog 99:119–122

Jiang H, Chen M, Li T, Liu H, Gong Y, Li M (2016) Molecular characterization of *Streptococcus agalactiae* causing community- and hospital-acquired infections in Shanghai, China. Front Microbiol 7:1308

Joubrel C, Tazi A, Six A, Dmytruk N, Touak G, Bidet P, Raymond J, Trieu Cuot P, Fouet A, Kerneis S, Poyart C (2015) Group B streptococcus neonatal invasive infections, France 2007–2012. Clin Microbiol Infect 21:910–916

Khodaei F, Najafi M, Hasani A, Kalantar E, Sharifi E, Amini A, Aghazadeh M (2018) Pilus–encoding islets in *S. agalactiae* and its association with antibacterial resistance and serotype distribution. Microb Pathog 116:189–194

Kobayashi M, Schrag SJ, Alderson MR, Madhi SA, Baker CJ, Meulen AS, Kaslow DC, Smith PG, Moorthy VS, Vekemans J (2016) WHO consultation on group B Streptococcus vaccine development: report from a meeting held on 27–28 April 2016. Vaccine. https://doi.org/10.1016/j.vaccine.2016.12.029

Kolar SL, Kyme P, Tseng CW, Soliman A, Kaplan A, Liang J, Nizet V, Jiang D, Murali R, Arditi M, Underhill DM, Liu GY (2015) Group B Streptococcus evades host immunity by degrading hyaluronan. Cell Host Microbe 18:694–704

Kwatra G, Cunnington MC, Merrall E, Adrian PV, Ip M, Klugman KP, Tam WH, Madhi SA (2016) Prevalence of maternal colonisation with group B Streptococcus: a systematic review and meta-analysis. Lancet Infect Dis 16:1076–1084

Lopez Y, Parra E, Cepas V, Sanfeliu I, Juncosa T, Andreu A, Xercavins M, Perez J, Sanz S, Vergara A, Bosch J, Soto SM (2018) Serotype, virulence profile, antimicrobial resistance and macrolide-resistance determinants in *Streptococcus agalactiae* isolates in pregnant women and neonates in Catalonia, Spain. Enferm Infecc Microbiol Clin 36(8):472–477

Lupo A, Ruppen C, Hemphill A, Spellerberg B, Sendi P (2014) Phenotypic and molecular characterization of hyperpigmented group B Streptococci. Int J Med Microbiol 304:717–724

Madhi SA, Dangor Z (2017) Prospects for preventing infant invasive GBS disease through maternal vaccination. Vaccine 35:4457–4460

Medugu N, Iregbu KC, Parker RE, Plemmons J, Singh P, Audu LI, Etefie E, Davies HD, Manning SD (2017) Group B Streptococcal colonization and transmission dynamics in pregnant women and their newborns in Nigeria: implications for prevention strategies. Clin Microbiol Infect 23:673.e9–673.e16

Melin P (2011) Neonatal group B Streptococcal disease: from pathogenesis to preventive strategies. Clin Microbiol Infect 17:1294–1303

Moreno MG, Trampuz A, Di Luca M (2017) Synergistic antibiotic activity against planktonic and biofilm embedded *Streptococcus agalactiae*, *Streptococcus pyogenes* and *Streptococcus oralis*. J Antimicrob Chemother 72(11):3085–3092

Nie S, Lu X, Hu YW, Zheng L, Wang Q (2018) Influence of environmental and genotypic factors on biofilm formation by clinical isolates of group B Streptococci. Microb Pathog 121:45–50

Nobbs AH, Jenkinson HF, Everett DB (2015) Generic determinants of Streptococcus colonization and infection. Infect Genet Evol 33:361–370

Otaguiri ES, Morguette AEB, Tavares ER, dos Santos PMC, Morey AT, Cardoso JD, Perugini MRE, Yamauchi LM, Yamada-Ogatta SF (2013) Commensal *Streptococcus agalactiae* isolated from patients seen at University Hospital of Londrina, Paraná, Brazil: capsular types, genotyping, antimicrobial susceptibility and virulence determinants. BMC Microbiol 13:297

Perichon B, Guignot J, Szili N, Gao C, Poyart C, Trieu-Cuot P, Dramsi S (2019) Insights into *Streptococcus agalactiae* PI-2b pilus biosynthesis and role in adherence to host cells. Microbe Infect 21:99–103

Rajagopal L (2009) Understanding the regulation of Group B Streptococcal virulence factors. Future Microbiol 4(2):201–221

Rato MG, Bexiga R, Florindo C, Cavaco LM, Vilela CL, Santos-Sanches I (2013) Antimicrobial resistance and molecular epidemiology of Streptococci from bovine mastitis. Veter Microbiol 161:286–294

Reyes J, Chaffer M, Rodriguez-Lecompte JC, Sanchez J, Zadoks RN, Robinson N, Cardona X, Ramirez N, Keefe GP (2017) Short communication: molecular epidemiology of *Streptococcus agalactiae* differs between countries. J Dairy Sci 100:9294–9297

Rosa-Fraile M, Spellerberg B (2017) Reliable detection of group B Streptococcus in the clinical laboratory. J Clin Microbiol 55(9):2590–2598

Ruppen C, Hemphill A, Sendi P (2017) *In vitro* activity of gentamicin as an adjunct to penicillin against biofilm group B Streptococcus. J Antimicrob Chemother 72:444447

Sadaka SM, Aly HA, Meheissen MA, Orief YI, Arafa BM (2018) Group B Streptococcal carriage, antimicrobial susceptibility, and virulence related genes among pregnant women in Alexandria, Egypt. Alexandria J Med 54:69–76

Shelburne SA, Keith D, Horstmann N, Sumby P, Davenport MT, Graviss EA, Brennan RG, Musser JM (2008) A direct link between carbohydrate utilization and virulence in the major human pathogen group A *Streptococcus*. Proc Nat Acad Sci 105(5):1698–1703

Skolnik K, Nguyen A, Thornton CS, Waddell B, Williamson T, Rabin HR, Parkins MD (2017) Group B streptococcus (GBS) is an important pathogen in human disease- but what about in cystic fibrosis? BMC Infect Dis 17:660

Springman AC, Lacher DW, Waymire EA, Wengert SL, Singh P, Zadoks RN, Davies HD, Manning SD (2014) Pilus distribution among lineages of group b Streptococcus: an evolutionary and clinical perspective. BMC Microbiol 14:159

Vornhagen J, Waldorf KMA, Rajagopal L (2017) Streptococcal infections: virulence factors, immunity, and prevention strategies. Trend Microbiol 25(11):919–931

Wang P, Ma Z, Tong J, Zhao R, Shi W, Yu S, Yao K, Zheng Y, Yang Y (2015) Serotype distribution, antimicrobial resistance, and molecular characterization of invasive group B Streptococcus isolates recovered from Chinese neonates. Int J Infect Dis 37:115–118

Xia FD, Mallet A, Caliot E, Gao C, Trieu-Cuot P, Dramsi S (2015) Capsular polysaccharide of Group B Streptococcus mediates biofilm formation in the presence of human plasma. Microbes Infect 17:71–76

Zhang Z, Lan J, Li Y, Hu M, Yu A, Zhang J, Wei S (2018) The pathogenic and antimicrobial characteristics of an emerging *Streptococcus agalactiae* serotype IX in Tilapia. Microb Pathog 122:39–45

Zimmermann P, Gwee A, Curtis N (2017) The controversial role of breast milk in GBS late-onset disease. J Infect 74:S34–S40

Enterococcal Infections and Drug Resistance Mechanisms

9

Muthu Vijaya Sarathy, Sivaraman Balaji, and Tingirikari Jagan Mohan Rao

Contents

Abstract

Multidrug resistance of pathogens is a serious issue which is emerging these days and causing a serious threat to mankind. The species belonging to *Enterococcus* genera, namely *Enterococcus faecium* (*E. faecium*) and *Enterococcus faecalis* (*E. faecalis*) are next to *Staphylococcus aureus* in causing severe health issues. The

M. V. Sarathy
Department of Microbial Biotechnology, Bharathiar University, Coimbatore, India

S. Balaji
Department of Biotechnology, Bharathiar University, Coimbatore, India

T. Jagan Mohan Rao (✉)
Department of Biotechnology, NIT Andhra Pradesh, Tadepalligudem, India
e-mail: tjmr@nitandhra.ac.in

B. Siddhardha et al. (eds.), *Model Organisms for Microbial Pathogenesis, Biofilm Formation and Antimicrobial Drug Discovery*,
https://doi.org/10.1007/978-981-15-1695-5_9

emerging resistance of *Enterococcus* species towards various antibiotics is posing a serious challenge in the therapy of diseases. *E. faecium* and *E. faecalis* exhibit resistance (intrinsic, acquired, and tolerance) against antibiotics. Moreover, the exchange of genetic material between or within the genus is the major concern for the proliferation of antibiotic resistance. Apart from the above the strains isolated from different geological niches exhibited varying pattern of resistance. However, the combination of cell wall active drugs with aminoglycosides acts as an alternative therapy in very few cases like *E. faecalis* associated endocarditis. With this knowledge, this chapter highlights the infection caused by *Enterococcus* species and its consequences. The sensitivity and resistance of *Enterococcus* species against different types of antibiotics will be discussed. The underlying molecular mechanism, genes responsible for resistance, and the treatment approaches for susceptible versus resistant strains will also be emphasized.

Keywords
Enterococcus species · Pathogen · Infection · Antibiotics · Resistance

9.1 Introduction

Hospital-associated *Enterococci* species are the major cause of infections like urinary tract infections (UTI), bacteremia, and endocarditis (Goh et al. 2017). Systemic infection of *Enterococci* starts from dense colonization in the gastrointestinal tract (GI), and gains access to various sites through the bloodstream (Goh et al. 2017). *Enterococcus faecium* and *E. faecalis* are the two strains reported with increased frequency of human infection as compared to other *Enterococcus* species (Gordon et al. 1992). Intestinal colonization of drug-resistant *Enterococci* is favored due to the long-term exposure of antibiotics in hospitalized patients. Infectious *Enterococci* are capable of forming biofilm and are often resistant to antibiotics like ampicillin, vancomycin, and aminoglycosides (Sydnor and Perl 2011; Van Harten et al. 2017). Global epidemiology report demonstrates that *Enterococcal* infections are the major cause of nosocomial diseases worldwide due to their antibiotic-resistant nature (Goh et al. 2017; Sydnor and Perl 2011). The standard line of evidence suggests that *Enterococcus faecium* is the most common bacteria acquiring intrinsic resistant on exposure to various antibiotics as compared to other *Enterococcal* species (Hollenbeck and Rice 2012). Emerging resistance to specific antibiotic was controlled through synergistic therapy. But the mechanism of synergistic bactericidal activity due to a combination of antibiotic remains a mystery. This chapter reviews the literature behind the genetic transfer of antibiotic resistance, a summary of various risk factors involved in enterococcal infection, virulent determinant, and the mechanism of antibiotic resistance along with possible treatment options.

9.1.1 Historical Overview of *Enterococcal* Antibiotic Resistance

Enterococcus is a common microbial flora of human and animal intestine (Dubin and Pamer 2014). The discovery of *Enterococcus genus* comes to attention through bacterial isolates from infective endocarditis in 1899, until then kept under the subset of *Streptococci* species (Schleifer and Kilpper-Balz 1984). Only after 1984, *Enterococcus* strains were proved to exhibit genetic diversity from *Streptococcus* species and allocated in to a separate genus (Dubin and Pamer 2014; Schleifer and Kilpper-Balz 1984). It acts as early gut colonizers in many animals including humans. It is the most common genera of fecal content in birds, reptiles, mammals, and humans (Gilmore et al. 2013). In human feces, *E. faecalis* and *E. faecium* were the most predominant species reported so far (Fisher and Phillips 2009). The genus of *Enterococcus* are mostly commensal in nature and some of the species are the leading cause of nosocomial infections in hospitalized patients next to *Staphylococci* (Gilmore et al. 2013). The antibiotic penicillin was found to be remarkably effective against *Streptococcal* infectious (Williamson et al. 1983). However, penicillin was less effective against *Enterococci* infections due to the inherent resistance (Handwerger et al. 1993). Later, several research findings suggest that the synergistic effect of these antimicrobial agents with aminoglycosides results in significant improvement in endocarditis caused by *Enterococcal* infection (Eliopoulos and Moellering 1982; Serra 1977). Noteworthy, multiple courses of antimicrobials usage in clinical care seem to evolve *Enterococcus* with antibiotic-resistant determinants. Thus, by the twentieth century, there is an incredible increase in multidrug resistant (MDR) *Enterococci* just by acquiring drug-resistant genes (Van Harten et al. 2017). In addition, the modern-day MDR *Enterococci* exposed to multiple antibiotics come into sight of diverged genetic clade with increased transposable elements, rapid mutation, and metabolic alterations (Gilmore et al. 2013). For instance, enterococci recruit vancomycin resistance with improved biochemical machinery of various gene clusters (Van Harten et al. 2017). These resistances were transmitted to most pathogenic bacteria like methicillin resistance *Staphylococcus aureus* (MRSA), making very serious public health concern. It is well documented with emerging resistance of enterococci against all the available gram-positive agents like quinupristin, daptopristin, linezolid, and tigecycline (Van Hoek et al. 2011; Hollenbeck and Rice 2012; Binda et al. 2014).

9.2 Risk Factors Associated with *Enterococcal* Infection

9.2.1 Stable Colonization

It was reported that *Enterococcus* strains are less-abundant bacteria inhabited in the intestine of human (Van Harten et al. 2017). However, the density of colonization increases with nosocomial infection of resistant *Enterococci* species. It was reported that the administration of multiple antibiotics to the hospitalized patients induces substantial change in the gut microbiota, thereby facilitating the colonization of resistant strains of *Enterococci* species (Ubeda et al. 2010). For instance, gram-negative commensals lipopolysaccharide and flagellin induce the paneth cells to produce REGIIIγ which are active against gram-positive bacteria. Thus, the decrease in population of gram-negative commensals due to antibiotic treatment facilitates an increase in the population of resistant *Enterococci* (Kinnebrew et al. 2010). Notably, the use of higher minimal inhibitory concentration of antibiotics like neomycin, vancomycin, and metronidazole in hospitalized patients displayed preferential growth of vancomycin-resistant *Enterococci* (Brandl et al. 2008).

9.2.2 Nosocomial Transmission

The dissemination of resistant *Enterococci* among the hospital settings is widely due to its adaptation to extreme environmental conditions prevailing in the surroundings. Hence, it can remain active for longer periods in surgical units, including surfaces of medical devices, doorknobs, bed rails, and so on (Sydnor and Perl 2011). The major reasons for the development of nosocomial infection caused by *Enterococci* species were due to the multi-dose regimen, long-term hospitalization, transmission through physical proximity of infected patients, organ transplantation, and co-morbidities like catheter-associated infections, and hemodialysis (Bradley and Fraise 1996). In the hospitalized patients, spreading of resistant *Enterococci* is widely through the nurse handling different patients suffering from various diseases (Sydnor and Perl 2011; Bradley and Fraise 1996).

9.2.3 Virulence Determinants of Enterococcal Infections

9.2.3.1 Secreted Virulence Factor

Secretions of virulent proteins were more prevalent in clinical isolates of *Enterococci* (Benachour et al. 2009). Cytolysin, one of the important virulent factors released outside the cell, is divided into two subunits, CylL-L and CylL-S. Cytolysins are capable of lysing RBC and WBC of humans, rabbits, and horses (Shokoohizadeh et al. 2018). Moreover, the virulence nature of cytolysin was due to membrane damage in host tissue. Gelatinase-E (gelE) is another important virulent factor with protease activity mainly degrading the misfolded proteins (Benachour et al. 2009). GelE gene expression is controlled by Fsr quorum sensing system, as GelE is regarded as the most frequent virulent determinants expressed (48.5%) in *E. faecalis* (Shokoohizadeh et al. 2018). Perhaps, GelE activates the peptidoglycan-degrading enzymes such as autolysin, in turn, the release of extracellular DNA and so mediates pathogenesis through biofilm formation. SagA is a major secreted antigen from *E. faecium* accomplishing broad-spectrum extracellular matrix binding (ECM) like fibrinogen, fibronectin, laminin, type-I and type-II collagen (Kropec et al. 2011).

9.2.3.2 Cell Surface Determinants

Cell surface determinants are another putative virulent factor comprising of an array of components which endorse colonization and biofilm formation for

Enterococcal infection (Kang et al. 2013; Soheili et al. 2014). The surface protein in *Enterococcus* is encoded by esp gene which helps in cell adhesion and commonly found in clinical isolates of patients suffering from endocarditis or urinary tract infections (UTIs) (Soheili et al. 2014). The surface of *E. faecalis* is comprised of protein, namely aggregate substance (AS), which aids in binding to cultured renal epithelial cells, colon cell internalization, and tolerant to immune response (polymorphonuclear neutrophils) (Sava et al. 2010). In addition, at the early stage of infection, microbial surface components recognizing adhesive matrix molecules (MSCRAMMs) recognize the microbial surface which help in adhering to LPxTG (Leu-Pro-x-Thr-Gly) type surface anchored proteins and pili gene cluster (PGC) proteins helps in *Enterococcal* binding to host extracellular matrix (Kang et al. 2013; Hendrickx et al. 2009). In this way, two reputed adhesins such as Ace and Acm which help in binding to collagen produced by *E. faecalis* and *E. faecium* were broadly studied in *Enterococcus* species (Kang et al. 2013). Ace adhesin provisionally expressed in clinical isolates of *E. faecalis* during its growth conditions in collagen was found to be a major contributor in endocarditis (Kang et al. 2013). On the other hand, Acm facilitates the binding of *E. faecium* to the collagen (Kang et al. 2013; Soheili et al. 2014). Acm gene expression was more frequently found in hospital-acquired *E. faecium* compared to the community (animal and human) acquired strains (Soheili et al. 2014). In *E. faecalis*, ElrA, an enterococcal leucine-rich protein-A, sustains the strain from evasion of macrophage recognition and adhesion (Nunez et al. 2018). Two important enterococcal proteins origin from pilus loci proficiency in biofilm formation of *E. faecalis* such as *Ebp* (endocarditis biofilm-associated pili) and *Bee* (Sava et al. 2010). *Ebp* helps in causing enterococcal pathogenesis in experimental endocarditis and UTI. Similarly, *Bee* an *Enterococcal* biofilm enhancer from pilus loci of *E. faecalis* mediates biofilm-associated pathogenesis (Hendrickx et al. 2009). Critical components of membrane structures like capsular polysaccharide, lipoteichoic acid (LTA) can promote skirting of *Enterococci* from phagocytosis (Hendrickx et al. 2009). For instance, Epa (Enterococcal polysaccharide antigen) a rhamnose containing polysaccharide has the ability to enhance the biofilm formation and translocation across enterocyte monolayer. The adhesion and biofilm formation is higher in mutant strains of *E. faecalis* as compared to wild-type strain due to higher production of lipoteichoic acid (LTA) and glycolipids in cell membrane (Hendrickx et al. 2009; Sava et al. 2010). Distribution of various virulent determinants expressed in *Enterococcal* cell wall is mentioned in Fig. 9.1.

9.2.3.3 Phosphotransferase System (PTS)

The role of PTS system is to translocate the phosphorylated carbohydrate across the bacterial cell membrane (Peng et al. 2017). Mostly in clinical samples of *E. faecium*, PTS system is associated with virulence and stress response (Paganelli et al. 2016). Virulent types of PTS system were widely distributed in clinical isolates and lack in commensal strains (Peng et al. 2017; Paganelli et al. 2016). Among the PTS system, the most notable one is PTS permease BepA, which plays a significant role in endocarditis (Paganelli et al. 2016).

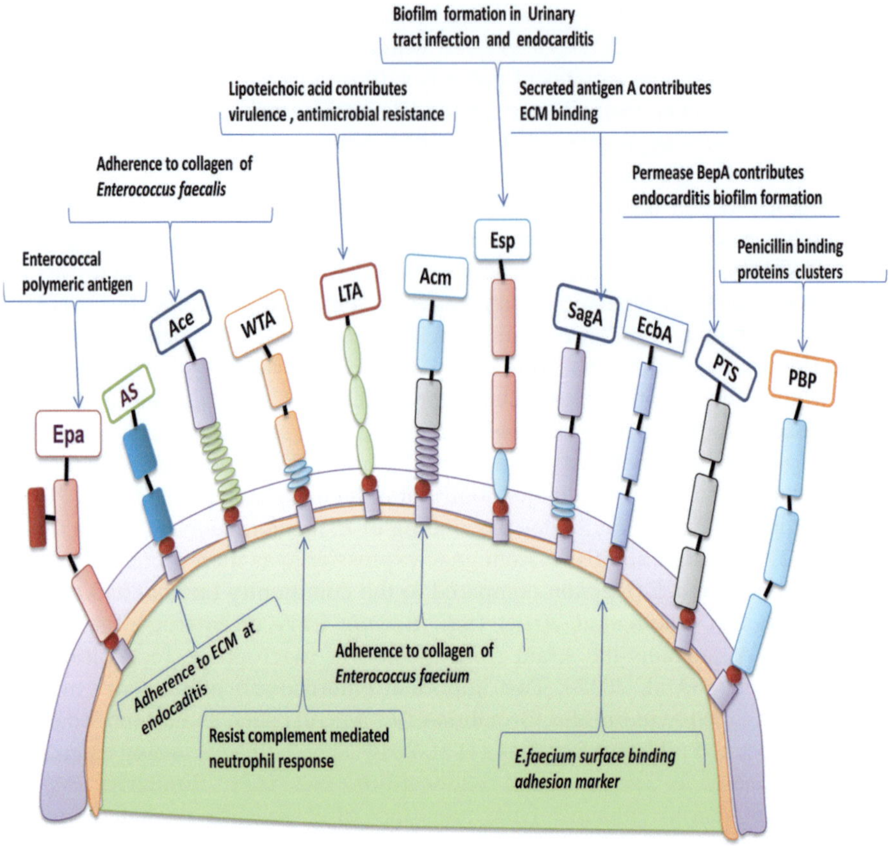

Fig. 9.1 Overview of enterococcal cell surface determinants responsible for virulence expression

9.2.4 Dissemination of Antibiotic Resistance Among Enterococci Species Through Horizontal Gene Transmission

The unrestrained practice of broad-spectrum antibiotics instigates the survival mechanism of certain gut microbial flora by promoting multidrug resistance in them (Chajęcka-Wierzchowska et al. 2019). The gastrointestinal tract of humans is the hot spot for antibiotic resistance. The gastrointestinal tract is colonized with diverse bacterial flora through biofilm formation and regarded as the hot spot region for acquisition of resistance (Schjorring and Krogfelt 2011). The contemporary disquiet in clinical microbiology is the acquired antibiotic resistance among *Enterococci* that colonizes gut lumen of human, which is highly difficult to medicate (Manson et al. 2010; Palmer et al. 2010). Consuming the fermented food, and ready to eat edible foods from various sources rich in bacterial population will act as a donor and/or recipient of antibiotic resistance (Togay et al. 2010; Chajęcka-Wierzchowska et al. 2019). The transition of *Enterococci* from harmless commensals to pesky pathogen has emerged over decades through various mechanisms such as the acquisition of resistant genes against several antibiotics from other organisms, gene mutations, and modification of bacterial surface molecules (Bourgeois-Nicolaos et al. 2006; Schjorring and Krogfelt 2011). However, these characteristics can be divided into internal and acquired resistance through horizontal gene transfer (HGT). Among these, HGT is the leading cause of infections in hospitals by gaining resistance against commercially available antibiotics (Chajęcka-Wierzchowska et al. 2019). Surprisingly Polidori et al. (2013) found that about 25% genome of *E. faecalis* V583 is integrated from other organisms which made this strain to resist the last-line antibiotic vancomycin. Few years before to this study, the total genome analysis of *E. faecalis* OG1X strain showed that there was no trace of foreign DNA in it (Bourgeois-Nicolaos et al. 2007). This clearly indicated that how selective pressure of antibiotics would coerce commensal to acquire genetic elements from their surroundings through an array of mechanisms. Moreover, *Enterococcus* species resistant to various antibiotics are emerging from diverse environmental sources which could probably enter our system and cause adverse effects (Conwell et al. 2017).

9.2.4.1 Antibiotic Resistance Mechanism in *Enterococci* Species

It is proven that gram-negative bacteria of the gut flora are more prone to accomplish antibiotic-resistant gene and might transfer it to other members of the microbiome (Bourgeois-Nicolaos et al. 2006). There is a more chance of this acquisition in the hospital environment where antibiotics are targeted frequently against gram-negative bacteria (Schjorring and Krogfelt 2011; Lerminiaux and Cameron 2019). Acquired resistance in *Enterococci* generally takes place by the exchange of pheromone-responsive genes or plasmids, or transposons. The most pathogenic species in *Enterococci* are *E. faecalis* and *E. faecium* which possess resistance towards several antibiotics (Palmer et al. 2010; Quinones et al. 2018). Most of the *Enterococci* strains hold multiple plasmids and transposons with the ability to resist antibiotics and exhibit virulence factors (Palmer et al. 2010). Some common virulent genes encoded in plasmids and transposons are enlisted in Table 9.1.

9.2.4.2 Plasmids

The antibiotic-resistant genes are present in the extra chromosomal elements called plasmids, which is transferred to neighboring species through bacterial conjugation (Clewell 2011). Among several genera that colonizes the gut lumen, plasmid characterizations are studied extensively in *Enterococci* bacteria (Di Sante et al. 2017). Based on the mode of transfer and host organisms the plasmids are categorized into two major classes: (a) pheromone-related plasmid and (b) broad host range plasmids (Weaver 2019).

9.2.4.2.1 Pheromone-Responsive Plasmids

These plasmids are transformed into recipient cells from donor cells in response to pheromone signals received from the environment (Clewell 2011). It was detected in *Streptococci* that pheromone signals are transferred through plasmids by conjugation (Tomura et al. 1973). In the entire *Enterococci* family, *E. faecalis*

Table 9.1 Plasmids and transposons in *Enterococcal* antibiotic resistance

S. No	Plasmid/ chromosome type	Antibiotic resistance gene	Resistance mechanism	References
1	pAD1	*vanB, ermB*	Carrier of resistant determinant	Clewell (2007)
2	pCF9	*tetM, uvrA*	Carrier of tetracycline A-resistance protein	Hirt et al. (2018)
3	pAMβ1	*ermB*	B-subunit, a carrier of resistant determinant	Clewell (2011)
4	pIP501	*cat, ermB*	Carrier of resistant determinant	Palmer et al. (2010)
5	pLG1	*vanA*	Carrier of vancomycin A-resistance protein	Laverde Gomez et al. (2011)
6	Tn5281	*aac-6′-aph-2″*	Aminoglycoside modifying enzyme	Feizabadi et al. (2008)
7	Tn5384	*ermB*	*β-lactamase* production	Bonafede et al. (1997)
8	Tn5385	*tetM, ermAM, aadE, Bla*	*β-lactamase* production	Rice and Carias (1998)
9	Tn*917*	*ermB, tnPR, mefE, tnPA*	Ribosome methylase activity, efflux action	Rice (2014)
10	Tn*552*	*Bla, aacA-aphD*	*β-lactamase*	Sarti et al. (2012)
11	Tn*1546*	*aac6′-aph2″*	Aminoglycoside modifying enzyme	Simjee et al. (2002)
12	Tn916	*tetM*	Ribosomal protection	Santoro et al. (2014)

predominantly possesses these pheromone-responsive plasmids and transforms it into other strains (Clewell 2011). It was witnessed that these kinds of plasmids could not replicate in the host cells other than *Enterococcal* species. Because, *Enterococci* chromosome expresses several peptides that could act as pheromones which stimulates specific pheromone-responsive plasmids in the target cells (Dunny 2013). Initially, the extra cellular signals produced by recipients will induce donors to initiate conjugation. During this transfer, recipient cells synthesize the chromosomally encoded lipoprotein fragment, which will be detected by donors. This process initiates the cascade reaction and stimulates the production of aggregation substances like *Asa1* and *PrgB* in the plasmids of the donor (Hollenbeck and Rice 2012). Then, these peculiar substances bind with the enterococcal binding element on the recipient cell surface and instigate conjugation. Experiments suggest that co-cultivation of donor and receipt strains enhances the conjugal exchange of pheromone-related plasmids (Dunny et al. 1978; Dunny and Clewell 1975). Though the transfer rate was minimal in the initial hours, the ratio of transconjugants and the parental strain was increased within a few hours from the startup and visible cell clumping started appearing in the medium. The efficiency of genetic transformation between *E. faecalis* strains is 10^{-3} donor cells during 4 h of matting (Kristich et al. 2014). More evidently, when the cell-free filtrate of recipient cells which presumed to contain the clumping inducing agent (heat resistance and enzyme sensitive compound) was added to the medium containing donor cells carrying conjugative plasmids induced cell aggregation (Dunny 2013). These signaling molecules were later called as bacterial sex pheromones. The widely considered pheromone-related plasmids among *Enterococcus* are pAD1 and pCF10 which encode for antibiotic resistance and virulence factor (Kristich et al. 2014). Transfer of plasmids among species was majorly controlled by two competitive molecules, namely pheromones and inhibitors.

9.2.4.2.2 pCF10

pCF10 is a 67.7 kb, a plasmid that habitually carries antibiotic resistance gene (tetracycline) and certain virulence genes (hemolysin and bacteriocins) which is efficiently disseminated among *Enterococci* of the different environment (Hirt et al. 2018). Over 25 genes from pCF10 are responsible for cell aggregation; matting channel formation and conjugal plasmid transfer have been reported. Conjugal transfer of this plasmid is majorly controlled by two peptides, i.e., cCF10 (c-pheromone, released by recipient cells) and iCF0 (i-inhibitor, expressed by donor cells), where conjugation is induced by cCF10 and inhibited by counteracting peptide iCF0. cCF10 peptide from recipient cell induces the pCF10 plasmid of contributor cells to initiate the conjugal transfer. In the absence of receiver, so as to avert the conjugal transfer, pCF10-encoded iCF10 will be expressed and halt the plasmid transmission (Kohler et al. 2019).

Upon external stress, bacteria will produce cCF10 in search of donor cells for conjugal transfer of plasmids. Thus, this cCF10 peptide is imported into a donor cell by surface receptor *PrgZ* and accumulates in the cytosol (Hirt et al. 2018; Breuer et al. 2018). There it competes with iCF10 for transcriptional factor *PrgX* in order to regulate the transcription of pCF10 plasmid by modifying the PQ promoter of the operon PrgQ and downstreaming genes (Kozlowicz et al. 2004; Chen et al. 2017). The negative regulator of PQ is reported to be *PrgX* which is derepressed by cCF10 and enhanced by iCF10. *PrgX* is a tetramer protein which could bind both cCF10 and iCF10 and interact with the operator sites of the PQ region contrarily. The structural difference in between these two peptide complex molecules either suppresses by inhibiting the RNA polymerase binding (iCF10) or enhances (cCF10) the expression of conjugation genes (Dunny 2013). The schematic representation for the spreading of antibiotic resistance in *Enterococcus* species is demonstrated in Fig. 9.2.

PrgQ operon (26 kb) encodes *PrgA*, *PrgB*, *PrgC* cell wall binding proteins which are responsible for cell aggregation, the formation of mating junctions, plasmid transfer into the recipient cell (Bhatty et al. 2015). In addition to this, the current results show that *PrgA* is the significant component involved in biofilm formation and binding of *Enterococci* into abiotic surfaces in the symbiotic environment (Bhatty et al. 2015). In some mammalian infection models these genes also expressed virulence factors. Regulation of *PrgQ* operon is majorly controlled by the relative concentration of peptides in the donor cells, which in turn, enhance the plasmid transfer in recipient cells (Chen et al. 2017; Kohler et al. 2019). In the GI tract, beyond plasmid transfer, acquisition of antibiotic, virulence genes, and other factors that aid in the interaction of enterococci with its mammalian host might lead

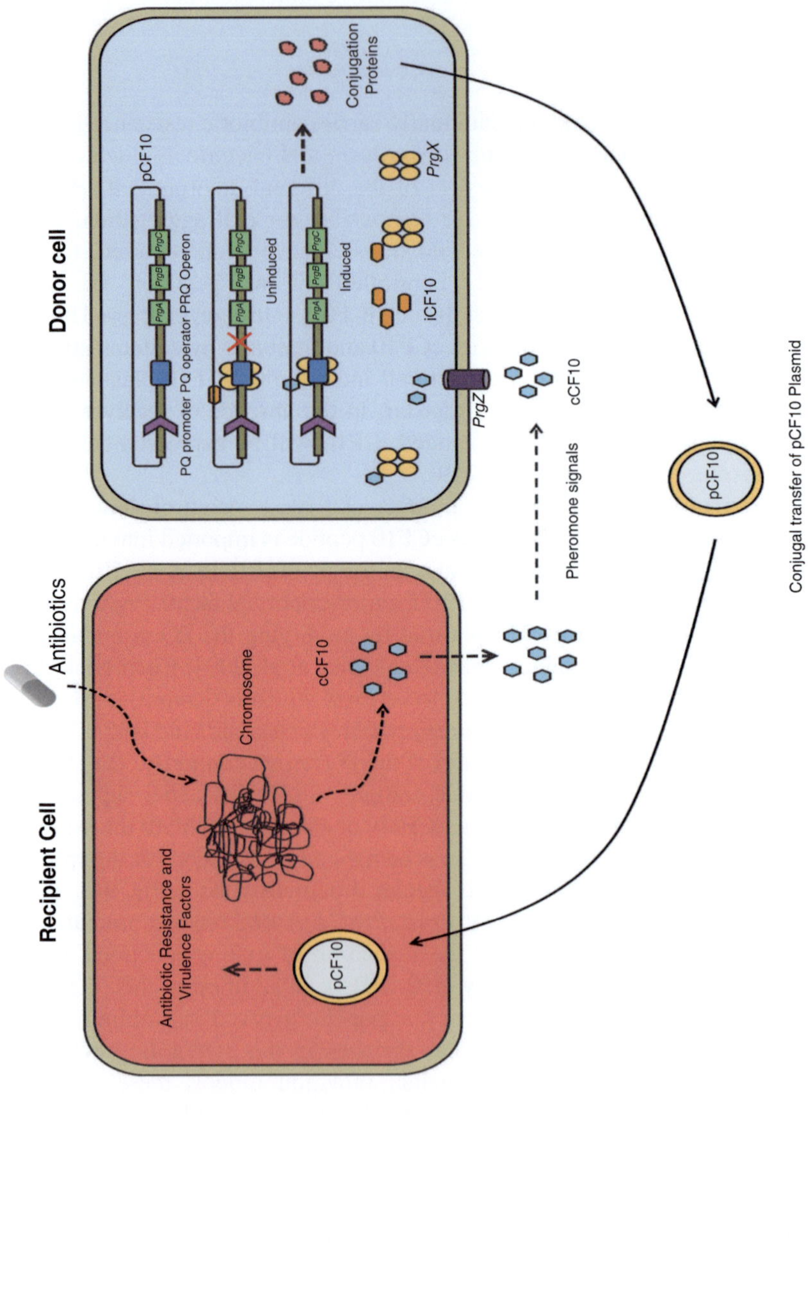

Fig. 9.2 Pheromone-responsive conjugal transfer of Pcf10 plasmid from donor to recipient cells

to the unsolicited effects. For instance, the expression of surface adherence molecules could negatively influence the host innate immune response and lead to adverse effects.

9.2.4.2.3 pAD1

It was reported that pAD1 is a 60 kb pheromone-related plasmid which is commonly reported in clinical samples of *E. faecalis* (Francia et al. 2004; Weaver 2019). In reply to the sex pheromone peptide cAD1 formed by receiver cell, pAD1 plasmid will encode mating responsive elements for its efficient own transfer reported in *E. faecalis* DS16 (Dunny and Berntsson 2016; An and Clewell 2002). pAD1 encodes various factors such as hemolysin and bacteriocins which trigger virulence in the mammalian host and develop resistance against several gram-positive bacterial species (Clewell 2007). There are 87 open reading frames (ORF) in the pAD1 plasmid among 9.2 kb segment that codes the genes responsible for cytolysin and two origins of replication, namely *oriT1* and *oriT2* responsible for conjugal transfer (Clewell 2011). In addition, this strain holds 28 kb pAD2 plasmid that encodes antibiotic resistance (erythromycin, streptomycin, tetracycline, and kanamycin). However, pAD2 plasmid does not encode any conjugative elements that make its own transfer (Showsh et al. 2001).

9.2.4.2.4 Wide Host Range Plasmids

These plasmids are involved in transferring genetic information to other gram-positive and gram-negative organisms. Among them, the extensively studied is mismatched group 18 also called as Inc18 (Kohler et al. 2018; Zhu et al. 2010). However, the transmission efficiency of this plasmid is very low when compared to the other type. These Inc18 plasmids are copious in *E. faecium* and their well-characterized prototypes are pAMβ1 (encodes resistance against macrolides, a class of antibiotics that includes erythromycin, roxithromycin, azithromycin, and clarithromycin) and pIP501 (encodes resistance against chloramphenicol in addition to macrolides) (Kohler et al. 2019). The *Enterococcal* strains that contain these plasmids transfer antibiotic-resistant genes to *Streptomyces*, *Leuconostoc*, *Listeria*, *Lactococcus,* and *E. coli* species, respectively (Kohler et al. 2018). In *Enterocccus* species tra genes encodes the plasmid transfer genes which help in easy transfer of plasmids. Tuohy et al. (2002) reported the transmission of wide host range plasmid pAMβ1 from *Lactobacillus lactis* to *Enterococcus* species in an animal model which contains the genetic material that could resist erythromycin and lincomycin. Interestingly this team observed that the bacterium could not transfer it into other bacteria like *Lactobacillus* and *Bifidobacterium*.

9.2.4.2.5 Transposons

Mobile genetic elements, i.e., transposons have been reported to hold and spread virulence factors and antibiotic resistance in *Enterococci* (Hegstad et al. 2010). The sequencing results reveal that mobile genetic elements are scattered within the genome of *Enterococci*. Transfer of these elements will lead to the emergence of numerous phenotypes, such as pathogenicity, ability to colonize the host cells, and communication between cells. So far, three types of transposons have been reported to transfer genetic materials in *Enterococci* such as Tn3, composite, and conjugative transposons (Hegstad et al. 2010). Complementation is difficult in transposons.

9.2.4.2.6 Tn3 Family Transposons

These are the widespread group of replicative transposons which is associated with dissemination of multidrug resistance. There are several Tn3 family transposons that have been reported in *Enterococci*, including Tn*917*, Tn*552*, and Tn*1546* (Nicolas et al. 2017). Tn*917* is the first Tn*3* family transposon identified in the *Enterococci* that confer resistance to macrolides, streptogramin B, and lincosamides antibiotics (Shaw and Clewell 1985). The *ermB* gene (RNA methylase), *mefE* (macrolide efflux genes *mef*), and mel that are found within this 5 kb element regulate their resistant mechanism. These elements have a broad host range such as *Bacillus*, *Enterococcus*, *Lactococcus*, *Listeria*, *Paenibacillus*, *Staphylococcus*, and *Streptococcus*. In addition, mutant libraries have been successfully developed using this transposon in *S. aureus* and *S. epidermidis* (Rice 2014).

Tn*552* is a β-lactamase encoding transposon identified from *S. aureus* (Rowland and Dyke 1990). Penicillin exposure among *Staphylococci* induces the production of β-lactamase to confer resistance. It is a very rare type of Tn*3* transposon that was detected in *Enterococci* where *bla* gene encodes the resistant element. In the majority of the *E. faecalis* strains, the regulative elements are present in the upstream region of the β-lactamase gene that is either truncated or deleted, as a result the expression might affect the isolates negatively. Since the regulatory genes are non-functional, the expression of β-lactamase was found to be constitutive in these strains and expressed at a very low level (Ono et al. 2005; Tomayko et al. 1996). This could be the reason why Tn*552* transposons are rare in *Enterococci* which exhibits a low level of resistance. However, β-lactamase flanking the complete regulatory genes is recognized in several *E. faecium* strains (Sarti et al. 2012).

Tn*1546* is an 11 kb transposon element regularly seen in the clinical samples of *E. faecium.* These elements confer resistance to vancomycin and are called vancomycin-resistant enterococci (VRE) (Novais et al. 2008; Sletvold et al. 2010; Freitas et al. 2013). Among the genotypes *vanA* and *vanB*, *vanA* is predominant and leads to the development of glycopeptide-resistant *Enterococci* worldwide. The *vanA* cluster comprises of seven genes which include *vanA*, *vanH*, *vanR*, *vanS*, *vanX*, *vanY*, and *vanZ* (Arthur et al. 1993). This transposon is often associated with plasmids and chromosomes, for instance, pIP816, a 4-kb conjugative plasmid present in VRE is known to have this transposon (Kristich et al. 2014). The occurrence of Tn*1546* element isolated from *E. faecium* of the distinct geographical region suggests its rapid spread through HGT. Interestingly, several mutations (deletions and insertions) have been described within Tn*1546*. These alterations result in the multi-resistant *Enterococci* (Kristich et al. 2014). For instance, the insertion of *fosB* gene within the Tn*1546* made the *E. faecium* to resist fosfomycin in addition to the vancomycin (Sun et al. 2017). In contrast, such mutation might cause nil effects, sequence result of Tn*1546* from the clinical isolates of *E. faecium* revealed several modifications in the cluster including 889 bp deletions in the ORF1. Sequence analysis of Tn1546 has revealed that, ORF1 had an 889-bp deletion and an IS1216V insertion at the 5′ end and an IS1251 insertion between *vanS* and *vanH* (Simjee et al. 2002). In the future, the diversity of Tn*1546* transposon will further increase by the sequencing of several *Enterococci* from various environmental sources.

9.2.4.2.7 Composite Transposons

These transposons are flanked by insertion sequences (IS) elements as their mobility is determined by its presence. These are flanked at both the ends and are oriented as direct or as inverted repeats (Hegstad et al. 209). Tn*1547* is the composite transposon that confers the VanB resistance in *Enterococcal* species (Garnier et al. 2000). This 64 kb element is transferred through the chromosome of *E. faecalis* to the hemolysin plasmid. This transposon is bordered by two addition sequences in straight alignment, namely IS256-like and IS16 (Quintiliani and Courvalin 1996).

Tn*5281* element is the next commonly observed composite transposon, which provides resistance against all aminoglycoside antibiotics excluding streptomycin in *E. faecalis* and *E. faecium* (Feizabadi et al. 2008). However, this transposon containing *Enterococci* strains prevalently resist the antibiotic gentamycin. The gene responsible for aminoglycoside modifying enzyme *aac (6′)-aph (2″)* is bordered by inverted replicas of IS*256* insertion sequence (Behnood et al. 2013). Interestingly, this transposon is similar to the composite transposon (Tn*4001*) of *Staphylococcus*, bordered by an inverted copy of IS256.

Tn*5385* is a multi-resistant composite element that is frequently found in the hospital strains of *E. faecalis* (Rice and Carias 1998). Tn*5385* could resist wide variety of antibiotics such as erythromycin, gentamicin, streptomycin, tetracycline, minocycline and penicillin. It is a 65 kb transposable element flanked by directly repeated copies of IS1216. Interestingly, Tn*5385* is composed of various transposons and transposon-like elements which include conjugative transposon Tn*5381*, composite transposon Tn*5384*, and indistinguishable elements belonging to *Staphylococcus* transposons Tn*4001* and Tn*552* (Rice et al. 2007, 2009). Importantly, various regions of Tn*5385* are linked by several IS elements like IS256, IS257, and IS1216. The genetic exchange of these transposons takes place through homologous recombination by the involvement of either neighboring sequences or inner portable elements.

9.2.4.2.8 Conjugative Transposons

Tn*916* is one of the most extensively characterized conjugative transposon from *E. faecalis* which is about 18 kb in size (Roberts and Mullany 2011; Rice 1998). This has the ability to resist tetracycline and minocycline by the production of *tet (M)* gene. Interestingly, Tn*1545* from *S. pneumonia* is similar and is even identical in many respects to this Tn*916* transposon (Santaro et al. 2014). These conjugative elements transpose to their target sites from their origin without any replication. Simply, they excised from the genomic material of donor cells to form a circular structure and transferred via conjugation finally integrated into the target genome by site-specific recombination (Celli and Trieu-Cuot 1998). The recent whole-genome sequencing results reveal that Tn*916* transposons are distributed among numerous strains of *Enterococci* family. The first completely described Tn*916* family transposon is Tn*5382* which confers resistance to *VanB* glycopeptide-resistance gene cluster (Garnier et al. 2000). This transposon could be transferred among different *Enterococcal* strains. However, emerging genomic sequences of *Enterococci* from diverse geographical regions would further widen the diversity of transposon across the *Enterococci* species and lead to further detailed classification.

9.3 Mechanism of Antibiotic Resistance in *Enterococcus* Species

Extensive research study on *Enterococcus* genus opens up the distribution of antibiotic resistance over the commensals and pathogenic forms. The diverse range of antibiotic-resistant mechanism observed in enterococci is depicted in Fig. 9.3.

9.3.1 Cell Wall Active Agents

9.3.1.1 Ampicillin/Penicillin Resistance

For β-lactam antimicrobial agents, inhibition of peptidoglycan synthesis has been a primary target in many gram-positive bacteria. Moreover, these kinds of antibiotics were highly approved in the initial stage because of no evidence of toxicity profile in the eukaryotic cell wall which lacks peptidoglycan (Fair and Tor 2014). However, clinical isolates of *Enterococci* species possess intrinsic tolerance to β-lactam antibiotics due to production of penicillin-binding proteins (PBPs) or due to the release of β-lactamases (Hollenbeck and Rice 2012). PBPs is a repository of a cluster of enzymes involved in cell wall synthesis which are categorized into two groups. One with bifunctional activity (transglycosylase) in "Class A" group while the another one with monofunctional activity (D, D-transpeptidase) in "Class B" group (Kong et al. 2010). Six putative PBPs genes were confirmed through DNA sequencing of *E. faecalis* and *E. faecium* with three genes each in class A and class B (Duez et al. 2004). The antibiotic resistance Pbp5 gene with independent operon structure is present in *E. faecalis* and *E. faecium*. PBP5 enzyme poses less affinity to ampicillin, besides higher antibiotic concentration is required to saturate enzyme active site (Fontana et al. 1996). Additionally, a mutation in the protein sequence with the substitution of Met → Ala at position 485 near the active serine residue provides resistance to ampicillin binding. Therefore in *E. faecium*, PBP5 sequence variations were observed between hospital-acquired variants with higher resistance (Pbp5-R) and from community-associated variants (Pbp5-S) with lower MIC results (Rice et al. 2004). Hence, in *Enterococci*, the mechanism of antibiotic resistance is coupled with overexpression and mutation in PBP5 thus shows greater degree of resistance towards ampicillin (Duez et al. 2001; Ono et al. 2005). In accordance with β-lactamases, both *E. faecalis* and *E. faecium* promote β-lactam ring cleavage (Murray 1992). *Enterococci* gene coding β-lactamases were found as *blaZ*, which is predicted to be originated from *Staphylococci*. But unlike *Staphylococci*, expression of *blaZ* in *Enterococci* is at lower level which is correlated with inoculum

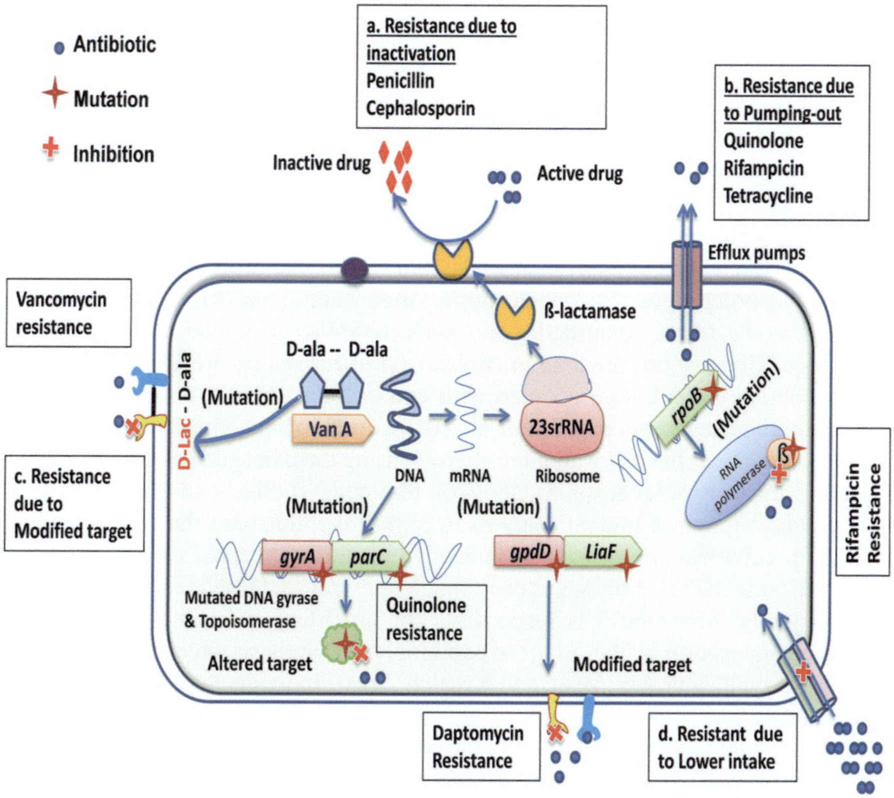

Fig. 9.3 Mechanism of antibiotic resistance in *Enterococcal* infections: (**a**) drug inactivation *Enterococci* were intrinsically resistance to certain antibiotics due to the production of modifying enzymes like β-lactamases. (**b**) Efflux pumps. Those pumps with resistance modulator will secrete the antibiotic outside the cell. (**c**) Resistance mutation. Mutation in the specific gene capable of altering the antibiotic binding site. For example, mutation in gyrA and parC inhibits binding of quinolone to DNA gyrase and topoisomerase. Similarly, rpoB mutation protects RNA polymerase from Rifampicin and gpdD LiaF mutations inhibit cell binding of daptomycin

volume (Hackbarth and Chambers 1993). The responsible enzyme for antibiotic-resistant came into view with the higher bacterial inoculums in *Enterococci*. On the other hand, the addition of β-lactamases inhibitor restores the efficiency of antibiotic activity against *Enterococci* (Sarti et al. 2012).

9.3.1.2 Cephalosporin Resistance

Intrinsic resistance to cephalosporin is a most familiar trait in enterococci which occurs through low binding affinity with Pbp5 (Rice et al. 2009). Some of the research findings suggest that *E. faecalis* and *E. faecium* resistance towards cephalosporin was linked to either PonA or PbpZ of class-A PBPs system (Arbeloa et al. 2004; Rice et al. 2009). This phenomenon illustrates that there is ideal cooperation between pbp5 and one or more class-A PBPs system to convey resistance mechanism against cephalosporins. Study with sequential deletion of class-A PBPs in both the strains reveals that there is an activation of the differential regulatory pathway in response to a cephalosporin (Rice et al. 2009). The intrinsic resistance to cephalosporin is controlled by two-component regulatory system (TCS) like CroR/S (Le Breton et al. 2007). Correspondingly, IreK, MurAA were also found to be an important regulatory system associated with cephalosporin resistance. Cognate response sensor (CroS) can phosphorylate itself by histidine kinase and has the ability to transfer the phosphate group to cognate response regulator (CroR) (Comenge et al. 2003). Activated CroR response regulator is capable of modifying DNA transcription in order to adapt cephalosporin stress conditions (Muller et al. 2006). However, some of the identified genes such as SalB, glutamine transporter with CroR regulation are not obvious in cephalosporin resistance. IreK is a eukaryotic type serine/threonine kinase coupled with extracellular PASTA domain through a transmembrane segment (Kristich et al. 2014). IreK exhibit kinase activity which help in maintaining the cell wall integrity by sensing the damaged peptidoglycan and signaling through autophosphorylation of threonine (Labbe and Kristich 2017). Subsequent deletion of mutant analyses in *E. faecalis* illustrates the significant role of IreK in cell envelop stress is an indicative of intrinsic cephalosporin resistance (Hall et al. 2013). MurAA gene (encoding for UDP-N-acetylglucosamine 1-carboxyvinyl transferase) is responsible for providing resistance to *E. faecalis* against cephalosporin by initiating the peptidoglycan synthesis which is not due to its homolog protein Mur AB (Vesic and Kristich 2012). Individual deletion mutants of either Mur AA or Mur AB capable of executing the enzyme activity for converting UDP-N-acetylglucosamine to UDP-N-acetyl glucosamine-enolpyruvate in the presence of UDP-N-acetylglucosamine 1-carboxyvinyl transferase is the major step in peptidoglycan synthesis. However, with the underlying physiology of *E. faecalis* Mur AA phenotype is more expressive in cephalosporin resistance rather than Mur AB with least.

9.3.1.3 Resistance Against Glycopeptide

Glycopeptide antibiotics like teicoplanin and vancomycin avoids the interaction between the peptidoglycan chain by adhering to D-alanine-D-alanine (D-Ala-D-Ala) terminal moiety of peptidoglycan precursor (Hollenbeck and Rice 2012). The pattern of resistance is highly relevant to the type of amino acid change occurred in precursor moiety. For instance, change in terminal pentapeptide amino acids from D-Ala-D-Ala to D-alanine-D-Lactate (D-Ala-D-Lla) reduces the binding attraction of vancomycin to peptidoglycan chain by 1000-folds, thus conferring a high level of resistance (Courvalin 2006). Minimal level of resistance was due to the modification of amino acid terminal to D-alanine-D-serine (D-Ala-D-Ser) that can exhibit a seven-fold reduction in binding affinity (Hollenbeck and Rice 2012). So far nine different vancomycin resistance gene clusters were identified in *Enterococcus* species (Courvalin 2006). Based on the genes encoded by this cluster, the enzymes are categorized into three groups: (1) enzymes that destroy precursor of terminal amino acids moiety (D-Ala-D-Ala); (2) enzymes for remodeling the new peptidoglycan precursor; (3) enzymes for two component regulatory systems (TCS) (Binda et al. 2014).

Vancomycin-resistant in *Enterococci* is commonly mediated by *vanA* cluster under the control of two promoters. The first promoter controls the expression and function of VanA by initiating VanR/S transcription (Depardieu et al. 2007). VanS is a transmembrane protein comprising of histidine kinase domain that can sense the glycopeptide signal and phosphorylate the VanR regulator (Binda et al. 2014). The triggered VanR initiates the transcription of the second promoter situated near the upstream of resistant genes (Depardieu et al. 2007). Vancomycin resistance begins from the transcription of *vanH* encoded dehydrogenase involved in the conversion of pyruvate to D-Lactate (Hollenbeck and Rice 2012). Next, the addition of D-Lac to D-Ala with the help of *vanA* encoded ligase results in pentapeptide structure. Then this pentapeptide integrates into the peptidoglycan structure by cross-linking. VanX and VanY work together in removing the usual D-Ala-D-Ala structure from the growing cell wall (Binda et al. 2014). Hence, the crucial stage in glycopeptide resistance is D-Ala clearance. The other group of enzymes which are similar to *vanA* is *vanB* cluster, involved in the regulation of resistance in *Enterococci* species followed by $VanS_B$ (sensor kinase) and $VanR_B$ (response regulator) posing distinct homolog to v*anA* cluster (Depardieu et al. 2007). VanC is reported in chromosomal DNA of *Enterococcus gallinarum* and *E. casseliflavus*. VanC promotes resistance to vancomycin through the incorporation of D-Ala-D-Ser to the terminal dipeptide, thus confers low-level resistance pattern (Binda et al. 2014).

9.3.1.4 Resistance to Daptomycin

Daptomycin (DAP) is a lipopeptide antibiotic whose mode of action is similar to that of cationic antimicrobial peptides (CAMPS from the immune system). It requires calcium ions for insertion into the bacterial cell membrane (CM) (Steenbergen et al. 2005). DAP starts with preferential binding at division septal plane for membrane entry. At the inner membrane space, presence of phospholipid (phosphatidylglycerol) stimulates the oligomer formation in DAP at the surface of the CM (Muraih et al. 2012). The oligomers of DAP enter the inner membrane by pore formation by the action of cardiolipin (CL), a membrane-associated phospholipid (Hollenbeck and Rice 2012). Thus, it disrupts the membrane integrity and functionality leading to cell death (Zhang et al. 2014). However, some of the recent research findings provides evidence of genetic changes for resistance mechanism in *Enterococci* based on three gene expression such as *liaF*, *gpdD*, and *cls*. *LiaF* is a member of three-component regulatory system (LiaFSR) in many gram-positive bacteria that govern the organized cell envelop response to stress conditions (Munita et al. 2012). The other two genes for DAP-R are *gpdD* and *cls* that encode for glycerol-phosphodiester phosphodiesterase and cardiolipin synthase which are involved in phospholipid metabolism (Arias et al. 2011). The DAP resistance mechanism is varied among the strains of *Enterococci*. For instance, in *E. faecalis* the redistribution of CL domain from the septal to non-septal areas within the CM is found to be a major reason for resistance phenotype. Based on the experimental evidence it was hypothesized that remodeling of CL microdomain was associated with change in LiaFSR system (Miller et al. 2013). Moreover, an in vitro study performed in *E. faecalis* has proved that changes in phospholipids contents of CM also plays a significant role in providing resistance against DAP insertion (Mishra et al. 2012). Prevention of DAP oligomerization occurs at phosphotidylglycerol deficiency, thereby CL domain traps/prevents the moment of monomeric DAP substance from reaching the inner leaflet of CM (Arias et al. 2011). Instead of DAP resistance based on septal diversion, DAP repulsion from the cell surface is considered as a prominent mechanism in *E. faecium*. In case of *E. faecalis*, the genetic basis of DAP resistance is due to change in *liaFSR* system. Mutation in *liaFSR* gene cluster with corresponding changes in TCS mediates DAP resistance in enterococci (Miller et al. 2013).

9.3.2 Agents Interfering Protein Synthesis

9.3.2.1 Aminoglycosides

Enterococci mediate intrinsic tolerance to various aminoglycoside through two important factors such as poor uptake and covalent modification in aminoglycoside molecule. Lack of sufficient concentration of antibiotics inhibits the protein synthesis and promotes resistance against aminoglycoside (Galimand et al. 2011). Naturally occurring *Enterococci* enzymes are capable of decreasing the binding affinity of aminoglycoside molecule with the ribosomal target by inducing changes in hydroxyl or amino group of aminoglycoside by covalent modification (Hollenbeck and Rice 2012). Similarly, *Enterococci* with *EfmM* encoding ribosomal RNA methyltransferase can induce alteration in ribosomal targets against aminoglycosides. Mostly, the enzymes encoded by chromosomes such as 6′ acetyltransferase (AAC (6′)-li) and APH (3′)-IIIa with phosphotransferase action are capable of modifying antibiotics like kanamycin, amikacin in *E. faecium* (Galimand et al. 2011).

Hence, among the aminoglycosides, gentamycin and streptomycin which are unaffected by intrinsic enzymes were selected for treating the enterococcal infection. The resistance arises for this compound when there is an increase in minimum inhibitory concentration (MIC) value. The antibiotics kanamycin and streptomycin with MIC of 900 and 500 μg/ml were sufficient to provide resistance against *Enterococcus* species (Hollenbeck and Rice 2012). A greater degree of resistance is mediated by streptomycin inactivation through the acquisition of streptomycin adenyl transferase in *Enterococci*. However, a bifunctional modifying enzyme like both AAC (6′)-li and APH (3′)-IIIa acts synergistically in providing resistance to almost all kind of aminoglycosides (Galimand et al. 2011).

9.3.2.2 Oxazolidinones

Linezolid is the FDA approved antibiotic employed for curing drug-resistant *Enterococcus* species (Hollenbeck and Rice 2012). It is a bacteriostatic agent that inhibits the initiation complex for protein synthesis, thereby affects the mRNA translation process (Bourgeois-Nicolaos et al. 2007). It mainly interacts with 23S rRNA and inhibits the peptide elongation in the A-site of the ribosome (Van Harten et al. 2017). Hence, a mutation in 23S rRNA encoding gene develops resistance by preventing linezolid binding. Substitution of uracil to guanosine at 2567 domain of 23S rRNA indicates mutation in patients suffering from linezolid resistant enterococcal infection (Hollenbeck and Rice 2012). Similarly, methylation of adenine at 2503 domain by enzymatic modification of 23S rRNA is another form of mutation which confers resistance mechanism in *Enterococci* (Van Harten et al. 2017).

9.3.2.3 Tetracyclines

Tetracycline promotes antibacterial activity by 30S subunit ribosomal binding and interfering with the docking of aminoacyl-tRNA (Van Hoek et al. 2011). Based on two general strategies such as antibiotic efflux and ribosome protection, multiple genes in enterococci develop resistance against tetracycline. Efflux pump is a transmembrane domain that confers tetracycline resistance (Fluit et al. 2005). The resistance determining genes (tetM, tetO, and tetS) encoded by chromosomes poses similar homology to elongation factors (EFs) which can hydrolyze GTP, thereby promotes remodeling of ribosome conformation to display tetracycline resistance (Celli and Trieu-Cuot 1998).

9.3.3 Agents Interfering Nucleic Acid Synthesis

9.3.3.1 Quinolones

DNA topoisomerase IV and DNA gyrase are targeted by quinolone (Van Hoek et al. 2011). Generally, DNA gyrase induces negative supercoil in double strand DNA and allows initiation of replication followed by topoisomerase IV which unwinds the newly replicated DNA that allows segregation to occur (Hawkey 2005). Both the process requires enzyme/DNA stable complex which can be interrupted by quinolones, thereby arrest the replication process. DNA gyrase and DNA topoisomerase are tetrameric in nature comprised of two different subunits (GyrA and GyrB) and (ParC and ParE) (Van Hoek et al. 2011). *Enterococci* mediate a higher degree of resistance towards quinolones, which has been reported through several mechanisms. Mutation of genes *qyrA* and *parC* in *E. faecalis* and *E. faecium* changes the binding affinity of quinolones to the target site (Lopez et al. 2011). These sites were named as quinolones resistance determining regions. Quinolone externalization through efflux pump encoded in NorA is a well described mechanism in *E. faecium* (Hawkey 2005). In *E. faecalis* presence of pentapeptide repeats encoded by *qnr* decreases quinolone binding to DNA (Van Hoek et al. 2011), hence protects DNA gyrase from the complex formation with quinolone.

9.3.3.2 Rifampicin

The primary target of rifampicin is a β-subunit of DNA-dependent RNA polymerase inhibiting enterococcal mRNA transcription (Enne et al. 2004; Deshpande et al. 2007). Mutation in rpoB gene encoding β-subunit of RNA polymerase is the major reason for the development of resistance against rifampicin (Kristich and Little 2012). Specific mutation at rpoB loci H486Y in *E. faecalis* and *E. faecium* mediates differential transcription of genes capable of producing cell wall resistance agents against cephalosporins (Enne et al. 2004). Despite mutation in rpoB gene rifampicin resistance was observed through efflux pump or enzyme inactivation (Kristich and Little 2012).

9.3.3.3 Trimethoprim

Bacterial enzymes involved in folate synthesis are the primary target for trimethoprim (Hollenbeck and Rice 2012). In particular, trimethoprim acts on enzymes involved in dihydrofolate production; therefore, conversion of tetrahydrofolate is limited (Grayson et al. 1990). Folate plays a vital role in many cellular functions such as thymidine synthesis. Due to insufficient uptake of exogenous folate from the environment, several bacteria have been reported to synthesize folate from p-amino benzoic acid (Hollenbeck and Rice 2012). However, *Enterococci* bypass the inhibition effect of trimethoprim just by utilizing exogenous folate, which was confirmed by in vivo studies (Grayson et al. 1990).

9.3.4 Treatment for Drug Resistance Enterococci Infection

Many alternative therapeutic strategies were revealed to combat antibiotic-resistant enterococcal infections that are listed in Table 9.2. Some of the approaches were found to be more effective and efficient under in vivo studies: (a) A novel class of bacteriocins produced by intestinal microbiota exerting potential antimicrobial activity and specificity are of clinical importance which does not entail harmful effects against commensal microbiota (Millette et al. 2008). (b) Investigational antibiotics such as cadazolid exhibit the development of lower propensity of antibiotic resistance among enterococcal infections (Seiler et al. 2015). (c) Active molecules such as RegIIIγ, stimulate the pheromone response in commensal bacteria and works in concert with mucosal immunity to impede the vancomycin-resistant *Enterococci* (VRE) outgrowth (Brandl et al. 2007; Gilmore et al. 2015). (d) Moreover, phage endolysin is an important therapeutic approach for efficiently clearing the *E. faecalis* clearance (Duerkop et al. 2012).

9.4 Conclusion

With the constant upsurge in the resistance against *Enterococcal* species towards various antibiotics, it is now hard to treat the disease with conventional antibiotics. The genetic versatility and antibiotic resistance ability makes the enterococcus species to excel other bacteria in ecological region. Though with a diverse array of virulent determinants, genes encoding surface proteins such as *ace, epa, agg* are potential factors for enterococcal infections. Thus, it ensued as a primary target for vaccination against strains with drug resistance. Moreover, understanding the membrane physiology of resistance varieties brings a new avenue of antibiotic discovery for cell wall active agents. Bacteriocins are one of the best therapeutic options in clearing resistant Enterococci upon specific targets, and capable of showing synergistic action with traditional antibiotics. Moreover, steps should be taken to limit bacteriocins resistant development and transfer of resistant trait among the pathogens. Horizontal gene transfer mechanism played a leading role in spreading the antibiotic resistance among this genus in the diverse niche. Many plasmids and transposable elements are

Table 9.2 Strategies employed for the treatment of drug-resistant enterococcal infection

S. No	Active compounds	Mode of transfer	Activity	Treatment options for	References
1	Nisin and pediocin	Administration of bacteriocins encoded lactobacillus	Colonization resistance	Vancomycin-resistant enterococci	Millette et al. (2008)
2	Cadazolid	Oral administration	Colonization resistance	Vancomycin-resistant enterococci	Seiler et al. (2015)
3	Barnesiella species	Fecal microbiota transplantation	Colonization resistance	Prevent infection of enterococci	Ubeda (2013)
4	RegIIIγ, a C-type lectin from paneth or intestinal cells	Oral administration of LPS or flagellin stimulates the production	Bactericidal activity	Inhibits the growth of resistant enterococci	Brandl et al. (2007)
5	Intact mucin production	Stimulation of goblet cells by commensal bacteria	Maintaining intestinal barrier integrity	Prevents invasion of infectious enterococci over the intestinal epithelium	Johansson et al. (2015)
6	Bacteriocin (BCN) production	Oral administration of BCN producing *Lactococcus lactis* MM19	Antimicrobial activity	Elimination of MDR enterococci	Borrero et al. (2015)
7	Pheromones, responsive conjugative plasmid	Administration of commensal *E. faecalis* with bacteriocins coded plasmid	Pheromones mediated cell death in competitors	Elimination of MDR *E. faecalis* V583	Gilmore et al. (2015)
8	Endolysin of bacteriophage ϕV 1/7	Intraperitoneal injection in mice	Phage-mediated lysis of competitors	Narrow spectrum antimicrobial against *E. faecalis*	Duerkop et al. (2012)
9	Probiotics	Oral supplementation of L. rhamnosus GG	Colonization resistant	Elimination of growth of VRE through growth competitors	Manley et al. (2007)
10	ß-lactamases inhibitors	Production through *Bacteroides sartorii* and *Parabacteroides distasonis*	Inactivate ß-lactamases	Replenish the ß-lactamases producing ampicillin sensitive strains	Caballero et al. (2017)

recognized to transfer such traits among *Enterococci* species. In addition to the reported class of transposable elements, mutations involving addition and deletion of nucleotides further increase its classification and increase its resistance to new antibiotics, thus cause devastating effects in humans unless this natural genetic transmission is hammered. CRISPR system can be considered to remove the antibiotic resistance genes from *Enterococcus* species.

References

An FY, Clewell DB (2002) Identification of the cAD1 sex pheromone precursor in *Enterococcus faecalis*. J Bacteriol 184(7):1880–1887

Arbeloa A, Segal H, Hugonnet JE et al (2004) Role of class A penicillin-binding proteins in PBP5-mediated beta-lactam resistance in *Enterococcus faecalis*. J Bacteriol 186(5): 1221–1228

Arias CA, Panesso D, McGrath DM et al (2011) Genetic basis for in vivo daptomycin resistance in Enterococci. N Engl J Med 365(10):892–900

Arthur M, Molinas C, Depardieu F, Courvalin P (1993) Characterization of Tn1546, a Tn3-related transposon conferring glycopeptide resistance by synthesis of depsipeptide peptidoglycan precursors in *Enterococcus faecium* BM4147. J Bacteriol 175(1):117–127

Behnood A, Farajnia S, Moaddab SR, Ahdi-Khosroshahi S, Katayounzadeh A (2013) Prevalence of aac (6′)-Ie-aph (2″)-Ia resistance gene and its linkage to Tn5281 in *Enterococcus faecalis* and *Enterococcus faecium* isolates from Tabriz hospitals. Iran J Microbiol 5(3):203–208

Benachour A, Morin T, Hebert L, Budin-Verneuil A, Le Jeune A, Auffray Y, Pichereau V (2009) Identification of secreted and surface proteins from *Enterococcus faecalis*. Can J Microbiol 55(8):967–974

Bhatty M, Cruz MR, Frank KL, Gomez JA, Andrade F, Garsin DA, Dunny GM, Kaplan HB, Christie PJ (2015) *Enterococcus faecalis* pCF 10-encoded surface proteins PrgA, PrgB (aggregation substance) and PrgC contribute to plasmid transfer, biofilm formation and virulence. Mol Microbiol 95(4):660–677

Binda E, Marinelli F, Marcone GL (2014) Old and new glycopeptide antibiotics: action and resistance. Antibiotics (Basel) 3(4):572–594

Bonafede ME, Carias LL, Rice LB (1997) Enterococcal transposon Tn5384: evolution of a composite transposon through co-integration of enterococcal and staphylococcal plasmids. Antimicrob Agents Chemother 41(9):1854–1858

Borrero J, Chen Y, Dunny GM, Kaznessis YN (2015) Modified lactic acid bacteria detect and inhibit multi resistant Enterococci. ACS Synth Biol 4(3):299–330

Bourgeois-Nicolaos N, Moubareck C, Mangeney N, Butel MJ, Doucet-Populaire F (2006) Comparative study of vanA gene transfer from *Enterococcus faecium* to *Enterococcus faecalis* and to *Enterococcus faecium* in the intestine of mice. FEMS Microbiol Lett 254(1):27–33

Bourgeois-Nicolaos N, Massias L, Couson B et al (2007) Dose dependence of emergence of resistance to linezolid in *Enterococcus faecalis* in vivo. J Infect Dis 195(10):1480–1488

Bradley CR, Fraise AP (1996) Heat and chemical resistance of Enterococci. J Hosp Infect 34(3):191–196

Brandl K, Plitas G, Schnabl B, DeMatteo RP, Pamer EG (2007) MyD88-mediated signals induce the bactericidal lectin RegIIIγ and protect mice against intestinal *Listeria monocytogenes* infection. J Exp Med 204(8):1891–1900

Brandl K, Plitas G, Mihu CN et al (2008) Vancomycin-resistant enterococci exploit antibiotic-induced innate immune deficits. Nature 455(7214):804–807

Breuer RJ, Hirt H, Dunny GM (2018) Mechanistic features of the enterococcal pCF10 sex pheromone response and the biology of *Enterococcus faecalis* in its natural habitat. J Bacteriol 200(14):e00733-17

Caballero S, Kim S, Carter RA, Leiner IM, Sušac B, Miller L, Kim GJ, Lilan L, Pamer EG (2017) Cooperating commensals restore colonization resistance to vancomycin-resistant *Enterococcus faecium*. Cell Host Microbe 21(5):592–602

Celli J, Trieu-Cuot P (1998) Circularization of Tn916 is required for expression of the transposon-encoded transfer functions: characterization of long tetracycline-inducible transcripts reading through the attachment site. Mol Microbiol 28(1):103–117

Chajęcka-Wierzchowska W, Zadernowska A, Zarzecka U, Zakrzewski A, Gajewska J (2019) Enterococci from ready-to-eat food–horizontal gene transfer of antibiotic resistance genes and genotypic characterization by PCR melting profile. J Sci Food Agric 99(3):1172–1179

Chen Y, Bandyopadhyay A, Kozlowicz BK, Haemig HA, Tai A, Hu WS, Dunny GM (2017) Mechanisms of peptide sex pheromone regulation of conjugation in *Enterococcus faecalis*. Microbiol 6(4):e00492

Clewell DB (2007) Properties of *Enterococcus faecalis* plasmid pAD1, a member of a widely disseminated family of pheromone-responding, conjugative, virulence elements encoding cytolysin. Plasmid 58(3):205–227

Clewell DB (2011) Tales of conjugation and sex pheromones: a plasmid and enterococcal odyssey. Mob Genet Elem 1(1):38–54

Comenge Y, Quintiliani R Jr, Li L et al (2003) The CroRS two-component regulatory system is required for intrinsic beta-lactam resistance in *Enterococcus faecalis*. J Bacteriol 185(24):7184–7192

Conwell M, Daniels V, Naughton PJ, Dooley JSG (2017) Interspecies transfer of vancomycin, erythromycin and tetracycline resistance among Enterococcus species recovered from agrarian sources. BMC Microbiol 17:19

Courvalin P (2006) Vancomycin resistance in gram-positive cocci. Clin Infect Dis 42:S25–S34

Depardieu F, Podglajen I, Leclercq R et al (2007) Modes and modulations of antibiotic resistance gene expression. Clin Microbiol Rev 20:79–114

Deshpande L, Fritsche T, Moet G et al (2007) Antimicrobial resistance and molecular epidemiology of vancomycin-resistant enterococci from North America and Europe: a report from the SENTRY antimicrobial surveillance program. Diagn Microbiol Infect Dis 58(2):163–170

Di Sante L, Morroni G, Brenciani A et al (2017) pHTβ-promoted mobilization of non-conjugative resistance plasmids from *Enterococcus faecium* to *Enterococcus faecalis*. J Antimicrob Chemother 72(9):2447–2453

Dubin K, Pamer EG (2014) Enterococci and their interactions with the intestinal microbiome. Microbiol Spectr 5(6). https://doi.org/10.1128/microbiolspec.BAD-0014-2016

Duerkop BA, Clements CV, Rollins D, Rodrigues JLM, Hooper LV (2012) A composite bacteriophage alters colonization by an intestinal commensal bacterium. Proc Natl Acad Sci U S A 109(43):17621–17626

Duez C, Zorzi W, Sapunaric F et al (2001) The penicillin resistance of *Enterococcus faecalis* JH2-2r results from an overproduction of the low-affinity penicillin-binding protein PBP4 and does not involve a psr-like gene. Microbiol 147:2561–2569

Duez C, Hallut S, Rhazi N et al (2004) The ponA gene of *Enterococcus faecalis* JH2-2 codes for a low affinity class A penicillin-binding protein. J Bacteriol 186(13):4412–4416

Dunny GM (2013) Enterococcal sex pheromones: signaling, social behavior, and evolution. Annu Rev Genet 47:457–482

Dunny GM, Berntsson RP (2016) Enterococcal sex pheromones: evolutionary pathways to complex, two-signal systems. J Bacteriol 198(11):1556–1562

Dunny GM, Clewell DB (1975) Transmissible toxin (hemolysin) plasmid in *Streptococcus faecalis* and its mobilization of a noninfectious drug resistance plasmid. J Bacteriol 124(2):784–790

Dunny GM, Brown BL, Clewell DB (1978) Induced cell aggregation and mating in *Streptococcus faecalis*: evidence for a bacterial sex pheromone. Proc Natl Acad Sci U S A 75(7):3479–3483

Eliopoulos GM, Moellering RC (1982) Antibiotic synergism and antimicrobial combinations in clinical infections. Rev Infect Dis 4(2):282–293

Enne V, Delsol A, Roe J et al (2004) Rifampicin resistance and its fitness cost in *Enterococcus faecium*. J Antimicrob Chemother 53(2):203–207

Fair RJ, Tor Y (2014) Antibiotics and bacterial resistance in the 21st century. Perspect Med Chem 6:25–64

Feizabadi MM, Shokrzadeh L, Sayady S, Asadi S (2008) Transposon Tn 5281 is the main distributor of the aminoglycoside modifying enzyme gene among isolates of *Enterococcus faecalis* in Tehran hospitals. Can J Microbiol 54(10):887–890

Fisher K, Phillips C (2009) The ecology, epidemiology and virulence of Enterococcus. Microbiol 155:1749–1757

Fluit AC, Florijn A, Verhoef J, Milatovic D (2005) Presence of tetracycline resistance determinants and susceptibility to tigecycline and minocycline. Antimicrob Agents Chemother 49(4):1636–1638

Fontana R, Ligozzi M, Pittaluga F, Satta G (1996) Intrinsic pencillin resistance in *Enterococci*. Microb Drug Resist 2(2):209–213

Francia MV, Fujimoto S, Tille P, Weaver KE, Clewell DB (2004) Replication of *Enterococcus faecalis* pheromone-responding plasmid pAD1: location of the minimal replicon and oriV site and RepA involvement in initiation of replication. J Bacteriol 186(15):5003–5016

Freitas AR, Novais C, Tedim AP et al (2013) Microevolutionary events involving narrow host plasmids influences local fixation of vancomycin-resistance in enterococcus populations. PLoS One 8(3):e60589

Galimand M, Schmitt E, Panvert M et al (2011) Intrinsic resistance to aminoglycosides in *Enterococcus faecium* is conferred by the 16S rRNA m5C1404-specific methyltransferase EfmM. RNA 17:251–262

Garnier F, Taourit S, Glaser P, Courvalin P, Galimand M (2000) Characterization of transposon Tn1549, conferring VanB-type resistance in *Enterococcus spp.* Microbiology 146:1481–1489

Gilmore MS, Lebreton F, van Schaik W (2013) Genomic transition of enterococci from gut commensals to leading causes of multidrug-resistant hospital infection in the antibiotic era. Curr Opin Microbiol 16(1):10–16

Gilmore MS, Rauch M, Ramsey MM et al (2015) Pheromone killing of multidrug-resistant *Enterococcus faecalis* V583 by native commensal strains. Proc Natl Acad Sci U S A 112(23):7273–7278

Goh HMS, Yong MHA, Chong KKL, Kline KA (2017) Model systems for the study of Enterococcal colonization and infection. Virulence 8(8):1525–1562

Gordon S, Swenson JM, Hill BC et al (1992) Antimicrobial susceptibility patterns of common and unusual species of enterococci causing infections in the United States. J Clin Microbiol 30(9):2373–2378

Grayson M, Thauvin-Eliopoulos C, Eliopoulos G et al (1990) Failure of trimethoprim-sulfamethoxazole therapy in experimental Enterococcal endocarditis. Antimicrob Agents Chemother 34(9):1792–1794

Hackbarth CJ, Chambers HF (1993) blaI and blaR1 regulate beta-lactamase and PBP 2a production in methicillin-resistant *Staphylococcus aureus*. Antimicrob Agents Chemother 37(5):1144–1149

Hall CL, Tschannen M, Worthey EA, Kristich CJ (2013) IreB, a Ser/Thr kinase substrate, influences antimicrobial resistance in *Enterococcus faecalis*. Antimicrob Agents Chemother 57(12):6179–6186

Handwerger S, Raucher B, Altarac D et al (1993) Nosocomial outbreak due to *Enterococcus faecium* highly resistant to vancomycin, penicillin and gentamicin. Clin Infect Dis 16(6):750–755

Hawkey P (2005) Mechanisms of quinolone action and microbial response. J Antimicrob Chemother 51(1):29–35

Hegstad K, Mikalsen T, Coque TM, Werner G, Sundsfjord A (2010) Mobile genetic elements and their contribution to the emergence of antimicrobial resistant *Enterococcus faecalis* and *Enterococcus faecium*. Clin Microbiol Infect 16(6):541–554

Hendrickx AP, Willems RJ, Bonten MJ, van Schaik W (2009) LPxTG surface proteins of Enterococci. Trends Microbiol 17(9):423–430

Hirt H, Greenwood-Quaintance KE, Karau MJ, Till LM, Kashyap PC, Patel R, Dunny GM (2018) *Enterococcus faecalis* sex pheromone cCF10 enhances conjugative plasmid transfer *in vivo*. MBio 9:e00037-18

Hollenbeck BL, Rice LB (2012) Intrinsic and acquired resistance mechanisms in enterococcus. Virulence 3:421–569

Johansson ME, Jakobsson HE, Holmen-Larsson J et al (2015) Normalization of host intestinal mucus layers requires long-term microbial colonization. Cell Host Microbe 18(5):582–592

Kang M, Ko YP, Liang X, Ross CL, Liu Q, Murray BE, Hook M (2013) Collagen- binding microbial surface components recognizing adhesive matrix molecule (MSCRAMM) of gram-positive bacteria inhibit complement activation via the classical pathway. J Biol Chem 288(28):20520–20531

Kinnebrew MA, Ubeda C, Zenewicz LA, Smith N, Flavell RA, Pamer EG (2010) Bacterial flagellin stimulates toll-like receptor 5-dependent defense against vancomycin-resistant enterococcus infection. J Infect Dis 201(4):534–543

Kohler V, Vaishampayan A, Grohmann E (2018) Broad-host-range Inc18 plasmids: occurrence, spread and transfer mechanisms. Plasmid 99:11–21

Kohler V, Keller W, Grohmann E (2019) Regulation of gram-positive conjugation. Front Microbiol 10:1134

Kong KF, Schneper L, Mathee K (2010) Beta-lactam antibiotics: from antibiosis to resistance and bacteriology. APMIS 118(1):1–36

Kozlowicz BK, Bae T, Dunny GM (2004) *Enterococcus faecalis* pheromone-responsive protein PrgX: genetic separation of positive autoregulatory functions from those involved in negative regulation of conjugative plasmid transfer. Mol Microbiol 54(2):520–532

Kristich CJ, Little JL (2012) Mutations in the β subunit of RNA polymerase alter intrinsic cephalosporin resistance in *Enterococci*. Antimicrob Agents Chemother 56(4):2022–2027

Kristich CJ, Rice LB, Arias CA (2014) Enterococcal infection-treatment and antibiotic resistance. Massachusetts Eye and Ear Infirmary, Boston, 24649502

Kropec A, Sava IG, Vonend C, Sakinc T, Grohmann E, Huebner J (2011) Identification of SagA as a novel vaccine target for the prevention of *Enterococcus faecium* infections. Microbiol 157(12):3429–3434

Labbe BD, Kristich CJ (2017) Growth- and stress-induced PASTA kinase phosphorylation in *Enterococcus faecalis*. J Bacteriol 199(21):e00363–e00317

Laverde Gomez JA, van Schaik W, Freitas AR et al (2011) A multiresistance megaplasmid pLG1 bearing a hylEfm genomic island in hospital *Enterococcus faecium* isolates. Int J Med Microbiol 301(2):165–175

Le Breton Y, Muller C, Auffray Y, Rince A (2007) New insights into the *Enterococcus faecalis* CroRS two component system obtained using a differential-display random arbitrarily primed PCR approach. Appl Environ Microbiol 73:3738–3741

Lerminiaux NA, Cameron AD (2019) Horizontal transfer of antibiotic resistance genes in clinical environments. Can J Microbiol 65(1):34–44

Lopez M, Tenorio C, Del Campo R, Zarazaga M, Torres C (2011) Characterization of the mechanisms of fluoroquinolone resistance in vancomycin-resistant Enterococci of different origins. J Chemother 23(2):87–91

Manley KJ, Fraenkel MB, Mayall BC, Power DA (2007) Probiotic treatment of vancomycin- resistant enterococci: a randomised controlled trial. Med J Aust 186(9):454–457

Manson JM, Hancock LE, Gilmore MS (2010) Mechanism of chromosomal transfer of *Enterococcus faecalis* pathogenicity island, capsule, antimicrobial resistance, and other traits. Proc Natl Acad Sci U S A 107(27):12269–12274

Miller C, Kong J, Tran TT et al (2013) Adaptation of *Enterococcus faecalis* to daptomycin reveals an ordered progression to resistance. Antimicrob Agents Chemother 57(11):5373–5383

Millette M, Cornut G, Dupont C, Shareck F, Archambault D, Lacroix M (2008) Capacity of human nisin- and pediocin-producing lactic acid bacteria to reduce intestinal colonization by vancomycin-resistant enterococci. Appl Environ Microbiol 74(7):1997–2003

Mishra NN, Bayer AS, Tran TT et al (2012) Daptomycin resistance in enterococci is associated with distinct alterations of cell membrane phospholipid content. PLoS One 7(8):e43958

Muller C, Le Breton Y, Morin T et al (2006) The response regulator CroR modulates expression of the secreted stress-induced SalB protein in *Enterococcus faecalis*. J Bacteriol 188(7):2636–2645

Munita J, Panesso D, Diaz L et al (2012) Correlation between mutations in liaFSR of *Enterococcus faecium* and MIC of daptomycin: revisiting daptomycin breakpoints. Antimicrob Agents Chemother 56(8):4354–4359

Muraih JK, Harris J, Taylor SD, Palmer M (2012) Characterization of daptomycin oligomerization with perylene excimer fluorescence: stoichiometric binding of phosphatidylglycerol triggers oligomer formation. Biochim Biophys Acta 1818(3):673–678

Murray BE (1992) Beta-lactamase-producing enterococci. Antimicrob Agents Chemother 36(11):2355–2359

Nicolas E, Oger CA, Nguyen N et al (2017) Unlocking Tn3-family transposase activity in vitro unveils an asymmetric pathway for transposome assembly. Proc Natl Acad Sci U S A 114(5):E669–E678

Novais C, Freitas AR, Sousa JC, Baquero F, Coque TM, Peixe LV (2008) Diversity of Tn1546 and its role in the dissemination of vancomycin-resistant Enterococci in Portugal. Antimicrob Agents Chemother 52(3):1001–1008

Nunez N, Derre-Bobillot A, Gaubert S (2018) Exploration of the role of the virulence factor ElrA during *Enterococcus faecalis* cell infection. Sci Rep 8(1):1749

Ono S, Muratani T, Mastumoto T (2005) Mechanisms of resistance to imipenem and ampicillin in *Enterococcus faecalis*. Antimicrob Agents Chemother 49(7):2954–2958

Paganelli FL, Huebner J, Singh KV et al (2016) Genome-wide screening identifies phosphotransferase system permease BepA to be involved in *Enterococcus faecium* endocarditis and biofilm formation. J Infect Dis 214(2):189–195

Palmer KL, Kos VN, Gilmore MS (2010) Horizontal gene transfer and the genomics of enterococcal antibiotic resistance. Curr Opin Microbiol 13(5):632–639

Peng Z, Ehrmann MA, Waldhuber A et al (2017) Phosphotransferase systems in *Enterococcus faecalis* OG1RF enhance anti-stress capacity *in vitro* and *in vivo*. Res Microbiol 168(6):558–566

Polidori M, Nuccorini A, Tascini C, Gemignani G, Iapoce R, Leonildi A, Tagliaferri E, Menichetti F (2013) Vancomycin-resistant (VRE) bacteremia in infective endocarditis successfully treated with combination daptomycin and tigecycline. J Chemother 23(4):240–241

Quinones D, Aung MS, Martins JS, Urushibara N, Kobayashi N (2018) Genetic characteristics of VanA-type vancomycin-resistant *Enterococcus faecalis* and *Enterococcus faecium* in Cuba. New Microbes New Infect 21:125–127

Quintiliani R Jr, Courvalin P (1996) Characterization of Tn1547, a composite transposon flanked by the IS16 and IS256-like elements, that confers vancomycin resistance in *Enterococcus faecalis* BM4281 Gene 172(1):1–8

Rice KC (2014) Creation of *Staphylococcal* mutant libraries using transposon Tn917. Methods Mol Biol 1373:97–102

Rice LB (1998) Tn916 family conjugative transposons and dissemination of antimicrobial resistance determinants. Antimicrob Agents Chemother 42(8):1871–1877

Rice LB, Carias LL (1998) Transfer of Tn5385, a composite, multiresistance chromosomal element from *Enterococcus faecalis*. J Bacteriol 180(3):714–721

Rice LB, Bellais S, Carias LL et al (2004) Impact of specific pbp5 mutations on expression of beta-lactam resistance in *Enterococcus faecium*. Antimicrob Agents Chemother 48(8):3028–3032

Rice LB, Carias LL, Hutton-Thomas R, Rudin S (2007) Interaction of related Tn916-like transposons: analysis of excision events promoted by Tn916 and Tn5386 integrases. J Bacteriol 189(10):3909–3917

Rice LB, Carias LL, Rudin S, Hutton R, Marshall S, Hassan M (2009) Role of class A penicillin-binding proteins in the expression of beta-lactam resistance in *Enterococcus faecium*. J Bacteriol 191(11):3649–3656

Roberts AP, Mullany P (2011) Tn 916-like genetic elements: a diverse group of modular mobile elements conferring antibiotic resistance. FEMS Microbiol Rev 35(5):856–871

Rowland SJ, Dyke KG (1990) Tn552, a novel transposable element from *Staphylococcus aureus*. Mol Microbiol 4(6):961–975

Santoro F, Vianna ME, Roberts AP (2014) Variation on a theme; an overview of the Tn916/Tn1545 family of mobile genetic elements in the oral and nasopharyngeal Streptococci. Front Microbiol 20(5):535

Sarti M, Campanile F, Sabia C, Santagati M, Gargiulo R, Stefani S (2012) Polyclonal diffusion of beta-lactamase-producing *Enterococcus faecium*. J Clin Microbiol 50(1):169–172

Sava IG, Heikens E, Huebner J (2010) Pathogenesis and immunity in enterococcal infections. Clin Microbiol Infect 16(6):533–540

Schjorring S, Krogfelt KA (2011) Assessment of bacterial antibiotic resistance transfer in the gut. Int J Microbiol 2011:312956

Schleifer KH, Kilpper-Balz R (1984) Transfer of *Streptococcus faecalis* and *Streptococcus faecium* to the genus enterococcus nom. Rev. as *Enterococcus faecalis* comb. Nov. and *Enterococcus faecium* comb. nov. Int J Syst Bacteriol 34(1):31–34

Seiler P, Enderlin-Paput M, Pfaff P et al (2015) Cadazolid does not promote intestinal colonization of vancomycin-resistant enterococci in mice. Antimicrob Agents Chemother 60(1):628–631

Serra P (1977) Synergistic treatment of enterococcal endocarditis. Arch Intern Med 137(11):1562

Shaw JH, Clewell DB (1985) Complete nucleotide sequence of macrolide-lincosamide-streptogramin B resistance transposon Tn917 in *Streptococcus faecalis*. J Bacteriol 164(2):782–796

Shokoohizadeh L, Ekrami A, Labibzadeh M, Ali L, Alavi SM (2018) Antimicrobial resistance patterns and virulence factors of Enterococci isolates in hospitalized burn patients. BMC Res Notes 11(1):1

Showsh SA, De Boever EH, Clewell DB (2001) Vancomycin resistance plasmid in *Enterococcus faecalis* that encodes sensitivity to a sex pheromone also produced by *Staphylococcus aureus*. Antimicrob Agents Chemother 45(7):2177–2178

Simjee S, White DG, McDermott PF, Wagner DD, Zervos MJ, Donabedian SM, English LL, Hayes JR, Walker RD (2002) Characterization of Tn1546 in vancomycin-resistant *Enterococcus faecium* isolated from canine urinary tract infections: evidence of gene exchange between human and animal enterococci. J Clin Microbiol 40(12):4659–4665

Sletvold H, Johnsen PJ, Wikmark OG, Simonsen GS, Sundsfjord A, Nielsen KM (2010) Tn 1546 is part of a larger plasmid-encoded genetic unit horizontally disseminated among clonal *Enterococcus faecium* lineages. J Antimicrob Chemother 65(9):1894–1906

Soheili S, Ghafourian S, Sekawi Z et al (2014) Wide distribution of virulence genes among *Enterococcus faecium* and *Enterococcus faecalis* clinical isolates. Sci World J 2014:623174

Steenbergen JN, Alder J, Thorne GM, Tally FP (2005) Daptomycin: a lipopeptide antibiotic for the treatment of serious Gram-positive infections. J Antimicrob Chemother 55(3):283–288

Sun L, Zhang P, Qu T, Chen Y, Hua X, Shi K, Yu Y (2017) Identification of novel conjugative plasmids with multiple copies of fosB that confer high-level fosfomycin resistance to vancomycin-resistant Enterococci. Front Microbiol 8:1541

Sydnor ER, Perl TM (2011) Hospital epidemiology and infection control in acute-care settings. Clin Microbiol Rev 24(1):141–173

Togay SO, Keskin AC, Acik L, Temiz A (2010) Virulence genes, antibiotic resistance and plasmid profiles of *Enterococcus faecalis* and *Enterococcus faecium* from naturally fermented Turkish foods. J Appl Microbiol 109(3):1084–1092

Tomayko JF, Zscheck KK, Singh KV, Murray BE (1996) Comparison of the ß-lactamase gene cluster in clonally distinct strains of *Enterococcus faecalis*. Antimicrob Agents Chemother 40(5):1170–1174

Tomura T, Hirano T, Ito T, Yoshioka M (1973) Transmission of bacteriocinogenicity by conjugation in group D S*treptococci*. Jpn J Microbiol 17(6):445–452

Tuohy KM, Ziemer CJ, Klinder A, Knöbel Y, Pool-Zobel BL, Gibson GR (2002) A human volunteer study to determine the prebiotic effects of lactulose powder on human colonic microbiota. Microb Ecol Health Dis 14(3):165–173

Ubeda C, Taur Y, Jenq RR et al (2010) Vancomycin-resistant enterococcus domination of intestinal microbiota is enabled by antibiotic treatment in mice and precedes bloodstream invasion in humans. J Clin Invest 120(12):4332–4341
Ubeda C, Bucci V, Caballero S et al (2013) Intestinal microbiota containing Barnesiella species cures vancomycin-resistant *Enterococcus faecium* colonization. Infect Immun 81(3):965–973
Van Harten RM, Willems RJL, Martin NI, Hendrickx APA (2017) Multidrug-resistant *Enterococcal* infections: new compounds, novel antimicrobial therapies? Trends Microbiol 25(6):467–479
Van Hoek AH, Mevius D, Guerra B et al (2011) Acquired antibiotic resistance genes: an overview. Front Microbiol 2:203
Vesic D, Kristich CJ (2012) MurAA is required for intrinsic cephalosporin resistance of *Enterococcus faecalis*. Antimicrob Agents Chemother 56(5):2443–2451
Weaver KE (2019) Enterococcal genetics. Microbiol Spectrum 7(2). https://doi.org/10.1128/microbiolspec.GPP3-0055-2018
Williamson R, Calderwood SB, Moellering RC et al (1983) Studies on the mechanism of intrinsic resistance to beta-lactam antibiotics in group D Streptococci. J Gen Microbiol 129(3):813–822
Zhang T, Muraih JK, Tishbi N et al (2014) Cardiolipin prevents membrane translocation and permeabilization by daptomycin. J Biol Chem 289(17):11584–11591
Zhu W, Murray PR, Huskins WC et al (2010) Dissemination of an *Enterococcus* Inc18-like vanA plasmid associated with vancomycin-resistant *Staphylococcus aureus*. Antimicrob Agents Chemother 54(10):4314–4320

Pathogenic *Escherichia coli:* Virulence Factors and Their Antimicrobial Resistance

10

Indu Sharma, Sagolsem Yaiphathoi, and Parijat Hazarika

Contents

Abstract

Escherichia coli (*E. coli*) falls under the family Enterobacteriaceae with the genus *Escherichia*. They are facultative anaerobic bacteria mostly at 37 °C and due to the presence of peritrichous flagella; they can be motile or nonmotile. They are enteric bacteria belonging to naturally beneficial flora of humans as well as gastrointestinal tract of warm-blooded animals. The pathotypes of pathogenic enteric *E. coli* strains have been classified into six groups which include Enterotoxigenic *E. coli* (ETEC), Enteropathogenic *E. coli* (EPEC), Enterohemorrhagic *E. coli* (EHEC), Enteroaggregative *E. coli* (EAEC), Enteroinvasive *E. coli* (EIEC), and diffusely adhering *E. coli* (DAEC). Another pathotype, adherent-invasive *E. coli* (AIEC)

I. Sharma (✉) · S. Yaiphathoi · P. Hazarika
Department of Microbiology, Assam University, Silchar, India

B. Siddhardha et al. (eds.), *Model Organisms for Microbial Pathogenesis, Biofilm Formation and Antimicrobial Drug Discovery*,
https://doi.org/10.1007/978-981-15-1695-5_10

which is the seventh group is observed along with Crohn's disease. *E. coli* infection in humans is usually observed while consuming food products that are contaminated that may belong to either animal origin foods, fresh vegetables, water meant for drinking contaminated by animal or human wastes. Some of the infections include UTI, gastrointestinal infections, neonatal meningitis, colorectal cancer, etc. Besides, multidrug resistance in *E. coli* is being increasingly observed across the world, which leads to a fearing issue both in humans and also in veterinary medicine field. The highly resistant drugs include tetracycline, quinolones, and carbapenemase. The ability of *E. coli* to form biofilm also makes the organisms resistant to different antimicrobial drugs.

Keywords
Escherichia coli · Virulence · Multidrug resistance · Biofilm · UTI

10.1 Introduction

Escherichia coli (*E. coli*) was identified first in the year 1885 and since then they have been considered as the important species of bacteria which is being studied (Clements et al. 2012). They fall under the family Enterobacteriaceae with the genus *Escherichia*. *E. coli* is facultative anaerobic bacteria mostly at 37 °C and due to the presence of peritrichous flagella, they can be motile or nonmotile (Croxen et al. 2013). They are enteric bacteria where most of them belong to naturally beneficial flora of humans as well as gastrointestinal tract of warm-blooded animals (Kabiru et al. 2015). Strains that acquire mechanisms for causing disease become pathogenic to humans (Barbau-Piednoir et al. 2018). In the environment, they are largely distributed in contaminated food and their major route for spreading is water. Some selective strains cause multiple infectious diseases both in community and hospital such as infection in the urinary tract, meningitis, sepsis, wound infections, and nosocomial pneumonia.

The pathotypes of pathogenic enteric *E. coli* strains classified into six groups include Enterotoxigenic *E. coli* (ETEC), Enteropathogenic *E. coli* (EPEC), Enterohemorrhagic *E. coli* (EHEC), Enteroaggregative *E. coli* (EAEC), Enteroinvasive *E. coli* (EIEC), and diffusely adhering *E. coli* (DAEC). Another pathotype, adherent-invasive *E. coli* (AIEC) is the seventh group, which is observed along with Crohn's disease (Servin 2014).

Determinations of the acute figure for the occurrence of *E. coli* infections are problematic throughout the world because of unidentification of the causative agents of diarrheagenic infections. Diarrhea usually occurring in children in the developing countries are due to Enterotoxigenic *E. coli*, Enteropathogenic *E. coli*, and Enteroaggregative *E. coli* and can lead to potentially fatal when kept untreated, while in developed countries of the world diarrhea does not seem to be serious as it is self-controlled. Incidence of food-borne infection, in the developed world is mainly caused by the *E. coli* pathotypes EHEC and more recently, EAEC and STEAEC (produced shiga toxin) have been observed to be involved. Many researchers have reported the persistence of EAEC in human beings since animal reservoirs have not yet been described. DAEC has been suggested to be an important source for diarrheagenic illness in children but still, controversy exists whether it is a causative agent or symptomatic disease in the combination of AIEC with Crohn's disease which has recently found (Abigail et al. 2012).

E. coli infection to humans mainly occurs due to the feeding on food products that are contaminated, which include: animal origin foods, fresh vegetables, water meant for drinking contaminated by animal or human wastes. The portal of spreading infection may be directly through the infected patient with a healthy person or infectious animals (Abigail et al. 2012).

10.1.1 Pathotypes of *E. coli* (Virulence Properties)

The infections of enteric *E. coli* can be seen in six pathotypes traditionally based on the pathogenicity profiles (Abigail et al. 2012), which includes: Enteropathogenic *E. coli* (EPEC), Enterohemorrhagic *E. coli* (EHEC), Enteroinvasive *E. coli* (EIEC), which includes *Shigella* sp., Enteroaggregative *Escherichia coli* (EAEC), Enterotoxigenic *Escherichia coli* (ETEC), and Diffusely Adherent *Escherichia coli* (DAEC) (Nataro and Kaper 1998).

Recently, two other pathotypes have emerged are as follows:

1. Adherent invasive *E. coli* (AIEC) that are related to Crohn's disease only and not with infections which causes diarrhea.
2. Shiga Toxin-producing Enteroaggregative *E. coli* (STEAEC) which caused *E. coli* outbreaks in Germany in 2011 (Abigail et al. 2012).

In the developing countries occurred diarrhea in children, is considered to be Enteropathogenic *E. coli* (EPEC), but its occurrence in developed countries is very low due to strict follow-up of high sanitation and hygienic practices (Clarke et al. 2002). Worldwide source of mortality in a child is due to childhood diarrhea and in developing countries alone, the estimation of affected children is 2.2 million (Ramana and Tamanna 2012). The mechanism of diarrhea caused by EPEC is found to be different from other types. The pathogenic phenomenon of EPEC includes fixing and destroying the lesion, followed by a number of changes in the intestinal cells physiologically (Vidal et al. 2007). Based on bundle forming pili (*BFP*) which is present in a particular *E. coli*, EPEC was divided into two groups; namely atypical EPEC and typical EPEC. Certain properties of EPEC such as virulence factor, antigenic property, self-formation of clusters, and restricted attachment to epithelial cells are due to adherence factor plasmid (*pEAF*), which encodes bundle forming pili (Trabulsi et al. 2002). Earlier, it was defined exclusively based on the existence of serotype O and H but presently it has been defined based on certain other pathogenic characteristics. The infection caused by EPEC is the histopathology that causes fixing and destroying of the lesion; which can be seen in the biopsy of intestinal specimens from infected patients or animals (Andrade et al. 1989).

Second type *E. coli*, EHEC causes hemorrhagic colitis (HC). Ingestion of undercooked food items is responsible for colonizing the humans gastrointestinal tract (Phillips 2000). A cytotoxin known as *stx* (Shiga toxin) attaches to the endothelial cells and expresses *Gb3* which allows absorption of this *stx* to bloodstream and helps in spreading the toxins to different parts of the body (Sandvig 2001). These virulence factors define EHEC, causing death of infected patients.

Since EIEC strains strictly resemble *Shigella* sp., they are responsible for causing watery diarrhea that leads to scanty dysenteric stool along with the presence of mucus and blood (Nataro and Kaper 1998). They are the causative source of bacillary dysentery in humans, especially those countries that are thriving on a low income (Croxen et al. 2013; Gomes et al. 2016). The pathogenic effect of EIEC is usually due to the attachment of the bacteria in the mucosal area of human colon, which leads to the expression of plasmid-borne and chromosomal genes (Harris et al. 1982; Sansonetti et al. 1982; Hale et al. 1983; Kaper et al. 2004). After penetrating the epithelial colonic cells, intracellular replication of EIEC takes place and begins to spread to nearby cells, which leads to the inflammatory crushing of the barrier of intestinal epithelial cells and dysentery occurs where blood mucus and leukocytes are accompanied in stool (DuPont et al. 1971; O'Brien et al. 1979; Taylor et al. 1988).

In 1987, a study was reported by Nataro et al. on the initial depiction of EAEC where different adherence patterns of *E. coli* were examined in the culture of Hep-2 cells. Till now, several studies have shown the relation of EAEC with diarrhea in developing countries, where persistent diarrhea for continuous 14 days or more is observed. These strains further enhance secretion of mucus from the mucosa and the stacked-brick binding manner to epithelial cells of bacteria is termed as Aggregative Adherence (AA) pattern. Researchers carried out many studies in countries that are still developing which prove that EAEC associated with diarrhea but the determination of the particular mechanism of its pathogenicity is difficult to understand, thus making the situation hard to conclude the clinical relevance of this microorganism. The infection caused by any subtype of EAEC will follow three stages of pathogenesis namely:

(a) Initial attachment to the mucosal surface
(b) Formation of biofilms
(c) Production of inflammation and toxins (Okeke et al. 2011; Da Re et al. 2013).

In the progressing world, ETEC is the utmost accepted cause of diarrhea (Qadri et al. 2005) and responsible for the cause of 400 million diarrheal cases every year, which involves children below the age group of five years resulting in three lakhs to five lakhs death where travelers diarrhea is considered as one of the main source (Walker et al. 2007; WHO 2006). Enterotoxigenic *Escherichia coli* contains two defined groups of enterotoxins that are heat-stable enterotoxin and heat labile enterotoxin. The organism colonizes the mucosal membrane surface of the small bowel by secreting the enterotoxins at higher level. Heat labile enterotoxins (LTs) of

E. coli which are oligomeric toxins are closely related to the cholera enterotoxin (CT) and is secreted by *Vibrio cholerae* (Sixma et al. 1993) whereas STs are monomeric toxins that is composed of multiple cysteine residues and their disulfide bonds leads to stability of heat of these toxins.

Among the *E. coli* strains that causes diarrhea, the diffusely adherent *E. coli* (DAEC) have been included which is subdivided into two classes, namely:

1. DAEC expressing Afa/Dr adhesins (*Afa/Dr* DAEC) and
2. DAEC not expressing Afa/Dr adhesins (Estrada-Garcia and Navarro-Garcia 2012).

The strain *Afa/Dr* DAEC is considered to be the causative agent of urinary tract infections leading to dangerous issues in pregnancy and diarrhea in children between 18 months and 5 years age group, but sometimes they may be in the form of asymptomatic intestinal microbiota strains in children (Estrada-Garcia and Navarro-Garcia 2012; Ricci et al. 2006). The diffuse attachment pattern of DAEC in HeLa or Hep 2 of epithelial cells that are cultured is considered as their characterization (Servin 2005). The adherence phenotype is shown because of the adhesins harbored by 75% of DAEC from the *Afa/Dr* family and hence much importance has been given to this strain (Kaper et al. 2004).

10.2 Pathogenicity

10.2.1 Urinary Tract Infection (UTI)

UTI a frequently occurring infection and the causative agent of *E. coli* leads to a significant amount of death and has been observed in both the sexes (male and female) but most commonly in women excluding her early childhood days (Raeispour and Ranjbar 2018). The infections among young sexually active women have been intended to surpass 0.5 episodes per annum (Stamm 2006; Hooton et al. 1996) that involves disorders like pyelonephritis and cystitis in the urinary tract (Momtaz et al. 2013). The bladder infection (cystitis) occurs when the lower portion of the urinary tract is affected and the kidney infection (pyelonephritis) takes place when the upper portion of the urinary tract is affected (Lane and Takhar 2011). Some of the commonly occurred symptoms which includes frequent urination, pain while passing urine, frequent urination urge even though the bladder is empty; while the symptoms of infection in kidney include fever and flank pain accompanied with the symptoms of infection in the lower urinary tract (Lane and Takhar 2011). In the year 2003, Foxman pointed out that for the whole life about 40–50% of women suffer from UTIs and most of the cases are usually ignored due to societal influences.

The UPEC strains cause severe UTI due to its virulence genes, which includes aerobactin (*aer*), proteinaceous appendages (P fimbriae), hemolysin (*hly*), type 1 fimbriae, a fimbrial adhesin (*afaI*), cytotoxic necrotizing factor (*cnf 1*), S fimbriae (*sfa*), adhesins and fimbriae; while few other virulence genes that play a role in disease causing in organism are capsule formation group II (*kpsMT*), protein T in outer membrane (*ompT*), uropathogenic specific protein (*usp*), iron protein (iroN), adhesin (*iha*), *set 1*, arginine N-succinyltransferase (*astA*), capsule synthesis group II the adhesin genes of S fimbriae and F1C fimbriae, S and F1C fimbriae; *iutA* (Siderophores included aerobactin), *traT*, serum resistance, and *fimH* (Soto et al. 2009; Bauer et al. 2002).

Various antimicrobial therapies have been used as remedies for UTI but due to an increase in antimicrobial-resistant frequency (Farshad et al. 2012; Tajbakhsh et al. 2015), UTI remains untreated. A complete antibiotic dose as prescribed by the doctors for the infected patients goes in vain due to ignorance on behalf of the patient, thus leading to antibiotic resistance development among such patients. Above all, the bacterial antibiotic-resistant patterns have been observed to be different among different geographical settings and such studies based on antibiotic resistance has become an important one. In a healthy society, different strains of *E. coli* are the main origin of dangerous infections caused by bacteria and certain reports have shown the effects of different patterns of antibiotic resistance (Ejrnaes et al. 2006; Anderson et al. 2005).

10.2.2 Gastrointestinal Infections

Food-borne infections in humans, especially diarrhea occurring in infants due to EPEC, are more frequent in countries that are still developing (da Silva et al. 2001). Genetically mutative strains can cause certain types of diarrhea in humans associated with enterohemorrhagic *E. coli* that may lead to systemic infections (Maity et al. 2010). The EPEC strains produce Vero toxins, which are different from the toxins produced by ETEC, but since the pathogenicity of EPEC strains is still not completely known (Leila et al. 1994), thus the uniqueness of infection caused by EPEC is the formation of histopathological lesion at the mucosal membrane of intestine surface which is similar to a structure of pedestal-like known as fixing and destroying of lesion (Vidal and Navarro-Garcia 2006). Moreover, the production of food in huge amounts and distribution of it can lead to infections which include a number of people in a short duration. Besides this, SLTEC strains also lead to major foodborne implications such as diarrhea, hemorrhagic colitis and hemolytic ureamic syndrome world widely. This strain belongs to a different range of serotypes, of which O157: H7 leads to large outbreaks (Xu et al. 1999).

10.2.3 Neonatal Meningitis (NMEC)

Neonatal meningitis (NMEC) is related to mortality and morbidity. Even after the advancement of intensive care, those who survive, suffer from neurologic complexity like hearing loss, delay in developmental and impairment of cognitive (Aruna Chandran et al. 2011, Theodoridou et al. 2013). NMEC common cause in newborns occurs following hematogenous spread (Zhao et al. 2018). The unfavorable step for the progress of meningitis is the occupation of spreading *E. coli* into the brain (Central Nervous System) (Kalliopi 2013).

Sometimes *E. coli* clubbed with group B *Streptococcus*, is found responsible for the most important cause of neonatal bacteremia in premature newborns (Cole et al. 2017). Another pathotype, ExPEC has also been observed as the source of causing neonatal meningitis *Escherichia coli* (NMEC). They can live in blood and cause meningitis to infants by invading the meninges. The virulence traits identifying diarrheagenic *E. coli* and other ExPEC pathovars have already been distinctly studied but NMEC pathotypes are yet to be elaborately viewed and discussed (Wijetunge et al. 2015).

10.2.4 Role of *E. coli* in Colorectal Cancer

Though tremendous advancements in therapeutic care have been commenced, still colorectal cancer (CRC) is considered as the third cause of cancer to both the sexes and also the fourth important cause of death by cancer worldwide currently. CRC is highly related to exposure to an environment that contributes notably to the colonic environment. In various studies, a relation within adherent *E. coli* in mucosal and CRC is clearly demonstrated (Buc et al. 2013) where the strain invades the epithelial cells of the intestine in culture and replicates within the macrophages (Maelle et al. 2014). Besides this, the second cause of death because of cancer remains to be CRC after lung neoplasia which is the cause of deaths for more than 6,00,000 lakhs every year. *E. coli* commensally can live as well as cause disease when carrying virulence factors. They produce toxins which include cyclomodulins that cause differentiation, apoptosis, and cell proliferation by disturbing the eukaryotic cell cycle of or damaging the DNA. The chromosomal instability involved in CRC is due to a particular toxin called colibactin coded by complex multienzyme polyketide synthase (pks genomic island) that can help to the formation of double-stranded DNA (Veziant et al. 2016).

10.2.5 Animal Diseases Caused by *E. coli*

In poultry, *E. coli* inhabits the lower portion of the digestive tract and colonizes within 24 h after hatching. Extraintestinal pathogenic *E. coli* strains colonize in the intestine and produce virulence factors that are responsible for causing disease in the extraintestinal sites. Other than intestine, poultry houses are found to serve as a reservoir for APEC (avian pathogenic *E. coli*) where such types of the environment allow the strain for several months to flourish over successive flocks. Severe diseases or poultry death can be due to the transmission of ExPEC from chicken to chicken through inhalation of contaminated dust or maybe pecks infected particles, which contain feces (Stromberg et al. 2017). The infection caused by ExPEC affects human health and is also the main reason for the heavy losses of the economy to the poultry industries worldwide (Stromberg et al. 2017).

E. coli is isolated most frequently from canine UTIs and some of the clonal types isolated from dogs are found to be related to human UTIs (LeCuyer et al. 2018). Besides this, ETEC in the pig's environment is also an important source for the bacterial transmission since these are protected by the manure and hence can survive for a minimum of 6 months and thereby cause neonatal diarrhea in certain animal species such as piglets, calves, sheep, and dogs (Dubreuil et al. 2016).

E. coli strains for the first time were linked to humans in 1982 due to consumption of undercooked meat that lead to outbreaks causing gastrointestinal disturbances. Since then Cattles and other ruminants have been recognised as reservoirs that act as sources of infections to human often consuming animal based food products such as beef, milk, and milk products that are undercooked or from manure which is contaminated (Gomes et al. 2016; Campos et al. 1994). Besides these, *E. coli* is also recognized as the leading and causative agent of acute mastitis and mastitis in cows of dairy farms and the disease symptoms can be completely subclinical to severe and life threatening. Even though mastitis caused due to *E. coli* is transient periodically, continual infections of intramammary glands can be seen. Certain findings have suggested that mammary-pathogenic *E. coli* (MPEC) infections have been persistent to about 5–20% (Lippolis et al. 2017).

10.3 Antimicrobial Resistance of *E. coli*

Over the last decades, antimicrobial drugs were widely used to treat infectious bacterial diseases in human beings, aquaculture, and horticulture and as enhancement in the growth of agriculture, high-yielding efficacy in livestock and poultry industries. However, antibiotics showed better results in limiting the infectious disease but their increased uses have led wide spread of antibiotic resistance gene.

Multidrug resistance *E. coli* increasingly observed across the world which leads to a fearing issue both in humans and also in veterinary medicine field due to their ability to acquire virulence factors, which make them overcome the host defense and resistance toward antibiotics and causes deleterious disease on a human host.

E. coli that acts as an indicator of antimicrobial resistance may be of its particular ecology since they are a part of commensal bacteria of human gut and animals; and is also found in the environment, Aarestrup et al. 2008. *E. coli* possesses resistance toward TEM-1 β-lactamase, extended-spectrum β-lactamases (ESBLs), through the mechanisms of carbapenemase or plasmid-mediated quinolone, ribosomal methylase affecting aminoglycoside, or plasmid-mediated fosfomycin.

Later, the organisms acquired newer resistance mechanisms which made them disseminate resistance traits throughout the world and are now world widely known as MDR (multidrug resistance) microorganisms (Coque et al. 2008). Recently, in 2014, WHO has reported *E. coli* as top nine international concerning microorganism which causes infectious disease in different sites.

10.3.1 Tetracycline Resistance in *E. coli*

The tetracycline antibiotics were first extracted from actinomycetes and discovered in the year 1940. These antibiotics have a wide range spectrum and showed high-level effectiveness against bacteria and are widely used as a growth enhancer in animal feeds (Chopra and Roberts 2001). *E. coli* possesses resistance toward tetracycline through various mechanisms by enzymatic inactivation, specific efflux pumps, ribosomal protection proteins (Burdett et al. 1982; Burdett 1986, 1991, 1996), ribosomal mutation, and through reduced permeability. Several researchers carried out and reported that *E. coli* acquires resistance genes such as tetracycline A (*tet A*), tetracycline B (*tet B*), tetracycline C (*tet C*), tetracycline D (*tet D*), and tetracycline G (*tet G*), which makes them resistance toward tetracycline. Other than these, many researchers have also reported that antibiotics tigecycline eravacycline, and omadacycline overcome resistance via efflux and ribosome protection (Jenner et al. 2013; Zhanel et al. 2016; Tanaka et al. 2016). However, new-generation tetracycline has emerged as new challenges by conferring resistance by deactivating the enzymatic action (Moore et al. 2005; Grossman et al. 2012, 2017).

Tetracycline possesses some enzymes that promote changing configuration of the covalent bond of an antibiotic scaffold that helps the bacteria to inactivate antibiotics (Davies 1994; Wright 2005) and the enzymes are widely known as Tetracycline destructases. These Tetracycline destructases have the ability to lower or destroy the antimicrobial activities by imparting resistance to the periplasm in the extrcellular space of the microorganisms.

10.3.2 Quinolone Resistance in *E. coli*

Relatively newly developed quinolone antibiotic agents produce their antibacterial effects by inhibiting two type's of topoisomerase II activity found in the bacteria but unfortunately, these organisms had already conferred resistance to quinolones many years ago. *E. coli* possesses resistance toward these antibiotics by a mutation in the chromosomes and plasmids and by a decrease in active efflux pump or chemical concentration (Jacoby 2005).

In quinolone resistance actions, DNA gyrase and topoisomerase IV enzymes play a very important function (Drlica and Zhao 1997). The DNA gyrase can remove both the terminals of supercoils (i.e., positive and negative terminals) by introducing negative into DNA (Jacoby 2005).

10.3.3 Carbapenemase Producing *E. coli*

Carbapenem is a broad-spectrum drug and is considered as the last choice of antibiotics against life-threatening diseases. Bacteria acquire resistance toward carbapenem by changing the structural gene within their PBPs (penicillin-binding proteins) and acquire metallo-β-lactamases, which makes them degrade carbapenems or by changing the configuration of membrane permeability (Zhanel et al. 2007).

Carbapenemases are divided into three types:

1. Class A carbapenemase (KPC)
2. Class B metalloenzymes and
3. Class D enzymes (OXA-48 type) (Nordmann et al. 2011).

The utility of this carbapenem was reduced due to the origin of resistant gene world widely where the carbapenem-resistant bacteria lead to great challenges for the treatment of infectious disease since these carbapenemase-resistant genes could live together with the β-lactamase and other resistance genes (Nordmann et al. 2011). *E. coli* isolates from clinical sources expressed carbapenem resistance through the wide range of expression of the plasmid-mediated class β-lactamase with deficiency of protein, which is located in the membrane (Stapleton et al. 1999).

10.4 Biofilm Formation in *E. coli*

Biofilms are aggregated by some groups of organisms constituting natural polymerase (extracellular polymeric substance) which are enclosed and help in function and structural formation of biofilms. Human infectious diseases around 60% are due to biofilms. They can be attached on different surfaces, such as living tissue, in medical dwelling devices, factories, and on the natural aquatic system (Arampatzi et al. 2011). Their formation is mainly enhanced by solid–liquid interface surface for their attachment and growth and the group of cells within the biofilm microcolonies makes an optimum environment for nutrient gradient, gene exchange, and quorum sensing (Arampatzi et al. 2011). They are mainly made of the microbial community of cells and complex natural polymerase substances (extracellular polymeric substance) (Flemming et al. 2000) and based on this composition they consist of a total of 50–90% organic material.

Recently in the year 2001, Sutherland IW and Flemming et al. 2000 reported that Gram-negative bacteria in the presence of D-glucuronic, D-galacturonic, and mannuronic acid linked with pyruvates linked ketal and acquired anionic properties which help in biofilms aggregation. For biofilms formation, three stages are necessary that is transition stages attachment (reversible to irreversible), formation, and maturation of microcolony. Other than this, many other factors and different bacterial appendages also help in biofilm formation. Further, it has been observed that bacterial flagella help in the reversible attachment (Prüß et al. 2006). Curli fimbriae, type I fimbriae which are encoded by *csg* operon and *fim* genes, respectively, promote for irreversible attachment (Prüß et al. 2006) while proteins, DNA, polysaccharides, and others are excreted at the time of maturation of biofilms formation by bacteria which is entrapped in the biofilms.

Biofilm aggregation exhibited by *E. coli* encloses the extracellular biochemical matrix which makes the organisms resistant to different antimicrobial drugs. Antibiotic resistance of bacteria in biofilms is easily reproduced in vitro, and they do not require host factors for the manifestation of biofilm defense. Other than these, many genes also help in biofilms formation.

Bacterial biofilm also possesses antimicrobial agent tolerant cells that have the ability to exhibit multidrug resistance by changing the biological integrate protein membrane. A change in protein structure reduces the effect of antimicrobial drugs. Alteration of the regulator gene structure of biological proteins (outer membrane protein B and outer membrane protein F) makes the *E. coli* resistant toward β-lactam antibiotics (Jaffe et al. 1982). In the year 1978, Pugsley and Schnaitman reported that due to absent of alteration of outer membrane protein F is observed to be less susceptible toward the antibiotics chloramphenicol and tetracycline.

10.5 Conclusion

E. coli truly versatile and diverse group of bacteria are virulent pathogenic organisms causing various diseases by exhibiting various antimicrobial resistance determinants and biofilms formation. Changes in human behavior (changes in food habits, food growing, and harvesting, etc.) and the environment helps the origin of new pathotypes which allows the strain to survive within the population. Recently, researchers involved on multidrug resistance of *E. coli* to determine the key role pathogens involved, prevalence of its resistance traits, mechanisms of acquisition, and spread of resistance genes as well as drug-resistance test; have suggested various strategies to reduce the effectiveness of resistance against existing drugs of public health concern since multidrug resistance in *E. coli* causes serious threat on public health due to the decline in the effectiveness of antibiotics. Thus, research on *E. coli* has become more challenging because it is constantly evolving, and therefore much more research is needed to puzzle out the contributions and relation between virulent properties and commensal microbiota.

References

Aarestrup FM, Wegener HC, Collignon P (2008) Resistance in bacteria of the food chain: epidemiology and control strategies. Expert Rev Anti-Infect Ther 6:733–750. https://doi.org/10.1586/14787210.6.5.733

Abigail F, Macdonald-Wallis C, Tilling K, Boyd A, Golding J, Smith GD, Henderson J et al (2012) Cohort profile: the Avon Longitudinal Study of Parents and Children: ALSPAC mothers cohort. Int J Epidemiol 42(1):97–110

Anderson KL, Whitlock JE, Harwood VJ (2005) Persistence and differential survival of fecal indicator bacteria in subtropical waters and sediments. Appl Environ Microbiol 71:3041–3048. https://doi.org/10.1128/AEM.71.6.3041-3048.2005

Andrade JRC, Daveiga VF, De santa rosa M, Suassuna I (1989) An endocytic process in HEp-2 cells induced by enteropathogenic *Escherichia coli.* J Med Microbiol 28:49–57. https://doi.org/10.1099/00222615-28-1-49

Arampatzi SI, Giannoglou G, Diza E (2011) Aristotle University medical journal scientific annals of the Medical School of Thessaloniki. Aristotle Univ Med J 38:21–29

Barbau-Piednoir E, Denayer S, Botteldoorn N et al (2018) Detection and discrimination of five E. coli pathotypes using a combinatory SYBR® green qPCR screening system. Appl Microbiol Biotechnol 102:3267–3285. https://doi.org/10.1007/s00253-018-8820-0

Bauer RJ, Zhang L, Foxman B et al (2002) Molecular epidemiology of 3 putative virulence genes for Escherichia coli urinary tract infection— usp, iha, and iroN E. coli. J Infect Dis 185:1521–1524. https://doi.org/10.1086/340206

Buc E, Dubois D, Sauvanet P et al (2013) High prevalence of mucosa-associated E. coli producing cyclomodulin and genotoxin in colon cancer. PLoS One 8:e56964. https://doi.org/10.1371/journal.pone.0056964

Burdett V (1986) Streptococcal tetracycline resistance mediated at the level of protein synthesis. J Bacteriol 165:564–569. https://doi.org/10.1128/jb.165.2.564-569.1986

Burdett V (1991) Purification and characterization of Tet(M), a protein that renders ribosomes resistant to tetracycline. J Biol Chem 266:2872–2877

Burdett V (1996) Tet(M)-promoted release of tetracycline from ribosomes is GTP dependent. J Bacteriol 178:3246–3251. https://doi.org/10.1128/jb.178.11.3246-3251.1996

Burdett V, Inamine J, Rajagopalan S (1982) Heterogeneity of tetracycline resistance determinants in Streptococcus. J Bacteriol 149:995–1004

Campos LC, Whittam TS, Gomes TA et al (1994) Escherichia coli serogroup O111 includes several clones of diarrheagenic strains with different virulence properties. Infect Immun 62:3282–3288

Chandran A, Herbert H, Misurski D, Santosham M (2011) Long-term sequelae of childhood bacterial meningitis. Pediatr Infect Dis J 30:3–6. https://doi.org/10.1097/INF.0b013e3181ef25f7

Chopra I, Roberts M (2001) Tetracycline antibiotics: mode of action, applications, molecular biology, and epidemiology of bacterial resistance. Microbiol Mol Biol Rev 65:232–260. https://doi.org/10.1128/MMBR.65.2.232-260.2001

Clarke SC, Haigh RD, Freestone PPE, Williams PH (2002) Enteropathogenic Escherichia coli infection: history and clinical aspects. Br J Biomed Sci 59:123–127

Clements A, Young JC, Constantinou N, Frankel G (2012) Infection strategies of enteric pathogenic Escherichia coli. Gut Microbes 3:71–87. https://doi.org/10.4161/gmic.19182

Cole BK, Scott E, Ilikj M et al (2017) Route of infection alters virulence of neonatal septicemia Escherichia coli clinical isolates. PLoS One 12:e0189032. https://doi.org/10.1371/journal.pone.0189032

Coque TM, Novais Â, Carattoli A et al (2008) Dissemination of clonally related Escherichia coli strains expressing extended-Spectrum β-lactamase CTX-M-15. Emerg Infect Dis 14:195–200. https://doi.org/10.3201/eid1402.070350

Croxen MA, Law RJ, Scholz R et al (2013) Recent advances in understanding enteric pathogenic Escherichia coli. Clin Microbiol Rev 26:822–880. https://doi.org/10.1128/CMR.00022-13

Da Re S, Valle J, Charbonnel N et al (2013) Identification of commensal Escherichia coli genes involved in biofilm resistance to pathogen colonization. PLoS One 8:e61628. https://doi.org/10.1371/journal.pone.0061628

Davies J (1994) Inactivation of antibiotics and the dissemination of resistance genes. Science 264:375–382. https://doi.org/10.1126/science.8153624

Drlica K, Zhao X (1997) DNA gyrase, topoisomerase IV, and the 4-quinolones. Microbiol Mol Biol Rev 61:377–392

Dubreuil JD, Isaacson RE, Schifferli DM (2016) Animal enterotoxigenic Escherichia coli. EcoSal Plus 7:ESP-0006-2016. https://doi.org/10.1128/ecosalplus.ESP-0006-2016

DuPont HL, Formal SB, Hornick RB et al (1971) Pathogenesis of Escherichia coli Diarrhea. N Engl J Med 285:1–9. https://doi.org/10.1056/NEJM197107012850101

Ejrnaes K, Sandvang D, Lundgren B et al (2006) Pulsed-field gel electrophoresis typing of Escherichia coli strains from samples collected before and after pivmecillinam or placebo treatment of uncomplicated community-acquired urinary tract infection in women. J Clin Microbiol 44:1776–1781. https://doi.org/10.1128/JCM.44.5.1776-1781.2006

Estrada-Garcia T, Navarro-Garcia F (2012) Enteroaggregative Escherichia coli pathotype: a genetically heterogeneous emerging foodborne enteropathogen. FEMS Immunol Med Microbiol 66:281–298. https://doi.org/10.1111/j.1574-695X.2012.01008.x

Farshad S, Ranjbar R, Japoni A et al (2012) Microbial susceptibility, virulence factors, and plasmid profiles of uropathogenic Escherichia coli strains isolated from children in Jahrom, Iran. Arch Iran Med 15:312–316

Flemming H-C, Wingender J, Griebe T, Mayer C (2000) Physico-chemical properties of biofilms. In: Evans LV (ed) Biofilms: recent advances in their study and control. Harwood Academic Publishers, Amsterdam, pp 19–34

Gomes TAT, Elias WP, Scaletsky ICA et al (2016) Diarrheagenic Escherichia coli. Braz J Microbiol 47:3–30. https://doi.org/10.1016/j.bjm.2016.10.015

Grossman TH, Fyfe C, O'Brien W et al (2017) Fluorocycline TP-271 is potent against complicated community-acquired bacterial pneumonia pathogens. mSphere 2:e00004-17. https://doi.org/10.1128/mSphere.00004-17

Grossman TH, Starosta AL, Fyfe C et al (2012) Target- and resistance-based mechanistic studies with TP-434, a novel fluorocycline antibiotic. Antimicrob Agents Chemother 56:2559–2564. https://doi.org/10.1128/AAC.06187-11

Hale TL, Sansonetti PJ, Schad PA et al (1983) Characterization of virulence plasmids and plasmid-associated outer membrane proteins in Shigella flexneri, Shigella sonnei, and Escherichia coli. Infect Immun 40:340–350

Harris JR, Wachsmuth IK, Davis BR, Cohen ML (1982) High-molecular-weight plasmid correlates with Escherichia coli enteroinvasiveness. Infect Immun 37:1295–1298

Hooton TM, Scholes D, Hughes JP et al (1996) A prospective study of risk factors for symptomatic urinary tract infection in young women. N Engl J Med 335:468–474. https://doi.org/10.1056/NEJM199608153350703

Jacoby GA (2005) Mechanisms of resistance to quinolones. Clin Infect Dis 41:S120–S126. https://doi.org/10.1086/428052

Jaffe A, Chabbert YA, Semonin O (1982) Role of porin proteins OmpF and OmpC in the permeation of beta-lactams. Antimicrob Agents Chemother 22:942–948. https://doi.org/10.1128/AAC.22.6.942

Jenner L, Starosta AL, Terry DS et al (2013) Structural basis for potent inhibitory activity of the antibiotic tigecycline during protein synthesis. Proc Natl Acad Sci 110:3812–3816. https://doi.org/10.1073/pnas.1216691110

Kabiru L, Bello M, Kabir J et al (2015) Detection of pathogenic Escherichia coli in samples collected at an abattoir in Zaria, Nigeria and at different points in the surrounding environment. Int J Environ Res Public Health 12:679–691. https://doi.org/10.3390/ijerph120100679

Kalliopi M (2013) Quality of life in chronic disease patients. Health Psychol Res 1:3

Kaper JB, Nataro JP, Mobley HLT (2004) Pathogenic Escherichia coli. Nat Rev Microbiol 2:123–140. https://doi.org/10.1038/nrmicro818

Lane DR, Takhar SS (2011) Diagnosis and management of urinary tract infection and pyelonephritis. Emerg Med Clin North Am 29:539–552. https://doi.org/10.1016/j.emc.2011.04.001

LeCuyer TE, Byrne BA, Daniels JB et al (2018) Population structure and antimicrobial resistance of canine uropathogenic Escherichia coli. J Clin Microbiol 56:e00788-18. https://doi.org/10.1128/JCM.00788-18

Leila C, Whittam TS, Gomes TA, Andrade JR, Trabulsi LR (1994) Escherichia coli serogroup O111 includes several clones of diarrheagenic strains with different virulence properties. Infect Immun 62(8):3282–3288

Lippolis JD, Holman DB, Brunelle BW et al (2017) Genomic and transcriptomic analysis of Escherichia coli strains associated with persistent and transient bovine mastitis and the role of colanic acid. Infect Immun 86:e00566-17. https://doi.org/10.1128/IAI.00566-17

Maelle PH, Friswell MK, Alswied A, Roberts CL, Song F, Flanagan PK, Knight P et al (2014) Colonic mucosa-associated diffusely adherent afaC+ Escherichia coli expressing lpfA and pks are increased in inflammatory bowel disease and colon cancer. Gut 63(5):761–770

Maity TK, Kumar R, Misra AK (2010) Prevalence of enteropathogenic Escherichia coli isolated from Chhana based Indian sweets in relation to public health. Indian J Microbiol 50:463–467. https://doi.org/10.1007/s12088-011-0102-9

Momtaz H, Karimian A, Madani M et al (2013) Uropathogenic Escherichia coli in Iran: Serogroup distributions, virulence factors and antimicrobial resistance properties. Ann Clin Microbiol Antimicrob 12:8. https://doi.org/10.1186/1476-0711-12-8

Moore IF, Hughes DW, Wright GD (2005) Tigecycline is modified by the flavin-dependent monooxygenase TetX. Biochemistry 44:11829–11835. https://doi.org/10.1021/bi0506066

Nataro JP, Kaper JB (1998) Diarrheagenic Escherichia coli. Clin Microbiol Rev 11:142–201. https://doi.org/10.1128/CMR.11.1.142

Nordmann P, Naas T, Poirel L (2011) Global spread of carbapenemase-producing enterobacteriaceae. Emerg Infect Dis 17:1791–1798. https://doi.org/10.3201/eid1710.110655

O'Brien AD, Gentry MK, Thompson MR et al (1979) Shigellosis and Escherichia coli diarrhea: relative importance of invasive and toxigenic mechanisms. Am J Clin Nutr 32:229–233. https://doi.org/10.1093/ajcn/32.1.229

Okeke IN, Macfarlane-Smith LR, Fletcher JN, Snelling AM (2011) IS3 profiling identifies the enterohaemorrhagic Escherichia coli O-island 62 in a distinct enteroaggregative E. coli lineage. Gut Pathog 3:4. https://doi.org/10.1186/1757-4749-3-4

Phillips AD (2000) Enterohaemorrhagic Escherichia coli O157:H7 target peyer's patches in humans and cause attaching/effacing lesions in both human and bovine intestine. Gut 47:377–381. https://doi.org/10.1136/gut.47.3.377

Prüß BM, Besemann C, Denton A, Wolfe AJ (2006) A complex transcription network controls the early stages of biofilm development by Escherichia coli. J Bacteriol 188:3731–3739. https://doi.org/10.1128/JB.01780-05

Qadri F, Svennerholm A-M, Faruque ASG, Sack RB (2005) Enterotoxigenic Escherichia coli in developing countries: epidemiology, microbiology, clinical features, treatment, and prevention. Clin Microbiol Rev 18:465–483. https://doi.org/10.1128/CMR.18.3.465-483.2005

Raeispour M, Ranjbar R (2018) Antibiotic resistance, virulence factors and genotyping of uropathogenic Escherichia coli strains. Antimicrob Resist Infect Control 7:118. https://doi.org/10.1186/s13756-018-0411-4

Ramana J, Tamanna (2012) dbDiarrhea: the database of pathogen proteins and vaccine antigens from diarrheal pathogens. Infect Genet Evol 12:1647–1651. https://doi.org/10.1016/j.meegid.2012.08.002

Ricci KA, Girosi F, Tarr PI et al (2006) Reducing stunting among children: the potential contribution of diagnostics. Nature 444:29–38. https://doi.org/10.1038/nature05444

Sandvig K (2001) Shiga toxins. Toxicon 39:1629–1635. https://doi.org/10.1016/S0041-0101(01)00150-7

Sansonetti PJ, Kopecko DJ, Formal SB (1982) Involvement of a plasmid in the invasive ability of Shigella flexneri. Infect Immun 35:852–860

Servin AL (2005) Pathogenesis of Afa/Dr diffusely adhering Escherichia coli. Clin Microbiol Rev 18:264–292. https://doi.org/10.1128/CMR.18.2.264-292.2005

Servin AL (2014) Pathogenesis of human diffusely adhering Escherichia coli expressing Afa/Dr Adhesins (Afa/Dr DAEC): current insights and future challenges. Clin Microbiol Rev 27:823–869. https://doi.org/10.1128/CMR.00036-14

da Silva ZN, da Cunha AS, Lins MC et al (2001) Isolation and serological identification of enteropathogenic Escherichia coli in pasteurized milk in Brazil. Rev Saude Publica 35:375–379. https://doi.org/10.1590/S0034-89102001000400007

Sixma TK, Kalk KH, van Zanten BAM et al (1993) Refined structure of Escherichia coli heat-labile enterotoxin, a close relative of cholera toxin. J Mol Biol 230:890–918. https://doi.org/10.1006/jmbi.1993.1209

Soto SM, Guiral E, Bosch J, Vila J (2009) Prevalence of the set-1B and astA genes encoding enterotoxins in uropathogenic Escherichia coli clinical isolates. Microb Pathog 47:305–307. https://doi.org/10.1016/j.micpath.2009.09.007

Stamm WE (2006) Theodore E. Woodward award: host-pathogen interactions in community-acquired urinary tract infections. Trans Am Clin Climatol Assoc 117:75–83; discussion 83–4

Stapleton PD, Shannon KP, French GL (1999) Carbapenem resistance in Escherichia coli associated with plasmid-determined CMY-4 beta-lactamase production and loss of an outer membrane protein. Antimicrob Agents Chemother 43:1206–1210

Stromberg ZR, Johnson JR, Fairbrother JM et al (2017) Evaluation of Escherichia coli isolates from healthy chickens to determine their potential risk to poultry and human health. PLoS One 12:e0180599. https://doi.org/10.1371/journal.pone.0180599

Tajbakhsh E, Khamesipour F, Ranjbar R, Ugwu IC (2015) Prevalence of class 1 and 2 integrons in multi-drug resistant Escherichia coli isolated from aquaculture water in Chaharmahal Va Bakhtiari province, Iran. Ann Clin Microbiol Antimicrob 14:37. https://doi.org/10.1186/s12941-015-0096-y

Tanaka SK, Steenbergen J, Villano S (2016) Discovery, pharmacology, and clinical profile of omadacycline, a novel aminomethylcycline antibiotic. Bioorg Med Chem 24:6409–6419. https://doi.org/10.1016/j.bmc.2016.07.029

Taylor DN, Echeverria P, Sethabutr O et al (1988) Clinical and microbiologic features of Shigella and enteroinvasive Escherichia coli infections detected by DNA hybridization. J Clin Microbiol 26:1362–1366

Theodoridou K, Vasilopoulou VA, Katsiaflaka A et al (2013) Association of treatment for bacterial meningitis with the development of sequelae. Int J Infect Dis 17:e707–e713. https://doi.org/10.1016/j.ijid.2013.02.009

Trabulsi LR, Keller R, Gomes TAT (2002) Typical and atypical enteropathogenic Escherichia coli. Emerg Infect Dis 8:508–513. https://doi.org/10.3201/eid0805.010385

Veziant J, Gagnière J, Jouberton E et al (2016) Association of colorectal cancer with pathogenic Escherichia coli: focus on mechanisms using optical imaging. World J Clin Oncol 7:293. https://doi.org/10.5306/wjco.v7.i3.293

Vidal JE, Canizález-Román A, Gutiérrez-Jiménez J, Navarro-García F (2007) Molecular pathogenesis, epidemiology and diagnosis of enteropathogenic Escherichia coli. Salud Publica Mex 49:376–386

Vidal JE, Navarro-Garcia F (2006) Efficient translocation of EspC into epithelial cells depends on enteropathogenic Escherichia coli and host cell contact. Infect Immun 74:2293–2303. https://doi.org/10.1128/IAI.74.4.2293-2303.2006

Walker RI, Steele D, Aguado T (2007) Analysis of strategies to successfully vaccinate infants in developing countries against enterotoxigenic E. coli (ETEC) disease. Vaccine 25:2545–2566. https://doi.org/10.1016/j.vaccine.2006.12.028

Wijetunge DSS, Gongati S, DebRoy C et al (2015) Characterizing the pathotype of neonatal meningitis causing Escherichia coli (NMEC). BMC Microbiol 15:211. https://doi.org/10.1186/s12866-015-0547-9

World Health Organization (WHO) (2006) Weekly epidemiological record. http://www.who.int/wer/2006/wer8111.pdf

Wright G (2005) Bacterial resistance to antibiotics: enzymatic degradation and modification. Adv Drug Deliv Rev 57:1451–1470. https://doi.org/10.1016/j.addr.2005.04.002

Xu J-G, Cheng B-K, Jing H-Q (1999) Escherichia Coli O157 H7 and Shiga-like-toxin-producing Escherichia Coli in China. World J Gastroenterol 5:191–194. https://doi.org/10.3748/wjg.v5.i3.191

Zhanel GG, Cheung D, Adam H et al (2016) Review of eravacycline, a novel fluorocycline antibacterial agent. Drugs 76:567–588. https://doi.org/10.1007/s40265-016-0545-8

Zhanel GG, Wiebe R, Dilay L et al (2007) Comparative review of the carbapenems. Drugs 67:1027–1052. https://doi.org/10.2165/00003495-200767070-00006

Zhao W-D, Liu D-X, Wei J-Y et al (2018) Caspr1 is a host receptor for meningitis-causing Escherichia coli. Nat Commun 9:2296. https://doi.org/10.1038/s41467-018-04637-3

Enterobacter Infections and Antimicrobial Drug Resistance

11

V. T. Anju, Busi Siddhardha, and Madhu Dyavaiah

Contents

Abstract

Recently, *Enterobacter* spp. are encountered as significant clinical pathogens. Most of them are naturally resistant to older and few newer antimicrobial agents. They have the inherent ability to develop antibiotic resistance to novel antibiotics.

V. T. Anju · M. Dyavaiah (✉)
Department of Biochemistry and Molecular Biology, School of Life Sciences, Pondicherry University, Puducherry, India

B. Siddhardha
Department of Microbiology, School of Life Sciences, Pondicherry University, Puducherry, India

B. Siddhardha et al. (eds.), *Model Organisms for Microbial Pathogenesis, Biofilm Formation and Antimicrobial Drug Discovery*,
https://doi.org/10.1007/978-981-15-1695-5_11

This genus is prevalent in nosocomial infections. Among the Enterobacteriaceae family, two well-known clinically important opportunistic pathogens emerged recently are *E. aerogenes* and *E. cloacae*. These are versatile pathogens known to cause nosocomial infections in intensive care unit patients. Recently, *Enterobacter* spp. are emerged in community other than the hospital settings causing life-threatening infections. This bacteria possesses intrinsic resistance to broad-spectrum beta-lactam drugs owing to the presence of beta lactamases. Majority of the *Enterobacter* resistance is due to the production of chromosomal AmpC beta-lactamases. AmpC is responsible for the resistance profiles of *Enterobacter* spp. toward first-, second-, and third-generation cephalosporins. Due to the acquisition of plasmids harboring genes that encode extended spectrum beta-lactamases and cephalosporinases, they often exhibit multiple drug resistance and are common in hospitalized patients. Emergence of plasmid-mediated quinolone resistance, aminoglycoside resistance, and carbapenem resistance appears to be rare cases. However, these resistant strains are troublesome during the antibiotic therapy as these drugs known as last line of treatment.

Keywords
Enterobacter spp. · Enterobacteriaceae · Multiple drug resistance · AmpC beta-lactamases

11.1 Introduction

Recently, *Enterobacter* spp. are rapidly establishing as deadly nosocomial pathogens. Hospital-acquired infections are one of the worse outcome associated with the multidrug-resistant *Enterobacter* spp. Recently, third-generation antibiotic, cephalosporin resistance among *Enterobacter* is important nosocomial pathogen associated with high mortality and infection morbidity (Ye et al. 2006). *Enterobacter* spp. causes several nosocomial infections such as bloodstream infections (BSIs), urinary tract infections (UTIs), ophthalmic infections, central nervous system infections, and skin and soft tissue infections. Among all, BSIs are the most invasive nosocomial infection caused by *Enterobacter* spp. These bacteria are ranked as the seventh most common infectious agent associated with nosocomial pneumonia and ninth most common pathogen causing nosocomial BSIs. The severity of this pathogen in hospital settings is high, where bloodstream infections are the second most invasive disease due to these bacteria (Kus 2014) (Fig. 11.1).

The genus *Enterobacter* was first suggested by Hormaeche and Edwards in 1960 (Grimont and Grimont 2006). *Enterobacter* spp. are having similar phenotypic and biochemical features of the genus *Klebsiella* but differ in their motility. Colonies of this genus are slightly mucoid in nature and less fermentative than *Klebsiella*. Clinically important and highly pathogenic one is known as *E. cloacae*. An anaerobic and yellow pigmented, *E. agglomerans* formerly called as *Erwinia herbicola* is

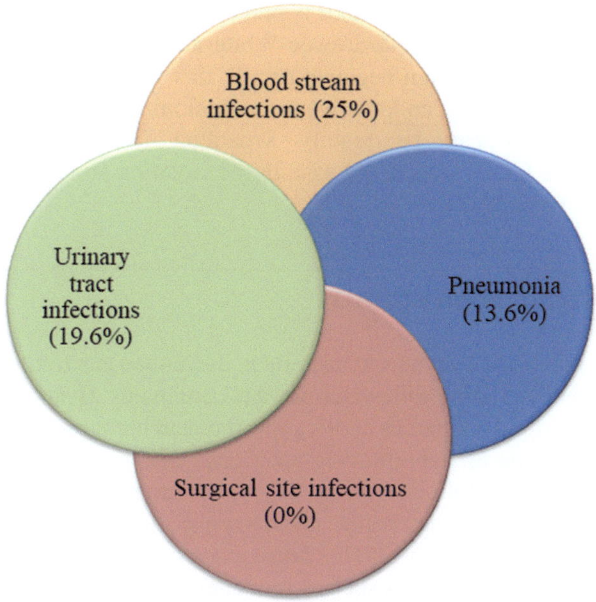

Fig. 11.1 Prevalence of *Enterobacter* species in different nosocomial infections

encountered occasionally in clinical settings (Greenwood 2012). This genus of bacteria is facultative anaerobes and straight Gram-negative bacilli having an approximate size of 0.6–1 mm × 1.2–3 mm. They are motile by peritrichous flagella and are nonspore formers. Bacteria are having capsules for protection from unfavorable conditions. *Enterobacter* colonies are pigmented or nonpigmented. Biochemical features of bacteria include mannitol fermenter, Voges-Proskauer test positive, methyl red test negative, ornithine positive, and citrate positive. They can grow on Moller's potassium cyanide medium at 30 °C and are lysine decarboxylase negative, gelatin test positive, indole negative, and oxidase test negative. Some of the species exhibit different biochemical properties. *E. agglomerans* do not deaminate phenylalanine and are ornithine decarboxylase negative. Also, they cannot produce hydrogen sulfide in triple sugar iron agar. Gelatin liquefaction test, indole test, oxidase test, and lysine decarboxylase test are some of the different tests for *E. aerogenes* and *E. gergoviae* (Kus 2014).

Enterobacter spp. belongs to the family Enterobacteriaceae. These are ubiquitous in the environment and are able to survive on dry and skin surfaces. This genus was recognized as pathogens after a nationwide septicemia outbreak at 25 hospitals in 1976 from unsterile intravenous solutions. These bacteria can cause sporadic outbreak owing to their ability to divide in the glucose-containing parental fluids (Maki et al. 1976). Several outbreaks of enterobacterial infections have been reported due to the use of unsterile humidifiers, respiratory therapy equipment, hydrotherapy water in burn unit, and enteral feedings. This genus causes a wide variety of nosocomial infections (infections of lung, urinary tract, abdominal cavity, and intravascular devices). *E. sakazakii* is one of the bacteria that causes neonatal sepsis along with meningitis (Nazarowec-White and Farber 1997; Bar-Oz et al. 2007). Like other enteric Gram-negative bacilli, this genera is gifted with several factors such as siderophores, endotoxins, and adhesions capable of initiating pathogenesis. This genus can be readily separated from other members of Enterobacteriaceae family due to the ease in the isolation from the clinical specimens (Patel and Patel 2016).

11.2 *Enterobacter* Infections

These bacteria inhabit as commensal bacteria in the human gastrointestinal tract and in the environment including soil, water, sewage, and plants. Historically, this genus was categorized as mild pathogens causing low threat to humans. Now, *Enterobacter* spp. acquired attention as opportunistic pathogens causing infections in immunocompromised patients within the clinics. *E. cloacae* are responsible for most of the nosocomial infections in human beings and known as the most medically significant species (Davin-Regli and Pagès 2015). *Enterobacter* species are associated with a wide variety of clinical infections such as lower respiratory tract infections, urinary tract infections, bacteremia, skin and soft tissue infections, endocarditis, central nervous system infections, bone and joint infections, and gastrointestinal tract infections. In case of animals, these species are rarely linked with pneumonia, peritonitis, intravenous catheter infections, wound infections, dermatitis, otitis media infections, and urinary tract infections (Weese 2008). The arrival of antibiotic resistance profiles among this genus is of great threat in global health care system, which affects both humans and companion animals. The failure of antibiotic therapy emerges with new risks in humans and animals. This may further end up with several significant public health consequences. Hence, alternatives strategies for the elimination of antimicrobial-resistant strains among *Enterobacter* spp. are important from the view of veterinary medicine and public health care system (Harada et al. 2017; Weese 2008).

This genus has been associated with several clinical syndromes and occasionally some of the syndromes mimics with the disease patterns of easily treatable pathogens such as *Staphylococcus aureus* and group A *Streptococci*. In recent times, high rates of coinfections with other pathogens are detected in the liver and lung transplant areas with *Enterobacter* infections. This includes the growing dominance in a variety of clinical syndromes and their etiologic part in cotton fever (Sanders and Sanders 1997). Overall, infections by *Enterobacter* spp. are mostly similar to those by other facultative Gram-negative bacilli. A broad range of infections of this genus include bacteremia, infections of urinary tract, lower respiratory tract, central nervous system, skin, soft tissue, bone, gastrointestinal tract, and other organs (Fig. 11.2). Recently, an *Enterobacter* spp., *E. bugandensis* is isolated from neonates and immunocompromised patients with sepsis. This is known as one of the newly isolated and highly pathogenic species of *Enterobacter* (Pati et al. 2018).

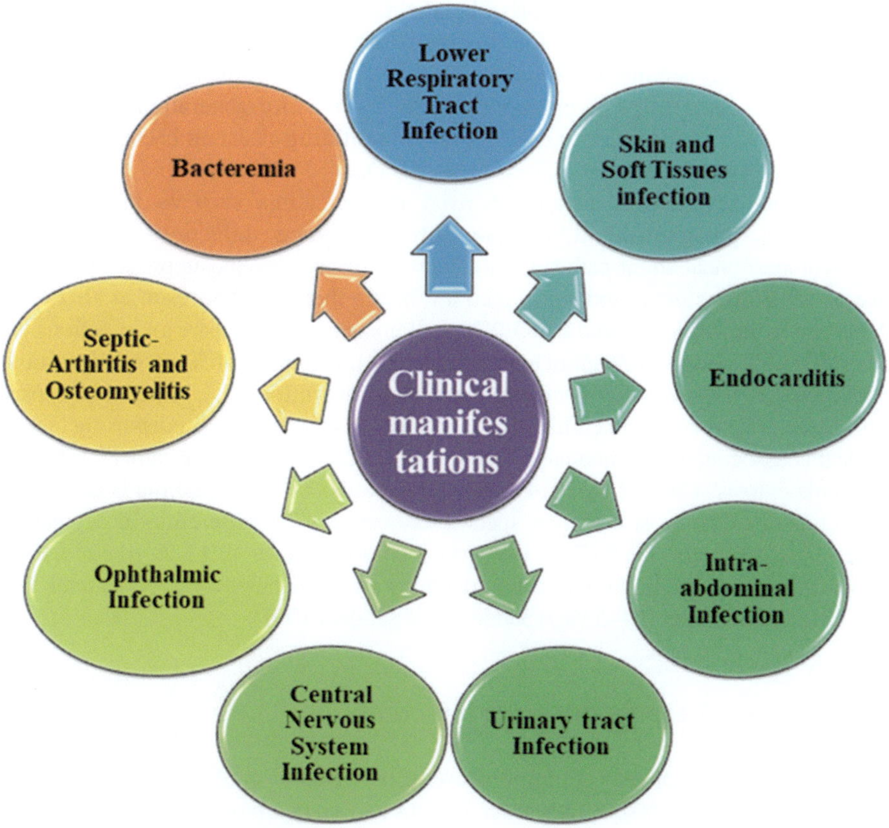

Fig. 11.2 Clinical manifestations of different *Enterobacter* infections

11.2.1 *Enterobacter bacteremia*

Often *Enterobacter* infections are complicated by their resistance to antibiotics of choice such as cephalosporins. In a study, 36% *Enterobacter* infections in intensive care units (ICUs) showed resistance to broad-spectrum cephalosporins. During the course of antibiotic therapy, these bacteria are adapted to the drugs through the increased production of beta-lactamase. Generally, cephalosporin-resistant *Enterobacter* bacteremia is significantly higher than those associated with the susceptible bacteremia. Species associated with *Enterobacter* bacteremia are *E. aerogenes*, *E. agglomerans*, *E. cloacae*, *E. asburiae*, *E. sakazakii*, *E. amnigenus*, *E. gergoviae*, and *E. hormaechei* (Kang et al. 2004). In all cases, bacteremia occurs in debilitated patients like those who had recently hospitalized, or received corticosteroid therapy, or previously received antibiotics or admitted to ICU. The mortality rate associated with *Enterobacter* bacteremia was 20 and 24% at 14 and 28 days after the diagnosis of bacteremia (Blot et al. 2003). Combination therapy is suggested for preventing the emergence of *Enterobacter* resistant strains. Combination therapy includes the administration of beta-lactam drugs other than cephalosporins and aminoglycosides. Combination therapy is more effective than monotherapy as the organism will become resistant to only one of the two drugs given and remain susceptible to other drugs. Even, more sensible use of third-generation cephalosporins may reduce the prevalence of nosocomial multidrug-resistant bacteremia associated with *Enterobacter* spp. (Chow et al. 1991).

There are several risk factors involved in bacteremia such as malignancies, prematurity, gastrointestinal disease, life-threatening infections, use of a ventriculoperitoneal shunt catheter, parenteral nutrition, immunosuppressive therapy, use of ventriculostomy, and prolonged antibiotic therapy (Andresen et al. 1994). A survey in the USA reported that 3.9% of all nosocomial bloodstream infections are caused by *E. cloacae* (Wisplinghoff et al. 2004). Around 5–6% of bacteremia is contributed by *Enterobacter* bacteremia and also known for 8.7% of neonatal sepsis. This bacteremia is severe in children younger than 18 months. The predisposing factors of deaths associated with neonatal bacteremia are prematurity, respiratory problems, and leukocytosis (Chen et al. 2014). Generally, 56–100% of bacteremia is developed institutionally. The most common species associated with bacteremia in decreasing order are *E. cloacae* (46–91% of isolates), *E. aerogenes* (9–43%), *E. agglomerans*, *E. sakazakii*, and others. Some of the infections are contributed by *Enterobacter* spp. that are polymicrobial (14–53%) (Sanders and Sanders 1997).

11.2.2 Lower Respiratory Tract Infections

Like *Enterobacter* spp. are common cause of bacteremia, most of these species are implicated in lower respiratory tract infections. They involved in lung abscess, pneumonia, emphysema, asymptomatic colonization in respiratory secretions, and purulent bronchitis (John et al. 1982). These species surpassed *Klebsiella* spp. and known as the third most cause of nosocomial respiratory tract infections in the USA. It is recognized as a cause of community-acquired pneumonia (Pareja et al. 1992). In last decades, prevalence of lower respiratory tract infections by *Enterobacter* spp. increased continuously. Statistics suggested that only 2–9% cases of respiratory tract infections were reported in 1970s. This rate was increased from 9.5% in the 1980s to 11% in 1990 (Jarvis and Martone 1992). There was a prevalence of *Enterobacter* infections in lung transplant recipients. It is reported that around 40% of lung transplant patients developed acute bacterial pneumonia exactly following the transplantation. Manifestations of clinical pneumonia caused by other Gram-negative bacilli differ from those caused by *Enterobacter* spp. (Sanders and Sanders 1997).

11.2.3 Endocarditis

Infective endocarditis is usually caused by Gram-positive bacteria and accounts for high risk of mortality in renal failure patients. Occasionally Gram-negative bacteria are encountered as one of the etiologic agents. *Enterobacter* is known to be a rare etiology of endocarditis. In a meta analyis of 2761 patients, two were confirmed with *Enterobacter* endocarditis (Karasahin et al. 2018). Another case series reviewed 37.7% mortality rate with *Enterobacter.* In a study, *Enterobacter* endocarditis was treated through monotherapy using carbapenem (Moon et al. 2012). Generally vancomycin along with meropenem is administered for critically ill patients and has risk factors for ESBL producing *Enterobacter* spp. (Gould et al. 2012). A case report showed that multidrug-resistant *E. cloacae* can be another possible pathogen of infective endocarditis and treated by continuous administration of beta-lactam drug and aminoglycoside (Yoshino et al. 2015).

11.2.4 Urinary Tract Infections

E. cloacae are one of the chief pathogens involved in urinary tract infection. Extended-spectrum beta-lactamase (ESBL) producing *Enterobacter* spp. are major in hospital-acquired urinary tract pathogens. One of the effective antibiotics against ESBL producing strains were carbapenems but the expression of carbapenemase provided resistance to all beta-lactam drugs (Xu and He 2019). In a case report, New Delhi metallo-β-lactamase 1 (NDM-1) producing *E. aerogenes* was reported to cause UTI after the insertion of JJ ureteric stents in areas of Croatia and Europe. Urinary pathology and even urosepsis are the most common complications developed after JJ stent insertion due to the antibiotic resistance determinants of bacteria (Franolić et al. 2019). Pathogens enter the lower urinary tract, urethra, and then spread to upper urinary tract. Sometimes, nosocomial UTI is acquired through bloodstream infections. Incidence of UTI is 50 times higher in women than men (Lipsky et al. 1980). Another resistant UTI causing species that express *AmpC* beta-lactamase is *E. cloacae*. They confer resistance to a wide range of antibiotics such as cephalosporins, penicillins, clavulanic acid, and quinolones (Pallett and Hand 2010). The known risk factors associated with ESBL producing UTI are recent hospitalization, presence of comorbidities, bladder catheterization, and prolonged stay in health care facility (Vardi et al. 2012).

11.3 Versatile *Enterobacter* spp. Challenging Antibiotic Treatment

Vast majority of *Enterobacter* infections in humans are caused by four important versatile species such as *E. cloacae*, *E. aerogenes*, *E. agglomerans* (*P. agglomerans*), and *E. sakazakii.* These bacteria are lactose fermenters, motile, and form mucoid colonies. These strains arise from endogenous intestinal flora of hospitalized patients and occur as cause of common source outbreaks or transmit from patient to patient. Especially those patients who had undergone antibiotic therapy and those admitted in intensive care units encounters more infections. These species cause variety of nosocomial infections (UTIs, pneumonia, wound, and burn infections, infections of intravascular devices, meningitis, and prosthetic devices) (Sanders and Sanders 1997). Among these, meningitis is primarily observed in neonates associated with contaminated powdered milk products (Bowen and Braden 2006).

E. aerogenes is isolated from clinical samples of urinary, blood, respiratory, or gastrointestinal tract. Epidemiology of these species was common in nosocomial infection outbreaks in Western Europe since 1993. *E. aerogenes* was regarded as a significant multidrug-resistant pathogen in intensive care units till 2003. International spread and transmission of extended-spectrum beta-lactam carrying epidemic plasmid in Europe resulted in *Enterobacter* infections in European hospitals and health care facilities. Also, antibiotic therapy using extended-spectrum cephalosporins and carbapenems caused an increase in infections caused by them in clinical wards. As a consequence of this therapy, pandrug-resistant *E. aerogenes* strains resistant to last line of drugs (colistin and carbapenems) are emerged. Identification of efflux pump mechanism in drug-resistant *E. aerogenes* highlighted the new methods in adaptive evolution of bacteria (Davin-Regli and Pagès 2015) (Fig. 11.3).

11.3.1 *Enterobacter cloacae*

E. cloacae are Gram-negative facultative bacteria widely found as saprophytes (sewage and soil) in nature. They are also found as commensals in human gastrointestinal tract. These are known as one of the important nosocomial pathogens causing UTIs, wound infections, sepsis in ICUs, and pneumonia with clinical significance (Wang et al. 2018). It is a known clinically significant species in *Enterobacter* genus owing to the presence of several antibiotic-resistant genes (Liu et al. 2013). This situation has become worse as there is a need for novel therapeutic drug discovery for broad-spectrum antibiotic-resistant *E. cloacae* (resistant to quinolones, carbapenems, and aztreonam). There are reports of combating *E. cloacae*

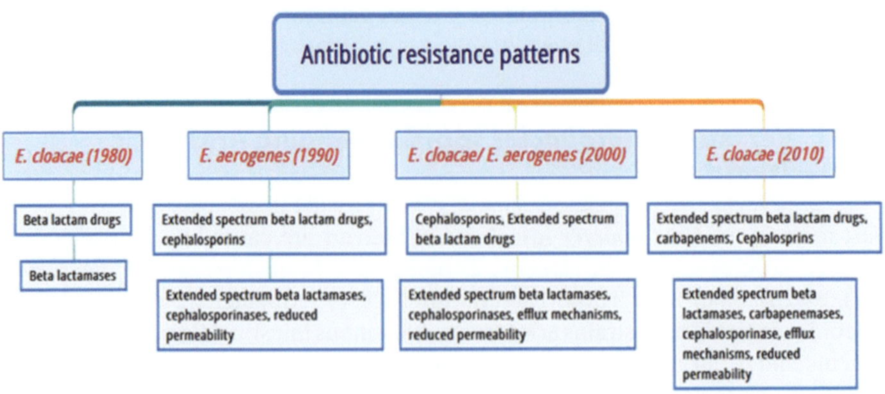

Fig. 11.3 Emergence of antibiotic resistance patterns in clinically important *Enterobacter* species with their mechanisms of antibiotic resistance

ceftazidime-resistant and cefotaxime-susceptible strains with a triple combination of colistin with amikacin and cefepime (Lima et al. 2017). This bacterium has also been isolated from rice, meat, vegetables, and food processing plants. They lead to food spoilage and food safety problems due to the production of putrescine and cadaverine (Liu et al. 2018).

Various virulence-associated genes are present in the pathogenic islands and codes for type IV or III secretion system that is acquired by horizontal gene transfer. Different antagonistic mechanisms present in *E. cloacae* allow them to endure in diverse environments (Liu et al. 2013). A case study reported these bacteria as unusual cause of necrosis of nasal mucosa. These species are intrinsically resistant to amoxicillin, ampicillin, and cephalosporins. Many studies revealed that the disproportionate use of broad-spectrum drugs (e.g., cephalosporins) developed *Enterobacter* sp. as significant nosocomial pathogens (Binar et al. 2015). Monotherapy or combination therapy using aminoglycosides was suggested for treatment of carbapenem or beta-lactam drug-resistant *E. cloacae*. Later, draft genome analysis showed that bacteria acquired high-level aminoglycoside resistance through *rmtD-2* gene (Martins et al. 2017). In an outbreak of bacteremia, *E. cloacae* were found to be the causative agent for 4.5% of all cases and remain endemic at medical center for 5 years (John et al. 1982). Infections caused by *E. cloacae* accounts for 5% hospital-acquired sepsis, 4% nosocomial pneumonia, 10% postsurgical peritonitis, and 4% nosocomial urinary tract pneumonia. Pathogenic mechanism in disease development includes the production of hemolysin, enterotoxins, and thiol activated pore-forming cytotoxins. Antibiotic-resistant biofilms are produced with the help of curli fimbriae. In general, these are widely found in nature but can also act as pathogens (Mezzatesta et al. 2012).

11.3.2 *Enterobacter aerogenes*

There is a greater concern associated with the nosocomial infections in immunocompromised individuals caused by *Enterobacter* sp. Among them, *E. aerogenes* and *E. cloacae* are the most important opportunistic pathogens, especially in patients on ventilation. *E. aerogenes* possess multiple resistance and virulence genes that contribute to increased pathogenesis. They produce extended-spectrum lactamases such as *AmpC* lactamase and have acquired resistance to cefoxitin, ampicillin, first-generation cephalosporins and amoxicillin (Azevedo et al. 2018). These bacteria were responsible for several outbreaks of nosocomial infections in Europe. Various redundant regulatory cascades present in bacteria efficiently allow the bacterial dissemination through control of membrane permeability during bacterial protection and expresses several detoxifying enzymes that codes for antibiotic resistance. Different factors involved in conferring antibiotic resistance include activation of OmpX porin membrane protein and other drug transporters such as AcrAB-TolC system, MacA, MdfA, OqxAB, Mar, Ram, Sox, and EmrE (Davin-Regli and Pagès 2015). The improved resistance toward broad-spectrum antibiotics was linked with the alterations in outer membrane that resulted in porin decrease and modification in the lipopolysaccharide components. To circumvent the emergence of beta-lactam-resistant strains, combination therapy of imipenem and colistin are suggested (Thiolas et al. 2005). Another mechanism involved in multiple drug resistance (MDR) is the overexpression of efflux pumps. These efflux pumps exclude the antibiotics before they reach to their target. Two efflux pumps, such as ABC transporter- and proton motive force-dependent types, were reported as active in MDR strains of *E. aerogenes* (Martins et al. 2010). *E. aerogenes* is known as the fourth most regularly isolated bacteria from hospital. This is due to the emergence of extended-spectrum cephalosporins and carbapenems. After that, there was an emergence of pan drug-resistant *E. aerogenes* isolates, which are resistant to last line of drugs. It has been studied that around 40% of the MDR strains possess active efflux pump systems (McCusker et al. 2019).

11.3.3 *Enterobacter sakazakii*

E. sakazakii is a known food-borne pathogen causing necrotizing enterocolitis, bacteremia, and meningitis in preterm and full-term immunocompromised infants (Hu et al. 2013). Now, *E. sakazakii* is called as *Cronobacter sakazakii* and in 2002, International Commission on Microbiological Specification for Foods classified this species as a severe threat for restricted populations. It is associated with life-threatening food-borne disease in infants where powdered infant formula acts as source of infection (Akineden et al. 2017). This organism was first characterized as yellow-pigmented coliform causing septicemia in infants. Antibiotics effective against this pathogen are acyl ureidopenicillins, carbapenems, aminoglycosides, aztreonam, antifolates, cephalosporins, chloramphenicol, quinolones, nitrofurantoin, and tetracyclines. They show resistance to benzylpenicillin oxacillin, some macrolides, and clindamycin (Abdesselam and Pagotto 2014). Fatal rate of *E. sakazakii* infection in infants can range up to 80% (Shukla et al. 2018).

Severe reported outcomes of this bacterial infection are brain abscess, seizures, hydrocephalus, developmental delay, and death in 40–80% cases. People at greater risk rate are premature infants rather than mature infants, children, and adults (Bowen and Braden 2006). According to Center for Disease Control and Prevention in the USA, four to six infection cases related to these bacteria are reported per year. Furthermore, immunocompromised elderly patients have also been reported with infections (Lou et al. 2014). Virulence traits of bacteria were found through complete genome sequencing. These virulence factors associated with pathogenesis include hemolysin, plasminogen activator (cpa), and siderophore interacting protein. Other proteins found in bacteria such as outer membrane protein (Omp) A and X plays vital part in the adhesion and internalization to the cells. They are resistant to antibiotic treatment as *E. sakazakii* form biofilms (Holý et al. 2019).

11.3.4 *Enterobacter agglomerans*

E. agglomerans are now classified into a new taxon of Enterobacteriaceae family; Pantoea based numerical phonotypical analysis. These are isolated widely from humans (wounds, internal organs, urine, and blood), animals, plant parts, water, and seeds. Some of them cause stalk and leaf necrosis on onions and some develop galls on *Wisteria japonica*, *Gypsophila paniculata,* etc. (Gavini et al. 1989). *P. agglomerans* are known as obligate infectious agents of plants and opportunistic human pathogens. In humans, mostly, infection by *P. agglomerans* is acquired through wound infection by plant material or by nosocomial infection in immunocompromised individuals. Possible clinical outcomes of *P. agglomerans* infection include synovitis, septic arthritis, peritonitis, osteomyelitis, endophthalmitis, and endocarditis. Epidemics of nosocomial septicemia caused by *P. agglomerans* have been reported in both adult and pediatric patients. Generally, nosocomial infections were mild and proper antibiotic therapy led to complete recovery. Reports of infections by *P. agglomerans* in vertebrate animals are few compared to humans (Dutkiewicz et al. 2016). The pathogenic determinants of *P. agglomerans* are described as pathogenicity island containing plasmid (150 kb pPATH) (Barash and Manulis-Sasson 2009), large Pantoea Plasmid (De Maayer et al. 2012) and the type III secretion system (T3SS) (Nissan et al. 2006). These bacteria cause soft tissue or bone joint infections followed by the penetration of trauma by vegetation (Völksch et al. 2009).

E. agglomerans causes a variety of nosocomial infections such as UTIs, septic arthritis, intra-abdominal infections, ophthalmic infections, central nervous system (CNS) infections, bacteremia, endocarditis, lower respiratory tract infections, skin infections, and osteomyelitis in individuals associated with intravenous lines and immunocompromised patients. This bacterium is an unusual cause of spondylodiscitis, which is one of the manifestations of osteomyelitis. Only 31 suspected cases of *E. agglomerans* spondylodiscitis reported previously (Jayaweera et al. 2016). A report showed *P. agglomerans* as an infrequent cause of peritonitis in peritoneal dialysis patients (Sastre et al. 2017). It is also known to be an unknown cause of infections in children. Often, these bacteria in association with other conventional pathogens cause bacteremia with indwelling central access in children. *P. agglomerans* is suspected as an etiologic agent of penetrating trauma through vegetation or by soil coated objects that persist to be resistant to the conventional therapy (Cruz et al. 2007). The relative contribution of most common infections of *P. agglomerans* includes UTIs (21.4%), wound infections (35.7%), and pneumonia (21.4%). Infections by these bacteria may cause serious morbidity and mortality, especially in children with pneumonia. The drug-resistant patterns of community and hospital-acquired strains may vary with their different pathogenic and clinical features. Even though, 21.4% of the *P. agglomerans* isolates obtained from hospital setting showed resistance to carbapenem and caused infection (Büyükcam et al. 2018). One report showed nosocomial outbreaks of *P. agglomerans* in a tertiary care center associated with in-house prepared anticoagulant dextrose solution which was used for the priming of the plasmapheresis machine and for hemodialysis in acute care. Nosocomial outbreaks reported due to the contaminated parenteral nutrition, transference tubes (used for intravenous purposes), blood and blood products (Boszczowski et al. 2012).

11.4 Antimicrobial Resistance Mechanisms and Associated Factors

Enterobacter spp. among Enterobacteriaceae family are documented as a foremost pathogen in hospital-acquired pathogen and continuously involved in several nosocomial outbreaks worldwide. The emergence of MDR strains is found to be responsible for the life threatening and more expensive outbreaks, which constitute a greater threat to the infection control teams. Increase in the number of carbapenemase and extended-spectrum beta-lactamase (ESBL) producing *Enterobacter* spp. remain as a concern for the physicians and scientists (Noël et al. 2019).

11.4.1 Antibiotic Resistance and Mechanism

Major mechanisms involved in *Enterobacter* spp. antibiotic resistance are alteration in the target of drug, production of an inactivating enzyme and modulating the ability of the drug to enter the cells. *Enterobacter* spp. acquires resistance to beta-lactam antibiotics and aminoglycosides through the production of an inactivating enzyme, whereas resistance to quinolones and trimethoprim is acquired by altering drug targets and their ability to accumulate in the cell. Studies have shown that all species of *Enterobacter* possess chromosomally encoded Bush group type 1 beta-lactamases (Bush et al. 1995; Cohen et al. 1993; Conus and Francioli 1992). In certain strains of *E. gergoviae*, *E. aerogenes*, and *E. sakazakii*, lactamases are produced in very low and noninducible concentrations. These strains have greater sensitivity toward ampicillin, cefoxitin, and older cephalosporins. Mostly, uniform resistance pattern is observed in wild-type strains of *E. sakazakii*, *E. cloacae*, *E. taylorae*, *E. asburiae*, and *E. aerogenes* owing to the presence of Bush type 1 beta-lactamases (Pitout et al. 1997).

Resistance among wild-type strains arises mainly from great specificity of drug to this enzyme or from the drug that acting as inducer of enzyme. There is an ampD gene in *Enterobacter* spp. which on mutation causes the emergence of resistance to extended-spectrum cephalosporins, aztreonam, and broad-spectrum penicillins. Usually, this gene prevents high level expression of beta-lactamases. Hence, such a mutation to this gene has been referred as stable depression mutation. These stable depressant mutants are also resistant to beta-lactam inhibitor beta-lactam drug combinations. Among beta-lactam drugs, only carbapenems and newer expanded spectrum cephalosporins (cefepime) maintain their activity toward *Enterobacter* infections (Schaberg et al. 1991; Ehrhardt and Sanders 1993; Bonten et al. 1994). Wild-type strains may become resistant to broad-spectrum penicillins like piperacillin through the achievement of plasmids encoding Bush group 2 TEM1, TEM2 or SHV lactamases (Huovinen et al. 1989; Liu et al. 1992). Recently scientists reported chromosomally encoded carbapenemases in *E. cloacae* which exhibited carbapenem resistance. Like Bush group 1 lactamases, this enzyme can also be induced by cefoxitin and carbapenems. *Enterobacter* spp. with aminoglycoside resistance are thought to produce one or multiple aminoglycoside inactivating enzymes. Acetylating enzymes such as AAC (6′), AAC (3) II, AAC (3) I, AAC (3) V and AAC (3) III, and nucleotidylating enzymes such as ANT (2″) have been investigated in *Enterobacter* spp. which exhibits aminoglycoside resistance (Huovinen et al. 1989; Maes and Vanhoof 1992) (Table 11.1).

11.4.2 Extended-Spectrum Beta-Lactamases (ESBL) and AmpC Beta-Lactamases

Beta-lactamases are bacterial enzymes produced to cleave beta-lactam antibiotics which results in the generation of inactive molecules. ESBLs have the ability to

Table 11.1 Different antibiotic resistance profiles exhibited by *Enterobacter* species

Enterobacter species	Antibiotics acquired resistance	References
Multidrug-resistant *Enterobacter* spp.	Tetracycline Amoxicillin Cephalosporins	Millar et al. (2008)
E. cloacae complex	Intrinsic resistance to penicillins, first- and second-generation cephalosporins Third-generation cephalosporins Aztreonam	Annavajhala et al. (2019)
E. cloacae	Extended-spectrum cephalosporins Plasmid-mediated quinolone Resistance	Harada et al. (2017)
E. aerogenes *E. cloacae* *E. cloacae* complex	Carbapenems	Khajuria et al. (2014)
E. bugandensis	Cefazolin Cefoxitin Oxacillin Penicillin Rifampin	Singh et al. (2018)
E. cloacae *E. aerogenes* *E. agglomerans*	Third-generation cephalosporins	Cosgrove et al. (2002)
Enterobacter spp.	Imipenem	Marchaim et al. (2008)
E. cloacae *E. aerogenes*	Sulfamethoxazole Cotrimoxazole Gentamicin Tobramycin Ampicillin Piperacillin Cefuroxime	Leverstein-van Hall et al. (2003)

inactivate oxyimino aminothiazolyl cephalosporins such as monobactam, aztreonam, ceftazidime, cefepime, and cefotaxime (Pitout and Laupland 2008). As they are resistant to extended-spectrum cephalosporins, ESBL producers are difficult to detect through zone diameters or MICs. Also, detection of ESBL producers is difficult due to the presence of inducible *AmpC* chromosomal enzymes. It has been shown that clavulanate induces AmpC beta-lactamases and hydrolyze the cephalosporins. *Enterobacter* spp. are resistant to third-generation (cefepime) and fourth-generation cephalosporins that act as poor substrates for beta-lactamases (Crowley and Ratcliffe 2003). Mostly, resistance to beta-lactam drugs is mediated by the hyperproduction of chromosomal AmpC beta-lactamase which resulted by induction or by selection of depressed mutant strains (Barnaud 2001). It has been reported a plasmid-mediated ESBL production among resistant *Enterobacter* spp. Most common beta-lactamases found in *Enterobacter* spp. belong to SHV-, CTX-M-, and TEM-derived lactamases. There are several other lactamases reported in different geographical areas. IBC-1 is one of the recently reported enzymes in *E. cloacae* in Greece. Another VEB-1 was reported in clinical samples containing *E. cloacae* and *E. sakazakii* in Bangkok and Thailand. SFO-1 reported in Japan and associated with *E. cloacae* (Schlesinger et al. 2005).

There are different classes of beta-lactamases based on their substrate and inhibitor specificity. Group 1 describes cephalosporinases which are not inhibited well by clavulanate. Group 2 describes enzymes such as penicillinase, broad-spectrum beta-lactamase, and cephalosporinase activity inhibited by beta-lactamase. Group 3 are metallo beta-lactamase hydrolyzing penicillins, carbapenems, and cephalosporins poorly inhibited by most of the beta-lactamase inhibitors (Bush 2013). Ambler class C (Bush-Jacoby group 1) enzymes do not belong to the ESBL-type enzymes but hydrolyze third-generation cephalosporins. Several Gram-negative bacteria possess chromosomally located genes coding *AmpC* and have been recognized in 1960s. In some species *AmpC* gene is chromosomally located and is intrinsic, especially in *E. aerogenes* and *E. cloacae*. This is due to the inducible expression of *AmpC* gene controlled by transcription factors in those species (Corvec et al. 2007; Macdougall 2011; Harris and Ferguson 2012). Some species are found with *AmpC* gene located on plasmids which can transfer between species. Plasmid-mediated AmpC gene is found in clinical species associated with community onset, nosocomial and health care-related infections. There are ESBL producing strains with plasmid-mediated *AmpC* gene that are frequently resistant to quinolones or trimethoprim sulfamethoxazole. These plasmid-mediated *AmpC* genes are not inducible in nature but there are reports of plasmid-mediated inducible *AmpC* gene transmitting into new hosts (Alvarez et al. 2004).

Further, there are no reports describing the superior action of any antimicrobial agent over carbapenems to combat or treat infections by ESBL or AmpC producers. Overuse of carbapenem has raised a new alarming threat of carbapenem-resistant strains that developing on a global scale. Beta-lactam/beta-lactamase inhibitors (BLBIs) such as piperacillin and tazobactam can be suggested for treatment of ESBL producers. But theoretical studies limited the use of piperacillin against the *AmpC* producers. For urinary infections caused by ESBL producers, amoxicillin–clavulanate is the choice of treatment. Some uncommon antibiotics such as temocillin, pivmecillinam, and fosfomycin are used to treat less critical infections (Harris 2015). A study reported intrinsic chromosomal resistance of *E. cloacae* to first-generation cephalosporins, penicillins, cephamycins, and beta-lactam or lactamase inhibitors due to the *AmpC* gene. One of the reasons for the resistance to cephalosporins by *Enterobacter* spp. is due to the overexpression of beta-lactamases. This class I beta-lactamases are encoded by chromosomal *AmpC* gene (Uzunović et al. 2018).

11.5 Conclusion and Future Perspectives

Recently, *Enterobacter* spp. are considered as significant clinical pathogens causing nosocomial infections. Most of the species associated with *Enterobacter* infections are innately resistant to older antibiotics. They have the ability to develop resistance to newer antibiotics also. Multiple drug-resistant strains are emerged in hospitals using beta-lactam drugs and cephalosporins. They have high prevalence of nosocomial infections in neonates and immunocompromised individuals. Among all isolates, *E. aerogenes* and *E. cloacae* are the most versatile opportunistic pathogens associated with nosocomial outbreaks. The saga of *Enterobacter* infections is associated with a logarithmic increase in the expression of beta-lactamases and extended-spectrum cephalosporinases. There are many fundamental questions to be answered for understanding the pathogenic mechanisms relevant to clinical infections caused by *Enterobacter* spp. that are different from other Gram-negative enteric bacilli. Strategies should be developed to suppress the expression of multiple drug-resistant strains and to minimize new emergence of resistance.

References

Abdesselam K, Pagotto F (2014) Bacteria: *Cronobacter* (*Enterobacter*) *sakazakii* and other Cronobacter spp. In: Encyclopedia of food safety. Elsevier, Cambridge, pp 424–432

Akineden Ö, Heinrich V, Gross M, Usleber E (2017) Reassessment of *Cronobacter* spp. originally isolated as *Enterobacter sakazakii* from infant food. Food Microbiol 65:44–50. https://doi.org/10.1016/j.fm.2017.01.021

Alvarez M, Tran JH, Chow N, Jacoby GA (2004) Epidemiology of conjugative plasmid-mediated AmpC β-lactamases in the United States. Antimicrob Agents Chemother 48:533–537. https://doi.org/10.1128/AAC.48.2.533-537.2004

Andresen J, Asmar BI, Dajani AS (1994) Increasing *Enterobacter* bacteremia in pediatric patients. Pediatr Infect Dis J 13:787–791. https://doi.org/10.1097/00006454-199409000-00007

Annavajhala MK, Gomez-Simmonds A, Uhlemann A-C (2019) Multidrug-resistant *Enterobacter cloacae* complex emerging as a global, diversifying threat. Front Microbiol 10:1–8. https://doi.org/10.3389/fmicb.2019.00044

Azevedo PAA, Furlan JPR, Oliveira-Silva M, Nakamura-Silva R, Gomes CN, Costa KRC, Stehling EG, Pitondo-Silva A (2018) Detection of virulence and β-lactamase encoding genes in *Enterobacter aerogenes* and *Enterobacter cloacae* clinical isolates from Brazil. Brazilian. J Microbiol 49:224–228. https://doi.org/10.1016/j.bjm.2018.04.009

Barash I, Manulis-Sasson S (2009) Recent evolution of bacterial pathogens: the gall-forming *Pantoea agglomerans* case. Annu Rev Phytopathol 47:133–152. https://doi.org/10.1146/annurev-phyto-080508-081803

Barnaud G (2001) Extension of resistance to cefepime and cefpirome associated to a six amino acid deletion in the H-10 helix of the cephalosporinase of an *Enterobacter cloacae* clinical isolate. FEMS Microbiol Lett 195:185–190. https://doi.org/10.1016/S0378-1097(01)00010-6

Bar-Oz B, Preminger A, Peleg O, Block C, Arad I (2007) *Enterobacter sakazakii* infection in the newborn. Acta Paediatr 90:356–358. https://doi.org/10.1111/j.1651-2227.2001.tb00319.x

Binar M, Arslan F, Tasli H, Karakoc O, Kilic A, Aydin U (2015) An unusual cause of necrosis and nasal septum perforation after septoplasty: *Enterobacter cloacae*. New Microbes New Infect 8:150–153. https://doi.org/10.1016/j.nmni.2015.07.002

Blot SI, Vandewoude KH, Colardyn FA (2003) Evaluation of outcome in critically ill patients with nosocomial *Enterobacter* bacteremia: results of a matched cohort study. Chest 123:1208–1213. https://doi.org/10.1378/chest.123.4.1208

Bonten MJ, Gaillard CA, van Tiel FH, Smeets HG, van der Geest S, Stobberingh EE (1994) The stomach is not a source for colonization of the upper respiratory tract and pneumonia in ICU patients. Chest 105:878–884. https://doi.org/10.1378/chest.105.3.878

Boszczowski I, Nóbrega de Almeida Júnior J, Peixoto de Miranda EJ, Pinheiro Freire M, Guimarães T, Chaves CE, Cais DP, Strabelli TM, Risek CF, Soares RE, Rossi F, Costa SF, Levin AS (2012) Nosocomial outbreak of *Pantoea agglomerans* bacteremia associated with contaminated anticoagulant citrate dextrose solution: new name, old bug? J Hosp Infect 80:255–258. https://doi.org/10.1016/j.jhin.2011.12.006

Bowen AB, Braden CR (2006) Invasive *Enterobacter sakazakii* disease in infants. Emerg Infect Dis 12:1185–1189. https://doi.org/10.3201/eid1208.051509

Bush K (2013) Proliferation and significance of clinically relevant β-lactamases. Ann N Y Acad Sci 1277:84–90. https://doi.org/10.1111/nyas.12023

Bush K, Jacoby GA, Medeiros AA (1995) A functional classification scheme for beta-lactamases and its correlation with molecular structure. Antimicrob Agents Chemother 39:1211–1233. https://doi.org/10.1128/AAC.39.6.1211

Büyükcam A, Tuncer Ö, Gür D, Sancak B, Ceyhan M, Cengiz AB, Kara A (2018) Clinical and microbiological characteristics of *Pantoea agglomerans* infection in children. J Infect Public Health 11:304–309. https://doi.org/10.1016/j.jiph.2017.07.020

Chen HL, Lu JH, Wang HH, Chen SJ, Chen CJ, Wu KG, Tang RB (2014) Clinical analysis of *Enterobacter* bacteremia in pediatric patients: a 10-year study. J Microbiol Immunol Infect 47:381–386. https://doi.org/10.1016/j.jmii.2013.03.016

Chow JW, Fine MJ, Shlaes DM, Quinn JP, Hooper DC, Johnson MP, Ramphal R, Wagener MM, Miyashiro DK, Yu VL (1991) *Enterobacter* bacteremia: clinical features and emergence of antibiotic resistance during therapy. Ann Intern Med 115:585–590

Cohen SP, Yan W, Levy SB (1993) A multidrug resistance regulatory chromosomal locus is widespread among enteric bacteria. J Infect Dis 168:484–488. https://doi.org/10.1093/infdis/168.2.484

Conus P, Francioli P (1992) Relationship between ceftriaxone use and resistance of *Enterobacter* species. J Clin Pharm Ther 17:303–305. https://doi.org/10.1111/j.1365-2710.1992.tb01308.x

Corvec S, Prodhomme A, Giraudeau C, Dauvergne S, Reynaud A, Caroff N (2007) Most *Escherichia coli* strains overproducing chromosomal AmpC beta-lactamase belong to phylogenetic group A. J Antimicrob Chemother 60:872–876. https://doi.org/10.1093/jac/dkm284

Cosgrove SE, Kaye KS, Eliopoulous GM, Carmeli Y (2002) Health and economic outcomes of the emergence of third-generation cephalosporin resistance in *Enterobacter* species. Arch Intern Med 162:185–190

Crowley B, Ratcliffe G (2003) Extended-spectrum β-lactamases in *Enterobacter cloacae*: underestimated but clincially significant. J Antimicrob Chemother 51:1316–1317

Cruz AT, Cazacu AC, Allen CH (2007) *Pantoea agglomerans*, a plant pathogen causing human disease. J Clin Microbiol 45:1989–1992. https://doi.org/10.1128/JCM.00632-07

Davin-Regli A, Pagès JM (2015) *Enterobacter aerogenes* and *Enterobacter cloacae*; versatile bacterial pathogens confronting antibiotic treatment. Front Microbiol 6:1–10. https://doi.org/10.3389/fmicb.2015.00392

De Maayer P, Chan WY, Blom J, Venter SN, Duffy B, Smits TH, Coutinho TA (2012) The large universal *Pantoea* plasmid LPP-1 plays a major role in biological and ecological diversification. BMC Genomics 13:625. https://doi.org/10.1186/1471-2164-13-625

Dutkiewicz J, Mackiewicz B, Kinga Lemieszek M, Golec M, Milanowski J (2016) *Pantoea agglomerans*: a mysterious bacterium of evil and good. Part III. Deleterious effects: infections of humans, animals and plants. Ann Agric Environ Med 23:197–205. https://doi.org/10.5604/12321966.1203878

Ehrhardt AF, Sanders CC (1993) Lactam resistance amongst *Enterobacter* species. J Antimicrob Chemother 32:1–11. https://doi.org/10.1093/jac/32.suppl_B.1

Franolić I, Bedenić B, Beader N, Lukić-Grlić A, Mihaljević S, Bielen L, Zarfel G, Meštrović T (2019) NDM-1-producing *Enterobacter aerogenes* isolated from a patient with a JJ ureteric stent in situ. CEN Case Rep 8:38–41. https://doi.org/10.1007/s13730-018-0360-z

Gavini F, Mergaert J, Beji A, Mielcarek C, Izard D, Kersters K, De Ley J (1989) Transfer of *Enterobacter agglomerans* (Beijerinck 1888) ewing and fife 1972 to *Pantoea* gen. nov. as *Pantoea agglomerans* comb. nov. and description of *Pantoe*a dispersa sp. nov. Int J Syst Bacteriol 39:337–345. https://doi.org/10.1099/00207713-39-3-337

Gould FK, Denning DW, Elliott TS, Foweraker J, Perry JD, Prendergast BD, Sandoe JA, Spry MJ, Watkin RW (2012) Guidelines for the diagnosis and antibiotic treatment of endocarditis in adults: a report of the Working Party of the British Society for Antimicrobial Chemotherapy. J Antimicrob Chemother 67:269–289. https://doi.org/10.1093/jac/dkr450

Greenwood D (2012) Medical microbiology, a guide to microbial infections: pathogenesis, immunity, laboratory investigation and control. Churchill Livingstone, London, p 795

Grimont F, Grimont PAD (2006) The genus *Enterobacter*. In: The prokaryotes. Springer, New York, pp 197–214

Harada K, Shimizu T, Mukai Y, Kuwajima K, Sato T, Kajino A, Usui M, Tamura Y, Kimura Y, Miyamoto T, Tsuyuki Y, Ohki A, Kataoka Y (2017) Phenotypic and molecular characterization of antimicrobial resistance in *Enterobacter* spp. isolates from companion animals in Japan. PLoS One 12:e0174178. https://doi.org/10.1371/journal.pone.0174178

Harris P (2015) Clinical management of infections caused by Enterobacteriaceae that express extended-spectrum β-lactamase and AmpC enzymes. Semin Respir Crit Care Med 36:056–073. https://doi.org/10.1055/s-0034-1398387

Harris PNA, Ferguson JK (2012) Antibiotic therapy for inducible AmpC β-lactamase-producing gram-negative bacilli: what are the alternatives to carbapenems, quinolones and aminoglycosides? Int J Antimicrob Agents 40:297–305. https://doi.org/10.1016/j.ijantimicag.2012.06.004

Holý O, Cruz-Córdova A, Xicohtencatl-Cortes J, Hochel I, Parra-Flores J, Petrželová J, Fačevicová K, Forsythe S, Alsonosi A (2019) Occurrence of virulence factors in *Cronobacter sakazakii* and *Cronobacter malonaticus* originated from clinical samples. Microb Pathog 127:250–256. https://doi.org/10.1016/j.micpath.2018.12.011

Hu X, Dou W, Fu L, Zhao G (2013) A disposable immunosensor for *Enterobacter sakazakii* based on an electrochemically reduced graphene oxide-modified electrode. Anal Biochem 434:218–220. https://doi.org/10.1016/j.ab.2012.11.013

Huovinen S, Klossner M, Katila M, Huovinen P (1989) Plasmid-mediated beta-lactamases among aminoglycoside resistant gram-negative bacilli. Scand J Infect Dis 21:303–309. https://doi.org/10.3109/00365548909035700

Jarvis WR, Martone WJ (1992) Predominant pathogens in hospital infections. J Antimicrob Chemother 29:19–24. https://doi.org/10.1093/jac/29.suppl_A.19

Jayaweera JAAS, Kothalawala M, Devakanthan B, Arunan S, Galgamuwa D, Rathnayake M (2016) Spondylodiscitis caused by *Enterobacter agglomerans*. Case Rep Infect Dis 2016:1–4. https://doi.org/10.1155/2016/8491571

John JF, Sharbaugh RJ, Bannister ER (1982) *Enterobacter cloacae*: Bacteremia, epidemiology, and antibiotic resistance. Rev Infect Dis 4:13–28. https://doi.org/10.1093/clinids/4.1.13

Kang CI, Kim SH, Park WB, Lee KD, Kim HB, Oh MD, Kim EC, Choe KW (2004) Bloodstream infections caused by *Enterobacter* species: predictors of 30-day mortality rate and impact of broad-Spectrum cephalosporin resistance on outcome. Clin Infect Dis 39:812–818. https://doi.org/10.1086/423382

Karasahin O, Yildiz Z, Unal O, Arslan U (2018) A rare cause of healthcare-associated infective endocarditis: *Enterobacter cloacae*. IDCases 12:18–20. https://doi.org/10.1016/j.idcr.2018.03.004

Khajuria A, Praharaj AK, Kumar M, Grover N (2014) Carbapenem resistance among *Enterobacter* species in a tertiary care hospital in Central India. Chemother Res Pract 2014:1–6. https://doi.org/10.1155/2014/972646

Kus JV (2014) Infections due to *Citrobacter and Enterobacter*. In: Reference module in biomedical sciences. Elsevier, Amsterdam, pp 1–12

Leverstein-van Hall MA, Blok HEM, Donders R et al (2003) Multidrug resistance among Enterobacteriaceae is strongly associated with the presence of integrons and is independent of species or isolate origin. J Infect Dis 187:251–259. https://doi.org/10.1086/345880

Lima TB, Silva ON, de Almeida KC, Ribeiro SM, Motta DO, Maria-Neto S, Lara MB, Filho CR, Ombredane AS, de Faria Junior C, Parachin NS, Magalhães BS, Franco OL (2017) Antibiotic combinations for controlling colistin-resistant *Enterobacter cloacae*. J Antibiot (Tokyo) 70:122–129. https://doi.org/10.1038/ja.2016.77

Lipsky BA, Hook EW, Smith AA, Plorde JJ (1980) Citrobacter infections in humans: experience at the Seattle veterans administration medical center and a review of the literature. Clin Infect Dis 2:746–760. https://doi.org/10.1093/clinids/2.5.746

Liu F, Wang F, Du L, Zhao T, Doyle MP, Wang D, Zhang W, Suna Z, Xu W (2018) Antibacterial and antibiofilm activity of phenyllactic acid against *Enterobacter cloacae*. Food Control 84:442–448. https://doi.org/10.1016/j.foodcont.2017.09.004

Liu PYF, Gur D, Hall LMC, Livermore DM (1992) Survey of the prevalence of βlactamases amongst 1000 gram-negative bacilli isolated consecutively at the Royal London Hospital. J Antimicrob Chemother 30:429–447. https://doi.org/10.1093/jac/30.4.429

Liu WY, Wong CF, Chung KM, Jiang JW, Leung FC (2013) Comparative genome analysis of *Enterobacter cloacae*. PLoS One 8:e74487. https://doi.org/10.1371/journal.pone.0074487

Lou X, Si G, Yu H, Qi J, LiuT FZ (2014) Possible reservoir and routes of transmission of *Cronobacter* (*Enterobacter sakazakii*) via wheat flour. Food Control 43:258–262. https://doi.org/10.1016/j.foodcont.2014.03.029

Macdougall C (2011) Beyond susceptible and resistant, part I: treatment of infections due to gram-negative organisms with inducible β-lactamases. J Pediatr Pharmacol Ther 16:23–30

Maes P, Vanhoof R (1992) A 56-month prospective surveillance study on the epidemiology of aminoglycoside resistance in a belgian general hospital. Scand J Infect Dis 24:495–501. https://doi.org/10.3109/00365549209052636

Maki DG, Rhame FS, Mackel DC, Bennett JV (1976) Nationwide epidemic of septicemia caused by contaminated intravenous products. Am J Med 60:471–485. https://doi.org/10.1016/0002-9343(76)90713-0

Marchaim D, Navon-Venezia S, Schwaber MJ, Carmeli Y (2008) Isolation of imipenem-resistant *Enterobacter* species: emergence of KPC-2 carbapenemase, molecular characterization, epidemiology, and outcomes. Antimicrob Agents Chemother 52:1413–1418. https://doi.org/10.1128/AAC.01103-07

Martins A, Spengler G, Martins M, Rodrigues L, Viveiros M, Davin-Regli A, Chevalier J, Couto I, Pagès JM, Amaral L (2010) Physiological characterisation of the efflux pump system of antibiotic-susceptible and multidrug-resistant *Enterobacter aerogenes*. Int J Antimicrob Agents 36:313–318. https://doi.org/10.1016/j.ijantimicag.2010.06.036

Martins ER, Casella T, Bueno MFC, Francisco GR, Tolentino FM, de Freitas ACT, Cerdeira L, Costa MN, Cevada C, Lincopan N, Garcia DO, Nogueira MCL (2017) Draft genome sequence of an aminoglycoside-resistant RmtD2-producing *Enterobacter cloacae* subsp. *cloacae* ST395 in Brazil. J Glob Antimicrob Resist 10:308–309. https://doi.org/10.1016/j.jgar.2017.07.012

McCusker MP, Alves Ferreira D, Cooney D, Martins Alves B, Fanning S, Pagès JM, Martins M, Davin-Regli A (2019) Modulation of antimicrobial resistance in clinical isolates of *Enterobacter aerogenes*: a strategy combining antibiotics and chemosensitisers. J Glob Antimicrob Resist 16:187–198. https://doi.org/10.1016/j.jgar.2018.10.009

Mezzatesta ML, Gona F, Stefani S (2012) *Enterobacter cloacae* complex: clinical impact and emerging antibiotic resistance. Future Microbiol 7:887–902. https://doi.org/10.2217/fmb.12.61

Millar M, Philpott A, Wilks M, Whiley A, Warwick S, Hennessy E, Coen P, Kempley S, Stacey F, Costeloe K (2008) Colonization and persistence of antibiotic-resistant Enterobacteriaceae strains in infants nursed in two neonatal intensive care units in East London, United Kingdom. J Clin Microbiol 46:560–567. https://doi.org/10.1128/JCM.00832-07

Moon J, Smith T, Sahud AG, Bhanot N (2012) An unusual etiology of infective endocarditis: *Enterobacter cloacae*. J Infect Chemother 18:925–930. https://doi.org/10.1007/s10156-012-0376-9

Nazarowec-White M, Farber JM (1997) *Enterobacter sakazakii*: a review. Int J Food Microbiol 34:103–113. https://doi.org/10.1016/S0168-1605(96)01172-5

Nissan G, Manulis-Sasson S, Weinthal D, Mor H, Sessa G, Barash I (2006) The type III effectors HsvG and HsvB of gall-forming *Pantoea agglomerans* determine host specificity and function as transcriptional activators. Mol Microbiol 61:1118–1131. https://doi.org/10.1111/j.1365-2958.2006.05301.x

Noël A, Vastrade C, Dupont S, de Barsy M, Huang TD, Van Maerken T, Leroux-Roels I, Delaere B, Melly L, Rondelet B, Dransart C, Dincq AS, Michaux I, Bogaerts P, Glupczynski Y (2019) Nosocomial outbreak of extended-spectrum β-lactamase-producing *Enterobacter cloacae* among cardiothoracic surgical patients: causes and consequences. J Hosp Infect 102:54–60. https://doi.org/10.1016/j.jhin.2019.01.001

Pallett A, Hand K (2010) Complicated urinary tract infections: practical solutions for the treatment of multiresistant gram-negative bacteria. J Antimicrob Chemother 65:iii25–iii33. https://doi.org/10.1093/jac/dkq298

Pareja A, Bernal C, Leyva A, Piedrola G, Maroto MC (1992) Etiologic study of patients with community-acquired pneumonia. Chest 101:1207–1210. https://doi.org/10.1378/chest.101.5.1207

Patel KK, Patel S (2016) *Enterobacter* spp.: an emerging nosocomial infection. Conf Proc 2:532–538

Pati NB, Doijad SP, Schultze T, Mannala GK, Yao Y, Jaiswal S, Ryan D, Suar M, Gwozdzinski K, Bunk B, Mraheil MA, Marahiel MA, Hegemann JD, Spröer C, Goesmann A, Falgenhauer L, Hain T, Imirzalioglu C, Mshana SE, Overmann J, Chakraborty T (2018) *Enterobacter bugandensis*: a novel enterobacterial species associated with severe clinical infection. Sci Rep 8:5392. https://doi.org/10.1038/s41598-018-23069-z

Pitout JD, Laupland KB (2008) Extended-spectrum β-lactamase-producing Enterobacteriaceae: an emerging public-health concern. Lancet Infect Dis 8:159–166

Pitout JD, Moland ES, Sanders CC, Thomson KS, Fitzsimmons SR (1997) Beta-lactamases and detection of beta-lactam resistance in *Enterobacter* spp. Antimicrob Agents Chemother 41:35–39

Sanders WE, Sanders CC (1997) *Enterobacter* spp.: pathogens poised to flourish at the turn of the century. Clin Microbiol Rev 10:220–241. https://doi.org/10.1128/CMR.10.2.220

Sastre A, González-Arregoces JE, Romainoik I, Mariño S, Lucas C, Monfá E, Stefan G, de León B, Prieto M (2017) Peritonitis causada por *Pantoea agglomerans* en diálisis peritoneal. Nefrologia 37:108–109. https://doi.org/10.1016/j.nefro.2016.09.006

Schaberg DR, Culver DH, Gaynes RP (1991) Major trends in the microbial etiology of nosocomial infection. Am J Med 91:S72–S75. https://doi.org/10.1016/0002-9343(91)90346-Y

Schlesinger J, Navon-Venezia S, Chmelnitsky I, Hammer-Münz O, Leavitt A, Gold HS, Schwaber MJ, Carmeli Y (2005) Extended-spectrum beta-lactamases among *Enterobacter* isolates obtained in Tel Aviv, Israel. Antimicrob Agents Chemother 49:1150–1156. https://doi.org/10.1128/AAC.49.3.1150-1156.2005

Shukla S, Haldorai Y, Bajpai VK, Rengaraj A, Hwang SK, Song X, Kim M, Huh YS, Han YK (2018) Electrochemical coupled immunosensing platform based on graphene oxide/gold nanocomposite for sensitive detection of *Cronobacter sakazakii* in powdered infant formula. Biosens Bioelectron 109:139–149. https://doi.org/10.1016/j.bios.2018.03.010

Singh NK, Bezdan D, Checinska Sielaff A, Wheeler K, Mason CE, Venkateswaran K (2018) Multi-drug resistant *Enterobacter bugandensis* species isolated from the international Space Station and comparative genomic analyses with human pathogenic strains. BMC Microbiol 18:175. https://doi.org/10.1186/s12866-018-1325-2

Thiolas A, Bollet C, La Scola B, Raoult D, Pagès JM (2005) Successive emergence of *Enterobacter aerogenes* strains resistant to imipenem and colistin in a patient. Antimicrob Agents Chemother 49:1354–1358. https://doi.org/10.1128/AAC.49.4.1354-1358.2005

Uzunović S, Ibrahimagić A, Bedenić B (2018) Antibiotic resistance in *Enterobacter cloacae* strains with derepressed/partly derepressed/inducible *AmpC* and extended-spectrum beta-lactamases in Zenica-Doboj Canton, Bosnia and Herzegovina. Med Glas 15:37–45. https://doi.org/10.17392/925-18

Vardi M, Kochavi T, Denekamp Y, Bitterman H (2012) Risk factors for urinary tract infection caused by enterobacteriaceae with extended-spectrum beta-lactamase resistance in patients admitted to internal medicine departments. Isr Med Assoc J 14:115–118

Völksch B, Thon S, Jacobsen ID, Gube M (2009) Polyphasic study of plant- and clinic-associated *Pantoea agglomerans* strains reveals indistinguishable virulence potential. Infect Genet Evol 9:1381–1391. https://doi.org/10.1016/j.meegid.2009.09.016

Wang YQ, Xiao D, Li J, Zhang HF, Fu BQ, Wang XL, Ai XM, Xiong YW, Zhang JZ, Ye CY (2018) Rapid identification and subtyping of *Enterobacter cloacae* clinical isolates using peptide mass fingerprinting. Biomed Environ Sci 31:48–56. https://doi.org/10.3967/bes2018.005

Weese JS (2008) Investigation of *Enterobacter cloacae* infections at a small animal veterinary teaching hospital. Vet Microbiol 130:426–428. https://doi.org/10.1016/j.vetmic.2008.02.009

Wisplinghoff H, Bischoff T, Tallent SM, Seifert H, Wenzel RP, Edmond MB (2004) Nosocomial bloodstream infections in US hospitals: analysis of 24,179 cases from a prospective nationwide surveillance study. Clin Infect Dis 39:309–317. https://doi.org/10.1086/421946

Xu J, He F (2019) Genomic analysis of two bacterial strains co-isolated from a urinary tract infection: NDM-1-producing *Enterobacter cloacae* accompanied by extended-spectrum β-lactamase-producing *Escherichia coli*. J Glob Antimicrob Resist 17:198–200. https://doi.org/10.1016/j.jgar.2019.04.007

Ye Y, Li JB, Ye DQ, Jiang ZJ (2006) *Enterobacter* bacteremia: clinical features, risk factors for multiresistance and mortality in a Chinese University Hospital. Infection 34:252–257. https://doi.org/10.1007/s15010-006-5038-3

Yoshino Y, Okugawa S, Kimura S, Makita E, Seo K, Koga I, Matsunaga N, Kitazawa T, Ota Y (2015) Infective endocarditis due to *Enterobacter cloacae* resistant to third- and fourth-generation cephalosporins. J Microbiol Immunol Infect 48:226–228. https://doi.org/10.1016/j.jmii.2012.07.015

Klebsiella pneumoniae Infections and Antimicrobial Drug Resistance

12

Ashwath Priyanka, Kotian Akshatha, Vijaya Kumar Deekshit, J. Prarthana, and Dharnappa Sannejal Akhila

Contents

A. Priyanka · K. Akshatha · V. K. Deekshit · D. S. Akhila (✉)
Nitte University Centre for Science Education and Research, NITTE (Deemed to be University), Mangaluru, Karnataka, India
e-mail: akhila@nitte.edu.in

J. Prarthana
SDM PG College, Ujire, Karnataka, India

B. Siddhardha et al. (eds.), *Model Organisms for Microbial Pathogenesis, Biofilm Formation and Antimicrobial Drug Discovery*,
https://doi.org/10.1007/978-981-15-1695-5_12

Abstract

Klebsiella species is ubiquitous in nature and is an important pathogen of humans and animals. *K. pneumoniae* is the most notorious opportunistic pathogen mainly affecting the hospitalized immunocompromised patients and accounts for urinary tract, respiratory tract, blood, and wound infection. They are the chief source of ventilator-associated pneumonia (VAP) and is accountable for 83% of hospital-acquired pneumonia (HAP). The antimicrobial resistance to human pathogens is being reported globally and has become a critical public health issue. In comparison to the susceptible pathogens, the rate of morbidity, and mortality is high with the infections caused by drug-resistant bacteria. Multidrug-resistant *K. pneumoniae* with diverse resistance determinants causing hospital and community acquired infections are of a major concern.

The mechanism of resistance known to date involves the production of beta-lactamases, such as extended-spectrum ß-lactamases, cephalosporinases, and carbapenemases. Constant horizontal transfer of antibiotic resistance genes via mobile elements essentially plasmids and transposons aid the production of extended spectrum beta-lactamases and other mechanisms of resistance that facilitate the survival of the *Klebsiella* in nosocomial environments. Currently, the emergence of *Klebsiella* strains that acquire plasmid-mediated resistance to ESBLs are of particular concern and are accountable for the failure of antibiotic treatment. The multidrug-resistant phenotype of the nosocomial *K. pneumoniae* caused by the presence of carbapenemases and extended-spectrum beta-lactamases make the therapeutic options limited.

The unending effort to develop new antibiotic has been outrun by the incidence of multidrug-resistant microbes and failed to replace the armamentarium required to combat the problem. Since discovering antibiotics takes a longer time, it is better to boost the activity of the existing antibiotic by inhibiting the mechanism that prevents the drug from acting will be an effective alternative. This chapter focuses on *K. pneumoniae* infections, pathogenicity, antibiotics, mode of action, antimicrobial resistance, therapies or alternative strategies for controlling drug resistance.

Keywords

K. pneumoniae · Multidrug resistance · Pathogenicity · Antimicrobial resistance · Nosocomial infections

12.1 Introduction

Klebsiella species belonging to Enterobacteriaceae are Gram-negative bacilli and are ubiquitous in nature. *Klebsiella* have two common habitats, one is the environment, where they are found in surface water, drinking water, sewage, industrial effluents, vegetation, soil and on plants, and the other being the mucosal surfaces of mammals such as humans, horses, or swine in which they colonize. The four-recognized species of the genus *Klebsiella* include *K. pneumoniae*, *K. oxytoca*, *K. terrigena,* and *K. planticola* (Donnenberg et al. 2005). Among these, *K. pneumoniae* and *K. oxytoca* are the most common human pathogen capable of causing neonatal infections of the bloodstream, urinary tract, central nervous system, lung, and soft tissue infections (Podschun and Ullmann 1998). *K. pneumoniae* was first isolated in the nineteenth century and were familiar as Friedlander's bacteria (Felson et al. 1949). *K. pneumoniae,* a nonmotile bacterium is a member of human gastrointestinal tract causes opportunistic infections such as urinary tract infections, wounds and respiratory infections, and causes ventilator-associated *pneumoniae* in hospitalized patients (Booker et al. 2008).

12.2 Classification of the Genus *Klebsiella*

As the taxonomy became increasingly refined due to the development of new methods such as numerical taxonomy, the species classification in this genus was continually revised. In the early 1980s, *Klebsiella* isolates from the environment, which was previously classified as "*Klebsiella*- like organisms" (groups J, K, L, and M), was later classified into provisional taxa. These groups led to the identification of four new species such as *K. terrigena* (Izard et al. 1981), *K. ornithinolytica* (Sakazaki et al. 1989), *K. planticola* (Bagley et al. 1981), and *K. trevisanii* (Ferragut et al. 1983). Since the DNA homology of the last two species was similar, they were combined as one and named *K. planticola* in the year 1986 (Gavini et al. 1986). Although *K. terrigena* and *K. planticola* are restricted to aquatic and soil environment, now are being reported in human clinical specimens. Since these isolates were obtained from polymicrobial specimens, it is not possible to distinguish them from other *Klebsiella* spp. In addition to *K. pneumoniae* and *K. oxytoca*, *K. granulata* is also now included in the list of human pathogens.

12.2.1 Differentiation of *Klebsiella* Species

Based on the biochemical reactions, *Klebsiella* species are identified and distinguished from other pathogens. It is a Gram-negative, nonmotile, encapsulated rod-shaped bacteria, produces lysine decarboxylase and shows a positive reaction for Voges-Proskauer test (Podschun and Ullmann 1992).

12.2.2 Typing of *Klebsiella* Isolates

It is essential in determining the clonality of the strains as it is mandatory in the management of endemic and epidemic nosocomial outbreaks of *Klebsiella* infections. Various methods adopted in the typing of *Klebsiella* are discussed below.

12.2.2.1 Biotyping

Biotyping is mainly based on the morphology of *Klebsiella* on a culture plate and the biochemical assay is performed in the laboratories as a gold standard method for the identification of *Klebsiella*. Biotyping can be performed by macro tube tests or by a combination of commercially available systems viz. API 20E with the additional macro tube methods. The biotyping of the *Klebsiella* is not much appreciated as an epidemiological tool due to the drawback of longer cultivation time of about 90 days for the confirmation of gelatinase production (Mitrophanov et al. 2008).

12.2.2.2 Serotyping

Serotyping is the most accepted technique for the typing of *Klebsiella* spp. and is based on the number of capsule antigens on the surface of the pathogen (Orskov and Orskov1984). The bacteria have a well-developed polysaccharide capsule and show a characteristic mucoid colony on the culture plate. Of the 82 capsule antigens described, 77 antigens are recognized and form the basis for the capsule antigen scheme. Regardless of the recognition of 12 different O-antigen types of *Klebsiella*, their classification is being held up by the heat-stable capsules (Ørskov and Fife-Asbury 1997). The method is very convenient as it aids in the differentiation of clinical isolates. The main drawback of the serotyping is the cross-reactions among the 77 capsule types. In addition, the procedure of typing is cumbersome due to the time-consuming longer procedures and difficulty in the interpretation of results due to the exhibition of weak reactions. The procedure is practiced in specialized laboratories since the anti-capsule antisera are not commercially available. However, neither the capsule typing, biochemical typing, phage typing, nor the biocin typing is singly sufficient to obtain results for epidemiological purpose. The combination of biotyping, as well as capsule typing, facilitates the differentiation of a large number of bio serotypes (Slopek et al. 1967).

12.2.2.3 Phage Typing

Discovered in the 1960s the phage typing is easy to read and is most accepted. The relative typing rate is as poor as 19–67% (Ślopek and Chapter 1978). Since it is just a method of typing and not an alternative to capsule typing, the procedure is not popular and may be used as an accessory method in addition to the serologic testing (Pieroni et al. 1994).

12.2.2.4 Bacteriocin Typing

Though capsule typing is suggested as a method for identifying *Klebsiella*, it is preferable to combine with a supplementary method independent of the capsule type for the epidemiological considerations. Several studies recommend typing of

Klebsiella via bacteriocins (Tomás et al. 1986; Bauernfeind 1984). Bacteriocins are bactericidal substances, made up of proteins with antimicrobial activity. Characterization of the isolates can be done by its competency in inhibiting specific indicator strains or by its susceptibility to bacteriocins. Instability of bacteriocin preparations, low type ability of strains and poor reproducibility are the major drawback of these methods (Bauernfeind et al. 1981). The limitations of these methods are summarized by the modification of the "scrape-and-point" procedure, thus avoiding the use of potentially unstable pre-produced and stored bacteriocins. The bacteriocins are synthesized on an agar medium prior before the strains to be typed and are inoculated by a multipoint inoculator. This procedure is proven effective as a bacteriocin typing for both clinical, environmental *Klebsiella* strain and also for nosocomial outbreaks of *Klebsiella* (Podschun and Ullmann 1993).

12.2.2.5 Molecular Typing Methods

Some of the most commonly used molecular typing methods for *Klebsiella* species are plasmid profiles (Combe et al. 1994; Nouvellon et al. 1994), ribotypes (Bingen et al. 1994), multilocus enzyme analysis, and pulsed-field gel electrophoresis (Poh et al. 1993; Gouby et al. 1992; Kitchel et al. 2009). The procedure followed varies from laboratory to laboratory and due to the lack of standardization it becomes laborious in the comparison.

12.2.2.6 Serotypes of *Klebsiella*

Klebsiella pneumoniae is recognized as an urgent threat to human health because of the multidrug-resistant strains and hypervirulent strains associated with hospital outbreaks. Virulence factors thought to be associated with invasive community-acquired infections include siderophores, specific polysaccharide capsule serotypes, and *rmpA* genes that are associated with hypermucoidy. Classic non-virulent *K. pneumoniae* (c-KP) strains are associated with pneumonia, urinary tract infection, and neonatal sepsis in immune compromised individuals. Being first recognized in Taiwan, the classic nonvirulent *K. pneumoniae* (c-KP) causes liver abscesses, meningitis, and endophthalmitis. K1 and K2 are the major capsular serotypes that cause liver abscesses. Genes like *mag*A and *K2*A are serotype specific genes for the K1 and K2 serotype, respectively.

The most important determinants of virulence in *K. pneumoniae* for liver abscesses are hyper mucoviscosity, rmpA (regulator of mucoid phenotype), aerobactin (an iron siderophore), kfu (an iron uptake system), alls (associated with allantoin metabolism), and K1/K2 capsules, respectively. The *rmpA* genes enhance the extracapsular polysaccharide (CPS) synthesis and thus confers highly mucoviscous phenotype of *K. pneumoniae* (Holt et al. 2015). It has been speculated that the structural genes for *Klebsiella* CPS synthesis are located near *his* on the chromosome (Liao et al. 2011).

K1 serotype is the major cause of primary liver abscesses and has greater potential for causing metastasis, whereas, K2 is a major cause of secondary liver abscesses (Yu et al. 2008). According to these capsular polysaccharides, *K. pneumoniae* can be classified into 77 serological K antigen types. High virulence of hv-KP is correlated with the enhanced capsule production often triggered by the regulator of the mucoid phenotype (*rmpA*) gene and mucoviscosity-associated gene A (*magA*) (Chung et al. 2007).

12.3 Transmission

The nosocomial pathogens are transmitted in the hospital either by direct or indirect contact. The direct contact requires physical contact between the infected individuals or the contaminated source and the susceptible host. Indirect contact is by the mechanical transfer of the pathogens between the hospitalized patients via hospital personnel or by contaminated medical equipment (Nazir and Kadri 2014).

12.3.1 Airborne Route

This route of transmission of the infectious agent is via dissemination of airborne droplet nuclei or the dust particles containing the infectious agents. The nosocomial pathogens carried in this fashion may be dispersed widely by the surrounding air and inhaled by the susceptible host in the same area or distant places depending on the environmental factors.

12.3.2 Droplet Route

The droplets produced as a result of sneezing, coughing, or even by talking can settle on the surrounding surfaces or the mucosal surfaces and can be transmitted to others (Beggs 2003).

12.3.3 Common Vehicle Transmission

This implies to the pathogens transmitted to the healthy host via contaminated sources such as food, water, medical equipments that include catheters, ventilators, and also hands of the hospitalized personnel (Reybrouck 1983).

12.3.4 The Four Most Common Nosocomial Infections Caused by *K. pneumoniae*

12.3.4.1 Urinary Tract Infections

UTI are the most common and frequent of all the nosocomial infections. The main reason being the indwelling urethral catheters. The effective, proven intervention in preventing nosocomial UTI is limiting the duration of catheter usage and maintaining aseptic insertion and maintenance of closed drainage. Usage of the catheters may be avoided unless and until there is a serious medical complication and are proven effective against nosocomial infections. Systemic antibiotic prophylaxis, bladder irrigation, the addition of antiseptics to the drainage bags, and antimicrobial coated catheters are proved to be ineffective (Warren 1997).

12.3.4.2 Surgical Wound Infections

There are several factors that influence frequency of surgical wound infection, to name a few techniques employed in the surgical course, the degree of endogenous contamination of the wound at the time of surgical procedure, duration of the surgical procedure, status of the underlying patient and the environment of the operative room (Owens and Stoessel 2008). Preventive measures in preventing surgical wound infections include optimal surgical technique, maintenance of a clean operative room environment with the restricted entry of staff members, sterile surgical equipment, preoperative preparation of the patients, and surgical wound surveillance program (Anderson et al. 2014).

12.3.4.3 Operative Room Environment

Airborne bacteria might be a cause of nosocomial respiratory tract diseases and hospital-acquired pneumonia (Johanson 1984). The infection can be minimized by maintaining clean operative and postoperative surroundings. Before any surgical interventions, all the surfaces must be disinfected thoroughly by using the recommended disinfectant. Unnecessary movement and conversations must be avoided within the operative and postoperative wards (Mangram et al. 1999).

12.3.4.4 Operative Room Attire

Usage of sterile gloves is recommended for all the operative staff. Hospital personnel entering the operative theater must and should wear the surgical attire restricted within the surgical ward. Coverage of the mouth and nose areas with the surgical mask within the operative ward must be mandatory. Therefore, maintenance of hygiene in the hospital and nursing settings are a prerequisite in the control and spread of any nosocomial infection.

12.4 Diagnosis

12.4.1 Conventional Methods

Usually, the identification of *K. pneumoniae* infection is confirmed by the culture of blood, sputum, urine, and aspirated body fluid, which includes pleural effusion, pericardial effusion, abscess material, and cerebrospinal fluid (Garner et al. 1988). In the identification of bacterial pneumonia, Gram staining may serve as presumptive identification, but the sensitivity is as less as 50% (Anitha 2012). Bacteria belonging to the genus *Klebsiella* appear as straight, short and Gram-negative encapsulated bacilli (Yu et al. 2008). Hence, capsule staining can be performed as one of the diagnostic methods in the confirmation of the species from the clinical specimens.

Apart from the conventional detection techniques, accurate, and rapid identification of the bacterium in the hospital settings is very crucial. Although real-time PCR is rapid, sensitive, and specific, the technique is not cost effective (Chen et al. 2011). Despite recent advances in the field of molecular biology and advancement in the availability of phenotypic identification kits, identifying the bacterial strains remains a difficult task in many routine microbiological laboratories (Vaneechoutte et al. 2009).

12.4.2 Molecular Methods of Diagnostics

12.4.2.1 Polymerase Chain Reaction

PCR detection of the genus-specific gene (*gyrA*) (Aly et al. 2014) and species detection by targeting the 16s rRNA for the identification of *K. pneumoniae* can be a basic rapid diagnostic tool in the molecular level. Other virulence genes of *Klebsiella* can also be used as an identification remark viz. *magA*, the mucoviscosity-associated gene; *kfu*, iron uptake system gene; *rmpA*, the extra polysaccharide synthesis regulator gene (Nassif et al. 1989); and *fimH*, fimbrial gene encoding type 1 fimbrial adhesion (Schembri et al. 2005).

PCR-ELISA can also be employed in the detection of *K. pneumoniae* clinical strains by using the 16S rDNA gene-based specific primers. PCR-ELISA is known to be accurate and a rapid method for the detection of infectious agents and has the advantage of being specific and a sensitive approach for the detection of *K. pneumoniae* strains (Mousavi et al. 2008).

12.4.3 Imaging Studies

12.4.3.1 Chest Radiography

Klebsiella usually affects the upper lobes, the involvement of the lower lobe is usually not common (Knight et al. 1975). The affected lobe is usually seen swollen producing a bulged fissure. But the clinical presentation is not exclusive for *Klebsiella* infections, *Haemophilus influenzae* also produces a similar radiographic appearance (Qureshi et al. 2014).

12.4.3.2 Chest Tomography

The patients under treatment for pneumonia, responding slowly are usually recommended for chest tomography. The findings help in excluding entities that are treatable with drainages such as empyema and respiratory tract obstruction caused by *K. rhinoscleromatis* (Qureshi et al. 2014).

12.4.3.3 DNA Microarray Technology

This method may be employed for the rapid detection of TEM, SHV, and CTX-M ESBLs, since the identification of the etiological agent is critical in the diagnosis of the disease (Fevre et al. 2011). Specific PCR assay was developed to discriminate the *K. pneumoniae* subspecies. This technique provides a platform for rapid and simple detection of rhinoscleroma.

12.4.3.4 Susceptibility Testing for ESBL-Producing Organisms

The increase in the incidence of ESBL-producing organisms has given rise for the effective screening methods for detection and has a sensitivity of as much as 98% for the detection of ESBL (Falagas and Karageorgopoulos 2009). The Vitek ESBL test is an automated broth microdilution method and has a sensitivity of 99.5% and a specificity of 100% (Spanu et al. 2006). It is the most reliable substitute.

It is the responsibility of all the individuals and the services providing health care in preventing the spread of nosocomial infections. Cooperation between the hospital personnel providing hospital care and hospital management is very necessary for reducing the risk of infections for the patients and the staff (Hawley 1985). Methods used for the diagnosis of *K. pneumoniae* are summarized in Table 12.1.

12.5 Virulence Factors Present in *Klebsiella pneumoniae*

K. pneumoniae harbors different virulence factors in order to grow and overcome the immune response by the host. *K. pneumoniae* uses pathogenic factors like capsule polysaccharide, lipopolysaccharide, fimbriae, outer membrane proteins, adhesins, and siderophores, for the survival and immune evasion during infection.

12.5.1 Capsule

The capsule is dense, approximately 160 nm in thickness consisting of a highly structured layer of surface-associated acidic polysaccharides, mainly composed of repeating three to six units of sugars. The composition is mainly dependent on the strain and is considered a dominant virulence factor (Shankar-Sinha et al.

Table 12.1 Methods used for diagnosis of *K. pneumoniae*

Diagnostic tools		
Conventional methods	Molecular methods	Imaging studies
Gram's staining	Polymerase chain reaction	Chest tomography
Capsule staining	PCR-ELISA	Chest radiography
Indole test	DNA microarray technology	Cystography
Citrate utilization test	LAMP assay	Computed tomography
Urease test	Real-time PCR assay	Magnetic resonance and imaging

2004; Li et al. 2014; Doorduijn et al. 2016). Capsule in the bacteria is known to show two pathogenic mechanisms: (1) protecting the bacteria from phagocytosis and (2) directly modifying the immune response. Production of the capsule in *K. pneumoniae* is important to cause infections in the host. The production of a capsule comprising of acidic polysaccharides by the Wzy-dependent polymerization pathway is the important characteristic of the genus *Klebsiella*. The complete ORF of *K. pneumoniae* capsule (*cps*) harbors around 16–25 genes with clusters ranging from 21 to 30 kb. Around 77 different types of capsule (K) antigen have been identified in *Klebsiella,* but only a few types have been systematically studied. Strains expressing K1 and K2 are highly virulent and the degree of virulence depends on the mannose content of capsular polysaccharides. K2 serotype is the prime serotype associated with UTI, pneumonia, or bacteremia and is hardly encountered in the environment sources. Presence of capsule protects the bacteria from opsonization and phagocytosis by macrophages during internalization into the host cell (Cortes et al. 2002). In addition, excess production of capsule helps them escape the neutrophil-mediated intracellular killing and leads to the development of abscess at different sites, such as the liver (Shankar-Sinha et al. 2004).

12.5.2 Lipopolysaccharides

Lipopolysaccharides (LPS) are considered a major and essential component of the cell membrane of every Gram-negative bacterium. Lipopolysaccharide is also an important pathogenic determinant in *K. pneumoniae* causing pneumonia, UTI, and bacteremia (Clements et al. 2007; Lugo et al. 2007; Llobet et al. 2011). Lipid A, core oligosaccharide (OS), and the O antigenic polysaccharide (O-PS) are three structural domains of LPS. O-antigen is the outermost component of LPS whereas core oligosaccharides anchor lipid A and O antigen. Capsular polysaccharides and the O-antigen portion of the lipopolysaccharides are the first molecules to come across the host immune system (Tomás et al. 1986).

Nine antigenic groups such as O1, O2, O2ac, O3, O4, O5, O7, O8, and O12 have been identified in *K. pneumoniae*. O1 is the common familiar among the invasive strains than noninvasive clinical strains (Dorman et al. 2018). *K. pneumoniae* O-antigen blocks the availability of complement components to activators and protects the bacteria against the complement-mediated killing process. Core polysaccharide of types 1 and 2 is identified in *Klebsiella* species. Lipid A and core polysaccharide protects mouse alveolar macrophages and contributes to resistance. Modification in the gene of lipid A and core polysaccharides leads to the attenuation of virulence of *Klebsiella* species and is proved in an animal model (Struve et al. 2009). LPS is known to have a dual effect on *K. pneumoniae,* infecting the host is beneficial and harmful because LPS also act as a strong immune activator.

12.5.3 Pili

Pili also familiar as fimbriae is filamentous projections on the bacterial surface. *K. pneumoniae* generally contains four types of fimbriae, namely type 1, type 3 fimbriae, Kpc fimbriae, and KPF-28 adhesin (Ong et al. 2008; Rosen et al. 2008). Types 1 and 3 pili are the predominant fimbrial adhesins present in *Klebsiella* species. Type 1 fimbriae are common in all the members of *Enterobacterial* species, but type 3 is specific to *Klebsiella* species though few studies have reported the expression of type 3 adhesins in *E. coli.* The genes required for structure and assembly of fimbriae are encoded on a gene cluster (*fim*). Types 1 and 3 fimbriae aids in the adherence to epithelial cells of the urogenital tract as well as intracellular biofilm formation within bladder umbrella cells in urinary tract infection and on abiotic surfaces. It has been assessed that most of the nosocomial infections (80%) are associated with an indwelling medical device. These devices offer a site for biofilm formation and the mechanical insertion of these devices further causes host cellular damage which provides attachment sites for the bacteria. Several studies have shown that types 1 and 3 fimbriae of *Klebsiella* are significant colonization factors required for biofilm formation (Miethke and Marahiel 2007).

12.5.4 Siderophores

Iron is an essential component in the growth of *K. pneumoniae*, required during the infection and is acquired from the environment (Raymond et al. 2003). As this metal is not available readily during the infection, the majority of the bacteria induce high-affinity iron-transport systems so that they can prevail over the low availability of the element. To acquire iron from the host during infection, these bacteria employ tactics by the secretion of siderophores, which have a higher affinity toward iron than the host transport proteins, which can steal iron from host iron-chelating proteins or obtain it from the environment (Bachman et al. 2009).

Most of the Gram-negative bacteria secrete small iron-chelating molecules called siderophores for their growth, virulence, and replication. The ability of the bacteria to produce siderophore depends on the iron and carbon contents in the culture medium. Different siderophores are expressed in *K. pneumoniae* viz. enterobactin, salmochelin, yersiniabactin, and aerobactin and have different roles in the infection process. Enterobactin produced by the members of Enterobacteriaceae has a greater affinity toward iron than host molecules (transferrin and lactoferrin). To overcome the effects, the host produces antimicrobial protein like lipocalin-2, which helps them in preventing the uptake of iron by the bacteria. To combat the effect of host antimicrobial protein, bacteria have acquired additional mechanism like stealth siderophores (salmochelin and yersiniabactin) which helps them to evade lipocalin-2-mediated iron starvation and proliferate within the host cell during overexpression of lipocalin-2 (Raffatellu et al. 2009). Existence of diversity in the siderophores impacts the replicative niche and pathogenesis of *K. pneumoniae* in the host (Bachman et al. 2011). Siderophores also modulate the host responses by activating the transcriptional factor HIF-1α in the Peyer's patches, epithelial and endothelial cells of human (Holden et al. 2016). It also promotes the dissemination of *K. pneumoniae* to the spleen by inducing stabilization of HIF-1α in lung epithelial cells. Hence, inactivation of siderophores may be helpful to the host not only by inhibiting pathogens from acquiring protein-bound iron but also by preventing pathogen dissemination through modulating the immune system.

12.5.5 Outer Membrane Proteins

Outer Membrane Proteins (OMPs) are considered as an important factor for the virulence in *K. pneumoniae,* which includes outer membrane protein A (OmpA), peptidoglycan-associated lipoprotein (Pal), and murein lipoprotein (LppA). OmpA protects the innate immune response of the host. These proteins contribute to the selective impermeability of the cell membrane in an LPS- and capsule-independent manner, integrity and also protect *K. pneumoniae* against certain antibiotics and anionic detergents (Sugawara et al. 2016). Outer membrane porins that exist as trimers act as water-filled protein channels allowing transportation of small hydrophilic molecules like iron, antibiotics and nutrients, and are also important in both virulence as well as antibiotic resistance in the bacteria. OmpK35 and OmpK36 are the two classical trimeric porins, which are produced by *K. pneumoniae.* Porin channels act as the route of penetration for the antimicrobial drugs, which should first penetrate the outer membrane to reach the periplasm (Tsai et al. 2011). Porins even serve as receptors for phages and bacteriocins also in combination with lipopolysaccharide and peptidoglycan, their structural role is to maintain the integrity of the cells.

12.5.6 Efflux System

Efflux system is proved to be vital mechanism of antibiotic resistance and found exclusively in Gram-negative bacteria. These systems allow the microorganisms to balance their internal environment by removing toxic substances, such as antimicrobial agents, metabolites and quorum-sensing-regulated expression of virulence determinant. Efflux pumps are grouped into resistance-nodulation-division (RND) family, the small multidrug resistance (SMR) family, the multi-antimicrobial extrusion (MATE) family, the major facilitator superfamily (MFS), and the ATP-binding cassette (ABC) superfamilies (Eswaran et al. 2004). To date, AcrAB, KexD, and OqxAB efflux systems have been identified to be involved in antibiotic resistance in *K. pneumoniae*. AcrAB efflux system encoded by *acrRAB* operon while *acrR* encodes the AcrAB repressor, where, *acrA* and *acrB* encode a periplasmic lipoprotein is attached to the inner membrane which bridges the outer, inner membranes, and an integral membrane protein situated in the cytoplasmic membrane. AcrB connects with outer membrane protein TolC, found in all Gram-negative bacteria is vital for the expulsion of dyes, detergents, and antimicrobial agents (Buckley et al.

2006). Several studies have proved that bacteria with deleterious efflux systems lost their pathogenicity in animal model experiments. The KpnEF efflux pump of *K. pneumoniae* mediates resistance to several dyes, detergents, and antimicrobial compounds such as SDS, deoxycholate, EtBr, cefepime, benzalkonium chloride ceftriaxone, colistin, erythromycin, rifampin, tetracycline, streptomycin, acriflavine, chlorhexidine, and triclosan. Thus, studies have shown that KpnEF efflux pump plays a key role in providing resistance toward broad-spectrum antimicrobial compounds (Bunikis et al. 2008).

12.6 Antimicrobial Resistance in *K. pneumoniae*

The discovery of antibiotics is an important milestone in the history of therapeutics, which led to effective control of infectious agents. Antibiotics are increasingly being thought the wonder drugs because of their selective inhibitory nature at minimum concentration. However, indiscriminate use of antibiotics in medical, veterinary, and agriculture sectors, has led to the emergence of antimicrobial-resistant strains. *Klebsiella* being an important human pathogen is found to be resistant to most of the present generation drugs and has been classified as one among the list of ESKAPE organisms (Boucher et al. 2009). The worldwide contribution of drug-resistant strains of *K. pnuemoniae* to the burden of antimicrobial resistance is increasing throughout the year. Multifactorial dissemination processes through mobile genetic elements play a very prominent role in the spread of multidrug-resistant *K. pnuemoniae* (Navon-Venezia et al. 2017).

12.6.1 Emergence of Antimicrobial Resistance in *K. pneumoniae* to Different Classes of Antibiotics

12.6.1.1 Cell Wall Synthesis Inhibitors

One of the most selective and potent classes of all the antibiotics is the cell wall synthesis inhibitors. Bacterial cell wall is synthesized in three different stages viz. synthesis of precursor molecules in the cytoplasm, transfer of precursor molecules through a lipid carrier and transpeptidation and carboxy peptidation reactions. All three stages can be inhibited by different groups of antimicrobial agents namely fosfomycin, bacitracin, and vancomycin or β–lactams, respectively. *K. pneumoniae* is known to display resistance toward most of the *β*-lactam drugs by producing *β*-lactamase enzymes such as cephalosporinases, extended-spectrum *β*-lactamases and carbapenemases. The existence of β-lactamase was first reported in 1940, which was before the commercial use of antibiotic penicillin (Abraham and Chain 1988). This could be a reason for the intrinsic resistance to penicillin in many of the environmental isolates. The chromosomally mediated *β*-lactamase such as penicillinase was first discovered in *K. pneumoniae* as a part of intrinsic resistance (Labia et al. 1979). The enzymes were later identified as class A group 2b *β*-lactamases (Petit et al. 1992). In addition to natural resistance, the massive use of *β*-lactam antibiotics against different groups of human bacterial pathogens, including *K. pneumoniae* initiated the emergence and spread of *β*-lactamase enzyme (Bush 2010). The first *β*-lactamase enzyme described was Temoneria (TEM)-1 followed by a sulfhydryl variable (SHV)-1 conferring resistance to penicillin but not to cephalosporins (Hæggman 2010). The derivatives of these enzymes were later classified as ESBL (extended-spectrum *β*-lactamases) shown to produce activity against oxyimino-*β*-lactam antibiotics through the modifications in the active site (Rawat and Nair 2010). Soon after the introduction of third-generation cephalosporins in 1982, resistance to oxyimino *β*-lactam was evident in *K. pneumoniae* and *Serratia marcescens* (Knothe et al. 1983). Since the first recorded outbreak of ESBL-producing strains in French hospitals, there was a rise in the outbreaks of *K. pneumoniae* and *E. coli* (Lewis et al. 2007). Later on, a shift has occurred from the initial predominance of TEM and SHV class of *β*-lactamases to the emergence of Cefotaximase-Munich (CTX-M)-type, which is the most commonly detected ESBL (van der Bij and Pitout 2012). To treat the infections caused by these ESBL producers, β-lactams of carbapenem group was more commonly used (Rawat and Nair 2010). This further caused the emergence of carbapenem resistance in Enterobacteriaceae worldwide (Tofteland et al. 2013; Storberg 2014). The widely distributed strain *K. pneumoniae* sequence type 258 harboring KPC (*K. pneumoniae* carbapenemase) belonging to the class A type *β*-lactamase caused national and international epidemics (Coetzee and Brink 2012). First reported in the late 1990s, to date more than ten different KPC variants have been identified (Walther-Rasmussen and Høiby 2007). These enzymes mainly provide resistance to cephalosporins, cephamycins, monobactams, carbapenems, and are weakly inhibited by clavulanic acid and tazobactam (Pitout 2012). KPC-producing *K. pneumoniae* isolates are the most important cause of nosocomial infections and are also endemic in certain parts of world such as Greece, northeastern USA, Colombia, Puerto Rico, Israel and China (Nordmann et al. 2009). Most of the resistance determinants of KPC were carried on transferable plasmids of variable size such as $bla_{\text{KPC-2}}$ on 100 Kb plasmid and $bla_{\text{KPC-3}}$ on 120 Kb plasmid (Samuelsen et al. 2009) associated with transposable element *Tn4401* which further supports the mobilization and spread of KPC among different human pathogens. The carbapenem-resistant strains also initiated the spread of NDM (New Delhi Metallo*β*-lactamase) belonging to class B β-lactamases and OXA-48 type belonging to class D *β*-lactamases showing resistance to virtually all class of β-lactams including present generation carbapenems (Hirsch and Tam 2010; NICD, 2013). In addition, *K. pneumoniae* occasionally harbors plasmid-mediated AmpC- *β*-lactamases. These enzymes are first reported in 1980s and are derived from the chromosomally encoded *AmpC* cephalosporinases, which are difficult to get inhibited by the first-generation β-lactamase inhibitors like clavulanic acid, sulbactam, and tazobactam (Philippon et al. 2002). These resistance determinants along with other *β*-lactamases bla_{NDM-1}, bla_{CMY-16}, bla_{OXA-48}, and $bla_{CTX-M-15}$ present on different plasmids in a single strain of *K. pneumoniae* pose a great challenge to the clinicians to treat the infection with the available antibiotics (Nordmann and Poirel 2014).

12.6.1.2 Protein Synthesis Inhibitors

Antimicrobial agents that inhibit several stages of protein synthesis are known to have a profound effect on cellular processes of bacteria. Tetracyclines, phenicols, aminoglycosides, MLSKs (macrolides, lincosamides, streptogramins, and ketolides), ansamycins, and oxazolidinones are the widely used classes of antibiotics under this category. Most of the antibiotics in this category work by inhibiting either transcription or translation initiation process during protein synthesis. Many affect either 50s or 30s ribosomal subunits during assembly and translation process. Aminoglycosides such as amikacin, tobramycin, and gentamicin in combination with β-lactams are the most frequently used antimicrobial agents against Gram-negative pathogens including *K. pneumoniae* until they were replaced by third-generation cephalosporins, fluoroquinolones, and carbapenems (Krause et al. 2016). Aminoglycosides irreversibly bind to the 30s-ribosomal subunit causing complete inhibition of protein synthesis leading to bacterial cell death (Shakil et al. 2008). Over clinical use of these antimicrobial agents has resulted in the emergence of resistance among *K. pneumoniae.* Bacteria can become resistant to aminoglycoside through a variety of ways including alterations of the ribosomal binding sites, drug inactivation by aminoglycoside modifying enzymes, reduced uptake of aminoglycoside due to the downregulation of porin proteins and overexpression of efflux pumps (Nasiri et al. 2018). Among these enzymatic modifications is the most commonly encountered mechanism. The aminoglycoside-modifying enzymes are classified into three major classes such as aminoglycoside acetyl transferase (*aac*), aminoglycoside nucleotidyl transferase (*ant*) and aminoglycoside phosphoryl transferase (*aph*) encoded either by chromosomal or plasmid-mediated resistance determinants (Kim et al. 2008). Of these, aminoglycoside acetyltransferase enzyme catalyzes the transfer of an acetyl group of acetyl coenzyme A to an amine group of aminoglycoside thereby making it inactive. While other two modifying enzymes (*aph* and *ant*) catalyze the transfer of γ-phosphate and nucleotide monophosphate to hydroxyl portions of aminoglycosides (Ramirez and Tolmasky 2010). Among these, aac(3)-II and aac(6′)-Ib group of modifying enzymes are most commonly reported in *K. pneumoniae* (Liang et al. 2015).

Decrease in the usage of aminoglycosides slowed down the emergence of new aminoglycoside resistance mechanisms until the discovery of plasmid-mediated enzyme 16S rRNA methyltransferase (16S RMTase) and is gaining importance due to its high prevalence in recent times (Liang et al. 2015). This enzyme methylates the binding site of the drug, thereby making the drug inactive, causing high-level resistance to aminoglycosides (Doi et al. 2016). Currently, eight 16s RMTase genes namely *armA*, *npmA*, *rmtA*, *rmtB*, *rmtC*, *rmtD*, *rmtE,* and *rmtF* have been identified among Gram-negatives with high prevalence of *rmtB* and *armA* (Xia et al. 2016). *armA* was the first 16sRMTase identified in *K. pneumoniae* in France in 2003 and is being increasingly reported worldwide along with β-lactam resistance among *Klebsiella* species and in other Gram-negative human pathogens (Nasiri et al. 2018; Costello et al. 2019). Recent studies have shown that the genes *rmtB* and $bla_{\text{CTX-M-55}}$ were carried on transferable plasmids harboring *IS26* and *Tn3* transposons in *K. pneumoniae* ST37 isolated from dogs and cats (Xia et al. 2017). However, further studies are needed to understand the transfer of these plasmid-mediated aminoglycoside genes from animals to humans. Chromosomal aminoglycoside resistance mechanisms include alterations in the cell membrane permeability due to the modifications in AcrAB-TolC and KpnEF efflux pumps or due to the loss of porin protein KpnO. However, different membrane apparatus exhibits variable affinities to different aminoglycosides. For instance, disruptions in AcrAB-TolC apparatus increased the sensitivity of *K. pnuemoniae* to tobramycin and spectinomycin (Padilla et al. 2010), while, mutant of KpnFF (Δ*kpnEF*) shows strong change in resistances to tobramycin and spectinomycin but exhibit low resistances to gentamycin and streptomycin (Srinivasan and Rajamohan 2013). Besides, it was also observed that the loss of porin protein caused resistances to tobramycin, spectinomycin and streptomycin (Srinivasan et al. 2012).

Although macrolides, lincosamides, streptogramins, ketolides and ansamycins are not considered for treating clinically relevant Gram-negative bacterial infections, a reasonable number of Gram-negative pathogens act as a reservoir of MLS resistance determinants. This can be successfully transferred to other pathogens of clinical importance through mobile genetic elements (Nguyen et al. 2009).

12.6.1.3 Cell Membrane Synthesis Inhibitors

The second most potent group of antimicrobial agents are cell membrane inhibitors. Although, not as selective as *β*-lactams, cell membrane inhibitors can work at stages where all other antimicrobial agents fail to control the infections caused by MDR strains. Polymyxins are the only drugs that are approved for human therapy under this category. Polymyxins are positively charged cations that are attracted toward the negatively charged LPS of the bacteria, modifies its structure and make the membrane more permeable. This further disrupts the osmotic balance through the displacement of cations (Ca^{2+}/Mg^{2+}) causing leakage of cellular molecules leading to cell death (Falagas and Kasiakou 2005). Colistin (polymyxin E) among polymyxins is considered as the drug of last reserve for treating Gram-negative MDR infections (Olaitan et al. 2014). The extensive use of colistin sulfate in veterinary medicine has led to the emergence of colistin-resistant strains. The first clinical case of colistin-resistant *K. pneumoniae* was reported in the late 1960s during the initial period of its use (Davis et al. 1969). However, during 1980s to 2000 use of polymyxin was restricted in human medicine due to its associated toxicity. Later in the early 2000s, with the increasing prevalence of XDR (extensively drug resistance) carbapenemase-producing *K. pneumoniae* (CPKP), the treatment mainly relied on polymyxins as the drug of last resort (Antoniadou et al. 2007). Soon after its comeback in clinical use, the first nosocomial outbreak of colistin-resistant MDR *K. pneumoniae* was reported from Greece in the year 2004 (Antoniadou et al. 2007) and since then there is an increasing trend of colistin-resistant clinical isolates (Marchaim et al. 2011). The main mechanism of polymyxin resistance in *K. pnuemoniae* is chromosomal-mediated target site modifications. This is named as LPS modification system. Thus, the altered the LPS structure in resistant strain decreases the anionic charge interfering with the binding of polymyxins (Navon-Venezia et al. 2017). The alteration in LPS is due to the mutations in certain core genes of lipid A such as *lpxM* and its regulator *ramA* (De Majumdar et al. 2015). Additional mechanisms like neutralization of lipid A, binding of amino arabinose (*pbgP* and *pmrE*), phosphoethanolamine (*pmrC*) to lipid A portion of cell membrane has also been reported (Llobet et al. 2011). Similarly, the enhanced activity of LPS-modifying gene regulators like *phoPQ*, *pmrA*, and *pmrD* could be another reason for resistance in *K. pneumoniae* (Navon-Venezia et al. 2017). Certain pathways such as CrrAB regulatory and TupA-like/glycosyltransferase system also involved in LPS modification leading to colistin resistance in *K. pneumoniae* (Wright et al. 2015).

The plasmid-mediated mcr (mobilized colistin resistance) genes can also lead to target site modification. *mcr* codes for phosphatidylethanolamine transferase, which is an enzyme that transfers phosphatidylethanolamine residue to the lipid A portion of Gram-negative bacteria, thereby modify its structure. This further results in reduced affinity of colistin and related polymyxins to lipid A portion of the cell membrane. The first acquired colistin resistance gene *mcr-1* has been detected in *E. coli* and *K. pneumoniae* in China and was present on an insertional element *Incl2* (Liu et al. 2016). Recently, three multidrug-resistant strains of *K. pneumoniae* isolated from chickens harboring *mcr 7.1* and $bla_{CTX\text{-}M\text{-}55}$ gene on an Incl2 conjugative plasmid was characterized (Yang et al. 2018). Among the three isolates, one isolate was found to carry other antibiotic resistance determinants such as oqxAB for quinolone resistance, *fosA* for fosfomycin resistance and *aph(3′)-Ia* for aminoglycoside resistance. Although mcr genes are of recent origin, the rapid spread of different classes of this gene could pose a serious threat if associated with pan-drug-resistant (PDR) strains (Karaiskos et al. 2017).

12.6.1.4 Nucleic Acid Synthesis Inhibitors

Most of the bacteriostatic agents belong to the category of nucleic acid synthesis inhibitors. One of the earliest nucleic acid synthesis inhibitor is a quinolone antibiotic, nalidixic acid, widely used for UTI infections and nosocomial infections. Later its potency was increased by the addition of fluorine group in its chemical structure. Nearly all modern quinolone antibiotics are fluoroquinolones with ciprofloxacin being most widely used. Fluoroquinolones mainly work by inhibiting the normal functioning of topoisomerase II (*gyrA* and *gyrB*) and topoisomerase IV enzymes (*parC* and *parE*) which are involved in DNA replication. Fluoroquinolones bind to the DNA–enzyme complex during replication and prevent the negative supercoiling of replication fork. However, other mechanisms of resistance such as presence of plasmid-mediated quinolone resistance (PMQR) genes and overexpression of efflux pumps have also been reported in certain Gram-negative pathogens. Quinolones have been in use since 1960s, but their clinical use increased rapidly after the introduction of fluoroquinolones in 1980s. The extensive use of fluoroquinolones led to the development of resistance in bacterial pathogens (Naeem et al. 2016). In *K. pneumoniae*, all the known mechanisms of quinolone/fluoroquinolone resistance have been reported including QRDR (quinolone resistance determining regions) mutations, PMQR, or MDR efflux pumps (Redgrave et al. 2014). The first chromosomal resistance mechanism by *K. pneumoniae* was observed against the first quinolone antibiotic, nalidixic acid and the first fluoroquinolone, norfloxacin (Davis et al. 1969). Like in other members of Enterobacteriaceae, the first QRDR mutation was seen in *gyrA* and *parC* subunits (Deguchi et al. 1997) followed by *gyrB* (Nam et al. 2013) and *parC* (Guillard et al. 2015) in MDR strains of *K. pneumoniae*. Another important resistance mechanism is the presence of PMQR genes. These genes are composed of Qnr pentapeptide proteins that protect topoisomerases II and IV from the action of quinolone/fluoroquinolone antibiotics. The first plasmid-mediated *qnr* was discovered in *K. pneumoniae* isolated from a clinical sample in the USA (Martínez-Martínez et al. 1998). Since then multiple genes (*qnrA1, qnrB1, qnrB4, qnrS1, oqxAB,* and *aac(6′)-1b-cr*) associated with PMQR have been discovered in MDR strains of *K. pneumoniae* (Yang et al. 2014). *aac(6′)-1b-cr* confer resistance to both aminoglycosides and fluoroquinolones and *qepA* encodes an efflux protein that recently identified in *K. pneumoniae* (Heidary et al. 2016). PMQR genes alone known to produce low or moderate level of resistance to fluoroquinolones (Fabrega et al. 2009). However, the combination of quinolone/fluoroquinolone-resistant mechanisms such as QRDR mutation and presence of PMQR determinants play a very important role in high-level fluoroquinolone resistance in *K. pneumoniae* (Azargun et al. 2019).

12.6.1.5 Antimetabolites

Antimetabolites are the chemicals which inhibits metabolites. They are also referred to as antifolates that inhibit folic acid synthesis in bacteria which otherwise required for the synthesis of adenine. The two important antimicrobial agents trimethoprim and sulfonamide belong to this category. Trimethoprim binds to dihydrofolate reductase enzyme and inhibits the reduction of dihydrofolic acid to tetrahydrofolic acid in the folate pathway. Similarly, sulfonamide inhibits dihydropteroate synthase thereby preventing the conversion of para aminobenzoic acid to dihydrofolic acid. Sulfonamide was first put into clinical use in the year 1935 (Sköld 2000). Since then, it has been extensively used although it resulted in serious side effects. Later, a new compound referred to as trimethoprim first used in the year 1962 in England for the treatment of urinary tract infection in combination with sulfonamide due to its synergistic effect in vitro (Bushby 1980). However, overuse of these drugs has resulted in the development of resistance among Enterobacteriaceae. The usual trimethoprim-resistance genes encountered in *K. pneumoniae* are *dfrA14*, *dfrA1dfrA5*, *dfrA8*, *dfrA12*, *dfrA13/21/22/23* family, *dfrA15*, *dfrA16*, and *dfrA17*. Similarly, *sul2* was found to be more common in *K. pneumoniae* followed by *sul2* and *sul3* conferring resistance to sulfonamide (Taitt et al. 2017). Resistance to trimethoprim and sulfonamide (co-trimoxazole) in *K. pneumoniae* is mostly associated with the plasmid IncA/C2 group that carries multiple resistance determinants conferring resistance to *β*-lactams (*bla*$_{NDM-1}$), chloramphenicols (*catA1*), aminoglycosides (*armA* or *rmtB* 16S RNA methylases) along with *AmpC* β-lactamase CMY-2 (Carattoli 2013). In general, most of the ESBL-producing *K. pneumoniae* usually show high-level resistance to co-trimoxazole (Somily et al. 2014; Stanley et al. 2018).

As a known cause of nosocomial infection, *K. pneumoniae* plays a prominent role in spreading the burden of antimicrobial resistance worldwide. The most recent

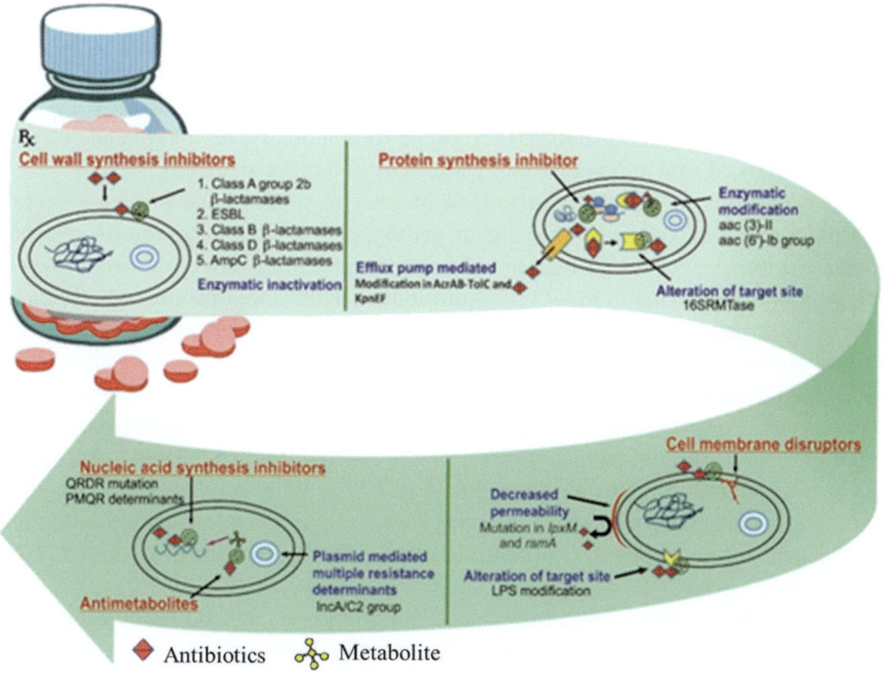

Fig. 12.1 Representative image of antimicrobial resistance mechanisms predominantly observed in *Klebsiella* toward different classes of antibiotics

emergence of XDR (resistant all drugs except cefepime, tigecycline, and ceftazidime-avibactam) and PDR (resistant to all drugs) strains of *K. pneumoniae* (Bi et al. 2017; Li et al. 2018; Krapp et al. 2018) would be a wakeup call for the world to contemplate on more strategic measures to control the spread of drug-resistant human pathogens like *K. pneumoniae.* Figure 12.1 represents an image of antimicrobial-resistant mechanisms predominantly observed in *Klebsiella* toward different classes of antibiotics.

Ideal therapeutic options for multidrug-resistant *K. pneumoniae* infections are not well established (Qureshi et al. 2014). Routine monitoring and prevention are a prerequisite in controlling any infection and their outbreaks.

12.7 Methods to Combat Antibiotic Resistance

12.7.1 Phage Therapy

Infections due to multidrug-resistant bacteria are increasing globally and is a critical issue. Bacteriophages are considered as an effective alternative for the treatment of bacterial infections. Phages are species and strain specific, some are polyvalent (Chibani-Chennoufi et al. 2004). The decline in the effectiveness of antibiotics has generated a need for a substitute therapy and hence, phage therapy can be one of the alternatives to combat bacterial infections. Frederick d' Herelle coined the term 'bacteriophages', convinced the use of phages as a therapeutic option and demonstrated the first clinical in the year 1919 in Paris at a hospital used to treat cases of pediatric dysentery (Chanishvili 2012). In the food industry, several phage preparations are marked safe and approved by the FDA (Monk et al. 2010).

Investigations of phage treatment on an animal model have proven effective against a range of clinically significant pathogens viz to treat antibiotic-resistant *Pseudomonas aeruginosa* infections of the skin, gastrointestinal tract, and lungs in mice model (Watanabe et al. 2007). Additional results show promising effects on *Vibrio parahaemolyticus, Staphylococcus aureus, Escherichia coli*, and *Acinitobacter baumani*. Human trails also have been proved effective against common pathogens such as *S. aureus, P. aeruginosa, E. coli, Proteus, Enterococcus*, and *Salmonella* species (Kutateladze and Adamia 2008). They also have concluded the therapeutics did not have any effect on the normal flora of the mice, thus proving it as a better alternative (Bogovazova et al. 1991). It has proven effective against pneumonia caused by multidrug-resistant *K. pneumoniae* experimented on animal model and thus considered a potent alternative to combat drug resistance (Fang et al. 2004).

Adverse reactions to antibiotics include anaphylaxis, cardiotoxicity, hepatoxicity, neurotoxicity, and gastrointestinal complications (Granowitz and Brown 2008). In contrast to this, phage therapy is considered safe since the translocation is across the epithelium and subsequently circulate within the blood (Górski et al. 2006). Phages are armed with enzymes on the exterior of the capsid that aid degradation of the extracellular polymeric substances (EPS) and disperse bacterial biofilms and thus allowing the phage to access the bacteria embedded within the EPS matrix (Abedon 2015). In contrast to the antibiotics, phages are more specific toward species and strain. Other complications of antibiotics include increased risk of asthma, diabetes, and obesity. The damage is less in case of phages and still is believed to reduce gut carriage of pathogens such as uropathogenic *E. coli* and *Shigella* (Mai et al. 2015). The available literature on the use of phage as an alternative therapy to combat bacterial infections, especially with respect to multidrug-resistant pathogens is a promising note. The combination of both phage therapy and antibacterial agents is significant in addressing the issue of antibiotic-resistant infections.

12.7.2 Gene Silencing/Knockout

Earlier days antibiotics were referred to as a magic bullet to combat-associated bacterial infection. Unfortunately, bacteria have devised a plethora of mechanisms that cause resistance to several antibiotics. Due to the pacing advent of new resistance mechanisms, there is a decline in the effectiveness of conventional antibiotic therapy, higher expenditures for health care, and immense risk of death. Modification in various drug enablers of bacteria viz increasing bacterial efflux pump, inactivation/

modification of drug, alteration of drug target site collectively contributes to the reduction in antibiotic potency (Tenover 2006). The unending effort to develop new antibiotic has been outrun by the incidence of multidrug-resistant microbes and also failed to replace the armamentarium required to combat this problem. Targeting drug resistance mechanism is the best option to tackle multidrug-resistant strain rather than a synthesis of a new antibiotic. Gene manipulation and gene editing is the best tool to modulate the antibiotic resistance but it failed to be used as therapeutics due to the ethical problem. Treatments of infectious disease by traditional antibiotic therapy have lagged behind the plethora of multidrug-resistant bacteria. Microbes acquire resistance to conventional antibiotic therapy by various drug resistance mechanisms. Efflux pump is considered to be the preferred route to expel diverse class of structurally unrelated drug and also prevent the emergence of drug-resistant mutant (Pérez et al. 2012; Ayhan et al. 2016). Reports have also shown that *Salmonella* Typhimurium strain lacking genes coding for efflux pump were totally avirulent in a mice infection model (Blair et al. 2015). Hence, it is better to tackle the drug-resistant mechanism in bacteria than scavenging for new antibiotics. RNA-mediated interference (RNAi) is an evolutionarily conserved natural phenomenon formerly discovered as an antiviral mechanism in plants and other organisms for the specific silencing of gene expression. Limited studies have used RNAi-based inhibition molecule to induce antibiotic sensitivity in drug-resistant bacteria (Yanagihara et al. 2005; Gong et al. 2013). Further study is required to extend the in vitro study to animal models of infection.

12.8 Prevention of Nosocomial Infections Caused by *Klebsiella*

A manual consolidating all the instructions and practices in the prevention of nosocomial infection prevention is mandatory (Chinn and Sehulster 2008). It is the duty of the infection control squad in developing, revision, and updating the manual.

12.8.1 Responsibility of the Infection Control Team

12.8.1.1 Role of the Hospital Administration

In the prevention of nosocomial infections, hospital administration and management play a crucial role in the establishment of the infection control board. It must also aid in the identification, monitoring, and implicating suitable methods in the control (Zingg et al. 2015). All the staff should be trained about the aspects of control of infection by techniques such as sterilization and disinfection. The hospital personnel including the nurse, housekeeping, laboratory technicians should be conveyed about hospital hygiene and their maintenance (Lówbúrý et al. 2013). Periodically the degree of the hospital-acquired infections and effective medical interventions should be reviewed (World Health Organization 2002).

12.8.1.2 Infection Control Team

The hospital management must have access to specialists in the field of epidemiology (Scheckler et al. 1998). However, the infection control team must ensure the appropriate management of infection control schedule. The infection control squad personnel also have the responsibility in invigilating the day-to-day functions, surveillance, evaluation, and supervision of the necessities viz. disinfectants and other sterilization agents required for the control of infections in the hospital settings (Scott et al. 2005).

12.8.1.3 Duty of the Nursing Staff

Implementing guidelines laid down by the infection control committee to the patient care service is the duty of the nursing staff (Grol et al. 2013). Knowledge of the nursing staff in preventing the spread of nosocomial infection, the practice of appropriate interventions for all the patients during the stay period is mandatory (Hooton et al. 2010).

Senior nursing heads are responsible for actively participating in the training program which includes supervising and implementing techniques in the prevention of the infections in the wards, operation theaters, intensive care units, and maternity units (Drachman 1981).

Maintaining hygiene and adopting good nursing practices in the wards. Aseptic conditions in the wards which include washing hands and reporting of any infections to the physicians in charge (Conly et al. 1989). Isolation of patients with any communicable diseases. Limiting patient visitors, hospital staffs, and equipment used for medical interventions (World Health Organization 2002).

12.8.1.4 Central Sterilization Service

A central sterilization facility should be provided in all the hospital settings. The responsibilities of the central sterilization facilities include cleaning, testing, and decontamination, storage of all the hospital equipment aseptically. This committee works hand in hand with the infection control team. They also monitor the cleaning procedures and decontamination of the contaminated equipment which include wrapping procedure and packing of the equipment according to the sterilization techniques. Sterilization conditions (temperature, humidity, and pressure). Higher authorities supervise the use of different physical, chemical, and biological methods in monitoring the sterilization process. Training the new staff members and periodic training to other staff members about the new technique employed.

There must be frequent training to the hospital personnel regarding hygiene and frequent cleaning and washing hands. Causes of contamination in the hospital premises and ways of minimizing it must be scrutinized by the hospital staff (Emori and Gaynes 1993).

12.8.1.5 Hospital Hygiene Services

They are responsible to overview and coordinate infection control activity and check the effectiveness of the methods employed in disinfection and sterilization, develop methods to improve hygiene in the hospital settings. Inspection and replacement of filters of all the equipment for ventilation. The hospital hygiene service may also assist and undertake research, which includes hospital hygiene and control of infections (Boyce and Pittet 2002).

12.9 Conclusion

Klebsiella is opportunistic nosocomial pathogen responsible for many hospital-acquired infections. The nosocomial infections are common among the hospitalized and immune compromised patients. The incidence of *Klebsiella* nosocomial infections ranges from 5 to 7% of the hospital-acquired infections and ranks them as the most significant nosocomial pathogen. The morbidity and mortality rates recorded as a result of *Klebsiella*-related infections are very high (as high as 50%). The prevention and control of nosocomial infections have resulted in considerable advancements in managing and controlling infections.

Alternative approaches for the prevention and control of nosocomial *Klebsiella*-related nosocomial infections is a necessity.

References

Abedon S (2015) Ecology of anti-biofilm agents I: antibiotics versus bacteriophages. Pharmaceuticals 8(3):525–558

Abraham EP, Chain E (1988) An enzyme from bacteria able to destroy penicillin 1940. Rev Infect Dis 10(4):677

Aly MM, Khalil S, Metwaly A (2014) Isolation and molecular identification of *Klebsiella* microbe isolated from chicks. Alex J Vet Sci 43(1):97–103

Anderson DJ, Moehring RW, Sloane R, Schmader KE, Weber DJ, Fowler JVG, Sexton DJ (2014) Bloodstream infections in community hospitals in the 21st century: a multicenter cohort study. PLoS One 9(3):e91713

Anitha TK (2012) Phenotypic characterization and antibiotic susceptibility pattern of *Klebsiella* species from blood cultures of neonates with septicemia. Doctoral dissertation

Antoniadou A, Kontopidou F, Poulakou G, Koratzanis E, Galani I, Papadomichelakis E, Kopterides P, Souli M, Armaganidis A, Giamarellou H (2007) Colistin-resistant isolates of Klebsiella pneumoniae emerging in intensive care unit patients: first report of a multiclonal cluster. J Antimicrob Chemother 59(4):786–790

Ayhan DH, Tamer YT, Akbar M, Bailey SM, Wong M, Daly SM, Greenberg DE, Toprak E (2016) Sequence-specific targeting of bacterial resistance genes increases antibiotic efficacy. PLoS Biol 14(9):e1002552

Azargun R, Barhaghi MH, Kafil HS, Oskouee MA, Sadeghi V, Memar MY, Ghotaslou R (2019) Frequency of DNA gyrase and topoisomerase IV mutations and plasmid-mediated quinolone resistance genes among *Escherichia coli* and *Klebsiella pneumoniae* isolated from urinary tract infections in Azerbaijan, Iran. J Glob Antimicrob Resist 17:39–43

Bachman MA, Miller VL, Weiser JN (2009) Mucosal lipocalin 2 has pro-inflammatory and iron-sequestering effects in response to bacterial enterobactin. PLoS Pathog 5(10):e1000622

Bachman MA, Oyler JE, Burns SH, Caza M, Lépine F, Dozois CM, Weiser JN (2011) *Klebsiella pneumoniae* yersiniabactin promotes respiratory tract infection through evasion of lipocalin. Infect Immun 79(8):3309–3316

Bagley ST, Seidler RJ, Brenner DJ (1981) *Klebsiella planticola* sp. nov.: a new species of Enterobacteriaceae found primarily in nonclinical environments. Curr Microbiol 6(2):105–109

Bauernfeind A (1984) Epidemiological typing of *Klebsiella* by Bacteriocins. In: Methods in microbiology, vol 16. Academic Press, Cambridge, pp 213–224

Bauernfeind AD, Petermüller C, Schneider RE (1981) Bacteriocins as tools in analysis of nosocomial *Klebsiella pneumoniae* infections. J Clin Microbiol 14(1):15–19

Beggs CB (2003) The airborne transmission of infection in hospital buildings: fact or fiction? Indoor Built Environ 12(1):9–18

Bi W, Liu H, Dunstan RA, Li B, Torres VV, Cao J, Chen L, Wilksch JJ, Strugnell RA, Lithgow T, Zhou T (2017) Extensively drug-resistant *Klebsiella pneumoniae* causing nosocomial bloodstream infections in China: molecular investigation of antibiotic resistance determinants, informing therapy, and clinical outcomes. Front Microbiol 8:1230

Bingen EH, Denamur E, Elion J (1994) Use of ribotyping in epidemiological surveillance of nosocomial outbreaks. Clin Microbiol Rev 7(3):311–327

Blair JM, Bavro VN, Ricci V, Modi N, Cacciotto P, Kleinekathöfer U, Ruggerone P, Vargiu AV, Baylay AJ, Smith HE, Brandon Y (2015) AcrB drug-binding pocket substitution confers clinically relevant resistance and altered substrate specificity. Proc Natl Acad 112(11):3511–3516

Bogovazova GG, Voroshilova NN, Bondarenko VM (1991) The efficacy of *Klebsiella pneumoniae* bacteriophage in the therapy of experimental *Klebsiella* infection. Zh. Mikrobiol. Epidemiol Immunobiol 4:5–8

Booker CW, Abutarbush SM, Morley PS, Jim GK, Pittman TJ, Schunicht OC, Perrett T, Wildman BK, Fenton RK, Guichon PT, Janzen ED (2008) Microbiological and histopathological findings in cases of fatal bovine respiratory disease of feedlot cattle in western Canada. Can Vet J 49(5):473

Boucher HW, Talbot GH, Bradley JS, Edwards JE, Gilbert D, Rice LB, Scheld M, Spellberg B, Bartlett J (2009) Bad bugs, no drugs: no ESKAPE! An update from the infectious diseases society of America. Clin Infect Dis 48(1):1–2

Boyce JM, Pittet D (2002) Guideline for hand hygiene in health-care settings: recommendations of the healthcare infection control practices advisory committee and the HICPAC/SHEA/APIC/IDSA hand hygiene task force. Infect Control Hosp Epidemiol 23(S12):S3–S40

Buckley AM, Webber MA, Cooles S, Randall LP, La Ragione RM, Woodward MJ, Piddock LJ (2006) The AcrAB–TolC efflux system of *Salmonella* enterica serovar typhimurium plays a role in pathogenesis. Cell Microbiol 8(5):847–856

Bunikis I, Denker K, Östberg Y, Andersen C, Benz R, Bergström S (2008) An RND-type efflux system in *Borrelia burgdorferi* is involved in virulence and resistance to antimicrobial compounds. PLoS Pathog 4(2):e1000009

Bush K (2010) Alarming ß-lactamase-mediated resistance in multidrug-resistant Enterobacteriaceae. Curr Opin Microbiol 13(5):558–564

Bushby SRM (1980) Sulfonamide and trimethoprim combinations. J Am Vet Med Assoc 176(10):1049–1053

Carattoli A (2013) Plasmids and the spread of resistance. Int J Med Microbiol 303(6-7):298–304

Chanishvili N (2012) Phage therapy—history from Twort and d'Herelle through Soviet experience to current approaches. Adv Virus Res 83:33–40

Chen L, Mediavilla JR, Endimiani A, Rosenthal ME, Zhao Y, Bonomo RA, Kreiswirth BN (2011) Multiplex real-time PCR assay for detection and classification of Klebsiella pneumoniae carbapenemase gene (bla KPC) variants. J Clin Microbiol 49(2):579–585

Chibani-Chennoufi S, Bruttin A, Dillmann ML, Brüssow H (2004) Phage-host interaction: an ecological perspective. J Bacteriol 186:3677–3686

Chinn RY, Sehulster L (2008) Guidelines for environmental infection control in health-care facilities; recommendations of CDC and Healthcare Infection Control Practices Advisory Committee (HICPAC). MMWR Recomm Rep 52(RR-10):1–42

Chung DR, Lee SS, Lee HR, Kim HB, Choi HJ, Eom JS, Kim JS, Choi YH, Lee JS, Chung MH, Kim YS (2007) Emerging invasive liver abscess caused by K1 serotype *Klebsiella pneumoniae* in Korea. J Infect 54(6):578–583

Clements A, Tull D, Jenney AW, Farn JL, Kim SH, Bishop RE, McPhee JB, Hancock RE, Hartland EL, Pearse MJ, Wijburg OL (2007) Secondary acylation of *Klebsiella pneumoniae* lipopolysaccharide contributes to sensitivity to antibacterial peptides. J Biol Chem 282(21):15569–15577
Coetzee J, Brink A (2012) The emergence of carbapenem resistance in Enterobacteriaceae in South Africa. S Afr Med J 26(4):239–240
Combe ML, Pons JL, Sesboue R, Martin JP (1994) Electrophoretic transfer from polyacrylamide gel to nitrocellulose sheets, a new method to characterize multilocus enzyme genotypes of *Klebsiella* strains. Appl Environ Microbiol 60(1):26–30
Conly JM, Hill S, Ross J, Lertzman J, Louie TJ (1989) Handwashing practices in an intensive care unit: the effects of an educational program and its relationship to infection rates. Am J Infect Control 17(6):330–339
Cortés G, Borrell N, de Astorza B, Gómez C, Sauleda J, Albertí S (2002) Molecular analysis of the contribution of the capsular polysaccharide and the lipopolysaccharide O side chain to the virulence of *Klebsiella pneumoniae* in a murine model of pneumonia. Infect Immun 70(5):2583–2590
Costello SE, Deshpande LM, Davis AP, Mendes RE, Castanheira M (2019) Aminoglycoside-modifying enzyme and 16S ribosomal RNA methyltransferase genes among a global collection of Gram-negative isolates. J Glob Antimicrob Resist 16:278–285
Davis B, Lilly HA, Lowbury EJ (1969) Gram-negative bacilli in burns. J Clin Pathol 22(6):634–641
De Majumdar S, Yu J, Fookes M, McAteer SP, Llobet E, Finn S, Spence S, Monaghan A, Kissenpfennig A, Ingram RJ, Bengoechea J (2015) Elucidation of the RamA regulon in *Klebsiella pneumoniae* reveals a role in LPS regulation. PLoS Pathog 11(1):e1004627
Deguchi T, Yasuda M, Nakano M, Ozeki S, Kanematsu E, Nishino Y, Ishihara S, Kawada Y (1997) Detection of mutations in the gyrA and parC genes in quinoloneresistant clinical isolates of Enterobacter cloacae. J Antimicrob Chemother 40(4):543–549
Doi Y, Wachino JI, Arakawa Y (2016) Aminoglycoside resistance: the emergence of acquired 16S ribosomal RNA methyltransferases. Infect Dis Clin 30(2):523–537
Donnenberg MS, Mandel GL, Bemet JE, John R, Mandel D (2005) Enterobacteriaceae: principles and practice of infectious (6th edn), 267–286
Doorduijn DJ, Rooijakkers SH, van Schaik W, Bardoel BW (2016) Complement resistance mechanisms of *Klebsiella pneumoniae*. Immunobiology 221(10):1102–1109
Dorman MJ, Feltwell T, Goulding DA, Parkhill J, Short FL (2018) The capsule regulatory network of *Klebsiella pneumoniae* defined by density-TraDISort. MBio 9(6):e01863
Drachman D (1981) A residential continuum for the chronically mentally ill: a Markov probability model. Eval Health Prof 4(1):93–104
Emori TG, Gaynes RP (1993) An overview of nosocomial infections, including the role of the microbiology laboratory. Clin Microbiol Rev 6(4):428–442
Eswaran J, Koronakis E, Higgins MK, Hughes C, Koronakis V (2004) Three's company: component structures bring a closer view of tripartite drug efflux pumps. Curr Opin Struct Biol 14:741–747
Fabrega A, Madurga S, Giralt E (2009) Mechanism of action of and resistance to quinolones. Microb Biotechnol 2:40–61
Falagas ME, Karageorgopoulos DE (2009) Extended-spectrum ß-lactamase-producing organisms. J Hosp Infect 73(4):345–354
Falagas ME, Kasiakou SK (2005) Colistin: the revival of polymyxins for the management of multidrug-resistant Gram-negative bacterial infections. Clin Infect Dis 40:1333–1341
Fang CT, Chuang YP, Shun CT (2004) A novel virulence gene in *Klebsiella pneumoniae* strains causing primary liver abscess and septic metastatic complications. J Exp Med 199:697
Felson B, Rosenberg LS, Hamburger M Jr (1949) Roentgen findings in acute Friedlander's pneumonia. Radiology 53:559–565
Ferragut C, Izard D, Gavini F, Kersters K, De Ley J, Leclerc H (1983) *Klebsiella trevisanii:* a new species from water and soil. Int J Syst Evol Microbiol 33(2):133–142

Fevre C, Passet V, Deletoile A, Barbe V, Frangeul L, Almeida AS, Sansonetti P, Tournebize R, Brisse S (2011) PCR-based identification of *Klebsiella pneumoniae* subsp. *rhinoscleromatis,* the agent of rhinoscleroma. PLoS Negl Trop Dis 5(5):e1052

Garner JS, Jarvis WR, Emori TG, Horan TC, Hughes JM (1988) CDC definitions for nosocomial infections. Am J Infect Control 16:128–140

Gavini F, Izard D, Pa G, Beji A, Ageron E, Leclerc H (1986) Priority of *Klebsiella planticol*a Bagley, Seidler, and Brenner 1982 over *Klebsiella trevisan*ii Ferragut, Izard, Gavini, Kersters, DeLey, and Leclerc 1983. Int J Syst Evol Microbiol 36(3):486–488

Gong L, Chen Y, Hu Z, Hu M (2013) Testing insecticidal activity of novel chemically synthesized siRNA against *Plutella xylostella* under laboratory and field conditions. PLoS One 8:e62990. https://doi.org/10.1371/journal.pone.0062990

Gouby A, Carles-Nurit MJ, Bouziges N, Bourg G, Mesnard R, Bouvet PJ (1992) Use of pulsed-field gel electrophoresis for investigation of hospital outbreaks of *Acinetobacter baumann*ii. J Clin Microbiol 30(6):1588–1591

Górski A, Wazna E, Dabrowska BW, Dabrowska K, Switala-Jelen K, Miedzybrodzki R (2006) Bacteriophage translocation. FEMS Immunol Med Microbiol 46(3):313–319

Granowitz EV, Brown RB (2008) Antibiotic adverse reactions and drug interactions. Crit care Clin 24(2):421–442

Grol R, Wensing M, Eccles M, Davis D (eds) (2013) Improving patient care: the implementation of change in health care. Wiley, Oxford

Guillard T, de Jong A, Limelette A, Lebreil AL, Madoux J, De Champs C, ComPath Study Group (2015) Characterization of quinolone resistance mechanisms in Enterobacteriaceae recovered from diseased companion animals in Europe. Vet Microbiol 94:23–29

Hawley JK (1985) Assessment of health risk from exposure to contaminated soil. Risk Anal 5(4):289–302

Hæggman S (2010) Evolution of beta-lactam: Resistance in *Klebsiella pneumoniae*. Institutionen för mikrobiologi, tumör-och cell biologi/Department of Microbiology, Tumor and Cell Biology; Apr 20

Heidary M, Bahramian A, Hashemi A, Goudarzi M, Omrani VF, Eslami G, Goudarzi H (2016) Detecon of acrA, acrB, aac (6′)-Ib-cr, and qepA genes among clinical isolates of Escherichia coli and Klebsiella pneumoniae. Acta Microbiol Immunol Hung 64(1):63–69

Hirsch EB, Tam VH (2010) Detection and treatment options for *Klebsiella pneumoniae* carbapenemases (KPCs): an emerging cause of multidrug-resistant infection. J Antimicrob Chemother 65(6):1119–1125

Holden VI, Breen P, Houle S, Dozois CM, Bachman MA (2016) Klebsiella pneumoniae siderophores induce inflammation, bacterial dissemination, and HIF-1a stabilization during pneumonia. MBio 7(5):e01397–16

Holt KE, Wertheim H, Zadoks RN, Baker S, Whitehouse CA, Dance D, Jenney A, Connor TR, Hsu LY, Severin J, Brisse S (2015) Genomic analysis of diversity, population structure, virulence, and antimicrobial resistance in Klebsiella pneumoniae, an urgent threat to public health. Proc Natl Acad Sci 112(27):E3574–E3581

Hooton TM, Bradley SF, Cardenas DD, Colgan R, Geerlings SE, Rice JC, Saint S, Schaeffer AJ, Tambayh PA, Tenke P, Nicolle LE (2010) Diagnosis, prevention, and treatment of catheter-associated urinary tract infection in adults: 2009 International Clinical Practice Guidelines from the Infectious Diseases Society of America. Clin Infect Dis 50(5):625–663

Izard D, Ferragut C, Gavini F, Kersters K, De Ley J, Leclerc H (1981) *Klebsiella terrigena*, a new species from soil and water. Int J Syst Evol Microbiol 31(2):16–127

Johanson WG Jr (1984) Prevention of respiratory tract infection. Am J Med Sci 76(5):69

Karaiskos I, Souli M, Galani I, Giamarellou H (2017) Colistin: still a lifesaver for the 21st century? Expert Opin Drug Metab Toxicol 13(1):59–71

Kim JY, Park YJ, Kwon HJ, Han K, Kang MW, Woo GJ (2008) Occurrence and mechanisms of amikacin resistance and its association with β-lactamases in *Pseudomonas aeruginosa*: a Korean nationwide study. J Antimicrob Chemother 62(3):479–483

Kitchel B, Rasheed JK, Patel JB, Srinivasan A, Navon-Venezia S, Carmeli Y, Brolund A, Giske CG (2009) Molecular epidemiology of KPC-producing *Klebsiella pneumoniae* isolates in the United States: clonal expansion of multilocus sequence type 258. Antimicrob Agents Chemother 53(8):3365–3370

Knight L, Fraser RG, Robson HG (1975) Massive pulmonary gangrene: a severe complication of *Klebsiella pneumonia*. Can Med Assoc J 112(2):196

Knothe H, Shah P, Krcmery V, Antal M, Mitsuhashi S (1983) Transferable resistance to cefotaxime, cefoxitin, cefamandole and cefuroxime in clinical isolates of *Klebsiella pneumoniae* and *Serratia marcescens*. Infection 11(6):315–317

Krapp F, Ozer EA, Qi C, Hauser AR (2018) Case report of an extensively drug-resistant Klebsiella pneumoniae infection with genomic characterization of the strain and review of similar cases in the United States. Open Forum Infect Dis 5:74

Krause KM, Serio AW, Kane TR, Connolly LE (2016) Aminoglycosides: an overview. CSH Perspect Med 6(6):a027029

Kutateladze Á, Adamia R (2008) Phage therapy experience at the Eliava institute. Med Mal Infect 38(8):426–430

Labia R, Fabre C, Masson JM, Barthelemy M, Heitz M, Pitton JS (1979) *Klebsiella pneumonia* strains moderately resistant to ampicillin and carbenicillin: characterization of a new β-lactamase. J Antimicrob Chemother 5(4):375–382

Lewis JS, Herrera M, Wickes B, Patterson JE, Jorgensen JH (2007) First report of the emergence of CTX-M-type extended-spectrum β-lactamases (ESBLs) as the predominant ESBL isolated in a US health care system. Antimicrob Agents Chemother 51(11):4015–4021

Li B, Zhao Y, Liu C, Chen Z, Zhou D (2014) Molecular pathogenesis of *Klebsiella pneumoniae*. Future Microbiol 9(9):1071–1081

Li L, Yu T, Ma Y, Yang Z, Wang W, Song X, Shen Y, Guo T, Kong J, Wang M, Xu H (2018) The genetic structures of an Extensively Drug Resistant (XDR) Klebsiella pneumoniae and its plasmids. Front Cell Infect Microbiol 8:446

Liang C, Xing B, Yang X, Fu Y, Feng Y, Zhang Y (2015) Molecular epidemiology of aminoglycosides resistance on *Klebsiella pneumonia* in a hospital in China. Int J Clin Exp Med 8(1):1381

Liao CH, Huang YT, Lai CC, Chang CY, Chu FY, Hsu MS, Hsu HS, Hsueh PR (2011) *Klebsiella pneumoniae* bacteremia and capsular serotypes, Taiwan. Emerg Infect Dis 17(6):1113

Liu YY, Wang Y, Walsh TR, Yi LX, Zhang R, Spencer J, Doi Y, Tian G, Dong B, Huang X, Yu LF (2016) Emergence of plasmid-mediated colistin resistance mechanism MCR-1 in animals and human beings in China: a microbiological and molecular biological study. Lancet Infect Dis 16(2):161–168

Llobet E, Campos MA, Giménez P, Moranta D, Bengoechea JA (2011) Analysis of the networks controlling the antimicrobial-peptide-dependent induction of *Klebsiella pneumoniae* virulence factors. Infect Immun 79(9):3718–3732

Lówbúrý ÉJL, Ayliffe GAJ, Geddes AM, Williams JD (2013) Control of hospital infection: a practical handbook. Springer, Berlin

Lugo JZ, Price S, Miller JE, Ben-David I, Merrill VJ, Mancuso P, Weinberg JB, Younger JG (2007) Lipopolysaccharide O-antigen promotes persistent murine bacteremia. Shock 27(2):186–191

Mai V, Ukhanova M, Reinhard MK, Li M, Sulakvelidze A (2015) Bacteriophage administration significantly reduces Shigella colonization and shedding by Shigella-challenged mice without deleterious side effects and distortions in the gut microbiota. Bacteriophage 5(4):e1088124

Mangram AJ, Horan TC, Pearson ML, Silver LC, Jarvis WR (1999) Hospital infection control practices advisory committee. Guideline for prevention of surgical site infection. Infect Control Hosp Epidemiol 20(4):247–280

Marchaim D, Chopra T, Pogue JM, Perez F, Hujer AM, Rudin S, Endimiani A, Navon-Venezia S, Hothi J, Slim J, Blunden C (2011) Outbreak of colistin-resistant, carbapenem-resistant *Klebsiella pneumoniae* in metropolitan Detroit, Michigan. Antimicrob Agents Chemother 55(2):593–599

Martínez-Martínez L, Pascual A, Jacoby GA (1998) Quinolone resistance from a transferable plasmid. Lancet 351(9105):797–799

Miethke M, Marahiel MA (2007) Siderophore-based iron acquisition and pathogen control. Mol Biol Rev 71(3):413–451

Mitrophanov AY, Jewett MW, Hadley TJ, Groisman EA (2008) Evolution and dynamics of regulatory architectures controlling polymyxin B resistance in enteric bacteria. PLoS Genet 4(10):e1000233

Monk AB, Rees CD, Barrow P, Hagens S, Harper DR (2010) Bacteriophage applications: where are we now? Lett Appl Microbiol 51(4):363–369

Mousavi SL, Nazarian S, Amani J, Karimi Rahgerdi A (2008) Rapid screening of toxigenic *Vibrio cholerae* O1 strains from South Iran by PCR-ELISA. Iran Biomed J 12(1):15–21

Naeem A, Badshah S, Muska M, Ahmad N, Khan K (2016) The current case of quinolones: synthetic approaches and antibacterial activity. Molecules 21(4):268

Nam YS, Cho SY, Yang HY, Park KS, Jang JH, Kim YT, Jeong JW, Suh JT, Lee HJ (2013) Investigation of mutation distribution in DNA gyrase and topoisomerase IV genes in ciprofloxacin-non-susceptible Enterobacteriaceae isolated from blood cultures in a tertiary care university hospital in South Korea, 2005–2010. Int J Antimicrob Agents 41(2):126–129

Nasiri G, Peymani A, Farivar TN, Hosseini P (2018) Molecular epidemiology of aminoglycoside resistance in clinical isolates of *Klebsiella pneumoniae* collected from Qazvin and Tehran provinces, Iran. Infect Genet Evol 64:219–224

Nassif XA, Fournier JM, Arondel J, Sansonetti PJ (1989) Mucoid phenotype of *Klebsiella pneumoniae* is a plasmid-encoded virulence factor. Infect Immun 57(2):546–552

National Institute for Communicable Diseases (NICD) (2013) Update on carbapenemase- producing Enterobacteriaceae. Communic Dis Communiq 12:5–6

Navon-Venezia S, Kondratyeva K, Carattoli A (2017) *Klebsiella pneumoniae*: a major worldwide source and shuttle for antibiotic resistance. FEMS Microbiol Lett 41(3):252–275

Nazir A, Kadri SM (2014) An overview of hospital acquired infections and the role of the microbiology laboratory. Int J Res Med Sci 2(1):21–27

Nguyen MC, Woerther PL, Bouvet M, Andremont A, Leclercq R, Canu A (2009) *Escherichia coli* as reservoir for macrolide resistance genes. Emerg Infect Dis 15(10):1648

Nordmann P, Poirel L (2014) The difficult-to-control spread of carbapenemase producers among Enterobacteriaceae worldwide. Clin Microbiol Infect 20(9):821–830

Nordmann P, Cuzon G, Naas T (2009) The real threat of *Klebsiella pneumoniae* carbapenemase-producing bacteria. Lancet Infect Dis 9(4):228–236

Nouvellon M, Pons JL, Sirot D, Combe ML, Lemeland JF (1994) Clonal outbreaks of extended-spectrum beta-lactamase-producing strains of *Klebsiella pneumoniae* demonstrated by antibiotic susceptibility testing, beta-lactamase typing, and multilocus enzyme electrophoresis. J Clin Microbiol 32(10):2625–2627

Olaitan AO, Morand S, Rolain JM (2014) Mechanisms of polymyxin resistance: acquired and intrinsic resistance in bacteria. Front Microbiol 5:643

Ong CL, Ulett GC, Mabbett AN, Beatson SA, Webb RI, Monaghan W, Nimmo GR, Looke DF, McEwan AG, Schembri MA (2008) Identification of type 3 fimbriae in uropathogenic *Escherichia coli* reveals a role in biofilm formation. J Bacteriol 190(3):1054–1063

Ørskov ID, Fife-Asbury MA (1997) New Klebsiella capsular antigen, K82, and the deletion of five of those previously assigned. Int J Syst Evol Microbiol 27(4):386–387

Owens CD, Stoessel K (2008) Surgical site infections: epidemiology, microbiology and prevention. J Hospital Infect 70:3–10

Padilla E, Llobet E, Doménech-Sánchez A, Martínez-Martínez L, Bengoechea JA, Albertí S (2010) *Klebsiella pneumoniae* AcrAB efflux pump contributes to antimicrobial resistance and virulence. Antimicrob Agents Chemother 54(1):177–183

Pérez A, Poza M, Fernández A, del Carmen Fernández M, Mallo S, Merino M, Rumbo-Feal S, Cabral MP, Bou G (2012) Involvement of the AcrAB-TolC efflux pump in the resistance, fitness, and virulence of *Enterobacter cloacae*. Antimicrob Agents Chemother 56(4):2084–2090

Petit A, Ben-Yaghlane-Bouslama H, Sofer L, Labia R (1992) Characterization of chromosomally encoded penicillinases in clinical isolates of *Klebsiella pneumoniae*. J Antimicrob Chemother 29(6):629–638

Philippon A, Arlet G, Jacoby GA (2002) Plasmid-determined AmpC-type β-lactamases. Antimicrob Agents Chemother 46(1):1–1

Pieroni P, Rennie RP, Ziola B, Deneer HG (1994) The use of bacteriophages to differentiate serologically cross-reactive isolates of *Klebsiella pneumoniae*. J Med Microbiol 41(6):423–429

Pitout JD (2012) Extraintestinal pathogenic *Escherichia coli*: an update on antimicrobial resistance, laboratory diagnosis and treatment. Expert Rev Anti-Infect Ther 10(10):1165–1176

Podschun R, Ullmann U (1992) Isolation of *Klebsiella terrigena* from clinical specimens. Eur J Clin Microbiol Infect Dis 11(4):349–352

Podschun R, Ullmann U (1993) Bacteriocin typing of environmental *Klebsiella* isolates. Int J Hyg Envir Health 195(1):22–26

Podschun R, Ullmann U (1998) *Klebsiella* spp. as nosocomial pathogens: epidemiology, taxonomy, typing methods, and pathogenicity factors. Clinical Microbiol Rev 11(4):589–603

Poh CL, Yap SC, Yeo M (1993) Pulsed-field gel electrophoresis for differentiation of hospital isolates of *Klebsiella pneumoniae*. J Hosp Infect 24(2):123–128

Qureshi H, Sharafkhaneh A, Hanania NA (2014) Chronic obstructive pulmonary disease exacerbations: latest evidence and clinical implications. Ther Adv Chronic Dis 5(5):212–227

Raffatellu M, George MD, Akiyama Y, Hornsby MJ, Nuccio SP, Paixao TA, Butler BP, Chu H, Santos RL, Berger T, Mak TW, Tsolis RM, Bevins CL, Solnick JV, Dandekar S, Baumler AJ (2009) Lipocalin-2 resistance confers an advantage to *Salmonella enterica* serotype typhimurium for growth and survival in the inflamed intestine. Cell Host Microbe 5:476–486

Ramirez MS, Tolmasky ME (2010) Aminoglycoside modifying enzymes. Microb Drug Resist 13(6):151–171

Rawat D, Nair D (2010) Extended-spectrum β-lactamases in gram negative Bacteria. J Glob Infect Dis 2(3):263

Raymond KN, Dertz EA, Kim SS (2003) Enterobactin: an archetype for microbial iron transport. Proc Natl Acad Sci 100(7):3584–3588

Redgrave LS, Sutton SB, Webber MA, Piddock LJ (2014) Fluoroquinolone resistance: mechanisms, impact on bacteria, and role in evolutionary success. Trends Microbiol 22(8):438–445

Reybrouck G (1983) Role of the hands in the spread of nosocomial infections. 1. J Hosp Infect 4(2):103–110

Rosen DA, Pinkner JS, Jones JM, Walker JN, Clegg S, Hultgren SJ (2008) Utilization of an intracellular bacterial community pathway in *Klebsiella pneumoniae* urinary tract infection and the effects of FimK on type 1 pilus expression. Infect Immun 76(7):3337–3345

Sakazaki R, Tamura K, Kosako Y, Yoshizaki E (1989) *Klebsiella ornithinolytica* sp. nov., formerly known as ornithine-positive *Klebsiella oxytoca*. Curr Microbiol 18(4):201–206

Samuelsen Ø, Naseer U, Tofteland S, Skutlaberg DH, Onken A, Hjetland R, Sundsfjord A, Giske CG (2009) Emergence of clonally related *Klebsiella pneumoniae* isolates of sequence type 258 producing plasmid-mediated KPC carbapenemase in Norway and Sweden. J Antimicrob Chemother 63(4):654–658

Scheckler WE, Brimhall D, Buck AS, Farr BM, Friedman C, Garibaldi RA, Gross PA, Harris JA, Hierholzer WJ, Martone WJ, McDonald LL (1998) Requirements for infrastructure and essential activities of infection control and epidemiology in hospitals: a consensus panel report. Infect Am J Med 19(2):114–124

Schembri MA, Blom J, Krogfelt KA, Klemm P (2005) Capsule and fimbria interaction in *Klebsiella pneumoniae*. Infect Immun 73(8):4626–4633

Scott AA, Hughes J, Hall D (2005) Views and understanding of senior nursing staff in relation to infection control. Am J Infect Control 6(2):23–26

Shakil S, Khan R, Zarrilli R, Khan AU (2008) Aminoglycosides versus bacteria–a description of the action, resistance mechanism, and nosocomial battleground. J Biomed Sci 15(1):5–14

Shankar-Sinha S, Valencia GA, Janes BK, Rosenberg JK, Whitfield C, Bender RA, Standiford TJ, Younger JG (2004) The *Klebsiella pneumoniae* O antigen contributes to bacteremia and lethality during murine pneumonia. Infec Immun 72(3):1423–1430

Sköld O (2000) Sulfonamide resistance: mechanisms and trends. Drug Resist Updat 3(3):155–160

Ślopek S, Chapter V (1978) Phage typing of Klebsiella. In: Methods in microbiology, vol 11. Academic Press, Cambridge, pp 193–222
Slopek S, Przondo-Hessek A, Milch H, Deak S (1967) A working scheme for bacteriophage typing of *Klebsiella* bacilli. Arch Immunol Ther Exp 15(4):589–599
Somily AM, Habib HA, Absar MM, Arshad MZ, Manneh K, Al Subaie SS, Murray TS (2014) ESBL-producing *Escherichia coli* and *Klebsiella pneumoniae* at a tertiary care hospital in Saudi Arabia. J Infect Dev Countr 8(9):1129–1136
Spanu T, Sanguinetti M, Tumbarello M, D'Inzeo T, Fiori B, Posteraro B, Santangelo R, Cauda R, Fadda G (2006) Evaluation of the new VITEK 2 extended-spectrum beta-lactamase (ESBL) test for rapid detection of ESBL production in Enterobacteriaceae isolates. J Clin Microbiol Infect 44(9):3257–3262
Srinivasan VB, Rajamohan G (2013) KpnEF, a new member of the *Klebsiella pneumoniae* cell envelope stress response regulon, is an SMR-type efflux pump involved in broad-spectrum antimicrobial resistance. Antimicrob Agents Chemother 57(9):4449–4462
Srinivasan VB, Venkataramaiah M, Mondal A, Vaidyanathan V, Govil T, Rajamohan G (2012) Functional characterization of a novel outer membrane porin KpnO, regulated by PhoBR two-component system in *Klebsiella pneumoniae* NTUH-K2044. PLoS One 7(7):e41505
Stanley IJ, Kajumbula H, Bazira J, Kansiime C, Rwego IB, Asiimwe BB (2018) Multidrug resistance among Escherichia coli and *Klebsiella pneumoniae* carried in the gut of out-patients from pastoralist communities of Kasese district, Uganda. PLoS One 13(7):e0200093
Storberg V (2014) ESBL-producing Enterobacteriaceae in Africa–a non-systematic literature review of research published 2008–2012. J Infect Dis Epidemiol 4(1):20342
Struve C, Bojer M, Krogfelt KA (2009) Identification of a conserved chromosomal region encoding *Klebsiella pneumoniae* type 1 and type 3 fimbriae and assessment of the role of fimbriae in pathogenicity. Infec Immun 77(11):5016–5024
Sugawara E, Kojima S, Nikaido H (2016) *Klebsiella pneumoniae* major porins OmpK35 and OmpK36 allow more efficient diffusion of β-lactams than their *Escherichia coli* homologs OmpF and OmpC. J Bacterial 198(23):3200–3208
Taitt CR, Leski TA, Erwin DP, Odundo EA, Kipkemoi NC, Ndonye JN, Kirera RK, Ombogo AN, Walson JL, Pavlinac PB, Hulseberg C (2017) Antimicrobial resistance of *Klebsiella pneumoniae* stool isolates circulating in Kenya. PLoS One 12(6):e0178880
Tenover FC (2006) Mechanism of antibiotic resistance in bacteria. Am J Infect Control 34:S3–S10
Tofteland S, Naseer U, Lislevand JH, Sundsfjord A, Samuelsen Ø (2013) A long-term low-frequency hospital outbreak of KPC-producing *Klebsiella pneumoniae* involving inter genus plasmid diffusion and a persisting environmental reservoir. PLoS One 8(3):e59015
Tomás JM, Benedi VJ, Ciurana B, Jofre J (1986) Role of capsule and O antigen in resistance of *Klebsiella pneumoniae* to serum bactericidal activity. Infect Immunol 54(1):85–89
Tsai YK, Fung CP, Lin JC, Chen JH, Chang FY, Chen TL, Siu LK (2011) *Klebsiella pneumoniae* outer membrane porins OmpK35 and OmpK36 play roles in both antimicrobial resistance and virulence. Antimicrob Agents Chemother 55(4):1485–1493
van der Bij AK, Pitout JD (2012) The role of international travel in the worldwide spread of multi resistant Enterobacteriaceae. J Antimicrob Chemother 67(9):2090–2100
Vaneechoutte M, Nemec A, Musílek M, van der Reijden TJ, van den Barselaar M, Tjernberg I, Calame W, Fani R, De Baere T, Dijkshoorn L (2009) Description of *Acinetobacter venetianus*ex Di Cello et al. 1997 sp. nov. Int J Syst Evol Microbiol 59(6):1376–1381
Walther-Rasmussen J, Høiby N (2007) Class A carbapenemases. J Antimicrob Chemother 60(3):470–482
Warren JW (1997) Catheter-associated urinary tract infections. Infect Dis Clin N Am 11(3):609–622
Watanabe R, Matsumoto T, Sano G, Ishii Y, Tateda K, Sumiyama Y, Uchiyama J, Sakurai S, Matsuzaki S, Imai S, Yamaguchi K (2007) Efficacy of bacteriophage therapy against gut-derived sepsis caused by *Pseudomonas aeruginosa* in mice. Antimicrob Agents Chemother 51(2):446–452
World Health Organization (2002) Prevention of hospital-acquired infections: a practical guide (No. WHO/CDS/CSR/EPH/2002.12). World Health Organization, Geneva

Wright MS, Suzuki Y, Jones MB, Marshall SH, Rudin SD, van Duin D, Kaye K, Jacobs MR, Bonomo RA, Adams MD (2015) Genomic and transcriptomic analyses of colistin-resistant clinical isolates of *Klebsiella pneumoniae* reveal multiple pathways of resistance. Antimicrob Agents Chemother 59(1):536–543

Xia J, Sun J, Cheng K, Li L, Fang LX, Zou MT, Liao XP, Liu YH, Liu YH (2016) Persistent spread of the rmtB 16S rRNA methyltransferase gene among Escherichia coli isolates from diseased food-producing animals in China. Vet Microbiol 188:41–46

Xia J, Fang LX, Cheng K, Xu GH, Wang XR, Liao XP, Liu YH, Sun J (2017) Clonal spread of 16S rRNA methyltransferase-producing *Klebsiella pneumoniae* ST37 with high prevalence of ESBLs from companion animals in China. Front Microbiol 8:529

Yanagihara K, Tashiro M, Fukuda Y, Ohno H, Higashiyama Y, Miyazaki Y, Hirakata Y, Tomono K, Mizuta Y, Tsukamoto K, Kohno S (2005) Effects of short interfering RNA against methicillin-resistant *Staphylococcus aureus* coagulase *in vitro* and *in vivo*. J Antimicrob Chemother 57(1):122–126

Yang HY, Nam YS, Lee HJ (2014) Prevalence of plasmid-mediated quinolone resistance genes among ciprofloxacin-nonsusceptible *Escherichia coli* and *Klebsiella pneumoniae* isolated from blood cultures in Korea. Can J Infect Dis Med Microbiol 25(3):163–169

Yang YQ, Li YX, Lei CW, Zhang AY, Wang HN (2018) Novel plasmid-mediated colistin resistance gene mcr-7.1 in *Klebsiella pneumoniae*. J Antimicrob Chemother 73(7):1791–1795

Yu WL, Ko WC, Cheng KC, Lee CC, Lai CC, Chuang YC (2008) Comparison of prevalence of virulence factors for *Klebsiella pneumoniae* liver abscesses between isolates with capsular K1/K2 and non-K1/K2 serotypes. Diagn Microbiol Infect Dis 62(1):1–6

Zingg W, Holmes A, Dettenkofer M, Goetting T, Secci F, Clack L, Allegranzi B, Magiorakos AP, Pittet D (2015) Hospital organisation, management, and structure for prevention of health-care-associated infection: a systematic review and expert consensus. Lancet Infect Dis 15(2):212–224

Pathogenesis and Drug Resistance of *Pseudomonas aeruginosa* 13

Paramanantham Parasuraman, Asad Syed, and Busi Siddhardha

Contents

P. Parasuraman · B. Siddhardha (✉)
Department of Microbiology, School of Life Sciences, Pondicherry University, Puducherry, India

A. Syed
Department of Botany and Microbiology, College of Science, King Saud University, Riyadh, Saudi Arabia

B. Siddhardha et al. (eds.), *Model Organisms for Microbial Pathogenesis, Biofilm Formation and Antimicrobial Drug Discovery*,
https://doi.org/10.1007/978-981-15-1695-5_13

Abstract

Pseudomonas aeruginosa is a prevalent pathogenic species reported with severe nosocomial infections. *P. aeruginosa* infections gain significant concern due to the greater morbidity and mortality coupled with the development of multidrug resistance phenomenon. Furthermore, they are potentially participating in the development of infections to different organs of the human and possibly evade from the host immune defence mechanism. *P. aeruginosa* involves in the productions several virulence factors which makes the bacterium remains as potential life-threatening pathogen. Additionally, biofilm formation behaviours of *P. aeruginosa* offer windows to evade different adverse environmental conditions and initiation of adaptation process according to the stress condition and deliver highly resistance strain. The current antibiotics used to combat the *Pseudomonas* infection are gradually losing their efficacy in controlling the bacterial infections due to the development of the resistance mechanism within the bacterial system. The different resistance mechanisms including intrinsic, acquired and adaptive resistance are supporting the *P. aeruginosa* for the development of high resistance against several antimicrobial agents. Recently, several alternative therapeutic options were reported such as interference in quorum sensing of *P. aeruginosa*, phage therapy and development of antimicrobial peptides alternative therapeutic options against antibiotics with limited incidence of resistance development. This chapter overviews the pathogenesis and drug resistance development in *P. aeruginosa* and also covers the recently available therapeutic alternatives to combat *Pseudomonas* infection.

Keywords

Pseudomonas aeruginosa · Pathogenesis · Biofilm · Antibiotics · Drug resistance · Novel therapeutic approaches

13.1 Introduction

The genus *Pseudomonas* belongs to the Pseudomonadaceae family, which is composed of a set of Gram-negative rods and flagellated bacteria. *Pseudomonas* species are ubiquitous in nature inhabiting different habitats including air, soil, and water ecosystems (Streeter and Katouli 2016). These habituation behaviours of *Pseudomonas* species are due to their versatile metabolic pathways, which subsequently lead to the development of adaptive behaviours against diverse ecological niches. They are capable in utilizing different substrates as carbon source for the synthesis of energy. Moreover, they could subsist in minimum nutrient condition, grow at temperature up to 42 °C and able to respire under anaerobic condition by employing nitrogen as terminal electron acceptor (Kung et al. 2010). They are widely regarded as opportunistic pathogens due to their ability to infect plants, animals and humans. These *Pseudomonas* strains are capable of producing several pigments including pyoverdine and pyocyanin, which help other way to detect their growth on the agar plates (Gellatly and Hancock 2013). Among the *Pseudomonas* strains, *P. aeruginosa* was found to be frequently associated strain in infections associated with humans. However, it is widely observed as opportunistic pathogen, which occasionally infects the normal hosts but capable to develop severe infections in immunocompromised individuals (Klockgether and Tümmler 2017). Furthermore, *P. aeruginosa* could able to cause wide spectrum of infections when the host fails to maintain its physiological condition including epithelial barriers, neutrophil production and mucociliary layer. Occasionally, even the contaminated medical devices are remains as a reason for the *P. aeruginosa* infections in humans (Salter 2015). The genome of *P. aeruginosa* was extensively characterized; initially it was investigated on PAO1 strain which was isolated from the wound infection. It was reported that *P. aeruginosa* holding a genome size approximately 6.3 Mb and holding 5570 predicted open reading frames (ORFs), which is comparatively larger when compared to other bacterial species (Silby et al. 2011). Due to the excess genome size of *P. aeruginosa*, they could harbour different genetic materials required for several biological processes including metabolism and efflux of different organic matters as well as putative chemotaxis system (Battle et al. 2009). These remarkable properties of *P. aeruginosa* facilitate the adaptive mechanism to the versatile environmental conditions. However, some of other bacterial genomes are also reported with larger genome size which occurred due to the duplication of genes unlike the genome of *P. aeruginosa* where it is composed of greater genetic and functional diversity (Lee et al. 2006). Interestingly, different isolates of *P. aeruginosa*, i.e. the different strains of *P. aeruginosa* isolated from different environmental conditions showed greater similarity in their genomes. For example, when the DNA of *P. aeruginosa* isolated from the soil sample was hybridized with the PAO1 microarray, the result showed the soil isolated shared 89 to 98% of genome similarity with the pathogenic strain (Wolfgang et al. 2003). These results indicated that *P. aeruginosa* possesses a highly conserved core set of the genetic material responsible for most of the virulence traits. The comparative study between the environmental isolates and the clinical isolates showed genome similarity among different genes responsible for the production virulence factors including pyocyanin, a type III system, lipase, protease, and rhamnolipid (Wiehlmann et al. 2007).

P. aeruginosa causes several infections in humans, including pneumonia that may lead to death when it was encountered by immune-compromised or cystic fibrosis (CF) individuals. *P. aeruginosa* is capable of producing both acute and chronic diseases, a person with CF arises life-long chronic *P. aeruginosa* disease that results in eventual death of the individual (Lorè et al. 2012). The disease CF is a genetic disorder where an individual is born with structurally normal lungs, but promotes slow progression of pulmonary malfunctioning with continuous susceptibility to chronic infections. These conditions provoke the development of bronchiectasis resulting in complete failure of the respiratory system, which subsequently leads to death (Bhagirath et al. 2016). An individual with CF is highly susceptible to different pathogenic microorganisms including *P. aeruginosa, Staphylococcus aureus, Burkholderia cepacia* and *Haemophilus influenzae* (da Silva Filho et al. 2013). The pulmonary infections in the individual with CF would be developing from their early stage of disease. The overall incidence states that pathogenic microorganisms *P. aeruginosa* and *S. aureus* are the common pathogens which are routinely isolated from the patients with CF. The epidemiological investigations evidencing that the rout of entry and transmission pathogen mainly by a direct contact in some cases it is occurring even through contaminated environment (Taccetti et al. 2008). Nevertheless, *P. aeruginosa* causes several other diseases in humans in which almost all the diseases are developed by different infectious stages including attachment, colonization, local invasion and dissemination to establish systemic infections. This bacterium is also capable of causing infections to major tissues of humans including dermal, ocular, cardiac, respiratory epithelial and transitional epithelial tissues of urinary tract (Kariminik et al. 2017; Azam and Khan 2019). The pathogenesis pattern of the *P. aeruginosa* are shown as multifactorial which occurs as results of expression of several genes and attacking multiple target on the host that makes *P. aeruginosa* infection remains as an potential threat in worldwide (Valentini et al. 2018). Although some of the virulence factors of the *P. aeruginosa* are known for a decade, some of the other significant virulence traits are recently identified owing to the emergence of whole-genome sequencing technology. As discussed earlier both the clinical and the environmental isolates of *P. aeruginosa* hold the genetic information for most of the virulence factors. However, the pathogenicity and type of infection are possibly altered by the host environment and state of its immune system, as a matter of fact that *P. aeruginosa* is highly capable to adapt to different lifestyles of the host (Huang et al. 2011).

Several therapeutic options are available and were practised to control the *P. aeruginosa* infection. However, still these infections remain a major threat leading to greater mortality due to the emergence of the multidrug-resistant strains. There have been several reports elucidating the risk factors associated with the emergence of multidrug-resistant strains. For instance, multidrug-resistant tuberculosis is found to be a serious concern worldwide. World Health Organization (WHO) and government global nations combined programme formulated several measures to control the disease associated with multidrug-resistant *Mycobacterium tuberculosis*. However, the success rate of this programme found to be very low as compared to the target aimed by the WHO 2015, 75–90% (Kibret et al. 2017). Unsurprisingly, some of the *P. aeruginosa* strains also showed resistance against third-generation antibiotics including carbapenem and cephalosporins suggesting for the treatment of multidrug-resistant phenomenon (Azam and Khan 2019). Furthermore, the recent antimicrobial agents prescribed for the treatment of pulmonary infections were result in disordering the normal function of the lungs, failure of respiratory system and premature death of the individual with CF. In case of individuals with *P. aeruginosa* infection, the quality and the life span are directly dependent on the success rate of the treatment process in clearing the initial infection (Stefani et al. 2017). The pathogens reside on the side or organ of infection are showing considerable variation, owing to intensive antimicrobial agent pressure and modification in the antimicrobial agent regimens. Due to these unfavourable conditions on the site of infection, the pathogens are pressured to acquired resistance genes or pretend as new pathogens (Pena et al. 2015; Nguyen et al. 2018). *P. aeruginosa* develops resistance to antimicrobial agents by altering several biological behaviours including reducing the membrane permeability, activation of efflux systems, synthesis of the enzyme which could able to degrade the antimicrobial agent and potentially modify the structure of the target (Bassetti et al. 2018). Interestingly, *P. aeruginosa* could develop resistance chromosomally (intrinsic resistance) as well as acquiring the resistance genes from the neighbouring microbes in the environment (Wagner et al. 2016). Till date, the reports on resistance development in *P. aeruginosa* to at least three antimicrobial agents from different classes of antibiotics including aminoglycosides, antipseudomonal penicillins, cephalosporins, carbapenems and fluoroquinolones (Hirsch and Tam 2010; Chatterjee et al. 2016).

The chronic infections of *P. aeruginosa* found to cause serious clinical complications due to its inherent ability to form a biofilm. Biofilm is an integrated community of sessile microorganisms protected within self-producing polymeric substance, provides homeostasis and stability during the progress of infection in host. This condition supports and promotes the resistance development by structurally limiting the permeability of the antimicrobial agents and facilitates the adaptive resistance mechanism in *P. aeruginosa*. It is extremely difficult to eradicate the biofilm state of *P. aeruginosa* by routine antimicrobial agents as biofilms are highly tolerant to most of the available antimicrobial agents as compared to its planktonic counterparts (Furiga et al. 2016). For example, the notorious biofilm-forming Gram-negative pathogen, *P. aeruginosa* has the inherent ability to develop both intrinsic and acquired drug resistance, as discussed earlier. It was evaluated that the biofilm and stationary phase of the planktonic cultures were exposed to the sublethal concentration of known antibiotics, ciprofloxacin and observed for the development of mutational resistance. It was reported that the bacteria in the biofilm are less probable to develop the mutational resistance as compared to the planktonic cultures, since biofilm residing bacteria are poorly exposed the antimicrobial agent (Ahmed et al. 2018).

In matured biofilm, bacteria accounts only for 10% while the rest 90% is composed of self-producing extracellular polymeric substances (EPS) matrix containing polysaccharides, proteins, extracellular DNA and lipids. The development of biofilm and stability of the biofilm is generally governed by different intrinsic and extrinsic factors such as genetic makeup of the isolate, hydrodynamic conditions, motility behaviours of pathogenic bacteria, nutrient availability, intracellular communication and the health of host immune system (Maurice et al. 2018). In healthcare sectors, biofilm infections remain highly problematic, since it could occur in host tissues or the surface of the host and medical instrumentals including joint and organ replacement, catheters and ventilators (Taylor et al. 2014). Indeed biofilm of *P. aeruginosa* being a sensation among the scientific community and potential efforts have been initiated to eradicate them. Three major strategies have been proposed as primary methods to eradicate biofilms including (1) controlling the microbial attachment to the surface, (2) interrupting the biofilm formation and architecture which enhance the internalization of antimicrobial agents and (3) Disruption of mature biofilm (Rasamiravaka et al. 2015; Rossi Gonçalves et al. 2017). There is an urgent call for the development of desired antimicrobial strategies in worldwide scientific community. This chapter presents key features of multidrug-resistant *P. aeruginosa*, biofilm formation, associated infection and current and future options which could help in eradicating the infection by multidrug-resistant *P. aeruginosa* strains.

13.2 *Pseudomonas aeruginosa* and Associated Infections

As a matter of fact, *P. aeruginosa* is a very common bacterium, which is frequently isolated from the human with a disease. Although it acts as an opportunistic pathogen, it has the potential enough to cause infection to any part of human organs or tissues. However, the incidence of disease progression is very common among the individual's underlying disease, age or impaired immune system. In recent years, the incidence of healthcare-associated infection by *P. aeruginosa* was increased worldwide which subsequently evolved as a major threat to the patients as well as the economy (Wu et al. 2019). The hospitalized patients are frequently encountered by the pathogenic microorganisms, since the environmental setup of the healthcare unit potentially contaminated by widespread pathogens. Although standard procedures are followed by the healthcare sectors to maintain the sterility, most of the pathogens have managed to persist. Inpatients are repeatedly infected with some pathogens through direct or indirect contact with the environmental surfaces of healthcare units which were contaminated by infected roommate or prior room occupant (Kohlenberg et al. 2010). It is interesting that *P. aeruginosa* resides in the environment of healthcare units are capable of thriving in the adverse condition with wide range of the temperature, relatively nutrient-limited conditions and continuous exposure to disinfectants. Especially, water system in the healthcare unit has a high tendency to be contaminated with *P. aeruginosa* infection and facilitates the successful establishment of biofilm that subsequently remains as a major reason for the transmission of the pathogen (Garvey et al. 2016). Inpatients are at high risk of acquiring different kinds of infections such as pneumonia, burn wound infection, acute and chronic pulmonary infection (Kerr and Snelling 2009).

13.2.1 Burn Wound Infection

Burn wounds could be specified as skin abrasion mediated by heat exposure, electrical shock, exposure to chemicals and radiations. Burn wounds could be categorized from moderate to severe based on their degree of damage on the skin. The incidence of burn wound infections (moderate to severe burns) have been increased every year worldwide and are very common in developing countries (Mofazzal Jahromi et al.

2018). As compared to earlier strategies, the success and the survival rate of the burn individual in the present scenario have been increased due to the advanced development in burn wound care units. As a result, significant percentage of the burn-related deaths were reduced, still complete eradication of burn-related death remains a dream due to the infection caused by the pathogenic microorganisms.

As per the recent trends in burn wound infections, more than 75% of the mortality is caused by microbial infections followed by burn shock and hypovolemia (Wang et al. 2018). *P. aeruginosa* and *S. aureus* are the most prevalent pathogens causing serious infections to burn wound patients and potentially reduced the success rate of burn wound treatment process (Gonzalez et al. 2016). The incidence between the burn wound individual and the pathogen could happen either by acquiring from the normal flora of the host or hospital environment. The drug-resistant strains make the condition even more worse and subsequently altering the therapeutic process. For instance, the recovering or less critical burn wounded individuals are also at high risk of acquiring chronic infections which makes the situation more complicated. This indicates the complication and the threat integrated with the prevention of burn wound infections inside the burn unit (Simões et al. 2018). Moreover, the antibiotic susceptibility pattern of normal microflora colonized in the burn wound is possibly changing their pattern as antibiotic resistance during the period of treatment process. Recently, serious investigations are executed in emergency basis for the development of potential drug candidates and therapeutic medication to address the problems associated with the burn wound infections. The following burn wound car unit protocol could possibly control the severity of the burn wound infection. These include surveillance on microbiota on the wound in needed basis, strict enforcement of patient and staff hygiene, isolation of infected individual and continuous monitoring of administration of antibiotics as well as gaining complete susceptibility profile against wound microbiota (Rafla and Tredget 2011).

13.2.2 Hospital- and Ventilator-Associated Pneumonia

Pneumonia is a respiratory disease characterized by inflammation of lungs with congestion caused by the most common nosocomial pathogen, *P. aeruginosa*. Pneumonia gains considerable attention in healthcare sectors where this disease is diagnosing repeatedly with hospitalized individuals whom have underlying medical condition or highly susceptible for *P. aeruginosa* infections (Fujii et al. 2014). Nosocomial pneumonia could be further classified based on the location and conditions of infection onset. If the development of disease observed and clinical manifestation results with inflammation in lung parenchymal after at least 48 h of hospitalization, it is termed as hospital-associated pneumonia (HAP) (Niederman 2010). Similarly, in ventilator-acquired pneumonia (VAP) where the pathogens are sourced from the ventilators. Ventilators are the mechanical device that was designed to breathe for patients where ventilator tube was inserted into the human through windpipe and facilitate the patient to breathe.

As discussed earlier, *P. aeruginosa* is the most common pathogen forming microcolonies on the medical device and infect the airway of an individual acquiring mechanical breath using ventilators. Ventilators acquiring pneumonia creates sever effect which leads to the death of the patient (Aarts et al. 2003). Despite the considerable development in antibiotics and its associated therapeutic practises, HAP and VAP remain a potential threat to the global healthcare sectors (Wilke et al. 2011; Bouglé et al. 2017). Diagnosing these HPA and VAP is possible by observing the clinical manifestation including fever, gradual increase in white blood cell count and microbiological procedures for pulmonary infections. The microbiological procedure confirms disease condition by bacterial loads in the samples including brush specimens, bronchoalveolar lavage fluid and endotracheal aspirate specimen. The recently multidrug-resistant strains of *P. aeruginosa* are prevalence in causing both HAP and VAP and remain reason for mortality associated with these diseases (Tumbarello et al. 2013; Micek et al. 2015).

13.2.3 *Pseudomonas aeruginosa* Infections in Cystic Fibrosis

Cystic fibrosis is caused due to the occurrence of mutation of cystic fibrosis transmembrane conductance regulator (CFTR) gene residing on chromosome VII and a cAMP-dependent chloride channel. These mutations resulted in the dehydrated and thickened airway surface liquid that hinders mucociliary clearance from the airway (Gellatly and Hancock 2013). The initial incidence of the pathogen in the modified airway surface liquid allows the establishment of an acute infection resulting in significant enhancement of inflammatory responses. This condition potentially impairs the immune system which consequently impairs the host to control the inflammation and results in chronic lung inflammation (Alhazmi 2015). Although *P. aeruginosa* not a causative agent for the CF since it being an autosomal receive genetic disorder but *Pseudomonas* plays an major role in increases of mortality rate in persons with CF. *P. aeruginosa* is capable to overcome from the treatment for acute infection and can able to adapt to the lung environment and potential establishing the biofilm resulting in chronic infection (Markou and Apidianakis 2014). Interestingly, *P. aeruginosa* strains isolated from the acute phase of lung infection are not similar to that of the strain observed in the chronic lung infection (Fig. 13.1). Herein, most of the virulent phenotypes observed in the isolates of acute lung infections are absent in their counterpart isolate of chronic infections (Sousa and Pereira 2014). For example, the inflammatory response stimulating virulence factors of *P. aeruginosa* including flagella and pili are not expressed in the isolates from chronic infection. Similarly, several other virulence factors are downregulated including type III secretion system. On the other hand, *P. aeruginosa* from chronic lung infections prevalently overexpresses the genes that take part in the successful establishment of biofilm, consequently production of the exopolysaccharide and alginate were drastically increased and the colonies become mucoid (Gellatly and Hancock 2013). The individual with CF disorder has a high tendency to be infected with chronic pseudomonal lung infections. Hyperactive inflammatory response remains

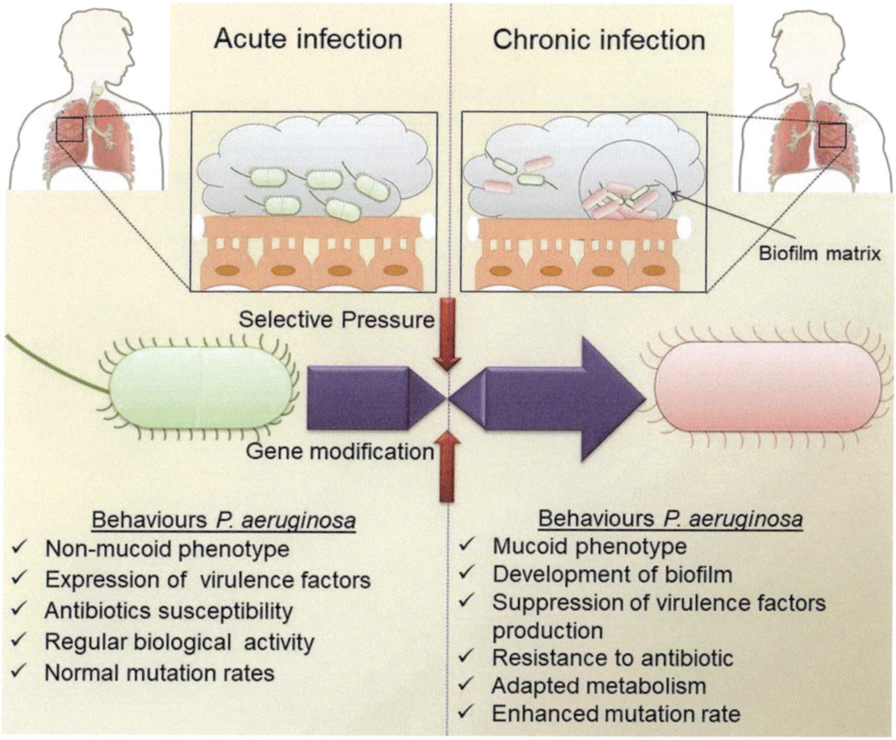

Fig. 13.1 Schematic representation of microevolution of *P. aeruginosa* during infection in CF lungs

a clinical manifestation of CF was potentially stimulated to worse by the toxins from the pathogens and result in gradual deterioration of lung function and subsequently leads to death of an individual (Furiga et al. 2016).

13.3 *Pseudomonas aeruginosa* Pathogenesis and Major Virulence Factors

Though the intensive advancement in the antimicrobial therapies addressed the answers to numerous deadly infections and diseased conditions, bacterial infections remain an uphill challenge for the scientific community which endorsed the ever-increasing burden of multidrug-resistant strains deteriorate the efficacy of the routine antibiotics. This brings a situation where the severely infected individuals continuously failed to respond to the appropriate antibiotic even when infected with susceptible microbial pathogens (Hauser 2011). To address upon the present problem, efforts are being made to gain appropriate knowledge about the bacterial pathogenesis and their adaptive mechanism toward antimicrobial agents. What

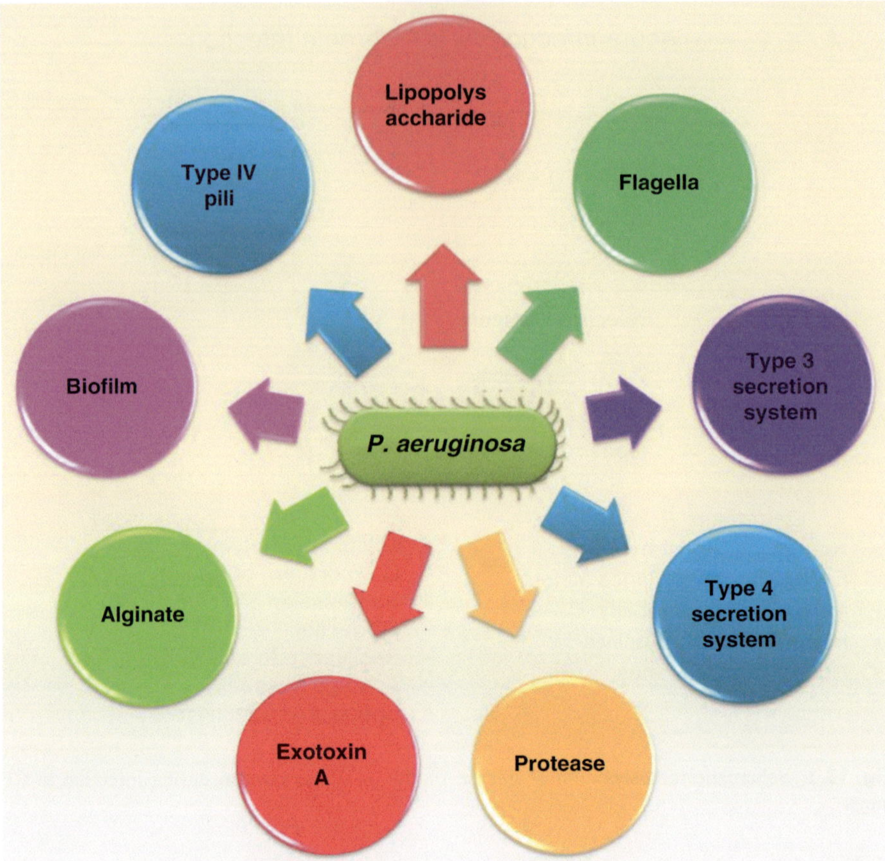

Fig. 13.2 Pathogenic factors of *P. aeruginosa*

follows is a maximum knowledge about known virulence determines suspected of contributing to *P. aeruginosa* pathogenesis (Fig. 13.2).

13.3.1 Lipopolysaccharide

Lipopolysaccharide (LPS) is widely accepted as a crucial virulence trait observed in *P. aeruginosa.* It is a complex glycolipid which is the major constituent of outer leaflet in the outer membrane of Gram-negative bacteria. The LPS plays a significant role in bacterial pathogenesis process by protecting the pathogen from the host defense mechanisms. Furthermore, it displays several other roles in bacterial pathogenicity including antigenicity, the inflammatory response, exclusion of external molecules and obstructing the interaction with antimicrobial agents (Gellatly and Hancock 2013). Since LPS gains biological importance in *P. aeruginosa* pathogenesis, intense study has been conducted to obtain vast knowledge about its biosynthesis pathways and other virulence impact of LPS. Enhancing the knowledge about the structure, function and participant genes of *P. aeruginosa* in the synthesis process of LPS could potentially increase the global understanding of the pathogenic profile of *P. aeruginosa* (Goldberg and Pier 1996). The LPS of *P. aeruginosa* constitutes three domains including lipid A which holds the disaccharide backbone that support the attachment of large amount of fatty acid chain subsequently facilitate a ground to the LPS to stable at the outer membrane. The structural difference of the lipid A in their number, position and properties of connected acyl groups is based on isolated and environmental conditions (Pier 2007). This lipid A glycolipid participates in the activation of host signalling pathways including TLR4 to NFκB by binding to coreceptors such as MD2 and CD14. Activation these signalling pathways leads to the synthesis of pro-inflammatory cytokines, chemokines, inflammation and subsequently induce endotoxic shock.

The variation that occurs in the lipid A could potentially alter the susceptibility nature of *P. aeruginosa* to different antimicrobial agents and also modify the inflammatory behaviours. For instance, the isolates from the chronically infected CF individual showed significantly enhanced inflammatory response and the extent of these alterations appears to gain high disease severity (King et al. 2009). Nevertheless, these modifications support *P. aeruginosa* to develop resistance against routinely administered antimicrobial agents. The second domain refers the polysaccharide core region, which is constructed by attaching nine or ten sugar molecules with lipid A and branched with oligosaccharide. The above-mentioned two domains are most common in the LPS molecules on the surface of the bacterial cells, which are also referred as lipid A-core. The third domain refers as O-specific polysaccharide, O antigen or O chain, which shows high variability due to its chemical constituent where it made up of continuous carbohydrate polymer, covalently attached to the core. Interestingly, *P. aeruginosa* is potential enough to synthesize two different O antigens which distinctly shows a variation in their structure, serology and their biosynthesis pathway in the same cells. Previously, these have been known to be called as A- and B-band O antigens, but recently according to the structural nomenclature and the O-serotyping pattern of *P. aeruginosa* the B-band O antigen is called as O-Specific Antigen, whereas the A-band O antigen is called as Common Polysaccharide Antigen (PIER 2007). The Common Polysaccharide Antigen production is very common, but not in all the strain of the *P. aeruginosa.* This antigens are the complex of homopolymer of d-rhamnose, which promote the weak antibody response. On the other hand, O-Specific Antigens are heteropolymer with continuous O unit with various sugar moieties. Unlike Common Polysaccharide Antigen, the O-Specific Antigens are potentially immunogenic and readily enhance the antibody response (Maldonado et al. 2016). The chemical structure, physiological behaviour and the number of sugar moieties in the O-Specific Antigen's structure could be varied from isolates to isolates and also based on their growth conditions (Werneburg et al. 2012).

13.3.2 Flagella and Type IV Pili

The attachment of bacterial pathogen to the host surface is an initial and critical step involved in the establishment of *P. aeruginosa* infection. Flagella and type IV pili are two major *P. aeruginosa* adhesins, which play a significant role in the initial attachment process. Furthermore, these two virulence phenotypes remain the major attributes for the spreading of disease from one organ to the other and also support the successful establishment of *P. aeruginosa* biofilm (Campodonico et al. 2010; Bucior et al. 2012). Flagella are one among organs of bacteria which is found to be highly complex and conserved even among diverse bacterial species. The synthesis process of flagella was encoded by more than 50 genes. The flagella in pathogen plays several roles in development of infection like maintaining the chemotaxis and mediating the motility. In some pathogen motility remains an important phenotype for survival, because the flagella provide an opportunity for the pathogen to acquire essential nutrients from the environment (Feldman et al. 1998).

Type IV pili are nanoscale protein filament with hair-like appendages decorated on the surface of *P. aeruginosa.* These type IV pili initiate and support the bacterial attachment process, facilitate the cell–cell aggregation, initiation and maturation of biofilm and providing twitching motility (Beaussart et al. 2014). *P. aeruginosa* has the inherent potential to exhibit different types of motility including swimming, twitching and swarming. The swimming motility frequently observed in aqueous conditions or in the medium plate with less agar concentration. This swimming motility is mediated by the flagella and hence called as flagellum-mediated motility. Similarly, twitching motility is mediated by type IV pili and mostly observed on the solid or interface medium. The third motility is termed as swarming motility that promotes coordinated and rapid movement of the *P. aeruginosa* on the semi-solid media. Recently, it was reported that swarming motility could be mediated by both flagella and type IV pili and in some cases even the amount of rhamnolipids present in the medium also affects the swarming motility of *P. aeruginosa* (Overhage et al. 2007). The role of the motility in the pathogenicity of the *P. aeruginosa* was investigated by several in vivo studies where the mutant strain with defective of both flagella and pili infected to burn wound mouse (Feldman et al. 1998).

13.3.3 Type III Secretion System

In bacteria, a secretion is a biological event in which the macromolecules of bacteria are transferred across the cellular envelope into the surrounding environment. This biological event believed to be more difficult in Gram-negative bacteria where macromolecules tasked to cross two membrane barriers. The evolution in Gram-negative bacteria addressed a path to overcome these complications by facilitating the development of specialized secretion systems. Till date, six secretion systems were reported in Gram-negative bacteria, *P. aeruginosa*. Type III secretion systems (T3SS) are widespread membrane-embedded nanomachinery that are found in different Gram-negative bacteria and facilitate transfer of toxin molecules also termed as effectors from bacterial cytosol to cytoplasm of the targeted eukaryotic cells (Puhar and Sansonetti 2014). This makes the difference among the secretion system that most of them secrete the effectors into the extracellular surroundings, where they bind to the distant target cell surface receptors. The effect of the toxin distinctly varies where in some cases it supports the symbiotic relationship between the host and the bacteria and in most cases it mediates the pathogenic features including membrane disruption, activation of apoptosis process and structural rearrangements in cytoskeleton (Burkinshaw and Strynadka 2014). However, T3SS apparatus are highly conserved at both structural and functional level among the bacterial species but potentially inject the effectors into the distinct hosts such as humans, plants and animals (Izoré et al. 2011). The apparatus of T3SS consists of three major parts including multiring basal body (made up of more than 25 different proteins and bridges all the layers of the bacteria including inner bacterial membrane), the peptidoglycan layer and the outer bacterial membrane; a hollow needle-like structure in which the bacteria mobilizes effector toxin from basal body in semi-unfolded form, it believed to have both inner and outer membrane component. Herein, the inner membrane established of the lipoprotein PscJ and outer membrane component made of oligomerized secretin PscC; and translocon, pore or the needle that inserted into the host cells and the release the toxin effectors and made up of two T3SS hydrophobic protein that play significant role the host immune system alteration (Anantharajah et al. 2016; Cascales 2017).

Most of the pathogenic strains of *P. aeruginosa* have the inherent ability to overcome the phagocytic clearance which is dependent upon the behaviour of T3SS (Burstein et al. 2015). Several clinical investigations also supported the statement that T3SS of *P. aeruginosa* has a crucial role in inducing pathogenicity and in most cases it leads to the death of the patient. Furthermore, in in vitro condition, T3SS mutant strain failed to establish the infection in most of the host unlike its counterpart wild strain (Galle et al. 2012). The T3SS of *P. aeruginosa* participates in the production and injection four effector toxins such as ExoS, ExoT, ExoY and ExoU. Recently, nucleoside diphosphate kinase was also reported to be transferred from pathogen to host cells via T3SS (Zhu et al. 2016). Although it seems T3SS involved in the production of limited numbers of effector toxins, but these limited effector toxins furnished *P. aeruginosa* to cause infection among wide range of hosts (Yamazaki et al. 2012). Interestingly, the strain nullified with all the four effector toxin genes is still capable to cause infections indicating the possibility of other effectors beyond these four effectors toxin which is yet be discovered (Galle et al. 2012). The activation of T3SS in *P. aeruginosa* occurred as responsive mechanism to the environmental signals including the attachment of pathogen on to the host and comparatively low concentration level of calcium in the growth medium or in the growing environment. However, the exact signalling mechanism is yet to be known (Anantharajah et al. 2016).

13.3.4 Type IV Secretion System (T4SS)

Similar to the T3SS, the T4SS also possesses core complex bridging both inner and outer bacterial membranes and pili that extends out into the extracellular surroundings (Depluverez et al. 2016). Based on the function, the T4SS could be classified into three distinct categories. The first category system involves the mobilization of single-stranded DNA into the host cells by conjugation process which potentially facilitates the pathogen to adapt to the host environmental condition, without surprising mediates the development of resistance mechanisms (Juhas et al. 2008). The first category T4SS system is activated only when the pathogen is in contact with the host. Similarly, second category system functions as transport to several protein molecules directly into the host cells, thus system involves successful establishment of infection by introducing pathogenic secretion into the host tissues from the pathogen (Christie et al. 2014). The third category of T4SS influences the adaptation of DNA from the external environment. It also involves in releasing both DNA and protein molecules to the exterior surface of bacteria. The ability to transfer both DNA and protein to the host cells make the T4SS system unique among the other secretion system (Trokter et al. 2014).

P. aeruginosa is one among the Gram-negative bacteria which employs T4SS for horizontal gene transfer that help to create a pathogenic island which subsequently plays a major role in its pathogenicity (Ma et al. 2003). In *P. aeruginosa*, pKLC and PAPI well-known pathogenicity islands that use T4SS for the spreading of pathogenic gene in *Pseudomonas*. The pathogenicity island, pKLC is constituted by combining phage and plasmid origin, which potentially transfer genetic information with increased excision rate from the chromosome. It being 103,532 bp long island with broad-spectrum open reading frames and genes encoding for type IV sex pili. Furthermore, it holds the gene encoded for *chvB*, which is found to be a major virulence trait in *P. aeruginosa* that commonly named as glucan synthetase (Juhas 2015). The other pathogenic island PAPI displayed as a larger island with 108 kb that encodes for several pathogenic traits. For the transfer genetic information from PAPI, it forms extrachromosomal circular intermediate before it is administrated into the recipient cells (Carter et al. 2010).

13.3.5 Proteases

Proteases are the enzymes synthesized by a microorganism which helps the microorganism, especially microbial pathogen in several ways including providing peptide nutrients and shows its effective contribution in the pathogenesis of infectious disease by different mechanism (Lantz 1997). These proteases are potentially capable of causing direct or indirect damage to host by lysing the cell surface and tissue protein or actively participates in inactivating the important proteins for host defence mechanism, respectively (Herwald and Egesten 2009; Musicki et al. 2009). Although microbial proteases are widely accepted as virulence factors due to their significant role in the pathogenesis of infectious diseases, limited reports are still available to confirm that microbial protease specifically participates in the microbial pathogenesis (Ingmer and Brøndsted 2009). Among the pathogenic microorganisms, *P. aeruginosa* was widely studied for the secretion of proteases and their participation in pathogenicity. In most cases, the ocular infection and sepsis condition by *P. aeruginosa* was mediated by proteases which potentially involved in the denaturation of immunoglobulins, fibrins and subsequently disrupt the epithelial cells (Alionte et al. 2001). *P. aeruginosa* is well known for its ability in producing alkaline protease via type I secretion system which readily participates in the degradation process of host complement protein and fibronectin. Nevertheless, they also produced two different elastolytic enzymes such as LasA and LasB. The production of these enzymes is mediated by the quorum sensing mechanisms of the *P. aeruginosa* and produced via type II secretion system. LasA proteases are serine protein and act on the cell wall of *Staphylococci*, hence often called as "Staphylolysin". LasB protease, often called as "elastase" which potentially acts on the lung surfactant proteins A and D (Matsumoto 2004; Gellatly and Hancock 2013). In recent years, different protease inhibitors including α_2-macroglobulin and specific elastase inhibitors were reported which control the *P. aeruginosa* infections and fatal septic shock which are mediated by bacterial proteases enzymes (Hobden 2002).

13.3.6 Exotoxin A

As discussed earlier, *P. aeruginosa* is capable of producing several extracellular virulence factors. Exotoxin A is one among them which attributes a significant role in the pathogenesis of *P. aeruginosa* infections (Al-Dahmoshi et al. 2018). *P. aeruginosa* utilizes a type II secretion system for the production of exotoxin A and the functional characters of this toxin are identical to Diphtheria toxin. The host receives this toxin through a surface receptor called α_2-macroglobulin receptor (Morlon-Guyot et al. 2009). Exotoxin A manages to internalize into a cytoplasm via previously mentioned transport mechanism and potentially affects the elongation factor, which subsequently inhibits protein synthesis in the host and suppresses the host immune response. Furthermore, the exotoxin A is also capable of inducing apoptosis process in host which finally leads to the death of the host cells (Pillar and Hobden 2002). This behaviour of exotoxin A allows the research community to conduct investigations to use these exotoxin A as apoptosis-inducing factor on cancer cells. The production of exotoxin A in *P. aeruginosa* mainly dependent on the availability of iron and the gene *regA* is directly involves in the transcriptional regulation in the production of exotoxin A. Similarly the genes *vfr* and *lasR* are indirectly participating in the transcriptional regulation process coupled with the synthesis of exotoxin in *P. aeruginosa* (Michalska and Wolf 2015).

13.3.7 Alginate

In *P. aeruginosa* infection, neutrophils remain an initial line of defence mechanism via phagocytosis. On the other hand, the alginate of *P. aeruginosa* protects the bacteria from several host defence mechanism including phagocytosis, oxygen radicals and other host immune defence mechanisms as well as from the treatment of antimicrobial agents (Colbert et al. 2018). The importance of clinical complications because of the presence of the alginate was reported that it is capable to create a condition of inefficient pulmonary clearance (Franklin et al. 2011). Moreover, they actively involved in spreading the infection from one organ to another, for instance alginate could able to mediate the spreading of lung infection to spleen (McCaslin et al. 2015). Interestingly, the pathogenic isolates of *P. aeruginosa*, PAO1 found to be non-mucoid. At the same time, isolates colonize on the host confronts several host immune feedback and exposed to different antimicrobial agents. These adverse conditions make the isolates to alter its transcriptional regulators and convert to a mucoid phenotype by synthesizing alginate. However, in some cases the mucoid phenotype resides in the lung infected area are failed to hold similar phenotype when they culture in laboratory conditions (Limoli et al. 2017).

13.3.8 Quorum Sensing and Biofilm Formations

Quorum sensing (QS) is a communication mechanism lies among the microbial community which facilitates a coordinated adaptation of microbial community to certain environmental condition, including adapting towards infection site within the host. This communication is mediated by a small membrane-diffusible signalling molecule also termed as autoinducers. These signals are coordinately synthesized by the microbial cells which act as cofactors for most of the transcriptional regulators coupled with virulent traits, when the concentration of this signalling molecules reaches its threshold level in the environment (Fila et al. 2018). However, the level of signalling molecules in the surrounding environment directly depends upon the number of microbial cells in the same environment. These signalling molecules coordinated the entire community to deliver a similar response. Furthermore, these communication networks, widely determined as a major cause for the evolution of multidrug resistance (Kalaiarasan et al. 2017).

P. aeruginosa utilizes this microbial communication to coordinate among their community. However, *P. aeruginosa* is well known for its ability to produce three signalling molecules including 3-oxo-dodecanoyl homoserine lactone (3-oxo-C_{12} HSL), butyryl homoserine lactone (C_4 HSL) and 2-heptyl-3-hydroxy-4-quinolone (PQS). The first signalling molecule 3-oxo-C_{12} HSL, is synthesized due the activation of LasI AHL synthase and binds with LasR transcriptional receptor protein. Similarly, the second signalling molecule, C_4 HSL, is produced by initiation of RhlI AHL synthase and bind with specific transcriptional receptor, RhlR. The synthesis process of the third signalling molecule is mediated by complex multistep biosynthesis pathway (Harmsen et al. 2010). This microbial communication network facilitates the *P. aeruginosa* several ways including survival of the bacteria, allows development of biofilm and controlling the production of different virulence factors. Several reports are supporting this context where the strain in absence of these signalling mechanisms was failed to produce different virulence determines and establishment of infections (Kariminik et al. 2017; Defoirdt 2018).

Biofilm could be simply defined as a biological gathering where microbial community attached with one another and develop a well-organized complex structure that is possible to happen under governs of quorum sensing system of bacteria (Palanisamy et al. 2014). In biofilms, all the bacterial cells are compactly packed inside the polymeric matrix also termed as extracellular polymeric substance (EPS). These polymeric matrixes contain several other biological substances including polysaccharides, nucleic acid, lipids and proteins (Koh et al. 2013). Among the total weight of biofilm more than 50–90% weight is shared by the polymeric matrix rest of the weights are shared by the bacterial cell in the biofilm. These thick complex matrixes facilitate the biofilm in several ways including protecting from both physical and chemical factors that subsequently affect the biofilm (Bai and Rai 2011). Due to the complex nature of the biofilm, potentially resist the penetration of any chemical from external environment to biofilm which makes most of the routine antibiotics inefficient to eradicate the biofilm. Interestingly, the bacterial cells show very slow growth than planktonic cells due to the establishment of the nutrient-limited environment inside the biofilm (Kalia et al. 2019). However, biofilm residing shows high resistance than its counterpart planktonic cells. These conditions made the *P. aeruginosa* infections remain as an unsolved problem in human healthcare sectors.

13.4 Antibiotic Resistance Mechanism of *Pseudomonas aeruginosa*

In recent years, treating the disease caused by *P. aeruginosa* becomes a potential issue worldwide due to the development of a resistance mechanism against available antibiotics. Based on recent WHO report, *P. aeruginosa* strain with carbapenem-resistant gene was found to cause most curial infection. Continuous instances of antimicrobial-resistant strains were observed due to the excessive use of antimicrobial agents in the infection treatment process (Chatterjee et al. 2016). *P. aeruginosa* demonstrates resistance to major classes of antibiotics such as aminoglycosides (e.g. gentamicin), quinolones (e.g. ciprofloxacin), β-lactams (e.g. cephalosporin) and polymyxins (e.g. colistin). The mechanisms underline with the antimicrobial resistance in *P. aeruginosa* lies with multifactorial processes like synthesis of antibiotic altering enzymes such as β-lactamases, metallo-β-lactamases and other enzymes which effectively alter the aminoglycoside; expression of the efflux pumps which potentially protect the bacterium from the internalization of antimicrobial agent into cytoplasm; capable to gain antibiotic resistance genes (Taylor et al. 2014). Furthermore, self-mutation also contribute to the successful establishment of the resistance mechanism in *P. aeruginosa*. Nevertheless, development of biofilm by the bacterium creates a micronutrient environment which potentially limits the penetration of the antimicrobial agent into the biofilm structure and direct multiresistance mechanism to cells reside in the biofilm matrix (Pang et al. 2019). The resistance mechanism could be majorly categorized into three different resistance mechanisms including intrinsic, acquiring and adaptive.

13.4.1 Intrinsic Resistance

The intrinsic resistance mechanism toward antimicrobial agents in *P. aeruginosa* could be defined as innate capability of a bacterium to suppress the activity of the antimicrobial agent by altering its structural and functional motifs. The intrinsic antibiotic resistance is found often in *P. aeruginosa* against different classes of antibiotics which could be achieved by the bacterium by controlling the permeability nature of an outer membrane, activation of efflux pump system which readily recognize the antibiotics and pumps out from the cell and expressing the genes encodes for the production of antibiotic inactivating enzymes (Fig. 13.3) (Valentini et al. 2018). The condition of lower permeability lies on the outer membrane of the bacterial cells making it difficult for antimicrobial agents to penetrate into the bacterial cytoplasm. Some of structurally small antimicrobial agents possessing the hydrophilic potential and readily dissolve in water molecules and pass thorough the water-filled porin channels (Breidenstein et al. 2011). The process involved in this

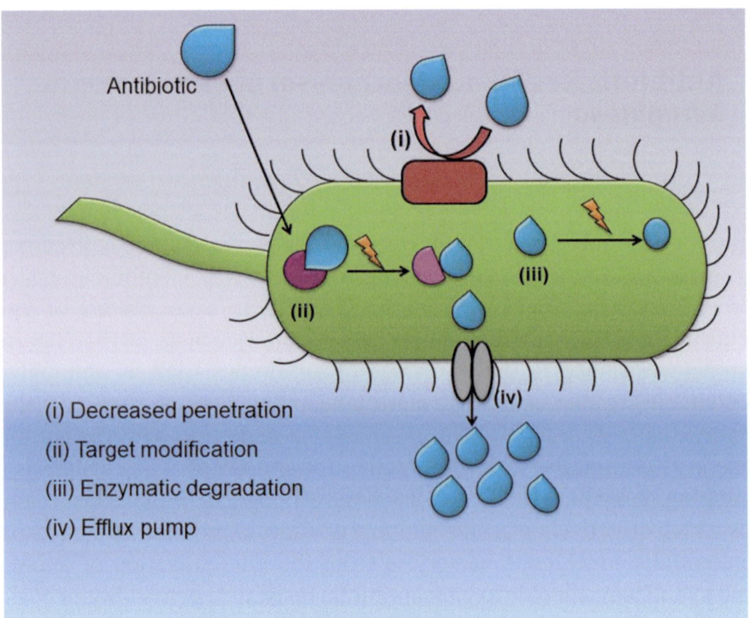

Fig. 13.3 Mechanisms and consequences of bacterial resistance to antibiotics

mode of penetration of the antimicrobial agents into bacterial cells found relatively slow which subsequently facilitate the option to develop more intrinsic resistance to different antimicrobial agents. Although the outer membrane plays a significant role in controlling the antibiotic entry, this mechanism alone is not sufficient to control the entry of the antibiotics. The other intrinsic resistance mechanism, efflux system provides a promising contribution that resists the internalization of the antimicrobial agents into the cells (Oliver et al. 2015). The synergizes mode of both discussed intrinsic mechanisms proceeding to the development of high level of the resistance to currently available antimicrobial agents relevant for *P. aeruginosa.*

13.4.2 Acquired Resistance

The major difference between the intrinsic and acquired resistance is the prior exposure to the antimicrobial agent. Wherein the intrinsic resistance it is not necessary that bacterium should previously exposed to antimicrobial agents, but in case of acquired resistance, it could be initiated once the bacterium experienced the antimicrobial agents (Friedman et al. 2016). In acquired resistance mechanism, once the bacteria experience the antimicrobial agent it undergoes several chromosomal gene modifications and develops resistance against same antimicrobial agent. In other way bacteria also undertake the acquired resistance via acquiring the genetic resistance information from different molecular elements including plasmids, transposons, interposons and integrons (Satpathy et al. 2016). The major mechanism of horizontal gene transfer necessitates transformation, transduction and conjugation. These genetic alterations on the bacterial system could potentially control the uptake of antimicrobial agents and changing the target of the antibiotics (Macia et al. 2014). Nevertheless, the gene modification also facilitates the functional behaviours of intrinsic resistance such as expression of efflux pump and production of antibiotic inactivating enzymes.

13.4.3 Adaptive Resistance

Adaptive resistance mechanism supports *P. aeruginosa* to overcome and continue to survive even after antibiotic treatment by altering the genetic information of the protein synthesis according to the environmental signals. Adaptive mechanism of the *P. aeruginosa* gains significant consideration in the clinical sectors due to its notable role in evading from the antimicrobial treatment process (Yelin and Kishony 2018). *P. aeruginosa* adaptive resistance was widely studied in which biofilm plays a significant role in creating persistent cells that subsequently case the persistent infection and result in poor prediction of the course of a disease. Unlike the other resistance mechanisms, the adaptive resistance mechanism is completely dependent upon the environmental circumstances which could regulate the different transcriptional factors of *P. aeruginosa* and surprisingly, the susceptibility nature of the bacterium most often reverted when the adverse environmental circumstances are normalized or removed (Bjarnsholt et al. 2005). Different environmental factors like temperature, pH, oxygen demand condition, DNA stress and nutrient deficiency as well as the motility of the organisms also remain as potential causes for the development of adaptive resistance.

Formation of biofilm remains an important factor for the initiation of an adaptive resistance mechanism. As discussed earlier, biofilm is an aggregation of microbial cells that attached one another and encapsulated in self-producing polymeric matrix and capable to attach to both biotic and abiotic surfaces (de la Fuente-Núñez et al. 2013). The *P. aeruginosa* cells reside in the biofilm are comparatively less sensitive to antibiotics and immunes defence mechanisms of host than its counterpart planktonic cells. Interestingly, the sensitivity of the *P. aeruginosa* against the antimicrobial agents would restore when the biofilm residing cells happen to grow outside of the biofilm (Jamal et al. 2018). This behavioural change is elucidating that biofilm-mediated resistance is independent of genomic modification and adaptive resistance mechanisms. The biofilm condition makes several physiological and phenotypic alterations in bacterium. For instance, the non-mucoid strain of *P. aeruginosa* happens to enter into the biofilm matrix would transcriptionally altered as from non-mucoid- to mucoid-producing strain (Macia et al. 2014).

13.5 Novel Strategies for Treating Drug-Resistant *P. aeruginosa* Infection

Although there are continuous instances of the antibiotic resistance strain, still antibiotics are used to control the infection caused by *P. aeruginosa.* Almost all the currently available antimicrobial agents are involved in either slowing down the organism's growth also termed as bacteriostatic agents or causing death which is also called as bactericidal agents (Christensen et al. 2012). Each antibiotic specifically interfered with some of the mechanisms underlined with the metabolic pathway notice in bacteria. For instance, certain class of antibiotic acts on the metabolic pathway coupled with bacterial cell wall synthesis, some of other class participates as an inhibitor in bacterial protein and nucleic acid synthesis metabolic pathways. Furthermore, certain group of also plays significant role in hindering nature of bacterial metabolism and cell membrane (Van Acker et al. 2014). As discussed earlier, due to indiscriminate use of antibiotics potentially causing emergence of resistance behaviours among the pathogen, which subsequently brings to the situation where the currently available antimicrobial agent no longer effective to control the bacterial infection (Zaidi et al. 2017). This makes for an urgent call to discover a novel antimicrobial agent to substitute the current antibiotics and potentially address the negative impact that lies with current antibiotics.

In recent years, several different approaches were initiated to improve the efficacy of the currently available antibiotics by means of different formulation procedures (Wagner et al. 2016; Smith et al. 2017). For example, antibiotics are prepared as an inhalation solution and used for long-term treatment processes which are most common in chronic infection of *P. aeruginosa.* However, this formulation process showed better recalcitrant efficacy, but still created a concern about increased antibiotic resistance. Similarly, some of the antibiotics are also formulated as a dry powder which is administered by inhalation. This produces significantly enhancing success rate of treatment procedure due to its ability to penetrate the sputum and potentially reach the infection site (Hurley et al. 2012). Furthermore, decoration of the antimicrobial agents on to the nanomaterial gains significant attention in recent years. As a result, the antibiotics like amikacin and ciprofloxacin are encapsulated in liposome and investigated for its outdo in controlling the bacterial infection as compared to routine administration procedure that lies with the utilization of antibiotics (Jeevanandam et al. 2016).

13.5.1 Anti-Quorum Sensing

Target the signalling system of the *P. aeruginosa* found to be an effective therapeutic option and the drug which involves in the interference of QS is also called as quorum sensing inhibitors (QSIs) (Bacha et al. 2016). The QSIs are novel class of antimicrobial agents because these drug molecules show less possibility to experience and development of bacterial resistance which remains a major drawback among antibiotics. The novel QSI should have some desired properties like should be small molecules which potentially inhibit the expression of QS-related genes; the drug should be more specific toward QS system of the bacteria; these QSI should not ant toxic symptom to the host cells during treatment process; the drug should not interfere with any other metabolic pathway like protein and DNA synthesis than QS circuit of the bacteria; the drug should be chemically stable enough to retain the host cell for sufficient time to enable the effective therapeutic actions (Chatterjee et al. 2016; Asfour 2018). Recently, several biological metabolites including phytochemicals and microbial metabolites are employed as potential anti-QS agents (Bhardwaj et al. 2013). For example, the phytochemical, mosloflavone was investigated for their ability in the hindering the QS of *P. aeruginosa* and suppression of virulence traits. The study release that the phytochemical, mosloflavone effectively blocks the QS system of *P. aeruginosa* and significantly reduces the production of different virulence factors. The study also reveals the toxicity profile of mosloflavone using in vivo model, *Caenorhabditis elegans* (Hnamte et al. 2019).

13.5.2 Bacteriophage Therapy

Bacteriophages are generally termed as bacterial viruses which are ubiquitous and found in diverse organisms and grow vigorously in the bacterial host (Rohde et al. 2018). The life cycle of bacteriophage is widely classified into virulent phage (lytic phage) which potentially causes death to the host and temperate phages (Pires et al. 2015). The lytic phages are initially attaching to the surface of the host cell and it extends its infection by injecting DNA into the host allows it DNA to replicate along with the host. After the complete establishment of infection, it induces the host to undergo for apoptosis process and subsequently kill the host. The releasing phages from the lysed host cell would potentially initiate the infection cycle in other host cells (Waters et al. 2017). On the other hand temperate phages also initiate the infection by attaching to the host cell and gradually integrate its genetic material into the host chromosome. In cases, they integrate its genetic material in bacterial plasmid and allows the transmission during the cell division without causing much adverse effect on the growth of host cell (Chan et al. 2018).

The idea of utilizing the phages as antimicrobial agent was reported in previous century but due to certain limitations and the phenomenal growth in antibiotics significantly reduced the popularity of the phage-mediated antimicrobial therapy. As a consequence of resistance mechanisms, alternative therapeutic options gained considerable interest in phages-mediated antimicrobial therapy (Hill et al. 2018; Law et al. 2019). The bacteriophage-mediated antimicrobial therapy found to be an potential alternative currently available antimicrobial agents in several ways: bacteriophages never initiate self-amplification process in absence of it susceptible bacteria; they are capable to penetrate into the biofilm matrix and infect the bacteria; their potential in killing even the persistent cells were found highly difficult using conventional antibiotics (Chan et al. 2018). A recent study showed that phage therapy could be an effective therapeutic aid against *P. aeruginosa* infection and study proved the result inference using zebrafish model system (Cafora et al. 2019).

13.5.3 Antibiofilm Peptides

As discussed earlier, the biofilm matrix of *P. aeruginosa* limits the efficacy of most of the antibiotics. The infections associated with biofilm bring the situation where the treatment such disease found utterly difficult to treat (Teerapo et al. 2019). There is an urgent call for the development of novel antimicrobial agent while the conventional antibiotic are increasingly inefficient in the controlling the bacterial infection. With this aspect, antimicrobial peptides are placed in the centre point of the attraction as an alternative approach for the treatment of bacterial infections (Pletzer and Hancock 2016). In general, antimicrobial peptides are the highly conserved molecules observed in diverse organisms and having greater concern with innate immunity of all species (Dostert et al. 2019). The antimicrobial peptides gained significant consideration due their specialized characteristic features including rapid action and wide spectrum of antimicrobial activity against different microorganisms such as bacteria, viruses, fungi and protozoa. Furthermore, antimicrobial peptides are less potential in the development resistance within the microbial cells, hence this has been displayed as promising antimicrobial agent to combat the multidrug-resistant strain-mediated infections (Grassi et al. 2019).

13.6 Conclusion

Prevention of individuals from *P. aeruginosa* infection remains significant burden for humans. Different resistances mechanisms such as intrinsic, acquired and adaptive possibly strengthen the multidrug-resistant behaviours of *P. aeruginosa,* which subsequently displays a potential tolerant to almost all the available antimicrobial agents. Moreover, the biofilm-forming nature of *P. aeruginosa* delivers several persistent cells which make the condition highly unfavourable for regular antimicrobial treatment process. Over several decades, several remedies were discovered to enhance the delivery and the efficacy of the antimicrobial agents to host that eventually control the microbial infection. However, *P. aeruginosa* is highly capable to adapt to most of the adverse conditions which lead to the repeated incidence of resistance development. In recent years, non-antibiotic therapies including anti-virulence, phage and antimicrobial peptide therapies are widely investigated for their antimicrobial potential which showed less potential in developing resistance. Still considerable knowledge about the host–microbial interactions is required to design innovative and effective antimicrobial agents to control infection caused by multidrug-resistant microorganisms.

References

Aarts M-A, Hancock JN, Heyland D, Marshall JC (2003) Antibiotics for ventilator-associated pneumonia. In: Aarts M-A (ed) The Cochrane database of systematic reviews. Wiley, Chichester

Ahmed MN, Porse A, Sommer MOA et al (2018) Evolution of antibiotic resistance in biofilm and planktonic *Pseudomonas aeruginosa* populations exposed to subinhibitory levels of ciprofloxacin. Antimicrob Agents Chemother 62:1–12. https://doi.org/10.1128/AAC.00320-18

Al-Dahmoshi HOM, Al-Khafaji NS, Abdulzahra Jeyad A et al (2018) Molecular detection of some virulence traits among *Pseudomonas aeruginosa* isolates, Hilla-Iraq. Biomed Pharmacol J 11:835–842. https://doi.org/10.13005/bpj/1439

Alhazmi A (2015) *Pseudomonas aeruginosa* – pathogenesis and pathogenic mechanisms. Int J Biol 7:44–67. https://doi.org/10.5539/ijb.v7n2p44

Alionte LG, Cannon BM, White CD et al (2001) *Pseudomonas aeruginosa* LasA protease and corneal infections. Curr Eye Res 22:266–271. https://doi.org/10.1076/ceyr.22.4.266.5509

Anantharajah A, Mingeot-Leclercq M-P, Van Bambeke F (2016) Targeting the type three secretion system in *Pseudomonas aeruginosa*. Trends Pharmacol Sci 37:734–749. https://doi.org/10.1016/j.tips.2016.05.011

Asfour H (2018) Anti-quorum sensing natural compounds. J Microsc Ultrastruct 6(1):1. https://doi.org/10.4103/JMAU.JMAU_10_18

Azam MW, Khan AU (2019) Updates on the pathogenicity status of *Pseudomonas aeruginosa*. Drug Discov Today 24:350–359. https://doi.org/10.1016/j.drudis.2018.07.003

Bacha K, Tariku Y, Gebreyesus F et al (2016) Antimicrobial and anti-quorum sensing activities of selected medicinal plants of Ethiopia: implication for development of potent antimicrobial agents. BMC Microbiol 16:139. https://doi.org/10.1186/s12866-016-0765-9

Bai AJ, Rai VR (2011) Bacterial quorum sensing and food industry. Compr Rev Food Sci Food Saf 10:183–193. https://doi.org/10.1111/j.1541-4337.2011.00150.x

Bassetti M, Vena A, Croxatto A et al (2018) How to manage *Pseudomonas aeruginosa* infections. Drugs Context 7:1–18. https://doi.org/10.7573/dic.212527

Battle SE, Rello J, Hauser AR (2009) Genomic islands of *Pseudomonas aeruginosa*. FEMS Microbiol Lett 290:70–78. https://doi.org/10.1111/j.1574-6968.2008.01406.x

Beaussart A, Baker AE, Kuchma SL et al (2014) Nanoscale adhesion forces of *Pseudomonas aeruginosa* type IV pili. ACS Nano 8:10723–10733. https://doi.org/10.1021/nn5044383

Bhagirath AY, Li Y, Somayajula D et al (2016) Cystic fibrosis lung environment and *Pseudomonas aeruginosa* infection. BMC Pulm Med 16:174. https://doi.org/10.1186/s12890-016-0339-5

Bhardwaj AK, Vinothkumar K, Rajpara N (2013) Bacterial quorum sensing inhibitors: attractive alternatives for control of infectious pathogens showing multiple drug resistance. Recent Pat Antiinfect Drug Discov 8:68–83. https://doi.org/10.2174/1574891X11308010012

Bjarnsholt T, Jensen PØ, Burmølle M et al (2005) *Pseudomonas aeruginosa* tolerance to tobramycin, hydrogen peroxide and polymorphonuclear leukocytes is quorum-sensing dependent. Microbiology 151:373–383. https://doi.org/10.1099/mic.0.27463-0

Bouglé A, Foucrier A, Dupont H et al (2017) Impact of the duration of antibiotics on clinical events in patients with *Pseudomonas aeruginosa* ventilator-associated pneumonia: study protocol for a randomized controlled study. Trials 18:37. https://doi.org/10.1186/s13063-017-1780-3

Breidenstein EBM, de la Fuente-Núñez C, Hancock REW (2011) *Pseudomonas aeruginosa*: all roads lead to resistance. Trends Microbiol 19:419–426. https://doi.org/10.1016/j.tim.2011.04.005

Bucior I, Pielage JF, Engel JN (2012) *Pseudomonas aeruginosa* pili and flagella mediate distinct binding and signaling events at the apical and basolateral surface of airway epithelium. PLoS Pathog 8:e1002616. https://doi.org/10.1371/journal.ppat.1002616

Burkinshaw BJ, Strynadka NCJ (2014) Assembly and structure of the T3SS. Biochim Biophys Acta, Mol Cell Res 1843:1649–1663. https://doi.org/10.1016/j.bbamcr.2014.01.035

Burstein D, Satanower S, Simovitch M et al (2015) Novel type III effectors in *Pseudomonas aeruginosa*. MBio 6:2–7. https://doi.org/10.1128/mBio.00161-15

Cafora M, Deflorian G, Forti F et al (2019) Phage therapy against *Pseudomonas aeruginosa* infections in a cystic fibrosis zebrafish model. Sci Rep 9:1527. https://doi.org/10.1038/s41598-018-37636-x

Campodonico VL, Llosa NJ, Grout M et al (2010) Evaluation of flagella and flagellin of *Pseudomonas aeruginosa* as vaccines. Infect Immun 78:746–755. https://doi.org/10.1128/IAI.00806-09

Carter MQ, Chen J, Lory S (2010) The *Pseudomonas aeruginosa* pathogenicity island PAPI-1 is transferred via a novel type IV pilus. J Bacteriol 192:3249–3258. https://doi.org/10.1128/JB.00041-10

Cascales E (2017) Inside the chamber of secrets of the type III secretion system. Cell 168:949–951. https://doi.org/10.1016/j.cell.2017.02.028

Chan BK, Turner PE, Kim S et al (2018) Phage treatment of an aortic graft infected with *Pseudomonas aeruginosa*. Evol Med Public Health 2018:60–66. https://doi.org/10.1093/emph/eoy005

Chatterjee M, Anju CP, Biswas L et al (2016) Antibiotic resistance in *Pseudomonas aeruginosa* and alternative therapeutic options. Int J Med Microbiol 306:48–58. https://doi.org/10.1016/j.ijmm.2015.11.004

Christensen LD, van Gennip M, Jakobsen TH et al (2012) Synergistic antibacterial efficacy of early combination treatment with tobramycin and quorum-sensing inhibitors against *Pseudomonas aeruginosa* in an intraperitoneal foreign-body infection mouse model. J Antimicrob Chemother 67:1198–1206. https://doi.org/10.1093/jac/dks002

Christie PJ, Whitaker N, González-Rivera C (2014) Mechanism and structure of the bacterial type IV secretion systems. Biochim Biophys Acta, Mol Cell Res 1843:1578–1591. https://doi.org/10.1016/j.bbamcr.2013.12.019

Colbert B, Kumari H, Piñon A et al (2018) Alginate-regulating genes are identified in the clinical cystic fibrosis isolate of *Pseudomonas aeruginosa* PA2192. bioRxiv:1–33. https://doi.org/10.1101/319004

da Silva Filho LVRF, Ferreira FDA, Reis FJC et al (2013) *Pseudomonas aeruginosa* infection in patients with cystic fibrosis: scientific evidence regarding clinical impact, diagnosis, and treatment. J Bras Pneumol 39:495–512. https://doi.org/10.1590/S1806-37132013000400015

de la Fuente-Núñez C, Reffuveille F, Fernández L, Hancock RE (2013) Bacterial biofilm development as a multicellular adaptation: antibiotic resistance and new therapeutic strategies. Curr Opin Microbiol 16:580–589. https://doi.org/10.1016/j.mib.2013.06.013

Defoirdt T (2018) Quorum-sensing systems as targets for antivirulence therapy. Trends Microbiol 26:313–328. https://doi.org/10.1016/j.tim.2017.10.005

Depluverez S, Devos S, Devreese B (2016) The role of bacterial secretion systems in the virulence of gram-negative airway pathogens associated with cystic fibrosis. Front Microbiol 7:1–8. https://doi.org/10.3389/fmicb.2016.01336

Dostert M, Belanger CR, Hancock REW (2019) Design and assessment of anti-biofilm peptides: steps toward clinical application. J Innate Immun 11:193–204. https://doi.org/10.1159/000491497

Feldman M, Bryan R, Rajan S et al (1998) Role of flagella in pathogenesis of *Pseudomonas aeruginosa* pulmonary infection. Infect Immun 66:43–51

Fila G, Krychowiak M, Rychlowski M et al (2018) Antimicrobial blue light photoinactivation of *Pseudomonas aeruginosa*: quorum sensing signaling molecules, biofilm formation and pathogenicity. J Biophotonics 11:e201800079. https://doi.org/10.1002/jbio.201800079

Franklin MJ, Nivens DE, Weadge JT, Lynne Howell P (2011) Biosynthesis of the *Pseudomonas aeruginosa* extracellular polysaccharides, alginate, Pel, and Psl. Front Microbiol 2:1–16. https://doi.org/10.3389/fmicb.2011.00167

Friedman ND, Temkin E, Carmeli Y (2016) The negative impact of antibiotic resistance. Clin Microbiol Infect 22:416–422. https://doi.org/10.1016/j.cmi.2015.12.002

Fujii A, Seki M, Higashiguchi M et al (2014) Community-acquired, hospital-acquired, and healthcare-associated pneumonia caused by *Pseudomonas aeruginosa*. Respir Med Case Rep 12:30–33. https://doi.org/10.1016/j.rmcr.2014.03.002

Furiga A, Lajoie B, El Hage S et al (2016) Impairment of *Pseudomonas aeruginosa* biofilm resistance to antibiotics by combining the drugs with a new quorum-sensing inhibitor. Antimicrob Agents Chemother 60:1676–1686. https://doi.org/10.1128/AAC.02533-15

Galle M, Carpentier I, Beyaert R (2012) Structure and function of the type III secretion system of *Pseudomonas aeruginosa*. Curr Protein Pept Sci 13:831–842. https://doi.org/10.2174/138920312804871210

Garvey MI, Bradley CW, Tracey J, Oppenheim B (2016) Continued transmission of *Pseudomonas aeruginosa* from a wash hand basin tap in a critical care unit. J Hosp Infect 94:8–12. https://doi.org/10.1016/j.jhin.2016.05.004

Gellatly SL, Hancock REW (2013) *Pseudomonas aeruginosa*: new insights into pathogenesis and host defenses. Pathog Dis 67:159–173. https://doi.org/10.1111/2049-632X.12033

Goldberg JB, Pier GB (1996) *Pseudomonas aeruginosa* lipopolysaccharides and pathogenesis. Trends Microbiol 4:490–494. https://doi.org/10.1016/S0966-842X(97)82911-3

Gonzalez MR, Fleuchot B, Lauciello L et al (2016) Effect of human burn wound exudate on *Pseudomonas aeruginosa* virulence. mSphere 1:1–14. https://doi.org/10.1128/mSphere.00111-15

Grassi L, Batoni G, Ostyn L et al (2019) The antimicrobial peptide lin-SB056-1 and its dendrimeric derivative prevent *Pseudomonas aeruginosa* biofilm formation in physiologically relevant models of chronic infections. Front Microbiol 10:1–14. https://doi.org/10.3389/fmicb.2019.00198

Harmsen M, Yang L, Pamp SJ, Tolker-Nielsen T (2010) An update on *Pseudomonas aeruginosa* biofilm formation, tolerance, and dispersal. FEMS Immunol Med Microbiol 59:253–268. https://doi.org/10.1111/j.1574-695X.2010.00690.x

Hauser AR (2011) *Pseudomonas aeruginosa*: so many virulence factors, so little time*. Crit Care Med 39:2193–2194. https://doi.org/10.1097/CCM.0b013e318221742d

Herwald H, Egesten A (2009) Bacterial proteases disarming host Defense. J Innate Immun 1:69–69. https://doi.org/10.1159/000181143

Hill C, Mills S, Ross RP (2018) Phages & antibiotic resistance: are the most abundant entities on earth ready for a comeback? Future Microbiol 13:711–726. https://doi.org/10.2217/fmb-2017-0261

Hirsch EB, Tam VH (2010) Impact of multidrug-resistant *Pseudomonas aeruginosa* infection on patient outcomes. Expert Rev Pharmacoecon Outcomes Res 10:441–451. https://doi.org/10.1586/erp.10.49

Hnamte S, Parasuraman P, Ranganathan S et al (2019) Mosloflavone attenuates the quorum sensing controlled virulence phenotypes and biofilm formation in *Pseudomonas aeruginosa* PAO1: in vitro, in vivo and in silico approach. Microb Pathog 131:128–134. https://doi.org/10.1016/j.micpath.2019.04.005

Hobden JA (2002) *Pseudomonas aeruginosa* proteases and corneal virulence. DNA Cell Biol 21:391–396. https://doi.org/10.1089/10445490260099674

Huang Z, Jiang Y, Liang J (2011) Pathogenesis could be one of the anti-cheating mechanisms for *Pseudomonas aeruginosa* society. Med Hypotheses 76:166–168. https://doi.org/10.1016/j.mehy.2010.09.007

Hurley MN, Cámara M, Smyth AR (2012) Novel approaches to the treatment of *Pseudomonas aeruginosa* infections in cystic fibrosis. Eur Respir J 40:1014–1023. https://doi.org/10.1183/09031936.00042012

Ingmer H, Brøndsted L (2009) Proteases in bacterial pathogenesis. Res Microbiol 160:704–710. https://doi.org/10.1016/j.resmic.2009.08.017

Izoré T, Job V, Dessen A (2011) Biogenesis, regulation, and targeting of the type III secretion system. Structure 19:603–612. https://doi.org/10.1016/j.str.2011.03.015

Jamal M, Ahmad W, Andleeb S et al (2018) Bacterial biofilm and associated infections. J Chin Med Assoc 81:7–11. https://doi.org/10.1016/j.jcma.2017.07.012

Jeevanandam J, Chan YS, Danquah MK (2016) Nano-formulations of drugs: recent developments, impact and challenges. Biochimie 128–129:99–112. https://doi.org/10.1016/j.biochi.2016.07.008

Juhas M (2015) Type IV secretion systems and genomic islands-mediated horizontal gene transfer in Pseudomonas and Haemophilus. Microbiol Res 170:10–17. https://doi.org/10.1016/j.micres.2014.06.007

Juhas M, Crook DW, Hood DW (2008) Type IV secretion systems: tools of bacterial horizontal gene transfer and virulence. Cell Microbiol 10:2377–2386. https://doi.org/10.1111/j.1462-5822.2008.01187.x

Kalaiarasan E, Thirumalaswamy K, Harish BN et al (2017) Inhibition of quorum sensing-controlled biofilm formation in *Pseudomonas aeruginosa* by quorum-sensing inhibitors. Microb Pathog 111:99–107. https://doi.org/10.1016/j.micpath.2017.08.017

Kalia VC, Patel SKS, Kang YC, Lee J-K (2019) Quorum sensing inhibitors as antipathogens: biotechnological applications. Biotechnol Adv 37:68–90. https://doi.org/10.1016/j.biotechadv.2018.11.006

Kariminik A, Baseri-Salehi M, Kheirkhah B (2017) *Pseudomonas aeruginosa* quorum sensing modulates immune responses: an updated review article. Immunol Lett 190:1–6. https://doi.org/10.1016/j.imlet.2017.07.002

Kerr KG, Snelling AM (2009) *Pseudomonas aeruginosa*: a formidable and ever-present adversary. J Hosp Infect 73:338–344. https://doi.org/10.1016/j.jhin.2009.04.020

Kibret KT, Moges Y, Memiah P, Biadgilign S (2017) Treatment outcomes for multidrug-resistant tuberculosis under DOTS-Plus: a systematic review and meta-analysis of published studies. Infect Dis Poverty 6:7. https://doi.org/10.1186/s40249-016-0214-x

King JD, Kocíncová D, Westman EL, Lam JS (2009) Review: Lipopolysaccharide biosynthesis in *Pseudomonas aeruginosa*. Innate Immun 15:261–312. https://doi.org/10.1177/1753425909106436

Klockgether J, Tümmler B (2017) Recent advances in understanding *Pseudomonas aeruginosa* as a pathogen. F1000Res 6:1261. https://doi.org/10.12688/f1000research.10506.1

Koh CL, Sam CK, Yin WF et al (2013) Plant-derived natural products as sources of anti-quorum sensing compounds. Sensors (Basel) 13:6217–6228. https://doi.org/10.3390/s130506217

Kohlenberg A, Weitzel-Kage D, van der Linden P et al (2010) Outbreak of carbapenem-resistant *Pseudomonas aeruginosa* infection in a surgical intensive care unit. J Hosp Infect 74:350–357. https://doi.org/10.1016/j.jhin.2009.10.024

Kung VL, Ozer EA, Hauser AR (2010) The accessory genome of *Pseudomonas aeruginosa*. Microbiol Mol Biol Rev 74:621–641. https://doi.org/10.1128/MMBR.00027-10

Lantz MS (1997) Are bacterial proteases important virulence factors? J Periodontal Res 32:126–132. https://doi.org/10.1111/j.1600-0765.1997.tb01393.x

Law N, Logan C, Yung G et al (2019) Successful adjunctive use of bacteriophage therapy for treatment of multidrug-resistant *Pseudomonas aeruginosa* infection in a cystic fibrosis patient. Infection 47:665–668. https://doi.org/10.1007/s15010-019-01319-0

Lee DG, Urbach JM, Wu G et al (2006) Genomic analysis reveals that *Pseudomonas aeruginosa* virulence is combinatorial. Genome Biol 7:R90. https://doi.org/10.1186/gb-2006-7-10-r90

Limoli DH, Whitfield GB, Kitao T et al (2017) *Pseudomonas aeruginosa* alginate overproduction promotes coexistence with Staphylococcus aureus in a model of cystic fibrosis respiratory infection. MBio 8:1–18. https://doi.org/10.1128/mBio.00186-17

Lorè NI, Cigana C, De Fino I et al (2012) Cystic fibrosis-niche adaptation of *Pseudomonas aeruginosa* reduces virulence in multiple infection hosts. PLoS One 7:e35648. https://doi.org/10.1371/journal.pone.0035648

Ma Q, Zhai Y, Schneider JC et al (2003) Protein secretion systems of *Pseudomonas aeruginosa* and P. fluorescens. Biochim Biophys Acta Biomembr 1611:223–233. https://doi.org/10.1016/S0005-2736(03)00059-2

Macia MD, Rojo-Molinero E, Oliver A (2014) Antimicrobial susceptibility testing in biofilm-growing bacteria. Clin Microbiol Infect 20:981–990. https://doi.org/10.1111/1469-0691.12651

Maldonado RF, Sá-Correia I, Valvano MA (2016) Lipopolysaccharide modification in gram-negative bacteria during chronic infection. FEMS Microbiol Rev 40:480–493. https://doi.org/10.1093/femsre/fuw007

Markou P, Apidianakis Y (2014) Pathogenesis of intestinal *Pseudomonas aeruginosa* infection in patients with cancer. Front Cell Infect Microbiol 3:1–5. https://doi.org/10.3389/fcimb.2013.00115

Matsumoto K (2004) Role of bacterial proteases in pseudomonal and serratial keratitis. Biol Chem 385:1007–1016. https://doi.org/10.1515/BC.2004.131

Maurice NM, Bedi B, Sadikot RT (2018) *Pseudomonas aeruginosa* biofilms: host response and clinical implications in lung infections. Am J Respir Cell Mol Biol 58:428–439. https://doi.org/10.1165/rcmb.2017-0321TR

McCaslin CA, Petrusca DN, Poirier C et al (2015) Impact of alginate-producing *Pseudomonas aeruginosa* on alveolar macrophage apoptotic cell clearance. J Cyst Fibros 14:70–77. https://doi.org/10.1016/j.jcf.2014.06.009

Micek ST, Kollef MH, Torres A et al (2015) *Pseudomonas aeruginosa* nosocomial pneumonia: impact of pneumonia classification. Infect Control Hosp Epidemiol 36:1190–1197. https://doi.org/10.1017/ice.2015.167

Michalska M, Wolf P (2015) Pseudomonas exotoxin A: optimized by evolution for effective killing. Front Microbiol 6:1–7. https://doi.org/10.3389/fmicb.2015.00963

Mofazzal Jahromi MA, Sahandi Zangabad P, Moosavi Basri SM et al (2018) Nanomedicine and advanced technologies for burns: preventing infection and facilitating wound healing. Adv Drug Deliv Rev 123:33–64. https://doi.org/10.1016/j.addr.2017.08.001

Morlon-Guyot J, Mere J, Bonhoure A, Beaumelle B (2009) Processing of *Pseudomonas aeruginosa* exotoxin A is dispensable for cell intoxication. Infect Immun 77:3090–3099. https://doi.org/10.1128/IAI.01390-08

Musicki B, Liu T, Lagoda GA et al (2009) Endothelial nitric oxide synthase regulation in female genital tract structures. J Sex Med 6:247–253. https://doi.org/10.1111/j.1743-6109.2008.01122.x

Nguyen L, Garcia J, Gruenberg K, MacDougall C (2018) Multidrug-resistant pseudomonas infections: hard to treat, but hope on the horizon? Curr Infect Dis Rep 20:23. https://doi.org/10.1007/s11908-018-0629-6

Niederman MS (2010) Hospital-acquired pneumonia, health care–associated pneumonia, ventilator-associated pneumonia, and ventilator-associated tracheobronchitis: definitions and challenges in trial design. Clin Infect Dis 51:S12–S17. https://doi.org/10.1086/653035

Oliver A, Mulet X, López-Causapé C, Juan C (2015) The increasing threat of *Pseudomonas aeruginosa* high-risk clones. Drug Resist Updat 21–22:41–59. https://doi.org/10.1016/j.drup.2015.08.002

Overhage J, Lewenza S, Marr AK, Hancock REW (2007) Identification of genes involved in swarming motility using a *Pseudomonas aeruginosa* PAO1 mini-Tn5-lux mutant library. J Bacteriol 189:2164–2169. https://doi.org/10.1128/JB.01623-06

Palanisamy N, Ferina N, Amirulhusni A et al (2014) Antibiofilm properties of chemically synthesized silver nanoparticles found against *Pseudomonas aeruginosa*. J Nanobiotechnology 12:2. https://doi.org/10.1186/1477-3155-12-2

Pang Z, Raudonis R, Glick BR et al (2019) Antibiotic resistance in *Pseudomonas aeruginosa*: mechanisms and alternative therapeutic strategies. Biotechnol Adv 37:177–192. https://doi.org/10.1016/j.biotechadv.2018.11.013

Pena C, Cabot G, Gomez-Zorrilla S et al (2015) Influence of virulence genotype and resistance profile in the mortality of *Pseudomonas aeruginosa* bloodstream infections. Clin Infect Dis 60:539–548. https://doi.org/10.1093/cid/ciu866

Pier G (2007) *Pseudomonas aeruginosa* lipopolysaccharide: a major virulence factor, initiator of inflammation and target for effective immunity. Int J Med Microbiol 297:277–295. https://doi.org/10.1016/j.ijmm.2007.03.012

Pillar CM, Hobden JA (2002) *Pseudomonas aeruginosa* exotoxin A and keratitis in mice. Investig Ophthalmol Vis Sci 43:1437–1444

Pires DP, Vilas Boas D, Sillankorva S, Azeredo J (2015) Phage therapy: a step forward in the treatment of *Pseudomonas aeruginosa* infections. J Virol 89:7449–7456. https://doi.org/10.1128/jvi.00385-15

Pletzer D, Hancock REW (2016) Antibiofilm peptides: potential as broad-spectrum agents. J Bacteriol 198:2572–2578. https://doi.org/10.1128/JB.00017-16

Puhar A, Sansonetti PJ (2014) Type III secretion system. Curr Biol 24:R784–R791. https://doi.org/10.1016/j.cub.2014.07.016

Rafla K, Tredget EE (2011) Infection control in the burn unit. Burns 37:5–15. https://doi.org/10.1016/j.burns.2009.06.198

Rasamiravaka T, Labtani Q, Duez P, El Jaziri M (2015) The formation of biofilms by *Pseudomonas aeruginosa*: a review of the natural and synthetic compounds interfering with control mechanisms. Biomed Res Int 2015:1–17. https://doi.org/10.1155/2015/759348

Rohde C, Wittmann J, Kutter E (2018) Bacteriophages: a therapy concept against multi-drug–resistant bacteria. Surg Infect 19:737–744. https://doi.org/10.1089/sur.2018.184

Rossi Gonçalves I, Dantas RCC, Ferreira ML et al (2017) Carbapenem-resistant *Pseudomonas aeruginosa*: association with virulence genes and biofilm formation. Braz J Microbiol 48:211–217. https://doi.org/10.1016/j.bjm.2016.11.004

Salter SJ (2015) Keeping an eye on P. aeruginosa. Nat Rev Microbiol 13:69–69. https://doi.org/10.1038/nrmicro3422

Satpathy S, Sen SK, Pattanaik S, Raut S (2016) Review on bacterial biofilm: an universal cause of contamination. Biocatal Agric Biotechnol 7:56–66. https://doi.org/10.1016/j.bcab.2016.05.002

Silby MW, Winstanley C, Godfrey SAC et al (2011) Pseudomonas genomes: diverse and adaptable. FEMS Microbiol Rev 35:652–680. https://doi.org/10.1111/j.1574-6976.2011.00269.x

Simões D, Miguel SP, Ribeiro MP et al (2018) Recent advances on antimicrobial wound dressing: a review. Eur J Pharm Biopharm 127:130–141. https://doi.org/10.1016/j.ejpb.2018.02.022

Smith WD, Bardin E, Cameron L et al (2017) Current and future therapies for *Pseudomonas aeruginosa* infection in patients with cystic fibrosis. FEMS Microbiol Lett 364:1–9. https://doi.org/10.1093/femsle/fnx121

Sousa A, Pereira M (2014) *Pseudomonas aeruginosa* diversification during infection development in cystic fibrosis lungs—a review. Pathogens 3:680–703. https://doi.org/10.3390/pathogens3030680

Stefani S, Campana S, Cariani L et al (2017) Relevance of multidrug-resistant *Pseudomonas aeruginosa* infections in cystic fibrosis. Int J Med Microbiol 307:353–362. https://doi.org/10.1016/j.ijmm.2017.07.004

Streeter K, Katouli M (2016) *Pseudomonas aeruginosa*: a review of their pathogenesis and prevalence in clinical settings and the environment. Infect Epidemiol Med 2:25–32. https://doi.org/10.18869/modares.iem.2.1.25

Taccetti G, Campana S, Neri AS et al (2008) Antibiotic therapy against *Pseudomonas aeruginosa* in cystic fibrosis. J Chemother 20:166–169. https://doi.org/10.1179/joc.2008.20.2.166

Taylor PK, Yeung ATY, Hancock REW (2014) Antibiotic resistance in *Pseudomonas aeruginosa* biofilms: towards the development of novel anti-biofilm therapies. J Biotechnol 191:121–130. https://doi.org/10.1016/j.jbiotec.2014.09.003

Teerapo K, Roytrakul S, Sistayanarain A, Kunthalert D (2019) A scorpion venom peptide derivative BmKn–22 with potent antibiofilm activity against *Pseudomonas aeruginosa*. PLoS One 14:e0218479. https://doi.org/10.1371/journal.pone.0218479

Trokter M, Felisberto-Rodrigues C, Christie PJ, Waksman G (2014) Recent advances in the structural and molecular biology of type IV secretion systems. Curr Opin Struct Biol 27:16–23. https://doi.org/10.1016/j.sbi.2014.02.006

Tumbarello M, De Pascale G, Trecarichi EM et al (2013) Clinical outcomes of *Pseudomonas aeruginosa* pneumonia in intensive care unit patients. Intensive Care Med 39:682–692. https://doi.org/10.1007/s00134-013-2828-9

Valentini M, Gonzalez D, Mavridou DA, Filloux A (2018) Lifestyle transitions and adaptive pathogenesis of *Pseudomonas aeruginosa*. Curr Opin Microbiol 41:15–20. https://doi.org/10.1016/j.mib.2017.11.006

Van Acker H, Van Dijck P, Coenye T (2014) Molecular mechanisms of antimicrobial tolerance and resistance in bacterial and fungal biofilms. Trends Microbiol 22:326–333. https://doi.org/10.1016/j.tim.2014.02.001

Wagner S, Sommer R, Hinsberger S et al (2016) Novel strategies for the treatment of *Pseudomonas aeruginosa* infections. J Med Chem 59:5929–5969. https://doi.org/10.1021/acs.jmedchem.5b01698

Wang Y, Beekman J, Hew J et al (2018) Burn injury: challenges and advances in burn wound healing, infection, pain and scarring. Adv Drug Deliv Rev 123:3–17. https://doi.org/10.1016/j.addr.2017.09.018

Waters EM, Neill DR, Kaman B et al (2017) Phage therapy is highly effective against chronic lung infections with *Pseudomonas aeruginosa*. Thorax 72:666–667. https://doi.org/10.1136/thoraxjnl-2016-209265

Werneburg M, Zerbe K, Juhas M et al (2012) Inhibition of lipopolysaccharide transport to the outer membrane in *Pseudomonas aeruginosa* by peptidomimetic antibiotics. Chembiochem 13:1767–1775. https://doi.org/10.1002/cbic.201200276

Wiehlmann L, Wagner G, Cramer N et al (2007) Population structure of *Pseudomonas aeruginosa*. Proc Natl Acad Sci 104:8101–8106. https://doi.org/10.1073/pnas.0609213104

Wilke M, Grube R, Bodmann K (2011) Guideline-adherent initial intravenous antibiotic therapy for hospital-acquired/ventilator-associated pneumonia is clinically superior, saves lives and is cheaper than non guideline adherent therapy. Eur J Med Res 16:315. https://doi.org/10.1186/2047-783X-16-7-315

Wolfgang MC, Kulasekara BR, Liang X et al (2003) Conservation of genome content and virulence determinants among clinical and environmental isolates of *Pseudomonas aeruginosa*. Proc Natl Acad Sci 100:8484–8489. https://doi.org/10.1073/pnas.0832438100

Wu Y-L, Yang X-Y, Ding X-X et al (2019) Exposure to infected/colonized roommates and prior room occupants increases the risks of healthcare-associated infections with the same organism. J Hosp Infect 101:231–239. https://doi.org/10.1016/j.jhin.2018.10.014

Yamazaki A, Li J, Zeng Q et al (2012) Derivatives of plant phenolic compound affect the type III secretion system of *Pseudomonas aeruginosa* via a GacS-GacA two-component signal transduction system. Antimicrob Agents Chemother 56:36–43. https://doi.org/10.1128/AAC.00732-11

Yelin I, Kishony R (2018) Antibiotic resistance. Cell 172:1136–1136.e1. https://doi.org/10.1016/j.cell.2018.02.018

Zaidi S, Misba L, Khan AU (2017) Nano-therapeutics: a revolution in infection control in post antibiotic era. Nanomedicine 13:2281–2301. https://doi.org/10.1016/j.nano.2017.06.015

Zhu M, Zhao J, Kang H et al (2016) Modulation of type III secretion system in *Pseudomonas aeruginosa*: involvement of the PA4857 gene product. Front Microbiol 7:1–13. https://doi.org/10.3389/fmicb.2016.00007

Acinetobacter baumannii: Infections and Drug Resistance

14

Jobina Rajkumari and Busi Siddhardha

Contents

J. Rajkumari · B. Siddhardha (✉)
Department of Microbiology, School of Life Sciences, Pondicherry University, Puducherry, India

B. Siddhardha et al. (eds.), *Model Organisms for Microbial Pathogenesis, Biofilm Formation and Antimicrobial Drug Discovery*,
https://doi.org/10.1007/978-981-15-1695-5_14

Abstract

Acinetobacter baumannii is currently considered a major clinical pathogen accountable for causing several nosocomial infections and hospital outbreaks all over the globe. *A. baumannii* is accountable for causing bloodstream infections, pneumonia, meningitis, wound infection, urinary tract infection, endocarditis, peritonitis, and keratitis especially in individuals with low immunity. The pathogen has been reported to exhibit resistance against broad classes of antibiotics including carbapenems, aminoglycoside, quinolones, tetracyclines, cephalosporins, and polymyxin. Various virulence mechanisms have been identified and proposed for the resistance of *A. baumannii* towards the different class of antibiotics including inactivation or alteration of antimicrobials, alteration in membrane proteins, efflux pumps system, changes in the drug target, and biofilm formation. Furthermore, the ability to acquire and express exogenous antibiotic resistance genes is another contributing factor in imparting drug resistance to the pathogen.

Keywords

A. baumannii · Carbapenems · Bacteremia · OmpA · Lipid A

14.1 Introduction

Acinetobacter baumannii is an opportunistic pathogen generally considered non-pathogenic to healthy individuals. However, in the past few decades, it has been extensively reported worldwide for its resistance to the conventional as well as recently developed antimicrobial drugs (Almasaudi 2018; Gokmen et al. 2016). There is an increasing prevalence of multidrug-resistant strains of *A. baumannii* especially in intensive care units (ICU) of the hospitals, causing infections in patients with invasive devices (ventilators and catheters) and low immunity (Singh et al. 2013). With its ability to develop antibiotic resistance and survive under dry conditions on inanimate objects for a long period, this gram-negative pathogen is responsible for causing numerous infections in the hospital environment (Inchai et al. 2015). *A. baumannii* is recognized as a major cause of life-threatening hospital-associated infections including pneumonia, meningitis, urinary tract infections, and severe bloodstream infection, especially in immune-compromised individuals (Santajit and Indrawattana 2016).

The indiscriminate administration of antibiotics along with various bacterial mechanisms has been attributed to the evolution of multiple drug-resistant strains of *A. baumannii* (Singh et al. 2013). This notorious pathogen also possesses a remarkable ability to rapidly acquire resistance foreign genetic material including a vast array of antibiotic resistance genes (Almasaudi 2018). Since the majority of the carbapenem-hydrolyzing enzymes are encoded in mobile genetic elements, it is proposed that the $bla_{\text{NDM-1}}$ gene in *A. baumannii* is apparently transmitted and acquired from carbapenem-resistant Enterobacteriaceae (*Escherichia coli* and *Klebsiella* spp.) via horizontal gene transfer (Jain et al. 2019). Furthermore, the ability to survive in severe conditions of temperature, pH, and in the presence of commonly used antimicrobial agents facilitates *Acinetobacter* to easily disseminate in the hospital settings (Chopra et al. 2013).

The remarkable property of *A. baumannii* to develop resistance against many antibiotics is a huge concern in the healthcare setting as these infections are difficult to treat. They are linked to increased mortality and morbidity in cases of ventilator-associated pneumonia (Inchai et al. 2015). *A. baumannii* is also responsible for causing community-acquired infections in patients with a history of alcoholism, cancer, diabetes, and obstructive pulmonary disorders (Harding et al. 2018).

Carbapenems are the most commonly recognized drugs for the management of *A. baumannii* associated infections. Hence, the emergence of carbapenem-resistant strains of *A. baumannii* has become a major challenge globally. In *A. baumannii,* numerous mechanisms including the production of β-lactamases, loss of outer membrane protein (OMPs), and efflux pump were found to cause in carbapenem resistance (Shoja et al. 2017). Hence, understanding the resistance mechanisms of *A. baumannii* is vital for the development of alternative tools and novel antimicrobial agents to combat the ever-increasing problem of drug resistance (Santajit and Indrawattana 2016). In the past few decades, the incidence of *A. baumannii* infections and outbreaks has drastically increased globally (Necati Hakyemez et al. 2013). Thus, the management of *A. baumannii* infections is currently one of the biggest challenges in clinical practice. In this chapter, the clinical implications of *A. baumannii* and its associated mechanisms of antimicrobial resistance have been presented in detail.

14.2 Virulence Factors

Recent studies have led to the identification of numerous virulence factors of *A. baumannii.* Outer membrane protein A (OmpA), efflux pump, phospholipase, and capsular polysaccharides were found to contribute to the pathogenicity and also impart antimicrobial resistance to *A. baumannii.* In addition, the lipopolysaccharide (LPS) of *A. baumannii* was found to induce the production of inflammatory cytokines, i.e., interleukin-8 and tumor necrosis factor. Other potential virulence factors including biofilm formation, penicillin-binding proteins, and alteration in outer membrane vesicles (OMVs) have also been implicated for the pathogenicity of *A. baumannii* (Wong et al. 2017). Targeted studies have also led to the identification of few virulence determinants of *A. baumannii,* including phospholipase D, metal (zinc and iron) acquisition system. Additionally, in the serum-resistant strain of *A. baumannii,* it was observed that the genes involved in competence, iron acquisition, type IV pili biogenesis, and efflux pumps get upregulated during the growth of the pathogen in human serum (Subashchandrabose et al. 2016).

OmpA coupled with the membrane efflux systems collectively participates in the expulsion of antimicrobial compounds from the periplasmic space (Lee et al.

2017). OmpA is also involved in invasion of epithelial cell and apoptosis. Phospholipases C and D aid in epithelial cell invasion and also help in the survival of the bacterial pathogen in the human serum (Antunes et al. 2014). Metal acquisition plays a vital role in the *A. baumannii* virulence. *A. baumannii* produces iron siderophores and acinetobactin in an iron limiting environment. Similarly, in Zn-limiting conditions, the zinc acquisition system (ZnuABC) is generally upregulated (Lee et al. 2017). In addition, the quorum-sensing system and biofilm-forming ability also aid in the survival of the cells on inert surfaces thereby contributing to drug resistance (Antunes et al. 2014). The above-mentioned virulence determinants serve as an attractive target for the development of new preventive strategies and antibiotics.

14.3 Clinical Manifestations

A. baumannii is recognized to cause fatal nosocomial infections, most commonly in patients with chronic disease or has undergone surgeries. The colonization and subsequent infection of *A. baumannii* commonly takes place in organ with higher fluid content such as urinary tract, respiratory tract, peritoneal cavity, and indwelling medical devices (Almasaudi 2018). It reported to cause secondary meningitis and infections to eye, skin and soft tissue, burn wound, and urinary tract. However, the most common clinical manifestation with the highest mortality rates is recorded in cases of ventilator-associated pneumonia and bacteremia. The pathogen usually gains entry into the host system via intravascular catheters, open wounds, and medical in-dwellings (Antunes et al. 2014).

14.3.1 Bacteremia

Bacteremia is considered as the most significant infection caused by *A. baumannii* in health care settings which is also associated with marked antimicrobial resistance and high mortality (Chopra et al. 2013). *A. baumannii* associated bacteremia is generally nosocomial and occurs commonly in ICUs (Wong et al. 2017). *A. baumannii* bacteremia commonly originates from the respiratory tract and intravascular catheters (Cisneros and Rodriguez-Bano 2002). It accounts for 52% of cases in critical care, followed by 13% in surgery, 10% in general medical, and 17% in cases of cancer (Wareham et al. 2008).

14.3.2 Acinetobacter Pneumonia

Ventilator-associated pneumonia (VAP) is a major nosocomial infection among critically immune-compromised individuals. VAP accounts for 25% of all types of ICU-acquired infections and exerts a huge burden on morbidity and health care costs (Balkhy et al. 2014). In recent years, *A. baumannii* has been widely documented and recognized as a major cause of VAP. The mortality rate in ICU patients with VAP ranged from 45.6% to 60.9% and VAP caused by extreme drug-resistant (XDR) *A. baumannii* has been found to be as high as 84.3%. Gokmen et al. (2016) reported three cases of VAP wherein the infection resulted due to the colonization of ICU-ventilator by carbapenem-resistant *A. baumannii* (CRA). All the isolated strains were carbapenem resistant and found to possess $bla_{OXA\text{-}23}$-like and $bla_{OXA\text{-}51}$-like gene (Gokmen et al. 2016).

A. baumannii can adhere and establish biofilms on the medical devices like the endotracheal tube and create a niche for the rapid dissemination of the bacterial cells, thus resulting in the increasing incidences of carbapenem-resistant *Acinetobacter* in the ICUs. *Acinetobacter* may also directly enter the alveoli of mechanically ventilated patients, thereby allowing the establishment of infection in lung tissue (Wong et al. 2017). In spite of the recent advancements in burns surgery and intensive care, VAP still remains the main cause of mortality in victims with major burn wounds. VAP is also accountable for significant morbidity and mortality in the intensive care pediatric ward (Rogers et al. 2014).

Community-acquired *A. baumannii* pneumonia (CAP-AB) is generally recorded during the rainy season especially in tropical parts of Australia and Asia. CAP-AB is particularly found in individuals with chronic obstructive pulmonary disease and cases of alcohol abuse (Almasaudi 2018). Oh et al. (2013) reported the first case of rapidly progressing and fatal CAP-AB in Korea wherein the patient died within 36 h of hospital admission (Oh et al. 2013).

14.3.3 Meningitis

Meningitis caused by multidrug-resistant *A. baumannii* usually arises post-neurosurgery or in the presence of a ventriculostomy (Wong et al. 2017). With mortality rate exceeding 15%, post-neurosurgical meningitis is a major clinical issue. The high-risk group for post-neurosurgical bacterial meningitis includes those with concomitant incision infection, cerebrospinal leakage, prolonged duration of surgery, a surgery that enters a sinus, prolonged external ventricular drainage, increased severity of illness, and need for repeat surgery (Kim et al. 2009).

14.3.4 Wound Infection

Patients with burn wounds are generally susceptible and associated with high risk for infection. *A. baumannii* infection represents a common cause of mortality in patients with burn wards. Shoja et al. (2017) reported a high population of carbapenem-resistant *A. baumannii* containing $bla_{OXA\text{-}23}$-like and $bla_{OXA\text{-}24}$-like genes among patients with burn wound (Shoja et al. 2017).

14.3.5 Urinary Tract Infection

Urinary tract infections are generally linked to the colonization of urinary catheters or percutaneous nephrostomy tubes by the pathogen (Wong et al. 2017). The cases of *A. baumannii* related UTI infections are rare with an estimated 1.6% being ICU-acquired UTIs. It is also unlikely for *A. baumannii* to cause severe UTI in outpatients (Almasaudi 2018).

14.3.6 Other Manifestations

A few reported cases of *A. baumannii* endocarditis exist, with the majority of these cases associated with prosthetic valves (Almasaudi 2018). Numerous reports from across the globe including India have documented the increasing prevalence of *Acinetobacter* spp. in lower respiratory tract infections (LRTI). In the recent study of Jain et al. (2019), *A. baumannii* was found to be among the most common pathogen responsible for LRTI in ICUs, accountable for almost 26.2% of the cases (Jain et al. 2019). Lagana et al. (2015) reported two cases of lethal and infective endocarditis on cardiac prostheses sustained by *A. baumannii* (Lagana et al. 2015).

Although not very frequent, MDR *A. baumannii* has been observed to cause peritonitis in peritoneal dialysis patients, thereby resulting in serious infection with a high possibility of mortality (Zhang et al. 2014). Ocular infections, particularly endophthalmitis and keratitis caused by *A. baumannii* have also been observed in recent years and mainly associated with prolonged use of contact lens or post-eye surgery. Chen et al. (2008) documented two cases of ocular infection caused by *A. baumannii*, one resulting in endogenous endophthalmitis and the other endophthalmitis following corneal transplant (Chen et al. 2008).

14.4 Mechanisms of Resistance

Antimicrobial resistance characteristics exhibited by the *A. baumannii* is contributed by multiple factors. The mechanisms of antimicrobial resistance in the clinical isolates of *A. baumannii* may be broadly divided into 4 categories: (1) inactivation or alteration of antimicrobials, (2) reduced entry and intracellular accumulation of antimicrobials, (3) alteration of the drug target, and (4) biofilm formation.

14.4.1 Inactivation or Alteration of Antimicrobials

A wide array of beta-lactamases produced by *Acinetobacter* species hydrolyze numerous class of drugs including β-lactams and cephalosporins, thereby conferring antibiotic resistance to *A. baumannii* (Manchanda et al. 2010). All the four classes of β-lactamases have been identified in *A. baumannii*. Some of the class A β-lactamases identified in *A. baumannii* includes CTX-M, GES, SCO, PER, KPC,

SHV, VEB, TEM, and CARB. Class β-lactamases are also known as metallo-β-lactamases (MBLs) and requires heavy metal like zinc or iron for catalysis (Lee et al. 2017). MBLs are known to impart resistance to carbapenems and to other β-lactams (Manchanda et al. 2010). Class C β-lactamases also known as AmpC cephalosporins are widely prevalent in *A. baumannii.* It is encoded by the *bla* gene and imparts broad-spectrum resistance to narrow and extended-spectrum cephalosporins, along with penicillin (Asif et al. 2018). Class D β-lactamases are also termed as oxacillinases (OXAs) due to their ability to hydrolyze isoxazolyl penicillin, oxacillin much faster than benzylpenicillin. The presence of class D β-lactamases and/or MBLs in *A. baumannii* is attributed to their high resistance towards carbapenems (Lee et al. 2017).

14.4.2 Reduced Entry and Intracellular Accumulation of Antimicrobials

The susceptibility towards a particular drug is determined by the balance in the uptake and elimination of antibiotics by the bacterial cell. Therefore, hampering the entry of antibiotic through the bacterial cell membrane is one of the strategies used by bacteria to develop antibiotic resistance (Santajit and Indrawattana 2016). The decrease in outer membrane permeability either due to alteration in porin channels and/or upregulation of multidrug efflux pumps significantly reduces the accessibility of the drugs towards the bacterial targets.

Porin channels along with other OMPs are known to facilitate the entry of antimicrobial agents inside the bacterial cells (Maragakis and Perl 2008). In *A. baumannii*, the presence of lesser porin channels with smaller size restricts the entry of the drug molecules to the intracellular bacterial targets (Singh et al. 2013). Many studies have also attributed carbapenem resistance to the reduction in the expression of OMPs (Asif et al. 2018). The loss of *A. baumannii* Omp29 which generates OXA-51-like or OXA-23-like carbapenemases is associated with high resistance towards imipenem. OmpA contributes to the resistance towards chloramphenicol, aztreonam, and nalidixic acid (Lee et al. 2017).

Acinetobacter species possess numerous efflux pumps system which actively removes a broad range of antimicrobial agents thereby preventing them from reaching the target site (Maragakis and Perl 2008). The most commonly expelled antimicrobials by the efflux pump system include tetracyclines, macrolides, and quinolones. The expulsion of the drug from the cell takes place at a high rate; hence, the drug concentrations are not sufficient enough to elicit an antibacterial effect (Santajit and Indrawattana 2016). The efflux system of *A. baumannii* is often associated with pathogenesis, virulence, and biofilm maturation. Furthermore, the overexpression of these efflux pumps plays a vital role in imparting antibiotic resistance (Ardehali et al. 2019).

The multidrug efflux systems may be categorized into (1) the ATP binding cassette (ABC), (2) major facilitator superfamily (MFS), (3) resistance nodulation division (RND), (4) multidrug and toxic compound extrusion (MATE), (5) SMR family,

Table 14.1 Efflux pumps of *Acinetobacter baumannii* belonging to different family and antimicrobial substrate

Efflux pump	Family	Antibiotics
Tet(A)	MFS	Tetracycline
Tet(B)	MFS	Minocycline, tetracycline
AbaF	MFS	Fosfomycin
AmvA	MFS	Erythromycin
MdfA	MFS	Ciprofloxacin, chloramphenicol
CmlA, CraA	MFS	Chloramphenicol
AdeABC	RND	β-lactams, chloramphenicol, fluoroquinolones, tetracycline, macrolide, aminoglycosides
AbeM	MATE	Aminoglycosides, chloramphenicol, fluoroquinolones, ethidium bromide, trimethoprim
AbeS	SMR	Chloramphenicol, erythromycin, nalidixic acid

and (6) the drug/metabolite transporter (DMT) superfamily. However, in the case of *A. baumannii,* the antimicrobial resistance is generally associated with MFS and RND family (Lee et al. 2011). Efflux pumps of *A. baumannii* belonging to different family and antimicrobial l substrate are summarized in Table 14.1.

Tet(A) and Tet(B) efflux systems belonging to the MFS confer resistance to tetracycline by exchanging a proton for a tetracycline-cation complex. In a recent study, Sharma et al. (2017) suggested that the AbaF efflux system mediates fosfomycin resistance to *A. baumannii* (Sharma et al. 2017). The MdfA efflux pumps mediate resistance to ciprofloxacin and chloramphenicol, while CmlA and CraA confer resistance to chloramphenicol (Vila et al. 2007). AmvA confers resistance to antibiotics (erythromycin), detergents (benzalkonium chloride), dyes (acriflavine), and disinfectants (methyl viologen) (Lee et al. 2017). Studies have also found that *A. baumannii* can actively pump disinfectant, chlorhexidine out of the cell using the *Acinetobacter* chlorhexidine transporter protein, AceI, thereby protecting the bacteria from external stress (Harding et al. 2018).

The RND-type efflux pump has been documented to impart resistance to aminoglycoside, chloramphenicol and macrolides (erythromycin), quinolones, tetracyclines, trimethoprim, and ethidium bromide (Almasaudi 2018). The AdeABC efflux pump in *A. baumannii* is also known to expel antimicrobial using proton motive force, thereby reducing the susceptibility of the pathogen towards β-lactams, fluoroquinolones, tetracycline, macrolides, chloramphenicol, and aminoglycosides (Asif et al. 2018). The overexpression of AdeABC also provides significant resistance to carbapenems, tigecycline, netilmicin, and gentamicin (Xu et al. 2019).

AbeM, a multidrug efflux pump of *A. baumannii* belonging to the MATE family, also utilizes the proton motive force to extrude aminoglycosides, chloramphenicol, fluoroquinolones, ethidium bromide, and trimethoprim out of the bacterial cell (Coyne et al. 2011). AbeS-efflux pump of the SMR family imparts resistance against chloramphenicol, erythromycin, and nalidixic acid (Lee et al. 2017).

14.4.3 Alteration of the Drug Target or Cellular Functions

The alteration in bacterial targets or functions as a result of point mutations decreases the affinity of antimicrobial drugs. In some cases, it may also up-regulate cellular functions such as efflux pumps or change the membrane proteins (Maragakis and Perl 2008). For example, mutations in the subunit of DNA gyrase, *gyrA* and topoisomerase IV, *parC* impart resistance against quinolones (Singh et al. 2013).

14.4.4 Biofilm Formation

Biofilms are complex microbial community predominantly attached to either a living or an inert surface. They are often encased by thick polysaccharide matrix constituting of self-produced polysaccharides, proteins, lipids, and extracellular DNA, which protects the biofilm structure from desiccation, immune system clearance, antibiotics, and other external stress (Maragakis and Perl 2008). Hence, it is difficult to eliminate biofilm encased bacterial pathogen using conventional antibiotics. The exceptionally high degree of antibiotic resistance of *Acinetobacter* and its ability to survive in the hospital setting may be attributed to its biofilm-forming ability (Babapour et al. 2016). A relationship between the biofilm-forming ability and drug resistance in the *A. baumannii* isolates due to the presence of extended-spectrum β-lactamase (ESBL) $bla_{\text{PER-1}}$ gene has been documented earlier (Badave and Kulkarni 2015).

Greene et al. (2016) provided evidence on the positive correlation between antibiotic resistance and tolerance to desiccation. It was observed that the high biofilm-forming MDR strains of *A. baumannii* were 50% less likely to die of desiccation as compared to the low biofilm-forming non-MDR strains (Greene et al. 2016). *Acinetobacter* is frequently isolated from indwelling devices, including cerebrospinal shunts, endotracheal tubes, vascular, and urinary catheters. Such colonization eventually leads to medical devices associated with infections like fever, pneumonia, bacteremia, and meningitis (Babapour et al. 2016). *A. baumannii* has also been reported to adhere to laryngeal, bronchial, and alveolar epithelial cells of the host (Doi et al. 2015).

Studies have shown the presence of high frequency of exogenous DNA in the genome of *A. baumannii* indicating the inherent ability of the pathogen to acquire resistance genes via horizontal gene transfer (Lee et al. 2017).

14.5 Resistance to Available Antimicrobial Drugs

The antimicrobial resistance of *A. baumannii* has progressively increased since the 1970s and in recent decades, it has acquired resistance to the majority of the commonly used antibiotics, thus leaving few choices of antimicrobial agents. Hence, the most serious concern in the management of *Acinetobacter* associated infection is its

Table 14.2 Mechanisms of resistance in *A. baumannii* towards different antibiotics

Class of antibiotic	Mechanisms of resistance
Carbapenems	Metallo-β-lactamases and class D OXA-type enzymes, altered outer membrane protein (OMPs) and porins channels, efflux pump-mediated expulsion
Aminoglycoside	Aminoglycoside modifying enzymes, alterations in the target ribosomal protein, altered OMPs and porins channels, efflux pump-mediated expulsion
Quinolones	Alteration in DNA gyrase and topoisomerase IV reduced expression of OMPs, efflux pump-mediated expulsion
Tetracyclines	Transposon-mediated efflux pumps, AdeABC efflux pump
Cephalosporins	AmpC β-lactamases, extended-spectrum β-lactamase (ESBL)
Polymyxin	Loss of lipopolysaccharides (LPS) production, alteration of cell membrane LPS

acquired MDR (Antunes et al. 2014). Table 14.2 represents the antibiotic resistance mechanisms observed in *A. baumannii* towards different classes of antibiotics.

14.5.1 Resistance to Carbapenems

A. baumannii also possesses AmpC-lactamase (bla_{ADC}) and OXA-51 serine-type oxacillinase ($bla_{\text{OXA-51}}$), which are known to naturally impart resistance to β-lactam antibiotics (Pagano et al. 2016). The carbapenemases includes the class D OXA-type enzymes and MBLs which contributes to the carbapenem resistance in *A. baumannii* (Lolans et al. 2006). Carbapenem resistance in *A. baumannii* involves either intrinsic or acquired $bla_{\text{OXA-23}}$, $bla_{\text{OXA-24/40}}$, and $bla_{\text{OXA-58}}$-like gene clusters (Singh et al. 2013). The second major mechanism is related to altered porins channels and penicillin-binding proteins. The smaller size and lower number of OMPs in *A. baumannii* causes a significant reduction in the permeability of the cell membrane to antimicrobial agents (Hsu et al. 2017). Finally, the upregulation of the RND-type efflux system (AdeABC, AdeFGH, and AdeIJK) also plays a considerable part in imparting resistance towards carbapenems (Hsu et al. 2017; Singh et al. 2013).

14.5.2 Resistance to Aminoglycoside

The major mechanism responsible for imparting resistance to *A. baumannii* towards aminoglycoside is the alteration in the amino or hydroxyl group of the drug by aminoglycoside modifying enzymes such as acetylases, adenylases, methyltransferases, and phosphotransferases (Asif et al. 2018). Other mechanisms include modification of the target ribosomal protein, efflux pump-mediated expulsion, and impaired transport of aminoglycosides like gentamicin, tobramycin, or amikacin into the bacterial cell (Almasaudi 2018).

14.5.3 Resistance to Quinolones

The most common mechanism involved in quinolones resistance is the mutations in the *gyrA* and *parC* genes which codes for DNA gyrase and topoisomerase IV. The resulting phenotypic alteration decreases the binding affinity of the quinolones to the enzyme-DNA complex. The reduction in intracellular drug accumulation due to the efflux systems and reduced expression of OMPs involved in drug influx also imparts resistance to quinolones (Doi et al. 2015). Plasmid-encoded genes found in *A. baumannii* such as *qnrA*, *qnrB,* and *qnrS* have also been identified to protect the DNA by preventing the interaction of quinolones to DNA gyrase and topoisomerase (Asif et al. 2018).

14.5.4 Resistance to Tetracyclines

A. baumannii exhibits two different resistance mechanisms towards tetracyclines. The first mechanism is regulated by transposon-mediated efflux pumps, TetA and TetB. TetA is responsible for the expulsion of tetracycline, while TetB ejects both tetracycline and minocycline out of the bacterial cell. The second mechanism is mediated by ribosomal protection protein encoded by *Tet(M)* gene. It protects the ribosome from the effect of doxycycline, tetracycline, and minocycline. The AdeABC efflux pump also aids in providing resistance to *A. baumannii* towards tigecycline (Almasaudi 2018).

14.5.5 Resistance to Cephalosporins

A majority of *A. baumannii* clinical isolates naturally produce AmpC β-lactamase which provides resistance against cephalosporins (ceftazidime or cefepime). Rezaee et al. (2013) studied the prevalence of various cephalosporin resistance mechanisms in the clinical isolates of *A. baumannii*. Out of the total 70 isolates screened, 98.5% of the cephalosporin-resistant isolates possess insertions upstream of the *ampC* gene, of which 69% and 8% were identified to be controlled by *ISAba1* and *ISAba125*, respectively (Rezaee et al. 2013). *A. baumannii* also produces extended-spectrum β-lactamase (ESBL) which also contributes to cephalosporin resistance (Doi et al. 2015).

14.5.6 Resistance to Polymyxin

The development of colistin (polymyxin E) resistance strains of *A. baumannii* is a matter of great concern (Almasaudi 2018). Till date, two primary mechanisms for colistin resistance have been recognized in *A. baumannii.* The first mechanism involves mutations in lipid A biosynthesis genes (*lpxA*, *lpxC*, and *lpxD*) which causes complete loss of lipopolysaccharides (LPS), an initial target of colistin (Cai et al. 2012). The second mechanism involves a mutation in *pmrA* and *pmrB* which controls the expression of genes involved in lipid A synthesis (Asif et al. 2018).

14.6 Global Incidence of Drug Resistant *Acinetobacter baumannii*

The increasing prevalence of antibiotic resistant *Acinetobacter* spp. in different geographical regions of the globe has been revealed in numerous reports. The association of OXA-58-like and OXA-24-like enzymes to the carbapenem resistance in *A. baumannii* was documented in 2008 in the Czech Republic. A few years later in 2011, an MDR strain of *A. baumannii* possessing the genes for NDM-1 and OXA-23 was identified again in the Czech Republic (Senkyrikova et al. 2013). Mesli et al. (2013) documented the first description of autochthonous *Acinetobacter* spp. in western Algeria with the ability to produce oxacillinases bla_{OXA-23}-like, bla_{NDM-1}-like, and bla_{OXA-24}-like metallo-β-lactamases (Mesli et al. 2013). In another report, Khorsi et al. (2015) highlighted the high prevalence of imipenem resistance in *A. baumannii* isolated from Algiers hospitals. The antibiotic resistance was found to be mediated predominantly by bla_{OXA-23}-like, bla_{OXA-24}-like, and bla_{NDM-1}-like genes (Khorsi et al. 2015).

Carvalho et al. (2009) reported the transmission of OXA-23 producing MDR clones of *A. baumannii* in hospitals throughout Rio de Janeiro, Brazil. A total of 110 imipenem-resistant *A. baumannii* isolates were identified in samples collected from eight hospitals between January 2006 and September 2007. All the isolated strains were MDR with 87.3% producing the carbapenemase OXA-23 (Carvalho et al. 2009). Al-Agamy et al. (2014) reported the high frequency of β-lactamase encoding genes in 40 carbapenem-resistant *A. baumannii* isolates obtained from two hospitals in Egypt. The carbapenem resistance was mediated by β-lactamase encoding genes, bla_{OXA-23}, bla_{OXA-24}, and bla_{OXA-58}, bla_{GES}, and *ISAba1–OXA* (Al-Agamy et al. 2014). Agodi et al. (2014) documented the emergence and dissemination of colistin- and carbapenem-resistant strains of *A. baumannii* in two hospitals in Sicily. The MDR strains were found to possess intrinsic bla_{OXA-51}-like carbapenemase gene and bla_{OXA-82}, which was flanked upstream by *ISAba1* (Agodi et al. 2014). Almaghrabi et al. (2018) reported the high frequency (69%) of MDR strains of *A. baumannii* in Aseer Region of Saudi Arabia. The isolates exhibited remarkable resistance to carbapenems and the drug resistance property was attributed to the presence of class D OXA-type enzymes (OXA-23 and OXA-24/40) (Almaghrabi et al. 2018).

The ever-increasing resistance of *A. baumannii* towards majority of the currently available antibiotics is a matter of great concern. This MDR pathogen is frequently associated with numerous hospital-associated infections resulting in high morbidity and mortality. Hence, the emergence and dissemination of these MDR strains of *A. baumannii* emphasize the urgent requirement to enforce infection control measures and develop effective treatment strategies to manage the further emergence and dissemination of these resistant *Acinetobacter* species.

References

Agodi A, Voulgari E, Barchitta M, Quattrocchi A, Bellocchi P, Poulou A, Tsakris A (2014) Spread of carbapenem- and colistin-resistant *Acinetobacter baumannii* ST2 clonal strain causing outbreaks in two Sicilian hospitals. J Hosp Infect 86(4):260–266

Al-Agamy MH, Khalaf NG, Tawfick MM, Shibl AM, El Kholy A (2014) Molecular characterization of carbapenem-insensitive *Acinetobacter baumannii* in Egypt. Int J Infect Dis 22:49–54

Almaghrabi MK, Joseph MRP, Assiry MM, Hamid ME (2018) Multidrug-resistant *Acinetobacter baumannii*: an emerging health threat in Aseer Region, Kingdom of Saudi Arabia. Can J Infect Dis Med Microbiol 2018:1

Almasaudi SB (2018) Acinetobacter spp. as nosocomial pathogens: epidemiology and resistance features. Saudi J Biol Sci 25(3):586–596

Antunes LC, Visca P, Towner KJ (2014) *Acinetobacter baumannii*: evolution of a global pathogen. Pathog Dis 71(3):292–301

Ardehali SH, Azimi T, Fallah F, Owrang M, Aghamohammadi N, Azimi L (2019) Role of efflux pumps in reduced susceptibility to tigecycline in *Acinetobacter baumannii*. New Microbes New Infect 30:100547

Asif M, Alvi IA, Rehman SU (2018) Insight into *Acinetobacter baumannii*: pathogenesis, global resistance, mechanisms of resistance, treatment options, and alternative modalities. Infect Drug Resist 11:1249–1260

Babapour E, Haddadi A, Mirnejad R, Angaji SA, Amirmozafari N (2016) Biofilm formation in clinical isolates of nosocomial *Acinetobacter baumannii* and its relationship with multidrug resistance. Asian Pac J Trop Biomed 6(6):528–533

Badave GK, Kulkarni D (2015) Biofilm producing multidrug resistant *Acinetobacter baumannii*: an emerging challenge. J Clin Diagn Res 9(1):DC08–DC10

Balkhy HH, El-Saed A, Maghraby R, Al-Dorzi HM, Khan R, Rishu AH, Arabi YM (2014) Drug-resistant ventilator associated pneumonia in a tertiary care hospital in Saudi Arabia. Ann Thorac Med 9(2):104–111

Cai Y, Chai D, Wang R, Liang B, Bai N (2012) Colistin resistance of *Acinetobacter baumannii*: clinical reports, mechanisms and antimicrobial strategies. J Antimicrob Chemother 67(7): 1607–1615

Carvalho KR, Carvalho-Assef AP, Peirano G, Santos LC, Pereira MJ, Asensi MD (2009) Dissemination of multidrug-resistant *Acinetobacter baumannii* genotypes carrying bla(OXA-23) collected from hospitals in Rio de Janeiro, Brazil. Int J Antimicrob Agents 34(1):25–28

Chen KJ, Hou CH, Sun MH, Lai CC, Sun CC, Hsiao CH (2008) Endophthalmitis caused by *Acinetobacter baumannii*: report of two cases. J Clin Microbiol 46(3):1148–1150

Chopra T, Marchaim D, Awali RA, Krishna A, Johnson P, Tansek R, Kaye KS (2013) Epidemiology of bloodstream infections caused by *Acinetobacter baumannii* and impact of drug resistance to both carbapenems and ampicillin-sulbactam on clinical outcomes. Antimicrob Agents Chemother 57(12):6270–6275

Cisneros JM, Rodriguez-Bano J (2002) Nosocomial bacteremia due to *Acinetobacter baumannii*: epidemiology, clinical features and treatment. Clin Microbiol Infect 8(11):687–693

Coyne S, Courvalin P, Perichon B (2011) Efflux-mediated antibiotic resistance in Acinetobacter spp. Antimicrob Agents Chemother 55(3):947–953

Doi Y, Murray GL, Peleg AY (2015) *Acinetobacter baumannii*: evolution of antimicrobial resistance-treatment options. Semin Respir Crit Care Med 36(1):85–98

Gokmen TG, Akcimen B, Kayar B, Marzi M, Koksal F (2016) The outbreak of Acinetobacter baumannii producing OXA-23 and OXA-51 type carbapenemases in a state hospital. J Exp Clin Med 33(3):157–161

Greene C, Vadlamudi G, Newton D, Foxman B, Xi C (2016) The influence of biofilm formation and multidrug resistance on environmental survival of clinical and environmental isolates of *Acinetobacter baumannii*. Am J Infect Control 44(5):e65–e71

Harding CM, Hennon SW, Feldman MF (2018) Uncovering the mechanisms of *Acinetobacter baumannii* virulence. Nat Rev Microbiol 16(2):91–102

Hsu LY, Apisarnthanarak A, Khan E, Suwantarat N, Ghafur A, Tambyah PA (2017) Carbapenem-resistant *Acinetobacter baumannii* and Enterobacteriaceae in South and Southeast Asia. Clin Microbiol Rev 30(1):1–22

Inchai J, Pothirat C, Bumroongkit C, Limsukon A, Khositsakulchai W, Liwsrisakun C (2015) Prognostic factors associated with mortality of drug-resistant *Acinetobacter baumannii* ventilator-associated pneumonia. J Intensive Care 3:9

Jain M, Sharma A, Sen MK, Rani V, Gaind R, Suri JC (2019) Phenotypic and molecular characterization of *Acinetobacter baumannii* isolates causing lower respiratory infections among ICU patients. Microb Pathog 128:75–81

Khorsi K, Messai Y, Hamidi M, Ammari H, Bakour R (2015) High prevalence of multidrug-resistance in *Acinetobacter baumannii* and dissemination of carbapenemase-encoding genes blaOXA-23-like, blaOXA-24-like and blaNDM-1 in Algiers hospitals. Asian Pac J Trop Med 8(6):438–446

Kim BN, Peleg AY, Lodise TP, Lipman J, Li J, Nation R, Paterson DL (2009) Management of meningitis due to antibiotic-resistant Acinetobacter species. Lancet Infect Dis 9(4):245–255

Lagana P, Melcarne L, Delia S (2015) *Acinetobacter baumannii* and endocarditis, rare complication but important clinical relevance. Int J Cardiol 187:678–679

Lee K, Yong D, Jeong SH, Chong Y (2011) Multidrug-resistant Acinetobacter spp: increasingly problematic nosocomial pathogens. Yonsei Med J 52(6):879–891

Lee CR, Lee JH, Park M, Park KS, Bae IK, Kim YB, Lee SH (2017) Biology of *Acinetobacter baumannii*: pathogenesis, antibiotic resistance mechanisms, and prospective treatment options. Front Cell Infect Microbiol 7:55

Lolans K, Rice TW, Munoz-Price LS, Quinn JP (2006) Multicity outbreak of carbapenem-resistant *Acinetobacter baumannii* isolates producing the carbapenemase OXA-40. Antimicrob Agents Chemother 50(9):2941–2945

Manchanda V, Sanchaita S, Singh N (2010) Multidrug resistant acinetobacter. J Global Infect Dis 2(3):291–304

Maragakis LL, Perl TM (2008) *Acinetobacter baumannii*: epidemiology, antimicrobial resistance, and treatment options. Clin Infect Dis 46(8):1254–1263

Mesli E, Berrazeg M, Drissi M, Bekkhoucha SN, Rolain JM (2013) Prevalence of carbapenemase-encoding genes including New Delhi metallo-beta-lactamase in Acinetobacter species Algeria. Int J Infect Dis 17(9):e739–e743

Necati Hakyemez I, Kucukbayrak A, Tas T, Burcu Yikilgan A, Akkaya A, Yasayacak A, Akdeniz H (2013) Nosocomial *Acinetobacter baumannii* infections and changing antibiotic resistance. Pak J Med Sci 29(5):1245–1248

Oh YJ, Song SH, Baik SH, Lee HH, Han IM, Oh DH (2013) A case of fulminant community-acquired *Acinetobacter baumannii* pneumonia in Korea. Korean J Intern Med 28(4):486–490

Pagano M, Martins AF, Barth AL (2016) Mobile genetic elements related to carbapenem resistance in *Acinetobacter baumannii*. Braz J Microbiol 47(4):785–792

Rezaee MA, Pajand O, Nahaei MR, Mahdian R, Aghazadeh M, Ghojazadeh M, Hojabri Z (2013) Prevalence of Ambler class A beta-lactamases and ampC expression in cephalosporin-resistant isolates of *Acinetobacter baumannii*. Diagn Microbiol Infect Dis 76(3):330–334

Rogers AD, Deal C, Argent AC, Hudson DA, Rode H (2014) Ventilator associated pneumonia in major paediatric burns. Burns 40(6):1141–1148

Santajit S, Indrawattana N (2016) Mechanisms of antimicrobial resistance in ESKAPE pathogens. Biomed Res Int 2016:2475067

Senkyrikova M, Husickova V, Chroma M, Sauer P, Bardon J, Kolar M (2013) *Acinetobacter baumannii* producing OXA-23 detected in the Czech Republic. SpringerPlus 2(1):296

Sharma A, Sharma R, Bhattacharyya T, Bhando T, Pathania R (2017) Fosfomycin resistance in *Acinetobacter baumannii* is mediated by efflux through a major facilitator superfamily (MFS) transporter-AbaF. J Antimicrob Chemother 72(1):68–74

Shoja S, Moosavian M, Rostami S, Farahani A, Peymani A, Ahmadi K, Ebrahimifard N (2017) Dissemination of carbapenem-resistant *Acinetobacter baumannii* in patients with burn injuries. J Chin Med Assoc 80(4):245–252

Singh H, Thangaraj P, Chakrabarti A (2013) *Acinetobacter baumannii*: a brief account of mechanisms of multidrug resistance and current and future therapeutic management. J Clin Diagn Res 7(11):2602–2605

Subashchandrabose S, Smith S, DeOrnellas V, Crepin S, Kole M, Zahdeh C, Mobley HL (2016) *Acinetobacter baumannii* genes required for bacterial survival during bloodstream infection. mSphere 1(1):e00013

Vila J, Marti S, Sanchez-Cespedes J (2007) Porins, efflux pumps and multidrug resistance in *Acinetobacter baumannii*. J Antimicrob Chemother 59(6):1210–1215

Wareham DW, Bean DC, Khanna P, Hennessy EM, Krahe D, Ely A, Millar M (2008) Bloodstream infection due to Acinetobacter spp: epidemiology, risk factors and impact of multi-drug resistance. Eur J Clin Microbiol Infect Dis 27(7):607–612

Wong D, Nielsen TB, Bonomo RA, Pantapalangkoor P, Luna B, Spellberg B (2017) Clinical and pathophysiological overview of Acinetobacter infections: a century of challenges. Clin Microbiol Rev 30(1):409–447

Xu C, Bilya SR, Xu W (2019) adeABC efflux gene in *Acinetobacter baumannii*. New Microbes New Infect 30:100549

Zhang W, Wu YG, Qi XM, Dai H, Lu W, Zhao M (2014) Peritoneal dialysis-related peritonitis with *Acinetobacter baumannii*: a review of seven cases. Perit Dial Int 34(3):317–321

Mechanism of Antibiotic Resistance and Pathogenicity of *Vibrio cholerae*

15

Subhasree Saha and Durg Vijai Singh

Contents

Abstract

Antibiotic resistance among the microorganisms is one of the major concerns in the field of infectious biology. With the discovery of advanced antibiotics, the microorganisms are also combating back by upgrading their resistance mechanisms. The antibiotic resistance mechanism in different bacteria not only enhances the virulence but also increases the mortality rate. In recent times, the phenomenon of antibiotic resistance is increasing in the microorganisms such as

S. Saha · D. V. Singh (✉)
Infectious Disease Biology, Institute of Life Sciences, Bhubaneshwar, India

Department of Biotechnology, School of Earth Biological and Environmental Sciences, Central University of South Bihar, Gaya, Bihar, India
e-mail: singhdv@ils.res.in; dvsingh@cusb.ac.in

B. Siddhardha et al. (eds.), *Model Organisms for Microbial Pathogenesis, Biofilm Formation and Antimicrobial Drug Discovery*,
https://doi.org/10.1007/978-981-15-1695-5_15

Pseudomonas aeruginosa, Mycobacterium tuberculosis, and *Streptococcus* spp., which can cause higher degree of health adversity, and the list will continue growing with the passing time. One such example is the *Vibrio cholerae,* a gram-negative bacterium belonging to family Vibrionaceae is responsible for causing the disease cholera which is characterized by profuse rice watery diarrhea. Cholera organism after reaching the upper small intestine adheres, colonizes, multiplies, and secretes cholera toxin. CT consists of five bindings (B) subunits and one active (A) subunit. The B subunits bind to the GM1 ganglioside receptors in the small-intestinal mucosa, and the A subunit enters into the cell where it activates adenylate cyclase. This leads to an up rise in cyclic AMP level, and an increase in chloride secretion and inhibition of neutral sodium chloride absorption, which in turn leads to a massive outpouring of fluid into the small intestine, thus causing severe watery diarrhea. The primary treatment for cholera is rehydration therapy. Antibiotic therapy shortens the duration of infection and adversity of the disease. However, with the increased incidence of drug resistance in *V. cholerae* strains, the treatment of infection became more challenging day by day. Antibiotic resistance mechanisms in *V. cholerae* include the expulsion of drugs via efflux pumps, SXT elements, conjugative plasmids, and sometimes random chromosomal mutations. In this chapter, we are describing different mechanism of antibiotic resistance and pathogenicity keeping *V. cholerae* in highlight.

Keywords
Antibiotic resistance · *Vibrio cholerae* · Pathogenicity · Efflux pumps · Biofilms · Quorum sensing

15.1 Introduction

The discovery of antibiotic was a remarkable event in the history of disease biology. Before 1928, when Alexander Fleming discovered penicillin, the rate of mortality was very high. The mechanisms of antibiotic resistance vary from microorganisms to microorganisms because of remodeling of drug binding targets, enzymatic cleavage of the drugs, modification of the cell wall to reduce the permeability, and the cell wall efflux pumps to exclude the drugs (Stewart and Costerton 2001) (Table 15.1). Specific antimicrobials target like bacterial protein synthesis which pioneered the production of ribosome protection proteins impedes the action of ribosome inhibiting drugs (Munita and Arias 2016). Apart from these, the bacteria produce biofilms, an extra-polymeric substance that forms a mesh-like structure which obstructs the drugs from acting upon the target sites allowing the microorganisms to become drug resistant (Singh et al. 2017b) (Figs. 15.1 and 15.2). With the growing times, the microorganisms have adopted several mechanisms to survive against the antibiotics, which are as follows.

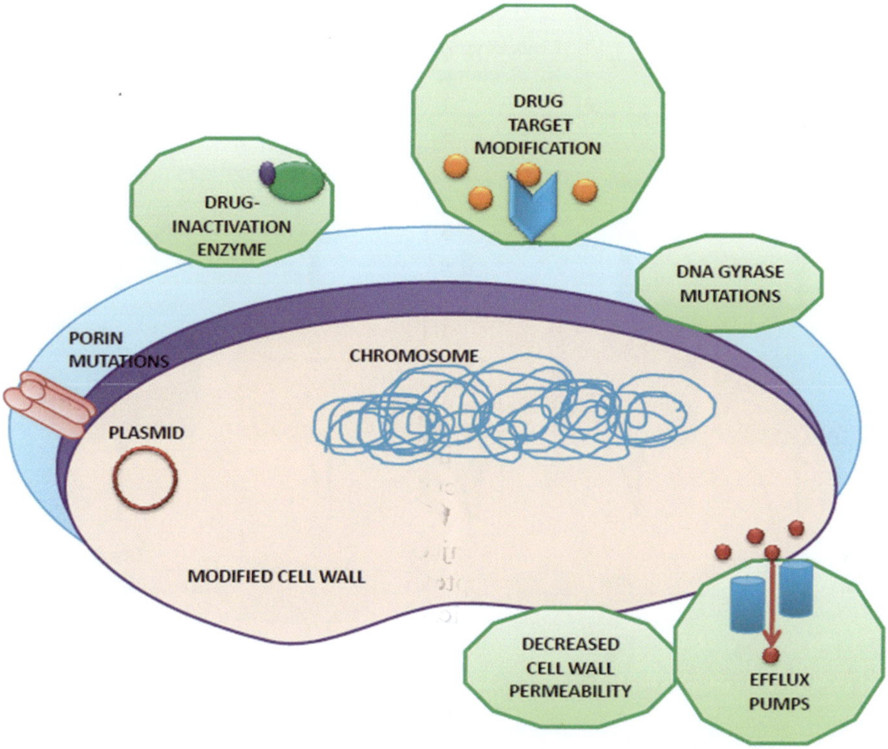

Fig. 15.1 Different drug resistant mechanisms in bacteria. Schematic diagram of various antibiotic resistance mechanisms employed by different microorganisms

15.1.1 Enzymatic Degradation

The degeneration of drugs aids the microorganisms in survival (Table 15.1). The secretion of enzymes such as aminoglycoside-modifying enzymes (AMES) or β-lactamases helps to nullify drugs like aminoglycoside and penicillin (Zhang and Dong 2004). The AMES are classified into three categories: aminoglycoside N-acetyltransferases (AACs), aminoglycoside O-nucleotidyltransferases (Bender et al. 2015), and aminoglycoside O-phosphotransferases (APIs) (Ramirez and Tolmasky 2010). AMES cause enzymatic alteration of the aminoglycoside at a particular position; however, the β-lactamases hydrolyze the cyclic ring present in the antibiotics, e.g., penicillin and cephalosporin (Zeng and Lin 2013).

15.1.2 Efflux Pumps

The efflux pumps function by excluding out the toxic compounds from the bacterial cell (Webber and Piddock 2003). The families of the efflux pumps range from the

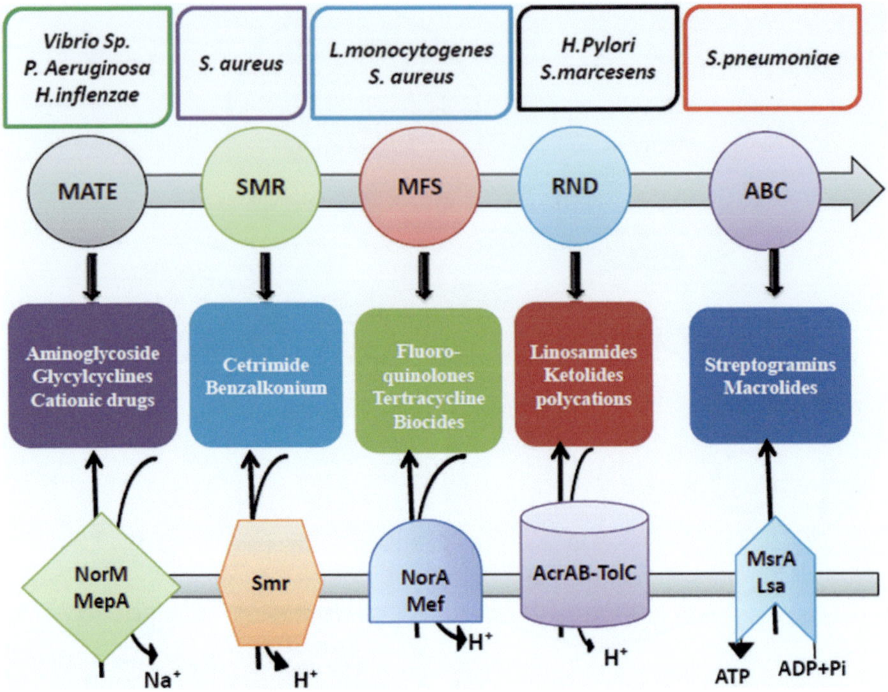

Fig. 15.2 Efflux mechanisms for drug resistance in microorganisms. *MATE* multidrug and toxin extrusion efflux, *SMR* small multidrug resistance, *MFS* major facilitator superfamily, *RND* resistance-nodulation division, *ABC* ATP-binding cassette

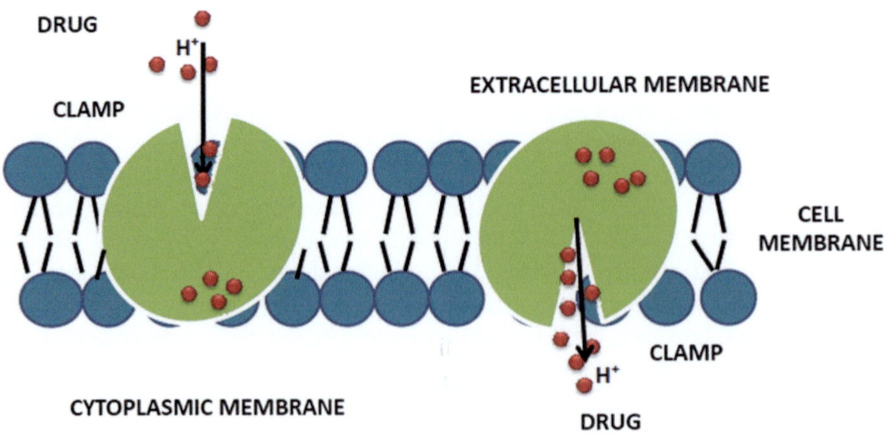

Fig. 15.3 "Clamp and Switch mechanism." Extrusion of drugs using the proton gradient drive

Table 15.1 Antibiotic resistance mechanisms in bacteria

Mode of antibiotics resistance	Examples of antibiotics	Bacterial targets responsible	Examples of bacteria with resistance mechanisms	References
Enzymatic degradation	Penicillin	β lactamases	ESBLs (extended spectrum β-lactamases) *Escherichia coli* and *Klebsiella pneumoniae*	Oli et al. (2017)
	Macrolides chloramphenicol	Aminoglycoside-modifying enzymes	*Staphylococcus spp., Streptococcus spp.*	Bush (2012)
Efflux pumps	Oxytetracycline	ABC transporters	*Serratia marcescens, Stenotrophomonas maltophila*	Greene et al. (2018)
	Erythromycin	MFS transporters	*Clostridium difficile E. coli*	Kumar et al. (2013)
Protein-based sequestration	Tetracycline	Ribosome Protection Proteins	*Camphylobacter jejuni Streptococcus spp.*	Connell et al. (2003a) Davis et al. (2011)
Drug target modification	Cephalosporin Carbepenem	Cell wall modification	*Streptococcus pneumonia Staphylococcus aureus*	Davis et al. (2011)
	Hygromycin Flambamycin	Mutation of 16s ribosomal RNA	*Propionibacterium acnes Helicobacter pylori*	Connell et al. (2003a)
SXT elements and integrons	Sulfamethoxazole Trimethopriin	Mobile genetic elements	*Vibrio cholerae*	Kitaoka et al. (2011)
Bacterial biofilms	Gentamycin Daptomycin	Drug target unavailability	*Psedomonas aeruginosa*	Hall and Mah (2017)

ATP driven transporter family, i.e., ATP-binding cassette (ABC) to the proton-based antiporter efflux pumps such as major facilitator superfamily (MFS) (Table 15.1). The secondary active transporters are multidrug and toxic compound extrusion (MATE), the resistance nodulation-cell division efflux family, the drug metabolism transporter family, and the small multidrug resistance (SMR) efflux pump family (Lin et al. 2015) (Fig. 15.3). Nowadays, the occurrence of multidrug resistance among the microorganisms is growing as a significant threat, and the efflux pumps aid them to thrive against the different groups of drugs by inhibiting the accumulation of drugs on the target side (Li et al. 2015).

The efflux pumps are distributed into three broad categories depending on the extrusion mechanisms. The primary active transporter utilizes the hydrolysis of ATP for the exclusion of compounds, the secondary carriers use ion gradient channeling for extruding out the drugs, and the third one is the phosphotransferase system (PTS) that utilizes phosphorylation of the drug substrate for inclusion and exclusion of toxic compounds from the cell (Andersen et al. 2015).

15.1.2.1 MATE

The multidrug and toxin extrusion efflux also known as MATE is the secondary active multidrug pumps that were recently determined as crucial factor for the removal of the cationic drugs through the cell membrane using Na^+ and H^+ gradient (Ogawa et al. 2015; Radchenko et al. 2015). MATE-like transporters are also present in the other host systems, but with different exclusion function, for example, the mammalian MATE efflux pumps helps in the excretion of xenobiotics compounds (Table 15.1). The MATE transporter family was first discovered in the *Vibrio parahaemolyticus* (NorM antiporter). The crystal complex of NorM shows an outward-open with two openings towards the outer layer of the cell inner membrane (IM) (Brown et al. 1999).

The MATE-type pumps phylogenetically divided into three broad clusters:

- The cluster 1 family mostly contains genes like NorM from *Vibrio* spp., which excludes the toxic compound like EtBr using the Na^+ pump.
- The cluster 2 families belong to eukaryotic MATE-family ranging from plants, animals, and fungi. The members of this cluster use the proton-driven pumps for the exclusion of toxic compounds.
- The cluster 3 belongs to MATE-family enhanced drug exclusion capacity. The DinF from the *E. coli* and MepA from *S. aureus* are part of this phylogenetic cluster (Miyamae et al. 2001).

V. cholerae has a series of efflux systems belonging to the MATE-family, e.g., VcmA, VcmB, VcmD, VcmH, VcmM, and VcrN. It utilizes these system arrays to exclude hydrophilic drugs like fluoroquinolones and norfloxacin (Taylor et al. 2012).

15.1.2.2 SMR

Small multidrug resistance drug pump also abbreviated as SMR are usually membrane-bound pumps that exclude the mono-cationic and dicationic compounds depending on the proton motive force driver (Misra and Bavro 2009). SMRs are the smallest of all the extruding pumps consisting of 110–150 amino acids in length that are found mostly in the prokaryotic system. SMRs specialize mainly in the exclusion of cationic toxic quaternary compounds (Srinivasan and Rajamohan 2013).

The NMR structure of various SMR pumps reveals that they are a consortium of four α-helical transmembrane strands, which also contains a conserved glutamate (Glu14) residue in the center of the first transmembrane strand. EmrE in *E. coli* and QacH-2 in *S. aureus* are the examples of SMR efflux pumps (Lorch et al. 2005). SMR aids the bacteria in survival by conferring high resistance against the antiseptics and dyes like safranin and crystal violet, which contain a high amount of quaternary ammonium compounds (Table 15.1). SMR-based drug resistance transferred to the other bacteria via plasmids and transposons (Shriram et al. 2018). The SMR family is subdivided into three categories (Bay et al. 2008).

15.1.2.2.1 Small Multidrug Pumps (SMP)

The SMPs subclass confers resistance to the lipophilic quaternary cationic compounds (QCC). SMPs are mostly grouped depending on the function, structure, and phylogenetic analysis (Table 15.1). The SMPs include transposons and integrons encoding genes like Hsmr from *Halobacterium salinarum,* EmrE from *E. coli* and QacF, QacG, etc. (Bay and Turner 2009).

15.1.2.2.2 The Suppressor of groEL Protein Subclass

The second category of the SMR is groEL mutation suppressor protein categorized depending on the phenotype and phylogeny (Paulsen et al. 1996). SUG actively excludes QACs and other toxic lipophilic cationic dyes using isogenic transportation. They might have a role even for the uptake of regulatory compounds. The sugE gene from *Citrobacter freundii* has been recognized to be in SUG category (Bay et al. 2017). The correlation between the GroEL and SUG protein is least understood; however, with advanced research, it is evident that the excessive accumulation of the SUG proteins inhibits the GroEL activity. The SUG protein has a homologous structure to the SMP pumps with α-helical transmembrane (Slipski et al. 2018).

15.1.2.2.3 Paired SMR Proteins

PSMRs are quite unusual in comparison to that of the SMP and SUG subclasses as the PSMR requires co-expression of both the SMR homologs. They are identified from the genome database survey using different genetic and biochemical standards (Bay et al. 2008). The PSMR consists of an extended hydrophilic loop present in the C-terminal and an additional longer hydrophilic loop (Sapula and Brown 2016). The PSMR members have a high sequence specificity among the bacterial genome. The *YvdR/YvdS* and *EbrA/EbrB* in bacilli family belong to the PSMR members. The drug resistance property of the PSMR against the toxic compounds depends on the co-expression of both PSMR proteins (Kolbusz et al. 2012).

15.1.2.3 MFS

The major facilitator superfamily, the abundant and ubiquitous efflux pump family, is present in most of the living forms. MFS belongs to the family of secondary active transporters, which excludes the toxic compounds using the proton gradient antiport system (Kumar et al. 2016). The MFS has a consortium of 12 or 14 α-helical segments, attached to the inner membrane (Arioli et al. 2014). The TetA was the primary efflux pump to be determined and well studied. The inference from the TetA crystal structure study shows that the MFS consists of two domains surrounding the central pore which help the MFS pump to "Clamp and switch" from the cytoplasmic side of the membrane to the periplasmic side of the layer which actively allows drugs and other toxic components to go out (Zárate et al. 2019) (Fig. 15.3).

MFS pump consists of α-helices and an inner loop that is made up of hydrophilic amino acids residues where both the C-terminal and N-terminal sides embedded in the inner loop. Recent research studies showed that the MFS efflux pump family is crucial for both acquired and inherit drug resistance among all bacteria. Some of the examples for MFS are *TetA, NorA, QacA*, and *EmrAB* (Kumar and Varela 2012). According to the "Clamp and Switch mechanism," the inner hydrophobic central pore is very crucial as it allows the drug to move through the cytoplasm and extra-cellular membrane (Fig. 15.4). The MFS efflux pump is tripartite system, which helps the inner membrane to span helping in the exclusion of toxic compounds (Singh et al. 2017b).

The MFS transporters present in *V. cholerae* EmrD-3 utilizes the proton-gradient-decoupling agent that causes exclusion of drugs like rifampicin, chloramphenicol, and bile salts (Singh et al. 2017a).

15.1.2.4 RND

The RND, stands for the resistance-nodulation-division, is predominant among efflux pump family in gram-negative bacteria, which releases the toxic compounds and antibiotics (Li and Nikaido 2009) and is crucial for providing antibiotic

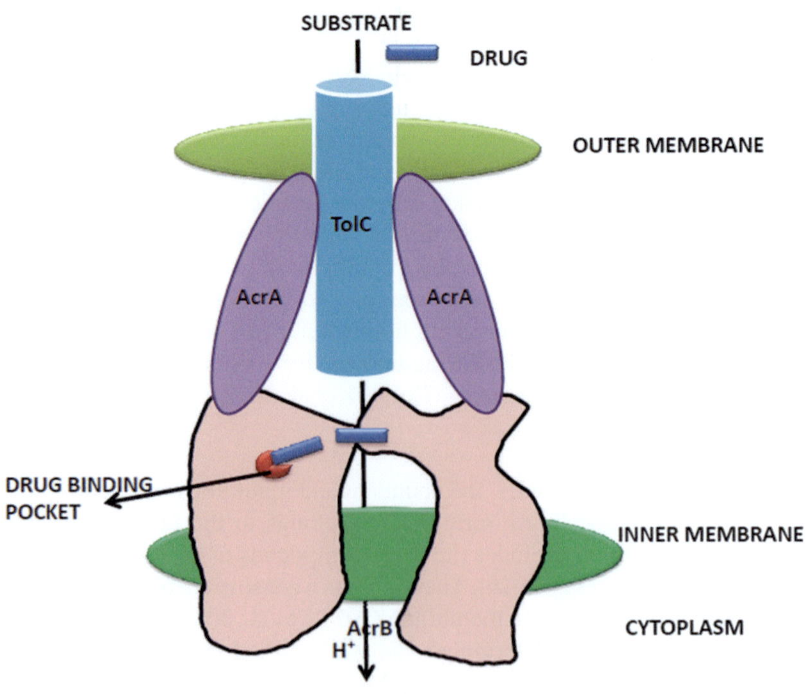

Fig. 15.4 The RND efflux pump mechanism. Schematic representation of tripartite RND multi-drug efflux system AcrAB–TolC of the gram-negative bacterium *Escherichia coli*. *AcrA* adapter protein, *TolC* outer membrane channel protein

resistance to the bacteria (Nikaido and Takatsuka 2009). AcrAB present in *E. coli* and MexB in *Pseudomonas aeruginosa* are known as a member of RND efflux pump proteins (Table 15.1) (Alvarez-Ortega et al. 2013).

RND pump is related to outer membrane channel protein like TolC and the adapter protein, AcrA in *E. coli* (Elkins and Nikaido 2002; Edward et al. 2003). The RND pumps are tripartite system just like MFS family, the three complex works together for the exclusion of compounds out of the cell, and the non-functionality of any component affects the efflux mechanism (Mao et al. 2002; Touzé et al. 2004) (Fig. 15.5). However, extrusion mechanism of RND differs from the MFS in terms of drug exclusion. In the case of RND, the drug excludes out directly from the inner membrane barrier rather than excluding the drug into the periplasm (Nikaido 1994). The RND efflux pumps are known for the exclusion of not only drugs like crystal violet, ethidium bromide, bile acids, and Triton-X (Yakushi et al. 2000) but also the lipophilic compounds by interacting with the membrane lipid bilayer (Yakushi et al. 2000) and capture the drug from the lipid bilayer or the aqueous interphase. The RND efflux pump mostly follows the three-step functional rotation for the expulsion of drugs (Davies et al. 2000).

The X-ray crystallography of RND efflux pump reveals that the RND pump is a tri-complex protein comprised of a large periplasmic head portion and a central pore domain (Zgurskaya 2002; Bolla et al. 2014). The RND efflux pumps consist of three promoters, which are essential for the drug exclusion mechanism. Among, the three promoters any of the promoters bind to the substrate that causes changes in the conformation (Bina et al. 2008; Taylor et al. 2012) (Fig. 15.4). Moreover, the change in conformation allows the exclusion of different compounds aiding the bacteria against drug resistance and survival (Murakami et al. 2002; Kim et al. 2004).

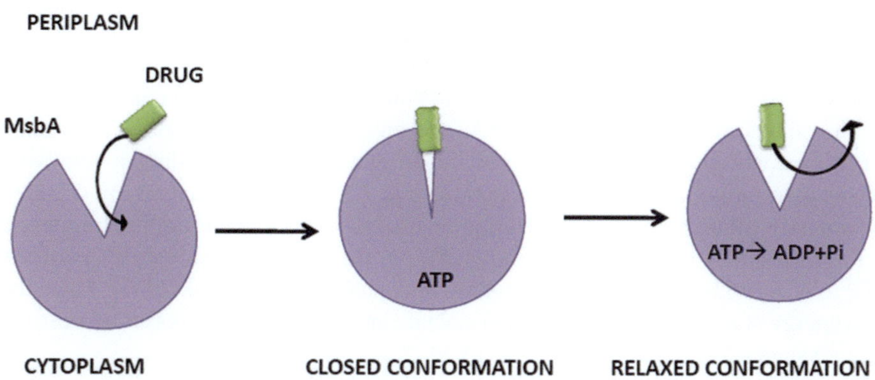

Fig. 15.5 The drug translocation mechanism of *Msb*A ABC transporters. Extrusion of drugs via conformation change

15.1.3 ABC

The ATP (adenosine triphosphate)-binding cassette is also referred to as the ABC efflux pump (Pohl et al. 2011). The ABC efflux pumps exclude the cytotoxic drugs entering the cell via passive diffusion (Ter Beek et al. 2014). The ABC pumps generally extrude a wide range of hydrophilic compounds from drugs to toxic organic substance (Lubelski et al. 2004). ABC transporters are the largest efflux pump family that is predominant in the eukaryotic organisms (Jones and George 2009). ABC transporters usually depend on hydrolysis of ATP, causing extrusion of toxic substances as the name itself signifies. Most of the ABC transporters exclude dyes, steroids, and different drugs (Zaidi et al. 2008).

The ABC transporter proteins are made up of two halves: one half is the transmembrane domain (TMD) that has a consortium of six α-helices, and the other half is the nucleotide binding domain (NBD) (Agustiandari et al. 2008; Davidson et al. 2008). X-ray crystallography of ABC transporter reveals a homodimer with a cone-shaped structure (Hürlimann et al. 2016). The TMD region made up of six α-helix monomers arranged in such a way that forms the peak of the cone which makes contact with the extracellular membrane, and NBD is a bit apart from the cone and the ATP binding sites projected towards the dimer center (Braibant et al. 2002). All six transmembrane helices are connected internally via loops towards the outer membrane (Moussatova et al. 2008). The collage of the transmembrane helices allows the formation of the entry and exit pores through which the passage of toxic compounds is difficult as the pores are very constrained (Zolnerciks et al. 2011). However, the cone structure contains a tunnel-like shape that allows the entry of compounds from the cytoplasmic membrane side (Dawson et al. 2007; Locher 2016). ABC pumps help the bacteria to survive against the ROS. LmrA in *Lactococcus lactis* and MsbA in *E. coli* are examples of ABC efflux transporters (Putman et al. 2000; Rice et al. 2014).

15.1.3.1 *V. cholerae* ABC Transporter

ABC transporter protein present in the *V. cholerae* is also referred to as MsbA (Chang 2003). It is a homolog to the *E. coli*-MsbA, with stark similarity but also has several differences as the VC-MsbA is a closed structure, with a conserved TMD and two NBDs interacting with each other (Ward et al. 2007). NBD consists of ATP binding site similar to that of Ec-MsbA, and once the substrate binds, it allows the conformational change and as a result of an enhanced ATP binding it relatively causes spanning of the NBD (Jones and George 2013). The two halves are oriented to face each other, causing dimerization of the NBDs producing the final structure of the MSB transporter and after dimerization, the NBD causes hydrolysis of the ATP to ADP with the release of inorganic phosphate (Borbat et al. 2007). Once the high energy state releases, the nBDs dissociate and allow the transporter to return to the free state (Dong et al. 2005) (Fig. 15.5).

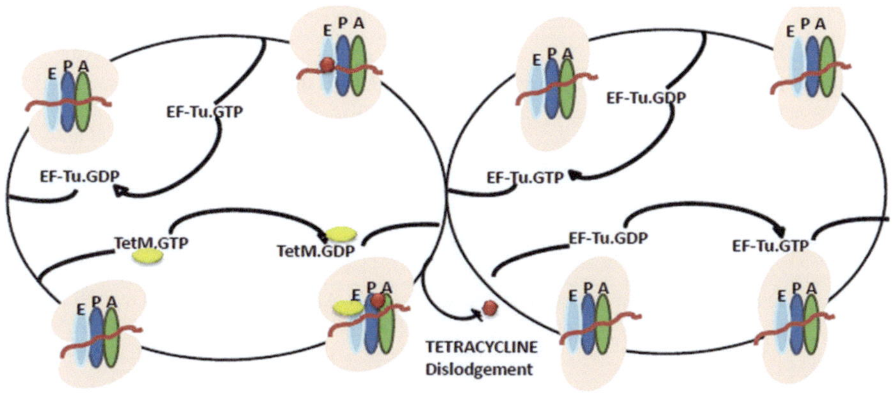

Fig. 15.6 The drug translocation mechanism by Tet(O)/(M) system. Dislodgment of the tetracycline by Tet (O) as Tet dissociates GTP to GDP allowing to the ribosome to continue with normal post-translation

15.1.4 Protein-Based Sequestration

The drug resistance mechanisms vary from microorganism to microbial species. Ribosome serves as a smooth and efficient target for antibiotics. However, microorganisms have employed sequestration of the ribosome targeting antibiotics by protecting the bacterial ribosome, thus inhibiting the antibiotics target (Poehlsgaard and Douthwaite 2005).

The Tet (O) and Tet (M) proteins from *Campylobacter jejuni* and *Streptococcus* sp. are good examples of bacterial resistance against the antibiotic tetracycline (Wang et al. 2014). The ribosome protection proteins protect all groups of bacteria (Amaral et al. 2014) (Fig. 15.7). Tetracycline commonly attaches itself to the ribosomal 30S subunit and then with the mRNA codon recognition site. Once, the drug binds with 16S rRNA in the presence of Mg^{2+} cation, a complex formation happens resulting in the disruption of binding of aminoacyl-tRNA, therefore, inhibition of the further protein translocation hence protein translation (Connell et al. 2003a, b).

Bacteria combat against tetracycline using several ways, including efflux pumps; however, the resistance via ribosome protection, providing protein like TetO/M, has higher efficiency (Bujard et al. 1996). The TetO/M proteins work via forming a bond with the elongation factor-G (EF-G) binding site that is present in the ribosome due to the similarity of TetO/M to the EF factor (Chopra and Roberts 2001) (Fig. 15.6). Moreover, as a result, the binding of TetO/M to EF-G hinders the post-translocation and causes the removal of bound complex from A-site. As a result, the bacterial translation continuously takes place (Deng et al. 2017).

Table 15.2 Van operon present in different microorganisms

Name	Function	References
VanR/ VanS	A two-component regulatory system which detects the vancomycin or may cause disruption of the cell	O'Driscoll and Crank (2015) Arthur et al. (1997) Faron et al. (2016)
VanX	D-dipeptidase that detects and cleaves D-Ala-D-Ala repeats	O'Driscoll and Crank (2015) McCafferty et al. (1997) Faron et al. (2016)
VanH	D-hydroxyacid dehydrogenase causes reduction of pyruvate to D-Lac for the aid VanA/B activity	O'Driscoll and Crank (2015) Faron et al. (2016)
VanA/B	Ligase that ligates the D-Ala-D-Lac helping the production of mutated pentapeptide that reduces the activity of vancomycin	Gutmann et al. (1992) Lai and Kirsch (1996) Baptista et al. (1997)
VanY	D, D-carboxypeptidase that cleaves the terminal D-Ala that cause reduction of peptapeptide that further reduces the affinity for the vancomycin	O'Driscoll and Crank (2015) Arthur et al. (1997)
VanZ	Unique but usually copresent with VanA, it confers resistance against teicoplanin. However, the mechanism is unknown	Arthur et al. (1997) Faron (2016)

15.1.5 Drug Target Modification

The bacterial cell wall is crucial for the survival and physiological maintenance and functioning (Munita and Arias 2016). Hence, it makes the cell wall a perfect target for the antibiotics to work (Table 15.2). However, with the emerging time, bacteria have developed several mechanisms to bypass the action of different antibiotics (Caddick et al. 2005).

Production of enzymes like β-lactamases and modification of peptidoglycan help the bacteria to survive against the cell wall targeting antibiotics (Reeve et al. 2015). The change helps the bacteria to survive against the antibiotics that target the bacterial cell wall (Džidić et al. 2008). For example, drugs like vancomycin interrupt the peptidoglycan formation by polymerization of the phospho-disaccharide-pentapeptide lipid complex is a crucial step, e.g., addition of ala-D-ala peptides present in the free carboxyl ends, and it stops the cross-linking of the penicillin binding protein (PBPs) hence, results in the death of the bacteria (Chaudhary et al. 2006). The seven different types of vancomycin resistance are known as VanA, VanB, VanC, VanD, VanE, VanG, and VanL (Périchon and Courvalin 2012).

15.1.6 Mutation of 16S Ribosomal RNA

The antibiotic/drug resistance mechanisms in different bacteria have a wide range; recently with in-depth studies, it was found that the mutation of the ribosomal RNA has added to the bacterial arsenal against the antibiotics (Sigmund et al. 1984). Drugs like tetracycline usually act against the bacteria by binding with 30S ribosomal subunit (Gerrits et al. 2002), and as a result, the protein synthesis will be inhibited. With the recent research studies, it revealed that for attachment of the drug, the interaction of 16s rRNA along with other ribosome related protein is crucial (De Stasio et al. 1989).

The ribosomal center consists of rRNA domain along with several ribosomal proteins like S7, S12, and other polypeptides (Marvig et al. 2012). Therefore, any mutation in the rRNA confers resistance to the bacteria against the protein synthesis targeting drug squad (Canu et al. 2002). However, the drug resistance ability of the bacteria comes at the expense of the bacterial growth rate; thereby, in normal conditions, the ribosome mutated bacteria have a low survival rate in comparison to the non-mutated bacteria (Gomez et al. 2017). However, the ribosomal mutation assists the bacteria for survival during antibiotics action (Limburg et al. 2004). One of the modifications that happen in nature is the methylation of the ribosome (Agustiandari et al. 2008), which is performed by the "house-keeping" methyltransferase enzymes that methylate the ribosome without hampering the normal functioning of the ribosome (Li et al. 2013).

One of the examples is *cfr* methyltransferase that targets the peptidyl transferase center (PTC), which is existent in the ribosomal subunit (Mankin and Polacek 2008). The PTC is the place where the addition of the nascent amino acids takes place, making PTC a precise target for the activity of several antibiotics. Therefore, methylation of the PTC will make the inefficient antibiotic activity (Long and Vester 2012). The *cfr* gene was discovered first in *S. aureus,* but with time, several homologs were found in several bacteria (Yang et al. 2009). The *cfr* methylates the A2503 position present in the PTC, and that the *cfr* provides resistance against five major classes of antibiotics (Bender et al. 2015).

Moreover, it is transferred among the microbial community using the transposons. Crystallography studies conferred that binding of drugs like linezolid to the ribosomal subunit (Long and Vester 2012) becomes ineffective due to the change in shape and also steric hindrance because of A2503 methylation (Kehrenberg et al. 2005). Apart from this, random mutation in the ribosome results in drug target binding affinity causing resistance against various antibiotics (Ippolito et al. 2008).

15.1.7 SXT Elements and Integrons

Mobile genetic factors are divided into three categories: plasmids, bacteriophages, and integrating and conjugative elements (ICEs). Both gram-positive and gram-negative bacteria contained ICE which had a role to play in horizontal gene transfer and genome plasticity. ICEs were identified as plasmids or transposons, curbing the knowledge and estimation of the actual existing elements (Wozniak et al. 2009). The SXT element was first discovered in *Vibrio cholerae* O139 strain MO10, isolated from an outbreak in Madras, India (Waldor et al. 1996; Toma et al. 2005). Also, such elements were reported from gram-positive bacteria *Enterococcus faecalis* as Tn916 (Gawron-Burke and Clewell 1982) and *Providencia rettgeri* isolated from human feces in South Africa in 1967 initially identified as IncJ plasmid (Böltner et al. 2002). SXT element provides resistance to *V. cholerae* against antibiotics like sulfamethoxazole, streptomycin, etc. (Wang et al. 2016). Also, the presence of SXT ICE, *dfr18*, *sulII*, *strAB*, and *aadA5* genes was reported in environmental *V. cholerae* O1 and non-O1, non-O139 strains which had been isolated before 1992, from Varanasi, India (Mohapatra et al. 2008, 2009), and in the *V. cholerae* serogroup, O1, and O139 (Burrus et al. 2006).

SXT carrying resistance genes encoding the antibiotics Sul-Tm, Sm, and Cm are integrated at 5′-end of *PrfC*, present in *V. cholerae* chromosome (Hochhut and Waldor 1999). Genetic analysis of R391 and SXT showed that the two elements shared the same attachment site *prfC* encoding for peptide chain release factor 3 (RF3) (Ahmed et al. 2005) and carried a closely related integrase gene (Hochhut et al. 2001). ICEs consist of four major modules: (1) the integration/excision module is allowing the mobility between the host and the recipient, (2) the replication/DNA processing module is assisting in the assembling and preparation for transfer; (3) the DNA secretion module allowing the migration of ICE, and (4) the regulation module which controls the expression. Based on functions and regulations, ICEs are classified into three major families (Carraro et al. 2014); however, recent studies revealed the presence of 428 ICEs found in 124 representative bacterial species showing high diversity and complexity (Bi et al. 2011).

ICEs carry three discrete functional modules that regulate its integration/excision, conjugative transfer, and regulation (Garris and Burrus 2013). SXT transfer requires three steps: (1) the excision of SXT from the host chromosome to form a circular intermediate, (2) the substrate DNA in a nicked form transfers to the recipient, and (3) integration of the transferred DNA takes place upon entry into the recipient strain. Integration of the SXT element at the *prfC* site involves site-specific recombination between *attP* present on the extrachromosomal form and the *attB* present in the host chromosome through a recA-independent process (Hochhut and Waldor 1999). Similarly, the excision of the SXT element occurs by site-specific recombination between *attL* and *attR* of the SXT element. This process results in the regeneration of the *attP* site on the excised circular form of the SXT/ICE and *attB* on the chromosome. Although SXT integration is regulated by Int (*int* gene), Xis (recombination directionality factor) (*xis* gene) assists in the process, promotes excision of the SXT, and inhibits its re-integration (Burrus and Waldor 2003). After excision, the single strand of ICE DNA acts as the substrate for conjugative transfer (Ceccarelli et al. 2008) and requires two clusters. The first cluster that acts as the substrate for the putative conjugative relaxase (*traI*) contains the origin of transfer (Morita et al. 2014; Ceccarelli et al. 2008). The second cluster requires the DNA processing module that includes the putative relaxase *traD* (a putative coupling factor) and *traJ* (encoding a protein has a weak homology with TraJ from plasmid F). The type IV secretion system facilitates the translocation of the ICE DNA which consists of three gene clusters: (1) *traLEKBVA,* (2) *s054/traC/trhF/traWUN,* and (3) *traFHG*, and involved in the formation of the pilus, the conjugative machinery, and the mating pair formation and stabilization. The *s063* region between *traN* and *traF* with no precise function identified, but its deletion resulted in a ~100-fold decrease in the frequency of transfer (Garriss et al. 2009). Wozniak et al. (2009) showed that most of the ICE core genes of unknown functions did not alter SXT transfer efficiency. The smallest element involved in SXT function was found to be of 29.7 kb consisting of 25 genes. Out of the 52 conserved genes of SXT/R391 ICEs, deletion of 27 genes resulted in little or no effect on SXT transfer efficiency and so was not included in the minimal functional backbone. It can be presumed that the conserved core genes not having any role in SXT function might enhance ICE fitness.

Integrons are gene-capturing systems incorporating exogenous open reading frames (ORFs) by site-specific recombination mediated by a tyrosine recombinase, converting the ORFs to functional genes by ensuring their correct expression (Mazel 2006). Integrons are classified into four class 1, 2, 3, or 4, and type 4 is referred to as the superintegron (Fluit and Schmitz 2004; Mazel 2006). Class I integrons consist of the integrase, *intI* gene, the attI recombination site, and a promoter that directs the transcription of the captured gene cassettes. Integron inserted gene cassettes mostly contain an imperfect repeat at the 3′ end of the gene, attC site, or 59 base element (Stokes et al. 1997). Class I integrons often harbor genes at their 3′ end encoding resistance to quaternary ammonium compounds and sulfonamides, *qacE*Δ*1* and *sul* (Dalsgaard et al. 2000; Recchia and Hall 1995).

15.2 Biofilms

Biofilms are the mesh-like structures made up of extra-polymeric substances (Kostakioti et al. 2013). The biofilms consist of cells, debris, proteins, polysaccharides, and extracellular DNA (Table 15.1). The biofilm serves several purposes for the bacteria; it helps the bacteria getting attached to the different surfaces and aids in survival against harsh environmental conditions (Davey and O'toole 2000).

For example, the *V. cholerae* survives the acidic pH of the intestine with the help of biofilm as the vibrios in the planktonic state are not able to withstand the acidic environment (Silva and Benitez 2016). The biofilm also helps the bacteria to have resistance against the antibiotics and provides a protective covering to the bacterial community so that drug cannot reach the target, thereby promotes the bacteria to thrive against the antibiotic (Higgins et al. 2007, Singh et al. 2017a). *V. cholerae* components like VPS (*Vibrio* polysaccharide) and the extracellular matrix protein help the bacteria to colonize inside the host (Teschler et al. 2015). The biofilm formation takes place in three steps: (1) planktonic state, (2) attachment of the cells to the surface, and (3) 3D mature biofilm and dispersal (Crouzet et al. 2014).

The biofilms in microbes secrete enzymes which degrade antibiotics, and other antimicrobials (Heithoff and Mahan 2004), for example, the biofilm in *Pseudomonas aeruginosa* produces a β-lactamase enzyme that causes cleavage of the β-lactam ring-containing antibiotics, making them inactive (Hall and Mah 2017). Apart from that, the extracellular DNA has been proven to be a pivotal part of the biofilm. The extracellular DNA released from the biofilm chelates the Mg^{2+} ion present in the matrix environment changing the pH condition of the environment and activates the different mechanism (Mulcahy et al. 2008). For example, in *Pseudomonas aeruginosa* phQ/pQ and pmrAB signaling pathways activated that helps in the deactivation of certain antibiotics (Lee and Zhang 2015).

15.3 Bacterial Quorum Sensing

Quorum sensing is the natural phenomenon of cell-based intercommunication among the microorganisms that aids in heightened virulence (Miller and Bassler 2001). Apart from that, the quorum sensing inducers cause enhancement of the expression of biofilm formation. Moreover, as mentioned earlier, biofilm helps the bacteria to survive against the antibiotics (Rutherford and Bassler 2012). Quorum

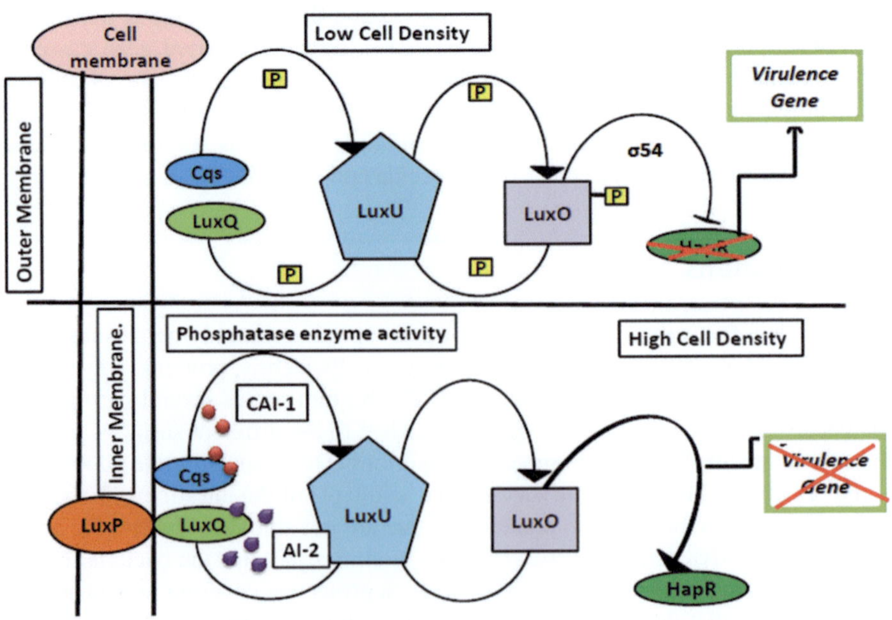

Fig. 15.7 The quorum sensing circuit of *Vibrio cholerae*

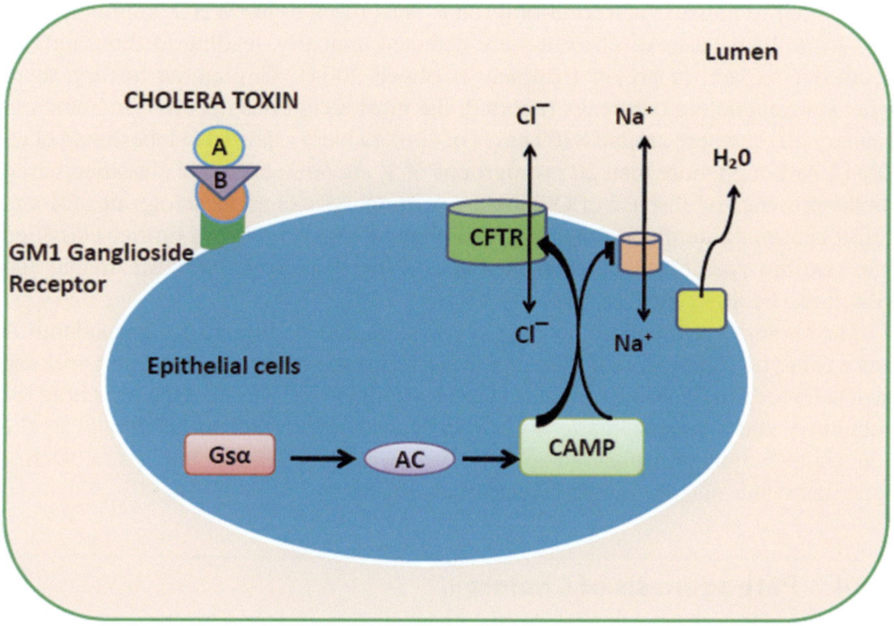

Fig. 15.8 Cholera toxin mode of action (AB) cholera toxin subunits, *GM1* GM1 ganglioside receptor, *Gsα* G protein, *Ac* adenylate cyclase, *cAMP* cyclic AMP, *CFTR* cystic fibrosis transmembrane conductor

sensing activates several signaling pathways that cause autolysis of the cells present in the biofilm (Abisado et al. 2018), and the extracellular DNA helps the bacteria thriving against the antibiotics (Lewenza et al. 2002). For example, in *Pseudomonas aeruginosa, vqs*M (Dong et al. 2007) gene, a master regulator, induces the antibiotic tolerance by activating the expression of *nfx*B, an antibiotic resistance regulator (Zhang and Dong 2004, Dong and Zhang 2005). It confers resistance against antibiotics like quinolones, kanamycin, and tetracycline (Morita et al. 2014).

Quorum sensing pathway in *V. cholerae* differs from the other pathogenic organisms because, in other organisms, the occurrence of high cell density is correlated to the expression of virulence gene and biofilm production (Antonova and Hammer 2011). However, in the case *V. cholerae*, the activation of quorum sensing signaling genes interferes with the expression of virulent genes (Freeman and Bassler 1999; Lenz et al. 2005). At low density LuxU gene in QS in *V. cholerae* represses the LuxO which is essential for the expression of virulence and activates the hapR gene (Jung et al. 2015; 2016) that is crucial for the production of protease thus helps the bacteria to come out from host (Cámara et al. 2002) (Fig. 15.7).

V. cholerae is a curve rod-shaped, non-obligatory anaerobic bacterium. This organism is the causative agent of cholera, a life-threatening disease leading to severe water loss, hypervolemic shock, and eventually death if left untreated (Kaper et al. 1995). This organism consists of a polar flagellum, which is crucial for the biofilm formation. *V. cholerae* is mostly found in the aquatic systems and transmitted to human via a fecal-oral route. According to the WHO, approximately 1.3–4.0 million cases of cholera were detected annually, leading to thousands of death due to lack of proper treatment (Colwell 2004). Throughout history, there were several cholera outbreaks reported; the most recent one occurred in Yemen in January 2019, where around 8369 cases of cholera were reported (Maheshwari et al. 2011). Although more than 205 serogroups of *V. cholerae* exist and classified based on the presence or absence of O-antigens in lipopolysaccharide, serogroups O1 and O139 caused epidemic and pandemic cholera. *V. cholerae* O1 is further classified into two biotypes: Classical and El Tor, and three serotypes: Ogawa, Inaba, and Hikojima (Feeley 1965).

The cholera toxin consists of two subunits: A and B (Fig. 15.8). The subunit A has a catalytic function, while the B subunit helps the holoenzyme to bind with the host cell receptor, GM1 ganglioside (van Heyningen et al. 1971). After attaching the subunit A starts its enzymatic action and causes enhanced production of cyclic adenosine-5-monophosphate (c-AMP) which eventually leads to the loss of electrolytes and water into the lumen (Pierce et al. 1971).

15.4 Pathogenesis of Cholera

The disease cholera is characterized by profuse rice watery diarrhea with symptoms such as anorexia, abdominal discomfort, and uncomplicated diarrhea. Cholera may lead rapidly to tachycardia, hypotension, and vascular collapse due to dehydration (Kaper et al. 1995; Morris and Black 1985). Ingested vibrios from contaminated water or food pass through the acid stomach and reach to the upper small intestine where it adheres, colonizes, multiplies, and secretes cholera toxin (Nelson et al. 2009). Colonization is facilitated by fimbria and toxin-coregulated pilus (TCP) that attaches to receptors on the mucosa (Taylor et al. 1987). CT has a molecular mass of 84,000 kDa and consists of five bindings (B) subunits and one active (A) subunit. The B subunits bind to the GM1 ganglioside receptors in the small-intestinal mucosa, and the A subunit is transported into the cell where it activates adenylate cyclase (Holmgren 1973; Van Heyningen et al. 1971). This activation leads to an increase in cyclic AMP, followed by an increase in chloride secretion in the crypt cells, and inhibition of neutral sodium chloride absorption in the villus cells, which in turn leads to a massive outpouring of fluid into the small intestine (Fig. 15.1). The diarrheal fluid contains large amounts of sodium, chloride, bicarbonate, and potassium, but little protein or blood cells. The loss of electrolyte-rich isotonic fluid leads to blood volume depletion with attendant low blood pressure and shock. Loss of bicarbonate and potassium leads to metabolic acidosis and potassium deficiency. The stools of cholera patients contain high concentrations of cholera vibrios (up to 10^8 bacteria per gram), and they are highly infectious. When passed into the environment, they can contaminate water sources and food and may seed an environmental reservoir.

15.5 Therapy

Genetic experiments showed the role of quorum sensing in biofilm formation and virulence gene regulation among *V. cholerae* (Hammer and Bassler 2003; Zhu et al. 2002). The increased drug resistance in *V. cholerae* has led to the urgency of development of new therapeutics by different pathogenesis mechanism. New therapeutics targeting *V. cholerae* quorum sensing would be helpful to decrease the quantitated release of toxins by the bacteria, which can ultimately reduce the severity of the infection. The nutrients concentration at surfaces are high and that attachment to the surface is a significant step in establishing infection by bacteria, therefore targeting biofilm would be to advantages. Cholera toxin is a hexameric A-B5 type toxin. The binding of CTX-B-subunits of the CT to monosialoganglioside (GM_1) receptor is considered as a recognition unit which acts as a delivery vehicle for A-subunit. Induction of immunogenicity was shown at mucosal surfaces after oral administration of CTB subunits of toxin. Therefore, an alternative approach is to use inhibitors against toxin binding receptor as a therapeutic agent for the treatment of infection.

There is a good relationship between cholera epidemics and the phage. The *ctx* gene is a part of the nonlytic CTXΦ prophage; however, there are many bacteriophages which can lyse this bacterium. Phage typing scheme classified vibrio phages into 27 types in O1 strains and 10 types in O139 strains (Sayamov 1963; Marcuk et al. 1971; Faloon et al. 2014). In the past, bacteriophages were used for the treatment of severe cases of cholera (Sayamov 1963; Marcuk et al. 1971;). Phage therapy can be used for the prevention of cholera and could successful in modulating epidemics by decreasing the required dose of infectious bacteria (Zahid et al. 2008). Apart from that, as discussed earlier quorum sensing in vibrio helps in the biofilm formation that increases the bacterial survival rate, making them resistant to harsh antibiotic treatment as well. According to recent studies, several QS inhibitors have been identified. For example, ML366 and ML370 which are identified as LuxO inhibitors in case of *V. cholerae* (Faloon et al. 2014) and Qstatin that targets the LuxR homologs and as a quorum sensing inhibitor causing inhibition of SmcR gene responsible for the virulence and biofilm formation (Kim et al. 2018). Compounds like ethanolamine have been identified as a cqsR gene regulator which is responsible for expression of virulence factors in *V. cholerae.*

15.6 Conclusions

V. cholerae is an autochthonous inhabitant of the aquatic environment and the causative agent of cholera. After infection cholera organism seeded into the environment where it acquires and transmits new genes, including antibiotic resistance. Cholera toxin consists of five B and one A subunits of which B subunits bind to the GM1 ganglioside, and the A subunit enters into the cell where it activates adenylate cyclase leading to an increase in cyclic AMP, thus causing an increase in water and chloride secretion, hence leads to severe watery diarrhea. The cholera treatment is dependent on oral rehydration therapy, but adjoint antibiotic treatment is essential for early recovery. With increase in antibiotic resistance, there is a need for modern-day alternative therapeutics. The bacteriophage cocktail treatment, usage of quorum sensing inhibitors, and anti-secretory drugs might help for the treatment of cholera. However, until developed a treatment strategy, it is essential to take proper prevention strategies like usage of hygiene, clean drinking water, and limited use of antibiotics.

Acknowledgments This study was supported by fund contributed by the Department of Biotechnology, New Delhi to the Institute of Life Sciences, Bhubaneswar. SS is grateful to the Department of Science and Technology, New Delhi for providing Inspire Junior Research Fellowship.

References

Abisado RG, Benomar S, Klaus JR, Dandekar AA, Chandler JR (2018) Erratum for Abisado et al., "bacterial quorum sensing and microbial community interactions". MBio 9(5):e01749–e01718

Agustiandari H, Lubelski J, Kuipers OP, Driessen AJ (2008) LmrR is a transcriptional repressor of expression of the multidrug ABC transporter LmrCD in Lactococcus lactis. J Bacteriol 190(2):759–763

Ahmed AM, Shinoda S, Shimamoto T (2005) A variant type of Vibrio cholerae SXT element in a multidrug-resistant strain of Vibrio fluvialis. FEMS Microbiol Lett 242(2):241–247

Alvarez-Ortega C, Olivares J, Martínez JL (2013) RND multidrug efflux pumps: what are they good for? Front Microbiol 4:7

Amaral L, Martins A, Spengler G, Molnar J (2014) Efflux pumps of gram-negative bacteria: what they do, how they do it, with what and how to deal with them. Front Pharmacol 4:168

Andersen J, He G-X, Kakarla P, KC R, Kumar S, Lakra W, Mukherjee M, Ranaweera I, Shrestha U, Tran T (2015) Multidrug efflux pumps from Enterobacteriaceae, Vibrio cholerae and Staphylococcus aureus bacterial food pathogens. Int J Environ Res Public Health 12(2):1487–1547

Antonova ES, Hammer BK (2011) Quorum-sensing autoinducer molecules produced by members of a multispecies biofilm promote horizontal gene transfer to Vibrio cholerae. FEMS Microbiol Lett 322(1):68–76

Arioli S, Guglielmetti S, Amalfitano S, Viti C, Marchi E, Decorosi F, Giovannetti L, Mora D (2014) Characterization of tetA-like gene encoding for a major facilitator superfamily efflux pump in Streptococcus thermophilus. FEMS Microbiol Lett 355(1):61–70

Arthur M, Depardieu F, Gerbaud G, Galimand M, Leclercq R, Courvalin P (1997) The VanS sensor negatively controls VanR-mediated transcriptional activation of glycopeptide resistance genes of Tn1546 and related elements in the absence of induction. J Bacteriol 179(1):97–106

Baptista M, Depardieu F, Reynolds P, Courvalin P, Arthur M (1997) Mutations leading to increased levels of resistance to glycopeptide antibiotics in VanB-type enterococci. Mol Microbiol 25(1):93–105

Bay DC, Rommens KL, Turner RJ (2008) Small multidrug resistance proteins: a multidrug transporter family that continues to grow. Biochim Biophys Acta 1778(9):1814–1838

Bay DC, Stremick CA, Slipski CJ, Turner RJ (2017) Secondary multidrug efflux pump mutants alter Escherichia coli biofilm growth in the presence of cationic antimicrobial compounds. Res Microbiol 168(3):208–221

Bay DC, Turner RJ (2009) Diversity and evolution of the small multidrug resistance protein family. BMC Evol Biol 9(1):140

Bender J, Strommenger B, Steglich M, Zimmermann O, Fenner I, Lensing C, Dagwadordsch U, Kekulé AS, Werner G, Layer F (2015) Linezolid resistance in clinical isolates of Staphylococcus epidermidis from German hospitals and characterization of two cfr-carrying plasmids. J Antimicrob Chemother 70(6):1630–1638

Bina XR, Provenzano D, Nguyen N, Bina JE (2008) Vibrio cholerae RND family efflux systems are required for antimicrobial resistance, optimal virulence factor production, and colonization of the infant mouse small intestine. Infect Immun 76(8):3595–3605

Bi D, Xu Z, Harrison EM, Tai C, Wei Y, He X et al (2011) ICEberg: a web-based resource for integrative and conjugative elements found in Bacteria. Nucleic Acids Res 40(database issue):D621–D626

Böltner D, MacMahon C, Pembroke JT, Strike P, Osborn AM (2002) R391: a conjugative integrating mosaic comprised of phage, plasmid, and transposon elements. J Bacteriol 184(18):5158-5169.

Bolla JR, Su C-C, Do SV, Radhakrishnan A, Kumar N, Long F, Chou T-H, Delmar JA, Lei H-T, Rajashankar KR (2014) Crystal structure of the Neisseria gonorrhoeae MtrD inner membrane multidrug efflux pump. PLoS One 9(6):e97903

Borbat PP, Surendhran K, Bortolus M, Zou P, Freed JH, Mchaourab HS (2007) Conformational motion of the ABC transporter MsbA induced by ATP hydrolysis. PLoS Biol 5(10):e271

Braibant M, Guilloteau L, Zygmunt MS (2002) Functional characterization of Brucella melitensis NorMI, an efflux pump belonging to the multidrug and toxic compound extrusion family. Antimicrob Agents Chemother 46(9):3050–3053

Brown MH, Paulsen IT, Skurray RA (1999) The multidrug efflux protein NorM is a prototype of a new family of transporters. Mol Microbiol 31(1):394–395

Burrus V, Waldor MK (2003) Control of SXT integration and excision. J Bacteriol 185(17):5045-5054.

Bujard H, Gossen M, Hillen W, Helbl V, Schnappinger D (1996) Tetracycline regulated transcriptional modulators with altered DNA binding specificities. Google Patents

Burrus V, Quezada-Calvillo R, Marrero J, Waldor MK (2006) SXT-related integrating conjugative element in New World Vibrio cholerae. Appl Environ Microbiol 72(4):3054–3057

Bush K (2012) Antimicrobial agents targeting bacterial cell walls and cell membranes. Rev Sci Tech 31(1):43–56

Caddick JM, Hilton AC, Rollason J, Lambert PA, Worthington T, Elliott TS (2005) Molecular analysis of methicillin-resistant Staphylococcus aureus reveals an absence of plasmid DNA in multidrug-resistant isolates. FEMS Immunol Med Microbiol 44(3):297–302

Cámara M, Hardman A, Williams P, Milton D (2002) Quorum sensing in Vibrio cholerae. Nat Genet 32(2):217

Canu A, Malbruny B, Coquemont M, Davies TA, Appelbaum PC, Leclercq R (2002) Diversity of ribosomal mutations conferring resistance to macrolides, clindamycin, streptogramin, and telithromycin in Streptococcus pneumoniae. Antimicrob Agents Chemother 46(1):125–131

Carraro N, Matteau D, Luo P, Rodrigue S, Burrus V (2014) The master activator of IncA/C conjugative plasmids stimulates genomic islands and multidrug resistance dissemination. PLoS Genet 10(10):e1004714.

Ceccarelli D, Daccord A, René M, Burrus V (2008) Identification of the origin of transfer (oriT) and a new gene required for mobilization of the SXT/R391 family of integrating conjugative elements. J Bacteriol 190(15):5328-5338.

Chang G (2003) Multidrug resistance ABC transporters. FEBS Lett 555(1):102–105

Chaudhary P, Kumar R, Verma AK, Singh D, Yadav V, Chhillar AK, Sharma G, Chandra R (2006) Synthesis and antimicrobial activity of N-alkyl and N-aryl piperazine derivatives. Bioorg Med Chem 14(6):1819–1826

Chopra I, Roberts M (2001) Tetracycline antibiotics: mode of action, applications, molecular biology, and epidemiology of bacterial resistance. Microbiol Mol Biol Rev 65(2):232–260

Colwell RR (2004) Infectious disease and environment: cholera as a paradigm for waterborne disease. Int Microbiol 7(4):285–289

Connell SR, Tracz DM, Nierhaus KH, Taylor DE (2003a) Ribosomal protection proteins and their mechanism of tetracycline resistance. Antimicrob Agents Chemother 47(12):3675–3681

Connell SR, Trieber CA, Dinos GP, Einfeldt E, Taylor DE, Nierhaus KH (2003b) Mechanism of Tet (O)-mediated tetracycline resistance. EMBO J 22(4):945–953

Crouzet M, Le Senechal C, Brözel VS, Costaglioli P, Barthe C, Bonneu M, Garbay B, Vilain S (2014) Exploring early steps in biofilm formation: set-up of an experimental system for molecular studies. BMC Microbiol 14(1):253

Davey ME, O'toole GA (2000) Microbial biofilms: from ecology to molecular genetics. Microbiol Mol Biol Rev 64(4):847–867

Davidson AL, Dassa E, Orelle C, Chen J (2008) Structure, function, and evolution of bacterial ATP-binding cassette systems. Microbiol Mol Biol Rev 72(2):317–364

Dalsgaard A, Forslund A, Serichantalergs O, Sandvang D (2000) Distribution and content of Class 1 integrons in different Vibrio cholerae O-serotype strains isolated in Thailand. Antimicrob Agents Chemother 44:1315-1321.

Davies JP, Chen FW, Ioannou YA (2000) Transmembrane molecular pump activity of Niemann-Pick C1 protein. Science 290(5500):2295–2298

Davis MA, Besser TE, Orfe LH, Baker KN, Lanier AS, Broschat SL, New D, Call DR (2011) Genotypic-phenotypic discrepancies between antibiotic resistance characteristics of Escherichia coli isolates from calves in management settings with high and low antibiotic use. Appl Environ Microbiol 77(10):3293–3299

Dawson RJ, Hollenstein K, Locher KP (2007) Uptake or extrusion: crystal structures of full ABC transporters suggest a common mechanism. Mol Microbiol 65(2):250–257

De Stasio E, Moazed D, Noller H, Dahlberg A (1989) Mutations in 16S ribosomal RNA disrupt antibiotic–RNA interactions. EMBO J 8(4):1213–1216

Deng Y, Sun C, Hunt DK, Fyfe C, Chen C-L, Grossman TH, Sutcliffe JA, Xiao X-Y (2017) Heterocyclyl tetracyclines. 1. 7-Trifluoromethyl-8-pyrrolidinyltetracyclines: potent, broad spectrum antibacterial agents with enhanced activity against Pseudomonas aeruginosa. J Med Chem 60(6):2498–2512

Dong J, Yang G, Mchaourab HS (2005) Structural basis of energy transduction in the transport cycle of MsbA. Science 308(5724):1023–1028

Dong Y-H, Wang L-H, Zhang L-H (2007) Quorum-quenching microbial infections: mechanisms and implications. Philos Trans R Soc B Biol Sci 362(1483):1201–1211

Dong Y-H, Zhang L-H (2005) Quorum sensing and quorum-quenching enzymes. J Microbiol 43(1):101–109

Džidić S, Šušković J, Kos B (2008) Antibiotic resistance mechanisms in bacteria: biochemical and genetic aspects. Food Technol Biotechnol 46(1):11–21

Edward WY, Aires JR, Nikaido H (2003) AcrB multidrug efflux pump of Escherichia coli: composite substrate-binding cavity of exceptional flexibility generates its extremely wide substrate specificity. J Bacteriol 185(19):5657–5664

Elkins CA, Nikaido H (2002) Substrate specificity of the RND-type multidrug efflux pumps AcrB and AcrD of Escherichia coli is determined predominately by two large periplasmic loops. J Bacteriol 184(23):6490–6498

Faloon P, Jewett I, Youngsaye W, Bennion M, Ng W-L, Hurley A, Lewis TA, Edwankar RV, Le H, Mosher CM (2014) Discovery of ML366, an inhibitor of Vibrio cholerae quorum sensing acting via the LuxO response regulator. In: Probe reports from the NIH molecular libraries program [internet]. National Center for Biotechnology Information, Bethesda

Faron ML, Ledeboer NA, Buchan BW (2016) Resistance mechanisms, epidemiology, and approaches to screening for vancomycin-resistant enterococcus in the health care setting. J Clin Microbiol 54(10):2436–2447

Feeley JC (1965) Classification of Vibrio cholerae (Vibrio comma), including El Tor vibrios, by infrasubspecific characteristics. J Bacteriol 89(3):665–670

Fluit AC, Schmitz FJ (2004) Resistance integrons and super integrons. Eur Soc Clin Microbiol Infect Dis 10:272-288.

Freeman JA, Bassler BL (1999) A genetic analysis of the function of LuxO, a two-component response regulator involved in quorum sensing in Vibrio harveyi. Mol Microbiol 31(2):665–677

Garriss G, Burrus V (2013) Integrating conjugative elements of the SXT/R391 family. Bacterial integrative mobile genetic elements. J Bacteriol 195(9):217-234.

Garriss G, Waldor MK, Burrus V (2009) Mobile antibiotic resistance encoding elements promote their own diversity. PLoS Genet 5(12):e1000775.
Gawron-Burke C, Clewell DB (1982) A transposon in Streptococcus faecalis with fertility properties. Nature 300(5889):281-284.
Gerrits MM, de Zoete MR, Arents NL, Kuipers EJ, Kusters JG (2002) 16S rRNA mutation-mediated tetracycline resistance in Helicobacter pylori. Antimicrob Agents Chemother 46(9):2996–3000
Gomez JE, Kaufmann-Malaga BB, Wivagg CN, Kim PB, Silvis MR, Renedo N, Ioerger TR, Ahmad R, Livny J, Fishbein S (2017) Ribosomal mutations promote the evolution of antibiotic resistance in a multidrug environment. eLife 6:e20420
Greene NP, Kaplan E, Crow A, Koronakis V (2018) Antibiotic resistance mediated by the MacB ABC transporter family: a structural and functional perspective. Front Microbiol 9:950
Gutmann L, Billot-Klein D, Al-Obeid S, Klare I, Francoual S, Collatz E, Van Heijenoort J (1992) Inducible carboxypeptidase activity in vancomycin-resistant enterococci. Antimicrob Agents Chemother 36(1):77–80
Hammer BK, Bassler BL (2003) Quorum sensing controls biofilm formation in Vibrio cholerae. Mol Microbiol 50:101-104.
Hall CW, Mah T-F (2017) Molecular mechanisms of biofilm-based antibiotic resistance and tolerance in pathogenic bacteria. FEMS Microbiol Rev 41(3):276–301
Heithoff DM, Mahan MJ (2004) Vibrio cholerae biofilms: stuck between a rock and a hard place. J Bacteriol 186(15):4835–4837
Higgins DA, Pomianek ME, Kraml CM, Taylor RK, Semmelhack MF, Bassler BL (2007) The major Vibrio cholerae autoinducer and its role in virulence factor production. Nature 450(7171):883
Holmgren J (1973) Comparison of the tissue receptors for Vibrio cholerae and Escherichia coli enterotoxins by means of gangliosides and natural cholera toxoid. Infect Immun 8:851-859.
Hochhut B, Waldor MK (1999) Site-specific integration of the conjugal Vibrio cholerae SXT element into prfC. Mol Microbiol 32(1):99–110
Hochhut B, Lotfi Y, Mazel D, Faruque SM, Woodgate R, Waldor MK (2001) Molecular analysis of antibiotic resistance gene clusters in *Vibrio cholerae* O139 and O1 SXT constins. Antimicrob Agents Chemother 45:2991–3000
Hürlimann LM, Corradi V, Hohl M, Bloemberg GV, Tieleman DP, Seeger MA (2016) The heterodimeric ABC transporter EfrCD mediates multidrug efflux in Enterococcus faecalis. Antimicrob Agents Chemother 60(9):5400–5411
Ippolito JA, Kanyo ZF, Wang D, Franceschi FJ, Moore PB, Steitz TA, Duffy EM (2008) Crystal structure of the oxazolidinone antibiotic linezolid bound to the 50S ribosomal subunit. J Med Chem 51(12):3353–3356
Jones PM, George AM (2009) Opening of the ADP-bound active site in the ABC transporter ATPase dimer: evidence for a constant contact, alternating sites model for the catalytic cycle. Proteins 75(2):387–396
Jones PM, George AM (2013) Mechanism of the ABC transporter ATPase domains: catalytic models and the biochemical and biophysical record. Crit Rev Biochem Mol Biol 48(1):39–50
Jung SA, Chapman CA, Ng W-L (2015) Quadruple quorum-sensing inputs control Vibrio cholerae virulence and maintain system robustness. PLoS Pathog 11(4):e1004837
Jung SA, Hawver LA, Ng W-L (2016) Parallel quorum sensing signaling pathways in Vibrio cholerae. Curr Genet 62(2):255–260
Kaper JB, Morris JG, Levine MM (1995) Cholera. Clin Microbiol Rev 8(1):48–86
Kehrenberg C, Schwarz S, Jacobsen L, Hansen LH, Vester B (2005) A new mechanism for chloramphenicol, florfenicol and clindamycin resistance: methylation of 23S ribosomal RNA at A2503. Mol Microbiol 57(4):1064–1073
Kim BS, Jang SY, Bang Y-J, Hwang J, Koo Y, Jang KK, Lim D, Kim MH, Choi SH (2018) QStatin, a selective inhibitor of quorum sensing in Vibrio species. MBio 9(1):e02262–e02217
Kim SH, Chang AB, Saier MH Jr (2004) Sequence similarity between multidrug resistance efflux pumps of the ABC and RND superfamilies. Microbiology 150(8):2493–2495
Kitaoka M, Miyata ST, Unterweger D, Pukatzki S (2011) Antibiotic resistance mechanisms of Vibrio cholerae. J Med Microbiol 60(4):397–407

Kolbusz MA, Slotboom DJ, Lolkema JS (2012) Role of individual positive charges in the membrane orientation and activity of transporters of the small multidrug resistance family. Biochemistry 51(44):8867–8876

Kostakioti M, Hadjifrangiskou M, Hultgren SJ (2013) Bacterial biofilms: development, dispersal, and therapeutic strategies in the dawn of the postantibiotic era. Cold Spring Harb Perspect Med 3(4):a010306

Kumar S, He G, Kakarla P, Shrestha U, KC R, Ranaweera I, Mark Willmon T, R Barr S, J Hernandez A, F Varela M (2016) Bacterial multidrug efflux pumps of the major facilitator superfamily as targets for modulation. Infect Disord Drug Targets 16(1):28–43

Kumar S, Mukherjee MM, Varela MF (2013) Modulation of bacterial multidrug resistance efflux pumps of the major facilitator superfamily. Int J Bacteriol 2013:204141

Kumar S, Varela MF (2012) Biochemistry of bacterial multidrug efflux pumps. Int J Mol Sci 13(4):4484–4495

Lai MH, Kirsch DR (1996) Induction signals for vancomycin resistance encoded by the vanA gene cluster in Enterococcus faecium. Antimicrob Agents Chemother 40(7):1645–1648

Lee J, Zhang L (2015) The hierarchy quorum sensing network in Pseudomonas aeruginosa. Protein Cell 6(1):26–41

Lenz DH, Miller MB, Zhu J, Kulkarni RV, Bassler BL (2005) CsrA and three redundant small RNAs regulate quorum sensing in Vibrio cholerae. Mol Microbiol 58(4):1186–1202

Lewenza S, Visser MB, Sokol PA (2002) Interspecies communication between Burkholderia cepacia and Pseudomonas aeruginosa. Can J Microbiol 48(8):707–716

Li W, Atkinson GC, Thakor NS, Allas Ü, Lu C-c, Chan K-Y, Tenson T, Schulten K, Wilson KS, Hauryliuk V (2013) Mechanism of tetracycline resistance by ribosomal protection protein Tet (O). Nat Commun 4:1477

Li X-Z, Nikaido H (2009) Efflux-mediated drug resistance in bacteria. Drugs 69(12):1555–1623

Li X-Z, Plésiat P, Nikaido H (2015) The challenge of efflux-mediated antibiotic resistance in gram-negative bacteria. Clin Microbiol Rev 28(2):337–418

Limburg E, Gahlmann R, Kroll HP, Beyer D (2004) Ribosomal alterations contribute to bacterial resistance against the dipeptide antibiotic TAN 1057. Antimicrob Agents Chemother 48(2):619–622. https://doi.org/10.1128/AAC.48.2.619-622.2004

Lin J, Nishino K, Roberts MC, Tolmasky M, Aminov RI, Zhang L (2015) Mechanisms of antibiotic resistance. Front Microbiol 6:34

Locher KP (2016) Mechanistic diversity in ATP-binding cassette (ABC) transporters. Nat Struct Mol Biol 23(6):487

Long KS, Vester B (2012) Resistance to linezolid caused by modifications at its binding site on the ribosome. Antimicrob Agents Chemother 56(2):603–612

Lorch M, Lehner I, Siarheyeva A, Basting D, Pfleger N, Manolikas T, Glaubitz C (2005) NMR and fluorescence spectroscopy approaches to secondary and primary active multidrug efflux pumps. Portland Press Limited, London

Lubelski J, Mazurkiewicz P, Van Merkerk R, Konings WN, Driessen AJ (2004) ydaG and ydbA of Lactococcus lactis encode a heterodimeric ATP-binding cassette-type multidrug transporter. J Biol Chem 279(33):34449–34455

Marcuk LM, Nikiforov VN, Scerbak JF et al (1971) Clinical studies of the use of bacteriophage in the treatment of cholera. Bull. World Health Organ 45:77-83.

Maheshwari M, Nelapati K, Kiranmayi B (2011) Vibrio cholerae-a review. Vet World 4(9):423

Mankin AS, Polacek N (2008) SPARK: a new peptidyl transferase activity assay. In: New antibiotic targets. Springer, Totowa, pp 107–116

Mao W, Warren MS, Black DS, Satou T, Murata T, Nishino T, Gotoh N, Lomovskaya O (2002) On the mechanism of substrate specificity by resistance nodulation division (RND)-type multidrug resistance pumps: the large periplasmic loops of MexD from Pseudomonas aeruginosa are involved in substrate recognition. Mol Microbiol 46(3):889–901

Marvig RL, Søndergaard MS, Damkiær S, Høiby N, Johansen HK, Molin S, Jelsbak L (2012) Mutations in 23S rRNA confer resistance against azithromycin in Pseudomonas aeruginosa. Antimicrob Agents Chemother 56(8):4519–4521

Mazel D (2006) Integrons: agents of bacterial evolution. Nature 4:608–620
McCafferty DG, Lessard IA, Walsh CT (1997) Mutational analysis of potential zinc-binding residues in the active site of the enterococcal D-Ala-D-Ala dipeptidase VanX. Biochemistry 36(34):10498–10505
Miller MB, Bassler BL (2001) Quorum sensing in bacteria. Annu Rev Microbiol 55(1):165–199
Misra R, Bavro VN (2009) Assembly and transport mechanism of tripartite drug efflux systems. Biochim Biophys Acta 1794(5):817–825
Miyamae S, Ueda O, Yoshimura F, Hwang J, Tanaka Y, Nikaido H (2001) A MATE family multidrug efflux transporter pumps out fluoroquinolones in Bacteroides thetaiotaomicron. Antimicrob Agents Chemother 45(12):3341–3346
Mohapatra H, Mohapatra SS, Mantri CK, Colwell RR, Singh DV (2008) *Vibrio cholerae* non-o1, non-o139 strains isolated before 1992 from Varanasi, India are multiple drug resistant, contain *int*sxt, *dfr18* and *aadA5* genes. Environ Microbiol 10(4):866–873
Mohapatra SS, Mantri CK, Harapriya M, Colwell RR, Singh DV (2009) Analysis of clonally related environmental *Vibrio cholerae* O1 El Tor isolated before 1992 from Varanasi, India reveals origin of SXT-ICEs belonging to O139 and O1 serogroups. Environ Microbiol Rep 2(1):50–57
Morris JG Jr, Black RE (1985) Cholera and other vibrios in the United States. New Eng J Med 312:343-350.
Morita Y, Tomida J, Kawamura Y (2014) Responses of Pseudomonas aeruginosa to antimicrobials. Front Microbiol 4:422
Moussatova A, Kandt C, O'Mara ML, Tieleman DP (2008) ATP-binding cassette transporters in Escherichia coli. Biochim Biophys Acta 1778(9):1757–1771
Mulcahy H, Charron-Mazenod L, Lewenza S (2008) Extracellular DNA chelates cations and induces antibiotic resistance in Pseudomonas aeruginosa biofilms. PLoS Pathog 4(11):e1000213
Munita JM, Arias CA (2016) Mechanisms of antibiotic resistance. Microbiol Spectr 4(2):VMBF-0016-2015. https://doi.org/10.1128/microbiolspec.VMBF-0016-2015
Murakami S, Nakashima R, Yamashita E, Yamaguchi A (2002) Crystal structure of bacterial multidrug efflux transporter AcrB. Nature 419(6907):587
Nelson EJ, Harris JB, Morris JG Jr, Calderwood SB, Camilli A (2009) Cholera transmission: the host, pathogen and bacteriophage dynamic. Nat Rev Microbiol 7(10):693
Nikaido H (1994) Prevention of drug access to bacterial targets: permeability barriers and active efflux. Science 264(5157):382–388
Nikaido H, Takatsuka Y (2009) Mechanisms of RND multidrug efflux pumps. Biochim Biophys Acta 1794(5):769–781
O'Driscoll T, Crank CW (2015) Vancomycin-resistant enterococcal infections: epidemiology, clinical manifestations, and optimal management. Infect Drug Resist 8:217
Ogawa W, Minato Y, Dodan H, Onishi M, Tsuchiya T, Kuroda T (2015) Characterization of MATE-type multidrug efflux pumps from Klebsiella pneumoniae MGH78578. PLoS One 10(3):e0121619
Oli AN, Eze DE, Gugu TH, Ezeobi I, Maduagwu UN, Ihekwereme CP (2017) Multi-antibiotic resistant extended-spectrum beta-lactamase producing bacteria pose a challenge to the effective treatment of wound and skin infections. Pan Afr Med J 27:66
Paulsen IT, Skurray RA, Tam R, Saier MH Jr, Turner RJ, Weiner JH, Goldberg EB, Grinius LL (1996) The SMR family: a novel family of multidrug efflux proteins involved with the efflux of lipophilic drugs. Mol Microbiol 19(6):1167–1175
Périchon B, Courvalin P (2012) Glycopeptide resistance. In: Antibiotic discovery and development. Springer, New York, pp 515–542
Pierce NF, Greenough W 3rd, Carpenter C Jr (1971) Vibrio cholerae enterotoxin and its mode of action. Bacteriol Rev 35(1):1
Poehlsgaard J, Douthwaite S (2005) The bacterial ribosome as a target for antibiotics. Nat Rev Microbiol 3(11):870
Pohl PC, Klafke GM, Carvalho DD, Martins JR, Daffre S, da Silva Vaz I Jr, Masuda A (2011) ABC transporter efflux pumps: a defense mechanism against ivermectin in Rhipicephalus (Boophilus) microplus. Int J Parasitol 41(13–14):1323–1333

Putman M, van Veen HW, Konings WN (2000) Molecular properties of bacterial multidrug transporters. Microbiol Mol Biol Rev 64(4):672–693

Radchenko M, Symersky J, Nie R, Lu M (2015) Structural basis for the blockade of MATE multidrug efflux pumps. Nat Commun 6:7995

Ramirez MS, Tolmasky ME (2010) Aminoglycoside modifying enzymes. Drug Resist Updat 13(6):151–171

Recchia GD, Hall RM (1995) Gene cassettes: a new class of mobile element. Microbiology 141:3015-3027.

Reeve SM, Lombardo MN, Anderson AC (2015) Understanding the structural mechanisms of antibiotic resistance sets the platform for new discovery. Future Microbiol 10(11):1727–1733

Rice AJ, Park A, Pinkett HW (2014) Diversity in ABC transporters: type I, II and III importers. Crit Rev Biochem Mol Biol 49(5):426–437

Rutherford ST, Bassler BL (2012) Bacterial quorum sensing: its role in virulence and possibilities for its control. Cold Spring Harb Perspect Med 2(11):a012427

Sayamov RM (1963) Treatment and prophylaxis of cholera with bacteriophage. Bull World Health Organ 28:361–367

Sapula SA, Brown MH (2016) Antimicrobial drug efflux pumps in Staphylococcus aureus. In: Efflux-mediated antimicrobial resistance in bacteria. Springer, Cham, pp 165–195

Shriram V, Khare T, Bhagwat R, Shukla R, Kumar V (2018) Inhibiting bacterial drug efflux pumps via phyto-therapeutics to combat threatening antimicrobial resistance. Front Microbiol 9:2990

Sigmund CD, Ettayebi M, Morgan EA (1984) Antibiotic resistance mutations in 16S and 23S ribosomal RNA genes of Escherichia coli. Nucleic Acids Res 12(11):4653–4664

Silva AJ, Benitez JA (2016) Vibrio cholerae biofilms and cholera pathogenesis. PLoS Negl Trop Dis 10(2):e0004330

Singh BR, Shoeb M, Sharma S, Naqvi A, Gupta VK, Singh BN (2017a) Scaffold of selenium nanovectors and honey phytochemicals for inhibition of Pseudomonas aeruginosa quorum sensing and biofilm formation. Front Cell Infect Microbiol 7:93

Singh S, Singh SK, Chowdhury I, Singh R (2017b) Understanding the mechanism of bacterial biofilms resistance to antimicrobial agents. Open Microbiol J 11:53

Slipski CJ, Zhanel GG, Bay DC (2018) Biocide selective TolC-independent efflux pumps in Enterobacteriaceae. J Membr Biol 251(1):15–33

Srinivasan VB, Rajamohan G (2013) KpnEF, a new member of the Klebsiella pneumoniae cell envelope stress response regulon, is an SMR-type efflux pump involved in broad-spectrum antimicrobial resistance. Antimicrob Agents Chemother 57(9):4449–4462

Stokes HW, O'Gorman DB, Recchia GD, Parsekhiam M, Hall RM (1997) Structure and function of 59-base element recombination sites associated with mobile gene cassettes. Mol Microbiol 26:731-745.

Stewart PS, Costerton JW (2001) Antibiotic resistance of bacteria in biofilms. Lancet 358(9276):135–138

Taylor RK, Miller VL, Furlong DB. Mekalanos JJ (1987) Use of phoA gene fusion to identify a pilus colonization factor coordinately regulated with cholera toxin. Proc Natl Acad Sci USA 84,:2833-2837.

Taylor DL, Bina XR, Bina JE (2012) Vibrio cholerae VexH encodes a multiple drug efflux pump that contributes to the production of cholera toxin and the toxin co-regulated pilus. PLoS One 7(5):e38208

Ter Beek J, Guskov A, Slotboom DJ (2014) Structural diversity of ABC transporters. J Gen Physiol 143(4):419–435

Teschler JK, Zamorano-Sánchez D, Utada AS, Warner CJ, Wong GC, Linington RG, Yildiz FH (2015) Living in the matrix: assembly and control of Vibrio cholerae biofilms. Nat Rev Microbiol 13(5):255

Toma C, Nakasone N, Song T, Iwanaga M (2005) Vibrio cholerae SXT element, Laos. Emerg Infect Dis 11(2):346

Touzé T, Eswaran J, Bokma E, Koronakis E, Hughes C, Koronakis V (2004) Interactions underlying assembly of the Escherichia coli AcrAB–TolC multidrug efflux system. Mol Microbiol 53(2):697–706

Van Heyningen WE, Carpenter CC, Pierce NF, Greenough WB III (1971) Deactivation of cholera toxin by ganglioside. J Infect Dis 124(4):415–418

Waldor MK, Tschape H, Mekalanos JJ (1996) A new type of conjugative transposon encodes resistance to sulfamethoxazole, trimethoprim, and streptomycin in Vibrio cholerae O139. J Bacteriol 178:4157–4165.

Wang R, Yu D, Yue J, Kan B (2016) Variations in SXT elements in epidemic Vibrio cholerae O1 El Tor strains in China. Sci Rep 6:22733

Wang W, Guo Q, Xu X, Sheng Z-k, Ye X, Wang M (2014) High-level tetracycline resistance mediated by efflux pumps Tet (A) and Tet (A)-1 with two start codons. J Med Microbiol 63(11):1454–1459

Ward A, Reyes CL, Yu J, Roth CB, Chang G (2007) Flexibility in the ABC transporter MsbA: alternating access with a twist. Proc Natl Acad Sci 104(48):19005–19010

Webber M, Piddock L (2003) The importance of efflux pumps in bacterial antibiotic resistance. J Antimicrob Chemother 51(1):9–11

Wozniak RA, Fouts DE, Spagnoletti M, Colombo MM, Ceccarelli D, Garriss G, Déry C, Burrus V, Waldor MK (2009) Comparative ICE genomics: insights into the evolution of the SXT/R391 family of ICEs. PLoS Genet 5(12):e1000786

Yakushi T, Masuda K, Narita S-i, Matsuyama S-i, Tokuda H (2000) A new ABC transporter mediating the detachment of lipid-modified proteins from membranes. Nat Cell Biol 2(4):212

Yang R, Cruz-Vera LR, Yanofsky C (2009) 23S rRNA nucleotides in the peptidyl transferase center are essential for tryptophanase operon induction. J Bacteriol 191(11):3445–3450

Zaidi AH, Bakkes PJ, Lubelski J, Agustiandari H, Kuipers OP, Driessen AJ (2008) The ABC-type multidrug resistance transporter LmrCD is responsible for an extrusion-based mechanism of bile acid resistance in Lactococcus lactis. J Bacteriol 190(22):7357–7366

Zahid MS, Udden SM, Faruque AS, Calderwood SB, Mekalanos JJ, Faruque SM (2008) Effect of phage on the infectivity of Vibrio cholerae and emergence of genetic variants. Infect Immun 76:5266–5273

Zárate SG, Morales P, Świderek K, Bolanos-Garcia VM, Bastida A (2019) A molecular modeling approach to identify novel inhibitors of the major facilitator superfamily of efflux pump transporters. Antibiotics 8(1):25

Zeng X, Lin J (2013) Beta-lactamase induction and cell wall metabolism in gram-negative bacteria. Front Microbiol 4:128

Zgurskaya HI (2002) Molecular analysis of efflux pump-based antibiotic resistance. Int J Med Microbiol 292(2):95–105

Zhu J, Miller MB, Vance RE, Dziejman M, Bassler BL, Mekalanos JJ (2002) Quorum-sensing regulators control virulence gene expression in Vibrio cholerae. Proc Natl Acad Sci USA 99:3129-3134.

Zhang LH, Dong YH (2004) Quorum sensing and signal interference: diverse implications. Mol Microbiol 53(6):1563–1571

Zolnerciks JK, Andress EJ, Nicolaou M, Linton KJ (2011) Structure of ABC transporters. Essays Biochem 50:43–61

Current Trends in *Mycobacterium tuberculosis* Pathogenesis and Drug Resistance

16

Vijay Kumar, Simranjeet Singh, Daljeet Singh, Shivika Datta, Sanjay Kumar, Shyam Babu Singh, and Joginder Singh

Contents

Vijay Kumar, Simranjeet Singh and Daljeet Singh contributed equally with all other contributors.

V. Kumar
Department of Chemistry, Regional Ayurveda Research Institute for Drug Development, Gwalior, M.P., India

S. Singh
Department of Biotechnology, Lovely Professional University, Jalandhar, Punjab, India

Punjab Biotechnology Incubators, Mohali, Punjab, India

Regional Advanced Water Testing laboratory, Mohali, Punjab, India

D. Singh · J. Singh (✉)
Department of Biotechnology, Lovely Professional University, Jalandhar, Punjab, India
e-mail: joginder.15005@lpu.co.in

S. Datta
Department of Zoology, Doaba College, Jalandhar, India

S. Kumar
Punjab Biotechnology Incubators, Mohali, Punjab, India

Regional Advanced Water Testing laboratory, Mohali, Punjab, India

S. B. Singh
Department of Ayurveda, Regional Ayurveda Research Institute for Drug Development, Gwalior, M.P., India

B. Siddhardha et al. (eds.), *Model Organisms for Microbial Pathogenesis, Biofilm Formation and Antimicrobial Drug Discovery*,
https://doi.org/10.1007/978-981-15-1695-5_16

Abstract

Tuberculosis (TB) is regarded as one of the highly infectious diseases which is caused by the species of Mycobacterium genus. Tuberculosis forms to be a major public health issue worldwide because it is anti-drug resistant; extensively drug-resistant (XDR) TB and multidrug-resistant (MDR) TB. Thus, there is an exigent need for the development of new anti-TB drugs. Various drugs are developed in the treatment of different ailments including chronic and TB related symptoms. The present study focuses on the evolution of drug resistance in *Mycobacterium tuberculosis*, the virulence of *Mycobacterium tuberculosis,* and preparation of model for evaluation of virulence caused by *Mycobacterium tuberculosis*. Efforts are also made to summarize the drug resistance mechanism in *Mycobacterium tuberculosis* including intrinsic and acquired drug resistance.

Keywords

Tuberculosis · Drug resistance · Anti-mycobacterial activity · Active compounds

16.1 Introduction

Tuberculosis (TB) is primarily caused by a single infectious agent, *M. tuberculosis,* which has remained the major cause of deaths worldwide (Glaziou et al. 2015). Even, the number of cases is subsequently increasing by the rate of 2% annually. In 1882, Robert Koch discovered *Mycobacterium tuberculosis* is responsible for TB, an airborne infection (Cambau and Drancourt 2014). *M. tuberculosis* is a pulmonary pathogen, but still it can exhibit dynamically from being asymptomatic to causing fatal disease (Smith 2003). Till date, a major pathogen of human TB is *M. tuberculosis* (Assam et al. 2013). Whereas, there are other causative agents of the same genus including *M. bovis, M. microti, M. leprae, M. canetti, M. africanum,* have also been found to cause of TB infection in humans (Banuls et al. 2015).

On the basis of public and clinical perspective, TB patients are categorized as LTBI (latent TB infection which is asymptomatic or non-transmissible state) or active TB (transmissible) (Lee 2016). Worldwide, about two billion people are suffering from LTBI. In the 17th report by WHO (World Health Organization), they cleared that there are 1.8 million death cases due to TB (Falzon et al. 2017). South Africa, India, China, and the Russian Federation are among the countries largely affected by TB (Jassal and Bishai 2010). Previously, primary drugs like para-aminosalicylic acid and streptomycin were thought to regulate the widespread disease. Furthermore, ethambutol, pyrazinamide, and rifampicin were also introduced (Murray et al. 2015). Due to this, nineteenth century was declared to be "Golden Age of TB Antibiotics." During this time, these affordable drugs were able to control and decline the TB cases globally. In the 1980s, the reemergence of drug-resistant form of TB during the epidemic of AIDS (acquired immune deficiency syndrome) led to the spread of TB to all corners of the world (Lange et al. 2014). At present, MDR-TB (multidrug resistance-TB) is widespread, nearly 5,80,000 new cases were recorded in 2015. Globally, 84 countries have been reported to be infected by XDR-TB (extensively drug-resistant tuberculosis) (Prasad et al. 2017). Thus, dealing with TB is challenging and therefore it requires targeted diagnosis, screening of drug resistance, and direct evaluation of patient under treatment for 6 months minimum. Moreover, there is requirement for the discovery and effective formulation of novel TB drugs for effective treatment of TB (Chetty et al. 2017).

In this chapter, some major points about virulence, pathogenesis, and drug resistance mechanism of *Mycobacterium tuberculosis* are incorporated and also provide an insight on the update on new drugs effective against TB.

16.2 *Mycobacterium tuberculosis* Virulence

Generally, it is not simple to understand what makes *M. tuberculosis* virulent, in spite of the information gathered in the last 100 years (Jagielski et al. 2016). As it does not involve the traditional factors of virulence like those found in the major disease caused by *E. coli* O157:H7, *Corynebacterium diphtheriae*, *Vibrio cholerae* and *Shigella dysenteriae* (Forrellad et al. 2013). A very limited information is available which elucidates the mechanism opted by *M. tuberculosis* to spread disease and how its virulence can be assessed (Pym et al. 2002). On understanding the literature content, it can open a new option that can be used to determine the effect of alteration of the bacterium during disease progression. The two terms "morbidity" and "mortality" have been mostly used to report about *M. tuberculosis* (Connell et al. 2011). Mortality signifies the percentage of animals died due to infection by calculating the time taken to die after the onset of infection (Hawn et al. 2014). Microbial load (i.e., numbers of microbes presented inside the infested host after the onset infection) is another factor which is associated with virulence. This knowledge enables us to compare the fitness of diverse microbial stains to endure host response when the host is infected (Hoff et al. 2011).

In addition to this, mutant strains of *M. tuberculosis* exhibit lower bacterial load on assessing their growth curve of infected animals during the process (Ribeiro et al. 2014). Mutants are broadly divided into three broad categories, i.e., persistence genes (*per*) as they grow normally in early stage but on the on-set of cell-mediated immunity the number gets declined; severe growth in vivo (sgiv) as these mutant do not multiply themselves but either they persist at same cell number or gets rapidly cleared and growth in vivo (giv) as in this case mutant initially get multiplied but multiplication rate is relatively less as compared to wild-type (Glickman and Jacobs 2001). This classification of mutation aids in understanding the genetic mechanism of bacterial genes in regulating the different stages of infection (McGrath et al. 2013). To confirm the standard genetic nomenclature, *M. tuberculosis* showing the reduced growth in mice is categorized with similar terminology, i.e., per, sgiv, and giv (Smith 2003). Morbidity is a primary factor analyzed during histopathology studies and is the important factor to characterize the mutant class of *M. tuberculosis* virulence (Sakamoto 2012). For example, *sigH* mutant genes of *M. tuberculosis* showed normal growth and high survival rate in mice and macrophages but histopathology analysis of lungs of infected mouse showed reduced virulence in comparison to that of wild-type species of *M. tuberculosis* (Kaushal et al. 2002).

There is a need for a better understanding of pathogenesis related to TB in order to effectively measure the mortality and morbidity induced by *M. tuberculosis* (Abbara and Davidson 2011). The unregulated developmental stage of *M. tuberculosis* in human cells at common site relates to lung damage which ultimately led to death because of oxygen scarcity. This anoxia occurs due to the damaging of parenchymal cells of lung that are usually involved in oxygen uptake, impediment of bronchiolar passages because of granulomatous growths, and due to the release of blood in adjacent lung tissue because of the bursting of liquefied granulomas (Delogu et al. 2013). Another form of TB, also known as tuberculomas, effects the brain by forming enlarged brain granulomas, which may result because of inflammatory response or seizures (Rock et al. 2008). Moreover, inflammatory responses are also responsible for extrapulmonary manifestations in TB patients, especially in bones (Lee 2015).

Inflammation response plays a key role here as they aid in controlling the infection but it also damages tissues of the host (Sasindran and Torrelles 2011). Various proteases have been found to be involved in tissue damage, especially cathepsin D that is majorly involved with granulomas liquefaction (Ehlers and Schaible 2013). Moreover, uptake by *M. tuberculosis* leads to the apoptosis of macrophages and damaging of adjacent tissues. TNF (tumor necrosis factor), is the key cytokine which gets elicited during inflammatory reaction triggered by the cellular immune system to restrict the widespread of infection (Dutta and Karakousis 2014). Mice which were unable to synthesize or trigger the TNF- did not form granulomas to restrict bacterial dissemination. But, during the presence of a large number of these cytokines, it causes severe inflammation in the lung and early death of mice (Shaler et al. 2011). TNF- is now considered to be the determinant factor of TB meningitis in a rabbit model, as it allows us to directly linked with the severity of disease caused by various strains of both *M. tuberculosis* and *M. bovis* as well as with cytokine level in the fluid of cerebrospinal portion (Tsenova et al. 2005). On analyzing the cytokine response and virulence in infected mice, it revealed that there are other factors other than TNF- involved with TB progression (Domingo-Gonzalez et al. 2016). The clinical strain, *M. tuberculosis* CDC1551, was previously considered to be highly virulent but recent studies revealed that CDC1551 induces cytokines synthesis along with TNF- at a higher level in comparison to other strains of *M. tuberculosis* in mice. Also, it was less virulent than other strains stated on behalf of mortality rate and bacterial load (Manca et al. 1999). Even comparative study conducted on the rabbit model shows a similar result for the virulence of H37Rv and

CDC1551. Another study evaluated the potential of two strains NHN5 and HN878 of *M. tuberculosis* to elicit the cytokine production and cause disease in the mouse model. For this, HN878 was found to be highly virulent in comparison to NHN5 (Manca et al. 2001).

Apoptosis is also one of the determinant factors, as infection of macrophages by *M. tuberculosis* depends on the TNF-. And, it was shown that the virulent strain of *M. tuberculosis* leads to less apoptosis (Behar et al. 2011). The result of above experiments highlights the complexity of the immune system as well as the effectors, but due to the inconsistency in the result, it is difficult to correlate the level of one or more cytokines like TNF- with the clinical model of the disease (Drain et al. 2018). It has become evident that the optimal balancing of these modulators of the immune system is very critical (Cooper 2009). In spite of the varied results, which makes the interpretation of data difficult but are valuable as they demonstrate that few species of *Mycobacterium* are highly virulent than other in clinical or animal models (Alvarez et al. 2009).

16.3 Model for Measuring *Mycobacterium tuberculosis* Virulence

Virulence of *Mycobacterium tuberculosis* is generally studied and measured on the animal or cell culture model. Therefore, different pathogenicity parameters are selected according to the model (Prozorov et al. 2014). The unique characteristic of *M. tuberculosis* to infect and survive in macrophages makes it the primary target; thus, cell lines and primary macrophages are used to check the effectiveness of *M. tuberculosis* as well as its mutants during the onset of infection (Pieters 2008). Thus, macrophages are chiefly targeted to assess the normal in vivo condition but difficult in propagating macrophages to a required number makes it incompetent for virulence experiment (Mehta et al. 2006). The immobilized cell lines like MH-S, THP, and J774 are most commonly used, whereas human macrophages from peripheral blood monocytes and murine bone-marrow derived macrophages are widely used macrophages to study the interaction among macrophages and *M. tuberculosis* (Majorov et al. 2003; Norris and Ernst 2018). Furthermore, besides the assessment of intracellular bacterial load, replication, and survival of *M. tuberculosis* in macrophage model, it can also be used to understand the mechanism of macrophage microbicide ability and how to counteract with it, like (a) generating resistance against reactive nitrogen/oxygen intermediates, (b) apoptosis inhibition, and (c) phagosome arresting (Bhat and Yaseen 2018).

Alternatively, the animal model aids in studying the diverse stages of TB infection. The most used animal models are rabbits, guinea pigs, and mice (Zhan et al. 2017). Most commonly used in vivo model is mice as it is genetically well-characterized; moreover, inbred strains and immunological reagents are also available (Singh and Gupta 2018). But species of mice are least susceptible to *M. tuberculosis* infection and their pathology is very different from humans (Kramnik and Beamer 2016). Similarly, guinea pigs are having high susceptible to infection of

M. tuberculosis and show similar ailments like disease dissemination, lung necrosis, and lymphadenopathy (Clark et al. 2015). Also, rabbit model on infecting with *M. bovis* develops granulomas in the lung which resembles the histology of human TB, but because of their size, cost, and very less number of immunological reagents makes it the less tractable model in comparison to mice (Chen et al. 2017). Due to the high similarity of *M. bovis* and *M. tuberculosis*, cattle have become an attractive model to study the pathogenicity of TB (Aguilar León et al. 2009). Even, the TB pathology in bovine shows close similarity with humans, results in the formation of caveating lung granulomas and exhibiting similar latent phase after prolong infection (Waters et al. 2011). The benefit of conducting the experiment on the cattle model allows us to conduct field trials and also make it an attractive model for vaccination studies (Buddle et al. 2018). Non-human primate models are the one which shows all the clinical states of the disease that are found in human TB and have given the invaluable contribution in TB research. But the high cost and ethical issues restrict their usage in research (Scanga and Flynn 2014). The bacterial load is one of the most important parameters for measuring the virulence in animal models other than morbidity and mortality (Dormans et al. 2004). Lastly, zebrafish model has also been found to be effective in elucidating the initial stages of mycobacterial infection, especially during the granuloma formation and its function in regulating the infection (Van Leeuwen et al. 2015). In a study, when zebrafish was infected with *M. marinum* it showed great resemblance with different stages of human tuberculosis; in reality, host genes, virulence factors, and immune cell types are conserved in this interacting model. This model revealed that RD1 locus of bacteria was involved during granuloma formation, whereas *ESX*-1 system was found to be accountable for the death of infested macrophages (Meijer 2016).

16.4 Drug Resistance in *Mycobacterium tuberculosis*

The primary mechanism which drives the drug-resistance mechanism in *M. tuberculosis* is due to the mutation of compensatory genes which encodes for drug-activating enzymes or drug targets (Palomino and Martin 2014). These mutations generally occur due to the deletion, insertion of SNPs (single nucleotide polymorphism), and very rarely due to the deletion of nucleotide in high number (Nguyen 2016). Contrasting other bacteria, *M. tuberculosis* does not develop mutation due to horizontal transfer of genetic material. Therefore, two mechanisms were reported to generate drug resistance mechanism in *M. tuberculosis*: first one is transmission and second one is acquired drug resistance (Almeida Da Silva and Palomino 2011).

Various studies conducted to assess the progressive development in drug resistance via WGS revealed that *M. tuberculosis* initially acquainted the resistance to isoniazid, followed by developing resistance against ethambutol or rifampicin, then against pyrazinamide, and lastly developed resistance against second as well as third-line drugs. This assessment has provided worthy insight into the evolution of *M. tuberculosis* pathogenicity (Gygli et al. 2017). Furthermore, recent studies have stated mutation leading to the development of drug resistance differs with respect to the lineage to the recipient strain (Ford et al. 2013). Thus, we have summarized the existing anti-TB and new drugs, with the action mechanism of drug and genes linked with resistance development (Table 16.1).

16.4.1 Intrinsic Drug Resistance

M. tuberculosis has been considered to evolve as well as develop various molecular mechanisms to neutralize the cytotoxic of various chemicals such as antibiotics (Davies and Davies 2010). These intrinsic resistance mechanisms have aided the *M. tuberculosis* to develop resistance against anti-TB agents, which has not only reduced the number of available drugs against TB but have made the exploration of novel anti-TB agents more difficult (Hameed et al. 2018). There are various mechanisms that are responsible for growth intrinsic resistance in strains of *M. tuberculosis* and other pathogenic strains.

One of the mechanisms is cell wall permeability which regulates the entry and exists of the chemical from the cell membrane. The reduction in the permeability of drug via cell wall of mycobacteria serves as the active barrier and hindrance for antibiotic therapy (Sarathy et al. 2012). For example, a report revealed that β-lactams penetration through cell walls of mycobacteria species to be 100 times slower than the cell wall of *E. coli*. The function of cell wall permeability in antibiotic resistance in mycobacterial strains has well comprehended by studying the mutant defects during cell wall biosynthesis (Smith et al. 2012). Mycobacterial cell wall regulates the penetration of antibiotic, and there are other specialized resistance mechanisms which detoxify antibiotic molecules that were able to enter into the cytoplasmic region (Mukhopadhyay et al. 2012). The specialized mechanism involves alteration of the target, mimicking of the target, drug modification, drug degradation, and drug efflux (Fig. 16.1).

Target alteration strategy is generally applied by bacteria to avoid the antibiotic action by modifying the target structure of the antibiotics and is usually adopted by strains of *M. tuberculosis* species and other mycobacterial strains to decrease the chances of binding of lincosamides and macrolides to ribosomes of *M. tuberculosis* (Fair and Tor 2014). Recent studies revealed that Erm37 gene has the protecting roles in mycobacterial strains from lincosamides and macrolides. *M. tuberculosis* uses a similar mechanism to neutralize the activity of capreomycin and viomycin drugs used for treating multiple drug resistance TB (Buriánková et al. 2004; Fu and Shinnick 2007). The studies conducted on *M. smegmatis* and *M. tuberculosis* revealed the association of tlyA gene with viomycin and capreomycin resistance (Maus et al. 2005). Another specialized method of intrinsic drug resistance, mimicking of the target, is effective in neutralizing the effect of fluoroquinolones. Fluoroquinolones are anti-TB drug which has bactericidal effect as it inhibits the action mechanism during DNA replication, transcription, as well as repair (Von Groll et al. 2009). Generally, these drugs bind with DNA topoisomerase or gyrase enzyme resulting in the complexation of DNA which prevents resealing of DNA strands and finally leads to DNA degradation and cell cessation (Ginsburg et al.

Table 16.1 List of the existing anti-TB and new drugs with their action mechanism and genes involved in resistance in *Mycobacterium tuberculosis*

Drug line	Drugs	Action mechanism of drug	Targeted genes involved in resistance	References
First-line drug	Amikacin	Inhibits protein synthesis	*eis, rrs* and tlyA	Jugheli et al. (2009)
	Capreomycin	Inhibits protein synthesis	eis, rrs and tlyA	Maus et al. (2005)
	Ethambutol	Inhibition of arabinogalactan biosynthesis	embB and ubiA	Somoskovi et al. (2001)
	Isoniazid	Inhibits mycolic acid synthesis	inhA, kasA and katG	Colangeli et al. (2007)
	Kanamycin	Inhibits protein synthesis	eis, rrs and tlyA	Jugheli et al. (2009)
	Pyrazinamide	Inhibits pantothenate and co-enzyme A synthesis, and reduction in membrane energy	panD, pncA, rpsA	Ramirez-Busby and Valafar (2015)
	Rifampicin	Inhibits RNA synthesis	rpoB	Kumar and Jena (2014)
	Streptomycin	Inhibits protein synthesis	gidB, rpsL and rrs	Spies et al. (2011)
Second-line drugs	Bedaquiline	Targets the ATP synthase of mycobacterial species and inhibits if functioning	atpE, pepQ and rv0678	Andries et al. (2014)
	Ethionamide	Inhibits mycolic acid synthesis	ethA, inhA, inhA promoter, mshA and ndh	Bollela et al. (2016)
	Fluoroquinolones	Inhibits DNA synthesis	gyrA and gyrB	Maruri et al. (2012)
	Para-aminosalicylic acid	Inhibits the functioning of thymine nucleotide and folic acid metabolism	folC, ribD and thyA	Rengarajan et al. (2004)
New drugs	Clofazimine	It ceases the mycobacterial growth by binding to its DNA and it also binds itself at potassium transporter in mycobacteria and inhibits its normal functioning	Ndh, pepQ, rv0678, rv1979c and rv2535c	Zhang et al. (2015)
	Delamanid	Selectively and specifically inhibits the mycolic acid biosynthesis	Ddn, fbiA, fbiB, fbic and fgd1	Fujiwara et al. (2018)
	Linezolid	Inhibits the synthesis of protein	rplC and rrl	Zhang et al. (2016)

Adapted from Dookie et al. (2018)

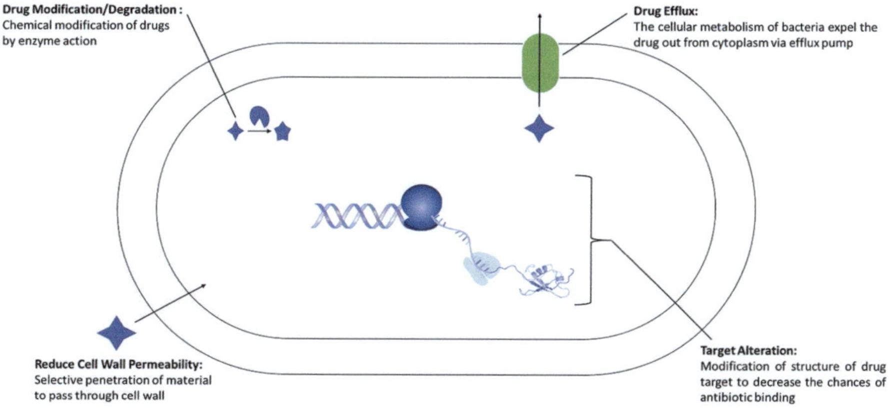

Fig. 16.1 Different intrinsic drug resistance mechanism by which mycobacteria become resistant

2003). Thus, on mapping, the acquired fluoroquinolone resistance revealed that protein MfpA attributes for intrinsic resistance (Hegde et al. 2005). Another mechanism which mycobacterial species employs to directly deactivate the active drug is drug modification (D'Ambrosio et al. 2015). Aminoglycosides drugs have always held the main position, even in the history of TB therapy. The target function of these drugs remained the same, i.e., to inhibit the synthesis of protein (Xie et al. 2011). Studies revealed that acetyltransferase plays a key role in the survival of mycobacterium species in macrophages of the host (Kim et al. 2012). Lately, it was discovered that it aids in changes in innate immunity of the host in contradiction to infection of mycobacterial species. These modifications in the host signaling molecule have suppressed the immune response like apoptosis, autophagy, and inflammation of the host infected by *M. tuberculosis* (Zhai et al. 2019).

Another strategy used by *M. tuberculosis* to subvert the action of the anti-TB drug is to degrade them via hydrolases (Nguta et al. 2015). These mechanisms have been broadly studied in β-lactams drugs, which have no effect on *M. tuberculosis* as well as on other mycobacterial strain. This action mechanism of these drugs is to inhibit the synthesis of cell wall synthesis by binding on penicillin-binding proteins (PBPs) region which leads to apoptosis (Kohanski et al. 2010). On analyzing the *M. tuberculosis* genome, it revealed that genome contains four sites which encode for PBPs, where β-lactams bind within detectable concentrations. This clears the fact that the least target affinity is not acceptable for β-lactam resistance in mycobacterial strain (Li et al. 2018). Hydrolytic enzyme, β-lactamases are considered to be the determinant for β-lactams resistant, as this enzyme hydrolyzes the β-lactam ring. This was confirmed by conducting experiment on *M. fallax* (highly susceptible to β-lactams drug), result of permeability assay revealed that rate of penetration in cell walls of *M. fallax* by β-lactams was similar to other mycobacterial species and permits the accumulation of β-lactam drugs to lethal concentration (Wang et al. 2006). But on engineering the *M. fallax* with gene expressing β-lactamase from *M.*

fortuitum showed the increase in resistance level similar with other species and revealed β-lactamases are the major cause of β-lactam susceptibility (Sauvage et al. 2006). BlaC is another β-lactamase which is effective against tuberculosis. Moreover, it has been found to have broad-substrate specificity because of flexible substrate binding nature. BlaI gene has been comprehended to regulate the function of BlaC in strains of *M. tuberculosis.* During β-lactams absence, there is the formation of homodimers of BlaI, which binds to the promoter region of BlaC by obstructing its transcription. But when this *M. tuberculosis* strain is subjected to β-lactams, it causes the dissociation of BlaI from DNA binding site and derepression of BlaC transcription, which results in the production of β-lactamase (Kurz and Bonomo 2012). Other than BlaC, *M. tuberculosis* also encodes other β-lactamases like BlaS, Rv3677c, and Rv0406c (Nampoothiri et al. 2008).

Lastly, the most commonly used method by microbes to avoid the action of drugs is to remove them from the cytoplasm via efflux mechanisms (Soto 2013). The trans-membrane proteins are the one which plays a key role in the mechanism. For example, there are 20 out of 36 genes which encode for membrane proteins in the genome of *E. coli*, which grant them the resistance to more than one drug (Niederweis et al. 2010). It is very improbable that now these transporter proteins have evolved themselves to act as specialized drug transporters (Feltcher et al. 2010). Various experiments revealed that mycobacteria contain 18 transporters which have conferred antibiotic resistance in them. Likewise, expression of EfpA and IniBAC is negatively regulated through Lsr2, which binds to AT-rich region of the sequence (Nguyen 2016). Significantly, the first-line drugs, isoniazid or ethambutol, have been found to have an inducible effect on Lsr2, which regulates the transcription of EfpA and IniBAC; thus, each transporter protein has evolved themselves to perform a specialized function in antibiotic resistant strain (Colangeli et al. 2007). Recent studies have also linked Lsr2 to changes in oxygen level involved in mycobacterial adaptation, thus providing us the connecting link between the pathogenesis and resistance of *M. tuberculosis* (Bartek et al. 2014). Another transporter protein that is effective in the efflux of anti-TB drugs like aminoglycosides, tetracycline, and spectinomycin is Tap. Some studies also confirmed the function of Tap in conferring the drug resistance to *M. tuberculosis* (Balganesh et al. 2012).

16.4.2 Acquired Resistance

The anti-TB drugs targeted binds to the target site with high affinity, as a result they obstruct the normal activity of the target molecule. But, modification in targeted site prevents the effective binding of the drug and generates resistant against the particular drug (Hoagland et al. 2016). In *M. tuberculosis* and other species, resistance occurs due to mutation (spontaneous) in the chromosomal genes encoding target molecules (Koch et al. 2018). Below, we have briefly discussed the point mutation allied with the resistance of strains of *M. tuberculosis* for first-line drugs like EMB, INH, PZA, and RIF as well as for second-line drugs like fluoroquinolones, bedaquiline, and macrolides.

16.5 Mutations Responsible for the Development of Acquired Resistance to First- and Second-Line TB Drugs

Recent studies have revealed that compensatory mutation in various genes like *ahpC, inhA, kasA, katG,* and *ndh* are all linked to INH resistance (Liu et al. 2018a). INH is one of the pro-drugs which uses peroxidase or catalase enzyme encoded by gene *katG* for its activation. A mutation in *katG* has been found to be linked with the reduced activity of catalase or peroxidase and is a common mechanism responsible for INH resistance (Cade et al. 2010). Another similar mechanism which confers low-level resistance towards INH occurs because of the mutation in inhA promoter (Bollela et al. 2016). Lately, Torres and his colleagues identified new mutation which comprehends for 98% of INH resistance induced by *fabG1*, *katG* mutation, or inhA promoter (Torres et al. 2015). Similar finding related to INH resistance and mutation is the discovery of harbinger mutation like *katG S315T*, which can serve as a valuable asset for reporting warning about the evolution of multidrug resistant. These results revealed the impact of these mutations on public health and have enabled to target treatment of the patients suffering from multiple-drug resistance TB (Pym et al. 2002), whereas 95% RIF resistant strains have reported about mutation in codons 507–533 of RNA polymerase beta-subunit gene (rpoB) (Van Deun et al. 2013).

Pyrazinamide (PZA) is an essential drug and considered to be short-term chemotherapy for TB as it is effective in reducing the treatment regimens (Chan et al. 2004). But now it has become ineffective due to the mutation in *pncA* gene, which reduces the activity of pyrazinamidase enzyme and becomes resistance against PZA (Ramirez-Busby and Valafar 2015). Various other studies also reported about the mutation in *clpC1, rpsA,* and *panD* which encodes for ATP-dependent ATPase, aspartate decarboxylase, and ribosomal protein S1, respectively, to be liable for PZA resistance (Zhang et al. 2017). Other first-line anti-TB drugs, such as EMB in combination with RIF, INH, and PZA, are used to treat TB and control the widespread of drug-resistant strains (Nasiri et al. 2016). On the contrary, various reports have documented showing that mutation in operon of embCAB especially in *embB* gene results in the development of resistance against EMB in *M. tuberculosis* making the treatment ineffective (Plinke et al. 2011). Although mutation of *ubiA* gene has also been established to be responsible for drug resistance in *M. tuberculosis strains* (Lingaraju et al. 2016), the other aminoglycosides drugs like amikacin and kanamycin are effective against TB. But, time strains of *M. tuberculosis* have developed the resistance to these aminoglycosides drug due to A1401G mutation in *rrs* gene which encodes for 16 s rRNA (Jugheli et al. 2009). In contrast to other bacteria, which contains multiple copies of genes, mycobacteria contain only one copy of this gene; hence, it defines that why mutation in this gene leads to aminoglycoside resistance (Garneau-Tsodikova and Labby 2016).

Capreomycin and viomycin are another set of drugs that have been used for the treatment of TB. But, due to similar function mechanism of these drugs with aminoglycosides drug they were found to be virulent against strains of *M. tuberculosis* having acquired resistance to kanamycin (Gualano et al. 2016). Fluoroquinolones, being the second-line anti-TB drugs, are used to treat infection caused by *M. fortuitum, M. kansasii,* and *M. simiae* (Ma et al. 2010). Generally, this drug targets the type II topoisomerases, DNA topoisomerase IV, and DNA gyrase enzyme which controls the functions like cell division, DNA replication, and supercoiling of DNA (Schluger 2013). The mutations in genes *gyrA* and *gyrB* are considered to be likely associated with fluoroquinolone resistance in mycobacterial strain. The substitution of codon 90 and 94 in *gyrA* gene is the mutation which is found to be involved with fluoroquinolone-resistant in *M. tuberculosis* (Maruri et al. 2012). Few studies also indicated that the efflux mechanism is also involved in fluoroquinolone resistance (Lu et al. 2014).

Another bacteriostatic drug, linezolid which inhibits the synthesis of protein by forming complexes on the 50S ribosomal subunit, is now clinically used for treating drug-resistant TB (Chetty et al. 2017). But, mutation in *rplC* and *rrl* gene has been discovered in linezolid resistant strain (Zhang et al. 2016). Recently approved diarylquinoline drug, bedaquiline, was also assessed for its resistance in mycobacterial strains. To our surprise, mutations in the *atpE* gene were found to be accountable for drug resistance (Andries et al. 2014). This was the brief discussion about the mutations responsible for the growth of acquired resistance in relation to TB drugs of the first or second line. And, we tried to highlight how these acquired mutations are making the situation difficult to regulate drug-resistance TB.

16.6 Mechanism of Drug Resistance

In 1948, the phenomenon of drug resistance was recorded while the first trial of TB therapy was being conducted (Gillespie 2002). As each novel anti-TB drug was discovered and brought into clinical trials, the prevalence of resistant strains was encountered within a decade (Rawal and Butani 2016). Genetic mutation is the key reason for the drug resistance in *M. tuberculosis* as there is no evidence or report for resistance development due to the acquisition of new DNA (Parida et al. 2015). Allelic exchange experiment has established the interconnection between drug resistance and mutation, which occurs due to a mutation in a subset of genes. There are two primary mechanisms involved in drug resistance: a) modification of the targeted molecules and b) due to defect in the enzyme function which changes its activity (Caminero et al. 2010).

Limitation in both genotypic and phenotypic drug-susceptibility test hampers the basic understanding of resistance mechanisms. Generally, the phenotypic test shows the dichotomous result, i.e., strain of *M. tuberculosis* is either resistant or susceptible to a specific set of drugs like ethambutol, rifampicin, and isoniazid (Ocheretina et al. 2014). Besides this, the genotypic test fails to detect the mutation present in the phenotypic resistant strain. Conclusively, identifying the mutation in phenotypic resistant strain with the help of genome or gene sequencing does not ensure to check the mutation responsible for the resistance (Yakrus et al. 2014). Hence, the phenotypic mutation could be any mutation from contemporary, intermediator, or causal mutation (Motiwala et al. 2010). This prompts them to design the diagnostic assay based on the causal mutation to identify the drug-resistant strains. That is why it is difficult to determine the mutation and categorized according to its type (Desjardins et al. 2016).

Till now, various groups have started to sequencing the whole genome of clinical isolates to find the novel mutation linked with resistance and long-term goal to develop a diagnostic test which could detect the resistant strain and can replace the culture-dependent drug susceptibility test (Iketleng et al. 2018). This approach has shown the feasibility in preliminary studies but lack of precision and high cost prevented its usage. Still, culture-based approach remained the reliable option for clinical care (Nahid et al. 2012).

16.7 Evolution of Drug Resistance in *Mycobacterium tuberculosis*

Bacterial epistasis and fitness are two main factors which influence the progress of drug resistance of *M. tuberculosis* strains (Al-Saeedi and Al-Hajoj 2017). Epistasis signifies genetic interaction of a certain set of genes, in which the phenotypic effect of the first mutation solely depends on the second mutations (Wong 2017). As observed, resistance strains carry the same resistant mutation which varies in their capacity during transmission from one to another patient, providing the evidence that genetic background of the strain can aid in determining the course of evolution to develop drug resistance (Trauner et al. 2014). On the contrary, bacterial fitness is the function of growth rate, transmissibility, and virulence. Thus, mutation results in the reduction of the bacterial number in contrast to wild-strain are considered to carry the "fitness cost" (Schulz et al. 2010). If one needs to immediately estimate the relative fitness of bacteria, he can determine it by measuring the growth rate of bacteria present in the culture (Ayabina et al. 2016). As evolution is a continuous process, various studies have provided the evidence in support that fitness of resistant mutant cannot be fixed (O'Neill et al. 2012). Another example of epistasis is the acquisition of compensatory mutation, which also plays vital role in the formation of drug-resistant strains imposing a great risk on human health (Müller et al. 2013). As of now, we do not have the adequate information to predict the epistasis interactions using bioinformatic tools; thus, we have to rely on the conventional approaches to gather knowledge about the genetics behind the development of drug resistance (Ngo and Teo 2019).

Recently, various research groups used WGS (whole genome sequencing) to gather information about molecular epidemiology, mutation frequency, and phylogeny to compare drug-resistant and drug-susceptible in *M. tuberculosis* strains. This approach also helps us to address the key contributor involved in the evolution of strains of *M. tuberculosis* (Ilina et al. 2013). Generally, there are various genes involved in transcriptional control, cell wall homeostasis, lipid metabolism, and purine metabolism during anti-TB therapy (Fonseca et al. 2015). Henceforth, these genes can assist us in understanding the drug-resistant mechanism. For example, ponA1 gene, whose actual function is unknown but is discovered to involve in the evolution of drug resistance in species of *M. tuberculosis* (Smith et al. 2012). Evidence from other studies prompted us to investigate the role of these genes and how these genes can be further used for diagnosis purpose. With time number of these genes is growing exponentially and we are also getting the supporting evidence to prove their role as epistasis from adaptation to resistance (Daya et al. 2015). *rpoC*, which act as a mediator from adaptation to RIF (rifampicin) resistance development and Rv3806c, which mediated the EMB (ethambutol) resistance are the examples of the genes identified in recent studies (Somoskovi et al. 2001).

Recently, RIF resistance is found to be induced due to the mutation of RNA polymerase by *rpoB* enzyme (Kumar and Jena 2014). Hence, mutation in multidrug-resistant strains of *M. tuberculosis* is nearly ubiquitous and is mostly found to be associated with compensatory mutations in *rpoA, rpoB,* and *rpoC* genes (RNA polymerase genes). Instead of this, the compensatory mutation also restores the baseline profile of cells (De Vos et al. 2013). Especially, mutants *rpoB* were found to improve the lipid profile and alter the expression of various proteins involved in lipid metabolism, specifically phthiocerol dimycocerosates (PDIMs). Therefore, lipid metabolism involving PDIMs has positive influence during the progress evolution of drug resistance (Lahiri et al. 2016). In addition to this, in vitro studies revealed that resistant mutant embB M306v contains a synonymous mutation in Rv3792 and non-synonymous mutation in Rv 3806c, which are a major contributor for developing EMB resistance (Safi et al. 2013). Thus, it is evident that epistasis does not depend on one drug. Therefore, different studies are being conducted to assess the interaction among the disparate drugs and mutations, and their eminent role in the growth and development of drug resistance. This proves the fact that positive epistasis can trigger multidrug resistance (Trauner et al. 2014).

Recently, it has been accorded that continuous exposure of drug imposes some constraint on the evolution of TB, which increase the chances of compensatory mutation in already resistant strains (Liu et al. 2018b). Once these strains get mutated, the strains possess the ability to transmit the mutated gene to the next generation alone (Banuls et al. 2015). Furthermore, continuous drug exposure is known to start accumulating the mutant which results in an increased level of resistant towards the particular drug. This is one of the factors, which influences isoniazid (INH) resistance in multidrug-resistant strains and also found to be contributing for high resistance against fluoroquinolones (FQ) (Dookie et al. 2018).

16.8 Conclusion

Due to the development of resistance to first- and second-line drugs, TB has remained the biggest concern worldwide. WHO has also issued some recommendations and guidelines for the proper care of TB patient in the public or private sector and are ensuring that precise diagnosis is being used and found effective in treating tuberculosis infection. With the failure of second-line drug like fluoroquinolones, which was effective in reducing the duration of chemotherapy and has limited drug treatment options for multiple drug-resistant tuberculosis. Therefore, the development of a new drug is urgently required along with that there is a need for exploring alternative treatments like host-directed therapy, personalized medicine, and more. Though we are still investigating basic biology as well as the pathogenesis of *M. tuberculosis* and exploring the different therapeutic options and new various anti-TB drugs. But there is a need to scale up various approaches, tools, and health care service in reliance with the government as it provides us to regulate the chaos induced by tuberculosis and the growing issue of drug-resistant tuberculosis.

References

Abbara A, Davidson RN (2011) Etiology and management of genitourinary tuberculosis. Nat Rev Urol 8(12):678–688

Aguilar León D, Zumárraga MJ, Jiménez Oropeza R, Gioffré AK, Bernardelli A, Orozco Estévez H, Cataldi AA, Hernández Pando R (2009) *Mycobacterium bovis* with different genotypes and from different hosts induce dissimilar immunopathological lesions in a mouse model of tuberculosis. Clin Exp Immunol 157(1):139–147

Almeida Da Silva PE, Palomino JC (2011) Molecular basis and mechanisms of drug resistance in *Mycobacterium tuberculosis*: classical and new drugs. J Antimicrob Chemother 66(7):1417–1430

Al-Saeedi M, Al-Hajoj S (2017) Diversity and evolution of drug resistance mechanisms in *Mycobacterium tuberculosis*. Infect Drug Resist 10:333–342

Alvarez AH, Estrada-Chávez C, Flores-Valdez MA (2009) Molecular findings and approaches spotlighting *Mycobacterium bovis* persistence in cattle. Vet Res 40(3):1–6

Andries K, Villellas C, Coeck N, Thys K, Gevers T, Vranckx L, Lounis N, de Jong BC, Koul A (2014) Acquired resistance of *Mycobacterium tuberculosis* to bedaquiline. PLoS One 9(7):e102135

Assam JP, Beng VP, Cho-Ngwa F, Toukam M, Ngoh AA, Kitavi M, Nzuki I, Nyonka JN, Tata E, Tedom JC, Skilton RA (2013) *Mycobacterium tuberculosis* is the causative agent of tuberculosis in the southern ecological zones of Cameroon, as shown by genetic analysis. BMC Infect Dis 13(1):431

Ayabina D, Hendon-Dunn C, Bacon J, Colijn C (2016) Diverse drug-resistant subpopulations of *Mycobacterium tuberculosis* are sustained in continuous culture. J R Soc Interface 13(124):20160745

Balganesh M, Dinesh N, Sharma S, Kuruppath S, Nair AV, Sharma U (2012) Efflux pumps of *Mycobacterium tuberculosis* play a significant role in antituberculosis activity of potential drug candidates. Antimicrob Agents Chemother 56(5):2643–2651

Banuls AL, Sanou A, Van Anh NT, Godreuil S (2015) *Mycobacterium tuberculosis*: ecology and evolution of a human bacterium. J Med Microbiol 64(11):1261–1269

Bartek IL, Woolhiser LK, Baughn AD, Basaraba RJ, Jacobs WR, Lenaerts AJ, Voskuil MI (2014) *Mycobacterium tuberculosis* Lsr2 is a global transcriptional regulator required for adaptation to changing oxygen levels and virulence. MBio 5(3):e01106–e01114

Behar SM, Martin CJ, Booty MG, Nishimura T, Zhao X, Gan HX, Divangahi M, Remold HG (2011) Apoptosis is an innate defense function of macrophages against *Mycobacterium tuberculosis*. Mucosal Immunol 4(3):279–287

Bhat KH, Yaseen I (2018) *Mycobacterium tuberculosis*: macrophage takeover and modulation of innate effector responses. In: Ribón W (ed) Mycobacterium-research and development. IntechOpen, London. https://doi.org/10.5772/intechopen.75003

Bollela VR, Namburete EI, Feliciano CS, Macheque D, Harrison LH, Caminero JA (2016) Detection of katG and inhA mutations to guide isoniazid and ethionamide use for drug-resistant tuberculosis. Int J Tuberc Lung Dis 20(8):1099–1104

Buddle BM, Vordermeier HM, Chambers MA, de Klerk-Lorist LM (2018) Efficacy and safety of BCG vaccine for control of tuberculosis in domestic livestock and wildlife. Front Vet Sci 5:259

Buriánková K, Doucet-Populaire F, Dorson O, Gondran A, Ghnassia JC, Weiser J, Pernodet JL (2004) Molecular basis of intrinsic macrolide resistance in the *Mycobacterium tuberculosis* complex. Antimicrob Agents Chemother 48(1):143–150

Cade CE, Dlouhy AC, Medzihradszky KF, Salas-Castillo SP, Ghiladi RA (2010) Isoniazid-resistance conferring mutations in *Mycobacterium tuberculosis* KatG: Catalase, peroxidase, and INH-NADH adduct formation activities. Protein Sci 19(3):458–474

Cambau E, Drancourt M (2014) Steps towards the discovery of *Mycobacterium tuberculosis* by Robert Koch, 1882. Clin Microbiol Infect 20(3):196–201

Caminero JA, Sotgiu G, Zumla A, Migliori GB (2010) Best drug treatment for multidrug-resistant and extensively drug-resistant tuberculosis. Lancet Infect Dis 10(9):621–629

Chan ED, Laurel V, Strand MJ, Chan JF, Huynh ML, Goble M, Iseman MD (2004) Treatment and outcome analysis of 205 patients with multidrug-resistant tuberculosis. Am J Respir Crit Care Med 169(10):1103–1109

Chen H, Liu X, Ma X, Wang Q, Yang G, Niu H, Li S, He B, He S, Dannenberg AM Jr, Zhu B (2017) A new rabbit-skin model to evaluate protective efficacy of tuberculosis vaccines. Front Microbiol 8:842

Chetty S, Ramesh M, Singh-Pillay A, Soliman ME (2017) Recent advancements in the development of anti-tuberculosis drugs. Bioorg Medicinal Chem Lett 27(3):370–386

Clark S, Hall Y, Williams A (2015) Animal models of tuberculosis: guinea pigs. Cold Spring Harb Perspect Med 5(5):a018572

Colangeli R, Helb D, Vilchèze C, Hazbón MH, Lee CG, Safi H, Sayers B, Sardone I, Jones MB, Fleischmann RD, Peterson SN (2007) Transcriptional regulation of multi-drug tolerance and antibiotic-induced responses by the histone-like protein Lsr2 in *M. tuberculosis*. PLoS Pathog 3(6):e87

Connell DW, Berry M, Cooke G, Kon OM (2011) Update on tuberculosis: TB in the early 21st century. Eur Respir Rev 20(120):71–84

Cooper AM (2009) Cell-mediated immune responses in tuberculosis. Annu Rev Immunol 27:393–422

D'Ambrosio L, Centis R, Sotgiu G, Pontali E, Spanevello A, Migliori GB (2015) New anti-tuberculosis drugs and regimens: 2015 update. ERJ Open Res 1(1):00010–02015

Davies J, Davies D (2010) Origins and evolution of antibiotic resistance. Microbiol Mol Biol Rev 74(3):417–433

Daya M, Van der Merwe L, Van Helden PD, Möller M, Hoal EG (2015) Investigating the role of gene-gene interactions in TB susceptibility. PLoS One 10(4):e0123970

De Vos M, Müller B, Borrell S, Black PA, Van Helden PD, Warren RM, Gagneux S, Victor TC (2013) Putative compensatory mutations in the rpoC gene of rifampin-resistant *Mycobacterium tuberculosis* are associated with ongoing transmission. Antimicrob Agents Chemother 57(2):827–832

Delogu G, Sali M, Fadda G (2013) The biology of *Mycobacterium tuberculosis* infection. Mediterr J Hematol Infect Dis 5(1):e2013070

Desjardins CA, Cohen KA, Munsamy V, Abeel T, Maharaj K, Walker BJ, Shea TP, Almeida DV, Manson AL, Salazar A, Padayatchi N (2016) Genomic and functional analyses of *Mycobacterium tuberculosis* strains implicate ald in D-cycloserine resistance. Nat Genet 48(5):544–551

Domingo-Gonzalez R, Prince O, Cooper A, Khader SA (2016) Cytokines and Chemokines in *Mycobacterium tuberculosis* Infection. Microbiol Spectr 4(5). https://doi.org/10.1128/microbiolspec.TBTB2-0018-2016

Dookie N, Rambaran S, Padayatchi N, Mahomed S, Naidoo K (2018) Evolution of drug resistance in *Mycobacterium tuberculosis*: a review on the molecular determinants of resistance and implications for personalized care. J Antimicrob Chemother 73(5):1138–1151

Dormans J, Burger M, Aguilar D, Hernandez-Pando R, Kremer K, Roholl P, Arend SM, Van Soolingen D (2004) Correlation of virulence, lung pathology, bacterial load and delayed type hypersensitivity responses after infection with different *Mycobacterium tuberculosis* genotypes in a BALB/c mouse model. Clin Exp Immunol 137(3):460–468

Drain PK, Bajema KL, Dowdy D, Dheda K, Naidoo K, Schumacher SG, Ma S, Meermeier E, Lewinsohn DM, Sherman DR (2018) Incipient and subclinical tuberculosis: a clinical review of early stages and progression of infection. Clin Microbiol Rev 31(4):e00021–e00018

Dutta NK, Karakousis PC (2014) Latent tuberculosis infection: myths, models, and molecular mechanisms. Microbiol Mol Biol Rev 78(3):343–371

Ehlers S, Schaible UE (2013) The granuloma in tuberculosis: dynamics of a host–pathogen collusion. Front Immunol 3:411

Fair RJ, Tor Y (2014) Antibiotics and Bacterial Resistance in the 21st Century. Perspect Medicin Chem 6:25–64

Falzon D, Schünemann HJ, Harausz E, González-Angulo L, Lienhardt C, Jaramillo E, Weyer K (2017) World Health Organization treatment guidelines for drug-resistant tuberculosis, 2016 update. Eur Respiratory Soc 49(3):1602308

Feltcher ME, Sullivan JT, Braunstein M (2010) Protein export systems of *Mycobacterium tuberculosis*: novel targets for drug development? Future Microbiol 5(10):1581–1597

Fonseca JD, Knight GM, McHugh TD (2015) The complex evolution of antibiotic resistance in *Mycobacterium tuberculosis*. Int J Infect Dis 32:94–100

Ford CB, Shah RR, Maeda MK, Gagneux S, Murray MB, Cohen T, Johnston JC, Gardy J, Lipsitch M, Fortune SM (2013) *Mycobacterium tuberculosis* mutation rate estimates from different lineages predict substantial differences in the emergence of drug-resistant tuberculosis. Nature Genet 45(7):784–790

Forrellad MA, Klepp LI, Gioffré A, Sabio y Garcia J, Morbidoni HR, Santangelo MD, Cataldi AA, Bigi F (2013) Virulence factors of the *Mycobacterium tuberculosis* complex. Virulence 4(1):3–66

Fu LM, Shinnick TM (2007) Genome-wide exploration of the drug action of capreomycin on *Mycobacterium tuberculosis* using Affymetrix oligonucleotide GeneChips. J Infect 54(3):277–284

Fujiwara M, Kawasaki M, Hariguchi N, Liu Y, Matsumoto M (2018) Mechanisms of resistance to delamanid, a drug for Mycobacterium tuberculosis. Tuberculosis 108:186–194

Garneau-Tsodikova S, Labby KJ (2016) Mechanisms of resistance to aminoglycoside antibiotics: overview and perspectives. Med Chem Comm 7(1):11–27

Gillespie SH (2002) Evolution of drug resistance in *Mycobacterium tuberculosis*: clinical and molecular perspective. Antimicrob Agents Chemother 46(2):267–274

Ginsburg AS, Grosset JH, Bishai WR (2003) Fluoroquinolones, tuberculosis, and resistance. Lancet Infect Dis 3(7):432–442

Glaziou P, Sismanidis C, Floyd K, Raviglione M (2015) Global epidemiology of tuberculosis. Cold Spring HarbPerspect Med 5(2):a017798

Glickman MS, Jacobs WR (2001) Microbial pathogenesis of *Mycobacterium tuberculosis*: dawn of a discipline. Cell 104(4):477–485

Gualano G, Capone S, Matteelli A, Palmieri F (2016) New Antituberculosis Drugs: From Clinical Trial to Programmatic Use. Infect Dis Rep 8(2):6569

Gygli SM, Borrell S, Trauner A, Gagneux S (2017) Antimicrobial resistance in *Mycobacterium tuberculosis*: mechanistic and evolutionary perspectives. FEMS Microbiol Rev 41(3):354–373

Hameed HM, Islam MM, Chhotaray C, Wang C, Liu Y, Tan Y, Li X, Tan S, Delorme V, Yew WW, Liu J (2018) Molecular targets related drug resistance mechanisms in MDR-, XDR-, and TDR-*Mycobacterium tuberculosis* strains. Front Cell Infect Microbiol 8:114

Hawn TR, Day TA, Scriba TJ, Hatherill M, Hanekom WA, Evans TG, Churchyard GJ, Kublin JG, Bekker LG, Self SG (2014) Tuberculosis vaccines and prevention of infection. Microbiol Mol Biol Rev 78(4):650–671

Hegde SS, Vetting MW, Roderick SL, Mitchenall LA, Maxwell A, Takiff HE, Blanchard JS (2005) A fluoroquinolone resistance protein from *Mycobacterium tuberculosis* that mimics DNA. Science 308(5727):1480–1483

Hoagland DT, Liu J, Lee RB, Lee RE (2016) New agents for the treatment of drug-resistant *Mycobacterium tuberculosis*. Adv Drug Deliv Rev 102:55–72

Hoff DR, Ryan GJ, Driver ER, Ssemakulu CC, De Groote MA, Basaraba RJ, Lenaerts AJ (2011) Location of intra-and extracellular *M. tuberculosis* populations in lungs of mice and guinea pigs during disease progression and after drug treatment. PLoS One 6(3):e17550

Iketleng T, Lessells R, Dlamini MT, Mogashoa T, Mupfumi L, Moyo S, Gaseitsiwe S (2018) *Mycobacterium tuberculosis* Next-generation whole genome sequencing: opportunities and challenges. Tuberc Res Treat 2018:1298542

Ilina EN, Shitikov EA, Ikryannikova LN, Alekseev DG, Kamashev DE, Malakhova MV, Parfenova TV, Afanas'ev MV, Ischenko DS, Bazaleev NA, Smirnova TG (2013) Comparative genomic analysis of *Mycobacterium tuberculosis* drug resistant strains from Russia. PLoS One 8(2):e56577

Jagielski T, Minias A, Van Ingen J, Rastogi N, Brzostek A, Żaczek A, Dziadek J (2016) Methodological and clinical aspects of the molecular epidemiology of *Mycobacterium tuberculosis* and other mycobacteria. Clin Microbiol Rev 29(2):239–290

Jassal MS, Bishai WR (2010) Epidemiology and challenges to the elimination of global tuberculosis. Clin Infect Dis 50:S156–S164

Jugheli L, Bzekalava N, de Rijk P, Fissette K, Portaels F, Rigouts L (2009) High level of cross-resistance between kanamycin, amikacin, and capreomycin among *Mycobacterium tuberculosis* isolates from Georgia and a close relation with mutations in the rrs gene. Antimicrob Agents Chemother 53(12):5064–5068

Kaushal D, Schroeder BG, Tyagi S, Yoshimatsu T, Scott C, Ko C, Carpenter L, Mehrotra J, Manabe YC, Fleischmann RD, Bishai WR (2002) Reduced immunopathology and mortality despite tissue persistence in a *Mycobacterium tuberculosis* mutant lacking alternative σ factor, SigH. Proc Natl Acad Sci U S A 99(12):8330–8335

Kim KH, An DR, Song J, Yoon JY, Kim HS, Yoon HJ, Im HN, Kim J, Lee SJ, Kim KH, Lee HM (2012) *Mycobacterium tuberculosis* Eis protein initiates suppression of host immune responses by acetylation of DUSP16/MKP-7. Proc Natl Acad Sci U S A 109(20):7729–7734

Koch A, Cox H, Mizrahi V (2018) Drug-resistant tuberculosis: challenges and opportunities for diagnosis and treatment. Curr Opin Pharmacol 42:7–15

Kohanski MA, Dwyer DJ, Collins JJ (2010) How antibiotics kill bacteria: from targets to networks. Nat Rev Microbiol 8(6):423–435

Kramnik I, Beamer G (2016) Mouse models of human TB pathology: roles in the analysis of necrosis and the development of host-directed therapies. Semin Immuno Pathol 38(2):221–237

Kumar S, Jena L (2014) Understanding rifampicin resistance in tuberculosis through a computational approach. Genomics Inform 12(4):276–282

Kurz SG, Bonomo RA (2012) Reappraising the use of β-lactams to treat tuberculosis. Expert Rev Anti-Infect Ther 10(9):999–1006

Lahiri N, Shah RR, Layre E, Young D, Ford C, Murray MB, Fortune SM, Moody DB (2016) Rifampin resistance mutations are associated with broad chemical remodeling of *Mycobacterium tuberculosis*. J Biol Chem 291(27):14248–14256

Lange C, Abubakar I, Alffenaar JW, Bothamley G, Caminero JA, Carvalho AC, Chang KC, Codecasa L, Correia A, Crudu V, Davies P (2014) Management of patients with multidrug-resistant/extensively drug-resistant tuberculosis in Europe: a TBNET consensus statement. The Eur Respir J 44(1):23–63

Lee JY (2015) Diagnosis and treatment of extrapulmonary tuberculosis. Tuberc Respir Dis 78(2):47–55

Lee SH (2016) Tuberculosis infection and latent tuberculosis. Tuberc Respir Dis 79(4):201–206

Li F, Wan L, Xiao T, Liu H, Jiang Y, Zhao X, Wang R, Wan K (2018) In vitro activity of β-lactams in combination with β-lactamase inhibitors against *Mycobacterium tuberculosis* clinical isolates. Biomed Res Int 2018:3579832

Lingaraju S, Rigouts L, Gupta A, Lee J, Umubyeyi AN, Davidow AL, German S, Cho E, Cho SN, Kim CT, Alland D (2016) Geographic differences in the contribution of ubiA mutations to high-level ethambutol resistance in *Mycobacterium tuberculosis*. Antimicrob Agents Chemother 60(7):4101–4105

Liu L, Jiang F, Chen L, Zhao B, Dong J, Sun L, Zhu Y, Liu B, Zhou Y, Yang J, Zhao Y (2018a) The impact of combined gene mutations in inhA and ahpC genes on high levels of isoniazid resistance amongst katG non-315 in multidrug-resistant tuberculosis isolates from China. Emerg Microbes Infect 7(1):1–10

Liu Q, Zuo T, Xu P, Jiang Q, Wu J, Gan M, Yang C, Prakash R, Zhu G, Takiff HE, Gao Q (2018b) Have compensatory mutations facilitated the current epidemic of multidrug-resistant tuberculosis? Emerg Microbes Infect 7(1):1–8

Lu J, Liu M, Wang Y, Pang Y, Zhao Z (2014) Mechanisms of fluoroquinolone monoresistance in *Mycobacterium tuberculosis*. FEMS Microbiol Lett 353(1):40–48

Ma Z, Lienhardt C, McIlleron H, Nunn AJ, Wang X (2010) Global tuberculosis drug development pipeline: the need and the reality. Lancet 375(9731):2100–2109

Majorov KB, Lyadova IV, Kondratieva TK, Eruslanov EB, Rubakova EI, Orlova MO, Mischenko VV, Apt AS (2003) Different innate ability of I/St and A/Sn mice to combat virulent *Mycobacterium tuberculosis*: phenotypes expressed in lung and extrapulmonary macrophages. Infect Immun 71(2):697–707

Manca C, Tsenova L, Barry CE, Bergtold A, Freeman S, Haslett PA, Musser JM, Freedman VH, Kaplan G (1999) *Mycobacterium tuberculosis* CDC1551 induces a more vigorous host response in vivo and in vitro, but is not more virulent than other clinical isolates. J Immunol 162(11):6740–6746

Manca C, Tsenova L, Bergtold A, Freeman S, Tovey M, Musser JM, Barry CE, Freedman VH, Kaplan G (2001) Virulence of a *Mycobacterium tuberculosis* clinical isolate in mice is determined by failure to induce Th1 type immunity and is associated with induction of IFN-α/β. Proc Natl Acad Sci U S A 98(10):5752–5757

Maruri F, Sterling TR, Kaiga AW, Blackman A, van der Heijden YF, Mayer C, Cambau E, Aubry A (2012) A systematic review of gyrase mutations associated with fluoroquinolone-resistant *Mycobacterium tuberculosis* and a proposed gyrase numbering system. J Antimicrob Chemother 67(4):819–831

Maus CE, Plikaytis BB, Shinnick TM (2005) Mutation of tlyA confers capreomycin resistance in *Mycobacterium tuberculosis*. Antimicrob Agents Chemother 49(2):571–577

McGrath M, Gey van Pittius NC, Van Helden PD, Warren RM, Warner DF (2013) Mutation rate and the emergence of drug resistance in *Mycobacterium tuberculosis*. J Antimicrob Chemother 69(2):292–302

Mehta PK, Pandey AK, Subbian S, El-Etr SH, Cirillo SL, Samrakandi MM, Cirillo JD (2006) Identification of *Mycobacterium marinum* macrophage infection mutants. Microb Pathog 40(4):139–151

Meijer AH (2016) Protection and pathology in TB: learning from the zebrafish model. Semin Immuno pathol 38(2):261–273

Motiwala AS, Dai Y, Jones-López EC, Hwang SH, Lee JS, Cho SN, Via LE, Barry CE III, Alland D (2010) Mutations in extensively drug-resistant *Mycobacterium tuberculosis* that do not code for known drug-resistance mechanisms. J Infect Dis 201(6):881–888

Mukhopadhyay S, Nair S, Ghosh S (2012) Pathogenesis in tuberculosis: transcriptomic approaches to unraveling virulence mechanisms and finding new drug targets. FEMS Microbiol Rev 36(2):463–485

Müller B, Borrell S, Rose G, Gagneux S (2013) The heterogeneous evolution of multidrug-resistant *Mycobacterium tuberculosis*. Trends Genet 29(3):160–169

Murray JF, Schraufnagel DE, Hopewell PC (2015) Treatment of tuberculosis. A historical perspective. Ann Am Thorac Soc 12(12):1749–1759

Nahid P, Kim PS, Evans CA, Alland D, Barer M, Diefenbach J, Ellner J, Hafner R, Hamilton CD, Iademarco MF, Ireton G (2012) Clinical Research and Development of Tuberculosis Diagnostics: Moving From Silos to Synergy. J Infect Dis 205:S159–S168
Nampoothiri KM, Rubex R, Patel AK, Narayanan SS, Krishna S, Das SM, Pandey A (2008) Molecular cloning, overexpression and biochemical characterization of hypothetical β-lactamases of *Mycobacterium tuberculosis* H37Rv. J Appl Microbiol 105(1):59–67
Nasiri MJ, Imani Fooladi AA, Dabiri H, Pormohammad A, Salimi Chirani A, Dadashi M, Houri H, Heidary M, Feizabadi MM (2016) Primary ethambutol resistance among Iranian pulmonary tuberculosis patients: a systematic review. Ther Adv Infect Dis 3(5):133–138
Ngo TM, Teo YY (2019) Genomic prediction of tuberculosis drug-resistance: benchmarking existing databases and prediction algorithms. BMC Bioinformatics 20(1):68
Nguta JM, Appiah-Opong R, Nyarko AK, Yeboah-Manu D, Addo PG (2015) Current perspectives in drug discovery against tuberculosis from natural products. Int J Mycobacteriol 4(3):165–183
Nguyen L (2016) Antibiotic resistance mechanisms in M. tuberculosis: an update. Arch Toxicol 90(7):1585–1604
Niederweis M, Danilchanka O, Huff J, Hoffmann C, Engelhardt H (2010) Mycobacterial outer membranes: in search of proteins. Trends Microbiol 18(3):109–116
Norris BA, Ernst JD (2018) Mononuclear cell dynamics in *M. tuberculosis* infection provide opportunities for therapeutic intervention. PLoS Pathog 14(10):e1007154
O'Neill MB, Mortimer TD, Pepperell CS (2012) Diversity of *Mycobacterium tuberculosis* across evolutionary scales. PLoS Pathog 11(11):e1005257
Ocheretina O, Escuyer VE, Mabou MM, Royal-Mardi G, Collins S, Vilbrun SC, Pape JW, Fitzgerald DW (2014) Correlation between genotypic and phenotypic testing for resistance to rifampin in *Mycobacterium tuberculosis* clinical isolates in Haiti: investigation of cases with discrepant susceptibility results. PLoS One 9(3):e90569
Palomino J, Martin A (2014) Drug resistance mechanisms in *Mycobacterium tuberculosis*. Antibiotics 3(3):317–340
Parida SK, Axelsson-Robertson R, Rao MV, Singh N, Master I, Lutckii A, Keshavjee S, Andersson J, Zumla A, Maeurer M (2015) Totally drug-resistant tuberculosis and adjunct therapies. J Intern Med 277(4):388–405
Pieters J (2008) *Mycobacterium tuberculosis* and the macrophage: maintaining a balance. Cell Host Microbe 3(6):399–407
Plinke C, Walter K, Aly S, Ehlers S, Niemann S (2011) *Mycobacterium tuberculosis* embB codon 306 mutations confer moderately increased resistance to ethambutol in vitro and in vivo. Antimicrob Agents Chemother 55(6):2891–2896
Prasad R, Singh A, Balasubramanian V, Gupta N (2017) Extensively drug-resistant tuberculosis in India: Current evidence on diagnosis & management. Indian J Med Res 145(3):271–293
Prozorov AA, Fedorova IA, Bekker OB, Danilenko VN (2014) The virulence factors of *Mycobacterium tuberculosis*: genetic control, new conceptions. Russ J Genet 50(8):775–797
Pym AS, Saint-Joanis B, Cole ST (2002) Effect of katG mutations on the virulence of *Mycobacterium tuberculosis* and the implication for transmission in humans. Infect Immun 70(9):4955–4960
Ramirez-Busby SM, Valafar F (2015) Systematic review of mutations in pyrazinamidase associated with pyrazinamide resistance in *Mycobacterium tuberculosis* clinical isolates. Antimicrob Agents Chemother 59(9):5267–5277
Rawal T, Butani S (2016) Combating tuberculosis infection: a forbidding challenge. Indian J Pharm Sci 78(1):8–16
Rengarajan J, Sassetti CM, Naroditskaya V, Sloutsky A, Bloom BR, Rubin EJ (2004) The folate pathway is a target for resistance to the drug para-aminosalicylic acid (PAS) in mycobacteria. Mol Microbiol 53(1):275–282
Ribeiro SC, Gomes LL, Amaral EP, Andrade MR, Almeida FM, Rezende AL, Lanes VR, Carvalho EC, Suffys PN, Mokrousov I, Lasunskaia EB (2014) *Mycobacterium tuberculosis* strains of the modern sublineage of the Beijing family are more likely to display increased virulence than strains of the ancient sublineage. J Clin Microbiol 52(7):2615–2624

Rock RB, Olin M, Baker CA, Molitor TW, Peterson PK (2008) Central nervous system tuberculosis: pathogenesis and clinical aspects. Clin Microbiol Rev 21(2):243–261

Safi H, Lingaraju S, Amin A, Kim S, Jones M, Holmes M, McNeil M, Peterson SN, Chatterjee D, Fleischmann R, Alland D (2013) Evolution of high-level ethambutol-resistant tuberculosis through interacting mutations in decaprenylphosphoryl-β-D-arabinose biosynthetic and utilization pathway genes. Nat Genet 45(10):1190–1197

Sakamoto K (2012) The pathology of *Mycobacterium tuberculosis* infection. Vet Pathol 49(3):423–439

Sarathy J, Dartois V, Lee E (2012) The role of transport mechanisms in *Mycobacterium tuberculosis* drug resistance and tolerance. Pharmaceuticals 5(11):1210–1235

Sasindran SJ, Torrelles JB (2011) *Mycobacterium tuberculosis* Infection and Inflammation: what is Beneficial for the Host and for the Bacterium? Front Microbiol 2:2

Sauvage E, Fonzé E, Quinting B, Galleni M, Frere JM, Charlier P (2006) Crystal structure of the *Mycobacterium fortuitum* class A β-lactamase: structural basis for broad substrate specificity. Antimicrob Agents Chemother 50(7):2516–2521

Scanga CA, Flynn JL (2014) Modeling tuberculosis in nonhuman primates. Cold Spring HarbPerspect Med 4(12):a018564

Schluger NW (2013) Fluoroquinolones in the Treatment of Tuberculosis: Which Is Best? Am J Respir Crit Care Med 188(7):768–769

Schulz zurWiesch P, Engelstädter J, Bonhoeffer S (2010) Compensation of fitness costs and reversibility of antibiotic resistance mutations. Antimicrob Agents Chemother 54(5):2085–2095

Shaler CR, Kugathasan K, McCormick S, Damjanovic D, Horvath C, Small CL, Jeyanathan M, Chen X, Yang PC, Xing Z (2011) Pulmonary mycobacterial granuloma: increased IL-10 production contributes to establishing a symbiotic host–microbe microenvironment. Am J Pathol 178(4):1622–1634

Singh AK, Gupta UD (2018) Animal models of tuberculosis: lesson learnt. Indian J Med Res 147(5):456–463

Smith I (2003) Mycobacterium tuberculosis pathogenesis and molecular determinants of virulence. Clin Microbiol Rev 16(3):463–496

Smith T, Wolff KA, Nguyen L (2012) Molecular biology of drug resistance in *Mycobacterium tuberculosis*. In: Pieters J, McKinney J (eds) Pathogenesis of *Mycobacterium tuberculosis* and its Interaction with the Host Organism. Springer, Berlin, pp 53–80

Somoskovi A, Parsons LM, Salfinger M (2001) The molecular basis of resistance to isoniazid, rifampin, and pyrazinamide in *Mycobacterium tuberculosis*. Respir Res 2(3):164

Soto SM (2013) Role of efflux pumps in the antibiotic resistance of bacteria embedded in a biofilm. Virulence 4(3):223–229

Spies FS, Ribeiro AW, Ramos DF, Ribeiro MO, Martin A, Palomino JC, Rossetti ML, da Silva PE, Zaha A (2011) Streptomycin resistance and lineage-specific polymorphisms in Mycobacterium tuberculosis gidB gene. J Clin Microbiol 49(7):2625–2630

Torres JN, Paul LV, Rodwell TC, Victor TC, Amallraja AM, Elghraoui A, Goodmanson AP, Ramirez-Busby SM, Chawla A, Zadorozhny V, Streicher EM (2015) Novel katG mutations causing isoniazid resistance in clinical *M. tuberculosis* isolates. Emerg Microbes Infect 4(1):1–9

Trauner A, Borrell S, Reither K, Gagneux S (2014) Evolution of drug resistance in tuberculosis: recent progress and implications for diagnosis and therapy. Drugs 74(10):1063–1072

Tsenova L, Ellison E, Harbacheuski R, Moreira AL, Kurepina N, Reed MB, Mathema B, Barry CE III, Kaplan G (2005) Virulence of selected *Mycobacterium tuberculosis* clinical isolates in the rabbit model of meningitis is dependent on phenolic glycolipid produced by the bacilli. J Infect Dis 192(1):98–106

Van Deun A, Aung KJ, Bola V, Lebeke R, Hossain MA, de Rijk WB, Rigouts L, Gumusboga A, Torrea G, de Jong BC (2013) Rifampin drug resistance tests for tuberculosis: challenging the gold standard. J Clin Microbiol 51(8):2633–2640

Van Leeuwen LM, Van der Sar AM, Bitter W (2015) Animal models of tuberculosis: zebrafish. Cold Spring HarbPerspect Med 5(3):a018580

Von Groll A, Martin A, Jureen P, Hoffner S, Vandamme P, Portaels F, Palomino JC, da Silva PA (2009) Fluoroquinolone resistance in *Mycobacterium tuberculosis* and mutations in gyrA and gyrB. Antimicrob Agents Chemother 53(10):4498–4500

Wang F, Cassidy C, Sacchettini JC (2006) Crystal structure and activity studies of the *Mycobacterium tuberculosis* β-lactamase reveal its critical role in resistance to β-lactam antibiotics. Antimicrob Agents Chemother 50(8):2762–2771

Waters WR, Palmer MV, Thacker TC, Davis WC, Sreevatsan S, Coussens P, Meade KG, Hope JC, Estes DM (2011) Tuberculosis immunity: opportunities from studies with cattle. Clin Dev Immunol 2011:768542

Wong A (2017) Epistasis and the evolution of antimicrobial resistance. Front Microbiol 8:246

Xie J, Talaska AE, Schacht J (2011) New developments in aminoglycoside therapy and ototoxicity. Hear Res 281(1–2):28–37

Yakrus MA, Driscoll J, Lentz AJ, Sikes D, Hartline D, Metchock B, Starks AM (2014) Concordance between molecular and phenotypic testing of *Mycobacterium tuberculosis* complex isolates for resistance to rifampin and isoniazid in the United States. J Clin Microbiol 52(6):1932–1937

Zhai W, Wu F, Zhang Y, Fu Y, Liu Z (2019) The Immune Escape Mechanisms of *Mycobacterium tuberculosis*. Int J Mol Sci 20(2):340

Zhan L, Tang J, Sun M, Qin C (2017) Animal models for tuberculosis in translational and precision medicine. Front Microbiol 8:717

Zhang S, Chen J, Cui P, Shi W, Zhang W, Zhang Y (2015) Identification of novel mutations associated with clofazimine resistance in *Mycobacterium tuberculosis*. J Antimicrob Chemother 70(9):2507–2510

Zhang S, Chen J, Cui P, Shi W, Shi X, Niu H, Chan D, Yew WW, Zhang W, Zhang Y (2016) *Mycobacterium tuberculosis* mutations associated with reduced susceptibility to linezolid. Antimicrob Agents Chemother 60(4):2542–2544

Zhang S, Chen J, Shi W, Cui P, Zhang J, Cho S, Zhang W, Zhang Y (2017) Mutation in clpC1 encoding an ATP-dependent ATPase involved in protein degradation is associated with pyrazinamide resistance in *Mycobacterium tuberculosis*. Emerg Microbes Infect 6(1):1–2

Pathogenesis of Fungal Infections and Drug-Resistance Phenomenon

17

Sudhakar Pola, Akella Vijayaramya, Pavani Sanapala, and V. A. Iswarya Deepthi

Contents

Abstract

Fungi the primitive eukaryotes are emerging as life-threatening pathogens of public health. Over a decade and ago, the frequency of fungal infections has been enormous, with an increased range of mortality and morbidity in immunocompromised patients. The risk of fungal infections is aggravated by random use of broad-spectrum antibacterial drugs, immunosuppressive agents, and various cancer chemotherapies. Most of the well-known fungal disease and the pathogenic fungi are Aspergillus, Blastomycosis, Candidiasis, Coccidioidomycosis,

S. Pola (✉) · A. Vijayaramya · P. Sanapala · V. A. I. Deepthi
Department of Biotechnology, Andhra University, Visakhapatnam, India
e-mail: sudhakar@andhrauniversity.edu.in

B. Siddhardha et al. (eds.), *Model Organisms for Microbial Pathogenesis, Biofilm Formation and Antimicrobial Drug Discovery*,
https://doi.org/10.1007/978-981-15-1695-5_17

Cryptococcus, and Dermatophytes. The resistance to antifungal medicines might be characteristic, acquired, or clinical. The comprehension of the mechanism of the clinical resistance effect is significant, unlike alternating treatment. In this chapter, after a concise overview of antifungal resistance, the molecular transport mechanism and mechanism of drugs will be detailed. It emerges that the main systems of resistance are necessarily appropriate to the deregulation of antifungal resistance effector genes. This deregulation in transcriptional regulators of the genes is due to the occurrence of point mutations. The study of antifungal its pathogenicity and resistance to drugs is essential for a better understanding of the human pathogenic fungal biology.

Keywords
Drug resistance mutation · PDR gene · Multi-drug resistance · ABC transporters · Antifungal resistance

17.1 Introduction

Fungi are ubiquitous in nature, contain membrane-bound cell organelles and also rigid cell wall encapsulated with a cell membrane. Usually, fungi occur in two basic forms, namely molds (vegetative multicellular organisms) and yeasts (unicellular filamentous organisms). Fungi subsist as a free-living saprobe with no apparent benefits from humans or animals. Characterization of fungi as a pathogen depends merely on the severity of the disease caused. Cause of disease from mild to severe range in the host system. The majority of fungal pathogens in mammals are so-called opportunistic since the disease is caused only when the host immune system lacks the defense response, one such example is the *Aspergillus*, the common mold. The infection of *Aspergillus* is known as aspergillosis with a few varied types present in the environment (Barnes and Marr 2006). Aspergillosis affects the respiratory system that is present in our normal indoor condition cause infections especially in those of weak immune system. Another parasitic fungal disease is blastomycosis, an ailment brought about by the parasite *Blastomyces dermatitidis* infecting the skin. The growth of Blastomyces lives in damp soil and in association with decomposing organic matter such as wooden leaves. The side effects of blastomycosis are frequently like influenza indications. The majority of this blastomycosis is found in midwestern, south-central, and south-eastern states in the USA (Furcolow et al. 1970; Bradsher et al. 2003). An additional example of fungal infection is Candidiasis caused by the organism *Candida* that is inhabitant of mouth, throat, gut, and vagina in the human body. Nearly 20 varieties of candida yeast organisms are reported to cause infections, the most common is *Candida albicans* (Nucci and Anaissie 2001). The other added examples are coccidioidomycosis and cryptococcosis. Coccidioidomycosis known as valley fever is a disease brought about by the growth of Coccidioides, which lives in the dirt of dry low precipitation territories. It is endemic in numerous zones of southwestern USA, Mexico, Central and South

America (Vugia et al. 2013). Cryptococcosis is a fungal infection caused by organisms that belong to the genus *Cryptococcus*, generally acquired by inhalation. The two species *Cryptococcus neoformans* and *Cryptococcus gattii* cause nearly all cryptococcal infections in both humans and animals (Knoke and Schwesinger 2009; Casadevall and Perfect 1998). Dermatophytes are organisms that trigger skin, hair, and nail infections. Diseases brought by these growths are additionally in some cases known as ringworm or tinea. The major recognized classes are Trichophyton rubrum and Trichophyton tonsurans (Hainer 2003).

Histoplasmosis is an infection triggered by the organism *Histoplasma capsulatum*. The fungus survives in surroundings for the most part together with a lot of bat or feathered creature droppings. The signs and symptoms of histoplasmosis are similar to that of pneumonia as the pathogen affects the lungs of human system (Manos et al. 1956). Mucormycosis otherwise called Zygomycosis is an uncommon disease caused by fungi that are related to a group of fungi known as Mucoromycotina in the type Mucorales. These parasites are commonly found in the dirt and along the decomposed biotic matter (Richardson 2009). Pneumocystis pneumonia (PCP) is a severe disease resulting from parasite *Pneumocystis jiroveci*. Pneumocystis pneumonia is one of the most frequent and severe opportunistic infections in people with weakened immune systems, particularly people with HIV and AIDS (Harris et al. 2010).

Sporotrichosis is a disease resulting from an organism called *Sporothrix schenckii*, which is a cutaneous disease and most frequently occurs when in contact with infected plants when the fungus penetrates through the skin (Bastos de Lima-Barros et al. 2011).

Exserohilum is another type of a fungus identified in soil and commonly on plants particularly grasses, and it flourishes in warm and moist atmospheres. It is an extremely uncommon disease in individuals, yet it has been distinguished as one of the dominating pathogens in an ongoing multistate episode of parasitic meningitis and other contagious contaminations related with debased steroid infusions (Revankar and Sutton 2010; Adler et al. 2006). Cladosporium is another uncommon reason for human ailment, yet it has been known to cause a few distinct sorts of diseases including skin, eye, sinus, and mind contaminations (Drabick et al. 1990; Gugnani et al. 2000; Sang et al. 2011). Cladosporium, in the same way as other sorts of organisms, has likewise been related to sensitivities and asthma (Sellart-Altisent et al. 2007).

In contemporary years, fungal infections have acquired significant/substantial prominence due to its enhancement in the immunocompromised population including HIV infected human beings, in which the patients receive immunosuppressive treatment for organ or bone marrow transplantation or cancer patients undergoing cytotoxic agents (Richardson and Lass-Flörl 2008).

A wide range of various treatment strategies are required to address this issue along with triumphing adverse effects and drug resistance. The main objective of this chapter is to review the latest phenomenon of fungi and importance of antifungal agents in clinical trials. Furthermore, countless extracts from various plants have also been demonstrated to combat with diverse fungi.

17.2 Chemical Classes

The treatment mainly depends on the accessibility of antifungal drugs. Numerous antifungal agents illustrate seven different chemical classes such as polyenes, pyrimidine analogs, azoles, candins, allylamines, thiocarbamates, and morpholines.

17.2.1 Polyenes

Polyenes are amphipathic (one hydrophilic charged site and another hydrophobic uncharged site) in nature corresponding to a class of natural compounds. Polyenes target ergosterol in the fungal membrane creating pores and permeating tiny molecules through the membrane causing cell death. Two major polyenes are amphotericin B and nystatin (Canuto and Rodero 2002).

17.2.2 Pyrimidine Analogs

Pyrimidine analogs consist of 5-fluorocytosine (5-FC) only. Susceptible fungi consist of cytosine deaminase that regenerates 5-fluorocytosine into 5-fluorouracil integrating into DNA and RNA, hindering cellular functions. Generally, 5-fluorocytosine has inadequate activity against the majority of filamentous fungi and dermatophytes rendering insufficient cytosine deaminase (Gehrt et al. 1995). Thus most often 5-FC is administered together with polyenes or other antifungal agents (Sanglard 2002).

17.2.3 Azoles

Azoles conjointly with allylamines, thiocarbamates, and morphlines obstruct ergosterol biosynthesis. Azoles hinder cytochrome P450 lanosterol demethylase, Erg11, or Cyp51, which is a fundamental phase in sterol biosynthesis, causing the substitution of ergosterol by methylated sterols in the plasma membrane (Sanglard 2002). Azoles might be responsible for sterol D22-desaturation (Erg5) by impeding cytochrome P450, as in sterol biosynthesis Erg11 takes precedence before Erg5 (Skaggs et al. 1996). Azole drugs are further classified into two classes: (1) imidazoles consisting of ketoconazole, miconazole, and clotrimazole mostly used for skin infections, and (2) triazoles consisting of fluconazole, voriconazole, itraconazole, and posaconazole commonly used for systemic infections (Sheehan et al. 1999).

17.2.4 Candins

Candins are the novel class of antifungal agents. Candins inhibits an enzyme complex b-1-3 glucan synthase, localized in the plasma membrane of fungi. Nowadays, compounds which are chemically related are available in three different types:

caspofungin, micafungin, and anidulafungin. Candins are mainly utilized in treating invasive *Candida* and *Aspergillus* infections (Perlin 2007). Candins repress the biosynthesis of ergosterol at various phases.

17.2.5 Allylamines, Thiocarbamates, and Morpholines

The allylamines (terbinafine) and thiocarbamates (tolnaftate) restrain a similar chemical, squalene epoxidase (Erg1), which illustrates to an early advance in ergosterol biosynthesis. The morpholines (fenpropimorph) hinder two various chemicals, Erg2 and Erg4, catalyzing sterol D14-reductase and D8-D7 isomerase, individually. Even though allylamines, thiocarbamates, and morpholines have an extensive activity spectra in contrast to fungal species, they are typically utilized as topical agents in treating dermatophyte diseases (Niewerth and Korting 2000).

The utilization of antifungal agents, particularly following recurrent or long-term treatments, results in unavoidable advancement of resistance. To establish the occurrence of resistance from these aggravates, the evaluation of antifungal susceptibility has been ordered utilizing various protocols. These protocols, in which fungal development is recorded in the presence of sequential medication dilutions over a limited period, produce a minimum inhibitory concentration (MIC) that is characterized as the least medication fixation bringing about a huge decrease of development. The MIC breakpoint values that are utilized to recognize resistant fungal isolates from susceptible isolates rely on few elements incorporating into in-vitro research facility or clinical perceptions.

17.3 Antifungal Resistance

Antifungal resistance by in vitro susceptibility testing in which MIC is estimated disproportionally in control cells and the organism, but that is yet helpless to drugs has the MIC exceeds the susceptibility breakpoint for that organism. Clinical resistance to fungal infection can be identified in vitro even though there is no microbial resistance from an antifungal agent (Espinel-Ingroff et al. 1997; Rex et al. 1997; Pfaller et al. 2008).

On the other hand, antifungal drug resistance is very critical because of the predetermined number of agents. Therefore, it is essential to comprehend the mechanisms of resistance to antifungal drug agents. In addition to this, the molecular comprehension of resistance mechanisms recognizes fungal genes, which would then be able to be utilized for resistance identification by molecular diagnostic tools. These genes and their related derivatives can encounter definite alternations in the advancement of resistance. Transcriptional administration of drug resistance genes is of remarkable concern since the modification of mechanism can be altered transitorily or constantly in fungal cells. Therefore, considering this, contemporary knowledge of molecular resistance mechanisms to antifungal agents was reviewed, however, by concentrating on transcriptional regulation of genes (Sanglard et al. 2009).

Exclusive therapeutic formulae for obtrusive fungal infections were amphotericin B deoxycholate and 5-fluorocytosine. Initially, the therapeutic substitutions originated with the establishment of itraconazole and fluconazole at the end of the 1980s. Progress in antifungal research over centuries evolved lipid formulation of amphotericin B, broad-spectrum triazoles, and exclusively novel antifungal agents. Adversely, excessive usage of triazoles in prophylactic and empiric treatments attributed drug-resistant pressure in both *Candida* and *Aspergillus* species (Perlin 2009). The consequences led to intrinsically resistant fungi or secondary resistant fungi, but the growth of the acquired fungi was not expedited in AIDS patients (Law et al. 1994). There was no horizontal resistance gene transfer technique acknowledged in fungi (Odds 2010). Presumptions have been made earlier that fungi were typically restricted to vertical gene transfer at a slower pace.

Antibiotic resistance in pathogenic fungi is a remarkable challenge in treating fungal infections triggered by these organisms. The equivalent exceptional preservation of the essential eukaryotic cell biology shown by fungal and animal cells that have enabled these littler eukaryotes to fill in as extraordinary model life form restricts the scope of growth explicit anti-infection agents that have been depicted. Moreover, mutant fungi are promptly isolated, both in the laboratory and in the clinic, that exhibit resistance from a broad scope of antibiotics besides the first utilized treatment. The wide choice of drug resistance is indicated as multidrug fight that shows up in microbes ranging from bacteria to the living beings. The restricted number of antifungal medications builds the phenotype an intense issue in the chemotherapeutic abolition of fungal parasites (Ling 1997). A large amount of multidrug resistance fungi have originated from research studies preferably from the yeast *Saccharomyces cerevisiae*, and the phenotype of this organism is indicated as PDR gene or pleiotropic drug resistance that influence the phenotype of PDR loci. With the advancement of ground-breaking, novel hereditary, and molecular biological procedures, specialists have conferred important understandings of information into the functioning of multidrug resistance from tests executed straightforwardly in infective organisms. This study concentrates on delivering an introduction to the various pathways poignant multidrug resistance in *S. cerevisiae* and by comparison of those pathways to infective growths, for instance, candida, fungus glabrata, and fungus genus fumigates.

The uncomplicated biology of *S. cerevisiae* persuaded the recognizable proof of a cistron arrangement of pleiotropic drug immune modification mapping to a cistron existing body VII that characterizes the PDR1 cistron (Rank et al. 1975). Pdr1p could be an atomic number 30 cluster containing positive transcriptional regulator known with the notable Gal4p interpretation issue (Balzi et al. 1987). Although Pdr1p was the primary multidrug resistance determinant recognized in *S. cerevisiae*, this issue is not an on the spot go-between of drug resistance. Pdr1p associated alternative transcriptional regulators change articulation of an assortment of proteins that demonstrate to counteract the virulent activity of medicine. Considering the proteins initially as an on the spot detoxifier of antifungal agents, we investigate the regulative transcription factors, and finally review the signals that regulate the expression of PDR genes. We are going to review a specific set of genes characterized by their regulation by Pdr1p and its homolog Pdr3p on the full because of the PDR pathway.

17.4 Membrane Transporters

The primarily recognized sequence that satisfied the factors through effector of drug resistance within the Pdr pathway was the PDR5 locus (Balzi et al. 1994; Bissinger and Kuchler 1994; Hirata et al. 1994). This sequence encodes associate degree of ATP-restricting transporter macromolecule that is private from the ABCG category of transporters (Dean and Allikmets 2001). Early investigations indicated that the loss of Pdr5p persuaded theatrical increment in drug sensitivity to a large scope of varied compounds (Leppert et al. 1990; Meyers et al. 1992). Moreover, direct organic chemistry tests showed that pdr5 cells were deficient within the flow of various dyes and radio-labelled probes (Kolaczkowski et al. 1996). Overproduction of Pdr5p by utilization of a high duplicate number of plasmids conveys the existence of hyperactive PDR1 alleles which is related to the PDR phenotype (Dexter et al. 1994; Katzmann et al. 1994; Leonard et al. 1994).

Together, this data firmly bolster the model that raised Pdr5p levels to feature the pleiotropic drug-resistant composition by increasing the movement of this multi-specific drug pump. Correlation of cells conveyancing hyperactive PDR1 alleles, and containing or lacking PDR5 contended that, although Pdr5p is a significant determinant within the PDR phenotype, the existence of this factor isn't adequate to clarify the whole range of drug resistance as seems. For example, overactive PDR1–6 mutants square measures soundproof to each cycloheximide and oligomycin. Expulsion of the PDR5 factor kills the enlarged cycloheximide resilience in every PDR1–6 cell, but does not decrease the abnormal state oligomycin opposition (Katzmann et al. 1994).

Screening a high-duplicate range cellular inclusion library for factors that impact oligomycin obstruction allows the convalescence of the YOR1 gene (Cui et al. 1996; Katzmann et al. 1994). YOR1 encrypts Associate in Nursing first principle transporter of the ABCC family that is essential for a normal oligomycin obstruction.

Loss of YOR1 from a PDR1–6 foundation diminishes oligomycin obstruction, but has no impact on the raised cycloheximide resilience given by this Pdr1p subsidiary (Katzmann et al. 1994). Hence, the evacuation of the PDR5 homolog SNQ2 factor from cells conveying a hyperactive gene of PDR1 diminished 4-nitroquinoline- N-oxide obstruction, however, did not impact cycloheximide or oligomycin opposition (Decottignies et al. 1995). This data serves to delineate a remarkable topic within the PDR composition in *S. cerevisiae*, even as contagious multidrug obstruction bushed all. Overrun of assorted first principle transporter proteins is needed for the statement of the complete scope of medication obstruction found in these multidrug-tolerant cells. This unremarkably happens due to Associate in Nursing adjustment within the movement of a translation issue and offers hit likeness with the multidrug obstruction found in class cells. Seclusion of the human MDR1 gene was cultivated by usage of cell lines that tremendously overproduce the multi-drug ABC transporter protein (Riordan et al. 1985; Roninson et al. 1986).

Most multidrug-safe mammalian cells seem to rise by means of enhancement of the gene cryptography of a specific ABC transporter gene (Gottesman et al. 1995 and Roninson 1992), instead of delivering a familiar interpretation factor that may, regardless of the careful mechanism, the two cells move towards becoming multidrug safe through the peak of ABC transporter articulation.

Alongside the first principle transporter proteins, proteins of the $64,000 assistant taxonomic group (MFS) likewise augment pleiotropic medicate opposition. Contrasted with the first principle transporter-encoding genes, our comprehension of the capability and guideline of the MFS proteins is at a previous stage. At any rate 20 distinctive MFS proteins show basic genes steady with or have simply been presented assume employment in drug resistance (Nelissen et al. 1997; Sá-Correia and Tenreiro 2002).

The large range of those film transporters proposes that their commitments to drug-resistant square measure doubtless unnoticed. The MFS proteins oft supply covering drug limpidity with first principle transporters, which can clarify to some extent why their better-plugged relatives have clouded MFS transporter inclusion in multidrug opposition, for example, the MFS supermolecule Flr1p adds to protection from the antifungal specialist fluconazole, even as cycloheximide (Alarco et al. 1997).

The first principle transporter Pdr5p to boot intervenes resilience to each of those mixes and may be a noteworthy determinant within the obstruction composition to those and completely different medications (Kontoyiannis 1999). Strikingly, whereas a pdr5p strain is utterly touchy to cycloheximide (Leppert et al. 1990), this affectability is often fully smothered if the interpretation issue Yap1p is overproduced (Dexter et al. 1994). This concealment is probably going due to the initiation of FLR1, a Yap1p target factor (Alarco et al. 1997). A lot of stays to be educated for the support of MFS proteins in eukaryotic multidrug opposition.

17.5 PDR Genes Regulated with Sphingolipid Homeostasis

The lipid synthesis of the plasma layer is the focal determinant managing entry of mixes from the outside condition to the inside of the cell. The conveyance of lipid segments in the inward and external handouts of the plasma film is unbalanced and inhibited by Pdr pathway to some extent. Together Pdr5p and Yor1p were discovered to upgrade the outward development (flop) of the phospholipid phosphatidylethanolamine (Decottignies et al. 1998; Pomorski et al. 2003). Phosphatidylethanolamine is regularly kept up at low dimensions in the external flyer by the fast internal development (flip) completed by aminophospholipid translocases, five of which can be found in the *S. cerevisiae* genome (Pomorski 2004).

Enigmatically, two phosphatidylinositol move protein homolog (PDR16 and PDR17) were additionally pragmatic to be the goal genes of the Pdr pathway and to impact phospholipid levels and medication obstruction (van den Hazel et al. 1999). These perceptions foresee that actuation of the Pdr pathway may trigger changes in phospholipid synthesis of the plasma layer, yet the results of these progressions stay dubious.

Eukaryotic film lipids incorporate sterols and sphingolipids into all the same phospholipids. In *S. cerevisiae*, steroid alcohol is the vital steroid within the cell and is delivered using the activity of work unit pathway (Sturley 2000). An excellent composition of various erg-invalid mutants is outrageous affectability to medication challenges that are reliable with a necessity for steroid alcohol in normal layer work. Examinations of work unit mutant strains demonstrate that Pdr5p transport action is unaffected in these mutants and suggest that loss of steroid alcohol could improve upstage dispersion of mixes over the changed layer (Emter et al. 2002; Kaur and Bachhawat 1999).

The last category of layer lipoid, sphingolipids, is assumed to attach with steroid alcohol to border microdomains known as lipoid rafts that area unit gathered within the external leaflet of the plasma film (Dickson and Lester 2002; Hannun and Obeid 2002). Investigation of the declaration of the IPT1 factor, secret writing of the last advance in sphingolipid synthesis (Dickson et al. 1997), designed up an instantaneous association between the Pdr pathway and synthesis of this category of film lipoid. IPT1 interpretation is unnatural by Pdr1p similarly as Pdr3p and reacts to sign notable to actuate the *S. cerevisiae* Pdr pathway. Strikingly, loss of IPT1 changed medication obstruction of the following mutants and proposes that normal sphingolipid substance is needed for wild-type dimensions of medication resistance (Hallstrom et al. 2001).

Later examinations gave proof that few biosynthetic advances upstream of the Ipt1p-catalyzed response were in addition accessible to increase the PDR pathway-intervened guidelines (Kolaczkowski et al. 2004). The hereditary associations between the Pdr and sphingolipid pathways were reached out by the finding that Pdr1p and Pdr3p manage the RSB1 factor (Kihara and Igarashi 2004; Panwar and Moye-Rowley 2006). Rsb1p is assumed to go about as Associate in the Nursing efflux of ceramide forerunners known as long-chain bases (LCBs) (Kihara and Igarashi 2002). LCBs will be cytotoxic whenever allowable to assemble, and Rsb1p will act to anticipate improper development of those sphingolipid intermediates. Shockingly, loss of Pdr5p and Yor1p from cells without ambiguity raises LCB obstruction (Kihara and Igarashi 2004) in Rsb1p-subordinate style, once more underlining the interconnections between sphingolipid synthesis and also the Pdr pathway. These varied interfaces between pleiotropic medicate opposition and also the equilibrium of layer lipids propose the probability that the physiological job of the Pdr pathway is to help guideline of the capability of the plasma film at the dimension of two lipids and film transporters.

17.6 Pleiotropic Drug Resistance Mutation

A focal determinant in the medication obstruction phenotype of *S. cerevisiae* is given by the guideline of trans-activation ability of a set number of authoritative proteins. Pdr1p and its homolog Pdr3p are Zn2Cys6-containing transcriptional administrative proteins that apply significant effects on the multidrug opposition phenotypes of cells. There are countless Zn2Cys6-containing interpretation factors in *S. cerevisiae* (p50), and a considerable lot of these have been appeared or are accepted to be engaged with pleiotropic drug-resistant (MacPherson et al. 2006).

We will concentrate on Pdr1p and Pdr3p as illustrative of the more significant number of Zn2Cys6-containing factors that add to medicate opposition. Progressively comprehensive contemplations of these and other transcriptional administrative proteins engaged with multidrug obstruction in *S. cerevisiae* are accessible in a few surveys (Fardeau et al. 2007; MacPherson et al. 2006; Moye-Rowley 2003a, b).

As referenced above, hyperactive mutant types of Pdr1p drove the underlying ID of the pleiotropic medicate obstruction phenotype in *S. cerevisiae* (Balzi et al. 1987; Carvajal 1997). Comparable mutant alleles of PDR3 have additionally been portrayed (Nourani et al. 1997). These single amino corrosive substitution types of Pdr1p and Pdr3p produce transcriptional administrative proteins that act as solid, constitutive activators of downstream gene articulation (Carvajal 1997; Nourani et al. 1997). Pdr1p and Pdr3p, both ties to components mentioned to as Pdr1p/Pdr3p reaction components (PDREs), found upstream of target genes (Delahodde et al. 1995; Katzmann et al. 1994). In vivo foot printing trials show that Pdr1p and Pdr3p are probably going to be constitutively bound to significant PDREs (Fardeau et al. 2007; Mamnun et al. 2002), a perception predictable with the constitutive atomic confinement of these proteins (Delahodde et al. 2001). Furthermore, both Pdr1p and Pdr3p can initiate the statement of another zinc bunch interpretation factor-encoding gene called YRR1 (Cui et al. 1998; Zhang et al. 2001). Expanded articulation of Yrr1p can enhance the transcriptional impacts of enactment of either Pdr1p or Pdr3p since this factor perceives a grouping unique about the PDRE (Le Crom et al. 2002).

Even though Pdr1p and Pdr3p share critical similarities, significant disparities are known. To start with, these elements are communicated at significantly various dimensions. Utilization of epitope-tagged types of the two proteins demonstrates that Pdr1p is available at almost multiple times the dimension of Pdr3p (Ghaemmaghami et al. 2003). Second, the guideline of these elements is receptive to various sign. Overproduction of the DnaK protein Ssz1p (Hallstrom et al. 1998) or the DnaJ Zuo1p persuades Pdr1p-subordinate gene interpretation yet has no impact on Pdr3p. Then again, cells coming up short on their mitochondrial genome (p0) enact Pdr3p work, however, have no impact on Pdr1p (Hallstrom and Moye-Rowley 2000a, b). PDR1 articulation levels are steady, though PDR3 is both auto-regulated and highly initiated in p0 cells (Delahodde et al. 2001; Hallstrom and Moye-Rowley 2000a, b). At long last, ongoing work from our research facility shows the Hsp70 protein Ssa1p can adversely direct Pdr3p, however not Pdr1p movement (107a). Even though these two translation elements share greater than 30% personality over their approximately 1000-amino-corrosive lengths (Delaveau et al. 1994), these distinctions show that Pdr1p and Pdr3p have non-identical tasks to carry out in the control of multidrug opposition.

The below-average of translation issue that has been connected with the rule of thumb of multidrug opposition is the basic region-leucine zipper (bZip) group of regulative macromolecules. Albeit a number of these bZip-containing variables are offered in *S. cerevisiae*, we are going to confine our exchange to Yap1p, the primary of those parts incontestable to be related to pleiotropic medicate opposition. Yap1p is healthier glorious for its important job in aerobic pressure resilience

(Moye-Rowley 2003a, b; Paget and Buttner 2003; Rodrigues-Pousada et al. 2004), and its guideline by oxidants has been the subject of escalated research. Quickly, Yap1p typically cycles between the core and the cytoplasm without stress (Kuge 1997).

Upon oxidant challenge, Yap1p quickly gathers in the core, where it actuates cancer prevention agent gene articulation (Coleman et al. 1999; Kuge and Jones 1994; Kuge 1997; Wu and Moye-Rowley 1994). Mutant types of Yap1p have been portrayed that are constitutively situated in the core and hyper-impervious to specific oxidants (Coleman et al. 1999; Kuge 1997; Wemmie et al. 1997).

Less is known about the reaction intervened by Yap1p upon medication challenge, yet the YAP1 gene was segregated as a high-duplicate number middle person of pleiotropic sedate obstruction alongside PDR5 (Leppert et al. 1990). As referenced above, Yap1p characterizes a pathway for drug resistance parallel to that of PDR5. Strikingly, Yap1p is known to initiate the outflow of any rate two unique MFS protein-encoding genes: ATR1 (Coleman et al. 1997) and FLR1 (Alarco et al. 1997). At any rate in *S. cerevisiae*, the control of multidrug opposition gene articulation has all the earmarks of being isolated between zinc group holding factors principally lashing the translation of ABC transporter-encoding genes, while bZip-containing proteins primarily act by controlling mRNA dimensions of MFS proteins. The method of reasoning hidden this division of transcriptional administrative circuits stays to be resolved.

17.6.1 *Candida albicans*

While the incredible hereditary genes related with the utilization of *S. cerevisiae* as a model parasite have permitted the rapid advancement in our comprehension of contagious multidrug opposition, this living being is certainly not a critical reason for human illness. The major parasitic wellspring of circulation system contaminations in people is from the family Candida which has developed in significance until candidemia now speaks to the fourth most regular nosocomial disease (Slavin et al. 2004). The essential *Candida* species related to candidemia is *Candida albicans* (Pfaller et al. 2001), and this life form has been the most seriously contemplated as far as multidrug opposition.

The first multidrug opposition gene from *C. albicans* was recuperated by choice of a section of *C. albicans* genomic DNA that modified medication obstruction in *S. cerevisiae* when carried on a proper plasmid (Ben-Yaacov et al. 1994). The primary gene confined in this style was assigned BENr and is presently cited to as MDR1 (Goldway et al. 1995). The protein encrypted by this particular gene is an individual from the significant facilitator superfamily of film transporters (Marger and Saier 1993). Not long after this discovering, two genes encoding homologs of the *S. cerevisiae* Pdr5p ABC transporter protein were distinguished (Prasad et al. 1995; Sanglard et al. 1997). These proteins were given the abbreviations CDR1 and CDR2 for Candida drug-resistant 1 and 2. Investigations in a few laboratories built up that mutant *C. albicans* strains coming up short on these ABC transporters were multidrug sensitive (Sanglard et al. 1996, 1997) and that various distinctive safe segregates were found to overproduce CDR1 and CDR2 transcripts (Sanglard et al. 1995; White 1997). Estimations of medication transport in strains designed to overproduce Cdr1p assistance with the possibility that this protein goes about as an ATP-subordinate medication efflux pump (Nakamura et al. 2001). Green fluorescent protein combinations to Cdr1p showed that this protein was fundamentally found in the plasma film, like the area of Pdr5p in *S. cerevisiae* (Shukla 2004).

Later work has concentrated on the control of translation of CDR1 and CDR2. Trials from two distinct research centers gave a portrayal of the advertiser area of CDR1 (de Micheli et al. 2002; Puri et al. 1999). CDR1 transcriptional control was observed to be intricate and the result of different DNA components of either positive or negative nature in the advertiser locale. Investigation of the CDR2 advertiser proposes that this gene is managed in parallel with CDR1 since azole-safe *C. albicans* disconnects regularly display raised mRNA levels relating to both of these ABC transporter-encoding genes (de Micheli et al. 2002).

A few distinctive transcriptional administrative proteins have been involved in the tweak of CDR1 and CDR2 articulation. The best described of these is the Zn2Cys6 group protein Tac1p; this factor ties to a solitary component in the CDR2 advertiser that contains pair rehashes of a CGG succession (Coste et al. 2006). These short trinucleotide rehashes are usually connected with the coupling locales of Zn2Cys6 bunch proteins (MacPherson et al. 2006). Tac1p shows the most noteworthy level of grouping likeness with a protein from *S. cerevisiae* assigned Hal9p, a figure included the control of articulation of the ENA1 sodium-potassium ATPase (Mendizabal et al. 1998).

Strikingly, the existence of a CDR2-lacZ combination in *S. cerevisiae* does not persuade the creation of critical beta-galactosidase action except if Tac1p is heterologous given (Coste et al. 2006), recommending that Hal9p cannot invigorate CDR2 articulation. Even though Pdr1p shares huge grouping similitude with Tac1p, there are clear contrasts between these two Zn2Cys6 bunch proteins at the dimension of DNA restricting particularity and protein arrangement. An intriguing shared characteristic somewhere often caused by gain-of-function mutations in the transcription factors of PDR1 and TAC1 result in profound responsive of target genes that these two significantly control multidrug efflux pumps and sway in their specific life forms.

In *S. cerevisiae*, PDR1 active mutants firmly persuade PDR5 and PDR15, even as another layer super molecule known as Rsb1p that is needed for cover from the long-chain base phytosphingosine (DeRisi et al. 2000). In *C. albicans*, TAC1 active alleles actuate CDR1 (Pdr5p homolog), CDR2 (Pdr15p homolog), and RTA3 (Rsb1p homolog).

A later optimistic controller of CDR1 was seen by viewing a *C. albicans* articulation library in a very *S. cerevisiae* cell conveyance of title a CDR1-lacZ journalist factor for clones that would raise articulation of this heterologous correspondent factor (Chen et al. 2004). This factor was assigned CaNDT80 since it had been found to cypher a homolog of the *S. cerevisiae* Ndt80p translation issue. ScNdt80p serves to enact genes engaged with monogenesis in *S. cerevisiae* (Chu and

Herskowitz 1998). CaNdt80p appears to possess much separated from the work of its *S. cerevisiae* connected super molecule since the *C. albicans* issue is important in the positive guideline of CDR1 throughout vegetative development.

A demonstration involving a significant job for a negative controller of CDR1 articulation has likewise gathered (Gaur et al. 2004). Until this point in time, the character of this factor is as yet dubious, however, its coupling site has been mapped to a component found downstream of the Tac1p acknowledgment component (Gaur et al. 2005). A 55-kDa protein has been demonstrated to be cross-connected to this negative administrative component, and decreases in the dimension of this factor are accepted to cause azole hyper-obstruction in some clinical disconnects. Together, this information proposes that control of articulation of CDR1 in *C. albicans* includes greater unpredictability at the dimension of the translation elements contrasted with the control of ScPDR5 articulation that is reliant on just Pdr1p and Pdr3p (Delaveau et al. 1994; Katzmann et al. 1994).

A few research facilities have additionally analyzed transcriptional control of CaMDR1. Point by point cancellation mapping analyses have been portrayed in a few productions (Harry et al. 2005; Hiller et al. 2006; Riggle and Kumamoto 2006; Rognon et al. 2006). These examinations have distinguished a few unique areas in the CaMDR1 advertiser that are engaged with basal, oxidant- or medication-incited articulation. There are contrasts in the exact jobs of these different components yet unmistakably the CaMDR1 advertiser is a complex transcriptional control district that coordinates various contributions to decide the correct articulation of CaMdr1p.

At any rate, two unique components have been connected to the trans-regulation of CaMDR1 articulation. Detachment of the *C. albicans* Yap1p homolog (Cap1p) demonstrated that CaMDR1 was a possible downstream focus of this translation factor (Alarco and Raymond 1999). This has been upheld by direct mutagenesis of putative Cap1p administrative components present in CaMDR1 (Rognon et al. 2006). Shockingly, different examinations exhibited that loss of CAP1 either had no impact on azole obstruction or contrarily affected resilience to this medication (Alarco and Raymond 1999). In any case, overproduction of a carboxy-terminal truncation mutant of Cap1p persuaded an emotional increment in azole resistance, a perception that corresponds with emphatically raised CaMDR1 translation (Alarco and Raymond 1999). This conduct is likely seen in *S. cerevisiae* when YAP1 is erased as yap1 cells display a humble (twofold) increment while overproduction of Yap1p created a striking increment in the MIC for fluconazole (Chen et al. 2007).

In conjunction with Cap1p, the *C. albicans* homolog of *S. cerevisiae* Mcm1p has additionally been involved responsible for CaMDR1 interpretation (Riggle and Kumamoto 2006; Rognon et al. 2006). Research facility strains of *C. albicans* that were chosen for high azole opposition were observed to be subject to the existences of a useful CaMcm1p restricting site for abnormal state generation of CaMDR1. Cap1p was not required for this impact contending that CaMcm1p is essential if not the sole factor that is upregulated in these azole-safe separates. Azole obstruction has likewise been demonstrated to be affected by changes in articulation of the *C. albicans* IPT1 gene (Prasad et al. 2005), proposing further protection with *S. cerevisiae*. The image that rises out of concentrates in *C. albicans* on the atomic premise of multidrug opposition is steady with this creature sharing a fundamentally the same as the scope of effector genes with *S. cerevisiae*, however with the guideline of gene articulation showing contrasts. This might be because of the altogether different milieus wherein these life forms ordinarily live. Further investigation of the opposition pathways in these yeasts will explain this thought.

17.6.2 *Candida glabrata*

C. glabrata is a haploid type of Candida that has risen to be mainly the second regular Candida living being related to fungemia (Pfaller et al. 2001). A reasonable contributing element to the quick development (2% during the 1970s to 20% currently) is the strong capacity of *C. glabrata* to get resistance to usually conveyed antifungal operators, for example, azoles. Long haul observing of *C. glabrata* related sickness shows that azole-safe disengages are expanding in recurrence, even from geographic districts in which *C. glabrata* was initially sensitive to these medications (Pfaller and Diekema 2007).

This is not found on account of *C. albicans* since clinical information show that azole medications hold their viability in controlling ailment related to this species (Hazen et al. 2003). A moment muddling factor for *C. glabrata* initiates from the routine multidrug opposition of azole-tolerant separates (Sanguinetti et al. 2005). Not exclusively does *C. glabrata* moderately and effectively convert to an azole-safe structure; however, it adds much of the time seems to turn out to be at the same time multidrug safe.

Protein and DNA progression likeness assessments demonstrate that *C. glabrata* and *S. cerevisiae* are immovably related living things (Wong et al. 2002). This similarity also speaks to this comfortable relationship in the multidrug deterrent pathways in these two yeasts. As depicted above for *S. cerevisiae* and *C. glabrata,* cells are very multidrug safe (Sanglard et al. 2001). This mitochondrial control of multidrug resistance proceeds through the authorization of the CgPdr1p protein (Tsai et al. 2006). An amazing complexity between *S. cerevisiae* and *C. glabrata* is that CgPdr1p addresses indisputably the best homolog of the Pdr1p/Pdr3p, a lot of proteins identified in *S. cerevisiae* (Vermitsky and Edlind 2004). CgPDR1 may contrast with a blend quality between the *S. cerevisiae* PDR1 and PDR3 loci. CgPDR1 shares greater course of action closeness with ScPdr1p; be that as it may, demonstrates the selection seen for ScPDR3 (Tsai et al. 2006).

Single amino destructive substitution freak kinds of CgPdr1p have been found that are associated with irregular state elucidation of both CgPDR1 and CgCDR1, similarly as incredible multidrug block (Vermitsky and Edlind 2004). This lead is clearly for all intents and purposes proportionate to that watched for the hyperactive alleles of ScPDR1 and ScPDR3.

Despite the fact that indisputably *C. glabrata* cells are exceedingly multidrug safe and happen a great part of the time, the activity of these cells in illness is more uncertain. *C. glabrata* clinical segregates that were changed over to petite status by either ethidium bromide or fluconazole treatment were found to have decreased damaging tendency in a mouse model of infectious pathogenesis (Brun et al. 2005). One stress with these assessments (noted by the makers of that audit) is that the evaluation of ruinous tendency was finished without the association of fluconazole. Since the nonappearance of drug assurance would clear the possible specific good position given by the unusual state verbalization of the CgCDR qualities and related multidrug resistance, the petite cells may be lost due to contention from the strong endogenous microbial verdure. There was proof of modest *C. glabrata* strains isolated from patients (Bouchara et al. 2000); in any case, a greater number of other clinical segregates are not clearly associated with a petite character (Shahi et al. 2007).

Despite the fact that during examination the modest strains of *C. glabrata* and *S. cerevisiae* show a multidrug-safe phenotype demonstrates the similar mitochondrial regulatory reason in these living things, an intriguing complexity has been uncovered by screening a collection of transposon-delivered *C. glabrata* freaks (Kaur et al. 2004). Expansion of a transposon into a couple of qualities was found to raise azole impediment yet, additionally, trigger petite course of action. Exactly when the specific weight was removed, the petite phenotype appeared to die down. Past examinations on both *C. glabrata* and *S. cerevisiae* petite freaks have not uncovered multidrug-safe freaks that show reversible direct (Sanglard et al. 2001; Zhang et al. 2001). One part of the *C. glabrata* petite multidrug-safe freaks that residual parts ill defined is the status of their mitochondrial genome.

The alleged high-repeat azole-safe freaks were displayed to do not have a mitochondrial genome (Sanglard et al. 2001) and eagerly take after the multidrug-safe freaks of *S. cerevisiae*. Three of the transposon-started *C. glabrata* petite freaks were reviewed for mitochondrial genome status (Kaur et al. 2004) and intriguingly found to at present hold cytoplasmic nucleoids. This is a critical potential refinement since different assorted *S. cerevisiae* mutants that are petite notwithstanding to express the multi-drug safe phenotype expression (Zhang et al. 2001).

The petite transposon expansion freaks in *C. glabrata* fuse consideration in the CgSHE9 quality. Aggravation freaks lacking SHE9 also called MDM33 (Messerschmitt et al. 2003) appear to bomb in the mitochondrial genome in neither *S. cerevisiae* nor *C. glabrata*; in any case, multidrug check has quite recently been studied for *C. glabrata* freaks. Further work is required to choose if the related *S. cerevisiae* freaks in like manner demonstrate the reversible multidrug-safe status of the three *C. glabrata* qualities. As suggested previously (Kaur et al. 2004), freaks that show a moderate reducing of mitochondrial work barring all out harm of mitochondrial genome could offer climb to this reversible multidrug.

Freaks of this sort would have clear central focuses in regard to the survival of chemotherapy in a patient since they could join the benefits of the overwhelming multidrug obstacle of a petite cell without always getting the touchy improvement of these mitochondrially inadequate with regard to strains. As depicted above for *S. cerevisiae*, multidrug restriction in *C. glabrata* is furthermore influenced by transcriptional control of MFS protein verbalization. The CgFLR1 quality is convinced by the *C. glabrata* Yap1p homolog (CgAP-1) and gives assurance from an extent of pros, including fluconazole (Chen et al. 2007). *C. glabrata* Cgap-1 freaks had run of the mill fluconazole obstruction, while Cgpdr1 freaks were unstable to this medicine.

The Yap1p-FLR1 fluconazole restriction pathway seems to serve a subordinate occupation in protection from this antifungal administrator in both *S. cerevisiae* and *C. glabrata*. Somewhat, incredibly, while Cgap-1 freaks are too much sensitive to oxidants, there is no perceptible effect on hurtfulness (Chen et al. 2007). Despite the fact that CgAP-1 is a noteworthy determinant of oxidative weight impediment, this does not appear to affect the limit of *C. glabrata* to colonize a mouse model.

The basic highlights between multidrug-safe separates of *S. cerevisiae* and *C. glabrata* have been extended by assessment of qualities that are transcriptionally open to a hyperactive freak sort of CgPdr1p by microarray (Vermitsky et al. 2006). While different qualities that are affected in *C. glabrata* address homologs that are in like manner brought up in light of *S. cerevisiae* hyperactive Pdr1p, the greatest social event of actuated transcripts contrasts with qualities that are strangely impelled in the pathogenic yeast (Vermitsky et al. 2006). These *C. glabrata* express qualities may address loci that are required for the living form to adequately withstand the troubles of hurtful blends, when revealed in an animal have.

17.6.3 *Aspergillus* Species

The fundamental human pathogen among the filamentous parasites is *Aspergillus fumigatus* (Richardson 2005). Pollutions identified with this living thing have a high distressingness and normally high security from the standard system of antifungal experts (Pfaller and Diekema 2004). As anyone might expect, the utilization of azole-based medications has the irksome consequence of raising the obstruction of ensuing disconnection from the patients. Fortunately, getting off a multidrug deterrent phenotype is commonly remarkable among *Aspergillus* species, nonetheless, this declaration must be qualified by the affirmation that assessment of this phenotype is tangled by specific inconveniences in setting up this phenotype (Moore et al. 2000). Additionally, the nuclear depiction of qualities related to *Aspergillus* multidrug restriction has been demolished by the nonappearance of an absolute genomic progression. This starting has late been developed (Nierman et al. 2005) and will stimulate examination of the loci related to medication opposition in this critical human pathogen.

Past examinations have given a few information concerning multidrug resistance qualities in *Aspergillus*. Degeneration of PCR cloning allowed the disconnection of two different ABC transporter-encoding qualities, and explanation of one of these in *S. cerevisiae* raised security from an antifungal drug of the echinocandin family (Tobin et al. 1997). Azole-safe withdraws picked in-vitro were found to have raised explanation of ABC transporter and MFS-encoding qualities (Nascimento et al. 2003). This data is enduring with similar systems for medicine detoxification existing in *Aspergillus* as have been portrayed in yeasts.

The availability of the genomic course of action of *Aspergillus* has quite recently given interesting bits of information into the comprehensible bit of multidrug resistance structures in this filamentous parasite. The *A. fumigatus* genome is half greater than that of *S. cerevisiae*, however then predicts the presence of 96 potential MFS multidrug transporters stood out from 24 from growing yeast (Da Silva Ferreira et al. 2005). Though *S. cerevisiae* contains 13 ABC transporter genes, the species look for a multi-drug obstacle, likewise, the Aspergillus is acknowledged to have 35 ABC transporter genes. Assessments of these proteins in *A. fumigatus* are required to avow that these proteins are related with medication obstruction in this living thing, notwithstanding, the presence of such a generally greater number proposes a hardening of the defensive drug limit of this pathogenic animal. It is striking to observe that the extended degree of MFS proteins diverged from ABC transporters found in the *Aspergillus* genome stood out from *S. cerevisiae* takes after the condition in pathogenic microorganisms. At present, there are no cases of ABC transporters in clinically material microorganism pathogens that are noteworthy in drug hindrance, while MFS protein collaboration in the shirking of antibacterial treatment is incredibly typical (Piddock 2006).

17.7 Conclusion

Multidrug restriction is an issue in chemotherapy in the conditions going from bacterial defilements to harmful development. The infectious prescription block is an especially extreme issue on account of the set number of antifungal blends (Kontoyiannis and Lewis 2002). Understanding the rule and limit of multidrug restriction pathways in parasites is still especially work in progression, be that as it may, its hugeness continues creating with the extending number of immunocompromised patients worldwide and their growing reliance on chemotherapy to control infectious maladies (Perlroth et al. 2007; Warnock 2006). A considerable amount of our present appreciation of the sub-nuclear reason for parasitic multidrug restriction springs from work in the ordinarily non-pathogenic yeast *S. cerevisiae*. Regardless, fast advances in the preliminary tractability of pathogenic parasites, for instance, the *Candida* and *Aspergillus* species will acquire the pathways with clinically huge life structures into center intrigue. This is a critical goal since the systems used for medicine impediment in these pathogens share some equivalence with those in *S. cerevisiae* yet have noteworthy differences that must be uncovered in the neighborhood living being.

References

Adler A, Yaniv I, Samra Z et al (2006) Exserohilum: an emerging human pathogen. Eur J Clin Microbiol Infect Dis 25:247–253. https://doi.org/10.1007/s10096-006-0093-3

Alarco AM, Raymond M (1999) The bZip transcription factor Cap1p is involved in multidrug resistance and oxidative stress response in Candida albicans. J Bacteriol 181:700–708

Alarco A-M, Balan I, Talibi D et al (1997) AP1-mediated multidrug resistance in Saccharomyces cerevisiae requires FLR1 encoding a transporter of the major facilitator superfamily. J Biol Chem 272:19304–19313. https://doi.org/10.1074/jbc.272.31.19304

Balzi E, Chen W, Ulaszewski S et al (1987) The multidrug resistance gene PDR1 from Saccharomyces cerevisiae. J Biol Chem 262:16871–16879

Balzi E, Wang M, Leterme S et al (1994) PDR5, a novel yeast multidrug resistance conferring transporter controlled by the transcription regulator PDR1. J Biol Chem 269:2206–2214

Barnes PD, Marr KA (2006) Aspergillosis: spectrum of disease, diagnosis, and treatment. Infect Dis Clin N Am 20:545–561. https://doi.org/10.1016/j.idc.2006.06.001

Bastos de Lima-Barros M, de Almeida-Paes M, Oliveira-Schubach A (2011) Sporothrix schenckii and Sporotrichosis. Clin Microbiol Rev 24:633–654. https://doi.org/10.1128/CMR.00007-11

Ben-Yaacov R, Knoller S, Caldwell GA et al (1994) Candida albicans gene encoding resistance to benomyl and methotrexate is a multidrug resistance gene. Antimicrob Agents Chemother 38:648–652. https://doi.org/10.1128/aac.38.4.648

Bissinger PH, Kuchler K (1994) Molecular cloning and expression of the *Saccharomyces cerevisiae* STS1 gene product. A yeast ABC transporter conferring mycotoxin resistance. J Biol Chem 269(6):4180–4186

Bouchara J-P, LE Boudouil S, Filmon R et al (2000) In-vivo selection of an azole-resistant petite mutant of Candida glabrata. J Med Microbiol 49:977–984. https://doi.org/10.1099/0022-1317-49-11-977

Bradsher RW, Chapman SW, Pappas PG (2003) Blastomycosis. Infect Dis Clin N Am 17:21–40, vii

Brun S, Dalle F, Saulnier P et al (2005) Biological consequences of petite mutations in Candida glabrata. J Antimicrob Chemother 56:307–314. https://doi.org/10.1093/jac/dki200

Canuto MM, Rodero FG (2002) Antifungal drug resistance to azoles and polyenes. Lancet Infect Dis 2(9):550–563

Carvajal E (1997) Molecular and phenotypic characterization of yeast PDR1 mutants that show hyperactive transcription of various ABC multidrug transporter genes. Mol Gen Genet MGG 256:406–415. https://doi.org/10.1007/s004380050584

Casadevall A, Perfect JR (1998) Cryptococcus neoformans. American Society of Microbiology, Washington

Chen C-G, Yang Y-L, Shih H-I et al (2004) CaNdt80 is involved in drug resistance in Candida albicans by regulating CDR1. Antimicrob Agents Chemother 48:4505–4512. https://doi.org/10.1128/AAC.48.12.4505-4512.2004

Chen K-H, Miyazaki T, Tsai H-F, Bennett JE (2007) The bZip transcription factor Cgap1p is involved in multidrug resistance and required for activation of multidrug transporter gene CgFLR1 in Candida glabrata. Gene 386:63–72. https://doi.org/10.1016/j.gene.2006.08.010

Chu S, Herskowitz I (1998) Gametogenesis in yeast is regulated by a transcriptional cascade dependent on Ndt80. Mol Cell 1:685–696. https://doi.org/10.1016/S1097-2765(00)80068-4

Coleman ST, Tseng E, Moye-Rowley WS (1997) Saccharomyces cerevisiae basic region-leucine zipper protein regulatory networks converge at the ATR1 structural gene. J Biol Chem 272:23224–23230. https://doi.org/10.1074/jbc.272.37.23224

Coleman ST, Epping EA, Steggerda SM, Moye-Rowley WS (1999) Yap1p activates gene transcription in an oxidant-specific fashion. Mol Cell Biol 19:8302–8313. https://doi.org/10.1128/mcb.19.12.8302

Coste A, Turner V, Ischer F et al (2006) A mutation in Tac1p, a transcription factor regulating CDR1 and CDR2, is coupled with loss of heterozygosity at chromosome 5 to mediate antifungal resistance in *Candida albicans*. Genetics 172(4):2139–2156

Cui Z, Hirata D, Tsuchiya E et al (1996) The multidrug resistance-associated protein (MRP) subfamily (Yrs1/Yor1) of Saccharomyces cerevisiae is important for the tolerance to a broad range of organic anions. J Biol Chem 271:14712–14716. https://doi.org/10.1074/jbc.271.25.14712

Cui Z, Shiraki T, Hirata D, Miyakawa T (1998) Yeast gene YRR1, which is required for resistance to 4-nitroquinoline N-oxide, mediates transcriptional activation of the multidrug resistance transporter gene SNQ2. Mol Microbiol 29:1307–1315. https://doi.org/10.1046/j.1365-2958.1998.01027.x

Da Silva Ferreira ME, Colombo AL, Paulsen I et al (2005) The ergosterol biosynthesis pathway, transporter genes, and azole resistance in Aspergillus fumigatus. Med Mycol 43:313–319. https://doi.org/10.1080/13693780400029114

de Micheli M, Bille J, Schueller C, Sanglard D (2002) A common drug-responsive element mediates the upregulation of the Candida albicans ABC transporters CDR1 and CDR2, two genes involved in antifungal drug resistance. Mol Microbiol 43:1197–1214. https://doi.org/10.1046/j.1365-2958.2002.02814.x

Dean M, Allikmets R (2001) Complete characterization of the human ABC gene family. J Bioenerg Biomembr 33(6):475–479

Decottignies A, Lambert L, Catty P et al (1995) Identification and characterization of SNQ2, a new multidrug ATP binding cassette transporter of the yeast plasma membrane. J Biol Chem 270:18150–18157. https://doi.org/10.1074/jbc.270.30.18150

Decottignies A, Grant AM, Nichols JW et al (1998) ATPase and multidrug transport activities of the overexpressed yeast ABC protein Yor1p. J Biol Chem 273:12612–12622. https://doi.org/10.1074/jbc.273.20.12612

Delahodde A, Delaveau T, Jacq C (1995) Positive autoregulation of the yeast transcription factor Pdr3p, which is involved in control of drug resistance. Mol Cell Biol 15:4043–4051. https://doi.org/10.1128/mcb.15.8.4043

Delahodde A, Pandjaitan R, Corral-Debrinski M, Jacq C (2001) Pse1/Kap121-dependent nuclear localization of the major yeast multidrug resistance (MDR) transcription factor Pdr1. Mol Microbiol 39:304–313. https://doi.org/10.1046/j.1365-2958.2001.02182.x

Delaveau T, Delahodde A, Carvajal E et al (1994) PDR3, a new yeast regulatory gene, is homologous toPDR1 and controls the multidrug resistance phenomenon. Mol Gen Genet 244:501–511. https://doi.org/10.1007/BF00583901

DeRisi J, van den Hazel B, Marc P et al (2000) Genome microarray analysis of transcriptional activation in multidrug resistance yeast mutants. FEBS Lett 470:156–160. https://doi.org/10.1016/S0014-5793(00)01294-1

Dexter D, Moye-Rowley WS, Wu AL, Golin J (1994) Mutations in the yeast PDR3, PDR4, PDR7 and PDR9 pleiotropic (multiple) drug resistance loci affect the transcript level of an ATP binding cassette transporter encoding gene, PDR5. Genetics 136:505–515

Dickson RC, Lester RL (2002) Sphingolipid functions in Saccharomyces cerevisiae. Biochim Biophys Acta Mol Cell Biol Lipids 1583:13–25. https://doi.org/10.1016/S1388-1981(02)00210-X

Dickson RC, Nagiec EE, Wells GB et al (1997) Synthesis of mannose-(inositol-P) 2-ceramide, the major sphingolipid in Saccharomyces cerevisiae, requires the IPT1 (YDR072c) gene. J Biol Chem 272:29620–29625. https://doi.org/10.1074/jbc.272.47.29620

Drabick JJ, Gomatos PJ, Solis JB (1990) Cutaneous cladosporiosis as a complication of skin testing in a man positive for human immunodeficiency virus. J Am Acad Dermatol 22:135–136. https://doi.org/10.1016/S0190-9622(08)80019-9

Emter R, Heese-Peck A, Kralli A (2002) ERG6 and PDR5 regulate small lipophilic drug accumulation in yeast cells via distinct mechanisms. FEBS Lett 521(1-3):57–61

Espinel-Ingroff A, Bartlett M, Bowden R et al (1997) Multicenter evaluation of proposed standardized procedure for antifungal susceptibility testing of filamentous fungi. J Clin Microbiol 35(1):139–143

Fardeau V, Lelandais G, Oldfield A et al (2007) The central role of PDR1 in the foundation of yeast drug resistance. J Biol Chem 282:5063–5074. https://doi.org/10.1074/jbc.M610197200

Furcolow ML, Busey JF, Menges RW, Chick EW (1970) Prevalence and incidence studies of human and canine blastomycosis II. Yearly incidence studies in three selected states, 1960–1967. Am J Epidemiol 92:121–131. https://doi.org/10.1093/oxfordjournals.aje.a121184

Gaur N, Puri N, Karnani N et al (2004) Identification of a negative regulatory element which regulates basal transcription of a multidrug resistance gene of. FEMS Yeast Res 4:389–399. https://doi.org/10.1016/S1567-1356(03)00204-6

Gaur NA, Manoharlal R, Saini P et al (2005) Expression of the CDR1 efflux pump in clinical Candida albicans isolates is controlled by a negative regulatory element. Biochem Biophys Res Commun 332:206–214. https://doi.org/10.1016/j.bbrc.2005.04.113

Gehrt A, Peter J, Pizzo PA et al (1995) Effect of increasing inoculum sizes of pathogenic filamentous fungi on MICs of antifungal agents by broth microdilution method. J Clin Microbiol 33(5):1302–1307

Ghaemmaghami S, Huh W-K, Bower K et al (2003) Global analysis of protein expression in yeast. Nature 425:737–741. https://doi.org/10.1038/nature02046

Goldway M, Teff D, Schmidt R et al (1995) Multidrug resistance in Candida albicans: disruption of the BENr gene. Antimicrob Agents Chemother 39:422–426. https://doi.org/10.1128/aac.39.2.422

Gottesman MM, Hrycyna CA, Schoenlein PV et al (1995) Genetic analysis of the multidrug transporter. Annu Rev Genet 29:607–649. https://doi.org/10.1146/annurev.ge.29.120195.003135

Gugnani HC, Sood N, Singh B, Makkar R (2000) Case report. Subcutaneous phaeohyphomycosis due to Cladosporium cladosporioides. Mycoses 43:85–87. https://doi.org/10.1046/j.1439-0507.2000.00545.x

Hainer BL (2003) Dermatophyte infections. Am Fam Physician 67(1):101–110

Hallstrom TC, Moye-Rowley WS (2000a) Hyperactive forms of the Pdr1p transcription factor fail to respond to positive regulation by the Hsp70 protein Pdr13p. Mol Microbiol 36:402–413. https://doi.org/10.1046/j.1365-2958.2000.01858.x

Hallstrom TC, Moye-Rowley WS (2000b) Multiple signals from dysfunctional mitochondria activate the pleiotropic drug resistance pathway in Saccharomyces cerevisiae. J Biol Chem 275:37347–37356. https://doi.org/10.1074/jbc.M007338200

Hallstrom TC, Katzmann DJ, Torres RJ et al (1998) Regulation of transcription factor Pdr1p function by an Hsp70 protein in Saccharomyces cerevisiae. Mol Cell Biol 18:1147–1155. https://doi.org/10.1128/mcb.18.3.1147

Hallstrom TC, Lambert L, Schorling S et al (2001) Coordinate control of sphingolipid biosynthesis and multidrug resistance in Saccharomyces cerevisiae. J Biol Chem 276:23674–23680. https://doi.org/10.1074/jbc.M101568200

Hannun YA, Obeid LM (2002) The Ceramide-centric universe of lipid-mediated cell regulation: stress encounters of the lipid kind. J Biol Chem 277:25847–25850. https://doi.org/10.1074/jbc.R200008200

Harris JR, Balajee SA, Park BJ (2010) Pneumocystis jirovecii pneumonia: current knowledge and outstanding public health issues. Curr Fungal Infect Rep 4:229–237. https://doi.org/10.1007/s12281-010-0029-3

Harry JB, Oliver BG, Song JL et al (2005) Drug-induced regulation of the MDR1 promoter in Candida albicans. Antimicrob Agents Chemother 49:2785–2792. https://doi.org/10.1128/AAC.49.7.2785-2792.2005

Hazen KC, Baron EJ, Colombo AL et al (2003) Comparison of the susceptibilities of Candida spp. to fluconazole and voriconazole in a 4-year global evaluation using disk diffusion. J Clin Microbiol 41:5623–5632. https://doi.org/10.1128/JCM.41.12.5623-5632.2003

Hiller D, Stahl S, Morschhauser J (2006) Multiple cis-acting sequences mediate upregulation of the MDR1 efflux pump in a fluconazole-resistant clinical Candida albicans isolate. Antimicrob Agents Chemother 50:2300–2308. https://doi.org/10.1128/AAC.00196-06

Hirata D, Yano K, Miyahara K, Miyakawa T (1994) Saccharomyces cerevisiae YDR1, which encodes a member of the ATP-binding cassette (ABC) superfamily, is required for multidrug resistance. Curr Genet 26:285–294. https://doi.org/10.1007/BF00310491

Katzmann DJ, Burnett PE, Golin J et al (1994) Transcriptional control of the yeast PDR5 gene by the PDR3 gene product. Mol Cell Biol 14:4653–4661. https://doi.org/10.1128/MCB.14.7.4653

Kaur R, Bachhawat AK (1999) The yeast multidrug resistance pump, Pdr5p, confers reduced drug resistance in erg mutants of Saccharomyces cerevisiae. Microbiology 145:809–818. https://doi.org/10.1099/13500872-145-4-809

Kaur R, Castano I, Cormack BP (2004) Functional genomic analysis of fluconazole susceptibility in the pathogenic yeast Candida glabrata: roles of calcium signaling and mitochondria. Antimicrob Agents Chemother 48:1600–1613. https://doi.org/10.1128/AAC.48.5.1600-1613.2004

Kihara A, Igarashi Y (2002) Identification and characterization of a Saccharomyces cerevisiae gene, RSB1, involved in sphingoid long-chain base release. J Biol Chem 277:30048–30054. https://doi.org/10.1074/jbc.M203385200

Kihara A, Igarashi Y (2004) Cross talk between sphingolipids and glycerophospholipids in the establishment of plasma membrane asymmetry. Mol Biol Cell 15:4949–4959. https://doi.org/10.1091/mbc.e04-06-0458

Knoke M, Schwesinger G (2009) One hundred years ago: the history of crytococcosis in Greifswald. Medical mycology in the nineteenth century. Mycoses 37:229–233. https://doi.org/10.1111/j.1439-0507.1994.tb00418.x

Kolaczkowski M, Michel van der R, Cybularz-Kolaczkowska A et al (1996) Anticancer drugs, ionophoric peptides, and steroids as substrates of the yeast multidrug transporter Pdr5p. J Biol Chem 271:31543–31548. https://doi.org/10.1074/jbc.271.49.31543

Kolaczkowski M, Kolaczkowska A, Gaigg B et al (2004) Differential regulation of Ceramide synthase components LAC1 and LAG1 in Saccharomyces cerevisiae. Eukaryot Cell 3:880–892. https://doi.org/10.1128/EC.3.4.880-892.2004

Kontoyiannis DP (1999) Genetic analysis of azole resistance by transposon mutagenesis in Saccharomyces cerevisiae. Antimicrob Agents Chemother 43:2731–2735. https://doi.org/10.1128/AAC.43.11.2731

Kontoyiannis DP, Lewis RE (2002) Antifungal drug resistance of pathogenic fungi. Lancet 359:1135–1144. https://doi.org/10.1016/S0140-6736(02)08162-X

Kuge S (1997) Regulation of yAP-1 nuclear localization in response to oxidative stress. EMBO J 16:1710–1720. https://doi.org/10.1093/emboj/16.7.1710

Kuge S, Jones N (1994) YAP1 dependent activation of TRX2 is essential for the response of Saccharomyces cerevisiae to oxidative stress by hydroperoxides. EMBO J 13:655–664. https://doi.org/10.1002/j.1460-2075.1994.tb06304.x

Law D, Moore CB, Wardle HM et al (1994) High prevalence of antifungal resistance in Candida spp. from patients with AIDS. J Antimicrob Chemother 34:659–668. https://doi.org/10.1093/jac/34.5.659

Le Crom S, Devaux F, Marc P et al (2002) New insights into the pleiotropic drug resistance network from genome-wide characterization of the YRR1 transcription factor regulation system. Mol Cell Biol 22:2642–2649. https://doi.org/10.1128/MCB.22.8.2642-2649.2002

Leonard PJ, Rathod PK, Golin J (1994) Loss of function mutation in the yeast multiple drug resistance gene PDR5 causes a reduction in chloramphenicol efflux. Antimicrob Agents Chemother 38:2492–2494. https://doi.org/10.1128/AAC.38.10.2492

Leppert G, McDevitt R, Falco SC et al (1990) Cloning by gene amplification of two loci conferring multiple drug resistance in Saccharomyces. Genetics 125:13–20

Ling V (1997) Multidrug resistance: molecular mechanisms and clinical relevance. Cancer Chemother Pharmacol 40:S3–S8. https://doi.org/10.1007/s002800051053

MacPherson S, Larochelle M, Turcotte B (2006) A fungal family of transcriptional regulators: the zinc cluster proteins. Microbiol Mol Biol Rev 70:583–604. https://doi.org/10.1128/MMBR.00015-06

Mamnun YM, Pandjaitan R, Mahé Y et al (2002) The yeast zinc finger regulators Pdr1p and Pdr3p control pleiotropic drug resistance (PDR) as homo- and heterodimers in vivo. Mol Microbiol 46:1429–1440. https://doi.org/10.1046/j.1365-2958.2002.03262.x

Manos NE, Ferebee SH, Kerschbaum WF (1956) Geographic variation in the prevalence of histoplasmin sensitivity. Dis Chest 29:649–668. https://doi.org/10.1378/chest.29.6.649

Marger MD, Saier MH (1993) A major superfamily of transmembrane facilitators that catalyse uniport, symport and antiport. Trends Biochem Sci 18:13–20. https://doi.org/10.1016/0968-0004(93)90081-W

Mendizabal I, Rios G, Mulet JM et al (1998) Yeast putative transcription factors involved in salt tolerance. FEBS Lett 425:323–328. https://doi.org/10.1016/S0014-5793(98)00249-X

Messerschmitt M, Jakobs S, Vogel F et al (2003) The inner membrane protein Mdm33 controls mitochondrial morphology in yeast. J Cell Biol 160:553–564. https://doi.org/10.1083/jcb.200211113

Meyers S, Schauer W, Balzi E et al (1992) Interaction of the yeast pleiotropic drug resistance genes PDR1 and PDR5. Curr Genet 21:431–436. https://doi.org/10.1007/BF00351651

Moore CB, Sayers N, Mosquera J et al (2000) Antifungal drug resistance in Aspergillus. J Infect 41:203–220. https://doi.org/10.1053/jinf.2000.0747

Moye-Rowley WS (2003a) Regulation of the transcriptional response to oxidative stress in fungi: similarities and differences. Eukaryot Cell 2:381–389. https://doi.org/10.1128/EC.2.3.381-389.2003

Moye-Rowley WS (2003b) Transcriptional control of multidrug resistance in the yeast saccharomyces. Prog Nucleic Acid Res Mol Biol 73:251–279

Nakamura K, Niimi M, Niimi K et al (2001) Functional expression of Candida albicans drug efflux pump Cdr1p in a Saccharomyces cerevisiae strain deficient in membrane transporters. Antimicrob Agents Chemother 45:3366–3374. https://doi.org/10.1128/AAC.45.12.3366-3374.2001

Nascimento AM, Goldman GH, Park S et al (2003) Multiple resistance mechanisms among Aspergillus fumigatus mutants with high-level resistance to itraconazole. Antimicrob Agents Chemother 47:1719–1726. https://doi.org/10.1128/AAC.47.5.1719-1726.2003

Nelissen B, Wachter R, Goffeau A (1997) Classification of all putative permeases and other membrane plurispanners of the major facilitator superfamily encoded by the complete genome of Saccharomyces cerevisiae. FEMS Microbiol Rev 21:113–134. https://doi.org/10.1111/j.1574-6976.1997.tb00347.x

Nierman WC, Pain A, Anderson MJ et al (2005) Genomic sequence of the pathogenic and allergenic filamentous fungus Aspergillus fumigatus. Nature 438:1151–1156. https://doi.org/10.1038/nature04332

Niewerth M, Korting HC (2000) The use of systemic antimycotics in dermatotherapy. Eur J Dermatol 10(2):155–160

Nourani A, Papajova D, Delahodde A, Jacq C (1997) Clustered amino acid substitutions in the yeast transcription regulator Pdr3p increase pleiotropic drug resistance and identify a new central regulatory domain. Mol Gen Genet MGG 256:397–405. https://doi.org/10.1007/s004380050583

Nucci M, Anaissie E (2001) Revisiting the source of Candidemia: skin or gut? Clin Infect Dis 33:1959–1967. https://doi.org/10.1086/323759

Odds FC (2010) Editorial: resistance to antifungal agents. Fungal Genet Biol 47:190. https://doi.org/10.1016/j.fgb.2009.12.005

Paget MSB, Buttner MJ (2003) Thiol-based regulatory switches. Annu Rev Genet 37:91–121. https://doi.org/10.1146/annurev.genet.37.110801.142538

Panwar SL, Moye-Rowley WS (2006) Long chain base tolerance in Saccharomyces cerevisiae is induced by retrograde signals from the mitochondria. J Biol Chem 281:6376–6384. https://doi.org/10.1074/jbc.M512115200

Perlin DS (2007) Resistance to echinocandin-class antifungal drugs. Drug Resist Updat 10(3):121–130

Perlin DS (2009) Antifungal drug resistance: do molecular methods provide a way forward? Curr Opin Infect Dis 22:568–573. https://doi.org/10.1097/QCO.0b013e3283321ce5

Perlroth J, Choi B, Spellberg B (2007) Nosocomial fungal infections: epidemiology, diagnosis, and treatment. Med Mycol 45:321–346. https://doi.org/10.1080/13693780701218689

Pfaller MA, Diekema DJ (2004) Rare and emerging opportunistic fungal pathogens: concern for resistance beyond Candida albicans and Aspergillus fumigatus. J Clin Microbiol 42:4419–4431. https://doi.org/10.1128/JCM.42.10.4419-4431.2004

Pfaller MA, Diekema DJ (2007) Epidemiology of invasive candidiasis: a persistent public health problem. Clin Microbiol Rev 20:133–163. https://doi.org/10.1128/CMR.00029-06

Pfaller MA, Diekema DJ, Jones RN et al (2001) International surveillance of bloodstream infections due to *Candida* species: frequency of occurrence and in vitro susceptibilities to fluconazole, ravuconazole, and voriconazole of isolates collected from 1997 through 1999 in the SENTRY antimicrobial surveillance program. J Clin Microbiol 39(9):3254–3259

Pfaller MA, Boyken L, Hollis RJ et al (2008) *In vitro* susceptibility of invasive isolates of Candida spp. to anidulafungin, caspofungin, and micafungin: six years of global surveillance. J Clin Microbiol 46(1):150–156

Piddock LJV (2006) Clinically relevant chromosomally encoded multidrug resistance efflux pumps in bacteria. Clin Microbiol Rev 19:382–402. https://doi.org/10.1128/CMR.19.2.382-402.2006

Pomorski T (2004) Tracking down lipid flippases and their biological functions. J Cell Sci 117:805–813. https://doi.org/10.1242/jcs.01055

Pomorski T, Lombardi R, Riezman H et al (2003) Drs2p-related P-type ATPases Dnf1p and Dnf2p are required for phospholipid translocation across the yeast plasma membrane and serve a role in endocytosis. Mol Biol Cell 14:1240–1254. https://doi.org/10.1091/mbc.e02-08-0501

Prasad R, De Wergifosse P, Goffeau A, Balzi E (1995) Molecular cloning and characterization of a novel gene of Candida albicans, CDR1, conferring multiple resistance to drugs and antifungals. Curr Genet 27:320–329. https://doi.org/10.1007/BF00352101

Prasad T, Saini P, Gaur NA et al (2005) Functional analysis of CaIPT1, a sphingolipid biosynthetic gene involved in multidrug resistance and morphogenesis of Candida albicans. Antimicrob Agents Chemother 49:3442–3452. https://doi.org/10.1128/AAC.49.8.3442-3452.2005

Puri N, Krishnamurthy S, Habib S et al (1999) CDR1, a multidrug resistance gene from Candida albicans, contains multiple regulatory domains in its promoter and the distal AP-1 element mediates its induction by miconazole. FEMS Microbiol Lett 180:213–219. https://doi.org/10.1111/j.1574-6968.1999.tb08798.x

Rank GH, Robertson AJ, Phillips KL (1975) Modification and inheritance of pleiotropic cross resistance and collateral sensitivity in Saccharomyces cerevisiae. Genetics 80:483–493

Revankar SG, Sutton DA (2010) Melanized fungi in human disease. Clin Microbiol Rev 23:884–928. https://doi.org/10.1128/CMR.00019-10

Rex JH, Pfaller MA, Galgiani JN et al (1997) Development of interpretive breakpoints for antifungal susceptibility testing: conceptual framework and analysis of *in vitro - in vivo* correlation data for fluconazole, itraconazole, and Candida infections. Clin Infect Dis 24(2):235–247

Richardson MD (2005) Changing patterns and trends in systemic fungal infections. J Antimicrob Chemother 56:i5–i11. https://doi.org/10.1093/jac/dki218

Richardson M (2009) The ecology of the Zygomycetes and its impact on environmental exposure. Clin Microbiol Infect 15:2–9. https://doi.org/10.1111/j.1469-0691.2009.02972.x

Richardson M, Lass-Flörl C (2008) Changing epidemiology of systemic fungal infections. Clin Microbiol Infect 14:5–24

Riggle PJ, Kumamoto CA (2006) Transcriptional regulation of MDR1, encoding a drug efflux determinant, in fluconazole-resistant Candida albicans strains through an Mcm1p binding site. Eukaryot Cell 5:1957–1968. https://doi.org/10.1128/EC.00243-06

Riordan JR, Deuchars K, Kartner N et al (1985) Amplification of P-glycoprotein genes in multidrug-resistant mammalian cell lines. Nature 316:817–819. https://doi.org/10.1038/316817a0

Rodrigues-Pousada CA, Nevitt T, Menezes R et al (2004) Yeast activator proteins and stress response: an overview. FEBS Lett 567:80–85. https://doi.org/10.1016/j.febslet.2004.03.119

Rognon B, Kozovska Z, Coste AT et al (2006) Identification of promoter elements responsible for the regulation of MDR1 from Candida albicans, a major facilitator transporter involved in azole resistance. Microbiology 152:3701–3722. https://doi.org/10.1099/mic.0.29277-0

Roninson IB (1992) From amplification to function: the case of the MDR1 gene. Mutat Res Genet Toxicol 276:151–161. https://doi.org/10.1016/0165-1110(92)90005-T

Roninson IB, Chin JE, Choi KG et al (1986) Isolation of human mdr DNA sequences amplified in multidrug-resistant KB carcinoma cells. Proc Natl Acad Sci 83:4538–4542. https://doi.org/10.1073/pnas.83.12.4538

Sá-Correia I, Tenreiro S (2002) The multidrug resistance transporters of the major facilitator superfamily, 6 years after disclosure of Saccharomyces cerevisiae genome sequence. J Biotechnol 98:215–226. https://doi.org/10.1016/S0168-1656(02)00133-5

Sang H, Zheng XE, Zhou WQ et al (2011) A case of subcutaneous phaeohyphomycosis caused by Cladosporium cladosporioides and its treatment. Mycoses. https://doi.org/10.1111/j.1439-0507.2011.02057.x

Sanglard D (2002) Resistance of human fungal pathogens to antifungal drugs. Curr Opin Microbiol 5(4):379–385

Sanglard D, Kuchler K, Ischer F et al (1995) Mechanisms of resistance to azole antifungal agents in Candida albicans isolates from AIDS patients involve specific multidrug transporters. Antimicrob Agents Chemother 39:2378–2386. https://doi.org/10.1128/AAC.39.11.2378

Sanglard D, Ischer F, Monod M, Bille J (1996) Susceptibilities of Candida albicans multidrug transporter mutants to various antifungal agents and other metabolic inhibitors. Antimicrob Agents Chemother 40:2300–2305. https://doi.org/10.1128/AAC.40.10.2300

Sanglard D, Ischer F, Monod M, Bille J (1997) Cloning of Candida albicans genes conferring resistance to azole antifungal agents: characterization of CDR2, a new multidrug ABC transporter gene. Microbiology 143:405–416. https://doi.org/10.1099/00221287-143-2-405

Sanglard D, Ischer F, Bille J (2001) Role of ATP-binding-cassette transporter genes in high-frequency acquisition of resistance to azole antifungals in Candida glabrata. Antimicrob Agents Chemother 45:1174–1183. https://doi.org/10.1128/AAC.45.4.1174-1183.2001

Sanglard D, Coste A, Ferrari S (2009) Antifungal drug resistance mechanisms in fungal pathogens from the perspective of transcriptional gene regulation. FEMS Yeast Res 9(7):1029–1050

Sanguinetti M, Posteraro B, Fiori B et al (2005) Mechanisms of azole resistance in clinical isolates of Candida glabrata collected during a hospital survey of antifungal resistance. Antimicrob Agents Chemother 49:668–679. https://doi.org/10.1128/AAC.49.2.668-679.2005

Sellart-Altisent M, Torres-Rodríguez JM, Gómez de Ana S, Alvarado-Ramírez E (2007) Nasal fungal microbiota in allergic and healthy subjects. Rev Iberoam Micol 24:125–130. https://doi.org/10.1016/S1130-1406(07)70027-X

Shahi P, Gulshan K, Moye-Rowley WS (2007) Negative transcriptional regulation of multidrug resistance gene expression by an Hsp70 protein. J Biol Chem 282:26822–26831. https://doi.org/10.1074/jbc.M704772200

Sheehan DJ, Hitchcock CA, Sibley CM (1999) Current and emerging azole antifungal agents. Clin Microbiol Rev 12(1):40–79

Shukla S (2004) Substitution of threonine-1351 in the multidrug transporter Cdr1p of Candida albicans results in hypersusceptibility to antifungal agents and threonine-1351 is essential for synergic effects of calcineurin inhibitor FK520. J Antimicrob Chemother 54:38–45. https://doi.org/10.1093/jac/dkh308

Skaggs BA, Alexander JF, Pierson CA et al (1996) Cloning and characterization of the *Saccharomyces cerevisiae* C-22 sterol desaturase gene, encoding a second cytochrome P-450 involved in ergosterol biosynthesis. Gene 169(1):105–109

Slavin M, Fastenau J, Sukarom I et al (2004) Burden of hospitalization of patients with Candida and Aspergillus infections in Australia. Int J Infect Dis 8:111–120. https://doi.org/10.1016/j.ijid.2003.05.001

Sturley S (2000) Conservation of eukaryotic sterol homeostasis: new insights from studies in budding yeast. Biochim Biophys Acta Mol Cell Biol Lipids 1529:155–163. https://doi.org/10.1016/S1388-1981(00)00145-1

Tobin MB, Peery RB, Skatrud PL (1997) Genes encoding multiple drug resistance-like proteins in Aspergillus fumigatus and Aspergillus flavus. Gene 200:11–23. https://doi.org/10.1016/S0378-1119(97)00281-3

Tsai H-F, Krol AA, Sarti KE, Bennett JE (2006) Candida glabrata PDR1, a transcriptional regulator of a pleiotropic drug resistance network, mediates azole resistance in clinical isolates and petite mutants. Antimicrob Agents Chemother 50:1384–1392. https://doi.org/10.1128/AAC.50.4.1384-1392.2006

van den Hazel HB, Pichler H, Matta MA do V et al (1999) PDR16 and PDR17, two homologous genes of Saccharomyces cerevisiae, affect lipid biosynthesis and resistance to multiple drugs. J Biol Chem 274:1934–1941. https://doi.org/10.1074/jbc.274.4.1934

Vermitsky J-P, Edlind TD (2004) Azole resistance in Candida glabrata: coordinate upregulation of multidrug transporters and evidence for a Pdr1-like transcription factor. Antimicrob Agents Chemother 48:3773–3781. https://doi.org/10.1128/AAC.48.10.3773-3781.2004

Vermitsky J-P, Earhart KD, Smith WL et al (2006) Pdr1 regulates multidrug resistance in Candida glabrata: gene disruption and genome-wide expression studies. Mol Microbiol 61:704–722. https://doi.org/10.1111/j.1365-2958.2006.05235.x

Vugia DJ, Tabnak F, Newton AE et al (2013) Impact of 2003 state regulation on raw oyster-associated Vibrio vulnificus illnesses and deaths, California, USA. Emerg Infect Dis 19:1276–1280. https://doi.org/10.3201/eid1908.121861

Warnock DW (2006) Fungal diseases: an evolving public health challenge. Med Mycol 44:697–705. https://doi.org/10.1080/13693780601009493

Wemmie JA, Steggerda SM, Moye-Rowley WS (1997) The Saccharomyces cerevisiae AP-1 protein discriminates between oxidative stress elicited by the oxidants H2O2 and diamide. J Biol Chem 272:7908–7914. https://doi.org/10.1074/jbc.272.12.7908

White TC (1997) Increased mRNA levels of ERG16, CDR, and MDR1 correlate with increases in azole resistance in Candida albicans isolates from a patient infected with human immunodeficiency virus. Antimicrob Agents Chemother 41:1482–1487. https://doi.org/10.1128/AAC.41.7.1482

Wong S, Butler G, Wolfe KH (2002) Gene order evolution and paleopolyploidy in hemiascomycete yeasts. Proc Natl Acad Sci 99:9272–9277. https://doi.org/10.1073/pnas.142101099

Wu AL, Moye-Rowley WS (1994) GSH1, which encodes gamma-glutamylcysteine synthetase, is a target gene for yAP-1 transcriptional regulation. Mol Cell Biol 14:5832–5839. https://doi.org/10.1128/MCB.14.9.5832

Zhang X, Cui Z, Miyakawa T, Moye-Rowley WS (2001) Cross-talk between transcriptional regulators of multidrug resistance in Saccharomyces cerevisiae. J Biol Chem 276:8812–8819. https://doi.org/10.1074/jbc.M010686200

Pathogenesis and Drug Resistance Profile of Food-Borne Pathogens 18

Preeti C. Sangave

Contents

Abstract

The World Health Organization (WHO) identifies food safety as one of the top ten priorities. Food-borne illnesses caused by food-borne pathogens are a major cause of morbidity and mortality globally. As a serious challenge of our generation, this issue is further exacerbated by the phenomenon of antimicrobial resistance among pathogens, which is another WHO listed top ten threats to global public health. The microbes have developed capabilities to resist the action of

P. C. Sangave (✉)
School of Pharmacy and Technology Management, SVKM's NMIMS, MPTP, Shirpur, India
e-mail: preeti.sangave@nmims.edu

B. Siddhardha et al. (eds.), *Model Organisms for Microbial Pathogenesis, Biofilm Formation and Antimicrobial Drug Discovery*,
https://doi.org/10.1007/978-981-15-1695-5_18

antibiotics—once called a wonder medicine, and are threatening to send us back to pre-antibiotic era. Pathogenesis and drug resistance profile of food-borne pathogens is complex subject with a vast expanse and spans diverse segments of food, food animals, the ecological niches, human populations, the pharmaceutical industry, and the clinic. New antibiotics or novel treatment options for tackling food-borne infections could only be put to clinical practice if the association of pathogenesis, development, and spread of antimicrobial resistance among causative pathogens is well understood. Thus, this chapter reviews the trends and factors contributing to pathogenesis along with mechanisms of observed antimicrobial resistances among the food-borne pathogens. The severity of the issue has also been discussed in the context of global spread of antibiotic use. Pathogenic mechanisms adopted by few food-borne pathogens have been illustrated at the end.

Keywords
Food safety · Antimicrobial resistance · Antibiotics · Pathogenesis · Food-borne pathogens

18.1 Introduction

Food safety happens to be one of the leading challenges of today's generation (Uyttendaele et al. 2015). According to the World Health Organization, over two million deaths arise from food-borne diseases annually (WHO 2019). These figures are further staggered by the appearance of pathogens resistant to antimicrobials. The recent data and predictions are indicating antimicrobial resistance to pose a serious risk to global public health and economy (Acar and Röstel 2001). Infections due to resistant pathogens have serious implications and can lead to multiple treatment failures. Patients thus may have to undergo longer hospital stays and would need expensive antibiotics to treat the resistant pathogens, adding to the overall health-care costs (Walsh and Duffy 2013).

The incidence of antibiotic resistance was first brought to public notice when the Swann Committee, UK, through their report published in 1969 raised a concern that an inappropriate use of antibiotics in food animals and veterinary medicine could be harmful to human health. The Committee recommended to ban the use of therapeutic antibiotics intended for humans as growth promoters for animals (McDermott et al. 2002). However, it was not until 1976 when clear evidence was observed, a report demonstrated a rise of tetracycline resistance in chickens soon after its introduction in animal feeds and later transmittance of resistance to microbiota of animal handlers. Subsequent to these first observations, many studies were published indicating the spread of antimicrobial resistance through the community, in the environment, and across the food chain, as major risk to human health and drew increased public and scientific attention (McDermott et al. 2002; Hamilton and Wenlock 2016).

The food animals and/or products derived from them have been found positive for antimicrobial resistant organisms, which are pathogenic to humans. For instance, strains showing resistance to vancomycin, belonging to *Enterococci* of food-animal origin, particularly have been associated with microbes found in the gastrointestinal tract of humans (Swartz 2002). *Listeria*, *Salmonella*, *Campylobacter*, *Escherichia coli* O157 (and other enterotoxin- and Shiga toxin-producing strains of *E. coli*), *Vibrio,* and *Yersinia* represent the major classes of intestinal pathogens of food-animal origin capable of causing serious illnesses. Most of them are commensals in cattle, poultry, and swine that also cause invasive infections both in humans and animals (except for *E. coli* O157, a microbe common in cattle and *Vibrio*, located in seawater and shellfish). The illnesses caused by food-borne pathogens have been reviewed in further detail by Swartz (2002). The spread of antimicrobial resistance in different food sources spanning fresh produce, milk and milk products, raw and processed meat, aquatic products, and retail foods has been summarized in Table 18.1.

A pathogen or pathogenic microorganism is described as an etiological agent that is able to cause damage to its host during, or as a consequence of, the host–microorganism interaction. The harm to the host may be imposed directly by the microorganism (e.g., by toxins or other so-called virulence factors) or indirectly through the activity of the host immune responses (Casadevall and Pirofski 2000). Pathogenicity is the capability of the pathogen to infect. Microorganisms express their pathogenicity by means of their virulence, a term that refers to the relative, measurable quantity of pathogenicity (Casadevall and Pirofski 2000). This chapter reviews the factors contributing to antimicrobial resistance in food-borne pathogens including transmission routes, mechanisms and illustrates examples of pathogens with their pathogenesis. Before proceeding, few relevant terminologies pertaining to microbial pathogenesis and resistance have been described in brief in Table 18.2.

Given the complexities of host–microbe/pathogen interactions (van Baarlen et al. 2007), other terminologies like infection, commensalism, colonization (Casadevall and Pirofski 2000), food-borne illnesses (Acar and Röstel 2001), and standards introduced for determining antimicrobial resistance of pathogen (like ISO, EUCAST) (Rodloff et al. 2008), with their detailed explanation have been provided elsewhere in literature.

18.2 Antibiotic Exposure/Consumption and Food Safety

In a white paper published by Center for Science in the Public Interest, USA (Walsh and Duffy 2013), findings related to a survey between 1973 and 2011 of food borne outbreaks caused by bacteria that were resistant to at least one antibiotic were presented. In a total of 55 outbreaks reported, over 56% of cases were related to food items which included dairy products, beef, and poultry. While in over 87% of the cases, *Salmonella* spp. was identified as most common causative agent of antibiotic

Table 18.1 Spread of antimicrobial resistance in the food sources

Food-borne pathogen	Food/food animal source	Reference
Drug-resistant *Salmonella* spp.	Retail meats—pork, chicken, beef, swine, Turkey and livestock, egg-shells, vegetables, fresh produce, fruit/fruit pulp, cereals, sea-food, cow's milk	VT Nair et al. (2018) and Mąka et al. (2015)
Methicillin-resistant *Staphylococcus aureus* (MRSA)	Livestock, pigs, broiler chickens and infecting dairy cows, ready-to-eat food, food-contact surface	Sergelidis and Angelidis (2017), Vanderhaeghen et al. (2010) and Aung et al. (2017)
Multidrug resistant (MDR) *Campylobacter* spp.	Feedlot cattle, food-producing animals, poultry meat, Turkey meat	Wang et al. (2014), Du et al. (2018) and Ge et al. (2003)
Enterococcus faecium (MDR)	Red meat, chickens, pork, Turkish cheese	Golob et al. (2019), Sanlibaba and Senturk (2018) and Daniel et al. (2015)
Resistant *Enterobacteriaceae*	Aquatic products, ready-to-eat (RTE) foods, raw meat products, mushrooms, frozen foods	Ye et al. (2018)
ß-lactam-resistant *E. coli*, *Staphylococcus aureus*-resistant type	Ready-to-eat food	Al-Humam (2019)
Shiga toxin-producing *E. coli* (STEC)	Semi-domesticated reindeer	Laaksonen et al. (2017)
Multidrug resistant *L. monocytogenes*	Retail ready-to-eat food samples	Chen et al. (2014)
Multidrug resistant—*Salmonella*, *Listeria Monocytogenes*, *Staphylococcus aureus*	Raw meats and cooked meat, raw milk	Gousia et al. (2011)
Multidrug resistant—*E. coli*, *Enterococcus spp.*, *Salmonella*, *Staphylococcus aureus*	Raw and processed meat products	Gwida (2015)
Multidrug resistant—*E. coli*, *Enterococcus spp.*, *Salmonella*, *Staphylococcus spp.*	Raw retail chicken	Ezepchuk (2017)

resistant outbreaks, 56% of pathogens exhibited multidrug resistance to five or more antibiotics (Walsh and Duffy 2013). With the growing severity and persistence of antimicrobial resistance, two aspects are followed, namely (1) use of antibiotics as essential medicines and (2) use of antibiotics for management of livestock productivity and warrant immediate attention.

Table 18.2 Some relevant terminologies related to microbial pathogenesis and resistance

Term	Explanation	Reference
Virulence	A extent of pathogenicity measured by the in vivo LD50 (lethal dose, 50%) and in vitro ID50 (infectious dose, 50%) tests	Ezepchuk (2013, 2017)
Pathogenic/ pathogenicity factors	The molecules produced by pathogen that are responsible for host cell specific interactions	Ezepchuk (2017)
Antimicrobial resistance	Determined on the basis of in vitro experimental parameters, and is related to the ability of microorganism to survive in the presence of a defined concentration of an antimicrobial compound	Acar and Röstel (2001)
Degrees of antimicrobial action	Explained in terms of the minimum inhibitory concentration (MIC) of an antimicrobial compound required that visibly inhibits bacterial growth *Susceptible*: Bacteria labeled as susceptible to a given antibiotic when it is inhibited in vitro by a concentration of this drug having a greater probability of therapeutic success *Resistant*: Bacteria can be designated as resistant to an antimicrobial compound, when it is inhibited in vitro by a concentration of this drug having a higher probability of therapeutic failure. This effect would be reflected when its MIC is higher than its wild-type counterpart	Acar and Röstel (2001), Rodloff et al. (2008)

18.2.1 Antibiotics as Essential Medicines

To enhance treatment outcomes through judicious use and to reduce resistance to antibiotics, WHO released revised model list of Essential Medicines, in 2017, and classified antibiotics under three categories—"Access, Watch, and Reserve" (WHO 2017).

In this list, the key 'Access' antibiotics are those which are available all the time to treat a wide range of common infections at affordable price and assured quality. For example, they include beta-lactams like amoxicillin, first- and second-generation cephalosporins like cefotaxime, ciprofloxacin, and aminoglycosides like meropenem. The second "Watch" group antibiotics are those recommended as first- or second-line treatments for limited, specific infections and possess high potential for developing resistance. For example, they include quinolones and fluoroquinolone like ciprofloxacin, macrolides like azithromycin, glycopeptides like vancomycin, penems like faropenem, carbapenems, and meropenem, and third-generation cephalosporins like cefixime. The third "Reserve" antibiotics group happens to be the "last resort" for life-threatening situations, when all alternatives have failed. For example, they include fourth- and fifth-generation cephalosporins, aztreonam, polymyxins like colistin, oxazolidones like linezolid, and daptomycin (WHO Advisory Group on Integrated Surveillance of Antimicrobial Resistance (AGISAR) 2019; EMA 2019).

In a recent detailed report, using existing consumption patterns in 76 countries over a period 2000–2015, the global antibiotic consumption in 2030 is projected to be 200% elevated than the 42 billion daily defined doses reported in 2015 (Klein et al. 2018). In 2000, the high income economies reported the highest antibiotic consumption rates, but by 2015, they were later replaced by low-middle income countries. Between 2000 and 2015, Asian countries like India recorded 103% antibiotic consumption, while China and Pakistan recorded 79% and 65%, respectively (Klein et al. 2018). Similarly, from 2000 to 2015, broad-spectrum penicillins (the Access group), widely used antibiotics globally, had a higher consumption rate of 39%. Although the antibiotic consumption rates increased overall for the next three widely consumed Watch group antibiotics cephalosporins, quinolones, and macrolides to 20%, 12%, and 12%, respectively, lower rates were reported in high income economies. In contrast by 2015, the antibiotic consumption rate in developing countries increased to 399%, 125%, and 119% respectively for these three classes (Klein et al. 2018). Utilization of newer and "last-resort" antibiotic classes increased globally between 2000 and 2015. In 2012, India became the largest consumer of oxazolidinones (the Reserve-group) in the world. Moreover by 2012, the sales in 2011–2012 of major Access group antibiotics rose by 13%, the Watch group by 24%, and the Reserve group by 69% in India (McGettigan et al. 2017). The global consumption of "Reserve group" antibiotics has been increasing concurrently with increasing evidences of resistances to colistin and carbapenems (the Watch group) thus calling an immediate attention (Klein et al. 2018; Founou et al. 2016).

18.2.2 Antibiotics Exposure to Food Animals

The major concern over the use of antibiotics for agricultural purposes arises due to the fact that residues of antibiotics remain on farm selects and in food products which breeds and in some cases act as reservoir of antibiotic resistant zoonotic bacteria. The consumption of contaminated food and/or water transfers resistant strains of bacteria to humans that further transmit and spread resistant determinants to bacterial species of human gut or other bacterial inhabitants of host (McDermott et al. 2002). There are many possible sources of contamination of foods by pathogens. For instance, fruits and vegetables can become contaminated during production on farms; seafoods and water-grown vegetables can become contaminated in their aquatic environment; eggs can become contaminated during formation (Founou et al. 2016); during animal slaughter or preparation food can become contaminated (McDermott et al. 2002), through workers during handling and preparation in food processing plants, and also at homes (Gwida 2015). Bacterial spores particularly resist heat processing and can thus possibly make heat-processed foods as vehicles of food-borne pathogens. Thus, the point at which contamination occurs depends on the pathogen or the source and on the opportunities for transfer during the food production-processing-preparation stages (Founou et al. 2016).

In recent years, studies are increasingly focusing on food as carriers of antibiotic resistance genes (Mathur and Singh 2005). For instance, vancomycin resistant enterococcus (VRE) has emerged as a recurring cause of nosocomial infections. One of the possibilities being that the resistant strains are being spread via the food chain, which got selected and augmented due to the use of avoparcin (having cross resistance to vancomycin) as a feed-additive/growth promoter in food animal (Marshall and Levy 2011). A study reported raw and processed meat products to be contaminated with resistant pathogens. Highest rate of resistance (62.5%) was exhibited to ciprofloxacin by *E. coli* from chicken. Isolates from pork showed 68.2% resistance to tetracycline while from chicken of 62.5%, and resistance to aminoglycosides was predominant in lamb/goat meat isolates. In almost 96% of the isolates of *Y. enterocolitica*, resistance to ampicillin was observed while all of the *C. jejuni* isolates were reported to be resistant to ampicillin, cefuroxime, and cephalothin (Gousia et al. 2011). In a Chinese study, partial antimicrobial resistance was observed in virulent *L. monocytogenes*, and listeriosis causing serotypes were frequently recovered from retail ready-to-eat foods (Chen et al. 2014). The food-borne transmission route was assessed to be one of the most important for spread of *Campylobacter* spp., *Salmonella* spp., and Shiga toxin-producing *E. coli*, then was via animal contact, next was direct human-to-human contact, and followed by waterborne transmission. For *Norovirus*, *Shigella spp*., and *E. histolytica,* human-to-human transmission was identified as the main exposure route, whereas for *V. cholerae* infections, waterborne transmission was estimated to be the main route for spread of resistance (Hald et al. 2016). Findings for *Bacteroides* indicated a 100-fold increase in gene transfer in bacteria possessing conjugative transposons (all encoding tetracycline resistance) on exposure to low concentrations of tetracycline (Wilson 2002).

The spread on antimicrobial resistance in food sources (Founou et al. 2016) and food pathogens (White et al. 2002) has been summarized in published literature. Table 18.3 illustrates representative evidences of multidrug resistances observed in different food sources.

In the food processing environment, a reversible adhesion occurs between the two surfaces involving the cell and the environmental matrix mainly through electrostatic, van der Waals, and/or hydrophobic interactions (Cappitelli et al. 2014). In many cases, the pathogens adhere to and are able to create biofilms on most surfaces they come in contact with, under most of the conditions prevalent in food manufacturing plants (Bridier et al. 2015). If the food contains moisture and nutrients in sufficient quantity or the pH, the redox potential, the temperatures are favorable or if they compete with the mixed microbial flora in the foods, the pathogenic bacteria in the contaminated food would multiply rapidly (Bryan 1978). For example, it was reported that *Listeria monocytogenes* could bind to different food-grade materials like rubbers, metals, and polymers (Di Bonaventura et al. 2008). Bacterial strains *L. monocytogenes*, Shiga toxin-producing *E. coli* (STEC), and *Salmonella* were found to persist in fresh produce fields (Strawn et al. 2013). Brooks et al. 2014 confirmed the prevalence of genes conferring resistance to methicillin, erythromycin, and tetracycline in swine manure wastewater (Brooks et al. 2014). *Salmonella* and *L. monocytogenes* were detected predominantly in water samples, while Shiga toxin-producing *Escherichia coli* (STEC) was detected in soil, water, and feces, reported

Table 18.3 Isolates from food sources with evidence of multidrug resistance to antibiotics

Organism name	Food source	Antimicrobial resistance/ multidrug resistant	Reference
Listeria monocytogenes	Food and food-processing environments	Clindamycin, ciprofloxacin, and rifampicin, linezolid, ampicillin, tetracycline trimethoprim/ sulfamethoxazole, and vancomycin	Conter et al. (2009)
Shiga toxin-producing *Escherichia coli* (*STEC*)	Poultry, cattle and swine	Cefoxitin, trimethoprim/ sulfamethoxazole, tetracycline, streptomycin, cefalothin, ampicillin, chloramphenicol, gentamicin	Singh et al. (2005)
Staphylococcus aureus	Animal origin food	Tetracycline, levofloxacin, tobramycin, methicillin, gentamicin, ciprofloxacin, moxifloxacin, and norfloxacin	Yucel et al. (2011)
Escherichia coli, *Campylobacter jejuni*, *Enterococcus* spp., *Salmonella* spp., *Yersinia enterocolitica*, and *Staphylococcus aureus*	Raw meat—Goat, sheep, chicken, pork and processed meat, roast beef, salami, mincemeat	Ciprofloxacin, ampicillin, cephalothin, and cefuroxime	Gousia et al. (2011)
Staphylococcus aureus and MRSA	Dairy products	Vancomycin, kanamycin, tetracycline, methicillin, and chloramphenicol	Sasidharan et al. (2011)
E. coli, *Salmonella*, thermophilic *Campylobacter*	Meat species—pork, beef, and poultry	Quinolones and tetracycline	Pesavento et al. (2007)

out of a longitudinal study carried out in five farms in State of New York (Strawn et al. 2013).

The antimicrobials are widely used for disease prevention and treatment among food animals and/or sources (Doyle et al. 2006). In the USA, these figures go up to 80% of the country's yearly antimicrobials that are being used in food animals, many of which are prescribed as human medicine. The antimicrobial agents widely used in food animals mainly belong to sulfonamides, macrolides, beta-lactams, aminoglycosides, and tetracyclines (McDermott et al. 2002). A survey projected, by 2030, the global consumption of antimicrobials in food animal production to rise by 67%, i.e., around 107,000 tons. In Asia alone, the rise is expected to be 46% and in the BRICS countries by 99%, in contrast to only 13% increase in human population over the same period (van Boeckel et al. 2015). In a developing nation like India that shows highest bacterial disease burden in the world and an overall increase use of antibiotics in food animals, the ever increasing prevalence and spread of antimicrobial resistance could drastically increase the mortality and morbidity (Van Boeckel et al, 2015).

The wide use of antibiotics as growth promoters for food animals has further influenced the prevalence and spread of antibiotic resistance (Aarestrup et al. 2008; Founou et al. 2016). Other factors which also contribute to the growing concern of antibiotic resistance include improper prescription practices, insufficient patient education, inadequate diagnostic facilities, unauthorized sale of antimicrobials, and inappropriate functioning of drug regulatory mechanisms (McGettigan et al. 2017). In a retrospective study, a connection between antibiotic resistance in coliform isolates from urine samples and the use of antibiotics by general practice was observed and study was also indicative of geographically contained effects from use of antibiotics to ensue in communities (Magee et al. 1999). Another study carried out in Cape Town concluded that an irrational use of antibiotics in cases due to incorrect dose, incorrect duration of therapy, undocumented diagnosis, and/or no need of antibiotic attributed to observed rise in resistance to antibiotics within community (Gasson et al. 2018). These factors have been discussed further in detail elsewhere (Leung et al. 2011; Odeyemi and Sani 2016; McGettigan et al. 2017). The selection pressure on bacteria increases with regular use of antimicrobials in modern food animal production and/or processing and prescription practices, to transform them into resistant-form (van Boeckel et al. 2015).

18.3 Genetic Basis of Pathogenesis and Virulence and Possible Relationship with Antibiotic Resistance

The evolution of pathogenicity is inextricably linked to the host as the gathered evidences suggest (Casadevall and Pirofski 2000). So, whenever evolution in pathogenicity is reported, the evolution of host sensitivity is implied as well (Casadevall and Pirofski 2000). In order to colonize and thus render a host susceptible to disease, the genetic framework of both host and pathogen need to enable molecular interactions. Genetic variations, assortment of the best-fitting mutants, and retention (or progression into disease) are the fundamental processes for tweaking the interactions (Mazel and Davies 1999). An integrated overview between mechanisms of virulence and antibiotic resistance was reported highlighting probability of genetic linkage between the two (Martínez and Baquero 2002). Recent findings indicated that very low antibiotic concentrations can select both low-level resistance (Murray et al. 2018) and also high-level resistance (Blázquez et al. 2018). Increasing antimicrobial resistance has been attributed to two main aspects, the presence of genes for resistance (Swartz 2002) and the selective pressure by the use of antibiotics (Martínez and Baquero 2002). Furthermore, the prevalence of resistance genes and their role in contributing virulence to a pathogen could be a pre-existing phenotype (an intrinsic phenomenon) or gathered in due course under antibiotic selection pressure (an acquired mechanism) (Founou et al. 2016; Martínez and Baquero 2002; Munita and Arias 2016). Figure 18.1 depicts the inter-relation of these mechanisms.

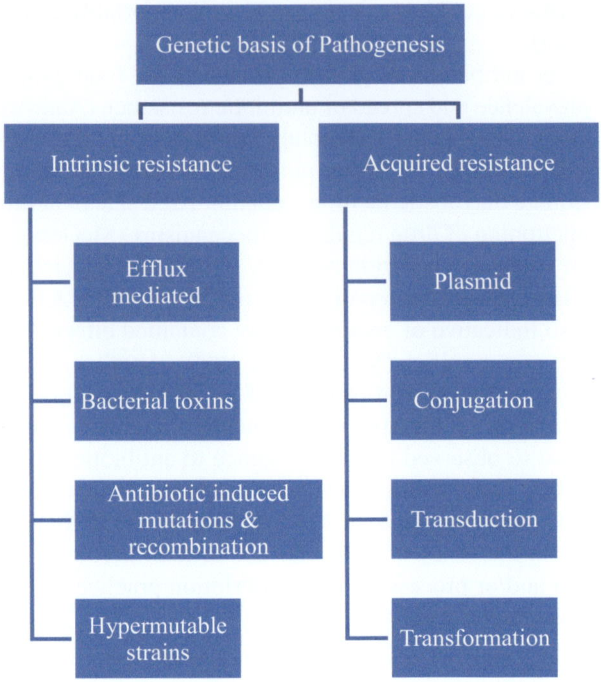

Fig. 18.1 Genetic basis of pathogenesis

18.3.1 Intrinsic Resistance Mechanisms

Bacterial species or genus can show characteristic resistance to a given antibiotic inherently, i.e., intrinsic or natural resistance, that confers an ability to the organism to survive in the presence of an antimicrobial agent (Mathur and Singh 2005). When the metabolic activity remains unaltered even in the presence of antimicrobial compound in surrounding, due to distinct genetic makeup of the microorganism, this phenomenon can be co-related as intrinsic or natural resistance of that microorganism (Wilson 2002; Walsh 2010). Resistance can be contributed by an innate phenotype in natural bacterial populations. Over a period of time, the microorganisms gather mutations in existing genes (in the chromosome or plasmid) and transmit the resistant genes to daughter cells via vertical gene transfer, contributing to a *natural*, *inherent,* or *intrinsic* resistance (Founou et al. 2016). Generally, intrinsic resistances are chromosomally encoded, and attribute unique properties and varying degree of differences observed in resistances between genera, species, and strains of bacteria (Martínez and Baquero 2002; Mathur and Singh 2005). For instance, intrinsic resistance can be associated with the ability to use efflux pumps to export antimicrobial compounds out of the bacterial cell, complexity in cell wall structures (Doyle et al. 2006; Dougherty and Pucci 2012), or the production of antibiotic degrading enzymes that can inactivate antimicrobial compounds within the bacterial cell (Founou et al.

2016). An oxazolidinone antibiotic, linezolid, used in treatment of infections caused by resistant Gram-positive bacteria acts by binding to the peptidyl transferase center on the ribosome subunit and inhibiting protein synthesis. Most of the known mechanisms of resistance to this antibiotic involve minor modifications to the linezolid binding site on ribosome (Long and Vester 2012).

18.3.1.1 Efflux Pump Mediated

The efflux mechanism, an ATP-dependent export of antimicrobials and other substances out of microbial cells, was first identified in the early 1980s and is now considered as a significant factor imparting resistance in microorganisms. Since then, many efflux transporters with an ability to export single or structurally diverse compounds like antibiotics and biocides have been reported. The efflux pumps contributing to antibiotic resistance have been extensively reviewed (Dougherty and Pucci 2012; Doyle et al. 2006). The bacterial efflux systems have been grouped into five classes based on the structural differences. Of these, one class of the major facilitator (MF) transporters is predominant in Gram-positive bacteria and contributes resistance to few antibiotics like macrolide, ketolide, chloramphenicol, or tetracycline (Dougherty and Pucci 2012). On the other hand, the resistance nodulation-division (RND) transporters are predominant in Gram-negative bacteria and in some cases offer resistance against erythromycin, tetracycline, or fluoroquinolones (Nikaido 1996). The RND pumps are encoded on the chromosome and are multidrug efflux pumps, and any errors like mutation in their local or global regulatory sequences can induce their over-expression. The drug-specific efflux pumps are a family of single component proteins and are encoded by mobile genetic elements. These mobile genetic elements (such as integrons, transposons, and plasmids) can further get transferred to and/or acquired from other organisms, thus conferring resistance (Dougherty and Pucci 2012).

18.3.1.2 Bacterial Toxins

By releasing toxins in the environment or injecting them into the host, some pathogens can cause damage from a distance. Such toxins are the hallmark factors for many pathogens that are able to create infectious conditions and symptoms leading to tissue damage, when they reach within the host body (Munita and Arias 2016). For instance, cholera, shiga, and botulinum toxins are classical examples, unique to few food-borne pathogens, each capable of disrupting major cellular activities and causing morbidity in afflicted individuals (Cegelski et al. 2008). For example, *Vibrio cholerae*, a causative agent of severe dehydration and diarrhea which can become life threatening if not treated in time. *V. cholerae* secretes cholera toxin that interacts with G proteins and cAMP in the intestinal lining of the host to disrupt the ion transport, triggering severe fluid loss and dehydration (Cegelski et al. 2008). *H. pylori* secretes VacA, a pore-forming toxin that disturbs the cell polarity, promoting apoptosis in epithelial cells and inhibiting T-cell proliferation and effector functions (Palframan et al. 2012).

18.3.1.3 Antibiotic Induced Mutations and Recombination

Antibiotics induce metabolic perturbations, target cellular processes, and are capable of inducing stress responses and genetic instability in bacteria. Induction of stress response regulatory sequences, production of reactive oxygen species, leading to enhanced mutation and recombination rates or horizontal gene transfer, can be induced by antibiotics. All these phenomena greatly impact the evolution and spread of antibiotic resistance among pathogens (Blázquez et al. 2018). In a study cefotaxime-resistant, *S. typhi* strains produced stronger biofilm when exposed to sub-MIC levels of cefotaxime as compared to sensitive strains as control group (Narasanna et al. 2017). As environmental factors, antibiotics can increase rate of mutation. They can trigger mutations not only at lethal concentrations of antibiotics, but also at sub-inhibitory concentrations in many species of bacteria and are capable of being spread via horizontal transfer (Blázquez et al. 2018; Marshall and Levy 2011). Not only antibiotics which directly affect DNA replication and integrity of bacterial genome like fluoroquinolone can increase mutation rates, but also antibiotics that do not affect DNA directly, such as the β-lactam ampicillin which disrupts peptidoglycan synthesis or the amino-glycoside streptomycin which interferes with protein synthesis, can also cause increase in mutation rates (Blázquez et al. 2018). The mechanisms of antibiotic-induced mutagenesis can be grouped in three largely overlapping classes: oxidative metabolism, the SOS, and the RNA polymerase sigma factor (RpoS)-regulated general stress responses (Blázquez et al. 2018). Apart from that, differences in recombination rates between bacterial species and also between commensal and pathogenic strains of the same species indicate recombination to be under strong selective pressure in bacterial populations. Recombination thus seems to contribute crucially in bacterial evolution (Blázquez et al. 2018). These phenomena have been reviewed in detail in published literature (Blázquez et al. 2018).

18.3.1.4 Antibiotics as Selectors of Hypermutable Bacteria

Bacterial cells with increased frequency of mutations called hypermutators or mutators have an enhanced probability of acquiring favorable mutations, including those that confer antibiotic resistance. By the process of natural selection of a single mutation, the proportion of mutated strain in the population can increase to as much as 0.5% from the original 0.001% (a frequency normally found in a typical *E. coli* population). Further with successive selection steps can increase this proportion up to 100% (LeClerc et al. 1996). In clinical settings like those during chronic infections, bacterial populations can face successive or alternative antibiotic challenges, mutant strains or mutators become more frequent in the challenged population. A given antibiotic can therefore select not only for resistance to itself but also, indirectly, for a higher capacity to acquire resistance to unrelated antimicrobials. Generally, inheritable hypermutation in bacteria is primarily due to mutations in genes of the mismatch repair (MMR) system (*mutS*, *mutL*, *uvrD*), which corrects mismatches produced during DNA replication (LeClerc et al. 1996). Hypermutable strains from pathogenic *E. coli* and *Salmonella enterica* isolates (with mutations over 1%) were identified to contain defective *mutS* allele with altered mismatch repair function. This mutant allele was found to increase the mutation rate and recombination frequency in different microbes, thus might be a factor also contributing towards emergence and spread of antibiotic resistance (LeClerc et al. 1996).

18.3.2 Acquired Resistance

A feature which would involve intra- and inter-species exchange of genetic material in bacteria would mean acquired resistance (Founou et al. 2016). Acquired resistance is a property in which an earlier sensitive bacterium transforms to resistant. This can happen as a result of (1) the genetic material undergoing mutation, or (2) the pathogen acquiring one or more antimicrobial resistance genes as a result of horizontal gene transfer through intra- or inter-species exchanges (Blázquez et al. 2018). Horizontal gene transfer offers new traits to the existing bacterial phenotype, and generates newer bacterial populations possessing a blend of new resistant genes and ways, thus conferring diverse resistance profiles (Founou et al. 2016).

For an organism to be a pathogen, the factors contributing pathogenicity should be transmitted in another organism that is ecologically compatible with the natural host. Pathogenic bacteria show their phenotypic traits which gathered over an extended period of time through evolution remaining in close vicinity with their potential hosts. In majority of cases, the virulence factors are located either in chromosomal gene clusters also known as pathogenicity islands or encoded in genetic accessory elements such as plasmids, or other transmissible elements and bacteriophages. This suggests that the evolution from an avirulent phenotype to a virulent type commonly implies the transfer of foreign fragments of DNA. The use of antimicrobial compounds may further co-select virulence and pathogenicity in these cases (Martínez and Baquero 2002).

Once genes imparting resistance to antibiotic have gathered by mutations, they can be exchanged among species by the mechanism of horizontal gene transfer. The mechanism can be executed by use of bacteriophage via transduction or through plasmids via transformation or through conjugation via conjugative transposons and integrons (Walsh 2010). Horizontal gene transfer has further identified in ecologies, such as water, manure, and soil, generating various levels of antibiotic resistant bacterial communities (Founou et al. 2016).

18.3.2.1 Plasmid

A plasmid is an extra-chromosomal genetic material, which carries various genes and performs their expression, similar to the chromosomal DNA. Such genes present on a plasmid are not generally found on the bacterial chromosome and they can attribute a specific function for improved adaptability, survival, and growth of the cell in the surrounding, for example, an R-plasmid (Martínez and Baquero 2002). Antibiotic resistance is generally present on regions on plasmids called jumping genes or transposons. These transposons are capable of jumping out of one plasmid and into another, thus carrying antibiotic resistance with them (Martínez and

Baquero 2002). This feature would cause the plasmids to build up resistance to a varied number of antibiotics, which are then subsequently available to be mobilized between intra- and inter-bacterial species (Martínez and Baquero 2002).

18.3.2.2 Conjugation

For conjugation to occur, the donor can only mobilize the fragment of DNA if it possesses a specialized type of transposon or a plasmid, later to bind to or recognize a suitable recipient cell and transfer the gene of interest. Genetic elements called as integrons have unique ability to capture genes carrying antibiotic resistance via site-specific recombination (Dougherty and Pucci 2012). The plasmid-borne integrons that encode for antibiotic resistance genes have been classified into four classes. Each integron has a conserved structure with an integrase gene (*int*), a distal recombination region (*attl*), and a promoter region (*Pant*). For example, class 1 integrons, the most common type, have been found to impart sulfonamide resistance (*sul1*), while class 2 integrons encode for trimethoprim resistance (Dougherty and Pucci 2012). Transfer of integrons by the mechanism of conjugation was observed between strains of *E. coli*. Class 1 integrons were identified in 16% of the isolates. In 95% of these cases, an adenyl acetyltransferase gene, aadA, conferring resistance to streptomycin, was detected, while in 19% of the isolates, a dfrA12 gene, conferring resistance to trimethoprim, was identified in integrons. The *sat1* gene was found in integrons from 7% of the isolates and conferred resistance to a not in use antibiotic—streptothricin (Singh et al. 2005).

18.3.2.3 Transduction

Transduction involves the transfer of DNA via bacteriophage from one bacterial species to another. When a virus collects a fragment of DNA conferring resistance from a bacterial cell and inserts it into another, the recipient cell thus can acquire resistance (Colavecchio et al. 2017). Genome sequences of *E. coli* O157:H7, *S. dysenteriae,* and Shiga toxin confirm the role in evolution of these human pathogens through bacteriophage mediated gene transfer (Colavecchio et al. 2017). The mechanisms involved in dissemination of resistance gene from one strain to another have been further reviewed (Colavecchio et al. 2017; Martínez and Baquero 2002).

18.3.2.4 Transformation

During transformation foreign free DNA is taken up by competent cell and is generally mediated by chromosomal genes. This mechanism is widely spread among bacterial population (Walsh 2010). However, the use of transformation mechanism to transfer genes for antimicrobial resistance, as intact functional piece, in food products has been less frequent (Verraes et al. 2013). The transferability of an oxazolidinone/phenicol resistance gene optrA in *Staphylococcus* isolates optrA was confirmed by transformation and conjugation assays (Li et al. 2016).

18.3.3 Invasion into the Host

18.3.3.1 Intestinal Mucus Lining

The intestinal mucus layer acts as primary barrier to limit invasion by commensal bacteria of the microflora or by food-borne pathogens (Kim et al. 2010; Ribet and Cossart 2015). The mucus is composed primarily of digestive enzymes, antimicrobial peptides, immunoglobulins, and glycoproteins called mucins. To cross the host mucus barrier is a challenge for the pathogenic bacteria during infection. Accordingly, few pathogens have developed mechanisms to invade this mucus layer to be able to reach epithelial cells. Such pathogens possess capabilities to either produce enzymes like proteases that can degrade target host mucins, or mobilize via flagella-based motility or offer resistance to antimicrobial products. These features have been observed in the case of pathogenic bacterium *Helicobacter pylori* colonized in gastric mucus (Ribet and Cossart 2015).

18.3.3.2 Adhesion

Adhesion is the prerequisite to initiate the process of interaction of the bacterial pathogen with the host (Ezepchuk 2017). Due to the mechanism of adhesion, the pathogen attempts to colonize to avoid mechanical clearing and offers selection advantage over the host endogenous flora (Katsikogianni et al. 2004). Interestingly, pathogens have evolved diverse molecular mechanisms assisting them to reach and adhere to host cells. For instance, by use of hair-like structures called as pili that present on the bacterial surface, by remodeling of host cytoskeleton through different effector proteins, or by production of cytoskeleton degrading enzymes (Ribet and Cossart 2015). Other structures involved in adhesion include lipopolysaccharides (LPS) and lipoteichoic acids (Ezepchuk 2017). Bacterial adhesins have been reviewed (Kline et al. 2009; Katsikogianni et al. 2004).

Pili Pili are polymeric hair-like appendages that project out from the outer surface of bacteria. They represent a first category of structures used by bacteria to bind to the host cells. These pili are composed of the major protein called pilin. These proteins are synthesized in the bacterial cytoplasm and are proteolytically processed followed by their translocation across the inner membrane. Once processed, pilin then only becomes competent for polymerization. A channel created by the secretin protein transports the assembled pili to the outer membrane (Ribet and Cossart 2015).

Remodeling of Host Cytoskeleton Few human pathogens like entero-hemorrhagic *Escherichia coli* (EHEC) and entero-pathogenic *E. coli* (EPEC) have elaborate mechanisms for creating their own receptors that can anchor the bacterium tightly to the host cell. During the process, bacterial effector protein Tir is delivered into the target cell via type III secretion system (T3SS) that form sharp needle-like structure spanning from the adhering bacterial cell to the host-cell plasma membrane. Tir thus provides direct connect between bacterial cell to the target cell through binding to bacterial intimin. Tir further recruits host cell cytoskeleton modulators such as the actin-related protein 2/3 (Arp2/3) complex which later extensively remodels the actin cytoskeleton, at the point of attachment of the pathogen (Kim et al. 2010; Ribet and Cossart 2015). Few bacterial pathogens can alter the cytoskeleton to invade and/or to become mobile in the host cell. To do so, they interact with G protein to modulate the actin filaments. This feature found in the bacterium *Salmonella enterica* injects the T3SS effector molecules, the proteins *SopE* and *SopE2*, into the host cell. These effectors activate the G-proteins in the target cell and further induce major alterations in the membrane leading to engulfment and internalization of the bacteria (Ezepchuk 2017). *Shigella flexneri* and *Listeria* spp. achieve their intracellular motility by making use of the bacterial effector proteins IcsA and ActA, respectively. IcsA is able to interact directly with the host cytoskeleton protein N-WASP (neural Wiskott–Aldrich syndrome protein), which then recruits a complex—the Arp2/3 complex capable of remodeling the actin filaments. Whereas the other effector protein ActA binds directly to both the Arp2/3 complex and the actin-associated protein VASP (vasodilator-stimulated phosphoprotein) (Bhavsar et al. 2007).

Cytoskeleton Degrading Enzymes Many pathogens are also able to produce different enzymes which have invasive function, in order to gain strong foothold in the host. They synthesize enzymes which have the ability to breakdown the cellular matrix, and penetrate the intercellular spaces to establish infection in the host (Ezepchuk 2017). For example, *Clostridium* synthesizes enzymes such as lecithinase, hyaluronidase, glycopeptidases, and proteases, which are able to digest extracellular matrix and invade the host tissue (Vedantam et al. 2012).

18.3.3.3 Intracellular Life of Pathogen in Host

Once the pathogen gains entry and get internalized as a transient membrane-enclosed compartment in the host cell, it can reside in, or escape from and survive to further establish the infection by multiplying in possible three host cellular environments: (1) lysosomes-like structures, possessing acidic pH and hydrolytic enzymes, (2) phagosomes-intracellular non-acidic vacuoles, which are usually modified by the pathogen, or (3) the cytoplasm (Ribet and Cossart 2015). For example, both *Salmonella* and *Legionella* spp. can occupy and reside in protected intracellular vacuole, after internalization, thus contributing to their pathogenesis (Ezepchuk 2017). In many cases, interaction of bacterial proteins with host cadherins or integrins can assist bacteria to be get internalized into non-phagocytic cells. The small size of bacteria aids its engulfment, by firmly attaching to extracellular matrix or neighboring cells. As observed in *Listeria*, the interaction between the surface protein, InlA, with the host receptor E-cadherin resulted into successful intersection of the intestinal barrier (Ribet and Cossart 2015; Ezepchuk 2017). After internalization in host post-entry, intracellular bacteria can survive and replicate in the host in three cellular environments: (1) vacuoles, those with acidic pH and hydrolytic enzymes, (2) intracellular non-acidic vacuoles, and (3) the cytosol, as a site to hide for some pathogens after escape from their internalization vacuole. For example, after internalization, Salmonella can hide in vacuoles that become acidic but are not lysosomes (Ezepchuk 2017). Only after a pathogen translocates through intestinal non-phagocytic cells and suppresses the host defense mechanisms, it is able to develop a stable infection in the host (Ezepchuk 2017; Ribet and Cossart 2015).

18.3.3.4 Interaction with Host's Immune Response

Host defense in the form of innate immunity get activated after its initial interaction with microbial antigens. Various types of antigenic determinants such as bacterial flagellin, the peptidoglycan of Gram-positive bacteria, the lipopolysaccharides of Gram-negative bacteria, and microbial DNA get involved in this recognition through pathogen-associated molecular patterns (PAMPs) (van Baarlen et al. 2007). As they are found in both pathogenic and nonpathogenic microorganisms, they are also termed as microbial-associated molecular patterns (MAMPs). Required for the microbial fitness, they are however not produced by the host (van Baarlen et al. 2007). Sialic acid-specific lectins form another class of pattern recognition molecules involved in host–pathogen interactions (Ezepchuk 2017). A set of toxic molecules named as super-antigens comprise of a family of exotoxins, such as enterotoxins, toxic shock syndrome toxin (TSST), and pyrogenic toxins which possess indirect mechanism of damaging the host (Ezepchuk 2017). Endotoxins (LPS) of Gram-negative bacteria are found on the outer membrane of the bacterial cell

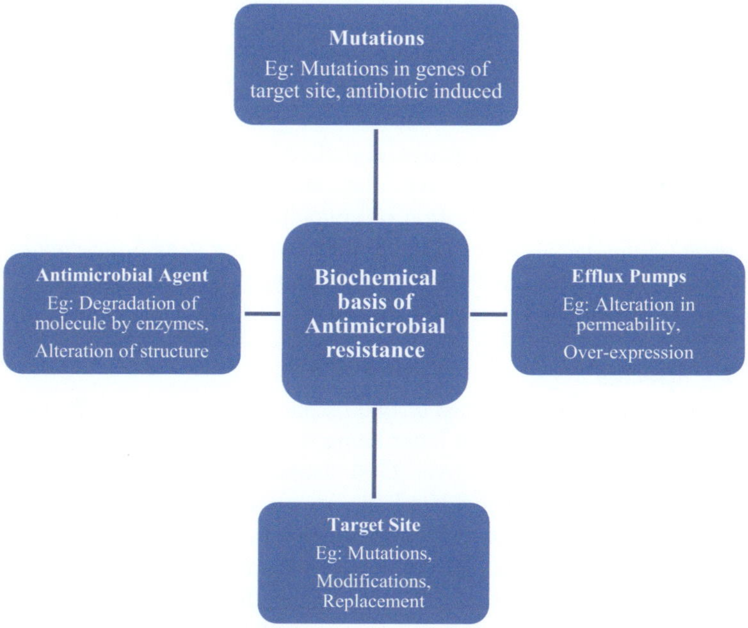

Fig. 18.2 Biochemical basis of antimicrobial resistance

wall. Endotoxins target the blood, and lymphoid cells along with components of immune and compliment systems (Ezepchuk 2017). Along with selective adhesion, the invasive and penetrative function is another feature, adopted by pathogens to survive into the new ecological spaces (Ezepchuk 2017). The biochemical mechanisms that form the basis of pathogenesis and virulence among different pathogens have been summarized in Fig. 18.2. The biochemical aspects of pathogenesis have also been reviewed in detail (Munita and Arias 2016; Mazel and Davies 1999; Martínez and Baquero 2002).

18.4 Food Borne Pathogens

In 2017, WHO published list of "priority pathogens"—families of bacteria that pose serious threat to human health, all of which are antibiotic resistant. Of which food-borne pathogens belonging to *Enterobacteriaceae* that are carbapenem-resistant and ESBL-producing have been grouped under "critical priority"; *Enterococcus faecium* that is vancomycin-resistant, *Staphylococcus aureus* that is methicillin-resistant, vancomycin-intermediate and resistant, *Helicobacter pylori* that is clarithromycin-resistant, *Campylobacter* spp. that is fluoroquinolone-resistant, *Salmonellae* spp. that is fluoroquinolone-resistant are labeled as "high priority" and *Shigella* spp.-fluoroquinolone-resistant grouped under "medium priority" (WHO Press 2013) The following section highlights epidemiology and pathogenicity of food-borne etiological agents few of which as well fall under the category of WHO priority pathogens.

18.4.1 Salmonella

Salmonella enterica is a Gram-negative, flagellated facultative anaerobe that causes salmonellosis with diarrhea, fever, and abdominal cramps as predominant symptoms. It mainly occurs after ingestion of contaminated food (Jo and Park 2019). Drug-resistant *Salmonella* causes 100,000 infections per year. *Salmonella* includes serotypes other than Typhi, as Paratyphi A, Paratyphi B, and Paratyphi C (United States Centers for Disease Control 2013).

18.4.1.1 Species

The genus *Salmonella* is comprised of two taxonomical species, *Salmonella bongori* and *Salmonella enterica*, with latter being the all medically relevant salmonellae. *Salmonella enterica* possess over 2500 distinct serovars. The pathogen can be host-adapted, host-restricted, or generalistic, depending on the broad range of hosts that it can infect. The pathogen is found all over the world in food chain, and is often associated with outbreaks of food-borne disease (VT Nair et al. 2018).

Salmonella strains associated with a high morbidity and mortality are grouped into typhoid *Salmonella*, and other which cause self-limited illness, as non-typhoid *Salmonella* (NTS). In humans, they are capable of causing bacteremia, gastroenteritis, enteric fever, and a carrier state (Eng et al. 2015). Based on their genomic and biochemical characteristics, *S. enterica* is further grouped into six subspecies. Over 2500 serotypes have been identified. Subspecies causing majority of *Salmonella* infections in mammals is *S. enterica* subsp. *enterica* (I) (Eng et al. 2015). Epidemiology and resistance: The contaminated water or foods are major causes of transmittance (Jo and Park 2019; VT Nair et al. 2018; Eng et al. 2015). In the early 1960s, *Salmonella* emerged resistant to chloramphenicol, a single antibiotic and since then, emergence of multi drug resistant strains has increased several folds (Jo and Park 2019). Multidrug-resistant *S. Typhimurium* DT104 is resistant to five antimicrobial drugs: streptomycin, sulfonamides, ampicillin, tetracycline, and chloramphenicol (Poppe et al. 1998). Animals (mainly swine and bovine) are the major reservoirs for observed antibiotic resistance patterns. The spread of antibiotic resistant strains from chicken and humans is alarming. Strains are mostly susceptible to fluoroquinolones (Wang et al. 2019). High incidences of mutators of over 1% have been reported among isolates of pathogenic *Salmonella enterica* (LeClerc et al. 1996). In 2000, out of 22 million cases of enteric fever, 200,000 cases were reported fatal worldwide. The mortality rate has been as high in spite of antibiotic therapy. Many Asian countries have high cases of occurrence of enteric fever, with over 100 cases per one lakh population annually out of which, India along with Pakistan have reported very high incidence rates of 214.2 cases and 451.7 cases per 100,000 population, respectively (Eng et al. 2015).

18.4.1.2 Transmission and Infection

The action mediated by flagella/flagellin fimbriae and/or non-fimbrial adhesion (lipopolysaccharide) offers *Salmonella* to adhere to the host, invade and enter intestinal epithelial cell. *Salmonella* pathogenicity island I and 2 (SPI-I and 2) express effector proteins called the Type III secretion system (T3SS) 1 and 2 and Type I secretion systems, respectively, which are used to invade the host. These secreted proteins are capable of suppressing the host immune system, for example, *PhoP/PhoQ* and *Inv*A genes (Andrews-Polymenis et al. 2010; Eng et al. 2015).

18.4.1.3 Pathogenicity Factors and Virulence

Two major groups of mechanism of resistance have been associated with *Salmonella* and they include the use of intrinsic mechanisms, such as inactivation of enzymes, alteration of target site and drug resistant efflux pumps, and acquired mechanisms, such as mutation, horizontal, and vertical gene transfers (Akinyemi and Ajoseh 2017). Many good reviews describe the mechanism of pathogenesis of *Salmonella* in much detail (Eng et al. 2015; Munita and Arias 2016).

18.4.2 Listeria monocytogenes

Listeria monocytogenes, a Gram-positive, non-spore-forming rod, can cause a severe food-borne disease characterized by meningitis, meningo-encephalitis, materno-fetal perinatal infections, and febrile gastroenteritis post-consumption of highly contaminated food products. *Listeria* species can sustain a broad pH and temperature range of 4.3–9.6 and 1–45 °C, respectively, and salt concentrations up to 10%. *Listeria* is considered as an opportunistic pathogen in immunocompromised individuals (Cossart and Toledo-Arana 2008).

18.4.2.1 Species

Within the genus *Listeria* there are seven species *L. monocytogenes*, *L. innocua*, *L. ivanovii*, *L. seeligeri*, *L. grayi*, *L. welshimeri*, and *L. murrayi*, of which *L. monocytogenes is* pathogenic, and capable of causing disease in both animals and humans, while *L. ivanovii* is capable of causing disease in sheep. In majority of cases, ingestion of contaminated food products causes listeriosis in humans. So far 13 serovars have been reported (Roberts and Wiedmann 2003).

18.4.2.2 Epidemiology

Listeriosis in humans caused by *L. monocytogenes* occurs at a mortality rate of 20–30% and occurs mainly as a mild noninvasive or an invasive gastrointestinal illness. *L. monocytogenes* is capable of crossing the blood–brain, the intestinal, and the placental barriers in a host. When phagocytosed, it is capable of escaping intracellular killing and can attack many types of non-phagocytic cells. *L. monocytogenes* can adhere to different surfaces, such as metal (stainless steel), glassware, plasticware, and rubber by forming biofilm (Roberts and Wiedmann 2003). A study reported occurrence, prevalence, and contamination levels due to a virulent *L. monocytogenes* having partial antimicrobial resistance, and serotypes associated with listeriosis in retail ready-to-eat foods in China (Chen et al. 2014).

18.4.2.3 Transmission and Infection

After entering into host tissues at the small intestine, the infection cycle begins. Following oral ingestion, *L. monocytogenes* survives gastric pH and accesses lymph nodes and blood vessels by invading intestinal epithelial cells. *L. monocytogenes* use cell surface proteins like internalin A (*InlA*) and internalin B (*InlB*) for attachment and entry (Cegelski et al. 2008). E-cadherin acts as cellular receptor for internalin, due to which *L. monocytogenes* invades into non-phagocytic cells or epithelial cells (Vivant et al. 2013). Once it gains entry in a host cell, *L. monocytogenes* use listeriolysin-O and zinc-dependent metalloprotease (*Mpl*) to sustain acidic pH in vacuole and evade lysis by getting enclosed in a membrane-bound vacuole. Later it breaks the vacuole and gets localized in the cytoplasm and migrates to neighboring cells to mitigate its effects. Pathogenicity factors of *L. monocytogenes* have been described in further detail (Roberts and Wiedmann 2003; Vivant et al. 2013).

18.4.2.4 Pathogenicity and Virulence Factors

Factors such as sugar uptake system, a peptidoglycan deacetylase, and a lipoate protein ligase along with other factors like a bile salt hydrolase and stress response factor are required for intracellular life in the host cell and identified to play important role during various stages of infection (Vivant et al. 2013). *L. monocytogenes* and *L. ivanovii* possess six virulence genes clustered pathogenicity island. The major regulator of gene expression is *PrfA*. The *prfA* gene functions as a transcriptional activator for all the other genes on the island, and thus is integral to *L. monocytogenes* virulence (Vivant et al. 2013).

18.4.3 Methicillin-Resistant S. aureus (MRSA)

Staphylococcus aureus, one of the first to be described as major human pathogen, possesses a diverse set of virulence factors and possesses ability to attain resistance to most antibiotics, makes *S. aureus* a "superbug." MRSA is *S. aureus* strain that has become resistant to beta-lactams, including methicillin (i.e., methicillin-resistant *S. aureus*). With over 80,000 cases per year due to severe MRSA infections and mortality over 11,000 per year, CDC describes MRSA as a serious threat to humans (Lakhundi and Zhang 2018; United States Centers for Disease Control 2013).

18.4.3.1 Species

Sta*phylococcus aureus* is a Gram-positive, spherical bacterium belonging to the *Staphylococcaceae* family. *S. aureus* is a commensal and present asymptomatically on the human body and guts of healthy individuals. It seldom causes life-threatening infections in healthy individuals. Wide clinical use of methicillin led to the appearance of MRSA in recent decades has witnessed the appearance of new MRSA clones (Sergelidis and Angelidis 2017).

18.4.3.2 Epidemiology

Not only humans, MRSA also inhabits in domesticated livestock, pets, aquatic, captive or free-living wild terrestrial species and acts as a permanent reservoir for human MRSA infections (Sergelidis and Angelidis 2017). *S. aureus* due to its diverse set of virulence factors and toxins can cause many toxin-mediated diseases like staphylococcal food-borne diseases and toxic shock syndrome. Patients in healthcare settings can also get MRSA infections. The origin and evolution of MRSA, details of their mobile genetic elements, molecular epidemiology, and global emergence of new MRSA clones have been recently reviewed in detail (Sergelidis and Angelidis 2017; Lakhundi and Zhang 2018).

18.4.3.3 Transmission and Infection

Infections due to MRSA cause higher mortality. In nosocomial infections, *S. aureus* is the causative agent in almost 25–50% cases further adding to longer hospital stays and associated health care costs. The strains principally are capable of causing skin and soft tissue infections. A wide repertoire of cell-associated virulence factors such as cytotoxins, enzymes, exotoxins, staphylococcal enterotoxins, and toxic shock syndrome toxin (TSST)-1 can cause damage to the host (Pinchuk et al. 2010).

18.4.3.4 Pathogenicity and Virulence Factors

Apart from above-mentioned pathogenicity factors, MRSA strains synthesize an altered penicillin-binding protein (PBP) coded by an acquired gene, *mecA* that has reduced affinity for most semisynthetic penicillins (Lakhundi and Zhang 2018; Sergelidis and Angelidis 2017). These properties put together make it an adaptable pathogen capable of causing wide range of infections (Lakhundi and Zhang 2018).

18.4.4 Campylobacter Jejuni

Campylobacter jejuni, a Gram-negative, motile, helical proteobacterium, is an etiological agent known to cause severe to life-threatening diarrhea. Other symptoms include fever, nausea and vomiting, watery to bloody stools that can be fatal in infants, elderly, and immune-compromised persons. In the developed countries, these infections outnumber combined infections caused by *E. coli* and *Salmonella* spp. and infect a large population in North American, Australian, and European each year (Dasti et al. 2010). *Campylobacter* can spread through contaminated food like raw, undercooked chicken. With over 300,000 deaths reported every year due to antibiotic-resistant *Campylobacter*, both the WHO and CDC have declared this infection a serious threat to public health (Karp et al. 2017; United States Centers for Disease Control 2013).

18.4.4.1 Species

C. jejuni is a harmless zoonotic organism that is present as commensal in the intestines of birds and other livestock (Dasti et al. 2010). Very diverse genetic makeup is found among isolates of *Campylobacter* in animals and humans contributing to variations in clinical manifestations. *Campylobacter* belongs to the family *Campylobacteriaceae* and comprises of 16 species. The strains, *C. jejuni* and *C. coli* are causes for over 90% of all human *Campylobacter* infections globally (Dasti et al. 2010).

18.4.4.2 Epidemiology

An infective dose as low as 5–800 organisms and ingestion of *C. jejuni* contaminated food especially poultry cause majority of human infections. The widespread occurrence, and emergence of strains resistant to the drugs used to treat human campylobacteriosis (fluoroquinolones and macrolides), is a cause of concern (Gilbert et al. 2007).

18.4.4.3 Transmission and Infection

C. jejuni is a distinct bacterium, capable to carry out N-linked glycosylation of more than 30 proteins related to adherence, invasion, and colonization. In spite of having a relatively small genome size, it lacks hallmark virulence factors such as type III secretion systems which are necessary in pathogenesis of other enteric organisms. It also lacks important stress response elements such as the sigma factor *RpoS*, but is capable to survive in diverse environmental niches (Dasti et al. 2010). After gaining entry, *C. jejuni* colonizes the lower intestinal tract often asymptomatically (Backert and Hofreuter 2013). Furthermore, invasion and translocation past the intestinal epithelial cell barrier are essential steps for progressing the infection (Dasti et al. 2010).

18.4.4.4 Pathogenicity and Virulence Factors

A multidrug efflux system, CmeABC, belonging to a resistance-nodulation-division (RND) type of MDR pump has been identified as an important mechanism for antimicrobial resistance. The efflux is coded by a three-gene operon (*cmeABC*) and confers resistance structurally diverse antimicrobials (Martinez and Lin 2006). Biofilm formation, bacterial flagella, amino acid uptake and utilization, ability to respire by use of lactate as carbon source have also been implicated in survival mechanisms and pathogenesis (Martinez and Lin 2006). Reviews on pathogenicity and disease mechanisms have been presented (Martinez and Lin 2006).

18.4.5 Shigella

Shigella is a Gram-negative bacterium belonging to the family *Enterobacteriaceae* and capable of causing shigellosis, a bacillary dysentery. It claims almost 27,000 deaths per year and CDC describes drug-resistant *Shigella* as serious threat (Marteyn et al. 2012; United States Centers for Disease Control 2013).

18.4.5.1 Species

Shigella includes four subgroups: *S. boydii*, *S. flexneri*, *S. sonnei*, and *S. dysenteriae*. Many serotypes have been reported among Shigella isolates: 15 serotypes have been reported for S. dysenteriae, 20 serotypes for *S. boydii*, 14 serotypes for *S. flexneri*, and *S. sonnei* with a single serotype. *S. dysenteriae* produces the Shiga toxins and causes severe bacillary dysentery, whereas *S. flexneri* and *S. sonnei* are endemically associated (Marteyn et al. 2012).

18.4.5.2 Transmission and Infection

Shigella spreads through contaminated surfaces, food, or water, or direct contact. After oral ingestion, *Shigella* sustains gastric acidic pH and crosses various parts of the GIT to finally reach the site of its colonization, i.e., colon and invades epithelial cells by use of secretion system. Furthermore, *Shigella* infection remains localized to the epithelial layer, where it may cause mucosal abscesses and ulcerations leading to extreme tissue damage and symptoms of dysentery (Marteyn et al. 2012).

18.4.5.3 Pathogenicity and Virulence Factors

Shigella virulence is attributed to the Type III-secretion-system (T3SS) present on plasmid, along with a group of secretory proteins which are the virulence effectors and to chaperone proteins. Many autotransporters which include Type V-secretion-system (T5SS), *SepA,* and *IcsA* are also encoded by the plasmid. The *RpoS*, *CpxA/cpxR* are the stress response component systems which offer resistance to the gastric pH and temperature, to *Shigella*. Several operons which encode the invasion genes are under the influence of *VirF*, a global transcriptional regulator (Marteyn et al. 2012).

18.5 Conclusion

Drug-resistance among food-borne pathogens is a matter of growing concern globally. It is evident that in order to acclimatize environmental challenges and survive in the ecological niches under the influence of antibiotic selection pressure, the food-borne pathogens are undergoing continuous irreversible evolution and in process display new virulence properties. Increased trade and traveling would further add to the rise and spread of resistance with large possibilities of gene exchange and modification among the bacteria at a scale beyond our imagination (Aarestrup et al. 2008).

To tackle this serious issue, consolidated efforts are required on two broad fronts. Firstly, experimental evidences gathered over previous few decades demonstrate multi-dimensional nature of intrinsic bacterial virulence, pathogenicity, and acquired resistances. Therefore, it is crucial that we determine the mechanisms of virulence and the consequences of host–pathogen interactions from both the pathogen and host perspective (Cegelski et al. 2008). Recent progress in the techniques used for identification of bacterial virulence along with growing knowledge of genomes (WHO 2018) of several pathogens should be used to identify novel targets to develop to design multidisciplinary and novel treatments.

Secondly, there have been several programs monitoring antibiotic consumption and the spread of antimicrobial resistance that have been implemented in countries such as EU, USA, Japan, Sweden, and Denmark and recently in countries such as India, China, Thailand, and South Africa (Founou et al. 2016). A consolidated action plan is still required to take proactive steps towards prevention of spread of food borne pathogens. This has been rightly pointed out by WHO, in its current Global Action Plan (Centers for Disease Control and Prevention 2018) for containment measures for antimicrobial resistances, and now it is the time for its execution.

Thus, different stake holders, i.e., scientists, medical professionals, governmental organizations, and general public have to come together to put a strong fight against food-borne and multidrug resistant pathogens for human existence.

Acknowledgement The author would like to express her deepest thanks to Dr. Nivedita M. Matkar for carefully reviewing the manuscript and valuable feedback.

References

Aarestrup FM, Wegener HC, Collignon P (2008) Resistance in bacteria of the food chain: epidemiology and control strategies. Expert Rev Anti-Infect Ther 6:733–750

Acar J, Röstel B (2001) Antimicrobial resistance: an overview antimicrobial resistance—a responsibility of the office international des epizooties why has the office international des epizooties. Rev Sci Tech 20:797–810

Akinyemi KO, Ajoseh SO (2017) Factors contributing to the emergence and spread of antibiotics resistance in *Salmonella* species. In: Current topics in *Salmonella* and Salmonellosis. InTech

Al-Humam NA (2019) Detection of *Escherichia coli*, *Salmonella* spp. and *Staphylococcus aureus* in ready-to-eat food in Al-Ahsa Province, Saudi Arabia. J Nutr Food Sci 9:754. https://doi.org/10.4172/2155-9600.1000754

Andrews-Polymenis HL, Bäumler AJ, McCormick BA, Fang FC (2010) Taming the elephant: *Salmonella* biology, pathogenesis, and prevention. Infect Immun 78:2356–2369. https://doi.org/10.1128/IAI.00096-10

Aung KT, Hsu LY, Koh TH et al (2017) Prevalence of methicillin-resistant *Staphylococcus aureus* (MRSA) in retail food in Singapore. Antimicrob Resist Infect Control 6:94. https://doi.org/10.1186/s13756-017-0255-3

Backert S, Hofreuter D (2013) Molecular methods to investigate adhesion, transmigration, invasion and intracellular survival of the foodborne pathogen campylobacter jejuni. J Microbiol Methods 95:8–23. https://doi.org/10.1016/j.mimet.2013.06.031

Bhavsar AP, Guttman JA, Finlay BB (2007) Manipulation of host-cell pathways by bacterial pathogens. Nature 449:827–834. https://doi.org/10.1038/nature06247

Blázquez J, Rodríguez-Beltrán J, Matic I (2018) Antibiotic-induced genetic variation: how it arises and how it can be prevented. Annu Rev Microbiol 72:209–230. https://doi.org/10.1146/annurev-micro-090817-062139

Bridier A, Sanchez-Vizuete P, Guilbaud M et al (2015) Biofilm-associated persistence of food-borne pathogens. Food Microbiol 45:167–178. https://doi.org/10.1016/j.fm.2014.04.015

Brooks JP, Adeli A, McLaughlin MR (2014) Microbial ecology, bacterial pathogens, and antibiotic resistant genes in swine manure wastewater as influenced by three swine management systems. Water Res 57:96–103. https://doi.org/10.1016/j.watres.2014.03.017

Bryan F (1978) Factors that contribute to outbreaks of food borne disease. J Food Prot 41:816–827

Cappitelli F, Polo A, Villa F (2014) Biofilm formation in food processing environments is still poorly understood and controlled. Food Eng Rev 6:29–42. https://doi.org/10.1007/s12393-014-9077-8

Casadevall A, Pirofski L (2000) Host-pathogen interactions: basic concepts of microbial commensalism, colonization, infection, and disease. Infect Immun 68:6511–6518. https://doi.org/10.1128/IAI.68.12.6511-6518.2000

Cegelski L, Marshall GR, Eldridge GR, Hultgren SJ (2008) The biology and future prospects of antivirulence therapies. Nat Rev Microbiol 6:17–27. https://doi.org/10.1038/nrmicro1818

Centers for Disease Control and Prevention (CDC) (2018) The core elements of human antibiotic stewardship programs in resource -limited settings: National and Hospital Levels. US Department of Health and Human Services, Atlanta, GA. https://www.cdc.gov/antibiotic-use/healthcare/implementation.html

Chen M, Wu Q, Zhang J et al (2014) Prevalence and characterization of *Listeria monocytogenes* isolated from retail-level ready-to-eat foods in South China. Food Control 38:1–7. https://doi.org/10.1016/j.foodcont.2013.09.061

Colavecchio A, Cadieux B, Lo A, Goodridge LD (2017) Bacteriophages contribute to the spread of antibiotic resistance genes among foodborne pathogens of the Enterobacteriaceae family—a review. Front Microbiol 8:1–13. https://doi.org/10.3389/fmicb.2017.01108

Conter M, Paludi D, Zanardi E et al (2009) Characterization of antimicrobial resistance of food-borne *Listeria monocytogenes*. Int J Food Microbiol 128:497–500. https://doi.org/10.1016/j.ijfoodmicro.2008.10.018

Cossart P, Toledo-Arana A (2008) *Listeria monocytogenes*, a unique model in infection biology: an overview. Microbes Infect 10:1041–1050. https://doi.org/10.1016/j.micinf.2008.07.043

Daniel DS, Lee SM, Dykes GA, Rahman S (2015) Public health risks of multiple-drug-resistant *Enterococcus* spp. in Southeast Asia. Appl Environ Microbiol 81:6090–6097. https://doi.org/10.1128/AEM.01741-15

Dasti JI, Tareen AM, Lugert R et al (2010) *Campylobacter jejuni*: a brief overview on pathogenicity-associated factors and disease-mediating mechanisms. Int J Med Microbiol 300:205–211. https://doi.org/10.1016/j.ijmm.2009.07.002

Di Bonaventura G, Piccolomini R, Paludi D et al (2008) Influence of temperature on biofilm formation by *Listeria monocytogenes* on various food-contact surfaces: relationship with motility and cell surface hydrophobicity. J Appl Microbiol 104:1552–1561. https://doi.org/10.1111/j.1365-2672.2007.03688.x

Dougherty TJ, Pucci MJ (2012) Antibiotic discovery and development. Springer, Berlin. https://doi.org/10.1007/978-1-4614-1400-1_10

Doyle MP, Busta F, Cords BR et al (2006) Antimicrobial resistance: implications for the food system: an expert report, funded by the IFT foundation. Compr Rev Food Sci Food Saf 5:71–137. https://doi.org/10.1111/j.1541-4337.2006.00004.x

Du Y, Wang C, Ye Y et al (2018) Molecular identification of multidrug-resistant *Campylobacter* species from diarrheal patients and poultry meat in Shanghai, China. Front Microbiol 9:1–8. https://doi.org/10.3389/fmicb.2018.01642

EMA (2019) European medicines agency (EMA). Committee for medicinal products for veterinary use (CVMP). Eur Med Agency 44:1–67

Eng S-K, Pusparajah P, Ab Mutalib N-S et al (2015) Salmonella: a review on pathogenesis, epidemiology and antibiotic resistance. Front Life Sci 8:284–293. https://doi.org/10.1080/21553769.2015.1051243

Ezepchuk YV (2013) Biology of pathogenicity (theoretical review). Annu Res Rev Biol 3:805–813

Ezepchuk YV (2017) Biological concept of bacterial pathogenicity (theoretical review). Adv Microbiol 07:535–544. https://doi.org/10.4236/aim.2017.77042

Founou LL, Founou RC, Essack SY (2016) Antibiotic resistance in the food chain: a developing country-perspective. Front Microbiol 7:1–19. https://doi.org/10.3389/fmicb.2016.01881

Gasson J, Blockman M, Willems B (2018) Antibiotic prescribing practice and adherence to guidelines in primary care in the Cape Town Metro District, South Africa. S Afr Med J 108:304. https://doi.org/10.7196/samj.2018.v108i4.12564

Ge B, White DG, McDermott PF et al (2003) Antimicrobial-resistant *Campylobacter* species from retail raw meats. Appl Environ Microbiol 69:3005–3007. https://doi.org/10.1128/AEM.69.5.3005-3007.2003

Gilbert JM, White DG, McDermott PF (2007) The US national antimicrobial resistance monitoring system. Future Microbiol 2:493–500. https://doi.org/10.2217/17460913.2.5.493

Golob M, Pate M, Kušar D et al (2019) Antimicrobial resistance and virulence genes in *Enterococcus faecium* and *Enterococcus faecalis* from humans and retail red meat. Biomed Res Int 2019:1–12. https://doi.org/10.1155/2019/2815279

Gousia P, Economou V, Sakkas H et al (2011) Antimicrobial resistance of major foodborne pathogens from major meat products. Foodborne Pathog Dis 8:27–38. https://doi.org/10.1089/fpd.2010.0577

Gwida M (2015) Prevalence and characterization of antibiotic resistance food borne pathogens isolated from locally produced chicken raw meat and their handlers. J Dairy Vet Anim Res 2:238–244. https://doi.org/10.15406/jdvar.2015.02.00062

Hald T, Aspinall W, Devleesschauwer B et al (2016) World Health Organization estimates of the relative contributions of food to the burden of disease due to selected foodborne hazards: a structured expert elicitation. PLoS One 11:1–35. https://doi.org/10.1371/journal.pone.0145839

Hamilton WL, Wenlock R (2016) Antimicrobial resistance: a major threat to public health. Cambridge Med J. https://doi.org/10.7244/cmj.2016.01.001

Jo SH, Park JM (2019) The dark side of organic vegetables: interactions of human enteropathogenic bacteria with plants. Plant Biotechnol Rep 13:105–110. https://doi.org/10.1007/s11816-019-00536-1

Karp BE, Tate H, Plumblee JR et al (2017) National Antimicrobial Resistance Monitoring System: two decades of advancing public health through integrated surveillance of antimicrobial resistance. Foodborne Pathog Dis 14:545–557. https://doi.org/10.1089/fpd.2017.2283

Katsikogianni M, Missirlis YF, Harris L, Douglas J (2004) Concise review of mechanisms of bacterial adhesion to biomaterials and of techniques used in estimating bacteria-material interactions. Eur Cells Mater 8:37–57. https://doi.org/10.22203/eCM.v008a05

Kim M, Ashida H, Ogawa M et al (2010) Bacterial interactions with the host epithelium. Cell Host Microbe 8:20–35. https://doi.org/10.1016/j.chom.2010.06.006

Klein EY, Van Boeckel TP, Martinez EM et al (2018) Global increase and geographic convergence in antibiotic consumption between 2000 and 2015. Proc Natl Acad Sci 115:E3463–E3470. https://doi.org/10.1073/pnas.1717295115

Kline KA, Fälker S, Dahlberg S et al (2009) Bacterial Adhesins in host-microbe interactions. Cell Host Microbe 5:580–592. https://doi.org/10.1016/j.chom.2009.05.011

Laaksonen S, Oksanen A, Julmi J et al (2017) Presence of foodborne pathogens, extended-spectrum β-lactamase -producing Enterobacteriaceae, and methicillin-resistant *Staphylococcus aureus* in slaughtered reindeer in northern Finland and Norway. Acta Vet Scand 59(2):2. https://doi.org/10.1186/s13028-016-0272-x

Lakhundi S, Zhang K (2018) Methicillin- resistant *Staphylococcus aureus*: molecular characterization, evolution, and epidemiology. Clin Microbiol Rev 31:e00020–e00018. https://doi.org/10.1128/CMR.00020-18

LeClerc JE, Li B, Payne WL, Cebula TA (1996) High mutation frequencies among *Escherichia coli* and *Salmonella* pathogens. Science (80-) 274:1208–1211. https://doi.org/10.1126/science.274.5290.1208

Leung E, Weil DE, Raviglione M, Nakatani H (2011) The WHO policy package to combat antimicrobial resistance. Bull World Health Organ 89:390–392. https://doi.org/10.2471/BLT.11.088435

Li D, Wang Y, Schwarz S et al (2016) Co-location of the oxazolidinone resistance genes optrA and cfr on a multiresistance plasmid from *Staphylococcus sciuri*. J Antimicrob Chemother 71:1474–1478. https://doi.org/10.1093/jac/dkw040

Long KS, Vester B (2012) Resistance to linezolid caused by modifications at its binding site on the ribosome. Antimicrob Agents Chemother 56:603–612. https://doi.org/10.1128/AAC.05702-11

Magee JT, Pritchard EL, Fitzgerald KA et al (1999) Antibiotic prescribing and antibiotic resistance in community practice: retrospective study, 1996-8. BMJ 319:1239–1240. https://doi.org/10.1136/bmj.319.7219.1239

Mąka Ł, Maćkiw E, Ścieżyńska H, Popowska M (2015) Occurrence and antimicrobial resistance of *Salmonella* spp. isolated from food other than meat in Poland. Ann Agric Environ Med 22:403–408. https://doi.org/10.5604/12321966.1167701

Marshall BM, Levy SB (2011) Food animals and antimicrobials: impacts on human health. Clin Microbiol Rev 24:718–733. https://doi.org/10.1128/CMR.00002-11

Marteyn BS, Gazi AD, Sansonetti PJ (2012) *Shigella*: a model of virulence regulation in vivo. Gut Microbes 3:104–120. https://doi.org/10.4161/gmic.19325

Martínez L, Baquero F (2002) Interactions among strategies associated with bacterial infection: pathogenicity, epidemicity, and antibiotic resistance. Clin Microbiol Rev 15:647–679. https://doi.org/10.1128/CMR.15.4.647

Martinez A, Lin J (2006) Effect of an efflux pump inhibitor on the function of the multidrug efflux pump CmeABC and antimicrobial resistance in *Campylobacter*. Foodborne Pathog Dis 3:393–402. https://doi.org/10.1089/fpd.2006.3.393

Mathur S, Singh R (2005) Antibiotic resistance in food lactic acid bacteria—a review. Int J Food Microbiol 105:281–295. https://doi.org/10.1016/j.ijfoodmicro.2005.03.008

Mazel D, Davies J (1999) Antibiotics, present and future. microbial biosystems: new frontiers. In: Bell CR, Brylinsky M, Johnson-Green P (eds) Proceedings of the 8th international symposium on microbial ecology. Atlantic Canada Society for Microbial Ecology, Halifax

McDermott PF, Zhao S, Wagner DD et al (2002) The food safety perspective of antibiotic resistance. Anim Biotechnol 13:71–84. https://doi.org/10.1081/ABIO-120005771

McGettigan P, Roderick P, Kadam A, Pollock AM (2017) Access, watch, and reserve antibiotics in India: challenges for WHO stewardship. Lancet Glob Health 5:e1075–e1076. https://doi.org/10.1016/S2214-109X(17)30365-0

Munita JM, Arias CA (2016) Mechanisms of antibiotic resistance. Microbiol Spectr 4(2). https://doi.org/10.1128/microbiolspec.VMBF-0016-2015

Murray AK, Zhang L, Yin X, Zhang T, Buckling A, Snape J, Gaze WH (2018) Novel insights into selection for antibiotic resistance in complex microbial communities. mBio 9:e00969–e00918. https://doi.org/10.1128/mBio.00969-18
Narasanna R, Chavadi M, Oli A, Chandrakanth K (2017) Effect of subinhibitory concentration of cefotaxime on biofilm formation. J Microbiol Infect Dis 7:67–74. https://doi.org/10.5799/ahinjs.02.2017.02.0257
Nikaido H (1996) Multidrug efflux pumps of gram-negative bacteria. J Bacteriol 178:5853–5859. https://doi.org/10.1128/jb.178.20.5853-5859.1996
Odeyemi OA, Sani NA (2016) Antibiotic resistance and burden of foodborne diseases in developing countries. Futur Sci OA 2:FSO139. https://doi.org/10.4155/fsoa-2016-0023
Palframan SL, Kwok T, Gabriel K (2012) Vacuolating cytotoxin A (VacA), a key toxin for *Helicobacter pylori* pathogenesis. Front Cell Infect Microbiol 2:1–9. https://doi.org/10.3389/fcimb.2012.00092
Pesavento G, Ducci B, Comodo N, Nostro AL (2007) Antimicrobial resistance profile of *Staphylococcus aureus* isolated from raw meat: a research for methicillin resistant *Staphylococcus aureus* (MRSA). Food Control 18:196–200. https://doi.org/10.1016/j.foodcont.2005.09.013
Pinchuk IV, Beswick EJ, Reyes VE (2010) Staphylococcal enterotoxins. Toxins (Basel) 2:2177–2197. https://doi.org/10.3390/toxins2082177
Poppe C, Smart N, Khakhria R et al (1998) *Salmonella typhimurium* DT104: a virulent and drug-resistant pathogen. Can Vet J 39:559–565
Ribet D, Cossart P (2015) How bacterial pathogens colonize their hosts and invade deeper tissues to cite this version: HAL Id: pasteur-01115526. Microbes Infect 17:173–183
Roberts AJ, Wiedmann M (2003) Pathogen, host and environmental factors contributing to the pathogenesis of listeriosis. Cell Mol Life Sci 60:904–918. https://doi.org/10.1007/s00018-003-2225-6
Rodloff A, Bauer T, Ewig S et al (2008) Susceptible, intermediate, and resistant—the intensity of antibiotic action. Dtsch Aerzteblatt Online 105:657–662. https://doi.org/10.3238/arztebl.2008.0657
Sanlibaba P, Senturk E (2018) Prevalence, characterization and antibiotic resistance of *Enterococci* from traditional cheeses in Turkey. Int J Food Prop 21:1955–1963. https://doi.org/10.1080/10942912.2018.1489413
Sasidharan S, Prema B, Yoga Latha L (2011) Antimicrobial drug resistance of *Staphylococcus aureus* in dairy products. Asian Pac J Trop Biomed 1:130–132. https://doi.org/10.1016/S2221-1691(11)60010-5
Sergelidis D, Angelidis AS (2017) Methicillin-resistant *Staphylococcus aureus*: a controversial food-borne pathogen. Lett Appl Microbiol 64:409–418. https://doi.org/10.1111/lam.12735
Singh R, Schroeder CM, Meng J et al (2005) Identification of antimicrobial resistance and class 1 integrons in Shiga toxin-producing *Escherichia coli* recovered from humans and food animals. J Antimicrob Chemother 56:216–219. https://doi.org/10.1093/jac/dki161
Strawn LK, Fortes ED, Bihn EA et al (2013) Landscape and meteorological factors affecting prevalence of three food-borne pathogens in fruit and vegetable farms. Appl Environ Microbiol 79:588–600. https://doi.org/10.1128/AEM.02491-12
Swartz MN (2002) Human diseases caused by foodborne pathogens of animal origin. Clin Infect Dis 34:S111–S122. https://doi.org/10.1086/340248
United States Centers for Disease Control (2013) Antibiotic resistance threats in the United States, 2013. Current 114:CS239559-B
Uyttendaele M, Franz E, Schlüter O (2015) Food safety, a global challenge. Int J Environ Res Public Health 13:2–7. https://doi.org/10.3390/ijerph13010067
van Baarlen P, van Belkum A, Summerbell RC et al (2007) Molecular mechanisms of pathogenicity: how do pathogenic microorganisms develop cross-kingdom host jumps? FEMS Microbiol Rev 31:239–277. https://doi.org/10.1111/j.1574-6976.2007.00065.x
van Boeckel TP, Brower C, Gilbert M et al (2015) Global trends in antimicrobial use in food animals. Proc Natl Acad Sci 112:5649–5654. https://doi.org/10.1073/pnas.1503141112

Vanderhaeghen W, Hermans K, Haesebrouck F, Butaye P (2010) Methicillin-resistant *Staphylococcus* aureus (MRSA) in food production animals. Epidemiol Infect 138:606–625. https://doi.org/10.1017/S0950268809991567

Vedantam G, Clark A, Chu M et al (2012) *Clostridium difficile* infection: toxins and non-toxin virulence factors, and their contributions to disease establishment and host response. Gut Microbes 3:121–134. https://doi.org/10.4161/gmic.19399

Verraes C, van Boxstael S, van Meervenne E et al (2013) Antimicrobial resistance in the food chain: a review. Int J Environ Res Public Health 10:2643–2669. https://doi.org/10.3390/ijerph10072643

Vivant A-L, Garmyn D, Piveteau P (2013) *Listeria monocytogenes*, a down-to-earth pathogen. Front Cell Infect Microbiol 3:1–10. https://doi.org/10.3389/fcimb.2013.00087

VT Nair D, Venkitanarayanan K, Kollanoor Johny A (2018) Antibiotic-resistant *Salmonella* in the food supply and the potential role of antibiotic alternatives for control. Foods 7:167. https://doi.org/10.3390/foods7100167

Walsh C (2010) The problem of antimicrobial resistance in the food chain. Teagasc, Dublin

Walsh C, Duffy G (2013) Antibiotic resistance in foodborne pathogens. Teagasc final report project RMIS no 5036

Wang Y, Zhang M, Deng F et al (2014) Emergence of multidrug-resistant *Campylobacter* species isolates with a horizontally acquired rRNA methylase. Antimicrob Agents Chemother 58:5405–5412. https://doi.org/10.1128/AAC.03039-14

Wang X, Biswas S, Paudyal N et al (2019) Antibiotic resistance in *Salmonella Typhimurium* isolates recovered from the food chain through National Antimicrobial Resistance Monitoring System between 1996 and 2016. Front Microbiol 10:1–12. https://doi.org/10.3389/fmicb.2019.00985

White DG, Zhao S, Simjee S et al (2002) Antimicrobial resistance of foodborne pathogens. Microbes Infect 4:405–412

WHO (2017) Report of the 21st WHO expert committee on the selection and use of essential medicines WHO headquarters, Geneva 27–31 March 2017 *Table of contents* annexes. 4:13

WHO (2018) Whole genome sequencing for foodborne disease surveillance: landscape paper. World Health Organization, Geneva

WHO Advisory Group on Integrated Surveillance of Antimicrobial Resistance (AGISAR) (2019) Critically important antimicrobials for human medicine 6th Revision 2018. Ranking of medically important antimicrobials for risk management of antimicrobial resistance due to non-human use

WHO Press (2013) Global priority list of antibiotic-resistant bacteria to guide research, discovery, and development of new antibiotics. WHO, Geneva, pp 1–7. https://doi.org/10.1590/S0100-15742013000100018

Wilson JW (2002) Mechanisms of bacterial pathogenicity. Postgrad Med J 78:216–224. https://doi.org/10.1136/pmj.78.918.216

Ye Q, Wu Q, Zhang S et al (2018) Characterization of extended-spectrum β-lactamase-producing enterobacteriaceae from retail food in China. Front Microbiol 9:1–12. https://doi.org/10.3389/fmicb.2018.01709

Yucel N, Citak S, Bayhün S (2011) Antimicrobial resistance profile of *Staphylococcus aureus* isolated from clinical samples and foods of animal origin. Foodborne Pathog Dis 8:427–431. https://doi.org/10.1089/fpd.2010.0707

Cell Lines as In Vitro Model for Studying Microbial Pathogenesis

19

Indranil Chattopadhyay

Contents

I. Chattopadhyay (✉)
Department of Life Sciences, Central University of Tamil Nadu,
Thiruvarur, Tamil Nadu, India
e-mail: indranil@cutn.ac.in

B. Siddhardha et al. (eds.), *Model Organisms for Microbial Pathogenesis, Biofilm Formation and Antimicrobial Drug Discovery*,
https://doi.org/10.1007/978-981-15-1695-5_19

Abstract

To understand the effect of microorganisms on host cells, in vitro cell line studies in addition to animal studies provide translational importance in human clinical trials. For better understanding the interaction between bacteria and host cells, cultured animal cell lines and cultured human epithelial cells have extensively used for understanding the mechanisms by which bacteria induce human diseases. Cell culture models define mechanisms about the mode of interactions between bacterial virulence factors and host epithelial cell receptors. Predatory bacteria do not produce cytotoxic and inflammatory response in human cells. We described the effect of chronic exposure of *Porphyromonas gingivalis* and *Fusobacterium nucleatum* on human oral cavity squamous cell carcinoma (SCC) cell. We discussed about the bacterial virulence factors interacting with cultured epithelial cells. This may permit research leading to innovative antimicrobial therapies.

Keywords

Cell line · Bacteria · Fungi · Virus · Microbial pathogenesis · Human diseases

19.1 Introduction

Studies of microorganisms in liquid broth or agar plates are well established. This will not address the issue about the effect of microorganism on host cells. Microorganisms grow in their natural habitats effect transcriptomic alterations in host cells (Palková 2004). Biofilm formation by bacteria has been recently focused to study the growth pattern of bacteria (Azeredo et al. 2017). Bacteria primarily invades epithelial mucous membranes of host. The bacteria having invasive property are evaluated by colony-forming unit (CFU)-enumeration on agar plates (Letourneau et al. 2011). Colonization of bacteria in epithelium of urinary bladder and intestine, and endothelial cell is quantified on the basis of fluorescence signals from green fluorescent protein (GFP) expressed bacteria. This will provide information about virulence property, immunomodulation, and antibacterial activity in the context of progression of infection (Pedersen et al. 2018). Pathogenic bacteria are used to grow on epithelial surfaces of host through irreversible attachment (Ribet and Cossart 2015). Bacteria-epithelial cell adhesion is studied in cultured cell lines at microtiter plate. Effect of bacteria on human cells in in vitro studies in addition to animal studies provides translation importance in human clinical trials. In this chapter, the application of cell line in microbial pathogenesis involved in human diseases was summarized.

19.2 Study the Effect of Shiga Toxin-Producing *Escherichia coli* on Colonization of Intestinal Cell Layers

To study the effect of Shiga toxin-producing *E. coli* O157:H7 (EDL933) on T84 cell line, formaldehyde-fixed T84 cell layers were infected with EDL933. Growth of EDL933 on the T84 cell layers was increased rapidly. STEC strain EDL933 is able to form biofilm on human intestinal epithelial layer effectively (Yu et al. 2014). Fixed cell layers are used as substratum surface for bacterial growth (Lai et al. 2013). Central virulence genes are upregulated by co-culture of STEC strain O157:H7 Sakai on adherent HeLa cells (Alsharif et al. 2015).

19.3 Colonization of *Staphylococcus aureus* on Endothelial Cell Layers

S. aureus ATCC strain 29213 was used as a bloodstream pathogen. To stimulate the opsonization, *S. aureus* ATCC 29213 was treated briefly with human blood plasma. DMEM+FBS was used as a culture medium to study the colonization of *S. aureus* in endothelial cells. Fibronectin in FBS was used for adhesion in *S. aureus* (Claes et al. 2014).

19.4 Test Susceptibility of *Chlamydia trachomatis* Serovar L2 to Antibiotics in Lymphoid Jurkat Cells

C. trachomatis L2 invasively enters lymphatic and sub epithelial layer of genital tract. It grows both in Jurkat and HeLa cells. The growth curve was analyzed by fluorescence staining and electron microscopy. To understand persistent infection of *C. trachomatis* in Jurkat cell and its susceptibility of doxycycline (DOX), azithromycin (AZM), and ofloxacin (OFLX), inclusion forming unit (IFU) assay was performed. Antibiotics showed effectiveness against *C. trachomatis* L2 growth in both Jurkat cells and HeLa cells. Bacteria showed more sensitivity in Jurkat cells than in HeLa cells. *C. pneumonia* induces infection in lymphocytes through reduced expression of CD3 and CD25 (Kubo et al. 2012; Yamaguchi et al. 2008; Hirai et al. 2010).

19.5 Biofilm Formation by *Candida tropicalis* on Catheter by Using Human Cells

Biofilm is defined as an attached microbial community on biological or inert surfaces. Biofilms are important virulence factor for pathogenicity of *Candida* species such as *C. albicans* and *C. tropicalis*. Biofilms formation by *C. tropicalis* ATCC 750 on blastoconidia composed small catheter fragments (SCF) were investigated to its interaction with human cells (HeLa and HUVEC). *C. tropicalis* reduces virulence property of *C. albicans* through reducing of cell number, metabolic activity, and hyphal growth (Capote-Bonato et al. 2018).

19.6 Effect of *S. aureus* on Human Dendritic Cells

Methicillin-sensitive and methicillin-resistant *S. aureus* kills human monocyte-derived DCs. *S. aureus* produce toxins such as leukocidins that target cells involved in innate immunity (Berends et al. 2019).

19.7 Effect of *C. albicans* and *S. aureus* on Different Types of Endothelial Cell

Endothelial cells are involved in the pathogenesis of infections induced by *C. albicans* and *S. aureus*. To understand the pathogenesis of interaction between microbes and endothelial cell, human umbilical vein endothelial cells (HUVECs) are used in vitro model. HMEC-1 cell line was developed by transfecting of simian virus 40A gene into human microvascular endothelial cells derived from human foreskin. These cell lines are applicable to study the multiple microorganisms such as Chlamydia pneumonia, *Brucella* spp. *Bartonella henselae*, *Mycobacterium tuberculosis*, *Rickettsia rickettsia*, *C. albicans*, and *S. aureus*. Wild-type *C. albicans* showed reduced adherence and invasive property to HMEC-1 cells as compared to HUVECs. Wild-type *S. aureus* showed similar strength of adhesion and invasion to HMEC-1 cells and HUVECs cells. Secretion of IL8 from HUVECs cells becomes higher as compared to HMEC-1 cells in response to both *C. albicans* and *S. aureus* (Seidl et al. 2012).

19.8 In Vitro Human Granuloma Model of Sarcoidosis and Latent Tuberculosis Infection

To understand the pathogenesis of formation of granuloma, an in vitro model of granuloma is formed from human peripheral blood mononuclear cells (PBMCs) of patients with active sarcoidosis and latent tuberculosis (TB) infection (LTBI). Incubation of PBMCs for 7 days was performed in the presence of uncoated polystyrene beads or beads coated with purified protein derivative (PPD) or human serum albumin. PBMCs of sarcoidosis and LTBI showed the formation of granulomas in response to PPD-coated beads and secrete cytokines which are involved in inflammation. This in vitro model provides information about molecular mechanisms and biomarker discovery for sarcoidosis, latent TB infection, and granuloma biology (Crouser et al. 2017).

19.9 In Vitro Study of Adhesion, Transmigration, Invasion of *Campylobacter jejuni*

C. jejuni lives in the gut of birds and domestic animals and it causes gastroenteritis in humans (Dasti et al. 2010). *C. jejuni* interacts with porcine small intestinal epithelial cell lines such as IPEC-1 and IPEC-J2, human cell lines such as HeLa cells, HepG2 cells, chicken hepatocellular carcinoma epithelial cells, and African green monkey kidney cell lines. *C. jejuni* strains NCTC 11168, 81-176, 81116, and F38011 are commonly used in vitro infection studies (Backert and Hofreuter 2013).

19.9.1 Effect of *P. gingivalis* on Oral Cancer Cell Lines

OSCC-derived cell lines such as SCC-25, BHY and human primary oral epithelial cells express B7-homologue receptors such as B7-H1 and B7-DC. B7-H1 receptors and B7-DC receptors are overexpressed on oral cancer cell lines after infection with *P. gingivalis* strains ATCC 33277 or W83. B7-H1 receptor regulates the development of regulatory T cells whereas B7-DC receptor which is expressed mainly on dendritic cells and macrophages is involved in survival of T cell, cytokine production, and proliferation (Groeger et al. 2011). Inaba et al. (2014) examined the effect of *P. gingivalis* ATCC 33277 on metastatic potential on two OSCC cell lines such as SAS cells and Ca9–22 cells. Invasive property was measured by the activation of human matrix metalloproteinase 9 (MMP9) in OSCC cell lines. In SAS cells, *P. gingivalis* ATCC 33277 induced the expression of proMMP9 enzyme and activation of MMP9. Upregulation of proMMP9 in response to *P. gingivalis* occurs through p38, ERK1/2, and NF-kB signalling pathway (Atanasova and Yilmaz 2014) (Fig. 19.1).

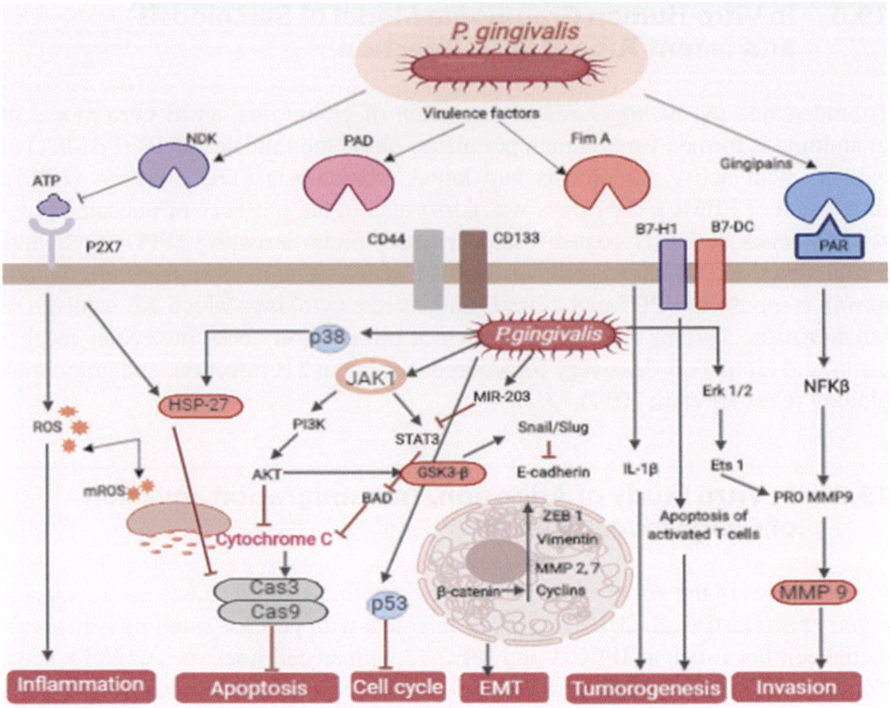

Fig. 19.1 Effect of *Porphyromonas gingivalis* on oral epithelial cell to drive oral cancer

19.9.2 Effect of *P. gingivalis* on Human Primary Oral Epithelial Cells (OEC)

P. gingivalis invades primary OECs in culture thorough β1-integrin (Yilmaz et al. 2008). *P. gingivalis* strain ATCC 33277 inhibits the apoptosis of OEC through depolarization of mitochondrial membrane, prevention of release of cytochrome-c, activation of phosphatidylinositol 3 kinase (PI3K), Akt and surviving and blocking of activation of caspase-3 and 9 (Yao et al. 2010). *P. gingivalis* also inhibits apoptosis of OEC cell through ATP coupled P2X7 signalling pathway (Yilmaz et al. 2008) (Fig. 19.1).

19.9.3 *P. gingivalis* Develop Resistance against Taxol in Oral Cancer Cells

P. gingivalis induces epithelial-mesenchymal-transition and develops resistance property against taxol in OSCC cells through overexpression of CD44 and CD133 (Woo et al. 2017).

19.9.4 Effect of *P. gingivalis* on Tumorigenic Properties of Human Immortalized Oral Epithelial Cells

Infection with *P. gingivalis* to human immortalized oral epithelial cells (HIOECs) for 5–23 weeks enhances proliferation, migration, and invasion of HIOECs. NNMT, FLI1, GAS6, lncRNA CCAT1, PDCD1LG2, and CD274 are differentially expressed in response to long-time exposure of *P. gingivalis*. These genes are considered as a potential biomarker for *P. gingivalis* induced OSCC with chronic periodontal infection (Geng et al. 2017).

19.9.5 Role of *P. gingivalis* in Epithelial-Mesenchyme-Transition of Human Primary Epithelial Cells

Infection of *P. gingivalis* induces EMT (epithelial-mesenchymal-transition) in primary oral epithelial cells (OECs) through overexpression of EMT-associated transcription factors, Slug, Snail, and Zeb1 at mRNA and protein level. It also induces overexpression of vimentin and downregulation of E-cadherin in OECs. Long-term infection of *P. gingivalis* induces the expression of matrix metalloproteinases (MMPs) 2, 7, and 9. Cellular migration of human primary OECs was enhanced in the presence of infection of *P. gingivalis* and *Fusobacterium nucleatum* (Lee et al. 2017). Epithelial-mesenchymal-transition (EMT) involves the loss of cell–cell adhesion in epithelial cells, cellular migration, invasion, and metastasis in cancer (Costa et al. 2015; Heerboth et al. 2015). MMP7, MMPs 2 and 9 are involved in metastasis of OSCCs (Hong et al. 2000) (Fig. 19.1).

19.9.6 Role of *F. nucleatum* Subspecies Animalis in Human Colorectal Tumors

F. nucleatum ssp. *animalis* ATCC 51191 strain was the most prevalent bacteria in human colorectal cancers. To identify the expression of CCL20 protein in *Fusobacterium* infected colorectal tumors, human cell–bacterium co-culture system under normoxic (optimal for human cell growth) and hypoxic (favorable for anaerobic growth of *Fusobacterium*) was established. CCL20 protein was overexpressed in *F. nucleatum* ssp. animalis infected low-passage colorectal cancer cell line HCP1 under normoxic conditions. CCL20 protein was expressed in high-passaged colorectal cancer cell lines (SW480, HT29, HCT116, and RKO) but CCL20 protein was not expressed in a noncancerous colonic cell line CCD841 infected with *F. nucleatum* ssp. *animalis*. CCL20 protein was also overexpressed in *F. nucleatum* ssp. *animalis* infected THP-1 monocytes. Hypoxic condition enhanced basal level expression of CCL20 protein in HCP1 and THP-1 cells which showed synergistic expression of CCL20 protein due to infection with *F. nucleatum* ssp. *animalis*. Infection of Fusobacterium and hypoxic microenvironment contribute the overexpression of CCL20 protein in colorectal cancer cells (Ye et al. 2017).

19.10 Effect of *P. gingivalis* on Human Monocyte-Derived Dendritic Cells (DCs)

P. gingivalis invades myeloid DCs through glycoprotein Mfa1 fimbriae. *P. gingivalis* with *Streptococcus gordonii* and *F. nucleatum* induces biofilm formation and invasion of DC cells through the expression of mfa-1. It has been reported that DC of periodontitis patients having abundant load of *F. nucleatum* and *P. gingivalis* showed increased expression of mfa-1 (El-Awady et al. 2019).

19.11 Study of Inflammatory Responses Induced by *Bdellovibrio bacteriovorus* Strains 109J and HD100, and *Micavibrio aeruginosavorus* Strain ARL-13 in Human Cell Lines

Some multi-drug resistant gram-negative bacteria are considered as predatory bacteria because they prey on other gram-negative bacteria. They are cytotoxic to human cell lines. *B. bacteriovorus* belonging to delta-proteobacterium and *M. aeruginosavorus* belonging to an alpha-proteobacterium are considered as predatory bacteria. High doses of these bacteria were challenged to five different human cell lines (HaCaT—human keratinocytes, HepG2—human liver epithelial cells, HK-2—human kidney epithelial cells, MD—loosely adherent human spleen monocytes, and THP-1—human blood monocytes) to study the inflammatory response. *B. bacteriovorus strains* 109J (ATCC® 43826™) and *M. aeruginosavorus* strain ARL-13 were used for the cytotoxicity and cytokine assay (GM-CSF, IFN-γ, IL-10, IL-12p70, IL-1β, IL-2, IL-6, IL-8, and TNF-α) to study the inflammatory response against lipopolysaccharide (LPS) and flagellum of these predatory gram-negative bacteria. *P. aeruginosa* strain PA14 and *E. coli* ATCC strain 43888 (serotype O157:H7) were used as positive controls for these studies. Cell viability was significantly higher for predatory bacterial strain in all five cell lines. *P. aeruginosa* strain PA14 was cytotoxic to all human cell lines. *P. aeruginosa* showed the reduction of cell viability of HaCaT cells, HepG2 cells, and HK-2 cells by 93.5%, 78.5%, and 90.5%, respectively. Levels of IL-1β and IL-6 were elevated in macrophages exposed to *B. bacteriovorus*. Levels of GMCSF, IL-10, and IL-6 were elevated in *M. aeruginosavorus* exposed macrophages. Levels of IL-1β, IL-6, GMCSF, IL-10, IL-12p70, and TNF-α were elevated in predatory bacteria exposed activated macrophages (THP-1). *B. bacteriovorus* and *M. aeruginosavorus* are not responsible for the significant elevation of the cytokines in four of the five human cell lines. Predatory bacteria did not produce cytotoxic effect and inflammatory response in different cell lines (Gupta et al. 2016).

19.12 Effect *Lactobacillus paracasei* on Human Intestinal Caco-2 Cell Line

Cell wall proteins of *L. paracasei* induce programmed cell death of Caco-2 cancer cell line and are considered as a chemotherapeutic agent. *L. paracasei* enhances annexin V and propidium iodide staining in Caco-2 cancer cell line (Nozari et al. 2019).

19.13 Effect of Metabiotic Derived from Probiotic *L. rhamnosus* MD 14 on Caco-2 and HT-29 Human Colon Cancer Cells

Metabiotic of probiotic *L. rhamnosus* MD 14 showed antitumorigenic property in colon cancer Caco-2 and HT-29 cells (Sharma et al. 2019).

19.14 Role of Proteases from Opportunistic Pathogen *Scedosporium aurantiacum* on Human Epithelial Cells

Peptidases of *S. aurantiacum* isolate (strain WM 06.482; CBS 136046) reduce cell viability of A549 human lung epithelial cells (Han et al. 2019).

19.15 Effect of Fungal Pathogen *S. aurantiacum* on Human Lung Epithelial Cells

S. aurantiacum strain WM 06.482 showed adherence property to human lung epithelial A549 cells which induced transcriptomic upregulation of genes involved in cell repair and inflammatory processes. Most of the differentially expressed genes induced by *S. aurantiacum* strain WM 06.482 in human lung epithelial A549 cells are involved in NF-kB pathway that drive the secretion of pro-inflammatory cytokines. *Scedosporium* species are involved in colonization in the lung of cystic fibrosis (CF) patients (Kaur et al. 2019).

19.16 Effect of Conidia of *Aspergillus fumigatus* on Human Bronchial Epithelial Cells

A. fumigatus is responsible for the development of allergic bronchopulmonary aspergillosis, aspergilloma, and invasive aspergillosis. *A. fumigatus* conidia infection induces transcriptomic alterations in human bronchial epithelial cell line (16HBE14o-). Genes involved in DNA repair (glutathione S-transferase) and inflammation (e.g., matrix metalloproteinases and chemokines) are significantly upregulated. Gene set enrichment analysis revealed that genes involved in chromatin assembly, G-protein-coupled receptor binding, chemokine activity, and glutathione metabolic process are overexpressed (Gomez et al. 2010).

19.17 Effect of *P. aeruginosa* on Alveolar Epithelial Cells to Understand Molecular Pathogenesis of Cystic Fibrosis

P. aeruginosa is responsible to develop cystic fibrosis (CF). *P. aeruginosa* induces genetic alterations in A549 cells. It shows high adherence to A549 cells. It induces reactive oxygen species (ROS) formation, production of pro-inflammatory cytokine and apoptosis in A549 cells (Hawdon et al. 2010).

19.18 Effect of *A. fumigatus* on Lung Epithelial Cells

In vitro cell line model was applied to study the interaction between lung epithelial cells such as BEAS-2B and HBE and *A. fumigatus*. Epithelial cells are incubated with culture filtrate (CF) secreted by mature mycelium of dormant *A. fumigatus* conidia. Adherent *A. fumigatus* conidia undergo internalization into endosomes through polymerization of actin surrounding the endosome. The adherent *A. fumigatus* conidia induce the activation of MyD-dependent NF-kB, PI3 kinase, and MAP kinase signalling pathway (MAPKs ERK1/2, JNK, and p38) that drive secretion of chemokine and cytokine. IL-6, IL-8, TNF-α, GM-CSF, and MCP1 are produced in A549 cells in response to fragments of mycelia (Osherov 2012).

19.19 Effect of Zika Virus Infections in Human Embryonic Kidney (HEK293) Cells

Zika virus (ZIKV) is responsible to develop neurological abnormalities in fetal stage. To understand the molecular pathogenesis in ZIKV-infected neuronal developmental, human embryonic kidney (HEK293) cell line was used. HEK293 cells were infected with ZIKV MR766 strain, PRVABC59 strain (PRV), and FLR strain. American PRV and FLR ZIKV isolates are able to infect human epithelial HEK293 cell lines derived from human embryonic kidney cells. PRV and FLR ZIKV strains induced cytopathic effects and cytolysis of HEK293 cells. ZIKV v-RNA genome is produced in ZIKV-infected HEK293 cell lines. ZIKV-infected HEK293 cells grow slowly. The persistently ZIKV-infected human cell lines will be helpful to study signal transduction pathway in host cell due to infection (Liu et al. 2019).

19.20 Effect of *S. epidermidis* in Human Alveolar Epithelial Cells

S. epidermidis is responsible to develop sepsis and bronchopulmonary dysplasia. To understand the molecular pathogenesis in neonatal inflammatory morbidities, A549 cells were infected by biofilm-positive and biofilm-negative strain of *S. epidermidis* for 24 h with infection ratio of 10. Both *S. epidermidis* strains are responsible to develop pro-inflammatory responses through the overexpression of tumor necrosis factor (TNF)-α, IL-1β, interleukin (IL)-6, IL-8, monocyte chemoattractant protein (MCP)-1, interferon γ-induced protein 10 (IP-10), and intercellular adhesion molecule (ICAM)-1in lung epithelial cells. Biofilm-positive *S. epidermidis* strains are more active to produce pro-inflammation than biofilm-negative strain (Dong et al. 2019).

19.21 Study of Viral Infection in Human Cell Lines

EBV, HTLV-1, HBV, B19V, HHV-6, and HHV-7 viruses were detected in 43 human cell lines. Virus-positive cell lines will be helpful to identify the viral integration site in host cell lines (Shioda et al. 2018).

19.22 Conclusion

The cell line study will provide information that may not be easily accessible clinically. Interactions between fungal pathogen and human lung epithelial cells provide the molecular mechanisms about the hyphal invasion of the human airway epithelial cells. In vitro study will provide molecular mechanisms of adaptive mutations of viruses and effect of persistence infection of viruses on host cells. In vitro study of effect of non-pathogenic predatory bacteria on host cell will provide information that can be helpful to develop live antibiotics or alternative to traditional antibiotics.

References

Alsharif G, Ahmad S, Islam MS et al (2015) Host attachment and fluid shear are integrated into a mechanical signal regulating virulence in Escherichia coli O157:H7. Proc Natl Acad Sci 112:5503–5508. https://doi.org/10.1073/pnas.1422986112

Atanasova KR, Yilmaz Ö (2014) Looking in the Porphyromonas gingivalis cabinet of curiosities: the microbium, the host and cancer association. Mol Oral Microbiol 29:55–66. https://doi.org/10.1111/omi.12047

Azeredo J, Azevedo NF, Briandet R et al (2017) Critical review on biofilm methods. Crit Rev Microbiol 43:313–351. https://doi.org/10.1080/1040841X.2016.1208146

Backert S, Hofreuter D (2013) Molecular methods to investigate adhesion, transmigration, invasion and intracellular survival of the foodborne pathogen Campylobacter jejuni. J Microbiol Methods 95:8–23. https://doi.org/10.1016/j.mimet.2013.06.031

Berends ETM, Zheng X, Zwack EE et al (2019) Staphylococcus aureus impairs the function of and kills human dendritic cells via the LukAB toxin. MBio 10. https://doi.org/10.1128/mBio.01918-18

Capote-Bonato F, Sakita KM, de Oliveira AG et al (2018) In vitro interaction of Candida tropicalis biofilm formed on catheter with human cells. Microb Pathog 125:177–182. https://doi.org/10.1016/j.micpath.2018.09.029

Claes J, Vanassche T, Peetermans M et al (2014) Adhesion of Staphylococcus aureus to the vessel wall under flow is mediated by von Willebrand factor-binding protein. Blood 124:1669–1676. https://doi.org/10.1182/blood-2014-02-558890

Costa LC, Leite CF, Cardoso SV et al (2015) Expression of epithelial-mesenchymal transition markers at the invasive front of oral squamous cell carcinoma. J Appl Oral Sci 23:169–178. https://doi.org/10.1590/1678-775720140187

Crouser ED, White P, Caceres EG et al (2017) A novel in vitro human granuloma model of sarcoidosis and latent tuberculosis infection. Am J Respir Cell Mol Biol 57:487–498. https://doi.org/10.1165/rcmb.2016-0321OC

Dasti JI, Tareen AM, Lugert R et al (2010) Campylobacter jejuni: a brief overview on pathogenicity-associated factors and disease-mediating mechanisms. Int J Med Microbiol 300:205–211. https://doi.org/10.1016/j.ijmm.2009.07.002

Dong Y, Glaser K, Schlegel N et al (2019) An underestimated pathogen: Staphylococcus epidermidis induces pro-inflammatory responses in human alveolar epithelial cells. Cytokine 123:154761. https://doi.org/10.1016/j.cyto.2019.154761

El-Awady A, de Sousa Rabelo M, Meghil MM et al (2019) Polymicrobial synergy within oral biofilm promotes invasion of dendritic cells and survival of consortia members. npj Biofilms Microbiomes 5:11. https://doi.org/10.1038/s41522-019-0084-7

Geng F, Liu J, Guo Y et al (2017) Persistent exposure to Porphyromonas gingivalis promotes proliferative and invasion capabilities, and tumorigenic properties of human immortalized oral epithelial cells. Front Cell Infect Microbiol 7. https://doi.org/10.3389/fcimb.2017.00057

Gomez P, Hackett TL, Moore MM et al (2010) Functional genomics of human bronchial epithelial cells directly interacting with conidia of Aspergillus fumigatus. BMC Genomics 11:358. https://doi.org/10.1186/1471-2164-11-358

Groeger S, Domann E, Gonzales JR et al (2011) B7-H1 and B7-DC receptors of oral squamous carcinoma cells are upregulated by Porphyromonas gingivalis. Immunobiology 216:1302–1310. https://doi.org/10.1016/j.imbio.2011.05.005

Gupta S, Tang C, Tran M, Kadouri DE (2016) Effect of predatory Bacteria on human cell lines. PLoS One 11:e0161242. https://doi.org/10.1371/journal.pone.0161242

Han Z, Kautto L, Meyer W et al (2019) Effect of proteases secreted by the opportunistic pathogen Scedosporium aurantiacum on human epithelial cells. Can J Microbiol 65(11):814–822. https://doi.org/10.1139/cjm-2019-0212

Hawdon NA, Aval PS, Barnes RJ et al (2010) Cellular responses of A549 alveolar epithelial cells to serially collected Pseudomonas aeruginosa from cystic fibrosis patients at different stages of pulmonary infection. FEMS Immunol Med Microbiol 59:207–220. https://doi.org/10.1111/j.1574-695X.2010.00693.x

Heerboth S, Housman G, Leary M et al (2015) EMT and tumor metastasis. Clin Transl Med 4:6. https://doi.org/10.1186/s40169-015-0048-3

Hirai I, Utsumi M, Yamaguchi H, Yamamoto Y (2010) Chlamydia pneumoniae infection suppresses Staphylococcus enterotoxin B-induced proliferation associated with down-expression of CD25 in lymphocytes. Can J Microbiol 56:289–294. https://doi.org/10.1139/W10-018

Hong S-D, Hong S-P, Lee J-I, Lim C-Y (2000) Expression of matrix metalloproteinase-2 and -9 in oral squamous cell carcinomas with regard to the metastatic potential. Oral Oncol 36:207–213. https://doi.org/10.1016/S1368-8375(99)00088-3

Inaba H, Sugita H, Kuboniwa M et al (2014) Porphyromonas gingivalis promotes invasion of oral squamous cell carcinoma through induction of proMMP9 and its activation. Cell Microbiol 16:131–145. https://doi.org/10.1111/cmi.12211

Kaur J, Kautto L, Penesyan A et al (2019) Interactions of an emerging fungal pathogen Scedosporium aurantiacum with human lung epithelial cells. Sci Rep 9:5035. https://doi.org/10.1038/s41598-019-41435-3

Kubo T, Ishida K, Matsuo J et al (2012) Chlamydia trachomatis serovar L2 infection model using human lymphoid Jurkat cells. Microb Pathog 53:1–11. https://doi.org/10.1016/j.micpath.2012.02.005

Lai Y, Rosenshine I, Leong JM, Frankel G (2013) Intimate host attachment: enteropathogenic and enterohaemorrhagic Escherichia coli. Cell Microbiol 15:1796–1808. https://doi.org/10.1111/cmi.12179

Lee J, Roberts JS, Atanasova KR et al (2017) Human primary epithelial cells acquire an epithelial-mesenchymal-transition phenotype during long-term infection by the oral opportunistic pathogen, Porphyromonas gingivalis. Front Cell Infect Microbiol 7:493. https://doi.org/10.3389/fcimb.2017.00493

Letourneau J, Levesque C, Berthiaume F et al (2011) In vitro assay of bacterial adhesion onto mammalian epithelial cells. J Vis Exp. https://doi.org/10.3791/2783

Liu H, Liao H-M, Li B et al (2019) Comparative genomics, infectivity and cytopathogenicity of American isolates of Zika virus that developed persistent infections in human embryonic kidney (HEK293) cells. Int J Mol Sci 20:3035. https://doi.org/10.3390/ijms20123035

Nozari S, Faridvand Y, Etesami A et al (2019) Potential anticancer effects of cell wall protein fractions from Lactobacillus paracasei on human intestinal Caco-2 cell line. Lett Appl Microbiol 69(3):148–154. https://doi.org/10.1111/lam.13198

Osherov N (2012) Interaction of the pathogenic mold Aspergillus fumigatus with lung epithelial cells. Front Microbiol 3:346. https://doi.org/10.3389/fmicb.2012.00346

Palková Z (2004) Multicellular microorganisms: laboratory versus nature. EMBO Rep 5:470–476. https://doi.org/10.1038/sj.embor.7400145

Pedersen RM, Grønnemose RB, Stærk K et al (2018) A method for quantification of epithelium colonization capacity by pathogenic bacteria. Front Cell Infect Microbiol 8:16. https://doi.org/10.3389/fcimb.2018.00016

Ribet D, Cossart P (2015) How bacterial pathogens colonize their hosts and invade deeper tissues. Microbes Infect 17:173–183. https://doi.org/10.1016/j.micinf.2015.01.004

Seidl K, Solis NV, Bayer AS et al (2012) Divergent responses of different endothelial cell types to infection with Candida albicans and Staphylococcus aureus. PLoS One 7:e39633. https://doi.org/10.1371/journal.pone.0039633

Sharma M, Chandel D, Shukla G (2019) Antigenotoxicity and cytotoxic potentials of metabiotics extracted from isolated probiotic, Lactobacillus rhamnosus MD 14 on Caco-2 and HT-29 human colon cancer cells. Nutr Cancer:1–10. https://doi.org/10.1080/01635581.2019.1615514

Shioda S, Kasai F, Watanabe K et al (2018) Screening for 15 pathogenic viruses in human cell lines registered at the JCRB Cell Bank: characterization of in vitro human cells by viral infection. R Soc Open Sci 5:172472. https://doi.org/10.1098/rsos.172472

Woo BH, Kim DJ, Choi JI et al (2017) Oral cancer cells sustainedly infected with Porphyromonas gingivalis exhibit resistance to taxol and have higher metastatic potential. Oncotarget 8:46981–46992. https://doi.org/10.18632/oncotarget.16550

Yamaguchi H, Matsuo J, Sugimoto S et al (2008) Inhibition of lymphocyte CD3 expression by Chlamydophila pneumoniae infection. Microb Pathog 45:290–296. https://doi.org/10.1016/j.micpath.2008.06.005

Yao L, Jermanus C, Barbetta B et al (2010) Porphyromonas gingivalis infection sequesters pro-apoptotic bad through Akt in primary gingival epithelial cells. Mol Oral Microbiol 25:89–101. https://doi.org/10.1111/j.2041-1014.2010.00569.x

Ye X, Wang R, Bhattacharya R et al (2017) Fusobacterium nucleatum subspecies animalis influences proinflammatory cytokine expression and monocyte activation in human colorectal tumors. Cancer Prev Res 10:398–409. https://doi.org/10.1158/1940-6207.CAPR-16-0178

Yilmaz Ö, Yao L, Maeda K et al (2008) ATP scavenging by the intracellular pathogen Porphyromonas gingivalis inhibits P2X 7-mediated host-cell apoptosis. Cell Microbiol 10:863–875. https://doi.org/10.1111/j.1462-5822.2007.01089.x

Yu Y, Sun X, Guan X et al (2014) Effects of hydroformylation treatment on the storage time and blood group antigen expressions of reagent red blood cells. Transfus Apher Sci 50:462–466. https://doi.org/10.1016/j.transci.2014.02.019

Animal Models to Understand Host–Pathogen Interactions

20

V. T. Anju, Busi Siddhardha, and Madhu Dyavaiah

Contents

Abstract

Infectious diseases are the outcome of molecular cross-communication between host and its pathogens. During the molecular cross talks, host–pathogen proteomics, genomics, and immunological responses are highly influenced. Host would respond to their pathogen through several mechanisms for the clearance of pathogens. It is always necessary to identify the underlying molecular mechanisms of pathogenicity. In general, host–pathogen cross talks are complex and dynamic in nature that exploits most of the host cell functions. Immune responses

V. T. Anju · M. Dyavaiah (✉)
Department of Biochemistry and Molecular Biology, School of Life Sciences, Pondicherry University, Puducherry, India

B. Siddhardha
Department of Microbiology, School of Life Sciences, Pondicherry University, Puducherry, India

B. Siddhardha et al. (eds.), *Model Organisms for Microbial Pathogenesis, Biofilm Formation and Antimicrobial Drug Discovery*,
https://doi.org/10.1007/978-981-15-1695-5_20

are initiated by host cell as a response to the pathogen attack. It was found to be very difficult, exclusive, and ethically inappropriate to evaluate the notorious pathogen interactions that cause adverse effects on human health at the organism level. Hence, the need for experimental animal models to understand host–pathogen interactions always emerges. Incorporation of a host animal model not only allows the identification of host–pathogen interactions but also lights into the phenotypic impacts and molecular mechanisms of pathogenicity. In ancient times, better understanding of virulence determinants and antimicrobial therapy has been hindered by the restrictions of adequate experimental models and necessary tools to measure the severity of infections. Laboratory hosts that have been employed as an alternative for mammal infection models are *Caenorhabditis elegans*, amoeba, *Drosophila melanogaster*, and *Danio rerio*. These models are used as infection models owing to their shorter generation times, flexibility, and affordability to study forward and reverse genetic analysis. Even though humans are excellent model to study human pathogens, their use in studies is limited due to the safety, ethical, and expense concerns. Among other primates, monkey, baboons, and chimpanzees are idyllic and mimic most of the infectious diseases. But rodents such as mice, rats, rabbits, hamsters, and guinea pigs are widely proposed model hosts due the limited application of other primate models. Here, we review the available animal models to study host–pathogen interactions with a focus to decipher pathogenicity mechanisms.

Keywords

Infectious disease · Host–pathogen cross talks · Host animal models · Pathogenicity

20.1 Introduction

Microorganisms were thought to be the primary invaders that involved in the host–pathogen interactions resulting in disease. Slowly, new insights into the host–pathogen interactions suggested that this does not always cause a disease. This has introduced new terms to enlighten the states in which host and pathogen do not result in a disease (Casadevall and Pirofski 2000). Host–pathogen interactions are also a type of interspecies interactions that result in infections occasionally but need to be figured out at the molecular level as early as possible. Generally, a pathogen protein attach to the receptors of human proteins and manipulates host biological processes. Deciphering the host–pathogen interactions is having an utmost role in developing suitable treatment methods. It has been reviewed that infections rose by HIV, plague, Ebola, cholera, other bacterial, and viral pathogens drag the circumstances more worse owing to the high mortality rate in each year. Infectious diseases not only affect human health but it has adverse effect on the economic status of country (Kösesoy et al. 2019). Host–pathogen interactions are followed by a cascade of cell signaling events. The early events involve the recognition of pathogen-associated molecular patterns (PAMPs) and the conserved microbial components by host cell pattern recognition receptors (PRRs). This binding is known to determine the rate of failure or success of an immune response. A complex cascade of cellular signaling events takes place followed by the PAMP–PPR interactions. Complex cell signaling events includes an early host response, pathogen clearance, activation of kinase pathways, production of effectors, activation of transcription factors, and modulation of innate and adaptive immune responses and finally leads to a pro-inflammatory or anti-inflammatory response (Bahia et al. 2018).

Emergence of coevolutionary dynamics between the host and pathogen is the most critical and well-studied interspecific interaction. Coevolution has been studied well among widespread ecosystems and found popularly in diverse set of host–parasitic interactions. Host–pathogen coevolution has a special role in shaping the diversity and population structure of hosts and pathogens. Coevolution of host and their pathogen helps in understanding the structure of communities, maintenance of sexual recombination, direction of species invasions, and population dynamics. The flow of coevolution is strongly influenced by the spatial structure of their populations and can occur in relatively short time. All cases of host–pathogen interactions harbor a genetic basis to infection. Size and genetic makeup of the pathogen and the density of susceptible host genotypes in the earlier generations will be the function of the frequency. Also, the chances of host becoming infected is a function of the frequency of pathogen genotypes and past genotypes of both populations. In general, each population can act as a dynamic target for others and hence these dynamics of one partner over the other helps to maintain the polymorphism. There is a pathogen specificity with an enhanced infection on a given host system but decreased infection at the community level due to the polymorphism (Morgan and Koskella 2017). Molecular cross talks between pathogen and host result in infectious diseases. There are several mechanisms underlying the pathogenic rewiring of host cells. Host–pathogen protein–protein interactions also mediate these molecular cross talks. Protein–protein interactions and protein complexes encompass the principal functional modules of the cell. Pathogenic hijacking or rewiring of host proteome involves the intervention at the signaling pathways and cellular functions to determine the strength of the virulent intervention. Phenotypic impact of a pathogen is directly related to its capacity to rewire the host interactome. This describes the impacts of each virulent protein that are linked to their number of interactions with the host proteins. Hence, mapping the host–pathogen protein interactions may offer valuable understandings of biological functions of virulent proteins that are critical to the progression and spread of pathogens. It also provides insights on the molecular basis of pathogenicity and possibly single out the pharmacological intrusion targets (Nicod et al. 2017).

All organisms sense and reply to their external stimuli through the production of second messengers (cyclic nucleotides). A universal second messenger, cyclic di-adenosine monophosphate is synthesized by diverse life forms (mammals, fungi, protozoa, and bacteria). cAMP regulates virulence gene expression in host cells owing to their influence on the transcription factors that are dependent on environmental control of secondary messenger production (McDonough and Rodriguez

2012). It is known as the bacterial signaling nucleotide produced by several human pathogens. C-di-AMP has central role in catabolic repression and virulence determinants' expression. Mostly, an infected host cell recognizes the synthesized c-di-AMP and triggers an innate immune response to prevent the colonization and transmission of pathogens and ultimately to clear the pathogens. It has been reported that long-standing interaction of host and pathogens results in the coevolution of both and controls the activation of innate immunity by the signaling molecule. These second messengers will be produced in the host cell in such way that it modulates the host response to intensify the infection by circumventing immune recognition (Devaux et al. 2018).

Most of the knowledge of host–pathogen interactions and their pathogenic mechanisms have risen from the use of various model systems including cell lines and animal models. Model systems are preferred in host–pathogen studies to confirm their pathogenic role causing a disease and also evaluate their immune responses. Cell lines are defined as indispensable powerful tools for learning the molecular and cellular mechanisms of pathogenesis. Recently, animal models are employed to evaluate the pathogenicity of pathogens in host cells, immune responses and to analyze the efficacy of a vaccine. Altogether, both cell lines and animal models are an integral part in the study of host–pathogen interactions and one must be acquainted with the knowledge of these models and their applications (Bhunia 2018). There are several excellent reviews and trailblazing contributions available in this research domain to enhance our understanding in the host–pathogen interactions and to provide new insights for deciphering the interactions through animal models. This chapter summarizes the fascinating reviews addressing various facets of host–pathogen interactions studies in animal models.

20.2 Importance of Animal Models

It is not recommended to use humans to evaluate host pathogens and pathogenicity owing to their ethical concerns and safety. From the research point of view, human models are ideal for studying host–pathogen interactions. Some cases of nonfatal diseases human volunteers have been incorporated and studied. To overcome the boundaries of using humans, animal models are frequently used and applied as a substitute for these studies. Most regularly used model hosts are *Caenorhabditis elegans* (nematode), *Dictyostelium discoideum* (amoeba), *Drosophila melanogaster* (fruit fly), *Danio rerio* (Zebra fish), *Cavia porcellus* (guinea pigs), *Mus musculus* (mouse), *Rattus* (rat), *Cricetinae* (hamster), and *Oryctolagus cuniculus* (rabbit). Among them, the nematode, fruit fly, amoeba, and zebra fish have major role in host–pathogen studies as these models exhibit shorter generation times and also due to their amenability and affordability to forward and reverse genetic studies. Owing to the ethical and expense-related concerns, widespread use of some nonhuman primates is limited, which is ideal to mimic many diseases (Lemaitre and Ausubel 2008). Compared to small mammals, morphological and genetic similarities between humans and primates provided an instinctive feeling that they may deliver more reliable and trustworthy data. Now, clinical studies and data would provide more relevant initial research orientations if it is encouraged (Druilhe et al. 2002).

Immunity of host and their susceptible profile toward infectious agents are defined based on multiple factors such as immune experience, environment, and generics of host and its pathogen. Studying the host–pathogen interactions directly on human models are quite challenging and complex. Hence, host genetics in causing diseases are more depend on rodent model systems (Noll et al. 2019). Most commonly used rodents include mice, rabbits, rats, guinea pigs, and hamsters. However, pathogenic studies using mouse models occupy the central part of the research flow with a successful interpretation of host–pathogen interactions. There are several advantages of using animal models in biomedical research over other model systems which includes: (1) animal models are used in such studies from past many years, (2) features of animal models are well documented for the correct application in different studies, (3) studies on these models provide information regarding current paradigm, and (4) they are readily available from the local suppliers (Druilhe et al. 2002). Among all, vertebrate and invertebrate models are having specific advantages when compared to both. Invertebrate models provide greater advantages over the other one due to their economy of size and ethical concerns. Vertebrate models are vital for the cellular and molecular analysis of host–pathogen interactions (Naglik et al. 2008) (Fig. 20.1).

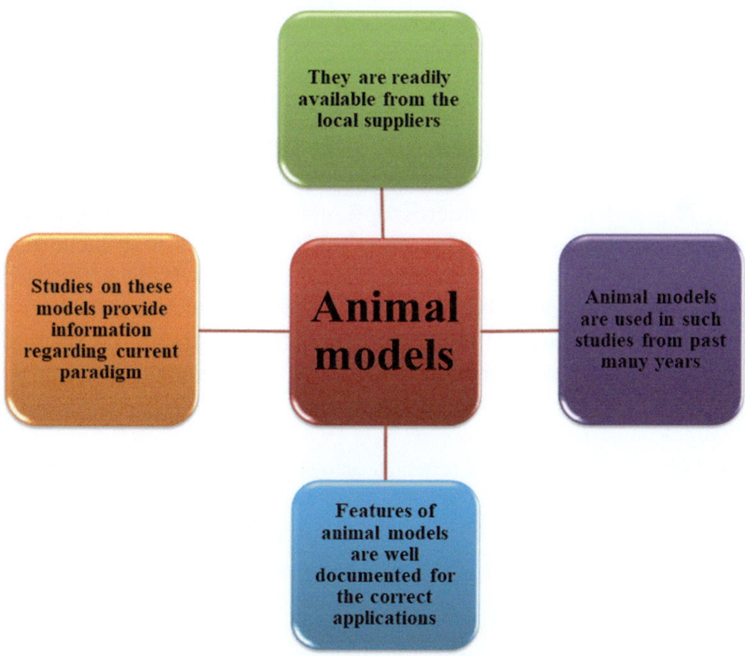

Fig. 20.1 Advantages of animal models used in different studies of host–pathogen interactions

Animal models also provide prospects to engineer and study the host–microbiota interactions with a level of experimental controls which is not possible with human models. Both the vertebrate and invertebrate models provide enough information regarding the microbial molecular patterns and host recognition receptors. These models are useful for studying the tractable genetics that are essential for enabling symbiosis by both the host and the pathogens. Model systems are extensively used in microbiome studies for revealing the host physiology, skeletal biology, and lipid metabolism. Ever increasing number of studies conducted in host–microbiome research area will prove the associations recognized between the human microbiota and disease (Kostic et al. 2013). Current scenario of biomedical research dealing with host–pathogen studies is dominated by the mouse, fruit fly, and nematode models. Researchers think that these models can be used to summarize the physiology and diseases in different species through manipulating some genes, which would actually make them as perfect models of human biology. However, still there are some limitations of using animal models, which have to be eliminated by the introduction of new or other unconventional model organisms. Most of mice models involved in different studies are young ones but most of diseases evaluated by different researchers are associated with old people such as cancer and neurological disorder. Recently, large scale collaborative research project results showed that genomic responses to acute inflammatory responses are greatly comparable to humans but are not portrayed by corresponding mouse models. New approaches have to be developed to gain more knowledge about the prevention and physiology of diseases. It has been found that animal model research focusing more often on laboratory species may weaken the chances of scientific progress in the forthcoming years (Conti et al. 2014) (Table 20.1).

20.3 The Nematode *Caenorhabditis elegans*

C. elegans is a soil inhabitant microscopic nematode having a length of 1 mm. Compatibility of *C. elegans* in laboratory work as models is influenced by their shorter generation time of 3 days and also their capacity to produce 300 progeny by a single animal. They naturally grow on agar plates containing *Escherichia coli* OP50, which is a uracil auxotroph facilitating controlled growth. These animals can be accommodated in laboratory conditions and grow quickly in large numbers.

Table 20.1 Different invertebrate and vertebrate animal models used in host–pathogen studies

Animal models	
Invertebrate models	Vertebrate models
Caenorhabditis elegans (nematode)	*Mus musculus* (mouse)
Dictyostelium discoideum (amoeba)	*Cavia porcellus* (guinea pig)
Drosophila melanogaster (fruit fly)	*Rattus* (rat)
Galleria mellonella (Greater wax moth)	*Danio rerio* (zebra fish)
Acanthamoeba castellanii (amoeba)	*Oryctolagus cuniculus* (rabbit)

Owing to their small size, around 10–20 infected worms can fit to the single well of a 384-well multititer plate. This nematode has become one of the preferred models for cell biologists and geneticists due to their ease of housing in laboratory conditions and simple body organization and hermaphroditic lifestyle. Infection studies using *C. elegans* can be profited from a host influenced circumstances in genetically characterized organisms (Lorenz et al. 2016). Another amazing benefit of using nematode is that adult worms are purely post-mitotic with the exception to germline. Adult worms are developed through the transition between four larval stages. At the end of transition from third to fourth larval stage only changes occur at the growth level of worms, not in their number of somatic cells. Many complexities raised while working with multicellular model systems whereas all cells turn over quickly are eliminated in the case of *C. elegans* model system. Number and identity of each cell vary from one worm to another worm is another advantage of using this model system (Marsh and May 2012).

C. elegans have been used as screening platforms for anti-infective molecules from long days. Nematode is an emerging powerful model system to study host–pathogen interactions and can be evaluated for multiple human pathogens for anti-infective development. The development of anti-infective agents is based on the fact that virulence determinants of pathogens causing disease in humans are also involved in killing nematode. *C. elegans* also mounts immune defense produced by the specific pathogen which involves the conserved innate immunity regulators. Pioneer work in *C. elegans* as animal models for screening of anti-infective agents is demonstrated by Frederick Ausubel (Peterson and Pukkila-Worley 2018). This well-known studied animal model for host–pathogen interactions is susceptible to *Enterococcus faecalis*, *Serratia marcescens*, *Salmonella typhimurium*, *Staphylococcus aureus*, and *Pseudomonas aeruginosa* by causing infection through their intestine. Many of the above-mentioned Gram-negative and Gram-positive bacteria cause infections in humans using the similar virulence factors (James et al. 2018).

There are some conserved immune signaling pathways in *C. elegans* which are always a center of attraction for studies. Immunity studies mainly focus on three approaches: (1) forward genetic screening for nematode mutants showing altered pathogen susceptibility, (2) application of reverse genetic approaches for evaluating conserved genes, and (3) assaying the gene expression that induced by pathogens or by regulation of signaling pathways (Kim 2008). Immune systems of worms are found to be comparably simple and evolutionarily predate those of higher organisms. Specifically, this organism lacks adaptive immunity with devoid of mobile immune cells. However, they carry three pairs of cells (coelomocytes) for detoxification processes that are not involved in any of the immune functions. They use only their innate immunity to mount a response toward pathogens for their removal and to resist them. Innate immune system regulates the signaling pathways of worms upon finding a pathogen at their transcriptional level and organized by several signaling cascades. Major pathways known to date are DAF-16 and DAF-2, p38/PMK-1, DBL-1, and ERK/MPK-1 signaling pathways (Williams and Schumacher 2018).

C. elegans–P. aeruginosa model is developed for host–pathogen interaction studies. *P. aeruginosa* strain PA14 causes disease in both animals and plants through a shared set of virulence factors. This bacterium kills the worms either by fast killing or by slow killing method. In slow killing mechanism, PA14 in a low salt medium kills worms within a period of 2–3 days through the accumulation of pathogen inside the intestine of worms. In addition, in a high salt medium PA14 adopts a fast killing mechanism within 4–24 h through the production of diffusible toxins. *P. aeruginosa* PA01 strain kills the worm through a rapid paralysis mechanism after 4 h (Tan 2000).

20.4 Amoeba

Amoeba is a eukaryotic microorganism with great diversity. They belong to different taxonomic group such as fungi, algae, and protozoa. They stand along among all microorganisms due to their amoeboid lifestyle. Amoeboid lifestyle is characterized by their capacity to change shape by forming pseudopods. Amoeba is considered as professional phagocytes owing to their ability to feed on bacteria and other microbes by phagocytosis. Even then, amoeba cannot degrade all microbes. Some of the microorganisms are able to resist the digestion by amoeba and even can use them as host cell for their activities. There are several studies of using amoebae as host models for analyzing the pathogenicity of pathogens. Most commonly used amoebae belong to the phylum Amoebozoa, which is closest to the phylum fungi and animals and the predominant representatives are from the genera *Dictyostelium* and *Acanthamoeba*. Several pathogenic fungi (*Cryptococcus*, *Candida*, and *Aspergillus*) and bacteria (*Mycobacterium*, *Legionella*, *Francisella*, and *Salmonella*) are known pathogens of these amoebae. Hence, amoeba models are useful for studying the complex interactions with the above pathogens (Thewes et al. 2019).

Acanthamoeba and *Dictyostelium* are identified as the natural and versatile host model for *Legionella* infection. Etiology and cellular host interactions of *L. pneumophila* have been particularly viewed in these amoebas. Owing to their similarity in causing infection in macrophages and amoebae, amoeba is used as a powerful model to study bacteria–macrophage interactions. Among these, *Acanthamoeba* are observed in habitats with *Legionella* positive isolates and are widely distributed. Yet, for laboratory purposes axenic growth of *Acanthamoeba* strains are mostly favored. This amoeba implements a biphasic life cycle comprising a trophozoite stage and cyst stage. A particular strain, *A. castellanii* adopts a diverse repository of pattern recognition receptors, which are thought to have orthologous roles in the innate immunity of higher organisms (Swart et al. 2018). *A. castellanii* can be used to study the molecular basis of different pathogen interactions as this amoeba interacts with wide variety of pathogens. The role and characterization of arsenal receptors utilized by this strain to engulf the pathogens would extend knowledge in how pathogenicity could be enhanced. Virulence gene expression in *A. castellanii* and mammalian cells would draw information regarding how these pathogens evolved and got adapted to different hosts (Guimaraes et al. 2016).

One of the merits of using this amoeba over other nonmammalian system is that they can be grown at 37 °C, which is the optimal temperature for most of the deadly pathogens. This will allow setting up of conditions in the laboratory that is more similar to the natural pathology of human pathogens and offers a great relevance. Complex mechanisms of host defense against pathogens can dissect from the *Acanthamoeba* due to their unicellular nature compared to the metazoal-nonvertebrate hosts. Information regarding the novel genes that are involved in mammalian pathogenesis caused by bacterial pathogens can be predicted using the amoeba model. Also, host cell components employed to respond to the pathogen attack can be identified using the unicellular amoeba compared to mammalian hosts (Sandstrom et al. 2011). There is some spine-like structures found on the surface of *Acanthamoeba* spp. known as acanthopodia. *A. castellanii* are known as simple, rapid, and low-cost model for studying host–pathogen interactions. Some of the pathogens are able to grow and internalize the amoeba. This may lead to the transmission to other susceptible hosts and exerts pathogenicity. These amoebas harboring the human pathogen act as Trojan horses and thus protect them from antimicrobial effectors and other environmental circumstances, which provide conditions for its survival and growth. Amoeba and macrophages are thought to share similar ability to ingest particles into the phagosomes. Presence of lysosomes makes both the cells hostile to infection by the pathogen. This model is attractive for phagocytosis of several pathogens such as *H. capsulatum*, *Sporothrix schenckii*, and *Blastomyces dermatitidis* (Singulani et al. 2018).

The interaction between *Legionella pneumophila* pathogen and the social amoeba has been explored using biochemical, cell biological, and genetical approaches with a focus on their small and large GTPases, autophagy components, phosphoinositide lipids, retromer complex, autophagy components, and bacterial effectors attacking these host factors. The genome of *D. discoideum* is having a size of 34 Mb and along with six chromosomes (size of 3.5–8.6 Mb). Studies related to genes can be performed using the model system where the genes in their mutant nature are able to cause disease in humans. In brief, complete information on genome of amoeba enhances the application of *D. discoideum* as an outstanding collection of genetic tools to evaluate their fundamental cellular functions. It is important to modulate various host cell processes through qualitative and quantitative approaches for an efficient replication and establishment of infection by *L. pneumophila* in *D. discoideum* and macrophages infection model. DNA microarray analysis comprising the half of the genome of amoeba identified around 371 genes that are regulated during an infection with *pathogenic L. pneumophila* Philadelphia-1 strain JR32 after 48 h of infection. Transcriptional analysis of *D. discoideum* infection model was revealed vital aspects of host–pathogen cross talk (Swart et al. 2018).

The genetically tractable, cooperative, and haploid social amoeba serves as a host for diverse pathogens such as *Legionella pneumophila*, *Mycobacterium* spp., *Pseudomonas aeruginosa*, and *Cryptococcus neoformans*. The studies on this amoeba enlighten more on host–pathogen interactions which include: (1) use of wild-type amoeba as a screening platform for extracting information regarding the virulence factors of intracellular, extracellular, and mutant pathogens; (2) mutants of this amoeba to classify the host susceptible and resistance determinants to infection; and (3) introduction of reporter strains of amoeba to understand in detail about the mechanism of host–pathogen cross talks (Steinert and Heuner 2005). This social amoeba is proven as a tool for finding several bacterial and fungal virulence factors. Specifically, tagging of genes with some markers such as green fluorescent protein allows the real and in vivo monitoring of unique virulence and host cell factors. Most of the assays that decipher host–pathogen interactions in amoeba integrate infection assay, phagocytosis assay, and confocal assay for in vivo monitoring of fluorescence (Ünal and Steinert 2006).

20.5 The Honeycomb Moth *Galleria mellonella*

Greater wax moth, *G. mellonella* are widely used for evaluating host–entomopathogenic microbe interactions. Not only for entamopathogenic microorganisms, *Galleria* can effectively employed as a reliable model to study the pathogenesis that exerted by many of the human pathogens. Vast opportunities open with the honeycomb wax host model to study the host–pathogen cross talks owing to their low rearing costs, ranking as an ethically acceptable model and their convenience in injection feasibility. Apart from this, growth of moth at 37 °C which is similar to that of human pathogens allows them to produce various pathogenic factors. Researchers have found a correlation between virulence of pathogen in mammals and this model. *Galleria* produces a complex innate immunity toward their pathogen. The multicomponent immune response produced in moth involves cellular phagocytosis, phenol oxidase-based melanization, and hemolymph coagulation. Pathogens will be destroyed by the production of lysozymes, antimicrobial peptide like defensins, and reactive oxygen species, which is similar to the mechanisms observed in mammals. *Galleria* can also recognize molecular patterns associated with nonself microbes through their germ line encoded receptors such as peptidoglycan and Toll recognition proteins. *Galleria* also employs danger signaling for detection of pathogens either through the sensing of peptides resulted from a protein cleavage process by metalloproteinase or nucleic acids produced by damaged cells (Mukherjee et al. 2010).

Galleria mellonella, a caterpillar of the wax moth are utilized in the host–pathogen interactions of *Burkholderia mallei*, *Pseudomonas aeruginosa*, *Proteus mirabilis*, *Burkholderia cepacia*, *Bacillus cereus*, *Francisella tularensis*, and several pathogenic fungi. *C. elegans* infection models for the study of *Acinetobacter* pathogenesis showed some limitations that recovered with the use of *Galleria*. This model is effectively used for the evaluation of efficacy of antimicrobial agents as the model is amenable to antibiotic treatment (Peleg et al. 2009). This model is considered as ethically acceptable owing to their ability to enhance reproducibility by introducing larger group sizes. This simple invertebrate animal is a promising infection model for *M. tuberculosis* complex. In the first time of *Galleria* infection model for tuberculosis pathogen showed replicate features of pathogenesis through the induction of granuloma-like structures and inclusion of lipid bodies, which are the unique features of infection. Use of this model has markedly reduced the use of more expensive and time-consuming mycobacterial infection models. Model is an effective tool for the assessment of unusual antimycobacterial drugs and novel vaccine entrants in vivo. Future studies of mycobacterium infection models with *Galleria* include the optimization studies with the pathogenic, nonpathogenic, drug-resistant, and drug-sensitive *M. tuberculosis* isolates (Li et al. 2018).

Greater wax moth belong to the Lepidoptera family has successfully used as a model to study the virulence of pathogenic fungi (*Candida albicans*, *Aspergillus fumigatus*, and *Cryptococcus neoformans*). Model is useful for the evaluation of antifungal drugs in the treatment of fungal infections. Hemocytes present in the hemolymph of *Galleria* presents a phagocytic effect against its pathogens. Another important role of immune system in pathogen defense is through the stimulation of melanization and encapsulation of foreign particles. Virulence of fungal pathogen can be assessed by the microbial burden, hemocyte density, and induction of microbial morphological changes in the moth. There are reports stating the killing effect of various *Candida* species in *G. mellonella*. Host–pathogen interactions of *C. tropicalis* and *Galleria* were fully characterized recently. Results indicated that *G. mellonella* is a nonconventional host to study the virulence of human fungal pathogen *C. tropicalis*. Also, this offers a feasible and simple model system for analyzing the antifungal drug efficacy and their protective role during *C. tropicalis* infection (Mesa-Arango et al. 2013).

20.6 The Fruit Fly—*Drosophila melanogaster*

Genetically tractable fruit fly, *D. melanogaster*, has delivered remarkable views into the host–pathogen interactions. This model provided that many of the aspects related to these host–pathogen interactions are conserved in higher organisms. Fruit fly possesses a well-established stand in the evaluation of host interactions with bacterial, viral, and fungal pathogens. Previously, host response of malarial parasite has been studied using *D. melanogaster* model prior to the sequencing of mosquito genome (Igboin et al. 2012). Fruit fly is used as a model of innate immune system owing to their simplicity and ease in which they can be applied in both forward and reverse genetics. Forward and reverse genetics allows the characterization and identification of innate immune responses produced against microbial pathogens that are preserved across evolution. Fruit fly generates the immune responses with three effector arms such as an inedible antimicrobial peptide response, a reactive oxygen response (by the enzyme phenoloxidase) and a cellular immune response through which foreign particles are phagocytized in fly hemocytes and accumulation of melanin pigment. Humoral antimicrobial peptide response is studied and controlled primarily by two pattern recognition pathways. The two pathways include Toll and Imd. Regulatory mechanisms of melanin deposition and cellular immunity are not fully explicated and studied widely only in recent years (Moule et al. 2010).

Drosophila recognize various Gram-negative and Gram-positive pathogens by sensing the specific forms of peptidoglycan present in the bacterial cell wall using peptidoglycan recognition proteins (PGRPs). These peptidoglycans found only in cell membrane of bacteria are essential glucopeptidic polymers (Lemaitre and

Hoffmann 2007). *D. melanogaster* is one the most and well-studied organism for a century of genetic work including RNA interference–hairpin constructs, reporter genes studies, and targeted gene expression to overexpress the recombinant proteins. This invertebrate organism provides an interesting alternative as host model for evaluating pathogenesis owing to their powerful genetics. Previously, genome-wide screen in fruit fly enabled us to identify the genes that are involved in the virulence of *Serratia marcescens* in the host infection model. A similar study was conducted to identify the genes associated with *P. aeruginosa* mutant virulence in fruit fly infection model. Unlike other invertebrate models, *D. melanogaster* is not useful for the high throughput screening of antimicrobial drugs but better designed for understanding the host–pathogen interactions in detail (Limmer et al. 2011).

Drosophila has been used for probing the mechanisms behind the interactions between *P. aeruginosa* virulence factors and host cells. It has been proven that Toll signaling pathway is induced in response to the *Pseudomonas* infection in the host cell that provided the insights how these virulence factors cause resistance in pathogen. One of the advantages of using *P. aeruginosa*–fruit fly infection model over human pathogenesis model is the manipulation of genome of host and pathogen (Lau et al. 2003). Recently, fruit fly is employed as a model system for host–symbiotic microbiota interactions other than the typical host–pathogen studies. Most commonly found gut microbiota of fruit fly are from the families of Lactobacillales, Enterobacteriaceae, and Acetobacteraceae. Gut microbiota communities are strongly dependent on the diet of the model system. Oxygen will be able to enter into the entire diameter of fruit fly gut as these gut flora are aero tolerant or obligate aerobes. It is possible to virtually culture all the gut microbiota in laboratory owing to the aerobic growth of this flora and their relative taxonomic simplicity. *Drosophila* possess a large potential to enable the better understanding of host–symbiont interactions due to the culturing of large proportions of fruit fly gut microbiome along with the rapid growth, wide collections of mutant flies, and their high reproductive capacity (Kostic et al. 2013).

Reports say that *P. aeruginosa* and *Plasmodium gallinaceum* can infect *D. melanogaster* but the former kills the model system where as latter one proliferate and develop within the fruit fly. Owing to the low cost of model system screening methodologies offer to unravel the mechanisms behind the host–pathogen interactions which could reduce the use of expensive or laborious vertebrate hosts. Also, the genetically tractable infection model allows a quick and possibly unbiased identification of host factors affected during the pathogenesis. In one study, *Mycobacterium marinum* causes systemic disease in fruit fly which is closely similar to the human tuberculosis. This bacterium had killed *Drosophila* with a lethal dose of 5 CFU. Also, adult flies or larvae can be easily infected with injected doses of bacteria for evaluating the pathogenesis. Genetic tools available in the fruit fly infection model are unparalleled and stand as a best studied model among all animal model. Like other vertebrates, *Drosophila* has bactericidal phagocytes called as hemocytes to fight with the pathogen attack (Dionne et al. 2003). *Drosophila* has been successively employed as a systemic infection model for evaluating antibacterial efficacy of phages toward the secondary opportunistic human pathogen, *P. aeruginosa*. Unlike antibacterial agents, phages can be easily and delicately counted by employing simple assays to study the pharmacokinetic properties of injected phages in the small-scale infection model. In order to address the bioactivity of antibacterial agents in this small-scale infection model, therapeutic phages would be transferred to the flies by placing starved flies in the media harboring appropriate number of phages. Routes of antibacterial administration can be consecutively exploited here to assess the antibacterial efficacy of bacteriophages against *P. aeruginosa* infection (Jang et al. 2019).

20.7 The Zebra fish *Danio rerio*

Zebra fish gained much interest as infection model for developmental biologists. It is a teleost fish which belongs to the family of Cyprinidae. They can breed easily and a single female can lay eggs up to a number of 200 per week. This fish was found as an amazing developmental model 30 years ago. Future expectations of zebra fish as model systems relies purely on different studies regarding the complete genome sequencing and expressed sequence tags sequencing projects to identify different zebra fish genes. Preliminary studies revealed that zebra fish shares many orthologous genes and conserved synteny with mammals (van der Sar et al. 2004). This model holds a position in the high throughput screening of drugs for inflammatory diseases, cancer, and infectious diseases. Zebra fish can be effectively used for genomic and mutant analysis with excellent opportunities of in vivo imaging. It is possible to study the different types of immune cell types synthesized toward disease progression with help of developing embryo immune system. In addition, zebra fish embryos and larvae are suitable for dichotomizing the innate immunity host factors related to the pathology owing to their temporal separation of innate responses from adaptive immunity responses. Moreover, immune systems of zebra fish and human share a remarkable similarity, which possess central role in biomedical applications. Current knowledge on the downstream signaling and signaling components involved in the innate immune responses of embryo are important to decipher the mechanism of host–pathogen interactions. Zebra fish larvae and embryo holds a position to fill the gap generated between the cell- and rodent-based model systems. This transparent model system also provides several advantages in drug trafficking and drug administration studies. Zebra fish has developed as an extremely powerful model over the past years for studying vertebrate host immune response and interaction with bacterial virulence factors, in vivo imaging, and genetic analysis and drug screening in fish larvae and embryos (Meijer and Spaink 2011).

In the present century, zebra fish larvae and embryo are accepted as genetically tractable and optically accessible with fully functional immune system that comprises macrophages and neutrophils, which mimic the mammalian counter parts. Several pathogenic interactions have been investigated to provide unprecedented resolution of cellular responses to the infections in vivo zebra fish model. This fish model has been proposed for several bacterial, viral, and fungal pathogens. Zebra fish and larvae models are utilized to understand the pathogenesis and cell biology rather than focusing the whole field of fish–human pathogen interactions. This model has effectively used for pathogen studies of several Gram-negative and Gram-positive bacteria such as *Salmonella typhimurium*, *Shigella flexneri*, *Pseudomonas aeruginosa*, *Burkholderia cenocepacia*, *Listeria monocytogenes*, *Staphylococcus aureus*, *Mycobacterium marinum*, *Mycobacterium abscessus*, and *Mycobacterium leprae* (Torraca and Mostowy 2018). Recently, infection course of pathogen studied using fluorescence-tagged pathogen allowing it to visualize through wild field epi-fluorescence microscopy due the transparency at larval stages. Further, recently, a method was introduced to monitor the *Salmonella typhimurium* infection progression using epi-fluorescence microscopy. Study also allowed visualizing the free-swimming bacteria through the circulatory system, phagocytosis of bacteria, and heterogeneous gene expression activation using a nontoxic inducer (Medina and Royo 2013).

Zebra fish embryos are well-known model for in vivo pathogenic studies of *P. aeruginosa*. There are two methods for zebra fish–pathogen studies where infection is achieved either by microinjection into the larvae or by static immersion method. A report was found with *P. aeruginosa* infection to zebra fish through both microinjection method and immersion method. Proteomics pathways affected by infection also evaluated both in pathogen and in host using non-isotopic metaproteomics methods. They found that metabolic pathways of fish such as hypoxia through HIF pathway was enriched by immersion method whereas inflammatory pathways mediated by chemokine and cytokine signaling molecules were enriched in infected larvae exposed to injection methods. They demonstrated the fitness of embryos as a model for assessing proteomic studies after infection (Díaz-Pascual et al. 2017).

Other than the conventional models, zebra fish xenografts are coming into the picture as a useful disease models and for translational research. Recently, scientists developed novel mouse–zebra fish hematopoietic tissue chimeric embryos for host–pathogen studies and hematopoiesis. Both the mouse and human hematopoietic tissues can be engrafted into the fish embryos for studies. Authors predicted that the chimeric embryos could be amended to study in vivo and real visualization of and analysis of host–pathogen interactions. Here, the zebra fish xenografts of murine tissue eliminates the generation of chimeric animals for different studies. Then, it also expands its area of studies that can be studied in zebra fish chimeras such as the murine cell behaviors (Parada-Kusz et al. 2018). Zebra fish continues to a model organism for disease and provides new insights into the disease mechanisms and its therapy (Patton and Tobin 2019).

20.8 Primates

Nonhuman primates used for host–pathogen studies are rhesus macaques (*Macaca mulatta*) and bonnet macaques (*Macaca radiata*). These nonhuman primates and humans are not studied widely for host–pathogen cross talks. These models are used only for evaluating specific pathogens (Burt et al. 2017). Early development and pathogenesis studies are performed in small mammals even though there is a constant pressure for the employment of nonhuman primates. Despite of their cost and relative scarcity, ethical concerns over their use often limits use in host–pathogen studies. There is always an intuitive feeling that only these models can provide more reliable data despite of their genetic and morphological similarities between primates and humans. Baboons are thought to be better primate model for evaluating course of pathology and disease. Exciting studies performed using baboons for evaluating schistosomiasis and periportal fibrosis (Druilhe et al. 2002). Diverse mammalian species are applied as experimental models to study infectious diseases caused by *P. aeruginosa* such as chinchilla to swine and up to nonhuman primates. These models served as suitable hosts to analyze the infections associated with *Pseudomonas* such as biofilm-associated infections (Lorenz et al. 2016).

20.8.1 Mouse

Based on a number of practical reasons, mice are preferred over other animal models. Major advantages of mouse animal models for host–pathogen studies are small size, cost effectiveness of maintenance in laboratory, availability of immunological tools for mice, and availability of genetically modified mouse strains. However, there are some criticisms of mouse models. Always, research with mice should be validated with other model systems to confirm the verdicts. Mouse model always stands as an important model of infection (Fonseca et al. 2017; Lowe et al. 2018). Inbred mouse strains are used past many years for studying the degree of susceptibility of different types of infectious agents. In order to gain more knowledge about the host responses toward these infections, a genetic approach in mice is adopted. Present advances such as germ line modification (BAS transgenics) which provided with positional cloning approach have made the studies easier. Quantitative trait loci and additional novel genetic loci, which play a vital role in host responses toward infections have been recognized. Thus, cloning and characterization of novel loci approaches would light the future years to unfold the story of genes and proteins involved in the host–pathogen interactions that eventually lead to onset and progression of an infection (Fortier et al. 2005).

Years ago, molecular and genetic toolbox created for mouse models empower the scientists manipulates and study the genes in vivo. Even though, mice are employed extensively to study the pathogenesis of human infections, these models summarizes many aspects of human infections as incorrect. Briefly, mouse is generally resistant to infections caused by HIV, *Plasmodium falciparum*, and *Shigella flexneri*. Host tropism or host restriction toward infections often stand as a hindrance for using mice as experimental model. In such situations, mice can be genetically engineered so that it strictly resembles to humans in all means of host–pathogen interactions (Coers et al. 2009). Another mouse pharyngeal colonization model is an inexpensive and available experimental model, which permits to evaluate broader pathogens. There is a great similarity between human and murine immune factors involved in pharyngeal colonization. In order to avail the pinpoint elements related to murine immune system responses towards pharyngeal colonization, humanized mice could be adopted. However, inbred mouse appears to be appropriate in those cases related to bacterial and host immune factors (Gogos and Federle 2019). Other than mouse models, humanized models are created through the reconstitution of immunocompromised mice with hematopoietic cells of different organs. In a study, humanized mice is employed to study pathogenesis of HIV/tuberculosis such that model could fully reflect its human immunity to tuberculosis pathogen (Fonseca et al. 2017).

Salmonella typhi is known to infect humans exclusively and owing to the lack of animal models host–pathogen studies related to typhoid fever has hampered. Currently, murine models with oral and systematic inoculation of streptomycin are used for evaluating intestinal pathology and inflammatory responses in patients with typhoid fever (de Jong et al. 2012).

Interaction of Toll-like receptor 4 and surface protein A during *Pseudomonas aeruginosa* lung infection in mouse model was evaluated. It has concluded that Toll-like receptor interaction with surface protein advances the host defense and relieves the tissue injury in a mouse model of bacterial lung infection. Also, pro-phagocytic and anti-inflammatory responses were studied in JAWS II dendritic cells and primary alveolar macrophages. Therapeutic potential of surface protein A-4 decreases bacterial burden, intracellular signaling, lactate levels, lung edema, and production of inflammatory cytokines and chemokines in infected mouse model. Altogether, this peptide may be helpful in reducing the bacterial load, tissue damage, and inflammation in bacterial infected murine model (Awasthi et al. 2019).

20.9 Large Models

Large animals such as swine, horse, cattle, sheep, and deer might be used as good experimental model for studying several human infectious diseases such as viral diarrhea, asthma, Crohn's disease, tuberculosis, and influenza. Use of large experimental animals provided numerous advances in developmental immunology studies. Over millennia, large animals and humans had established as out bred populations and their size also adds several advantages. Hence, it is credible that how their immune responses are sculpted by exposure to a similar range of pathogens (Conti et al. 2014).

20.10 Future Perspectives and Conclusion

Although humans are the best suitable model for evaluating host–pathogen interactions, other vertebrate and invertebrate models are preferred as laboratory animal models owing to the ethical concerns associated with humans. Animal models paved the way for much of the studies related to infectious disease to understand its pathogenicity. These models are one of the well-known factors involved in the current needs to study and develop antimicrobial therapy to combat human pathogens. There are some models which entirely reflect the host–pathogen cross talk in humans other than the models that based on the type of host and its specificity for pathogen. Yet, there are some poorly adapted models, which provide contradictory information regarding pathogenesis mechanisms. Hence, there is a need to develop advanced animal models for improved studies in biomedical research. Scientists are engaged in the development of novel model systems for the improved evaluation and understanding of pathogenicity in hosts. Novel models to be developed could better define the relevance of laboratory hosts in pathogen studies and to understand the mechanisms of virulence. In near future, combination of different animal models could provide new insights into the understanding of pathogenesis.

References

Awasthi S, Singh B, Ramani V et al (2019) TLR4-interacting SPA4 peptide improves host defense and alleviates tissue injury in a mouse model of *Pseudomonas aeruginosa* lung infection. PLoS One 14:e0210979. https://doi.org/10.1371/journal.pone.0210979

Bahia D, Satoskar AR, Dussurget O (2018) Editorial: cell signaling in host–pathogen interactions: the host point of view. Front Immunol 9:1–4. https://doi.org/10.3389/fimmu.2018.00221

Bhunia AK (2018) Animal and cell culture models to study foodborne pathogen interaction. In: Foodborne microbial pathogens. Springer, New York, pp 113–123

Burt FJ, Chen W, Miner JJ et al (2017) Chikungunya virus: an update on the biology and pathogenesis of this emerging pathogen. Lancet Infect Dis 17:e107–e117. https://doi.org/10.1016/S1473-3099(16)30385-1

Casadevall A, Pirofski L-A (2000) Host-pathogen interactions: basic concepts of microbial commensalism, colonization, infection, and disease. Infect Immun 68:6511–6518. https://doi.org/10.1128/IAI.68.12.6511-6518.2000

Coers J, Starnbach MN, Howard JC (2009) Modeling infectious disease in mice: co-adaptation and the role of host-specific IFNγ responses. PLoS Pathog 5:e1000333. https://doi.org/10.1371/journal.ppat.1000333

Conti F, Abnave P, Ghigo E (2014) Unconventional animal models: a booster for new advances in host-pathogen interactions. Front Cell Infect Microbiol 4:1–4. https://doi.org/10.3389/fcimb.2014.00142

de Jong HK, Parry CM, van der Poll T, Wiersinga WJ (2012) Host–pathogen interaction in invasive salmonellosis. PLoS Pathog 8:e1002933. https://doi.org/10.1371/journal.ppat.1002933

Devaux L, Kaminski P-A, Trieu-Cuot P, Firon A (2018) Cyclic di-AMP in host–pathogen interactions. Curr Opin Microbiol 41:21–28. https://doi.org/10.1016/j.mib.2017.11.007

Díaz-Pascual F, Ortíz-Severín J, Varas MA et al (2017) *In vivo* host-pathogen interaction as revealed by global proteomic profiling of zebrafish larvae. Front Cell Infect Microbiol 7:1–11. https://doi.org/10.3389/fcimb.2017.00334

Dionne MS, Ghori N, Schneider DS (2003) *Drosophila melanogaster* is a genetically tractable model host for *Mycobacterium marinum*. Infect Immun 71:3540–3550. https://doi.org/10.1128/IAI.71.6.3540-3550.2003

Druilhe P, Hagan P, Rook GA (2002) The importance of models of infection in the study of disease resistance. Trends Microbiol 10:s38–s46. https://doi.org/10.1016/S0966-842X(02)02437-X

Fonseca KL, Rodrigues PNS, Olsson IAS, Saraiva M (2017) Experimental study of tuberculosis: from animal models to complex cell systems and organoids. PLoS Pathog 13:e1006421. https://doi.org/10.1371/journal.ppat.1006421

Fortier A, Min-Oo G, Forbes J et al (2005) Single gene effects in mouse models of host: pathogen interactions. J Leukoc Biol 77:868–877. https://doi.org/10.1189/jlb.1004616

Gogos A, Federle MJ (2019) Modeling *Streptococcus pyogenes* pharyngeal colonization in the mouse. Front Cell Infect Microbiol 9:1–12. https://doi.org/10.3389/fcimb.2019.00137

Guimaraes AJ, Gomes KX, Cortines JR et al (2016) *Acanthamoeba* spp. as a universal host for pathogenic microorganisms: one bridge from environment to host virulence. Microbiol Res 193:30–38. https://doi.org/10.1016/j.micres.2016.08.001

Igboin CO, Griffen AL, Leys EJ (2012) The Drosophila melanogaster host model. J Oral Microbiol 4:10368. https://doi.org/10.3402/jom.v4i0.10368

James CG, Morah O, Panwala V, Yarmand A (2018) DFB1655 is not affected by a missense mutation in dop-1 or treatment with chlorpromazine hydrochloride. J Exp Microbiol Immunol 4:1–9

Jang H-J, Bae H-W, Cho Y-H (2019) Exploitation of Drosophila infection models to evaluate antibacterial efficacy of phages. Methods Mol Biol 1898:183–190

Kim D (2008) Studying host-pathogen interactions and innate immunity in *Caenorhabditis elegans*. Dis Model Mech 1:205–208. https://doi.org/10.1242/dmm.000265

Kösesoy İ, Gök M, Öz C (2019) A new sequence based encoding for prediction of host–pathogen protein interactions. Comput Biol Chem 78:170–177. https://doi.org/10.1016/j.compbiolchem.2018.12.001

Kostic AD, Howitt MR, Garrett WS (2013) Exploring host-microbiota interactions in animal models and humans. Genes Dev 27:701–718. https://doi.org/10.1101/gad.212522.112

Lau GW, Goumnerov BC, Walendziewicz CL et al (2003) The *Drosophila melanogaster* toll pathway participates in resistance to infection by the gram-negative human pathogen *Pseudomonas aeruginosa*. Infect Immun 71:4059–4066. https://doi.org/10.1128/IAI.71.7.4059-4066.2003

Lemaitre B, Ausubel FM (2008) Animal models for host–pathogen interactions. Curr Opin Microbiol 11:249–250. https://doi.org/10.1016/j.mib.2008.05.002

Lemaitre B, Hoffmann J (2007) The host defense of *Drosophila melanogaster*. Annu Rev Immunol 25:697–743. https://doi.org/10.1146/annurev.immunol.25.022106.141615

Li Y, Spiropoulos J, Cooley W et al (2018) *Galleria mellonella* - a novel infection model for the *Mycobacterium tuberculosis* complex. Virulence 9:1126–1137. https://doi.org/10.1080/21505594.2018.1491255

Limmer S, Quintin J, Hetru C, Ferrandon D (2011) Virulence on the fly: *Drosophila melanogaster* as a model genetic organism to decipher host-pathogen interactions. Curr Drug Targets 12:978–999. https://doi.org/10.2174/138945011795677818

Lorenz A, Pawar V, Häussler S, Weiss S (2016) Insights into host-pathogen interactions from state-of-the-art animal models of respiratory *Pseudomonas aeruginosa* infections. FEBS Lett 590:3941–3959. https://doi.org/10.1002/1873-3468.12454

Lowe DE, Robbins JR, Bakardjiev AI (2018) Animal and human tissue models of vertical *Listeria monocytogenes* transmission and implications for other pregnancy-associated infections. Infect Immun 86:1–15. https://doi.org/10.1128/IAI.00801-17

Marsh EK, May RC (2012) *Caenorhabditis elegans*, a model organism for investigating immunity. Appl Environ Microbiol 78:2075–2081. https://doi.org/10.1128/AEM.07486-11

McDonough KA, Rodriguez A (2012) The myriad roles of cyclic AMP in microbial pathogens: from signal to sword. Nat Rev Microbiol 10:27–38

Medina C, Royo JL (2013) Zebrafish as a model organism to study host–pathogen interactions. Methods 62:241–245. https://doi.org/10.1016/j.ymeth.2013.04.012

Meijer AH, Spaink HP (2011) Host-pathogen interactions made transparent with the zebrafish model. Curr Drug Targets 12:1000–1017. https://doi.org/10.2174/138945011795677809

Mesa-Arango AC, Forastiero A, Bernal-Martínez L et al (2013) The non-mammalian host *Galleria mellonella* can be used to study the virulence of the fungal pathogen *Candida tropicalis* and the efficacy of antifungal drugs during infection by this pathogenic yeast. Med Mycol 51:461–472. https://doi.org/10.3109/13693786.2012.737031

Morgan AD, Koskella B (2017) Coevolution of host and pathogen. In: Genetics and evolution of infectious diseases. Elsevier, Amsterdam, pp 115–140

Moule MG, Monack DM, Schneider DS (2010) Reciprocal analysis of *Francisella novicida* infections of a *Drosophila melanogaster* model reveal host-pathogen conflicts mediated by reactive oxygen and imd-regulated innate immune response. PLoS Pathog 6:e1001065. https://doi.org/10.1371/journal.ppat.1001065

Mukherjee K, Altincicek B, Hain T et al (2010) *Galleria mellonella* as a model system for studying *Listeria* pathogenesis. Appl Environ Microbiol 76:310–317. https://doi.org/10.1128/AEM.01301-09

Naglik JR, Fidel PL, Odds FC (2008) Animal models of mucosal *Candida* infection. FEMS Microbiol Lett 283:129–139. https://doi.org/10.1111/j.1574-6968.2008.01160.x

Nicod C, Banaei-Esfahani A, Collins BC (2017) Elucidation of host–pathogen protein–protein interactions to uncover mechanisms of host cell rewiring. Curr Opin Microbiol 39:7–15. https://doi.org/10.1016/j.mib.2017.07.005

Noll KE, Ferris MT, Heise MT (2019) The collaborative cross: a systems genetics resource for studying host-pathogen interactions. Cell Host Microbe 25:484–498. https://doi.org/10.1016/j.chom.2019.03.009

Parada-Kusz M, Penaranda C, Hagedorn EJ et al (2018) Generation of mouse-zebrafish hematopoietic tissue chimeric embryos for hematopoiesis and host-pathogen interaction studies. Dis Model Mech 11:dmm034876. https://doi.org/10.1242/dmm.034876

Patton EE, Tobin DM (2019) Spotlight on zebrafish: the next wave of translational research. Dis Model Mech 12:dmm039370. https://doi.org/10.1242/dmm.039370

Peleg AY, Jara S, Monga D et al (2009) *Galleria mellonella* as a model system to study *Acinetobacter baumannii* pathogenesis and therapeutics. Antimicrob Agents Chemother 53:2605–2609. https://doi.org/10.1128/AAC.01533-08

Peterson ND, Pukkila-Worley R (2018) *Caenorhabditis elegans* in high-throughput screens for anti-infective compounds. Curr Opin Immunol 54:59–65. https://doi.org/10.1016/j.coi.2018.06.003

Sandstrom G, Saeed A, Abd H (2011) *Acanthamoeba*-bacteria: a model to study host interaction with human pathogens. Curr Drug Targets 12:936–941. https://doi.org/10.2174/138945011795677845

Singulani JL, Scorzoni L, de Oliveira HC et al (2018) Applications of invertebrate animal models to dimorphic fungal infections. J Fungi 4:118. https://doi.org/10.3390/jof4040118

Steinert M, Heuner K (2005) *Dictyostelium* as host model for pathogenesis. Cell Microbiol 7:307–314. https://doi.org/10.1111/j.1462-5822.2005.00493.x

Swart AL, Harrison CF, Eichinger L et al (2018) *Acanthamoeba* and *Dictyostelium* as cellular models for *Legionella* infection. Front Cell Infect Microbiol 8:1–17. https://doi.org/10.3389/fcimb.2018.00061

Tan M (2000) *Caenorhabditis elegans*: a model genetic host to study *Pseudomonas aeruginosa* pathogenesis. Curr Opin Microbiol 3:29–34. https://doi.org/10.1016/S1369-5274(99)00047-8

Thewes S, Soldati T, Eichinger L (2019) Editorial: amoebae as host models to study the interaction with pathogens. Front Cell Infect Microbiol 9:1–3. https://doi.org/10.3389/fcimb.2019.00047

Torraca V, Mostowy S (2018) Zebrafish infection: from pathogenesis to cell biology. Trends Cell Biol 28:143–156. https://doi.org/10.1016/j.tcb.2017.10.002

Ünal C, Steinert M (2006) *Dictyostelium discoideum* as a model to study host–pathogen interactions. In: *Dictyostelium discoideum* protocols. Humana Press, Totowa, pp 507–516

van der Sar AM, Appelmelk BJ, Vandenbroucke-Grauls CMJE, Bitter W (2004) A star with stripes: zebrafish as an infection model. Trends Microbiol 12:451–457. https://doi.org/10.1016/j.tim.2004.08.001

Williams AB, Schumacher B (2018) Nematoda: the *Caenorhabditis elegans* model for innate immunity–interactions between worms and pathogens, and their responses to immunogenic damage. In: Cooper EL (ed) Advances in comparative immunology. Springer, Cham, pp 117–134

Caenorhabditis elegans as Pathogenesis Model to Understand Bacterial Virulence 21

Arun Kumar and Mojibur Khan

Contents

Abstract

Caenorhabditis elegans is an emerging pathogenesis model for studying conserved virulence mechanisms of pathogens and host immune-defense responses against the pathogenic infection. A mutation-based screening of pathogens can be successfully performed for investigating the potential genes of pathogens involved in encoding virulence factors that further kills *C. elegans*. On the other hand, the availability of genetic mutant libraries of *C. elegans* offers the advantage that genetic tractability is possible in both host and pathogen. Thus, the

A. Kumar · M. Khan (✉)
Molecular Biology and Microbial Biotechnology Laboratory, Division of Life Sciences, Institute of Advanced Study in Science and Technology (IASST), Guwahati, Assam, India

B. Siddhardha et al. (eds.), *Model Organisms for Microbial Pathogenesis, Biofilm Formation and Antimicrobial Drug Discovery*,
https://doi.org/10.1007/978-981-15-1695-5_21

high-throughput screening on *C. elegans* may identify new antimicrobials that suppress the virulence ability of pathogens.

Keywords
C. elegans · Pathogen · Probiotic · Commensal · Reactive oxygen species

21.1 Introduction

Caenorhabditis elegans is an invertebrate model organism used for studying longevity, fat storage, neurodegenerative diseases, pathogenesis, and immunity. Its candidature is supported by the fact that it has a short lifespan of 2–3 weeks, easily propagated through self-fertilization and safely used under laboratory conditions. Like larger animals, *C. elegans* consists few of the same organs, including a mouth, pharynx, intestine, collagenous cuticle, and gonad, while they lack respiratory or circulatory system. In nature, *C. elegans* is a microbe-eating nematode (i.e., feeding on bacteria and fungi) and found on different rotting fruits and compost (Félix et al. 2011). The abundance of nematodes on these decomposing plant materials suggests its exposure to different array of microorganisms. In support, researchers have characterized the intestinal microbiota of *C. elegans* based on their geographical locations. These studies showed that their gut microbiota provides resistance against stress conditions (i.e., temperature, pH, osmolarity, and nutrient) and pathogenic infections (Samuel et al. 2016). Further investigations support parallel results in rodents, suggesting the beneficial role of native gut microbiota on host health.

C. elegans is gaining acceptance as one of the powerful model to investigate the effect of their microbial interactions on host health, as they mimic several features of human physiology. Their intestine contains 20 epithelial cells, which are structurally and functionally similar to human intestine. The intestinal epithelial cells form microvilli structures that are composed of intermediate filaments and actin, as similar to humans (McGhee 2013). In addition to nutrient absorption, their intestine provides second largest surface area of nematode's body for host–microbe interactions (McGhee 2013). Several other features support its use for studying host–microbe interactions. Firstly, the available genetic mutants make it a great model to interrogate the effect of host genetics on physiology. Secondly, their transparency allows the visualization of internal tissues and colonizing microbes using microscopic techniques. Thirdly, the use of sodium hypochlorite as bleaching agent allows only the survival of nematode eggs while diminishing the possibility of microbial growth. Thus, axenic or monoaxenic conditions can be easily maintained (Stiernagle 2006). Fourthly, the fluorescent-tagged microbes allow the visualization of *C. elegans*–microbe interactions (Rezzoagli et al. 2019; Gomez et al. 2012). Lastly, the nematodes contain two sexes, i.e., hermaphrodite (XX) and males (XO). Self-fertile or hermaphrodite nematodes lay about 300 eggs and their shorter lifespan of approximately 3 weeks allows synchronous studies of multiple populations (Rezzoagli et al. 2019; Gomez et al. 2012).

About two-third genes of humans are conserved in these nematodes (Lai et al. 2000). Therefore, several human pathogens are able to infect them, which allows us to study immune responses triggered in response to pathogenic infections. Availability of gene mutant libraries of *C. elegans* in the "*C. elegans* Genetics Center" along with advanced technologies such as CRISPR/Cas9 system, transcriptomics, proteomics, advanced imaging technologies, and RNA interference technology may contribute to interrogate the effect of host–pathogen interactions on their candidate genes or mechanistic pathways and their resultant effect on host health.

21.2 Microbial Pathogenesis in *C. elegans*

Evidences suggest that conserved pathogenic and host-mediated defense mechanisms exist in *C. elegans*. They consist of innate immune system providing defense against pathogenic infections, as they lack adaptive immune system (Kim et al. 2002). Firstly, these nematodes avoid pathogens based on their previous encounter (Meisel and Kim 2014). Secondly, their pharynx bacterial grinders also provide protection against pathogens via degrading them (Labrousse et al. 2000). Thirdly, their gut epithelia produce antimicrobial effectors (lysozymes, caenacins, caenopores, neuropeptide-like proteins, and defensin-like peptides) upon pathogenic infection in the intestine (Dierking et al. 2016). Lastly, the innate immunity against microbial infections involves four mechanistic pathways involving mitogen-activated protein kinase (MAPK), a transforming growth factor-beta pathway (TGF-β), insulin-like signaling, and programmed cell death (Evans et al. 2008; Kim et al. 2002; Aballay and Ausubel 2001; Zhang and Zhang 2012). In addition, RNA interference (RNAi) confers resistance against viral infections (Gammon et al. 2017).

21.2.1 *C. elegans*–Bacteria Interactions

Colonization of bacterial pathogens into the intestine of nematodes has been evidenced as preferred mechanisms of infection, but they may release toxins after colonization. Several reports suggest the role of toxins in killing *C. elegans* faster than colonization (e.g., *Listeria monocytogenes*) within their intestine. For example, *B. thuringiensis* secrete pore-forming toxins, *S. aureus* release α-hemolysin, and *E. faecalis* produce cytolysin, all cause lysis of intestinal cells of *C. elegans* (Jiang and Wang 2018). Recognition of bacterial infections may involve many downstream signaling mechanisms. However, p38 MAPK serves as a core pathway involved in defending them against pathogens (Huffman et al. 2004). Further detailed investigations are still needed to know how their immune system is activated against bacterial infections. Reports suggest that damage caused by these bacterial infections may be sensed by their immune system (Zugasti et al. 2014).

Transcriptomics was used to reveal differential gene expression of *C. elegans* in response to their culturing on pathogens, including *Bacillus megaterium*, *Escherichia coli*, *Micrococcus luteus*, and *Pseudomonas sp*. Their results suggested an upregulation in the expression of 366 gene by twofold. Previously, 11.6% genes of 366 genes are known to be important in pathogen or defense-related pathways. *C. elegans* fed on *Pseudomonas sp*. and *B. megaterium* showed upregulation of 2 and 14 defense gene expressions, which extends the lifespan of nematodes as compared to *E. coli*. Their data suggest collagen and cuticle may play a pivotal role in host-specific defense mechanisms, while 23.1% genes of total unknown genes might play a role in an environment-specific phenotypic adaptation under laboratory conditions (Coolon et al. 2009). Co-colonization of defensive *E. faecalis* and pathogenic *S. aureus* within the gut of *C. elegans* revealed reciprocal adaptation via genetic diversification and local adaptation (Ford et al. 2016). An RNAi screening of *C. elegans* infected with *E. faecalis* has shown that knockdown of *nhr-49* shortened the lifespan via regulating their fat metabolism. Additionally, *ajm-1* or *dlg-1* gene

Table 21.1 Killing of *C. elegans* by bacterial and fungal pathogens of humans

Pathogen	Mechanism of interactions	Effect on *C. elegans*	References
P. aeruginosa	Induces p38 MAPK pathway	Fast and slow killing based on medium used	Tan et al. (1999)
S. aureus	Downregulates the components of p38 MAPK immunity	Mutants defective in esp-2/sek-1 and esp-8/nsy-1 increased their susceptibility to *S. aureus*-based killing	Sifri et al. (2003)
S. marcescens	Triggers the expression of antimicrobial peptides and proteins	Distended intestine	Mallo et al. (2002)
S. enterica	P38 MAPK-dependent programmed cell death	Persistent infection of intestine	Aballay and Ausubel (2001)
E. coli	Require virulence global regulator Ler and bacterial tryptophanase	Strain-based slow and fast killing	Mellies et al. (2006) Anyanful et al. (2005)
Listeria monocytogenes	*Require virulence factors such as prfA* and *degU*	Decreased their egg laying ability and survival	Thomsen et al. (2006)
E. faecalis	*Knockdown of nhr-49*, *dlg-1*, and *ajm-1 genes*	Persistent infection of intestine *Reduces survival*	Sim and Hibberd (2016)
C. albicans	Induces p38 MAPK and raises antibacterial responses, involving caenacin (*cnc-4* and *cnc-7*) and antibacterial factor *abf-2*	Both hyphal and yeast forms are pathogenic	Pukkila-Worley et al. (2011)
Cryptococcus neoformans	Raises immune response genes *abl-1* and *lys-7*	Not only live, but also heat-killed yeast can kill them	Marsh et al. (2011)

knockdown in *C. elegans* undergoing infection with *E. faecalis* showed their role in maintaining the intestinal epithelium layer, which serves as an immune barrier. Thus, *nhr-49*, *ajm-1*, and *dlg-1* mechanistically link the fat metabolism, epithelium junction integrity, and innate immune defense (Sim and Hibberd 2016) (Table 21.1). Interestingly, co-evolution of pathogenic *S. aureus* and beneficial *E. faecalis* within nematode's gut secreted superoxide anions by *E. faecalis* against the competing pathogen *S. aureus* (King et al. 2016). Therefore, *C. elegans* may be used to study the inter-microbial interaction and their interactions may be conserved in human as well, which may prove to be potentially therapeutic or diagnostic in future (summarized in Fig. 21.1).

21.2.2 *C. elegans*–Fungi Interactions

Host–fungal interaction studies have also been performed in *C. elegans*. The fungal pathogen *Candida albicans* infects the intestine of nematodes. Transcriptional profile of nematodes exposed to *C. albicans* suggested that host factors control their

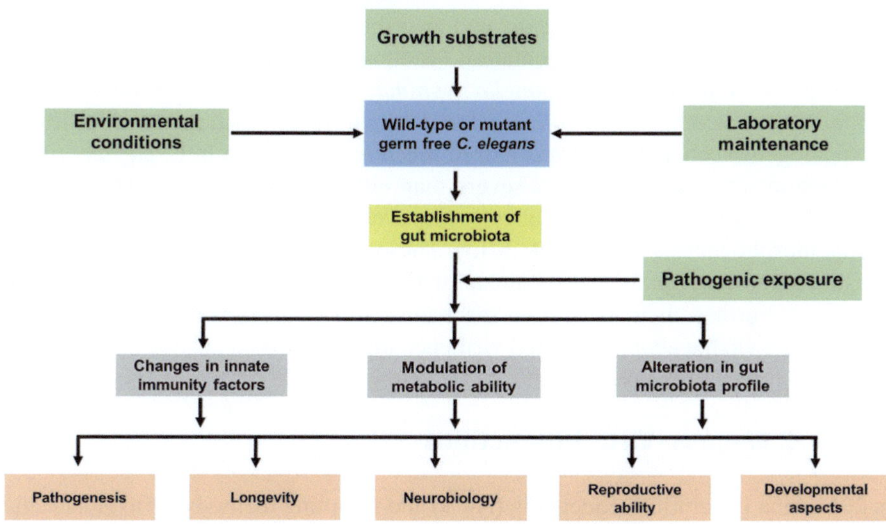

Fig. 21.1 Modeling of host–pathogen interactions in *C. elegans*. Adult N2 or mutant nematodes are transferred to a Petri plate containing nematode growth medium (NGM) agar seeded with *E. coli* OP50, a standard bacterial food source. After reaching sufficient number of eggs, the bleaching procedure can be followed with sodium hypochlorite to synchronize nematodes at egg stage. These synchronized eggs may be washed with M9 buffer and plated onto a fresh NGM Petri plate and maintained at environmental conditions to be studied. After establishment of their gut microbiota, these nematodes may be transferred to other plates for their exposure to pathogens and their effect on nematode's physiology may be studied, including pathogenesis, longevity, neurobiology, reproductive ability, and developmental aspects. Additionally, the effect of other microbial counterparts such as probiotics and commensals may be studied based on their pre-exposure and post-exposure treatment after pathogenic treatment.

susceptibility to this fungal infection. They found that a yeast form of this pathogen infected the intestine of these nematodes, but heat-killed *C. albicans* were unable to infect nematodes. In contrast, the transcriptional analysis of heat-killed and live *C. albicans* infected *C. elegans* showed overlapping with 56% of their host genes. On the other hand, transcriptional profile of this alive fungal pathogen infected nematodes showed less transcriptional overlap with bacterial pathogens (e.g., *S. aureus* (12%)) (Pukkila-Worley et al. 2011). Their results suggested that nematodes can differentiate between different microbial pathogens. The fungal infections are not only intestinal. Other fungal pathogens such as *Drechmeria coniospora* also infects the cuticle of *C. elegans* and penetrates all parts of its body. In response to fungal pathogens, *C. elegans* immune system is activated and results in host-mediated upregulation of antimicrobial peptides and proteins via p38 MAPK pathway (Table 21.1). However, another report suggests that dihydrocaffeic acid receptor-1 (DCAR-1) upregulates the expression of AMP genes against fungal pathogens (Lebrigand et al. 2016).

C. elegans can also be used to study interaction of microorganisms belonging to more than one kingdom. One interesting study shows that co-infection of fungal pathogen *C. albicans* with *Salmonella enterica* serovar *typhimurium* inhibited *C. albicans*-induced filamentation. This may be due to a secretary molecule released by *bacterial pathogen* in *C. elegans* and also their sterilized bacterial filtrate was also able to inhibit fungal filamentation (Tampakakis et al. 2009). Another study found that co-infection of *Salmonella typhimurium* and *E. coli* revealed that *S. typhimurium* outcompetes *E. coli* in colonizing gut. In contrast, the co-infection of two pathogens such as *C. albicans* and *E. faecalis* was studied in *C. elegans*. Their co-infection was found to be less severe than either pathogen alone. This may be due to *E. faecalis* derived anti-fungal compounds inhibiting *C. albicans*-induced hyphal morphogenesis (Cruz et al. 2013). These kinds of studies mimic the gastrointestinal tract environment of human in *C. elegans* as a host (summarized in Fig. 21.1). Further, their interaction mechanisms in *C. elegans* may provide more detailed knowledge about microbial virulence mechanisms.

21.2.3 *C. elegans*–Virus Interactions

In 2011, next generation sequencing (NGS) and sampling methods led to isolate and identify first natural infecting virus, i.e., Orsay virus (Franz et al. 2014). This virus was found to infect intestinal cells of *C. elegans*, i.e., *rde-1* gene which is conserved in *C. elegans*, *Drosophila melanogaster*, plants, and other vertebrate animals. Further investigation suggests that Orsay virus infection in mutant strains of *C. elegans,* i.e., *rde-1* led to increase in the replication of this virus by 100-fold than *C. elegans* strain N2, while other RNAi pathway mutant, i.e., rde-4 and drh-1 also showed to replicate Orsay virus by 100 times, suggested that RNAi mechanism is conserved in *C. elegans* as an authentic mechanism against viral infections (Guo et al. 2013). An alternative antiviral defense mechanism is also operated in *C. elegans,* i.e., STAT pathway. This pathway controls the Orsay virus infection at a particular level and involving kinase SID3 and transcription factor STA-1. STA-1 serves as an ortholog responsible for interferon signaling in mammalian system (Jose et al. 2012).

21.3 *C. elegans*-Mediated Protective Mechanisms Against Pathogen

21.3.1 Learning Behavior

The tempting delicious food-related smell can make us more crave and attract us towards the food. Likewise, different organisms also show an oriented movement towards the external stimulus in nature, such as light, odor, temperature, magnetic field, etc. This phenomenon is called as taxis. Several reports suggest that *C. elegans* also show this kind of taxis behavior towards different water-soluble chemoattractants (e.g., NaCl) and odorants. When a gradient of chemoattractants, i.e., NaCl, is provided to nematodes, they gradually migrate towards the higher concentration (Oda et al. 2011; Hirotsu and Iino 2005). Presently, the scientific community uses similar pathogenic interventions to study their avoidance behavior against pathogens and its underlying molecular mechanisms. If two bacterial food choices are there, *C. elegans* may firstly show their preference to pathogenic bacteria and once infected, it may shift preference to non-pathogenic bacteria (Fig. 21.2) (Zhang et al. 2005). This may be due to sensing of damaged tissues via serotonin signaling which is a gastrointestinal response against enteric pathogenic bacteria. Sensory neurons, ADF (amphid neuron, dual) and ASI (amphid neuron, single), transfer information from environmental cues to RIA interneuron via insulin-like signaling and serotonin (Zhang et al. 2005). Their intestine produces ins-11, an insulin-like

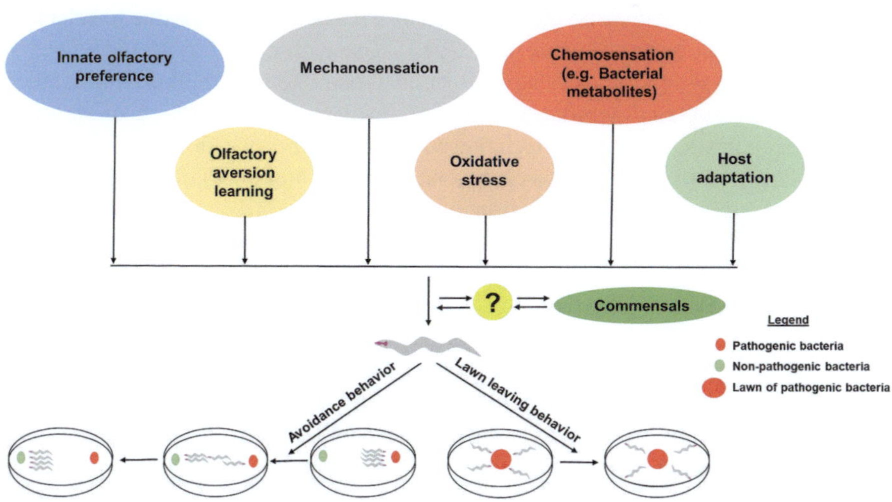

Fig. 21.2 Different mechanisms involved in *C. elegans* avoidance and lawn leaving behavior against pathogenic microorganisms

neuropeptide that controls the insulin pathways and secretion of serotonin. The pathogenic avoidance behavior is negatively controlled by ins-11 via MAPK pathway and a transcription factor hlh-30. The ins-11 attenuation increased their survival against pathogenic bacteria (Lee and Mylonakis 2017), while another neuropeptide ins-6 secreted from ASI (food-sensing neuron) controlled the preference of non-pathogenic bacteria over pathogens (Chen et al. 2013).

Toll-like receptor-1 (TLR-1) is an essential player of innate immunity in avoiding pathogens. Carbon dioxide (CO_2) produced by pathogens may be sensed by chemosensory neuron BAG via TLR-1 signaling resulting in their pathogen avoidance behavior. These results suggested that *C. elegans* showed their innate avoidance behavior by surveilling the metabolic activities of the surrounding microbes (Pujol et al. 2001; Brandt and Ringstad 2015). Also, superoxide dismutase-1 (SOD-1) modulated the gustatory neuron, amphid sensory neurons, right (ASER), thereby affected the avoidance behavior (Horspool and Chang 2017). Moreover, *C. elegans* exhibited lawn leaving behavior when propagated monoaxenically on pathogenic strains (Zhang et al. 2005). However, a mutant lacking *hecw-1* failed to exhibit lawn leaving behavior against pathogens due to reduced function of OLL chemosensory neuron. In contrast, npr-1 (neuropeptide receptor family-1) mutation suppressed hecw-1 mediated lawn leaving behavior that indicated that npr-1 has a vital role in lawn leaving behavior (Chang et al. 2011) (Fig. 21.2).

Pathogen-secreted metabolites also modulate avoidance behavior of *C. elegans* against pathogens. For example, *P. aeruginosa*-released pyochelin and phenazine-1-carboxamide activated chemosensory neuron signaling pathway in nematodes and led to upregulate the expression of DAF-7/TGF-β. This increase in DAF-7/TGF-β expression limits nematode's exposure to *P. aeruginosa* (Wang and Sherwood 2011; Zhang and Zhang 2012). Besides, natural selection also contributes to their avoidance behavior against pathogens. Exposure of *C. elegans* for 30 generations to *Serratia marcescens* increased their pathogenic avoidance behavior due to selection pressure like phenomenon (Penley et al. 2017). These observations provide clues of co-evolutionary (sympatric) mechanisms of host–pathogen interactions. Altogether, the pathogen-associated avoidance behavior involves the integration of multiple sensory mechanisms involving mechanosensation, chemosensation to environmental cues (bacterial metabolites, oxygen concentration, and odor), learned behavior, innate preferences, and host adaptation during co-evolution.

21.3.2 Pharyngeal Grinder

The pharynx is known as a chewing organ of *C. elegans*, as it breaks down the ingested microbes. It consists of 8 muscles and 20 neurons (Avery 1993). Firstly, they feed on microbe from outside environment and crushed by their pharyngeal grinder. This grinding of microbial cells provides nutrients to *C. elegans*. Reports suggest that pharynx grinder-defective *C. elegans* live shorter compared to control N2 strain. Aged *C. elegans* are more susceptible to bacterial infections, as their grinder ability decreases (Hsiao et al. 2013). In addition, the functioning of innate immune system also declines as they age (e.g., p38 MAPK kinase) (Youngman et al. 2011). The pharyngeal grinder mutant *phm-2* observed to increase bacterial accumulation in their gut if infected with *S. typhimurium* compared to the control N2 strain. Thus, age-related deterioration of pharynx grinder shows relation to the declined innate immune mechanisms which further results in the accumulation of bacterial load (Portal-Celhay et al. 2012).

21.3.3 Antimicrobial Proteins and Peptides

Innate immune mechanisms of nematodes are independent of canonical pathogen-associated molecular patterns of pathogens, as they lack Nod-like receptors (NLRs) and NF-kB ortholog. Toll-like receptors (TLRs) are the only pathogen recognizing receptors found in nematodes (Akira et al. 2006). Therefore, the recognition mechanisms of pathogens are not still very clear. Probably, the pathogen-induced damage may be recognized by their innate immunity, known as damage-associated molecular patterns (DAMPS) and leads to stress/innate immune response (insulin-like signaling, p38 MAPK, and TGF-β). The activated innate immunity may upregulate the expression of antimicrobial peptides (AMPs). These upregulated AMPs include caenacins or neuropeptide-like proteins, lysozymes, caenopores, and defensin-like peptides which are produced by nematode's gut epithelia (Dierking et al. 2016). Exposure of a pathogen *S. marcescens* to *C. elegans* overexpressed their intestinal lysozymes (*lys-7* and *lys-8*) (Mallo et al. 2002). While another study showed that exposure of *S. typhimurium* and *E. coli* to nematode's mutants defective in *lys-7* reduced their lifespan via insulin-like signaling (Portal-Celhay et al. 2012). In addition, *lys-1* mutant's exposure to the same pathogens shortened their lifespan via p38 MAPK and TGF-β mechanisms (Portal-Celhay et al. 2012). Future research on nematode's innate immunity may fill a gap between antimicrobial effectors and its resultant protection to the host.

21.3.4 Reactive Oxygen Species

Reactive oxygen species (ROS) modify or damage the genetic material, lipids, and proteins of pathogens. It exerts this effect due to generation of by-products of aerobic metabolic machinery (superoxide anion (O_2^-), hydrogen peroxide (H_2O_2), and hydroxyl radical (•OH)). ROS act in two ways, either activate the innate immunity (e.g., p38 MAPK) or act as microbicidal against pathogens, thereby considered as an essential effector of gut immunity in *C. elegans* (Ray et al. 2012; Roos et al. 2003). Exposure of a bacterial pathogen *E. faecalis* and further inhibiting nematode's oxidative stress defensive genes (i.e., *sod-3* and *ctl-2*) using RNAi decreased their survival against pathogen. These oxidative stress responses are induced intestinally via DAF-16 and/or SKN-1 transcription factors (Van Der Hoeven et al. 2011; Chávez et al. 2007). Thus, the role of ROS can be effectively studied in defending the host against pathogens.

21.4 Colonized Microbe-Associated Protection Against Pathogenic Infections

21.4.1 Commensal-Mediated Protection

Commensals residing in host's gut procure aid without affecting the physiology of the host. In other terms, they do not exert any beneficial or inimical role to the host. *C. elegans* is found in various ecological niches which exhibit a diverse spectrum of microbial commensal to pathogens, including bacteria, fungus, and viruses (Jiang and Wang 2018). The concept of disease ecology can be dissected after amalgamating experimental observations mimicking their natural host–pathogen interactions by culturing them on their natural environment substrates such as soil supplements with decaying organic matter such as rotten plant matter (Fig. 21.3). More recently, the deep sequencing of 16S rDNA V4 facilitated the investigation of gut microbiota profiling of *C. elegans* based on strain, diet selectivity, geographical locations, and laboratory maintenance approaches. Dirksen et al. (2016) collected the substrates (i.e., compost, rotting fruits and stems, invertebrate hosts used for transportation of their larvae) from a large set of diverse natural habitats from Northern Germany, France, and Portugal. They identified that their native gut microbiota falls into 29 bacterial genera which further belong to their respective phylum Proteobacteria, Firmicutes, Bacteroidetes, and Actinobacteria. While Samuel et al. (2016) characterized the *C. elegans* natural gut microbiota on rotting fruits, vegetation, and snail from different geographical locations of Paris, France, and Europe. They reported thousands of bacterial operational taxonomic units (OTUs), approximately 2400 OTUs belong to most abundant phyla Proteobacteria, Actinobacteria, Bacteroidetes, and Firmicutes. However, Proteobacteria is the most dominant phylum in the

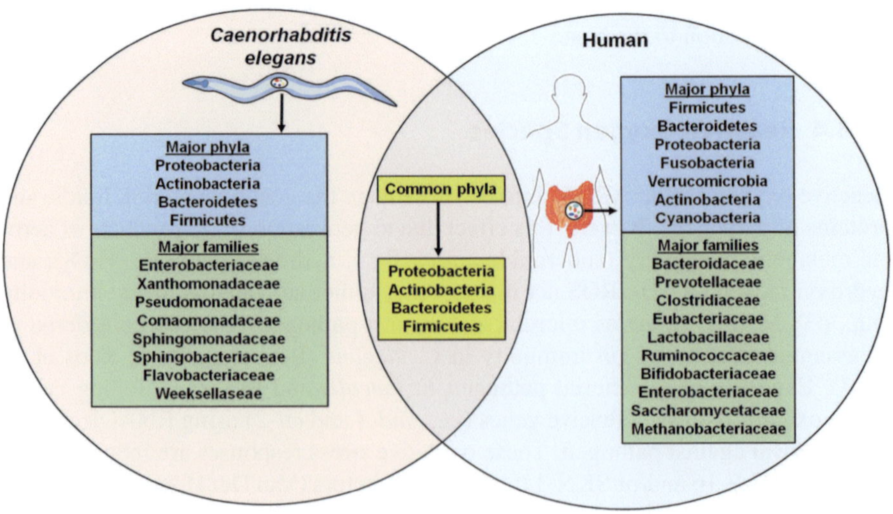

Fig. 21.3 A comparative presentation of major gut microbial profiles of *C. elegans* and human

nematode's gut microenvironment. The reconstructed microbiome composition consisting of alpha-Proteobacteria promotes nematode's proliferation. In contrast, Bacteroidetes and Gamma-Proteobacteria appear to negatively impact the physiology of nematodes (Samuel et al. 2016). Therefore, these studies suggest role of native gut microbiota in maintaining their physiology (Fig. 21.3).

In addition, the intestine of *C. elegans* may also contain such host-specific commensals that protect them from infections via releasing antimicrobial compounds, but these commensals do not have any protective effect on non-host defense mechanisms (Berg et al. 2016). An increase in these commensals may alleviate the drastic effects of pathogenic infection through depriving them from important nutrients (e.g., vitamins or iron) or downregulate the stress/immune responses. In contrast, harmful bacteria may be unable to use important nutrients resulting in limiting their survival (Samuel et al. 2016). The isolated beneficial microorganisms from their intestine, including *Enterobacter* sp., *Gluconobacter* sp., and *Providencia* sp. have shown their protective role against harmful bacteria such as *Chryseobacterium* sp., and *Serratia* sp. isolated from the same gut environment (Samuel et al. 2016). *B. subtilis* GS67, a commensal gut microbiota of *C. elegans* protected them against the *Bacillus thuringiensis* infection via releasing an antibiotic, namely fengycin (Iatsenko et al. 2014). A further exploration of commensal communities harboring in the gut of *C. elegans* may allow detailed investigation of their defensive role against infection, such as affecting the availability of nutrient and production of important metabolites.

21.4.2 Probiotic-Mediated Protection

The term "Probiotics" is used for a group of microorganisms to survive in the host's gut environment and provides beneficial health effects on the host (Fuller et al. 1989). The most common probiotic bacteria include *Lactobacillus*, *Bifidobacterium*, and *Bacillus* (Clark and Hodgkin 2014). Several probiotic mechanisms have been suggested which may improve host-defense responses against pathogenic infections: (1) induce the immune responses against pathogens (Kim and Mylonakis 2012), (2) interfere with the release of virulence factors (i.e., toxins) by reducing or inhibiting the expression of virulence coding genes (Fang et al. 2018), (3) inhibiting the expression of cell membrane receptors causing pathogens to adhere to gut epithelium (Ikeda et al. 2007; Zhou et al. 2014a), and (4) the direct bactericidal (via released compounds) or competitive exclusion of pathogenic microorganisms into their intestine (Zhou et al. 2014b). The probiotic-released pharmacological compounds are bacterial strain-specific and concentration-dependent, involving anti-pathogenic and physiological modulatory activities to the host. Shanahan et al. (2009) called these probiotic bioactive molecules as "pharmabiotics" (Shanahan et al. 2009).

C. elegans consist of various defense mechanisms to distinguish between probiotic and the pathogenic bacteria. The pathogen responses act via insulin-like signaling and DAF-16 pathways. In contrast, the probiotic bacteria act independent of insulin-like signaling. Other host response pathways are more strain dependent which includes TGF-β, MAPK, toll/IL-1 resistance (TIR), and nuclear hormone receptors (NHRs) (Clark and Hodgkin 2014). Recent reports suggest that *Lactobacillus acidophilus* strain NCFM improved their survival rate against intestinal infections with *E. faecalis* and *Staphylococcus aureus* through activation of p38 MAPK and β-catenin signaling pathways (Kim and Mylonakis 2012), while *Lactobacillus casei* protect the nematodes against the *Klebsiella pneumonia* infection via p38 MAPK pathway and also protect from oxidative stress by improving GSH and SOD levels (Kamaladevi and Balamurugan 2016). In contrast, heat-killed *Lactobacillus* sp. increases the expression of defense-related genes (acdh-1 and cnc-2) which suppresses the pathogenesis of *Salmonella* sp. and *Yersinia* sp. (Lee et al. 2015). *Lactobacillus zeae* also protect the nematodes against enterotoxigenic *E. coli* by affecting their enterotoxin gene expression (Zhou et al. 2014a). Further advances in the research of host–microbe interaction may reveal more conserved pathways (Fig. 21.4).

Additionally, a consortium consisting of *Lactobacillus reuteri*, *Lactobacillus salivarius*, and *Pediococcus acidilactici* has been shown to inhibit their development and show anti-tumor like activity. The inhibitory effect of this consortium can be used against parasitic nematodes (Fasseas et al. 2013). Moreover, different bacterial-induced mechanisms may operate in *C. elegans* propagated monoaxenically or in a consortium. Furthermore, the investigations of probiotics in ameliorating the intestinal status of nematodes may reveal their impact on the overall health status of *C. elegans*.

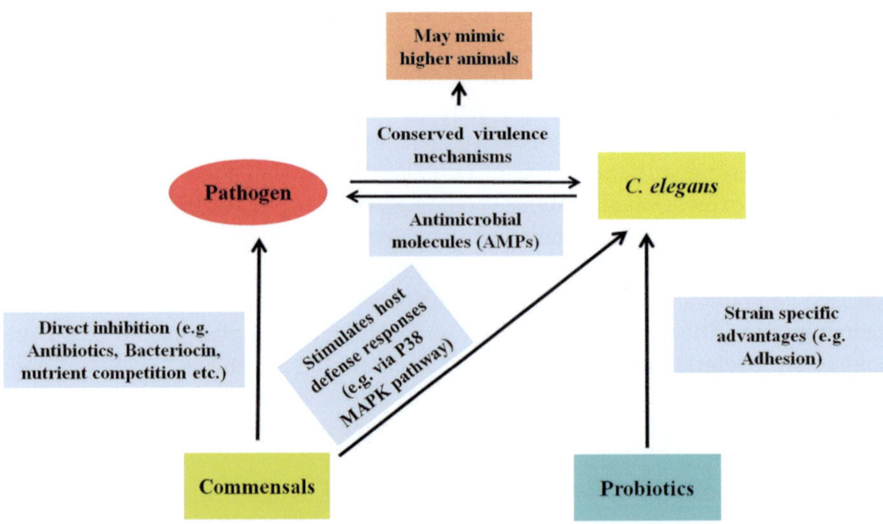

Fig. 21.4 Cross-talk of commensals, probiotic, and pathogens in the gut of *C. elegans*

21.5 Conclusion

C. elegans as a eukaryotic host offers a platform to understand fundamental host–pathogen interactions, and to reveal evolutionary conserved pathways operated in response to infection with bacteria, fungus, and viruses. Not only it allows us to study mono-infection, but also inter-microbial and inter-kingdom investigation can be performed.

References

Aballay A, Ausubel FM (2001) Programmed cell death mediated by ced-3 and ced-4 protects *Caenorhabditis elegans* from *Salmonella typhimurium*-mediated killing. Proc Natl Acad Sci 98(5):2735–2739

Akira S, Uematsu S, Takeuchi O (2006) Pathogen recognition and innate immunity. Cell 124(4):783–801

Anyanful A, Dolan-Livengood JM, Lewis T, Sheth S, DeZalia MN, Sherman MA, Kalman LV, Benian GM, Kalman D (2005) Paralysis and killing of *Caenorhabditis elegans* by enteropathogenic *Escherichia coli* requires the bacterial tryptophanase gene. Mol Microbiol 57(4):988–1007

Avery L (1993) The genetics of feeding in *Caenorhabditis elegans*. Genetics 133(4):897–917

Berg M, Zhou XY, Shapira M (2016) Host-specific functional significance of *Caenorhabditis* gut commensals. Front Microbiol 7:1622

Brandt JP, Ringstad N (2015) Toll-like receptor signaling promotes development and function of sensory neurons required for a *C. elegans* pathogen-avoidance behavior. Curr Biol 25(17): 2228–2237

Chang HC, Paek J, Kim DH (2011) Natural polymorphisms in *C. elegans* HECW-1 E3 ligase affect pathogen avoidance behaviour. Nature 480(7378):525

Chávez V, Mohri-Shiomi A, Maadani A, Vega LA, Garsin DA (2007) Oxidative stress enzymes are required for DAF-16-mediated immunity due to generation of reactive oxygen species by *Caenorhabditis elegans*. Genetics 176(3):1567–1577

Chen Z, Hendricks M, Cornils A, Maier W, Alcedo J, Zhang Y (2013) Two insulin-like peptides antagonistically regulate aversive olfactory learning in *C. elegans*. Neuron 77(3):572–585

Clark LC, Hodgkin J (2014) Commensals, probiotics and pathogens in the *Caenorhabditis elegans* model. Cell Microbiol 16(1):27–38

Coolon JD, Jones KL, Todd TC, Carr BC, Herman MA (2009) *Caenorhabditis elegans* genomic response to soil bacteria predicts environment-specific genetic effects on life history traits. PLoS Genet 5(6):e1000503

Cruz MR, Graham CE, Gagliano BC, Lorenz MC, Garsin DA (2013) *Enterococcus faecalis* inhibits hyphal morphogenesis and virulence of *Candida albicans*. Infect Immun 81(1):189–200

Dierking K, Yang W, Schulenburg H (2016) Antimicrobial effectors in the nematode *Caenorhabditis elegans*: an outgroup to the Arthropoda. Philos Trans R Soc Lond B Biol Sci 371(1695):20150299

Dirksen P, Marsh SA, Braker I, Heitland N, Wagner S, Nakad R, Mader S, Petersen C, Kowallik V, Rosenstiel P, Félix MA, Schulenburg H (2016) The native microbiome of the nematode *Caenorhabditis elegans*: gateway to a new host-microbiome model. BMC Biol 14(1):38

Evans EA, Chen WC, Tan MW (2008) The DAF-2 insulin-like signaling pathway independently regulates aging and immunity in *C. elegans*. Aging Cell 7(6):879–893

Fang K, Jin X, Hong SH (2018) Probiotic *Escherichia coli* inhibits biofilm formation of pathogenic *E. coli* via extracellular activity of DegP. Sci Rep 8(1):4939

Fasseas MK, Fasseas C, Mountzouris KC, Syntichaki P (2013) Effects of *Lactobacillus salivarius*, *Lactobacillus reuteri*, and *Pediococcus acidilactici* on the nematode *Caenorhabditis elegans* include possible antitumor activity. Appl Microbiol Biotechnol 97(5):2109–2118

Félix M-A, Ashe A, Piffaretti J, Wu G, Nuez I, Bélicard T, Jiang Y, Zhao G, Franz CJ, Goldstein LD (2011) Natural and experimental infection of *Caenorhabditis* nematodes by novel viruses related to nodaviruses. PLoS Biol 9(1):e1000586

Ford SA, Kao D, Williams D, King KC (2016) Microbe-mediated host defence drives the evolution of reduced pathogen virulence. Nat Commun 7:13430

Franz CJ, Renshaw H, Frezal L, Jiang Y, Félix M-A, Wang D (2014) Orsay, Santeuil and Le Blanc viruses primarily infect intestinal cells in *Caenorhabditis* nematodes. Virology 448:255–264

Fuller M, McWilliam R, Wang T, Giles L (1989) The optimum dietary amino acid pattern for growing pigs: 2. Requirements for maintenance and for tissue protein accretion. Br J Nutr 62(2):255–267

Gammon DB, Ishidate T, Li L, Gu W, Silverman N, Mello CC (2017) The antiviral RNA interference response provides resistance to lethal arbovirus infection and vertical transmission in *Caenorhabditis elegans*. Curr Biol 27(6):795–806

Gomez F, Monsalve GC, Tse V, Saiki R, Weng E, Lee L, Srinivasan C, Frand AR, Clarke CF (2012) Delayed accumulation of intestinal coliform bacteria enhances life span and stress resistance in *Caenorhabditis elegans* fed respiratory deficient *E. coli*. BMC Microbiol 12(1):300

Guo X, Zhang R, Wang J, Ding S-W, Lu R (2013) Homologous RIG-I–like helicase proteins direct RNAi-mediated antiviral immunity in *C. elegans* by distinct mechanisms. Proc Natl Acad Sci 110(40):16085–16090

Hirotsu T, Iino Y (2005) Neural circuit-dependent odor adaptation in *C. elegans* is regulated by the Ras-MAPK pathway. Genes Cells 10(6):517–530

Horspool AM, Chang HC (2017) Superoxide dismutase SOD-1 modulates *C. elegans* pathogen avoidance behavior. Sci Rep 7:45128

Hsiao J-Y, Chen C-Y, Yang M-J, Ho H-C (2013) Live and dead GFP-tagged bacteria showed indistinguishable fluorescence in *Caenorhabditis elegans* gut. J Microbiol 51(3):367–372

Huffman DL, Abrami L, Sasik R, Corbeil J, van der Goot FG, Aroian RV (2004) Mitogen-activated protein kinase pathways defend against bacterial pore-forming toxins. Proc Natl Acad Sci 101(30):10995–11000

Iatsenko I, Yim JJ, Schroeder FC, Sommer RJ (2014) B. subtilis GS67 protects *C. elegans* from gram-positive pathogens via fengycin-mediated microbial antagonism. Curr Biol 24(22):2720–2727

Ikeda T, Yasui C, Hoshino K, Arikawa K, Nishikawa Y (2007) Influence of lactic acid bacteria on longevity of *Caenorhabditis elegans* and host defense against *Salmonella enterica* serovar enteritidis. Appl Environ Microbiol 73(20):6404–6409

Jiang H, Wang D (2018) The microbial zoo in the *C. elegans* intestine: Bacteria, fungi and viruses. Viruses 10(2):85

Jose AM, Kim YA, Leal-Ekman S, Hunter CP (2012) Conserved tyrosine kinase promotes the import of silencing RNA into *Caenorhabditis elegans* cells. Proc Natl Acad Sci 109(36): 14520–14525

Kamaladevi A, Balamurugan K (2016) *Lactobacillus* casei triggers a TLR mediated RACK-1 dependent p38 MAPK pathway in *Caenorhabditis elegans* to resist *Klebsiella pneumoniae* infection. Food Funct 7(7):3211–3223

Kim DH, Feinbaum R, Alloing G, Emerson FE, Garsin DA, Inoue H, Tanaka-Hino M, Hisamoto N, Matsumoto K, Tan M-W (2002) A conserved p38 MAP kinase pathway in *Caenorhabditis elegans* innate immunity. Science 297(5581):623–626

Kim Y, Mylonakis E (2012) *Caenorhabditis elegans* immune conditioning with the probiotic bacterium *Lactobacillus acidophilus* strain NCFM enhances gram-positive immune responses. Infect Immun 80(7):2500–2508

King KC, Brockhurst MA, Vasieva O, Paterson S, Betts A, Ford SA, Frost CL, Horsburgh MJ, Haldenby S, Hurst GD (2016) Rapid evolution of microbe-mediated protection against pathogens in a worm host. ISME J 10(8):1915

Labrousse A, Chauvet S, Couillault C, Kurz CL, Ewbank JJ (2000) *Caenorhabditis elegans* is a model host for *Salmonella typhimurium*. Curr Biol 10(23):1543–1545

Lai C-H, Chou C-Y, Ch'ang L-Y, Liu C-S, Lin W-c (2000) Identification of novel human genes evolutionarily conserved in *Caenorhabditis elegans* by comparative proteomics. Genome Res 10(5):703–713

Lebrigand K, He LD, Thakur N, Arguel M-J, Polanowska J, Henrissat B, Record E, Magdelenat G, Barbe V, Raffaele S (2016) Comparative genomic analysis of *Drechmeria coniospora* reveals core and specific genetic requirements for fungal endoparasitism of nematodes. PLoS Genet 12(5):e1006017

Lee J, Choe J, Kim J, Oh S, Park S, Kim S, Kim Y (2015) Heat-killed *Lactobacillus spp.* cells enhance survivals of *Caenorhabditis elegans* against *Salmonella* and *Yersinia* infections. Lett Appl Microbiol 61(6):523–530

Lee K, Mylonakis E (2017) An intestine-derived neuropeptide controls avoidance behavior in *Caenorhabditis elegans*. Cell Rep 20(10):2501–2512

Mallo GV, Kurz CL, Couillault C, Pujol N, Granjeaud S, Kohara Y, Ewbank JJ (2002) Inducible antibacterial defense system in *C. elegans*. Curr Biol 12(14):1209–1214

Marsh EK, van den Berg MC, May RC (2011) A two-gene balance regulates *Salmonella typhimurium* tolerance in the nematode *Caenorhabditis elegans*. PLoS One 6(3):e16839

McGhee JD (2013) The *Caenorhabditis elegans* intestine. Wiley Interdiscip Rev Dev Biol 2(3):347–367

Meisel JD, Kim DH (2014) Behavioral avoidance of pathogenic bacteria by *Caenorhabditis elegans*. Trends Immunol 35(10):465–470

Mellies JL, Barron AM, Haack KR, Korson AS, Oldridge DA (2006) The global regulator Ler is necessary for enteropathogenic *Escherichia coli* colonization of *Caenorhabditis elegans*. Infect Immun 74(1):64–72

Oda S, Tomioka M, Iino Y (2011) Neuronal plasticity regulated by the insulin-like signaling pathway underlies salt chemotaxis learning in *Caenorhabditis elegans*. J Neurophysiol 106(1):301–308

Penley MJ, Ha GT, Morran LT (2017) Evolution of *Caenorhabditis elegans* host defense under selection by the bacterial parasite *Serratia marcescens*. PLoS One 12(8):e0181913

Portal-Celhay C, Bradley ER, Blaser MJ (2012) Control of intestinal bacterial proliferation in regulation of lifespan in *Caenorhabditis elegans*. BMC Microbiol 12(1):49

Pujol N, Link EM, Liu LX, Kurz CL, Alloing G, Tan M-W, Ray KP, Solari R, Johnson CD, Ewbank JJ (2001) A reverse genetic analysis of components of the toll signaling pathway in *Caenorhabditis elegans*. Curr Biol 11(11):809–821

Pukkila-Worley R, Ausubel FM, Mylonakis E (2011) Candida albicans infection of *Caenorhabditis elegans* induces antifungal immune defenses. PLoS Pathog 7(6):e1002074

Ray PD, Huang B-W, Tsuji Y (2012) Reactive oxygen species (ROS) homeostasis and redox regulation in cellular signaling. Cell Signal 24(5):981–990

Rezzoagli C, Granato E, Kuemmerli R (2019) In vivo microscopy reveals the impact of *Pseudomonas aeruginosa* social interactions on host colonization. ISME J 13(10):2403–2414

Roos D, van Bruggen R, Meischl C (2003) Oxidative killing of microbes by neutrophils. Microbes Infect 5(14):1307–1315

Samuel BS, Rowedder H, Braendle C, Félix M-A, Ruvkun G (2016) *Caenorhabditis elegans* responses to bacteria from its natural habitats. Proc Natl Acad Sci 113(27):E3941–E3949

Shanahan F, Stanton C, Ross P, Hill C (2009) Pharmabiotics: bioactives from mining host-microbe-dietary interactions. Funct Food Rev 1:1–6

Sifri CD, Begun J, Ausubel FM, Calderwood SB (2003) *Caenorhabditis elegans* as a model host for *Staphylococcus aureus* pathogenesis. Infect Immun 71(4):2208–2217

Sim S, Hibberd ML (2016) *Caenorhabditis elegans* susceptibility to gut *Enterococcus faecalis* infection is associated with fat metabolism and epithelial junction integrity. BMC Microbiol 16(1):6

Stiernagle T (2006) Maintenance of *C. elegans*. In: The *C. elegans* Research Community (ed) WormBook. WormBook, Pasadena. doi/10.1895/wormbook. 1.101. 1

Tampakakis E, Peleg AY, Mylonakis E (2009) Interaction of *Candida albicans* with an intestinal pathogen, *Salmonella enterica* serovar Typhimurium. Eukaryot Cell 8(5):732–737

Tan M-W, Rahme LG, Sternberg JA, Tompkins RG, Ausubel FM (1999) *Pseudomonas aeruginosa* killing of *Caenorhabditis elegans* used to identify *P. aeruginosa* virulence factors. Proc Natl Acad Sci 96(5):2408–2413

Thomsen LE, Slutz SS, Tan M-W, Ingmer H (2006) *Caenorhabditis elegans* is a model host for *Listeria monocytogenes*. Appl Environ Microbiol 72(2):1700–1701

Van Der Hoeven R, McCallum KC, Cruz MR, Garsin DA (2011) Ce-Duox1/BLI-3 generated reactive oxygen species trigger protective SKN-1 activity via p38 MAPK signaling during infection in *C. elegans*. PLoS Pathog 7(12):e1002453

Wang Z, Sherwood DR (2011) Dissection of genetic pathways in *C. elegans*. Methods Cell Biol 106:113–157. Elsevier

Youngman MJ, Rogers ZN, Kim DH (2011) A decline in p38 MAPK signaling underlies immunosenescence in *Caenorhabditis elegans*. PLoS Genet 7(5):e1002082

Zhang X, Zhang Y (2012) DBL-1, a TGF-β, is essential for *Caenorhabditis elegans* aversive olfactory learning. Proc Natl Acad Sci 109(42):17081–17086

Zhang Y, Lu H, Bargmann CI (2005) Pathogenic bacteria induce aversive olfactory learning in *Caenorhabditis elegans*. Nature 438(7065):179

Zhou M, Yu H, Yin X, Sabour PM, Chen W, Gong J (2014a) *Lactobacillus zeae* protects *Caenorhabditis elegans* from enterotoxigenic *Escherichia coli*-caused death by inhibiting enterotoxin gene expression of the pathogen. PLoS One 9(2):e89004

Zhou M, Zhu J, Yu H, Yin X, Sabour P, Zhao L, Chen W, Gong J (2014b) Investigation into in vitro and in vivo models using intestinal epithelial IPEC-J2 cells and *Caenorhabditis elegans* for selecting probiotic candidates to control porcine enterotoxigenic *Escherichia coli*. J Appl Microbiol 117(1):217–226

Zugasti O, Bose N, Squiban B, Belougne J, Kurz CL, Schroeder FC, Pujol N, Ewbank JJ (2014) Activation of a G protein–coupled receptor by its endogenous ligand triggers the innate immune response of *Caenorhabditis elegans*. Nat Immunol 15(9):833

Zebra Fish Infection Model: From Pathogenesis to Therapeutics

22

Brijesh Singh Sisodia, Vijay Kumar, Simranjeet Singh, Satyender Singh, Shivika Datta, and Joginder Singh

Contents

B. S. Sisodia (✉)
Department of Biochemistry, Regional Ayurveda Research Institute for Drug Development, Gwalior, M.P., India

V. Kumar
Department of Chemistry, Regional Ayurveda Research Institute for Drug Development, Gwalior, M.P., India

S. Singh
Department of Biotechnology, Lovely Professional University, Jalandhar, Punjab, India

Punjab Biotechnology Incubators, Mohali, Punjab, India

Regional Advanced Water Testing Laboratory, Mohali, Punjab, India

S. Singh
Regional Advanced Water Testing Laboratory, Mohali, Punjab, India

S. Datta
Department of Zoology, Doaba College, Jalandhar, Punjab, India

J. Singh (✉)
Department of Biotechnology, Lovely Professional University, Jalandhar, Punjab, India
e-mail: joginder.15005@lpu.co.in

B. Siddhardha et al. (eds.), *Model Organisms for Microbial Pathogenesis, Biofilm Formation and Antimicrobial Drug Discovery*,
https://doi.org/10.1007/978-981-15-1695-5_22

Abstract

Earlier, murine infection models were the preferred host for studying interaction between the host and the microbial pathogen. However, all pathogens do not infect mice and they do not always replicate human infections. Moreover, mammalian models are not very suitable for large scale screens due to high cost involved. Since Zebrafish (*Danio rerio*) are transparent in early development, genetically modifiable, and have functional innate immune system with neutrophils and macrophages similar to human counterparts, it has come to the forefront of biomedical research. This organism has been used to investigate wide varieties of viral, bacterial, and fungal pathogens illuminating our understanding of disease pathogenesis and host–pathogen interactions. Due to recent technical advancements, zebrafish has become a preferred model for infectious diseases for studying mechanisms of pathogenesis, virulence factors, response of the host immune system, immune evasion, drug resistance, and potential therapeutic applications. This chapter will describe zebrafish as a model host system for infectious diseases along with several advantages of using it to understand immune response, host–pathogen interactions and their therapeutic potential.

Keywords

Zebrafish · Infection · Host–microbes interactions · Immune response · Innate immunity · Fungal infections

22.1 Introduction

In human infectious diseases, thorough analysis of the communication that occurs between host defense systems and the pathogenic microbes is not practical. Thus, an appropriate host is needed that can mimic human immune response to pathogens. For this purpose, mammalian model hosts like murine models and nonhuman primates have been in discussion for many years; however, these are not ideal in many instances leading to development of new host models. Later, few models such as nematodes (*Caenorhabditis elegans*) and flies (*Drosophila melanogaster*) have been developed that allows analysis of innate response during microbial infection. These were successfully used to address specific questions while analyzing mechanism of infectious diseases (Sullivan and Kim 2008; Allen and Neely 2010; Torraca and Mostowy 2018).

Zebrafish (*Danio rerio*) embryos and larvae are widely accepted in vivo models to study for a variety of pathogenic diseases in humans. However, zebrafish model is becoming increasingly popular for addressing more complex questions pertaining to adaptive and innate immune responses as a vertebrate model host. Zebrafish is an attractive and model host to examine mechanism of infectious diseases, owing to the wealth of knowledge generated from decades of research done on vertebrate biology. Zebrafish have early emergence of innate immune system, i.e., primitive macrophages have been described in zebrafish at 22 h postfertilization (hpf) that are able to defend experimental infections, whereas primitive neutrophils were observed at 36 hpf that synchronize with macrophages to provide stronger immunity. Further, local and systemic infections in zebrafish can be achieved by microinjection. Such features have resulted in the establishment of *D. rerio* infection models for different varieties of viral, bacterial, and fungal pathogens. Recent studies of various pathogenic infections in zebrafish embryo model have illuminated our understanding of pathogenesis and cell biology (Phelps and Neely 2005; Sullivan and Kim 2008; Meeker and Trede 2008; Torraca and Mostowy 2018).

22.2 Methods for Investigation of Zebrafish–Pathogen Interactions

Zebrafish infections can be established via microinjection. For local infections, yolk sac and body cavities like hindbrain ventricle larvae can be injected. The local infections can also be pursed subcutaneous and intramuscular. Integral dissemination of pathogenic microbes throughout the host body can be achieved by using intravenous injections via duct of Cuvier or caudal blood/vein island. Since zebrafish are able to absorb compounds dissolved in bath water through skin, it is very useful for screening of large compound library including antimicrobials, for evaluation of toxicity and bioactivities. Intravital imaging is another powerful tool to study zebrafish–microbe interactions in vivo as it allows visualization of living cells in the true environment of complex multicellular organism. It is achieved by combining fluorescently labeled transgenic lines and microbial pathogens of zebrafish with selected fluorescent tags to specific cell or gene types (Torraca and Mostowy 2018).

Genetic manipulation in zebrafish is obtained through injection of genetic constructs and chemicals at one-cell stage of eggs. Transient knockdown of specific gene is done via morpholino oligonucleotide, RNA-binding oligomers and injected to block the maturation and translation of pre-mRNA. For achieving transient overexpression of genes, synthesized plasmid DNA or mRNA is injected in zebrafish egg resulting in ubiquitous and cell or tissue-specific expression, respectively. Stable integration of DNA is obtained by transgenesis and insertional mutagenesis. Transgenesis frequency can be enhanced using transposases like meganucleases or Tol2, i.e., l-SceI. Stable mutants of zebrafish can be obtained efficiently by techniques of targeted mutagenesis like transcription activator-like effector nuclease (TALEN), zinc-finger nuclease (ZFN), and CRISPR (associated protein 9/clustered regulatory interspaced short palindromic repeats, Cas9/CRISPR). In these techniques, site-directed double-stranded break is drafted which is fixed through an error-inclined non-homologous end joining system. Recently, CRISPR/Cas9 technology has been a method of choice for obtaining zebrafish mutants (Varshney et al. 2015).

22.3 Zebrafish as a Model Host for Infectious Diseases

Zebrafish has several advantages like low cost, small size, prolific breeding, ease of maintenance, and ex vivo embryo development that make it an ideal model host system for infectious diseases. Its embryos are apparently transparent for first few days postfertilization (dpf) and larvae are translucent for 21 days. Optically transparent properties allow analysis of ex vivo embryo development using simple microscope to examine how mutations affect tissue and organ development in vertebrates that is not possible with conventional mammalian models. This property is also very useful for research in infectious diseases as fluorescently labeled viruses or bacteria can be used to visualize embryos under fluorescent stereomicroscope for microbial load and its dissemination (Allen and Neely 2010; Torraca and Mostowy 2018). Furthermore, recently several zebrafish transgenic strains with tissue-specific fluorescent markers have been developed that are specifically useful in studying host–pathogen interactions. For visualizing macrophages in vivo fli1: enhanced GFP (EGFP) strain (Lawson and Weinstein 2002) or a lysC:EGFP strain (Hall et al. 2007) is useful, whereas live analysis of neutrophils or heterophils is possible with myeloid-specific peroxidase: EGFP strain (Mathias et al. 2006; Clay et al. 2007). Several important interactions between vertebrate host and pathogenic microbes have been discovered by using labeled fluorescent dyes in both bacteria and immune cells of zebrafish (Renshaw et al. 2006; Redd et al. 2006). However, this type of analysis can only be performed up to third week postfertilization as optical transparency and visualization of fluorescent markers in internal tissues/organs is disturbed by the development of melanization. This limitation has been overcome by growth of a clear adult zebrafish strain which enables visualization of interiors of adult fish stages in vivo (White et al. 2008).

The ease of obtaining hundreds of embryos per animal pair is a great advantage for high throughput pharmacological treatment in vivo via bathwater exposure, a feature which was available previously for in vitro culture models and invertebrates. The genome of zebra fish is almost similar to humans (Howe et al. 2013). Nearly about 70% immune associated genes are orthologous to zebrafish. Several genomic studies and other reports revealed that adaptive and innate immune system in zebrafish is highly similar with humans (Murayama et al. 2006; Stein et al. 2007; Sullivan and Kim 2008). Therefore, zebrafish is emerging as a perfect model system to explore development, function, and diseases of immune system (Sullivan and Kim 2008; Meeker and Trede 2008). Development of immune system occurs in stages as it gets functional innate immunity by 2 days postfertilization (2dpf), whereas adaptive immunity is delayed in zebrafish. This provides a great opportunity to investigate functional response of innate system during microbial infection in absence of adaptive immunity. This provides another advantage to zebrafish embryos as an infectious disease model (Trede et al. 2004; Lam et al. 2004; Meeker and Trede 2008; Sullivan and Kim 2008).

22.3.1 Zebrafish as a Model Host for Viral Infection

There is a need to discover effective vaccines, adjuvant therapies, and antiviral medicines. Suitable animal models of human viral diseases are required to achieve these goals. These models essentially help the replication of human viruses and produce immune response via conserved signaling cascades (Goody et al. 2014). The action response to viral infection solely depends on the recognition of various viral mechanisms by host cells. These receptors initiate antiviral defense mechanisms conceded by various signaling pathways. Both adult and larval zebrafish have been used to study several types of viral infections. Zebrafish model of hepatitis C virus (HCV) has also been reported to investigate for studying the mechanism of its replication. The replicon vector was microinjected into larvae of zebrafish. The virus was visualized by enhanced green fluorescent protein (EGFP). In order to check its use in drug screening two clinically used drugs, oxymatrine and ribavirin, were investigated and the drugs have negative influence on the replication of the HCV subreplicon at larval stage. This model has also been demonstrated to check the mechanisms replication of HCV and in the discovery of new drugs for its treatment (Ding et al. 2015). Study of nervous necrosis virus in zebrafish infection model showed that an active interferon-α (IFN-α) response is responsible for reduced viral load and mortality in acute and persistent infection. Further, IFN-α treatment inhibiting virus replication in susceptible zebrafish larvae suggests possible application antiviral therapies (Lu et al. 2008).

22.3.2 Zebrafish as a Model Host for Bacterial Infections

Zebrafish has been employed as model host to investigate numerous types of bacterial infections including *Shigella flexneri, Listeria monocytogenes, Staphylococcus aureus, Pseudomonas aeruginosa, Burkholderia cenocepacia, Salmonella Typhimurium, and Mycobacterium* species.

22.3.2.1 Understanding Host–Microbe Interaction

Zebrafish embryo model has been utilized to explore the host–microbe interactions for intracellular infection by *Listeria monocytogenes*. Macrophages driven rapid phagocytosis was also visualized in vivo of all the green fluorescent protein (GFP)-expressing bacteria within 1 h of infection. Further, formation of actin comet tail for invasion of adjoining cells by the bacteria was also observed using transmission electron microscopy and staining techniques (Levraud et al. 2009).

22.3.2.2 Revealing Mechanisms of Disease Pathogenesis and Virulence

In embryos of zebra fishes, development of immune system has been exploited to demonstrate that in order to counter *Pseudomonas aeruginosa* infection both macrophages and neutrophils are needed. Infection of zebrafish at 50 hpf leads to attenuated infection as both macrophages and neutrophils are functional. This observation is similar to what observed in murine model. However, infection gets lethal in the presence of macrophages at 28 hpf. To further investigate the mechanism, gene knockdown was achieved using morpholino oligonucleotide technology in which nonionic oligos are microinjected to inhibit RNA translation of gene of interest. Knockdown of *pu1* gene eliminates myeloid cells including macrophages and neutrophils resulting in high susceptibility to *P. aeruginosa* infection, whereas *gata1* expression knockdown enhanced myeloid cells led to increased survival of zebrafish (Clatworthy et al. 2009).

Zebrafish embryo is also a useful host for large scale mutagenesis screens in 96-well plate owing to its small size. Observations for persistence or clearance of fluorescently labeled pathogen can efficiently be achieved using fluorescent stereomicroscope. Adult zebrafish have been also used for screening mutagenesis at large scale. Such screens have been used to recognize virulence factors/genes which are involved in pathogenesis (Kizy and Neely 2009; Miller 2005).

22.3.2.3 Discovering Immunological Bottleneck and Bacterial Persistence

Staphylococcus aureus, a member of human microflora in skin and mucosa, may cause complications in systemic infection of zebrafish model like abscesses and septicemia (Prajsnar et al. 2008, 2012). *S. aureus* infection model has also been reported to show the role of nerve growth factor-β (NGF-β). Mutations in protein tropomyosin receptor kinase A (TRKA) and NGF-β receptor resulted in severe *S. aureus* infection. In zebrafish, macrophages and neutrophils are able to clear most of the systemic *S. aureus* infection through phagocytosis. However, few cells remain viable and continue to replicate. Here neutrophils appear as immunological bottleneck resulting in bacterial replication and clonal selection. Reduced clonal selection has been observed in neutrophils depleted zebrafish lines. *S. aureus* is also known for development of antibiotic resistance as sub-curative doses of antibiotic lead to growth of antibiotic-resistant strains (McVicker et al. 2014). Collectively, *S. aureus*–zebrafish infection model provides significant understanding for cellular mechanism of bacterial persistence.

22.3.2.4 Inflammatory Macrophages Act as Replicative Niche

Burkholderia cenocepacia infection is a serious health concern for cystic fibrosis patients as it results in various dreadful diseases like progressive decline of lung function, bacteremia, and abscesses. *B. cenocepacia*–zebrafish infection model was established to investigate *Burkholderia* pathogenesis in vivo. Studies have shown that only macrophages are essential for survival and replication of *B. cenocepacia* in vivo. Further, chemical induced disruption of macrophages results in reduction of bacterial replication which confers zebrafish to the infection. This is responsible to move from chronic to acute disease resulting in inflammation and death of zebrafish at next level (Mesureur et al. 2017; Vergunst et al. 2010).

22.3.2.5 Understanding Immune Evasion

A natural pathogen *Mycobacterium marinum* of aquatic species including fishes is a close relative of *M. tuberculosis* responsible for tuberculosis disease in humans. *M. marinum*–zebrafish infection model has been employed to study the mechanisms behind human tuberculosis (Cronan and Tobin 2014). Study of *M. marinum*–zebrafish infection model suggests that virulence determinant region of difference 1 (RD1) induces formation of mycobacterial granulomas by macrophages and adaptive immunity is not required for it (Davis et al. 2002). Neutrophils help in dissemination of bacteria and other materials from necrotic macrophages at advanced and inflammatory stages (Yang et al. 2012). The tumor necrosis factor (TNF) is known to control the product of reactive oxygen species (ROS) in macrophages, which significantly plays a role in mycobacterial restriction. Thus, TNF mediates both susceptibility and resistance to mycobacteria via mitochondrial ROS (Roca and Ramakrishnan 2013).

Signaling via toll-like receptor (TLR) is important to mediate antagonist effect against mycobacterial autophagy at inflammatory stages of infection; however, *M. marinum* is able to evade recognition of TLR at early stages of infection (Van der Vaart et al. 2014). Moreover, macrophages are reported to kill mycobacteria via phagolysosome maturation; however, *M. marinum* has a virulence factor (MarP) for acid tolerance enabling the pathogen to survive in these compartments (Levitte et al. 2016).

22.3.3 Zebrafish as a Model Host for Fungal Infections

Research in fungal pathogens–zebrafish infection model has also been focused as it causes fatal diseases globally, resulting in significant morbidity and mortality. Most of the fungal diseases are the results from impaired immunity or high-dose inoculum. Organ transplantation, cancer chemotherapy, and HIV/AIDS epidemic have created a large population of immune-compromised humans that frequently succumb to opportunistic fungal infections in zebrafish models (Brown et al. 2012; Yoshida et al. 2017).

Studies using zebrafish model for *Aspergillus fumigatus* infection revealed that neutrophils are unable to phagocytose fungal spores, and suggested macrophages as crucial for host defense (Knox et al. 2014, 2017). Another research using zebrafish infection demonstrated role for calcineurin in protection against *Aspergillus*, consistent with the findings from mouse models. Here, activation of calcineurin results in the dephosphorylation of NFAT cells. Further, FK506 treatment restricts recruitment of neutrophil due to decreased production of TNF-α by macrophages (Herbst et al. 2015). Such studies signify important role of calcineurin in macrophages for *Aspergillus* control in vivo.

Another fungal pathogen *Candida albicans* with huge clinical impact is an opportunistic fungal pathogen primarily affecting immunocompromised population. *C. albicans*–zebrafish infection models have also been employed to figure out virulence factors and suggest an important role for the formation of hyphae in *C.*

albicans in its pathogenicity (Chen et al. 2015; Brothers et al. 2011; Chao et al. 2010). Fungal dissemination can be observed by 24 h postinfection (hpi), followed by uncontrolled hyphal growth resulting in lethality. Here, macrophages inhibit germination, but not replication (Brothers et al. 2011). Another significant insight came from this model is a new insight role for NADPH oxidase in controlling hyphal growth, preventing germination, and recruiting macrophages through ROS (Brothers et al. 2011, 2013). In summary, zebrafish infection model enables novel discoveries in host pathogen interplay.

22.3.4 Zebrafish as a Model Host for Monitoring Chemical Hazards in Ecosystem

Excessive use of chemical in day to day ranging from agriculture purposes to manufacturing industries (Kumar et al. 2018a, 2019a; Kapoor et al. 2019; Bhati et al. 2019; Singh et al. 2019a) is of great concern to develop various tools to efficiently and accurately assess their risks and hazards to ecological and human receptors (Singh et al. 2019b, c; Kumar et al. 2019b, c; Sidhu et al. 2019). Such chemicals include different classes of pesticides, heavy metals, polychlorinated biphenyls, poly aromatic hydrocarbons, etc. (Kumar et al. 2018b, c). These chemicals are highly persistent in the ecosystem and have both acute and chronic responses (Datta et al. 2018; Singh et al. 2018; Kaur et al. 2018; Singh et al. 2017; Kumar et al. 2017; Wani et al. 2017). They result in ailment of various diseases such as Alzheimer's, reproductive dysfunction, birth defects, diabetes, Parkinson's diseases, and various types of cancers (Mishra et al. 2016; Singh et al. 2016; Kumar et al. 2014, 2015, 2016).

Various efforts are used to minimize animal testing by using in silico models or in vitro assays to screen various hazardous chemicals during developmental process for investigation. Embryos of zebra fish have emerged as a promising approach for evaluating toxicity of various chemicals. Various studies have reported that zebra fish embryos have a lower sensitivity against eurotoxicants. One major advantage to work with embryos of zebra fish is that they provide real time screening of various chemical toxicants in multi-well plates, thereby reducing cost, efforts, and time associated with its screening. Secondly, the offspring of zebra fish are developed rapidly in order to testify various contaminants during short span of time. The chorion of zebra fish is transparent, so it becomes easy to visualize developmental and morphological abnormalities during testing. It is a reliable indicator of acute toxicity to fish in general and shows promise as a more cost effective replacement for more resource intensive juvenile fish tests.

22.4 Limitations of Zebrafish Infection Model

Zebrafish have demonstrated its application as a model host for infectious diseases; however, there are few drawbacks or limitations such as lack of availability of zebrafish-specific antibodies and cell lines. Further, most of the human pathogens have optimum temperature of 37 °C, whereas zebrafish require 28 °C as optimal temperature. Some virulence genes are reported to be temperature regulated and may affect pathogenesis. However, several virulence genes have been reported by mutagenesis studies to behave in zebrafish model similar to the mammalian hosts (Kizy and Neely 2009; Miller 2005).

22.5 Conclusion and Future Prospects

The future of zebrafish as infectious disease model holds great promises. Recent technical advancements will further promote zebrafish model for better interactions with host pathogens. Transparent adult zebrafish may be utilized in great ways for observing dissemination and accumulation of microbial pathogens in real time generating new knowledge about its mechanism. Large scale mutagenesis screens of zebrafish embryos will be useful to identify strains susceptible and resistant to infection. Further, 96-well plate format will be used for efficacy screens for new drug discovery for several infectious diseases. Zebrafish are advantageous model host to checkout a wide range of microbial infections owing to the physiological and genetic similarities with humans and in higher vertebrates. Recent researches provide an insight on novel mechanism of cellular immunity and host–pathogen interactions. The new knowledge generated may be potentially utilized to generate various therapeutic strategies for working against humans diseases including infection and inflammation.

References

Allen JP, Neely MN (2010) Trolling for the ideal model host: zebrafish take the bait. Future Microbiol 5(4):563–569

Bhati S, Kumar V, Singh S, Singh J (2019) Synthesis, biological activities and docking studies of piperazine incorporated 1, 3, 4-oxadiazole derivatives. J Mol Str 1191:197–205

Brothers KM, Newman ZR, Wheeler RT (2011) Live imaging of disseminated candidiasis in zebrafish reveals role of phagocyte oxidase in limiting filamentous growth. Eukaryot Cell 10(7):932–944

Brothers KM, Gratacap RL, Barker SE, Newman ZR, Norum A, Wheeler RT (2013) NADPH oxidase-driven phagocyte recruitment controls *Candida albicans* filamentous growth and prevents mortality. PLoS Pathog 9(10):e1003634

Brown GD, Denning DW, Gow NA, Levitz SM, Netea MG, White TC (2012) Hidden killers: human fungal infections. Sci Transl Med 4:165rv113

Chao CC, Hsu PC, Jen CF, Chen IH, Wang CH, Chan HC (2010) Zebrafish as a model host for *Candida albicans* infection. Infect Immun 78(6):2512–2521

Chen YZ, Yang YL, Chu WL, You MS, Lo HJ (2015) Zebrafish egg infection model for studying *Candida albicans* adhesion factors. PLoS One 10(11):e0143048

Clatworthy AE, Lee JS, Leibman M, Kostun Z, Davidson AJ, Hung DT (2009) *Pseudomonas aeruginosa* infection of zebrafish involves both host and pathogen determinants. Infect Immun 77(4):1293–1303

Clay H, Davis JM, Beery D, Huttenlocher A, Lyons SE, Ramakrishnan L (2007) Dichotomous role of the macrophage in early *Mycobacterium marinum* infection of the zebrafish. Cell Host Microbe 2(1):29–39

Cronan MR, Tobin DM (2014) Fit for consumption: zebrafish as a model for tuberculosis. Dis Models Mech 7(7):777–784

Datta S, Singh J, Singh J, Singh S, Singh S (2018) Assessment of genotoxic effects of pesticide and vermicompost treated soil with Allium cepa test. Sus Environ Res 28(4):171–178

Davis JM, Clay H, Lewis JL, Ghori N, Herbomel P, Ramakrishnan L (2002) Real-time visualization of mycobacterium-macrophage interactions leading to initiation of granuloma formation in zebrafish embryos. Immunity 17(6):693–702

Ding CB, Zhao Y, Zhang JP, Peng ZG, Song DQ, Jiang JD (2015) A zebrafish model for subgenomic hepatitis C virus replication. Int J Mol Med 35(3):791–797

Goody MF, Sullivan C, Kim CH (2014) Studying the immune response to human viral infections using zebrafish. Dev Comp Immunol 46(1):84–95

Hall C, Flores MV, Storm T, Crosier K, Crosier P (2007) The zebrafish lysozyme C promoter drives myeloid-specific expression in transgenic fish. BMC Dev Biol 7:42

Herbst S, Shah A, Mazon-Moya M, Marzola V, Jensen B, Reed A (2015) Phagocytosis-dependent activation of a TLR9-BTK-calcineurin-NFAT path-way co-ordinates innate immunity to *Aspergillus fumigatus*. EMBO Mol Med 7(3):240–258

Howe K, Clark MD, Torroja CF, Torrance J, Berthelot C, Muffato M, Collins JE, Humphray S, McLaren K, Matthews L, McLaren S (2013) The zebrafish reference genome sequence and its relationship to the human genome. Nature 496(7446):498

Kapoor D, Singh S, Kumar V, Romero R, Prasad R, Singh J (2019) Antioxidant enzymes regulation in plants in reference to reactive oxygen species (ROS) and reactive nitrogen species (RNS). Plant Gene 19:100182

Kaur P, Singh S, Kumar V, Singh N, Singh J (2018) Effect of rhizobacteria on arsenic uptake by macrophyte Eichhornia crassipes (Mart.) Solms. Int J Phytoremediation 20(2):114–120

Kizy AE, Neely MN (2009) First *Streptococcus pyogenes* signature-tagged mutagenesis screen identifies novel virulence determinants. Infect Immun 77(5):1854–1865

Knox BP, Deng Q, Rood M, Eickhoff JC, Keller NP, Huttenlocher A (2014) Distinct innate immune phagocyte responses to *Aspergillus fumigatus* conidia and hyphae in zebrafish larvae. Eukaryot Cell 13(10):1266–1277

Knox BP, Huttenlocher A, Keller NP (2017) Real-time visualization of immune cell clearance of *Aspergillus fumigatus* spores and hyphae. Fungal Genet Biol 105:52–54

Kumar V, Upadhyay N, Kumar V, Kaur S, Singh J, Singh S, Datta S (2014) Environmental exposure and health risks of the insecticide monocrotophos—a review. J Biodivers Environ Sci 5:111–120

Kumar V, Singh S, Singh J, Upadhyay N (2015) Potential of plant growth promoting traits by bacteria isolated from heavy metal contaminated soils. Bull Environ Contam Toxicol 94(6):807–814

Kumar V, Kaur S, Singh S, Upadhyay N (2016) Unexpected formation of N′-phenyl-thiophosphorohydrazidic acid O, S-dimethyl ester from acephate: chemical, biotechnical and computational study. 3 Biotech 6(1):1

Kumar V, Singh S, Singh R, Upadhyay N, Singh J (2017) Design, synthesis, and characterization of 2, 2-bis (2, 4-dinitrophenyl)-2-(phosphonatomethylamino) acetate as a herbicidal and biological active agent. J Chem Biol 10(4):179–190

Kumar V, Singh S, Singh R, Upadhyay N, Singh J, Pant P, Singh R, Srivastava B, Singh A, Subhose V (2018a) Spectral, structural and energetic study of acephate, glyphosate, monocrotophos and phorate: an experimental and computational approach. J Taibah Uni Sci 12(1):69–78

Kumar V, Singh S, Singh A, Dixit AK, Shrivastava B, Kondalkar SA, Singh J, Singh R, Sidhu GK, Singh RP, Subhose V (2018b) Determination of phytochemical, antioxidant, antimicrobial, and protein binding qualities of hydroethanolic extract of *Celastrus paniculatus*. J Appl Biol Biotech 6(06):11–17

Kumar V, Singh S, Singh A, Dixit AK, Srivastava B, Sidhu GK, Singh R, Meena AK, Singh RP, Subhose V, Prakash O (2018c) Phytochemical, antioxidant, antimicrobial, and protein binding qualities of hydro-ethanolic extract of *Tinospora cordifolia*. J Biologically Active Products Nat 8(3):192–200

Kumar V, Singh S, Srivastava B, Bhadouria R, Singh R (2019a) Green synthesis of silver nanoparticles using leaf extract of *Holoptelea integrifolia* and preliminary investigation of its antioxidant, anti-inflammatory, antidiabetic and antibacterial activities. J Environ Chem Eng 7(3):103094

Kumar V, Singh S, Singh R (2019b) Phytochemical constituents of guggul gum and their biological qualities. Mini-Rev Org Chem 16

Kumar V, Singh S, Singh A, Subhose V, Prakash O (2019c) Assessment of heavy metal ions, essential metal ions, and antioxidant properties of the most common herbal drugs in Indian Ayurvedic hospital: for ensuring quality assurance of certain Ayurvedic drugs. Biocatal Agric Biotechnol 18:101018

Lam SH, Chua HL, Gong Z, Lam TJ, Sin YM (2004) Development and maturation of the immune system in zebrafish, *Danio rerio*: a gene expression profiling, *in situ* hybridization and immunological study. Dev Comp Immunol 28(1):9–28

Lawson ND, Weinstein BM (2002) *In vivo* imaging of embryonic vascular development using transgenic zebrafish. Dev Biol 248(2):307–318

Levitte S, Adams KN, Berg RD, Cosma CL, Urdahl KB, Ramakrishnan L (2016) Mycobacterial acid tolerance enables phagolysosomal survival and establishment of tuberculous infection in vivo. Cell Host Microbe 20(2):250–258

Levraud JP, Disson O, Kissa K, Bonne I, Cossart P, Herbomel P, Lecuit M (2009) Real-time observation of *Listeria monocytogenes*-phagocyte interactions in living zebrafish larvae. Infect Immun 77(9):3651–3660

Lu MW, Chao YM, Guo TC, Santi N, Evensen Ø, Kasani SK, Hong JR, Wu JL (2008) The interferon response is involved in nervous necrosis virus acute and persistent infection in zebrafish infection model. Mol Immunology 45(4):1146–1152

Mathias JR, Perrin BJ, Liu TX, Kanki J, Look AT, Huttenlocher A (2006) Resolution of inflammation by retrograde chemotaxis of neutrophils in transgenic zebrafish. J Leukoc Biol 80(6): 1281–1288

McVicker G, Prajsnar TK, Williams A, Wagner NL, Boots M, Renshaw SA, Foster SJ (2014) Clonal expansion during *Staphylococcus aureus* infection dynamics reveals the effect of antibiotic intervention. PLoS Pathog 10(2):e1003959

Meeker ND, Trede NS (2008) Immunology and zebrafish: spawning new models of human disease. Dev Comp Immunol 32(7):745–757

Mesureur J, Feliciano JR, Wagner N, Gomes MC, Zhang L, Blanco-Gonzalez M, van der Vaart M, O'Callaghan D, Meijer AH, Vergunst AC (2017) Macrophages, but not neutrophils, are critical for proliferation of *Burkholderia cenocepacia* and ensuing host-damaging inflammation. PLoS Pathog 13(6):e1006437

Miller J (2005) Large scale screen highlights the importance of capsule for virulence in the zoonotic pathogen, *Streptococcus iniae*. Infect Immun 73(2):921–934

Mishra V, Gupta A, Kaur P, Singh S, Singh N, Gehlot P, Singh J (2016) Synergistic effects of Arbuscular mycorrhizal fungi and plant growth promoting rhizobacteria in bioremediation of iron contaminated soils. Int J Phytoremediation 18(7):697–703

Murayama E, Kissa K, Zapata A, Mordelet E, Briolat V, Lin HF, Handin RI, Herbomel P (2006) Tracing hematopoietic precursor migration to successive hematopoietic organs during zebrafish development. Immunity 25(6):963–975

Phelps HA, Neely MN (2005) Evolution of the zebrafish model: from development to immunity and infectious disease. Zebrafish 2(2):87–103

Prajsnar TK, Cunliffe VT, Foster SJ, Renshaw SA (2008) A novel vertebrate model of *Staphylococcus aureus* infection reveals phagocyte-dependent resistance of zebrafish to non-host specialized pathogens. Cellular Microbiol 10(11):2312–2325

Prajsnar TK, Hamilton R, Garcia-Lara J, McVicker G, Williams A, Boots M, Foster SJ, Renshaw SA (2012) A privileged intraphagocyte niche is responsible for disseminated infection of *Staphylococcus aureus* in a zebrafish model. Cellular Microbiol 14(10):1600–1619

Redd MJ, Kelly G, Dunn G, Way M, Martin P (2006) Imaging macrophage chemotaxis *in vivo*: studies of microtubule function in zebrafish wound inflammation. Cell Motil Cytoskeleton 63(7):415–422

Renshaw SA, Loynes CA, Trushell DM, Elworthy S, Ingham PW, Whyte MK (2006) A transgenic zebrafish model of neutrophilic inflammation. Blood 108(13):3976–3978

Roca FJ, Ramakrishnan L (2013) TNF dually mediates resistance and susceptibility to mycobacteria via mitochondrial reactive oxygen species. Cell 153(3):521–534

Sidhu GK, Singh S, Kumar V, Dhanjal DS, Datta S, Singh J (2019) Toxicity, monitoring and biodegradation of organophosphate pesticides: a review. Critic Rev Environ Sci Tech 49:1–53

Singh S, Singh N, Kumar V, Datta S, Wani AB, Singh D, Singh K, Singh J (2016) Toxicity, monitoring and biodegradation of the fungicide carbendazim. Environ Chem Lett 14(3):317–329

Singh S, Kumar V, Upadhyay N, Singh J, Singla S, Datta S (2017) Efficient biodegradation of acephate by *Pseudomonas pseudoalcaligenes* PS-5 in the presence and absence of heavy metal ions [cu (II) and Fe (III)], and humic acid. 3 Biotech 7(4):262

Singh S, Kumar V, Chauhan A, Datta S, Wani AB, Singh N, Singh J (2018) Toxicity, degradation and analysis of the herbicide atrazine. Environ Chem Lett 16(1):211–237

Singh S, Kumar V, Singh J (2019a) Kinetic study of the biodegradation of glyphosate by indigenous soil bacterial isolates in presence of humic acid, Fe (III) and cu (II) ions. J Environ Chem Eng 7(3):103098

Singh S, Kumar V, Sidhu GK, Datta S, Dhanjal DS, Koul B, Janeja HS, Singh J (2019b) Plant growth promoting rhizobacteria from heavy metal contaminated soil promote growth attributes of *Pisum sativum* L. Biocatal Agri Biotech 17:665–671

Singh S, Kumar V, Singh S, Singh J (2019c) Influence of humic acid, iron and copper on microbial degradation of fungicide Carbendazim. Biocatal Agri Biotech 20:101196

Stein C, Caccamo M, Laird G, Leptin M (2007) Conservation and divergence of gene families encoding components of innate immune response systems in zebrafish. Genome Biol 8(11):R251

Sullivan C, Kim CH (2008) Zebrafish as a model for infectious disease and immune function. Fish Shellfish Immunol 25(4):341–350

Torraca V, Mostowy S (2018) Zebrafish infection: from pathogenesis to cell biology. Trends Cell Biol 28(2):143–156

Trede NS, Langenau DM, Traver D, Look AT, Zon LI (2004) The use of zebrafish to understand immunity. Immunity 20(4):367–379

Van der Vaart M, Korbee CJ, Lamers GE, Tengeler AC, Hosseini R, Haks MC, Ottenhoff TH, Spaink HP, Meijer AH (2014) The DNA damage-regulated autophagy modulator DRAM1 links mycobacterial recognition via TLR-MYD88 to autophagic defense. Cell Host Microbe 15(6):753–767

Varshney GK, Sood R, Burgess SM (2015) Understanding and editing the zebrafish genome. In: Advances in genetics, vol 92. Academic Press, San Diego, pp 1–52

Vergunst AC, Meijer AH, Renshaw SA, O'Callaghan D (2010) *Burkholderia cenocepacia* creates an intramacrophage replication niche in zebrafish embryos, followed by bacterial dissemination and establishment of systemic infection. Infect Imm 78(4):1495–1508

Wani AB, Chadar H, Wani AH, Singh S, Upadhyay N (2017) Salicylic acid to decrease plant stress. Environ Chem Lett 5(1):101–123

White RM, Sessa A, Burke C (2008) Transparent adult zebrafish as a tool for *in vivo* transplantation analysis. Cell Stem Cell 2(2):183–189

Yang CT, Cambier CJ, Davis JM, Hall CJ, Crosier PS, Ramakrishnan L (2012) Neutrophils exert protection in the early tuberculous granuloma by oxidative killing of mycobacteria phagocytosed from infected macrophages. Cell Host Microbe 12(3):301–312

Yoshida N, Frickel EM, Mostowy S (2017) Macrophage–microbe interactions: lessons from the zebrafish model. Frontiers Immunol 8:1703

Mycobacterium Pathogenesis and Drug Discovery: Looking Through the Zebrafish Keyhole

23

Nishant Gopalan, Vipin Gopinath, and K. Madhavan Nampoothiri

Contents

Abstract

Mycobacterium tuberculosis is like a fugitive parasite that predominantly attacks and manipulates the macrophages from the host's defense, while constantly escaping execution through various routes that are attributed to its biochemistry and genetic versatility. Even though, it is primarily classified as a respiratory disease, macrophages and fibroblasts that form granulomas at the site of infection, pathogenesis, persistence as well as dissemination. Understanding the

N. Gopalan (✉)
V.V.P. Engineering College, Rajkot, Gujarat, India

V. Gopinath
Malabar Cancer Centre, Kannur, Kerala, India

K. M. Nampoothiri
MPTD, CSIR-NIIST, Thiruvananthapuram, Kerala, India

B. Siddhardha et al. (eds.), *Model Organisms for Microbial Pathogenesis, Biofilm Formation and Antimicrobial Drug Discovery*,
https://doi.org/10.1007/978-981-15-1695-5_23

composition and the dynamics of the granuloma microenvironment becomes the best bet at understanding the pathogenesis of the diseases. The most used murine models used for studying the host–pathogen interactions between *Mycobacterium tuberculosis* and humans are not natural host–pathogen pairs. On the extreme end of the range of models available is the study involving the use of nonhuman primates as a model system for the study of *M. tuberculosis,* which though very accurate is expensive, time consuming, and can lead to ethical conundrums. *Danio rerio* or zebrafish embryos are susceptible to *Mycobacterium marinum,* a *Mycobacterium*, that is genetically very similar to the *Mycobacterium tuberculosis* complex. Even though zebrafish cannot be inoculated with *Mycobacterium tuberculosis*, however, they are natural hosts for *M. marinum.* Since *M. marinum* pathogenesis in zebrafish embryo is very similar to that of humans, with formation of caseating granulomas (a property that is not observed in murine models with diffuse granulomas), they serve as better in vivo models for study of host–pathogen interplay. Zebrafish embryo is optically clear; as a result events of pathogenesis or even drug discovery/efficacy can be studied without sacrificing the larvae through in vivo imaging, fluorescence microscopy in particular. In the past, this remarkable model has revealed groundbreaking revelations like the fact that granuloma formation, earlier thought to be advantageous to the host, for containment of the pathogen is actually mediated by the pathogen to enable its spread after early infection; akin to the assembly of early humans for sharing fires. The goal of this chapter is to shed light on various discoveries in the field of *M. tuberculosis* infection and treatment, which were made possible because of the versatility of the zebrafish disease model.

Keywords

Mycobacterium tuberculosis · Pathogenesis · *Danio rerio* · *Mycobacterium marinum* · Host–pathogen

23.1 Introduction

Tuberculosis as a disease is one of the prime causes of fatality among human populations. The causative agent, *M. tuberculosis* has evolved as an obligate human pathogen. The disease has also been a cause of major concern, due to the developing of multidrug-resistant tuberculosis strains (MDR TB) and extensively drug-resistant TB (XDR TB). The physiology and the biochemistry of the bacterium along with a highly choreographed evasion mechanism that the *Bacillus* spp applies ensure that the bacteria escape from new drugs also in an easy manner (Control and Prevention 2006). Since the discovery of the *M. tuberculosis* that is responsible for TB by Robert Koch in 1882, extensive research has been carried out regarding various aspects of tuberculosis, including infection pathogenesis and dissemination. Detrimental infectious agents that target the immune system are known to be some of the most difficult problems for modern sciences to circumvent. Even though causative microbes infect and compromise the health of the host, disease state is a consequence of the cross talk and interactions between the host defense systems and causative agent. Understanding this cross talk and recognizing targets that are detrimental to development of a disease state without the loss of life is something that researchers try to do so as to combat the disease. Understanding the pathogenesis of the disease through a correct disease model is required so that realistic aspects of the disease may be revealed. One gains true insights about the process of pathogenesis of a disease when one studies the process in the same organism, that the disease targets; however, that may not be always possible as many aspects of the pathogenesis occur prior to an individual approaching a medical personnel, and it is unethical to infect fresh individuals with the disease-causing agent so as to gain understanding of the pathogenesis. To overcome these issues, it is necessary to choose a model system of the disease that closely resembles that of the natural host and pathogen.

23.2 *Mycobacterium tuberculosis* and Classical Pathogenesis

As a pathogen, the *Mycobacterium tuberculosis* Complex (MBC), a group of *Mycobacteria* with very close resemblance (more than 99% similarity at 16s rRNA level) have coexisted with humans since more than 70,000 years and have migrated with the modern humans out of Africa as established by phylogenetic analysis (Comas et al. 2013). Even though the MBC is genetically close related, different species have different host specificities with a specific repository of genetic material that helps them survive in the respective hosts. *Mycobacterium tuberculosis* is an intracellular parasite of the human macrophage cells (only human host) and is one among the top ten reasons for high mortality rate in the world. About 10 million new cases of TB were reported as late as 2017 (World Health Organization 2018). The heavy burden of tuberculosis along with a very long treatment time involving anti-mycobacterial drugs leads to a high failure rate in the treatment of tuberculosis. If the anti-mycobacterial drugs are not taken by the patient at correct intervals and for the prescribed amount/time or if they have been misdiagnosed and given wrong drugs or dosages, it leads to MDR TB and if treatment failure happens multiple times, it may even lead to XDR TB (Ormerod 2005). Although many vaccines are being tested against TB, the only one that has been in active use is the BCG vaccine, using the *Mycobacterium bovis* bacillus Calmette–Guérin. The vaccine has been used for more than a hundred years now and works well against systemic spread of TB, but does not affect the dormant or persistent TB (Ottenhoff and Kaufmann 2012). The BCG vaccine during the early years of deployment may have been useful for protection against pulmonary TB, but now in more recent trials have proved to be futile in providing protection against pulmonary TB (Fine 1989). The higher number of deaths caused due to TB is clearly due to the mutability and selection of virulent strains of the pathogen, abuse, and mismanagement of drugs and inefficacy of the vaccines being used against the pathogen. To mitigate such successful pathogen, understanding its pathogenesis is absolutely necessary.

While most of the studies looked at *Mycobacterium tuberculosis* as a pathogen of the human macrophage, the broader picture, with the involvement of the granuloma, its microenvironment and the interaction of the pathogen with the macrophage within a granuloma was overlooked (Balcewicz-Sablinska et al. 1998; McDonough et al. 1993; Schlesinger et al. 1990). The studies limited to interactions between the pathogen and macrophages did shed light on various aspects of pathogenesis, including the variability displayed by *Mycobacterium tuberculosis* when it comes to usage of different cell surface receptors when gaining entry into a macrophage cell (Aderem and Underhill 1999); the various mechanisms by which the pathogen, after being engulfed by the macrophage prevents the maturation of the phagolysosome through the alteration of Rab GTPase composition of the phagosome and the omission of proton ATPase pump to prevent the reduction of pH of the phagosome (Clemens et al. 2000; Sturgill-Koszycki et al. 1994). The classical pathogenesis data suggested that the bacterium is transmitted through aerosol droplets inhaled and delivered to the alveoli in the lungs, where it is phagocytosed by the residing macrophages. The *Mycobacterium* then thrives and divides inside the macrophages while the host enlists the help of various immune cells to the site of infection resulting in the formation of an aggregate of cells referred to as the granuloma, which is the classical hallmark of the disease (Eum et al. 2010; Wolf et al. 2007). The phenomenon of formation of these granulomas was thought to be initiated by the host immune system to reduce the dissemination of the bacteria throughout the system (Bold and Ernst 2009; Ulrichs and Kaufmann 2006). The granuloma is made up of multiple types of immune cells including the cells of the humoral immunity in the periphery, with tightly interlocked epithelioid macrophages and dendritic cells. The granuloma also has neutrophils and natural killer cells. The necrotic center of the granuloma is filled with fatty acids (caseous/caseating) secreted by foamy cells and giant cells. The presence of such complex granulomas is a hallmark of tuberculosis (Ramakrishnan 2012). However, most of the studies on granuloma and its role in pathogenesis have been carried out on static models, with fixed slides, which limit the understanding of the utility of these complex structures in the pathogenesis of tuberculosis (Ramakrishnan 2012; Saunders and Cooper 2000). Tuberculosis may be active TB or latent TB depending on the condition of the host immunity; when the host immune system is undermined, latent TB can become active and granulomas break open into the alveoli of the host and the pathogen is disseminated as an aerosol ready to infect a new host. Mycobacteria are known to have PAMPs (pathogen-activated molecular patterns) on bacterial exterior. However, these PAMPs are masked by specific cell surface lipid components called phthiocerol dimycoceroserate (PDIM) and hence the employment of activated macrophages, which are detrimental to the pathogen, is prevented. The mycobacteria also use a surface lipid, phenolic glycolipid (PGL), which leads to the secretion of the macrophage chemokine CCL2, which in turn attracts macrophages that are not microbicidal for the mycobacteria (Cambier et al. 2014). With such complex interplay of the host immune system and the pathogen and a near-perfect strategy employed by *Mycobacterium tuberculosis* for evasion, it is essential that we study it thoroughly so as to find weak spots in its seemingly impenetrable and exhaustive armor.

23.3 Models Used for Studying Pathogenesis

23.3.1 In Vitro Models of Pathogenesis

One of the most common models that have been used for studying pathogenesis of *Mycobacterium tuberculosis* includes the use of macrophage cell cultures to understand the events of phagocytosis and evasion of phagosome maturation. The pathogen is evolved enough to be promiscuous about the cell surface receptor that it chooses for entry into the macrophage, these include different receptors like Fc receptor, mannose receptor, and even the complement receptor. Many of the aspects of mycobacterial pathogenesis initially were discovered by in vitro studies on fast-growing relatives of the MBC. The most common relative of the MBC that has been used extensively for understanding the MBC is *Mycobacterium smegmatis*. Studies on *Mycobacterium smegmatis* have revealed various aspects about sliding motility, biofilm formation, and even the ESX-1 secretory system, which are essential for escaping host immune system (absent in *M. bovis*) (Coros et al. 2008; Recht and Kolter 2001). *Mycobacterium marinum* and *Mycobacterium bovis* have been used along with *Mycobacterium smegmatis* to reveal various aspects that are conserved among these organisms and *Mycobacterium tuberculosis*.

23.4 In Vivo Models of Pathogenesis Used Apart from Zebrafish–*M. marinum*

Apart from the zebrafish–*M. marinum* host–pathogen pair as a disease model, other in vivo models include cell culture-based models with macrophage cultures infected with *M. tuberculosis*, mouse, rabbit, guinea pigs, and primate models infected with *M. tuberculosis*. Initially, rabbit and guinea pig models have helped to understand efficiency of vaccines (Baldwin et al. 1998; Dannenberg Jr 1994; McMurray 1994). These models do display some hallmarks of the pathophysiology of the disease similar to humans and have helped to advance our knowledge with specific details about cavitary lesions but without tools required for genetic and immunological manipulations for these systems. The mouse model for tuberculosis has customarily been used for most of the studies regarding pathogenesis because it is easy to rear mice in a Bio Safety Level 3 (BSL3) laboratory, and the exhaustive protocols that have been set for creating knockout mutants made this model as the first choice for many *Mycobacterium tuberculosis* researchers. The cornel murine model of tuberculosis could be used to study latent tuberculosis and activation of latent tuberculosis to the active form, which are quintessential human tuberculosis events (Scanga et al. 1999). Limitations with the mouse models included the presence of diffuse lesions with granuloma that did not possess a caseous necrotic centers and the absence of hypoxic conditions at the center of the granuloma (Aly et al. 2006; Tsai et al. 2006; Young 2009). Therefore, results obtained from persistent/latent tuberculosis murine models may not reflect the outcomes and situations in a human latent tuberculosis event, also drugs that may be active in the murine model with the absence of hypoxia may or may not function the way they were predicted to act due to the differences in the transcriptional state of the pathogen during hypoxia and its absence and due differences in the pH.

Nonhuman primates have also been successfully used as tuberculosis models, with great accuracy in reproducing the pathophysiology of the human tuberculosis; including infection through aerosol inhalation, establishment of a latent tuberculosis infection when the number of initial colony forming units used (no. of active cells) is ≤ 100 (Walsh et al. 1996; Zeng et al. 2011). The *Macaca mulatta* (rhesus macaques) model has shown remarkable similarity with the human tuberculosis phenomenon, with clear evidence of protection mediated by vaccination using the BCG strain (Verreck et al. 2009). The nonhuman primate model shows even more similarities with the presence of different types of granulomas that the human tuberculosis shows, which include the caseous necrotic granulomas, the calcified, fibrotic, and cavitary granulomas (Kaushal et al. 2012). The tuberculosis model using the nonhuman primate rhesus macaques can also be used to study the pathogenesis of reactivation of *M. tuberculosis* when infected with the HIV equivalent SIV_{mac} virus (Mehra et al. 2011). It makes the above model the closest possible model is to study the actual phenomenon of a human infected with *M. tuberculosis*. Then, why we cannot use this model for finding drug candidates and for studying the complex pathogenesis of *M. tuberculosis*? Why do we still rely on models that may not be completely accurate? Table 23.1 gives a comparative analysis of various systems and depending on the merits and demerits of the system model for understanding host–pathogen interactions can be selected. Usually as the complexity of the model increases, the information extractable from the interaction exchanges that occur between the host and the pathogen pair increases, and as we move up the ladder from fast-growing mycobacteria in broth, to cell culture-based models, to small mammals, and at the top nonhuman primate models, we are able to inquire about processes that are very specific to the pathogenesis in humans. But the zebrafish system, even if it is below the small mammal models as far as relatedness and complexity are concerned, gives us the opportunity to carry out live imaging of the whole process of host–pathogen interaction due to the optical transparency of the embryo. An adult zebrafish matures within three months compared to much longer time periods for small mammals and nonhuman primates; also the zebrafish can be kept at high densities (up to 5 individuals in 1 liter of water), while small mammals and nonhuman primates require cages or housings with large footprints. These features become beneficial with the inclusion of *M. marinum* as the pathogen for host–pathogen interaction makes the zebrafish–*M. marinum* pair a lucrative system for studying host–pathogen interactions.

Table 23.1 Comparative features of different pathogenesis and disease models for tuberculosis

Model	*Mycobacterium smegmatis* and other broth grown mycobacterial strains (apart from *M. tuberculosis*)	Macrophage cell culture models infected with *M. tuberculosis*	Small mammal model including murine, guinea pig, and rabbit models	Nonhuman primate models	*Danio rerio* infected with *M. marinum*
Type	In vitro	In vivo	In vivo	In vivo	In vivo
Economics	Very economical	Very economical	Expensive	Very expensive	Moderately economic
Ease of infection with mycobacterial strain.	Very easy	Comparatively easy	Requires expertise in the handling of small mammals	Technically intensive as it requires knowledge of primate handling and use of tools for intratracheal delivery of pathogen	Technically intensive as it requires precision injection into specific locations of the embryo without puncturing adjacent regions, using a micro-injector
Ease of handling/ biosafety concerns	Safest system to handle	Primary precautions and BSL-2 laboratory required	Relatively unsafe compared to cell culture models	Requires the highest level of safety	A safe system with simple BSL-1 laboratory required
Generation time of pathogen	3–4 h	20–24 h	20–24 h	20–24 h	4 h
Clinical similarity with human–*M. tuberculosis* system	Least related to actual human infection	Slightly similar to human system, but dynamics with other immune cells cannot be studied	Similar to human system, with interactive immune system, but dissimilar pathologies	Extremely similar to human system, with interactive immune system, and similar pathologies	Extremely similar to human system, with interactive immune system, and limited similarity to observed pathologies
Ethical restrictions	None	None	Minor	Major	Minor
Ease of genetic manipulations and/ or availability of tools and reagents	Comparatively very easy, with easily available tools for pathogen manipulation	Comparatively easy, with easily available tools for pathogen and host manipulation	Difficult with easily available tools for pathogen but host manipulation may not be easy	Difficult with easily available tools for pathogen but host manipulation is very difficult	Comparatively very easy, with easily available tools for pathogen and host manipulation with high reproduction rates

23.5 The Zebrafish *M. marinum* Duo as a Model for Pathogenesis

23.5.1 The Zebrafish as a Model Organism

Danio rerio has been used as a model organism for drug discovery and has the advantage of being tractable or amenable to genetic modifications. Hence can be inquired utilizing forward or reverse genetic screens (phenotype to gene position and gene position to phenotype, respectively) (Lawson and Wolfe 2011). There are protocols, which allow for creation of haploid zebrafish embryos as well as established the procedures for generation of diploid homozygous individuals that can be used for screening recessive mutants. These protocols are openly available and shared on the web by a group of dedicated curators who run ZFIN a very comprehensive website, that allows collaborations and information exchange regarding the zebrafish model organisms to come (Howe et al. 2012; Westerfield 2000). Zebrafish started out as a model organism for the study of vertebrate developmental biology, with heavy emphasis on the use of forward genetic screening techniques involving screening of large number of zebrafish embryos for mutant phenotypes, which were then studied for mapping of the responsible gene for the mutant phenotype. As technology advances, more and more emphasis was laid on reverse genetic screens, where the gene-specific mutations were carried out and the corresponding phenotypes were studied. Reverse genetic screens are becoming exceedingly viable options with the advent of tools that allow knockouts of genes with specific sequences using zinc finger proteins, TALENs, and recently with the use of CRISPR-CAS9-based genome editing (Lawson and Wolfe 2011; Xiao et al. 2013). Apart from these modern tools, classical methods for inactivating translation of targeted genes, by using Morpholino oligonucleotides that are complimentary to the target RNA that has been expressed by the organism; TILLING (Target-induced local lesions in genome) which involves the induction of germline mutation induction followed by screening of mutant progeny through either sequencing or PCR for mutation in a particular location or gene (Bill et al. 2009). These features make the zebrafish a genetically tractable model organism with a very active scientific community that makes update each other as soon as new discoveries in protocols are achieved. Zebrafish are more suitable for this job due to the availability of tools and services specific to zebrafish. A prime example is the Zebrafish International Resource Centre (ZIRC), which is a repository for mutant and transgenic zebrafish lines, that deals with research resources specifically for zebrafish (ESTs, health services, antibodies against specific proteins, etc.) and it is run by the same curators that run ZFIN. As a result ZIRC gives very relevant technical data about the services it provides (Murray et al. 2016).

Zebrafish early embryos are optically transparent, which allows the use of zebrafish lines that express tags like the green fluorescent protein (GFP) or the mCherry/DsRed protein, which can be visualized as green and red fluorescent probes when observed through a fluorescent microscope. These tags can be used for tracking a single protein by employing fusion proteins of the tag and the protein of interest.

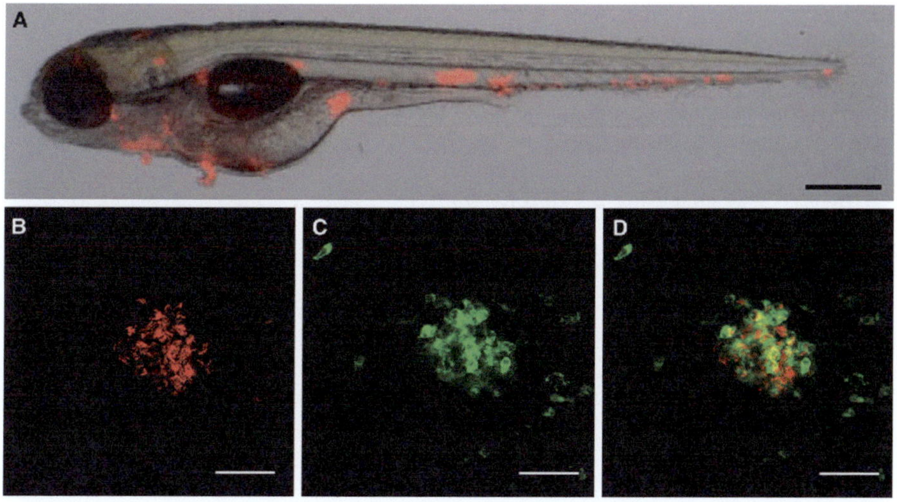

Fig. 23.1 (**A**) Merged image of visible light (bright field) microscopy and fluorescent microscopy of a zebrafish embryo infected with *M. marinum* tagged with a red fluorescent marker and photographed at 5 dpi. (Image (**A**) adapted with permission (creative commons license) from Stoop et al. (2011). Scale bar 500 μm. The act of formation of the granuloma with the gathering of mycobacteria can be visualized from panel (**B–D**). (**B**) *M. marinum E11* (in red), (**C**) phagocytes stained with anti-L-Plastin (in green), (**D**) merge of panel (**B**, **C**) reaffirms the co-localization of the phagocytes pathogen within the early granuloma formed in a zebrafish embryo/larva (Images (**B–D**) adapted with permission (creative commons license) from Van Leeuwen et al. (2014). Scale bar, 35 μm (**B–D**)

They can even be used to tag whole cells, by placing the tags downstream to a promoter that is exclusively expressed in those cells. Optical transparency coupled with cell-specific tagging allows for live imaging of cellular interactions through the use of fluorescent microscopes (Fig. 23.1). These features make the zebrafish system a favorable one for the study of host–pathogen interactions, because it facilitates real-time imaging of exchanges that take place between zebrafish immune cells and any pathogen (both tagged with different colored tags). A list of different lines of zebrafish that are available with tags for different types of cells of the immune system is displayed in Table 23.2.

23.5.2 *Mycobacterium marinum* and Its Resemblance with the *Mycobacterium tuberculosis* Complex

The *Mycobacterium tuberculosis* genome is 1.5 times smaller than the *Mycobacterium marinum*, the additional information in the genome of *M. marinum* is supposedly owing to the broad host specificity that the bacterium has compared to *M. tuberculosis*, which has specialized itself for the human host. The 16s rRNA analysis shows that both the organisms had a common ancestor; after the split, *M. tuberculosis* genome shrunk losing a whole lot of information required for survival

Table 23.2 Listing of various lines of zebrafish that can be used to visualize a specific immune cells for understanding their roles in pathogenesis

Marker	Transgenic line	Specificity
apoeb apolipoprotein Eb	*Tg(apoeb:lynEGFP)*	Microglial cells
coro1a coronin, actin-binding protein, 1A	*Tg(coro1a:EGFP)*	Macrophages, neutrophils, and thymocytes
csf1ra/fms colony stimulating factor 1 receptor, a	*TgBAC(csf1ra:Gal4-VP16)/ Tg(UAS-E1b:Eco.NfsBmCherry*	Macrophages (highly motile), and xantophore cells (immobile)
li1a friend leukemia integration 1a	*Tg(fli1a:EGFP)*	Primitive macrophages (dull), endothelial cells (bright) and subsets of erythrocytes (dull)
lyz/lysC lysozyme	*Tg(lyz:EGFP)* *Tg(lyz:DsRed2)* *Tg(-4.1lyz:EGFP)* *Tg(lyz:Gal4-VP16)*	Neutrophils
mhc2dab major histocompatibility complex class II DAB gene	*Tg(mhc2dab:EGFP)* *Tg(mhc2dab:mCherry)*	Antigen presenting cells (APCs)
mpeg1 macrophage expressed 1	*Tg(mpeg1:EGFP)* *Tg(mpeg1:mCherry)* *Tg(mpeg1:Gal4-VP16)* *Tg(mpeg1:mCherry-F)* *Tg(mpeg1:Dendra2)* *Tg(mpeg1:YFP)*	Macrophages
mpx/mpo myeloid-specific peroxidase	*TgBAC(mpx:EGFP)* *Tg(mpx:GFP)* *Tg(-8mpx:mCherry)* *Tg(-8mpx:DsRed-F)* *Tg(-8 mpx:EGFP-F)* *Tg(-8mpx:Dendra2)*	Neutrophils and macrophages
myd88 myeloid differentiation primary response gene (88)	*Tg(myd88:EGFP)* *Tg(myd88:DsRed2)*	Subsets of myeloid leukocytes, distal pronephric ducts and cloaca
ptprc/cd45 protein tyrosine phosphatase, receptor type, C	*Tg(ptprc:DsRed)*	Macrophages, granulocytes and T lymphocytes
spi1b/pu.1 spleen focus forming virus (SFFV) proviral integration oncogene spi1b	*Tg(-5.3spi1b:EGFP)* *Tg(-9.0spi1b:EGFP)* *Tg(-4spi1b:Gal4/UAS:EGFP)* *Tg(-4spi1:lynEGFP)*	Early myeloid cells

Table adapted with permission (creative commons license) from Torraca et al. (2014)

in environmental niches, while adapting and duplicating regions with respect to the specific host niche of the human macrophage. *M. marinum* genome has expanded acquiring new genetic material through horizontal gene transfer causing a gross difference of extra 2.2 Mb of DNA in *M. marinum* (Stinear et al. 2008; Tønjum et al.

1998; Van Pittius et al. 2006). Both the genomes are 85% similar and about 14% of the DNA from *M. tuberculosis* does not have any orthologous regions in the *M. marinum* genome. There is no evidence of *M. tuberculosis* exclusive region to be important for pathogenesis. Tobin and Ramakrishnan (2008), who are the pioneers in the study of *M. marinum* zebrafish system study who speculate that the genomic DNA is non-orthologous between *M. marinum* and *M. tuberculosis* and may be involved with organ specificity. *M. tuberculosis* prefers lungs over other organs; certain mutants like in the *VirS* locus leads to a phenotype with no change in pathogenesis, but differences in differential organ load have been identified (Singh et al. 2005). Table 23.3 adapted and modified from Tobin and Ramakrishnan (2008) displays a list of genes that have similar phenotypes for deletion mutants, indicating a similarity in functions.

The zebrafish infected with *M. marinum* model can be used to study the dynamics of the formation of caseating granulomas, and with a marked difference in the time period between the onset of innate immunity and humoral immunity. The role of innate as well as adaptive immunity can be selectively visualized. Functional macrophages start circulating in the optically clear zebrafish embryos at 30 h post-fertilization (3 dpf), while adaptive immunity truly becomes active after ~4 weeks postfertilization (Bertrand and Traver 2009; Volkman et al. 2010). The zebrafish embryos can then be injected before the 1021 cells phase into the yolk sac to induce a systemic infection, while hindbrain ventricle maybe a choice of injection of mycobacterial to study the effect of a local infection. The early stages of infection lead to formation of macrophage aggregates, with the formation of granulomas, and later stages lead to spread and dissemination of granulomas throughout the tissues of organism. The infection leads to wasting of tissue, and granulomas with a caseous center just like in the advanced stages of the human tuberculosis disease (Swaim et al. 2006).

23.5.3 Paradigm Shifting Findings Through the Use of Zebrafish Infected with *M. marinum* Model

There have been major discoveries that sent shockwaves throughout the tuberculosis research community, with all the information gained from the study of the Zebrafish–*M. marinum* model. The zebrafish model of the disease gave critical information regarding the dynamic nature of the granuloma. It was through the live imaging that the paradigm of a static granuloma was broken, with the evidence through fluorescent tagging of both macrophages and mycobacteria, and through forward genetic screening, it was understood that virulence determinant ESAT6 leads to secretion of mmp9 by the epithelial cells in the vicinity, which in turn leads to aggregation and enlistment of uninfected macrophages to the site of infection and leads to granuloma formation, where additional bacteria are phagocytosed (Volkman et al. 2010). It was observed that the newly infected bacteria even leave the granuloma, acting as a torchbearer for the formation of secondary granulomas. According to the study conducted by Davis and Ramakrishnan (2009), about 19% of the

Table 23.3 Virulence factors common among the two mycobacteria

Molecule/protein	Corresponding gene	Function of gene/protein	Role in virulence	Present in both pathogens	References
Cell wall lipids					
Mycolic acid	kasB	β-ketoacyl-acyl carrier protein synthases involved in mycolic acid synthesis	Required for cording phenotype associated with virulence	Yes	Gao et al. (2003)
Phthiocerol dimycocerosate (PDIM) and phenolic glycolipid (PGL)	fad28/acyl-CoA synthase	Involved in biosynthesis of PDIM and PGL, by synthesis of mycocerosates	Postulated to have a role in cell permeability modulation	Yes	Yu et al. (2012)
	Mas	Involved in biosynthesis of PDIM and PGL, synthesizes the methyl-branched lipids, mycocerosates	Postulated to inhibit proinflammatory cytokine response	Yes	Yu et al. (2012)
Trehalose dimycolate (TDM)	ffbpABC	Transport of trehalose monomycolate across the cell wall	Multiple postulated roles	Yes	Nobre et al. (2014)
ESX secretion (protein secretion across bacterial cell wall and membrane)					
ESX 1	Locus associated with secretion system	Locus associated with secretion system	Postulated in multiple roles viz., phagolysosome maturation stop/cytokine production attenuation in macrophage	Yes	Lewis et al. (2003), Lesley and Ramakrishnan (2008)
ESAT 6	esat6	Pore formation in host vacuole and induction of MMP9 protein in host epithelial cell	Required for vacuole/phagosome escape via pore formation and induction of MMP9 for recruitment of uninfected macrophages	Yes	Smith et al. (2008), Volkman et al. (2010)
ESX-5	esx5	Secretion of PE and PPE proteins of mycobacteria across cell wall	Regulation of persistent infection	Yes	Abdallah et al. (2009), Weerdenburg et al. (2012)

infected macrophages that leave the granuloma lead to the establishment of a new granuloma. They also found out experimentally that defective ESAT6 secretion greatly reduced the chemotaxis of macrophages free of the *Bacillus* toward the granuloma.

Earlier it was thought that an immunosuppressive state led to the activation and progress of tuberculosis, while an elevated immune response would be advantageous, but through experiments on the zebrafish–*M. marinum* model, it was discovered that balanced production of pro-inflammatory and anti-inflammatory eicosanoids is necessary for controlling the pathogen. Tobin et al. (2010) corroborated that the *lta4 h* locus was responsible for this balancing act in zebrafish and humans infected with *M. marinum* and *M. tuberculosis,* respectively (Tobin et al. 2012).

Through studies on zebrafish models, it was established that just by being inside macrophages; mycobacteria can develop a drug-tolerant subpopulation, and subsequently can expand this subpopulation as long as the selective pressure is applied. These mycobacteria appear to evolve into drug-tolerant mutants expressing drug efflux pumps to tolerate anti-mycobacterial drugs overnight, with a very small initial population, which has been perplexing, but the discovery has led to combinatorial therapy with molecules like verapamil that act as efflux pump inhibitors (Adams et al. 2011; Gupta et al. 2013).

23.6 High-Throughput Application of the Zebrafish Tuberculosis Disease Model

The drawbacks of the zebrafish model include tedious process of infection of embryos with the correct number of mycobacteria and at the precise location (caudal vein/hindbrain ventricle/yolk sac). This is followed by sorting a large number of embryos according to the expression of the tagged proteins and studying the process through live imaging in selected few embryos, which are ideal for the intended study with the correct amount of fluorescence markers and bacterial loads. These drawbacks, however, have been looked into and corrected by the use of robotic arm-based injection of embryos with precise volumes in precise 3D space, followed by sorting of the embryos based on fluorescence using a device called Complex Object Parametric Analyzer and Sorter (COPAS), which is similar to a FACS or fluorescence-activated cell sorter (Carvalho et al. 2011). The same group that developed the COPAS also developed an advanced imaging system that allows high-resolution imaging of zebrafish embryos (Veneman et al. 2014). All in the entire model has been used to shed light on various preexisting notions leading to a vigorous growth in the field of host–pathogen interaction not just about tuberculosis, but also other pathogens. The system has also sparked a host of studies with regard to drug discovery using the in vivo model.

References

Abdallah AM, Verboom T, Weerdenburg EM, Gey Van Pittius NC, Mahasha PW, Jiménez C, Parra M, Cadieux N, Brennan MJ, Appelmelk BJ (2009) PPE and PE_PGRS proteins of *Mycobacterium marinum* are transported via the type VII secretion system ESX-5. Mol Microbiol 73:329–340

Adams KN, Takaki K, Connolly LE, Wiedenhoft H, Winglee K, Humbert O, Edelstein PH, Cosma CL, Ramakrishnan L (2011) Drug tolerance in replicating mycobacteria mediated by a macrophage-induced efflux mechanism. Cell 145:39–53

Aderem A, Underhill DM (1999) Mechanisms of phagocytosis in macrophages. Annu Rev Immunol 17:593–623

Aly S, Wagner K, Keller C, Malm S, Malzan A, Brandau S, Bange FC, Ehlers S (2006) Oxygen status of lung granulomas in *Mycobacterium tuberculosis*-infected mice. J Pathol 210:298–305

Balcewicz-Sablinska MK, Keane J, Kornfeld H, Remold HG (1998) Pathogenic *Mycobacterium tuberculosis* evades apoptosis of host macrophages by release of TNF-R2, resulting in inactivation of TNF-α. J Immunol 161:2636–2641

Baldwin SL, D'souza C, Roberts AD, Kelly BP, Frank AA, Lui MA, Ulmer JB, Huygen K, McMurray DM, Orme IM (1998) Evaluation of new vaccines in the mouse and guinea pig model of tuberculosis. Infect Immun 66:2951–2959

Bertrand JY, Traver D (2009) Hematopoietic cell development in the zebrafish embryo. Curr Opin Hematol 16:243–248

Bill BR, Petzold AM, Clark KJ, Schimmenti LA, Ekker SC (2009) A primer for morpholino use in zebrafish. Zebrafish 6:69–77

Bold TD, Ernst JD (2009) Who benefits from granulomas, mycobacteria or host? Cell 136:17–19

Cambier C, Takaki KK, Larson RP, Hernandez RE, Tobin DM, Urdahl KB, Cosma CL, Ramakrishnan L (2014) Mycobacteria manipulate macrophage recruitment through coordinated use of membrane lipids. Nature 505:218

Carvalho R, De Sonneville J, Stockhammer OW, Savage ND, Veneman WJ, Ottenhoff TH, Dirks RP, Meijer AH, Spaink HP (2011) A high-throughput screen for tuberculosis progression. PLoS One 6:e16779

Clemens DL, Lee B-Y, Horwitz MA (2000) Deviant expression of Rab5 on phagosomes containing the intracellular pathogens *Mycobacterium tuberculosis* and *Legionella pneumophila* is associated with altered phagosomal fate. Infect Immun 68:2671–2684

Comas I, Coscolla M, Luo T, Borrell S, Holt KE, Kato-Maeda M, Parkhill J, Malla B, Berg S, Thwaites G (2013) Out-of-Africa migration and neolithic coexpansion of *Mycobacterium tuberculosis* with modern humans. Nat Genet 45:1176

Control and Prevention (2006) Emergence of *Mycobacterium tuberculosis* with extensive resistance to second-line drugs--worldwide, 2000-2004. MMWR 55:301

Coros A, Callahan B, Battaglioli E, Derbyshire KM (2008) The specialized secretory apparatus ESX-1 is essential for DNA transfer in *Mycobacterium smegmatis*. Mol Microbiol 69:794–808

Dannenberg AM Jr (1994) Rabbit model of tuberculosis. In: Tuberculosis. American Society of Microbiology, Washington, pp 149–156

Davis JM, Ramakrishnan L (2009) The role of the granuloma in expansion and dissemination of early tuberculous infection. Cell 136:37–49

Eum S-Y, Kong J-H, Hong M-S, Lee Y-J, Kim J-H, Hwang S-H, Cho S-N, Via LE, Barry CE (2010) Neutrophils are the predominant infected phagocytic cells in the airways of patients with active pulmonary TB. Chest 137:122–128

Fine PE (1989) The BCG story: lessons from the past and implications for the future. Rev Infect Dis 11:S353–S359

Gao LY, Laval F, Lawson EH, Groger RK, Woodruff A, Morisaki JH, Cox JS, Daffe M, Brown EJ (2003) Requirement for kasB in Mycobacterium mycolic acid biosynthesis, cell wall impermeability and intracellular survival: implications for therapy. Mol Microbiol 49:1547–1563

Gupta S, Tyagi S, Almeida DV, Maiga MC, Ammerman NC, Bishai WR (2013) Acceleration of tuberculosis treatment by adjunctive therapy with verapamil as an efflux inhibitor. Am J Respir Crit Care Med 188:600–607

Howe DG, Bradford YM, Conlin T, Eagle AE, Fashena D, Frazer K, Knight J, Mani P, Martin R, Moxon S a T (2012) ZFIN, the Zebrafish model organism database: increased support for mutants and transgenics. Nucleic Acids Res 41:D854–D860

Kaushal D, Mehra S, Didier P, Lackner A (2012) The non-human primate model of tuberculosis. J Med Primatol 41:191–201

Lawson ND, Wolfe SA (2011) Forward and reverse genetic approaches for the analysis of vertebrate development in the zebrafish. Dev Cell 21:48–64

Lesley R, Ramakrishnan L (2008) Insights into early mycobacterial pathogenesis from the zebrafish. Curr Opin Microbiol 11:277–283

Lewis KN, Liao R, Guinn KM, Hickey MJ, Smith S, Behr MA, Sherman DR (2003) Deletion of RD1 from *Mycobacterium tuberculosis* mimics bacille Calmette-Guerin attenuation. J Infect Dis 187:117–123

McDonough KA, Kress Y, Bloom BR (1993) Pathogenesis of tuberculosis: interaction of *Mycobacterium tuberculosis* with macrophages. Infect Immun 61:2763–2773

McMurray DN (1994) Guinea pig model of tuberculosis. In: Tuberculosis. American Society of Microbiology, Washington, pp 135–147

Mehra S, Golden NA, Dutta NK, Midkiff CC, Alvarez X, Doyle LA, Asher M, Russell-Lodrigue K, Monjure C, Roy CJ (2011) Reactivation of latent tuberculosis in *Rhesus macaques* by coinfection with simian immunodeficiency virus. J Med Primatol 40:233–243

Murray KN, Varga ZM, Kent ML (2016) Biosecurity and health monitoring at the Zebrafish International Resource Center. Zebrafish 13:S–30-S-38

Nobre A, Alarico S, Maranha A, Mendes V, Empadinhas N (2014) The molecular biology of mycobacterial trehalose in the quest for advanced tuberculosis therapies. Microbiology 160:1547–1570

Ormerod LP (2005) Multidrug-resistant tuberculosis (MDR-TB): epidemiology, prevention and treatment. Br Med Bull 73:17–24

Ottenhoff TH, Kaufmann SH (2012) Vaccines against tuberculosis: where are we and where do we need to go? PLoS Pathog 8:e1002607

Ramakrishnan L (2012) Revisiting the role of the granuloma in tuberculosis. Nat Rev Immunol 12:352

Recht J, Kolter R (2001) Glycopeptidolipid acetylation affects sliding motility and biofilm formation in *Mycobacterium smegmatis*. J Bacteriol 183:5718–5724

Saunders BM, Cooper AM (2000) Restraining mycobacteria: role of granulomas in mycobacterial infections. Immunol Cell Biol 78:334–341

Scanga CA, Mohan V, Joseph H, Yu K, Chan J, Flynn JL (1999) Reactivation of latent tuberculosis: variations on the Cornell murine model. Infect Immun 67:4531–4538

Schlesinger L, Bellinger-Kawahara C, Payne N, Horwitz M (1990) Phagocytosis of *Mycobacterium tuberculosis* is mediated by human monocyte complement receptors and complement component C3. J Immunol 144:2771–2780

Singh A, Gupta R, Vishwakarma R, Narayanan P, Paramasivan C, Ramanathan V, Tyagi AK (2005) Requirement of the mymA operon for appropriate cell wall ultrastructure and persistence of *Mycobacterium tuberculosis* in the spleens of guinea pigs. J Bacteriol 187:4173–4186

Smith J, Manoranjan J, Pan M, Bohsali A, Xu J, Liu J, Mcdonald KL, Szyk A, Laronde-Leblanc N, Gao L-Y (2008) Evidence for pore formation in host cell membranes by ESX-1-secreted ESAT-6 and its role in *Mycobacterium marinum* escape from the vacuole. Infect Immun 76:5478–5487

Stinear TP, Seemann T, Harrison PF, Jenkin GA, Davies JK, Johnson PD, Abdellah Z, Arrowsmith C, Chillingworth T, Churcher C (2008) Insights from the complete genome sequence of *Mycobacterium marinum* on the evolution of *Mycobacterium tuberculosis*. Genome Res 18:729–741

Stoop EJM, Schipper T, Rosendahl Huber SK, Nezhinsky AE, Verbeek FJ, Gurcha SS, Besra GS, Vandenbroucke-Grauls CMJE, Bitter W, van der Sar AM (2011) Zebrafish embryo screen for mycobacterial genes involved in the initiation of granuloma formation reveals a newly identified ESX-1 component. Dis Model Mech 4(4):526–536

Sturgill-Koszycki S, Schlesinger PH, Chakraborty P, Haddix PL, Collins HL, Fok AK, Allen RD, Gluck SL, Heuser J, Russell DG (1994) Lack of acidification in *Mycobacterium* phagosomes produced by exclusion of the vesicular proton-ATPase. Science 263:678–681

Swaim LE, Connolly LE, Volkman HE, Humbert O, Born DE, Ramakrishnan L (2006) *Mycobacterium marinum* infection of adult zebrafish causes caseating granulomatous tuberculosis and is moderated by adaptive immunity. Infect Immun 74:6108–6117

Tobin DM, Ramakrishnan L (2008) Comparative pathogenesis of *Mycobacterium marinum* and *Mycobacterium tuberculosis*. Cell Microbiol 10:1027–1039

Tobin DM, Vary JC Jr, Ray JP, Walsh GS, Dunstan SJ, Bang ND, Hagge DA, Khadge S, King M-C, Hawn TR (2010) The lta4h locus modulates susceptibility to mycobacterial infection in zebrafish and humans. Cell 140:717–730

Tobin DM, Roca FJ, Oh SF, Mcfarland R, Vickery TW, Ray JP, Ko DC, Zou Y, Bang ND, Chau TT (2012) Host genotype-specific therapies can optimize the inflammatory response to mycobacterial infections. Cell 148:434–446

Tønjum T, Welty D, Jantzen E, Small P (1998) Differentiation of *Mycobacterium ulcerans, M. marinum*, and *M. haemophilum*: mapping of their relationships to *M. tuberculosis* by fatty acid profile analysis, DNA-DNA hybridization, and 16S rRNA gene sequence analysis. J Clin Microbiol 36:918–925

Torraca V, Masud S, Spaink HP, Meijer AH (2014) Macrophage-pathogen interactions in infectious diseases: new therapeutic insights from the zebrafish host model. Dis Model Mech 7:785–797

Tsai MC, Chakravarty S, Zhu G, Xu J, Tanaka K, Koch C, Tufariello J, Flynn J, Chan J (2006) Characterization of the tuberculous granuloma in murine and human lungs: cellular composition and relative tissue oxygen tension. Cell Microbiol 8:218–232

Ulrichs T, Kaufmann SH (2006) New insights into the function of granulomas in human tuberculosis. J Pathol 208:261–269

Van Leeuwen LM, Van Der Kuip M, Youssef SA, De Bruin A, Bitter W, Van Furth AM, Van Der Sar AM (2014) Modeling tuberculous meningitis in zebrafish using *Mycobacterium marinum*. Dis Model Mech 7:1111–1122

Van Pittius NCG, Sampson SL, Lee H, Kim Y, Van Helden PD, Warren RM (2006) Evolution and expansion of the *Mycobacterium tuberculosis* PE and PPE multigene families and their association with the duplication of the ESAT-6 (esx) gene cluster regions. BMC Evol Biol 6:95

Veneman WJ, Marín-Juez R, De Sonneville J, Ordas A, Jong-Raadsen S, Meijer AH, Spaink HP (2014) Establishment and optimization of a high throughput setup to study *Staphylococcus epidermidis* and *Mycobacterium marinum* infection as a model for drug discovery. J Vis Exp 88:e51649

Verreck FA, Vervenne RA, Kondova I, Van Kralingen KW, Remarque EJ, Braskamp G, Van Der Werff NM, Kersbergen A, Ottenhoff TH, Heidt PJ (2009) MVA. 85A boosting of BCG and an attenuated, phoP deficient M. tuberculosis vaccine both show protective efficacy against tuberculosis in *Rhesus macaques*. PLoS One 4:e5264

Volkman HE, Pozos TC, Zheng J, Davis JM, Rawls JF, Ramakrishnan L (2010) Tuberculous granuloma induction via interaction of a bacterial secreted protein with host epithelium. Science 327:466–469

Walsh GP, Tan EV, Cruz ECD, Abalos RM, Villahermosa LG, Young LJ, Cellona RV, Narareno JB, Horwitz MA (1996) The Philippine cynomolgus monkey (*Macaca fasicularis*) provides a new nonhuman primate model of tuberculosis that resembles human disease. Nat Med 2:430

Weerdenburg EM, Abdallah AM, Mitra S, De Punder K, Van Der Wel NN, Bird S, Appelmelk BJ, Bitter W, Van Der Sar AM (2012) ESX-5-deficient *Mycobacterium marinum* is hypervirulent in adult zebrafish. Cell Microbiol 14:728–739

Westerfield M (2000) The zebrafish book: a guide for the laboratory use of zebrafish. http://zfin.org/zf_info/zfbook/zfbk.html

Wolf AJ, Linas B, Trevejo-Nuñez GJ, Kincaid E, Tamura T, Takatsu K, Ernst JD (2007) *Mycobacterium tuberculosis* infects dendritic cells with high frequency and impairs their function *in vivo*. J Immunol 179:2509–2519

World-Health-Organization (2018) Global tuberculosis report 2018. World Health Organization, Geneva

Xiao A, Wang Z, Hu Y, Wu Y, Luo Z, Yang Z, Zu Y, Li W, Huang P, Tong X (2013) Chromosomal deletions and inversions mediated by TALENs and CRISPR/Cas in zebrafish. Nucleic Acids Res 41:e141–e141

Young D (2009) Animal models of tuberculosis. Eur J Immunol 39:2011–2014

Yu J, Tran V, Li M, Huang X, Niu C, Wang D, Zhu J, Wang J, Gao Q, Liu J (2012) Both phthiocerol dimycocerosates and phenolic glycolipids are required for virulence of *Mycobacterium marinum*. Infect Immun 80:1381–1389

Zeng G, Chen CY, Huang D, Yao S, Wang RC, Chen ZW (2011) Membrane-bound IL-22 after de novo production in tuberculosis and anti-*Mycobacterium tuberculosis* effector function of IL-22+ CD4+ T cells. J Immunol 187:190–199

Murine Models to Study Acute and Chronic Bacterial Infections

24

Suparna Chakraborty and Santasabuj Das

Contents

S. Chakraborty · S. Das (✉)
Division of Clinical Medicine, ICMR National Institute of Cholera and Enteric Diseases, Kolkata, India
e-mail: dasss.niced@gov.in

B. Siddhardha et al. (eds.), *Model Organisms for Microbial Pathogenesis, Biofilm Formation and Antimicrobial Drug Discovery*,
https://doi.org/10.1007/978-981-15-1695-5_24

Abstract

Despite spectacular advances in gaining insights into the biology of bacterial pathogens and the human host and the availability of different classes of antibiotics, bacterial diseases continue to be a major killer around the world, affecting both the developing and developed nations. The latter half of the last century had witnessed a revolution in biomedical sciences and health science research, including research in infectious diseases. Although the discovery of antibiotics dates back to the late 1920s, little was known about the specific mechanisms underlying host–pathogen interactions that involve microbial pathogenesis and host immune responses before the introduction and widespread use of molecular techniques. In fact, the vast majority of knowledge in the field is generated in the last four decades, primarily through the experiments involving gene-modified host and pathogenic microorganisms. An expanding number of disease models have greatly helped the cause of better understanding of infectious diseases and development of novel therapeutics, including drugs and vaccines. However, murine models have remained invaluable to study human infections, including those caused by bacterial pathogens, because of the genetic, anatomical, and physiological similarities between the human and mouse and the ease of genetic and other manipulations (construction of transgenic, knockout, and humanized animals) of mouse. Laboratory-bred mice with homogeneous genetic backgrounds are cheap, easily available, and amenable to experimental interventions. This review discusses the current trends of the use of murine models to study different bacterial infections of the human and contrasts the traditional and more recent mouse models. A better understanding of various murine models, along with the advantages and limitations of each of them will offer new insights into the pathogenesis of acute and chronic bacterial infections and pave the way for better control of these diseases.

Keywords

Murine model · Acute infection · Chronic infection · Bacterial pathogen

24.1 Introduction

Human is exposed to numerous pathogenic bacteria in the environment throughout their course of life. But the infection only takes place when the host's immune system fails to contain the invading bacteria. Persistent infections may go asymptomatic or result in clinically noticeable symptoms. Certain bacterial pathogens, such as *Mycobacterium tuberculosis* and *Salmonella enterica* serovar Typhi (*S.* Typhi), adapt the art of escaping the host defense machinery and persist within the host for decades without showing any symptoms (Gomez and McKinney 2004; Monack et al. 2004). On the other hand, *Pseudomonas aeruginosa* and *Escherichia coli* develop symptomatic acute or chronic infections.

24.1.1 Importance of Animal Models to Study Bacterial Infections

The fundamental rationale behind using a model organism to study disease pathogenesis lies in the anatomical and physiological relatedness of the affected species. This holds true for infectious diseases, including bacterial infections as well where the same tissues are infected with similar disease manifestations for the model host and the natural hosts. Animal experimentations are extensively used to study human diseases, especially the bacterial infections to elucidate their interactions with the host immune system and for assessing novel therapeutics. Studies involving animal models help us to understand the disease pathogenesis, host immune responses against the invading pathogens and most importantly, to evaluate the efficacy of antibiotics and potential vaccine candidates. Hence, a suitable animal model bridges the gap between the characterization of a pathogen in vitro and the clinical disease to address cure of the infection. Finch and Crimmins (2004) proposed that early age infections burden the patients with "cohort morbidity phenotype," which persists throughout their lives (Finch and Crimmins 2004). Even after the acute bacterial infection is resolved by the standard treatment, an "immunologic scar" persists. Currently, heightened concerns about antimicrobial resistance have led the researchers to develop optimized dosing regimens, using animals. Animal models, especially mouse models have been widely employed for quantitative and qualitative study of the host immune response and also to translate this knowledge at the clinic.

24.1.2 Mouse Models of Bacterial Infections

The choice of proper animals is one of the most important things to be considered while studying any disease. *Mus musculus* has remained a much favored model to study infectious bacterial diseases and the subsequent host immune response. The similarities between the human and the mouse in their genetic makeup, anatomical structures, and physiological functions have made the murine model invaluable to study bacterial diseases that affect human beings. Numerous inbred and outbred mouse strains are currently available to establish infectious disease models. Inbred mouse strains provide a uniform response to infections as well as investigational treatments and novel vaccine candidates, whereas an outbred strain maintains heterogeneity. Inbred mouse strains include BALB/c, C57BL/6, DBA, and CBA, whereas MF1, CD-1(Swiss), Swiss-Webster, and NMRI fall under the category of outbred strains. With the help of genome engineering, mouse can be genetically manipulated with relative ease to create transgenic, knockout, and knockin animals. Immunodeficient mouse models are useful to dissect bacterial infections and their interactions with the host immune system. Nude mice (nu) have mutation at *Foxn1* (winged-helix/forkhead transcription factor) gene and lack T cells due to athymia, which blocks the development of thymus-derived T cells. Severe combined immunodeficiency (SCID) strains have mutation at *Prkdc/scid* (protein kinase, DNA activated, catalytic polypeptide) protein. The latter is essential to join the nonhomologous ends of double-stranded DNA. The *Rag1* or *Rag2* (recombination activating genes) gene deletions in mice result in the arrest of rearrangement of B and T cell receptors and differentiation of the cells. Although mice have been extensively used to study disease pathogenesis, host–pathogen interactions, identify the virulence factors critical for establishing infections, and test the potential of drug and vaccine candidates, there are limitations of using mouse models while studying human biology. Unlike humans, absence of functional TLR10 and the presence of other innate immune responders, such as TLR11, TLR12, and TLR13 in mice made mouse models somewhat incongruent. Some pathogenic bacteria are restricted to the humans and are difficult to mimic in mouse models. To overcome these limitations, immunodeficient mice were engrafted with human tissues or cells to generate mouse/human chimera, which became increasingly popular to study human restricted pathogens. In this review, we will discuss various mouse models used to study acute and chronic bacterial infections.

24.2 Mouse Models to Study Acute Bacterial Infections

Table 24.1 summarizes the mouse models that were used to study acute bacterial infections affecting humans.

24.2.1 Cholera

Cholera is manifested by acute watery diarrhea and caused by *Vibrio cholerae*, a Gram-negative bacterium. If the circulatory volume loss is not replenished immediately, dehydration, leading to hypovolemic shock rapidly ensues. Genes encoding the two major virulence determinants of *V. cholerae*, cholera toxin (CT) and toxin co-regulated pilus (TCP) are parts of the *ToxR* regulon. Their expressions are regulated by the transcriptional regulators, encoded by *toxRS*, *tcpPH*, and *toxT* genes (Bina et al. 2003). The rabbit ileal loop model has been extensively used to study the pathogenesis of toxigenic *V. cholerae* strains as well as purified CT (Ritchie et al. 2010). However, the only physiologically relevant model of cholera developed to date is the RITARD model. We had improvised a mouse ileal loop model to test the functions of CT as well as infection with the live organisms. We demonstrated that CT suppresses cathelicidin and human beta-defensin 2, two endogenously produced cationic antimicrobial peptides that are abundant at the intestinal mucosal surface and aid in immune evasion (Chakraborty et al. 2008). However, *V. cholerae* fails to colonize the adult mouse intestine. Instead, a suckling mouse model has been widely used to study intestinal colonization of *V. cholerae* and the role of TCP in this process. This model was tested for therapeutic interventions against *V. cholerae* infection. In an intriguing set of experiments, researchers had shown that *E. coli* Nissle, expressing the *V. cholerae* quorum sensing molecules, autoinducer-2 (AI-2) and cholera autoinducer-1 (CAI-1) suppresses virulence genes (CT and TCP) expression by *V. cholerae*. An 80% reduction in CT production led to >90% survival of the mice (Duan and March 2010). Phage

Table 24.1 Mouse models of acute bacterial infections

Disease	Organism	Mouse strain	Aim of study	Major questions addressed/ finding	References
Cholera	*Vibrio cholerae* 569B inaba	Swiss mice; ICR (outbred)	Disease pathophysiology	Fluid accumulation in ileal loops inoculated with bacteria and toxin	Basu and Pickett (1969), Sawasvirojwong et al. (2013)
	Vibrio cholerae	Suckling CD-1 mice	Disease model	Colonization studies and virulence factor	Klose (2000), Duan and March (2010)
	Vibrio cholerae	BALB/c	Streptomycin-pretreated mouse model	Colonization studies and protective antibody response	Nygren et al. (2009)
	Vibrio cholerae	Mouse ileal loop model	Disease pathogenesis	Immune evasion	Chakraborty et al. (2008)
	V. cholerae El tor INABA	Balb/c	Pulmonary infection model	Colonization studies and virulence factor	Fullner et al. (2002), Pineyro et al. (2010)
	Vibrio cholerae	Infant mouse model	Infant mouse model	Phage therapy	Yen et al. (2017)
Pneumonia	*S. pneumoniae*	C57BL/6	Disease pathogenesis	Efficacies of different antibiotics	Azoulay-Dupuis et al. (1991)
	S. pneumoniae	MF1	Disease pathogenesis	Evaluating the role of pneumolysin and autolysin	Canvin et al. (1995)
	S. pneumoniae	Infant NMRI mice	Disease model description	Early-life immunization	Jakobsen et al. (2002)
	L. pneumophila	A/J	Disease model description	Acute pneumonia	Yamamoto et al. (1988), Yoshida et al. (1991), Brieland et al. (1994)
	L. pneumophila	C57BL/6	Disease model description	Study of flagellin mutants	Archer et al. (2009), Archer et al. (2010), Berrington et al. (2010)

(continued)

Table 24.1 (continued)

Disease	Organism	Mouse strain	Aim of study	Major questions addressed/ finding	References
Acute bacterial meningitis	*S. pneumoniae*	C57BL/6	Elucidate the role of nitric oxide synthase	To study innate immune response	Koedel et al. (2001)
	S. pneumoniae	CD-1	Therapeutic efficacy of drugs	Infection induced via intracerebral route	Shapiro et al. (2000)
	S. pneumoniae	BALB/c, C57BL/6, CD-1	Disease pathogenesis	Model of brain damage in pneumococcal meningitis	Grandgirard et al. (2007)
	S. pneumoniae	Swiss mice, C57/129	Disease pathogenesis	Infection through intranasal route and intraperitoneal route	Tan et al. (1995), Zwijnenburg et al. (2001)
	Neisseria meningitidis	CD46 transgenic mouse expressing human CD46	Model to study disease pathophysiology	To study bacterial load, quantification of antibody titers and cytokine response	Johansson et al. (2003), Johansson et al. (2005)
	N. meningitidis	Transgenic mice expressing human factor(fH)	Model to study disease pathophysiology	Antibiotic susceptibility testing	Belkacem et al. (2016)
	L. monocytogenes	C57BL/6	Oral model description	Disease pathogenesis	Czuprynski et al. (2003), Bergmann et al. (2013)
	L. monocytogenes	C57BL/6	Intracerebral model description	Immunomodulation and study the function of TNF and IL10	Schluter et al. (1996), Deckert et al. (2001), Virna et al. (2006)
Shigellosis	*S. flexneri*	SCID-HU-INT chimeric mouse	Disease pathogenesis	Immune evasion	Zhang et al. (2001)
	S. flexneri	Newborn BALB/c mice	Model description and disease pathogenesis	Mucosal inflammatory process	Siegrist (2001), Fernandez et al. (2003)
	S. flexneri	C57BL/6	Intraperitoneal disease model	Evaluating vaccine candidates	Yang et al. (2013)
	S. flexneri	Balb/c	Intraperitoneal disease model	Therapeutic efficacy of Toll-like receptor 3	Ta et al. (2017)

Salmonellosis: systemic infection	*S.* Typhi	C57/BL6	tlr11−/− mice model	Disease pathogenesis	Mathur et al. (2012)
	S. Typhi	B6.C3Pde6b^{rd1} Hsp4^{le}/J	Model description	Colonization study and Virulence factor	Spano et al. (2016)
	S. Typhi	Swiss mice; C57BL/6	Disease pathogenesis	Protective antibody responses	Ghosh et al. (2011), Das et al. (2017)
Salmonella-associated enterocolitis	*S. enterica* typhimurium	C57BL/6	Streptomycin-pretreated colitis mouse model	Molecular mechanisms of enteric salmonellosis	Barthel et al. (2003), Coburn et al. (2005)
	S. enterica typhimurium	C57BL/6	Streptomycin-pretreated colitis mouse model	Colonization studies and virulence factor	Hapfelmeier et al. (2004), Stecher et al. (2004)
Campylobacter infection	*Campylobacter jejuni*	BALb/c	Model description and disease pathophysiology	Colonization studies	McCardell et al. (1986a, b), Stanfield et al. (1987)
	Campylobacter jejuni	C57BL/6	Model description	Influence of diet and antibiotics in alterations of gut microflora	Giallourou et al. (2018)
	Campylobacter jejuni	C3H mice with limited gut flora (LF)	Model description	Colonization determinants and host immune response	Chang and Miller (2006)
Sepsis	*Streptococcus pneumoniae*	CBA/N, DBA/2, BALB/c	Disease pathogenesis	Drug efficacy and virulence study	Briles et al. (1981)
	Streptococcus pneumoniae, Klebsiella pneumoniae	CD-1	Model description and disease pathogenesis	Immune response	Wang et al. (2001)

(continued)

Table 24.1 (continued)

Disease	Organism	Mouse strain	Aim of study	Major questions addressed/ finding	References
Skin infection	*Staphylococcus aureus, MSSA, MRSA*	ICR	Burnt skin model	Efficacy of antibiotics	Akiyama et al. (1994), Yamakawa et al. (2002)
	Staphylococcus aureus and *Streptococcus pyogenes*	MF1	Skin suture-wound model	Efficacy of antibiotics	Boon and Beale (1987), Gisby and Bryant (2000)
Polymicrobial infection	*Escherichia coli*, *Bacteroides fragilis*, and *Clostridium perfringens*	Diabetic mouse strain BKS.Cg-m +/+ Leprdb/J	Disease progression	Influence of diabetes on polymicrobial infection	Mastropaolo et al. (2005)
	P. gingivalis, *T. denticola*, and *T. forsythia*	ApoE$^{-/-}$ B6.129P2-*Apoe*^{tm1Unc}/J	Model description	Relations between periodontal disease and associated atherosclerosis	Rivera et al. (2013), Chukkapalli et al. (2015), Chukkapalli et al. (2017)
	Candida albicans, Pseudomonas aeruginosa	C3H/HeN	Neutropenic mouse model	*To study fungal secreted products that inhibit bacterial colonization*	Lopez-Medina et al. (2015)
	Candida albicans, Pseudomonas aeruginosa	Cr1 CF1 BR non-Swiss mice	Model description	Sequential infection of a burned mouse model and disease	Neely et al. (1986), Peters and Noverr (2013)
Anaerobic infection	*Fusobacterium necrophorum*	Swiss mice; Balb/c	Disease pathogenesis	Efficacy of antibiotics	Wilkins and Smith (1974), Maestrone et al. (1975), Nagaoka et al. (2013)
	Fusobacterium necrophorum	A/J; CF1	Model description	Disease pathogenesis, efficacy on antibiotics	Abe et al. (1976), Hill (1977)
	Escherichia coli	Streptomycin-treated CD1 mice	Model description	Colonization studies	Jones et al. (2011)

Colitis in mice	*C. rodentium*	C57BL/6	Disease pathophysiology	Role of MyD88 signaling	Gibson et al. (2008), Lebeis et al. (2008); Bhinder et al. (2014)
	C. rodentium	C57BL/6-*Igh-6*^{tm1Cgn} mice	Disease pathophysiology	Role of B cells and T cell subsets	Simmons et al. (2003), Maaser et al. (2004)
	C. rodentium	C57BL/6J	Disease pathophysiology	Host responses to A/E lesion	Higgins et al. (1999), Newman et al. (1999), Gonçalves et al. (2001)
	EPEC	C57BL/6J	Model description	Disease pathophysiology	Savkovic et al. (2005)
	EPEC and EHEC	BALB/c mice	Model description	Cross-reactive protection	Calderon Toledo et al. (2011)

therapy for cholera was also successfully tested in the infant mouse model. Orogastric administration of a cocktail of three lytic phages 3 h prior to *V. cholerae* infection resulted in 2-log decrease in the number of surviving *V. cholerae* in the small intestine (Yen et al. 2017). Innate immune response to *V. cholerae* infection was studied using both infant and adult mice. At the peak of infection (24–30 h post-infection), several pro-inflammatory mediators, such as KC, MIP-2, NOS-2, IL-6, and IL-17a, were upregulated in the neonatal intestinal tissues (Bishop et al. 2014). This was in contrast to adult mice (streptomycin treatment model), which showed only modest increase in KC, MIP-2, and IL-6 expression. In the ketamine-dependent adult mouse model, IL-6 and MIP-2 were increased, but not the levels of NOS-2 and IL-17a. While no changes in TNF-α and IL-1β expressions were observed in any of the above models, IL-10 was induced only in the adult mice. Intriguingly IL-10 expression displayed strain specificity to *V. cholerae* infection, with C57/BL6 being more efficiently colonized than BALB/c or CD-1 strain. Inflammation in the infant mouse model is dependent on CT, and the accessory toxins like RtxA and HlyA play a minor role in CTX positive *V. cholerae* strain. In contrast, RtxA appears to be more important than CT for inflammation in the adult mouse model. Innate immune response to *V. cholerae* is also characterized by the recruitment of neutrophils and appearance of neutrophil extracellular traps (NETs), but secreted DNAses from *V. cholerae* disrupt the NETs (Bishop et al. 2014).

24.2.2 Acute Bacterial Meningitis

Acute bacterial meningitis, a life-threatening bacterial infection, is characterized by inflammation of the meninges and the subarachnoid space. This infection is fatal in 5–40% of patients, while 30% of the survivors develop serious neurological sequelae, such as seizures, deafness, and impairment of learning and memory (Leib and Tauber 1999; Koedel et al. 2002). The most common bacterial pathogens involved are *Streptococcus pneumoniae*, *N. meningitidis,* and *Listeria monocytogenes*. After the introduction of the Hib vaccine, *S. pneumoniae* became the major bacteria responsible for acute bacterial meningitis in all age groups. After colonizing the nasopharynx, the bacteria reach the lungs and from there, invades the bloodstream to effectively cross the blood–brain barrier (Leib and Tauber 1999). Rat and rabbit models were previously used to study meningitis, but two different types of mouse models have been established recently. Direct inoculation of the bacterium into the CNS through the intracerebral or the intracisternal route is useful to study host–pathogen interactions after the infection is established. Blood and brain homogenate levels of KC, IL6, and MIP-2 were elevated. Brain histology uniformly showed meningeal inflammation, characterized by neutrophilic infiltration, microglial activation, and hippocampal apoptosis. Parenchymal, subarachnoidal, and cortical hemorrhages were also seen in some mice (Mook-Kanamori et al. 2012). On the other hand, intranasal and intraperitoneal inoculation aids to study disease progression from colonization to establishment of the disease (Tan et al. 1995; Zwijnenburg et al. 2001). A female transgenic mouse model of intraperitoneal infection with bioluminescent *N. meningitidis* was demonstrated to be particularly useful for antibiotic susceptibility testing (AST) (Belkacem et al. 2016).

24.2.3 *Salmonella* Infection

Salmonella serotypes can cause enteritis, enteric (typhoid and paratyphoid) fever, and bacteremia in humans. Enteric fever caused by *Salmonella enterica* serovar Typhi (*S.* Typhi) and *S. enterica* serovar Paratyphi (*S.* Paratyphi) A and B is a human restricted, potentially life-threatening illness. The infections spread by the fecal–oral route via contaminated food and water and result in similar clinical illnesses. Annually, 26.9 and 5.9 million cases of typhoid and paratyphoid fevers, respectively, are reported (Crump et al. 2004; Buckle et al. 2012). Upper small intestinal biopsy of patients reveals diffuse enteritis, caused by infiltration of mononuclear cells. Mucosal thickening of the ileum and enlargement of the mesenteric lymph nodes, spleen, and liver are also observed. Bacterial colonization often leads to capillary thrombosis in the Peyer's patches of the terminal ileum, which further results in hemorrhage, necrosis, ulceration, and intestinal perforation (Sprinz et al. 1966; Bitar and Tarpley 1985; Kraus et al. 1999). Acute enteritis due to *S. enterica* serotypes, Typhimurium and Enteritidis infection represents the second most common cause of bacterial food-borne disease in the USA (Mead et al. 1999) and is characterized by mucosal edema and infiltration of PMN leukocytes (Day et al. 1978). In the mouse, *S. enterica* serotypes Typhimurium (previously known as *Bacillus* Typhi) causes human typhoid-like illness. Since the tissue lesions of mice infected with *S.* Typhimurium are similar to those of the patients suffering from typhoid, murine typhoid model has been extensively used to study human typhoid fever. However, despite ~98% overall identity of the *S.* Typhimurium and *S.* Typhi genomes, the pathogenicity islands of the two serotypes contain ~11% non-homologous genes that account for different host specificity and disease manifestations. The use of knockout mice has offered new insights into the pathogenesis of typhoid. A study reported that deletion of TLR11, which is expressed in mice, but not in humans results in the establishment of *S.* Typhi infection in mice after oral administration (Mathur et al. 2012). Mice lacking Rab32 (Rab family GTPase) and BLOC-3 (Nucleotide exchange factor) was also shown to be susceptible to *S.* Typhi infection. *S.* Typhimurium blocks this pathway by delivering two effector molecules, namely SopD2 (Rab32 GAP) and GtgE (Rab32 protease). A mutant strain of *S.* Typhimurium devoid of these two effector molecules showed reduced virulence in mice (Spano et al. 2016). O'Brien (1982) first reported that mice become more prone to *S.* Typhi infection when administrated intraperitoneally with iron alone or with an iron chelator (O'Brien 1982). Based on this, we developed an iron overload mouse model for systemic *S.* Typhi infection through the natural route. Swiss Albino mice were intraperitoneally injected with Fe^{3+} and desferrioxamine 4 h before the bacterial challenge. We observed that outer membrane proteins of *S.* Typhi, such as T2544 and T2942 play a major role in bacterial adhesion and/or invasion of the intestinal epithelial cells and disease development in the mice (Ghosh et al. 2011;

Chowdhury et al. 2015). This model was also successfully used to study the protective immune responses induced by T2544 (Ghosh et al. 2011; Das et al. 2017). *S.* Typhimurium-induced enterocolitis was modeled in streptomycin-pretreated mice. Streptomycin eliminates the commensal microflora, allowing intestinal colonization of mice with *S.* Typhimurium. Mutant *S.* Typhimurium strains with defective SPI-1 type III secretion system fail to cause systemic infection, but successfully colonize the lower intestinal tract of streptomycin pretreated mice (Barthel et al. 2003; Hapfelmeier et al. 2004; Stecher et al. 2004; Coburn et al. 2005).

24.2.4 Shigellosis

Bacillary dysentery or shigellosis is an acute invasive disease caused by a Gram-negative, facultative bacteria that belong to the *Shigella* sp. Broadly, four different species of *Shigella,* namely *Shigella flexneri*, *Shigella sonnei*, *Shigella dysenteriae*, and *Shigella boydii* cause Shigellosis, which remains a major threat to public health in the developing world. The age group primarily affected is 0-5 years and accounts for more than one million deaths every year (Kotloff et al. 1999). Patients with shigellosis develop fever, intestinal cramps, and discharge of mucopurulent and bloody feces. *Shigella* invades the intestinal epithelial layer of the human and nonhuman primates and causes infiltration of the inflammatory cells into it, resulting in tissue edema and areas of epithelial erosions (Mathan and Mathan 1991). Numerous in vivo models, such as the macaque monkey, rabbit, guinea pig, and mouse models have been developed to mimic human shigellosis. Among these models, macaque monkey is the most relevant to assess immunogenicity and protection by vaccine administration (Phalipon and Sansonetti 2007). Guinea pigs, resistant to both natural and experimental infections of the intestine by *Shigella* were used to demonstrate acute inflammation, characterized by keratoconjunctivitis following bacterial infection of the eye (Sereny test) (Sereny 1957). Adult mouse fails to establish intestinal *Shigella* infection due to the lack of IL-8, a key player of the human immune system, which recruits neutrophils to the site of infection (Singer and Sansonetti 2004). However, a mouse model of pulmonary infection developed by intranasal administration of *Shigella* showed acute bronchopneumonia due to massive neutrophilic infiltration. This model is still used to assess immunization efficacy and protection against bacterial challenge (Sharma et al. 2016). Zhang et al. (2001) established successful colonization of the bacterium in human intestinal xenografts in severe combined immunodeficient mice (SCID-HU-INT mice) (Zhang et al. 2001). Another group reported a 4-days old neonatal mouse model to study the pathogenesis of *Shigella* infection (Fernandez et al. 2003). Yang et al. (2013) had established shigellosis in adult mice by intraperitoneal (i.p.) infection (Yang et al. 2013). The model mimics human shigellosis in many respects and was used to study protective efficacy of potential vaccine candidates. High doses (10^8 CFU) of *S. flexneri* cause severe dysentery with massive destruction of the mucosal tissues and reduced colon length, leading to 100% mortality within 18 h of infection. In contrast, low dose (~10^6 CFU) infection induces diarrhea and mild inflammatory response. CFU counts in the stool and infected colon support bacterial colonization of the intestinal tissues. We have used the intraperitoneal model of shigellosis to study therapeutic efficacy of Toll-like receptor 3 ligand, poly(I;C) that functions by inducing cationic antimicrobial peptide cathelicidin in the colonic epithelium (Ta et al. 2017).

24.2.5 *Citrobacter rodentium* Infection

Infection pattern of *Citrobacter rodentium,* a Gram-negative, mouse restricted, mucosal pathogen provides a robust model to study bacteria-induced colitis in mice. Upon infection, it injects several effector proteins into the host cells to subverts the innate immune system of the host and leads to the formation of characteristic attaching and effacing lesions. Studies reported the role of B cells (Maaser et al. 2004), T cell subsets (Simmons et al. 2003) and MyD88 signaling (Gibson et al. 2008; Bhinder et al. 2014) in the development of immunity toward *Citrobacter rodentium*-induced colitis in vivo. *Citrobacter rodentium* is also used as a surrogate model to study the disease pathogenesis of two diarrheagenic pathogens of the human, namely EPEC (enteropathogenic *E. coli*) and EHEC (enterohemorrhagic *E. coli*).

24.2.6 *Campylobacter jejuni* Infection

Absence of animal models, especially the mouse models that replicate the human disease has long been a major bottleneck in the study of *C. jejuni* pathogenesis. In chickens and most other animals, the bacteria reside as a commensal microbe within the intestine, while causing severe diarrhea and gastroenteritis in humans (Walker et al. 1986; Wassenaar and Blaser 1999; Enocksson et al. 2004). A large number of studies reported intestinal colonization in both the neonatal and adult mice, which increases after antibiotic pre-treatment. However, reports of diarrhea and deaths in non-manipulated adult mice are rare (Baqar et al. 1999). Vancomycin pre-treatment of mice deficient in Single IgG IL-1 Related Receptor (SIGIRR), a negative regulator of MyD88-dependent signaling resulted in widespread and extensive colonization of *C. jejuni,* accompanied by severe gastroenteritis involving high levels of Th1/Th17 cytokines. SIGIRR−/− mice showed increased susceptibility to infection by *Citrobacter rodentium* and *Salmonella enterica* serovar Typhimurium, the two natural enteric bacterial pathogens of mice. However, Tlr4−/−/Sigirr−/− mice were largely unresponsive to infection by *C. jejuni* despite heavy colonization, while Tlr2−/−/Sigirr−/− mice developed exaggerated inflammation and pathology. This indicates that TLR4 signaling is mainly responsible for enteritis seen in this model, whereas TLR2 signaling is protective (Stahl et al. 2014). A weaned mouse model of *C. jejuni* infection was developed that replicated the two widely recognized clinical manifestations of the human disease, namely, enteropathy with growth failure and overt bloody diarrhea. The animals were fed Zn-deficient or protein-deficient diet and treated with vancomycin to disrupt the commensal microflora before *C. jejuni* infection. The Zn-deficiency model showed severe enteropathy with bloody diarrhea, fecal shedding of bacteria for more than 2 weeks and growth failure. In contrast, protein-deficient diet modeled a milder disease with modest enteropathy, but no overt diarrhea as observed for childhood infection in the developing world (Giallourou et al. 2018).

24.2.7 Skin Infections

Human skin primarily provides a physical barrier to the invading organisms. It can secrete antimicrobial peptides, which aid the host defense system to fight off the pathogen. Methicillin-sensitive and -resistant *Staph. aureus* (MSSA and MRSA) and *Streptococcus pyogenes* are the major causal agents of primary skin and soft tissue infections, leading to a wide range of dermatological diseases, including atopic dermatitis, psoriasis, and skin cancers. Burnt skin model (Akiyama et al. 1994; Yamakawa et al. 2002) and skin suture-wound model (Boon and Beale 1987; Gisby and Bryant 2000) were used to assess the efficacy of the antibiotics.

24.2.8 Bacteremia and Sepsis

Bacteremia and sepsis pose major threats to public health. Bone et al. (1989) defined the term sepsis as a host systemic response to infection measured by several clinical parameters, such as blood leukocyte count, heart and respiratory rates, and the body temperature (Bone et al. 1989). In contrast, bacteremia denotes the bacterial load in the blood. *S. pneumoniae*, *P. aeruginosa*, and *L. penumophilia* are the most common bacteria associated with acute bacteremic pneumonia in humans (Whitney 2016). These pathogens frequently colonize the upper respiratory tract and often lead to the secondary pneumonia following a viral infection (Morens et al. 2008). Mouse models of pneumonia allow the evaluation of the bacterial load in the lungs and the blood, histological analysis of lung tissue inflammation, quantification of antibody titers, and screening of antimicrobials. Despite intensive studies, murine models of sepsis have remained a topic of debate. A comparison of the complex physiology of the human with a single genetic strain of mouse may be inappropriate. Inoculation of bacteria through the intraperitoneal or intravenous route induces sepsis in mice, but the disease is usually short lived, with either death of the animal or clearing of the bacteria. Inclusion of fibrin clots impregnated with standardized number of bacteria into the peritoneal cavity added an advantage to the bacterial injection model of sepsis. This fibrin clot, serving as a reservoir of inoculums, allows monitoring the rate of bacterial clearance and titrating the mortality rate. This model showed the importance of early antibiotic treatment over delayed administration in sepsis (Kumar et al. 2006). The cecal slurry method is based on injecting cecal contents intraperitoneally, which is quantified as a fecal mass and resuspended in liquid, whereas cecal ligation and puncture (CLP) model involves ligation of a part of the cecum in conjunction with cecal colotomies via needle puncture. Due to incomplete intestinal development of neonatal mice and small organ size, the cecal slurry model is advantageous over CLP method to study sepsis. Studies on cecal slurry model of sepsis showed distinctive genetic and cytokine responses (Gentile et al. 2014).

24.2.9 Polymicrobial and Anaerobic Infections

Animal models are the key to our current knowledge about the pathogenesis of human infections and the therapeutic advancement to combat them. However, the reductionist approach of using one infectious agent at a time has limitations and fails to capture the complexities of human infections that often occur with more than one pathogen, infecting simultaneously or sequentially. Concomitant infection(s) could change the outcome of infectious diseases, due to either synergistic or antagonistic interactions between the pathogens or by modulating the host immune response.

Periodontal infection is a classic example of polymicrobial infection caused by several early colonizer pathogens, including *Prevotella, Peptostreptococcus, Campylobacter*, and *F. nucleatum* and some late colonizers, such as *Treponema denticola, Porphyromonas gingivalis*, and *Bacteroides forsythus*. There are numerous reports of an association between periodontal infection and atherosclerotic vascular disease, and epidemiological studies have suggested that periodontal infection is a risk factor for coronary artery disease (CAD). Causative association between the two was addressed in two widely used mouse models of accelerated atherosclerosis, namely the ApoE−/− and LDLR−/− mice (Rivera et al. 2013; Chukkapalli et al. 2015; Chukkapalli et al. 2017). Simultaneous or sequential infection of the mouse oral cavity with periodontal pathogens proved polymicrobial synergism in the aggravation of the pre-existing aortic atherosclerotic lesions. This may be contributed by the activation of aortic TLR and inflammasome signaling pathways and increase in the local aortic oxidative stress (Velsko et al. 2015). The models demonstrated hematogenous dissemination of the periodontal pathogens to distant tissues, including the heart and the aorta. The ApoE−/− mice also showed unique changes in the inflammatory response to polymicrobial infection with the progression of the atherosclerotic disease (Chukkapalli et al. 2015). The intestine is host to a vast number and varieties of microorganisms, including bacteria and fungi. Thus, bacterial–fungal interactions are likely to have significant implications for human health and disease. Using polymicrobial infection of a neutropenic mouse strain, Lopez-Medina and colleagues showed that co-colonization with *Candida albicans* significantly inhibits *P. aeruginosa* virulence without interference of the colonization ability of either organism. *C. albicans* inhibits the expression of *P. aeruginosa* siderophores, pyochelin and pyoverdine through fungal secreted products (Lopez-Medina et al. 2015). In contrast, sequential infection of a burned mouse model with sublethal doses of elastase-producing *P. aeruginosa* strain and *C. albicans* exhibited a mortality rate of 60% as opposed to less than 10% mortality of unburned mice challenged in the same way or burned mice receiving a single organism or elastase-negative *P. aeruginosa* (Neely et al. 1986). While monomicrobial infection with *C.*

albicans or *Staphylococcus aureus* is nonlethal in a mouse peritonitis model, coinfection caused ~40% mortality that was perhaps precipitated by increased pro-inflammatory cytokine secretion (Peters and Noverr 2013). Similarly, nonlethal intravenous infection of *E. coli* or administration of *E. coli* lipopolysaccharide (LPS) accelerated death due to lethal doses of *C. albicans* given through the intravenous route (Akagawa et al. 1995).

24.3 Mouse Models to Study Chronic Bacterial Infections

More than half of the world population is affected by chronic bacterial infections. Chronicity may result from incomplete cure of an acute infection. However, certain bacterial pathogens have evolved with specific mechanisms to survive within the host for a long time despite antimicrobial treatment. Various mouse models used to study chronic bacterial infections are summarized in Table 24.2.

24.3.1 *Helicobacter pylori* Infection

Helicobacter pylori, a Gram-negative bacterium colonizes the gastroduodenal epithelium and persists over there for decades by escaping the host defense mechanisms. The infection may be asymptomatic or may cause mild gastritis to gastric and duodenal ulcers (Nomura et al. 1994) and gastric adenocarcinoma (Wotherspoon et al. 1991). Since the natural *H. pylori* infection is restricted to humans, establishing a mouse model was initially challenging. An earlier study reported colonization of *H. pylori* in the nude mice (Karita et al. 1991; Karita et al. 1995; Marchetti et al. 1995). Later on, *H. pylori* and *H. felis* infections were established in several wild-type and gene-modified mouse strains to model different outcomes of human infections (Zhang and Moss 2012). C57BL/6 mice were extensively used to study gastric carcinogenesis. However, high-grade dysplasia was observed with *H. felis*, but not with *H. pylori* infection (Kim et al. 2003). Recently, INS-GAS mouse that overexpress human gastrin gene under the control of an insulin promoter was used to investigate the synergistic role of *Helicobacter* infection and hypergastrinemia in gastric carcinogenesis (Wang et al. 2000) C57BL/6 mice also develop severe gastritis; but unlike human disease, which shows both neutrophil and mononuclear cell infiltration into the lesion, mouse gastritis is predominantly infiltrated by the mononuclear cells (Sakagami et al. 1996). In contrast, BALB/c mice exhibit a Th2-predominant response to *Helicobacter* infection, characterized by fewer epithelial lesions, but larger number of intralesional bacteria (Mohammadi et al. 1996). These mice were also used as a model for *Helicobacter*-induced MALT lymphoma (Thompson et al. 2004). *H. pylori* infection in cytokine knockout mice shed light on the underlying mechanisms of the development of gastritis and gastric carcinoma. IFN-γ was found to play a critical role in *Helicobacter*-induced gastric inflammation and epithelial cell damage (Sawai et al. 1999), while *Helicobacter* infection and IL-1β synergistically promote gastric neoplasia (Tu et al. 2008). Similarly, IL-10 deficiency in the 129/EvSv background leads to severe hyperplastic gastritis when infected with *H. felis* (Berg et al. 1998). Mice deficient in p27 tumor suppressor gene developed metaplasia, dysplasia, and finally gastric cancer after *H. pylori* infection (Kuzushit et al. 2005).

24.3.2 Chronic Respiratory Infections

A common mouse model for chronic lung infection is cystic fibrosis (CF), an autosomal recessive disorder. It is caused by mutations in the cystic fibrosis transmembrane conductance regulator (CFTR) gene that is located on human chromosome 7 and encodes a cAMP-regulated chloride channel (Guilbault et al. 2007). The multi-organ exocrine dysfunction observed in CF patients is difficult to replicate in mouse. However, several CF mouse models have been generated, either by disruption of the CFTR gene or by introduction of frequently occurring mutations, such as F508 and G551D mutations in the ES cells (Semaniakou et al. 2018). Human patients with

Table 24.2 Mouse models of chronic bacterial infections

Disease	Organism	Mouse strain	Aim of study	Major questions addressed/ finding	References
H. pylori infection	*H. pylori*	Athymic and euthymic BALB/c	Model description	Immune response	Karita et al. (1991), Karita et al. (1995), Marchetti et al. (1995), Rabelo-Goncalves et al. (2005)
	Mouse adapted *H. pylori*	CD1	Disease pathogenesis	Colonization and vaccine studies	Dorer et al. (2013)
Chronic respiratory infection	*S. pneumoniae*	CBA/J	Intranasal aerosol pneumonia model	Effectively mimic the episode of pneumonia in humans	Iizawa et al. (1996)
	Pseudomonas aeruginosa	C57BL/6	Chronic infection by mucoid strain	Dose-dependent response	Bragonzi et al. (2005)
	Mycobacterium tuberculosis	Chimeric IFNγR1-deficient mouse	Disease pathogenesis and cellular response	Role of IFNγ, cellular response during early life infection	Saito et al. (2001), Desvignes and Ernst (2009)

(continued)

Table 24.2 (continued)

Disease	Organism	Mouse strain	Aim of study	Major questions addressed/ finding	References
Chronic *Salmonella* infection	*S.* Typhimurium	129X1/SvJ Nramp1+/+ mouse	Chronic *Salmonella* infection model	Colonization and persistence inside host	Monack et al. (2004)
	S. Typhimurium	129X1/SvJ Nramp1+/+ mouse	Chronic *Salmonella* infection model	Bacterial colonization in the gall bladder tissue	Crawford et al. (2010)
	S. Typhi	Immuno-deficient Rag2(−/−) γc(−/−) mice	Chronic *Salmonella* infection model	Bacterial pathogenesis and therapeutic strategies	Song et al. (2010), Firoz Mian et al. (2011)
UTI	*E. coli*	BALB/c	Disease pathophysiology	Comparing the route of inoculation	Schaeffer et al. (1987); Hopkins et al. (1995)
	E. coli	C57BL/6	Disease pathophysiology	Host immune response	Gunther et al. (2001), Lane et al. (2007)
Chronic colitis infection	*C. difficile*	C57BL/6	Disease pathophysiology	Prophylactic effects of *B. fragilis in a CDI mouse model*	Deng et al. (2018)
	C. difficile	C57BL/6	Mouse relapse CDI model	Host immune response	Sun et al. (2011)
	C. difficile	C57BL/6	Mouse model of antibiotic-induced CDAD	Disease pathogenesis	Chen et al. (2008)
Chronic wound infection	*MRSA, Pseudomonas aeruginosa*	FVB/N	Disease model description with bioluminescent bacteria	Enhancement of antimicrobial photodynamic therapy	Fila et al. (2016)
	Staphylococcus aureus Staphylococcus epidermidis	C57Bl6/J	Murine cutaneous wound model	Wound healing and biofilm visualization	Schierle et al. (2009)
	Pseudomonas aeruginosa	Diabetic mice	Disease model description	Wound healing and antibiotic tolerance	Zhao et al. (2010), Watters et al. (2013)

CF are frequently colonized by *P. aeruginosa* and *S. aureus*, whereas spontaneous colonization with the typical CF pathogens was not found in CF mice. Administration of free *P. aeruginosa* causes acute, not chronic, lung infection, and specific techniques such as intranasal inoculation with agar beads laden with *P. aeruginosa* revealed defects in bacterial clearance, resulting an inflammatory response in CFTR mutant mice (Heeckeren et al. 1997). Shah et al. demonstrated that overexpression of ATP12A in CF mice airways leads to impaired airway host defense and increased bacterial load (Shah et al. 2016). Very recently, a G542XCF mouse model was generated with CRISPR/Cas9 gene editing method (McHugh et al. 2018). These mice have decreased CFTR expression and lack CFTR functions in the airways and intestine.

Inbred mouse strains CBA, C3HeB/FeJ, DBA/2, and 129SvJ are susceptible to chronic *Mycobacterium tuberculosis*, whereas C57BL/6 J and BALB/c are resistant (Medina and North 1998). In vivo studies help us to understand the mechanisms of disease progression as well as cellular response mediated by the host to combat the infection.

24.3.3 Chronic *Salmonella* Infection

Regardless of adequate treatment of acute typhoid fever, *S.* Typhi persists in the gallbladder of 3–5% of the affected individuals, resulting in an asymptomatic and chronic infection. They shed bacteria in their stools and urine, which can further infect other individuals, leading to a major public health hazard. There are cases of asymptomatic carriers who developed adenocarcinoma of the gall bladder (Gonzalez-Escobedo et al. 2011). Monack et al. (2004) studied the chronic infection of *S.* Typhimurium by infecting 129X1/SvJ Nramp1+/+ mouse through oral route and found detectable presence of the bacterium in the gall bladder for one year post-infection (Monack et al. 2004). Using this model, another study was conducted where mice were fed with lithogenic diet for 6–8 weeks, which facilitates the formation of cholesterol gallstones in mice. Subsequent *S.* Typhimurium infection resulted in enhanced bacterial colonization in the gall bladder tissue and bile of the mice. Electron microscopic analysis showed bacterial biofilms on the gallstones (Crawford et al. 2010). A separate study suggested that an immunodeficient Rag2(−/−)γc(−/−) mice, engrafted with human fetal liver hematopoietic stem and progenitor cells effectively support *S.* Typhi infection and is very useful to elucidate bacterial pathogenesis (Song et al. 2010). Another study with humanized mouse model, where a lymphoid RAG-2-(−/−) γ(c)(−/−) mouse was engrafted with human leukocytes demonstrated that intravenous administration of *S.* Typhi was able to establish successful infection. A significant bacterial load was found in the liver, spleen, blood, and bone marrow of these mice (Firoz Mian et al. 2011).

24.3.4 Chronic Colitis Model

An increased incidence of morbidity and mortality of patients infected with *Clostridium difficile* are observed due to the emergence of hypervirulent, antibiotic-resistant strains. This Gram-positive, spore-forming bacillus causes nosocomial antibiotic-associated diarrhea and colitis, which are collectively referred to as CDI. Numerous in vivo models including hamsters, rabbits, guinea pigs, and rats were used to study the pathogenesis of CDI. However, a recently developed mouse CDI model closely resembled the human infection (Chen et al. 2008). Invading bacterium releases two exotoxins, TcdA and TcdB to establish successful colonization of the host. One study reported that vancomycin treatment of mice delays disease recurrence and generate neutralizing polysera against bacterial exotoxins (Sun et al. 2011). Another study showed that short-chain fatty acids, such as butyrate protects intestinal epithelial cells of CDI mouse from tissue damage by the activation and stabilization of HIF-1 transcription factor (Fachi et al. 2019). Clinical cases reported significant decrease of *Bacteroides fragilis* in patients infected with *Clostridium difficile,* leading to the study prophylactic role of *Bacteroides fragilis* in CDI mouse model. *Bacteroides fragilis* confers protection by modulating the gut microbiota of the infected mice (Deng et al. 2018).

24.3.5 Urinary Tract Infection

Urinary tract infection is another severe bacterial infection, frequently caused by uropathogenic *Escherichia coli*. Though it primarily affects women, young boys and elderly men are also at risk. Upper UTI involves bladder and kidney, which are known as cystitis and pyelonephritis, respectively. There are many murine models to study the pathophysiology of UTI. Hopkins et al. (1995) effectively introduced bacteria into the bladder through intraurethral and transurethral routes, with the induction of vesicoureteral reflux (VUR) (Hopkins et al. 1995). Most studies with mouse models used polyethylene plastic catheter tubes, fitted over needles for the administration of bacteria through it. This method reduces induction of VUR. There are reports of similar mouse models to study several other uropathogenic bacteria, such as *Enterococcus faecalis* and *Klebsiella pneumoniae* (Rosen et al. 2008a, b).

24.3.6 Chronic Wound Infections

Chronic infections with multispecies biofilms are a major problem in the diabetic population. In an obese diabetic mouse (BKS.Cg-m+/+ Lepr(db)/J) model for human type 2 diabetes, inner thigh abscess was induced by subcutaneous injection of mixed cultures containing *Escherichia coli*, *Bacteroides fragilis*, and *Clostridium perfringens*. Synergy, as determined by statistically significant increase in the number of bacterial pathogens following coinfection as opposed to infection with the individual pathogens was found between *E. coli* and *B. fragilis*, but *C. Perfringens* was antagonistic (Mastropaolo et al. 2005). Biofilm-associated chronic wound infection was also modeled in *db/db* mice (diabetic mice that have a spontaneous mutation in the leptin receptor, resulting in insulin tolerance). A full thickness, excision wound was transferred and infected over time with bacterial species of the mouse normal skin flora, such as *S. aureus,* coagulase-negative *Staphylococcus* sp., *Enterococcus* sp., *Enterobacter cloacae*, and *Pseudomonas* sp. (Dhall et al. 2014). In an elegant experiment, Dalton et al. (2011) demonstrated "biofilm transplant" from an infected wound and showed that the transfer of planktonic cultures would instead allow growth of only one species (Dalton et al. 2011).

24.4 Bystander Infection

In most intances, mice being used in studying the host's immune system in response to bacterial infection are housed in specific pathogen-free conditions in the laboratories. Humans and mice both are exposed to the various pathogens in the environment throughout their lives. This has shaped their immune system to fight against those disease-causing pathogens. Specific pathogen-free conditions lead to immature immune system in mice. This could be less representative of the real life infections. Previous history of microbial exposure would be helpful to recapitulate the disease and immune response (Beura et al. 2016; Reese et al. 2016). A recent work reported that significant colonization of a novel protist, residing in the mouse intestine, protected mice from *Salmonella* Typhimurium infection (Chudnovskiy et al. 2016). Pathogen-free animals did not develop colitis. C57L/J mice infected with *Helicobacter* species and on lithogenic diet developed gallstones (Maurer et al. 2005) and enhanced colitis (Ward et al. 1996). Bystander infections could alter the basal immune activation of the host. A study reported significant increase in lymphatic leakage into the mesenteric adipose tissue and reduced number of migratory dendritic cells were observed with *Yersinia pseudotuberculosis* infection in mice. However, although the bacterial infection is resolved with antibiotic treatment, it results in alterations of the intestinal microbiota, which persists for long time after pathogen clearance. Similar observations were seen with *Yersinia enterocolitica* infection in Toll-like receptor 1-deficient mice. These studies explain that acute infections drive changes in intestinal microflora which eventually lead to chronic illness (Fonseca et al. 2015, Kamdar et al. 2016).

24.5 Conclusion

Murine models are attractive tools for researchers. They are extensively used in biomedical studies. They can be easily manipulated genetically to mimic human physiological conditions and disease states, which further aid in studying disease pathophysiology. Although mice and humans differ in the disease response to a particular pathogen, genetic engineering solves the issues by creating desired mutations to reflect the human infection and study disease progression. The mouse model indeed allows unbiased approaches to screen various drug targets and potential vaccine candidates.

References

Abe PM, Lennard ES, Holland JW (1976) *Fusobacterium necrophorum* infection in mice as a model for the study of liver abscess formation and induction of immunity. Infect Immun 13(5):1473–1478

Akagawa G, Abe S, Yamaguchi H (1995) Mortality of *Candida albicans*-infected mice is facilitated by superinfection of *Escherichia coli* or administration of its lipopolysaccharide. J Infect Dis 171(6):1539–1544

Akiyama H, Kanzaki H, Abe Y, Tada J, Arata J (1994) *Staphylococcus aureus* infection on experimental croton oil-inflamed skin in mice. J Dermatol Sci 8(1):1–10

Archer KA, Alexopoulou L, Flavell RA, Roy CR (2009) Multiple MyD88-dependent responses contribute to pulmonary clearance of *Legionella pneumophila*. Cell Microbiol 11(1):21–36

Archer KA, Ader F, Kobayashi KS, Flavell RA, Roy CR (2010) Cooperation between multiple microbial pattern recognition systems is important for host protection against the intracellular pathogen *Legionella pneumophila*. Infect Immun 78(6):2477–2487

Azoulay-Dupuis E, Bedos JP, Vallée E, Hardy DJ, Swanson RN, Pocidalo JJ (1991) Antipneumococcal activity of ciprofloxacin, ofloxacin, and temafloxacin in an experimental mouse pneumonia model at various stages of the disease. J Infect Dis 163(2):319–324

Baqar S, Burg EF, Murphy JR (1999) Chapter 26 - mouse models of *Campylobacter jejuni* infection. In: Zak O, Sande MA (eds) Handbook of animal models of infection. Academic, London, pp 223–239

Barthel M, Hapfelmeier S, Quintanilla-Martínez L, Kremer M, Rohde M, Hogardt M, Pfeffer K, Rüssmann H, Hardt W-D (2003) Pretreatment of mice with streptomycin provides a *Salmonella enterica* serovar Typhimurium colitis model that allows analysis of both pathogen and host. Infect Immun 71(5):2839–2858

Basu S, Pickett MJ (1969) Reaction of *Vibrio cholerae* and choleragenic toxin in ileal loop of laboratory animals. J Bacteriol 100(2):1142–1143

Belkacem N, Hong E, Antunes A, Terrade A, Deghmane A-E, Taha M-K (2016) Use of animal models to support revising meningococcal breakpoints of β-lactams. Antimicrob Agents 60(7):4023–4027

Berg DJ, Lynch NA, Lynch RG, Lauricella DM (1998) Rapid development of severe hyperplastic gastritis with gastric epithelial dedifferentiation in *Helicobacter felis*-infected IL-10(-/-) mice. Am J Pathol 152(5):1377–1386

Bergmann S, Beard PM, Pasche B, Lienenklaus S, Weiss S, Gahan CG, Schughart K, Lengeling A (2013) Influence of internalin A murinisation on host resistance to orally acquired listeriosis in mice. BMC Microbiol 13:90

Berrington WR, Iyer R, Wells RD, Smith KD, Skerrett SJ, Hawn TR (2010) NOD1 and NOD2 regulation of pulmonary innate immunity to *Legionella pneumophila*. Eur J Immunol 40(12):3519–3527

Beura LK, Hamilton SE, Bi K, Schenkel JM, Odumade OA, Casey KA, Thompson EA, Fraser KA, Rosato PC, Filali-Mouhim A, Sekaly RP, Jenkins MK, Vezys V, Haining WN, Jameson SC, Masopust D (2016) Normalizing the environment recapitulates adult human immune traits in laboratory mice. Nature 532:512

Bhinder G, Stahl M, Sham HP, Crowley SM, Morampudi V, Dalwadi U, Ma C, Jacobson K, Vallance BA (2014) Intestinal epithelium-specific MyD88 signaling impacts host susceptibility to Infectious colitis by promoting protective goblet cell and antimicrobial responses. Infect Immun 82(9):3753–3763

Bina J, Zhu J, Dziejman M, Faruque S, Calderwood S, Mekalanos J (2003) ToxR regulon of *Vibrio cholerae* and its expression in vibrios shed by cholera patients. Proc Natl Acad Sci U S A 100(5):2801–2806

Bishop AL, Patimalla B, Camilli A (2014) *Vibrio cholerae*-induced inflammation in the neonatal mouse cholera model. Infect Immun 82(6):2434–2447

Bitar R, Tarpley J (1985) Intestinal perforation in typhoid fever: a historical and state-of-the-art review. Rev Infect Dis 7(2):257–271

Bone RCC, Fisher CJ, Clemmer TP, Slotman GJ, Metz CA, Balk RA (1989) Sepsis syndrome: a valid clinical entity. Crit Care Med 17(5):389–393

Boon RJ, Beale AS (1987) Response of *Streptococcus pyogenes* to therapy with amoxicillin or amoxicillin-clavulanic acid in a mouse model of mixed infection caused by *Staphylococcus aureus* and *Streptococcus pyogenes*. Antimicrob Agents Chemother 31(8):1204–1209

Bragonzi A, Worlitzsch D, Pier GB, Timpert P, Ulrich M, Hentzer M, Andersen JB, Givskov M, Conese M, Doring G (2005) Nonmucoid *Pseudomonas aeruginosa* expresses alginate in the lungs of patients with cystic fibrosis and in a mouse model. J Infect Dis 192(3):410–419

Brieland J, Freeman P, Kunkel R, Chrisp C, Hurley M, Fantone J, Engleberg C (1994) Replicative *Legionella pneumophila* lung infection in intratracheally inoculated A/J mice. A murine model of human Legionnaires' disease. Am J Pathol 145(6):1537–1546

Briles DE, Nahm M, Schroer K, Davie J, Baker P, Kearney J, Barletta R (1981) Antiphosphocholine antibodies found in normal mouse serum are protective against intravenous infection with type 3 Streptococcus pneumoniae. J Exp Med 153(3):694–705

Buckle GC, Walker CLF, Black RE (2012) Typhoid fever and paratyphoid fever: systematic review to estimate global morbidity and mortality for 2010. J Glob Health 2(1):010401

Calderon Toledo C, Arvidsson I, Karpman D (2011) Cross-reactive protection against enterohemorrhagic *Escherichia coli* infection by enteropathogenic *E. coli* in a mouse model. Infect Immun 79(6):2224–2233

Canvin JR, Marvin AP, Sivakumaran M, Paton JC, Boulnois GJ, Andrew PW, Mitchell TJ (1995) The role of pneumolysin and autolysin in the pathology of pneumonia and septicemia in mice infected with a type 2 pneumococcus. J Infect Dis 172(1):119–123

Chakraborty K, Ghosh S, Koley H, Mukhopadhyay AK, Ramamurthy T, Saha DR, Mukhopadhyay D, Roychowdhury S, Hamabata T, Takeda Y, Das S (2008) Bacterial exotoxins downregulate cathelicidin (hCAP-18/LL-37) and human beta-defensin 1 (HBD-1) expression in the intestinal epithelial cells. Cell Microbiol 10(12):2520–2537

Chang C, Miller JF (2006) *Campylobacter jejuni* colonization of mice with limited enteric flora. Infect Immun 74(9):5261–5271

Chen X, Katchar K, Goldsmith JD, Nanthakumar N, Cheknis A, Gerding DN, Kelly CP (2008) A mouse model of *Clostridium difficile*-associated disease. Gastroenterology 135(6):1984–1992

Chowdhury R, Mandal RS, Das S (2015) An AIL family protein promotes type three secretion system-1-independent invasion and pathogenesis of *Salmonella enterica* serovar Typhi. Cell Microbiol 17(4):486–503

Chudnovskiy A, Mortha A, Kana V, Kennard A, Ramirez JD, Rahman A, Remark R, Mogno I, Ng R, Gnjatic S, Amir ED, Solovyov A, Greenbaum B, Clemente J, Faith J, Belkaid Y, Grigg ME, Merad M (2016) Host-protozoan interactions protect from mucosal infections through activation of the inflammasome. Cell 167(2):444–456

Chukkapalli SS, Velsko IM, Rivera-Kweh MF, Zheng D, Lucas AR, Kesavalu L (2015) Polymicrobial oral infection with four periodontal bacteria orchestrates a distinct inflammatory response and atherosclerosis in ApoE null mice. PLoS One 10(11):e0143291

Chukkapalli SS, Easwaran M, Rivera-Kweh MF, Velsko IM, Ambadapadi S, Dai J, Larjava H, Lucas AR, Kesavalu L (2017) Sequential colonization of periodontal pathogens in induction of periodontal disease and atherosclerosis in LDLR null mice. Pathog Dis 75(1):ftx003

Coburn B, Li Y, Owen D, Vallance BA, Finlay BB (2005) *Salmonella enterica* serovar Typhimurium pathogenicity island 2 is necessary for complete virulence in a mouse model of Infectious enterocolitis. Infect Immun 73(6):3219–3227

Crawford RW, Rosales-Reyes R, de Ramirez-Aguilar ML, Chapa-Azuela O, Alpuche-Aranda C, Gunn JS (2010) Gallstones play a significant role in *Salmonella* spp. gallbladder colonization and carriage. Proc Natl Acad Sci U S A 107(9):4353–4358

Crump J, Luby SP, Mintz ED (2004) The global burden of typhoid fever. Bull World Health Organ 82(5):346–353

Czuprynski CJ, Faith NG, Steinberg H (2003) A/J mice are susceptible and C57BL/6 mice are resistant to *Listeria monocytogenes* infection by intragastric inoculation. Infect Immun 71(2):682–689

Dalton T, Dowd SE, Wolcott RD, Sun Y, Watters C, Griswold JA, Rumbaugh KP (2011) An in vivo polymicrobial biofilm wound infection model to study interspecies interactions. PLoS One 6(11):e27317

Das S, Chowdhury R, Ghosh S, Das S (2017) A recombinant protein of *Salmonella Typhi* induces humoral and cell-mediated immune responses including memory responses. Vaccine 35(35 Pt B):4523–4531

Day DW, Mandal BK, Morson BC (1978) The rectal biopsy appearances in *Salmonella colitis*. Histopathology 2(2):117–131

Deckert M, Soltek S, Geginat G, Lütjen S, Montesinos-Rongen M, Hof H, Schlüter D (2001) Endogenous interleukin-10 is required for prevention of a hyperinflammatory intracerebral immune response in *Listeria monocytogenes* meningoencephalitis. Infect Immun 69(7):4561–4571

Deng H, Yang S, Zhang Y, Qian K, Zhang Z, Liu Y, Wang Y, Bai Y, Fan H, Zhao X, Zhi F (2018) *Bacteroides fragilis* prevents *Clostridium difficile* infection in a mouse model by restoring gut barrier and microbiome regulation. Front Microbiol 9:2976

Desvignes L, Ernst JD (2009) Interferon-gamma-responsive nonhematopoietic cells regulate the immune response to *Mycobacterium tuberculosis*. Immunity 31(6):974–985

Dhall S, Do DC, Garcia M, Kim J, Mirebrahim SH, Lyubovitsky J, Lonardi S, Nothnagel EA, Schiller N, Martins-Green M (2014) Generating and reversing chronic wounds in diabetic mice by manipulating wound redox parameters. J Diabetes Res 2014:18

Dorer MS, Cohen IE, Sessler TH, Fero J, Salama NR (2013) Natural competence promotes Helicobacter pylori chronic infection. Infect Immun 81(1):209–215

Duan F, March JC (2010) Engineered bacterial communication prevents *Vibrio cholerae* virulence in an infant mouse model. Proc Natl Acad Sci U S A 107(25):11260–11264

Enocksson A, Lundberg J, Weitzberg E, Norrby-Teglund A, Svenungsson B (2004) Rectal nitric oxide gas and stool cytokine levels during the course of infectious gastroenteritis. Clin Diagn Lab Immunol 11(2):250–254

Fachi JL, Felipe JDS, Pral LP, da Silva BK, Correa RO, de Andrade MCP, da Fonseca DM, Basso PJ, Camara NOS, de Sales e Souza EL, dos Santos Martins F, Guima SES, Thomas AM, Setubal JC, Magalhães YT, Forti FL, Candreva T, Rodrigues HG, de Jesus MB, Consonni SR, Farias ADS, Varga-Weisz P, Vinolo MAR (2019) Butyrate protects mice from *Clostridium difficile*-Induced colitis through an HIF-1-dependent mechanism. Cell Rep 27(3):750–761

Fernandez MI, Thuizat A, Pedron T, Neutra M, Phalipon A, Sansonetti PJ (2003) A newborn mouse model for the study of intestinal pathogenesis of shigellosis. Cell Microbiol 5(7):481–491

Fila G, Kasimova K, Arenas Y, Nakonieczna J, Grinholc M, Bielawski MP, Lilge L (2016) Murine model imitating chronic wound infections for evaluation of antimicrobial photodynamic therapy efficacy. Front Microbiol 7:1258

Finch CE, Crimmins EM (2004) Inflammatory exposure and historical changes in human lifespans. Science 305(5691):1736–1739

Firoz Mian M, Pek EA, Chenoweth MJ, Ashkar AA (2011) Humanized mice are susceptible to *Salmonella typhi* infection. Cell Mol Immunol 8(1):83–87

Fonseca DM, Hand TW, Han SJ, Gerner MY, Glatman Zaretsky A, Byrd AL, Harrison OJ, Ortiz AM, Quinones M, Trinchieri G, Brenchley JM, Brodsky IE, Germain RN, Randolph GJ, Belkaid Y (2015) Microbiota-dependent sequelae of acute infection compromise tissue-specific immunity. Cell 163(2):354–366

Fullner KJ, Boucher JC, Hanes MA, Haines GK, Meehan BM, Walchle C, Sansonetti PJ, Mekalanos JJ (2002) The contribution of accessory toxins of *Vibrio cholerae* O1 El Tor to the proinflammatory response in a murine pulmonary cholera model. J Exp Med 195(11):1455–1462

Gentile LF, Nacionales DC, Lopez MC, Vanzant E, Cuenca A, Szpila BE, Cuenca AG, Joseph A, Moore FA, Leeuwenburgh C, Baker HV, Moldawer LL, Efron PA (2014) Host responses to sepsis vary in different low-lethality murine models. PLoS One 9(5):e94404

Ghosh S, Chakraborty, Nagaraja T, Basak S, Koley H, Dutta S, Mitra U, Das S (2011) An adhesion protein of *Salmonella enterica* serovar Typhi is required for pathogenesis and potential target for vaccine development. Proc Natl Acad Sci U S A 108(8):3348–3353

Giallourou N, Medlock GL, Bolick DT, Medeiros PH, Ledwaba SE, Kolling GL, Tung K, Guerry P, Swann JR, Guerrant RL (2018) A novel mouse model of *Campylobacter jejuni* enteropathy and diarrhea. PLoS Pathog 14(3):e1007083

Gibson DL, Ma C, Bergstrom KS, Huang JT, Man C, Vallance BA (2008) MyD88 signalling plays a critical role in host defence by controlling pathogen burden and promoting epithelial cell homeostasis during *Citrobacter rodentium*-induced colitis. Cell Microbiol 10(3):618–631

Gisby J, Bryant J (2000) Efficacy of a new cream formulation of mupirocin: comparison with oral and topical agents in experimental skin infections. Antimicrob Agents Chemother 44(2):255–260

Gomez JE, McKinney JD (2004) *M. tuberculosis* persistence, latency, and drug tolerance. Tuberculosis 84(1):29–44

Gonçalves NS, Ghaem-Maghami M, Monteleone G, Frankel G, Dougan G, Lewis DJM, Simmons CP, MacDonald TT (2001) Critical role for tumor necrosis factor alpha in controlling the number of lumenal pathogenic bacteria and immunopathology in Infectious colitis. Infect Immun 69(11):6651–6659

Gonzalez-Escobedo G, Marshall JM, Gunn JS (2011) Chronic and acute infection of the gall bladder by *Salmonella Typhi*: understanding the carrier state. Nat Rev Microbiol 9(1):9–14

Grandgirard D, Steiner O, Tauber MG, Leib SL (2007) An infant mouse model of brain damage in pneumococcal meningitis. Acta Neuropathol 114(6):609–617

Guilbault C, Saeed Z, Downey GP, Radzioch D (2007) Cystic fibrosis mouse models. Am J Respir Cell Mol Biol 36(1):1–7

Gunther NWT, Lockatell V, Johnson DE, Mobley HL (2001) In vivo dynamics of type 1 fimbria regulation in uropathogenic *Escherichia coli* during experimental urinary tract infection. Infect Immun 69(5):2838–2846

Hapfelmeier S, Ehrbar K, Stecher B, Barthel M, Kremer M, Hardt WD (2004) Role of the *Salmonella* pathogenicity island 1 effector proteins *SipA*, *SopB*, *SopE*, and *SopE2* in *Salmonella enterica* subspecies 1 serovar Typhimurium colitis in streptomycin-pretreated mice. Infect Immun 72(2):795–809

Heeckeren A, Walenga R, Konstan MW, Bonfield T, Davis PB, Ferkol T (1997) Excessive inflammatory response of cystic fibrosis mice to bronchopulmonary infection with *Pseudomonas aeruginosa*. J Clin Invest 100(11):2810–2815

Higgins LM, Frankel G, Douce G, Dougan G, MacDonald TT (1999) *Citrobacter rodentium* infection in mice elicits a mucosal Th1 cytokine response and lesions similar to those in murine inflammatory bowel disease. Infect Immun 67(6):3031–3039

Hill GB (1977) Therapeutic evaluation of minocycline and tetracycline for mixed anaerobic infection in mice. Antimicrob Agents Chemother 11(4):625–630

Hopkins WJ, Hall JA, Conway BP, Uehling DT (1995) Induction of urinary tract infection by intraurethral inoculation with *Escherichia coli*: refining the murine model. J Infect Dis 171(2):462–465

Iizawa Y, Kitamoto N, Hiroe K, Nakao M (1996) *Streptococcus pneumoniae* in the nasal cavity of mice causes lower respiratory tract infection after airway obstruction. J Med Microbiol 44(6):490–495

Jakobsen H, Bjarnarson S, Del Giudice G, Moreau M, Siegrist CA, Jonsdottir I (2002) Intranasal immunization with pneumococcal conjugate vaccines with LT-K63, a nontoxic mutant of heat-Labile enterotoxin, as adjuvant rapidly induces protective immunity against lethal pneumococcal infections in neonatal mice. Infect Immun 70(3):1443–1452

Johansson L, Rytkonen A, Bergman P, Albiger B, Kallstrom H, Hokfelt T, Agerberth B, Cattaneo R, Jonsson AB (2003) CD46 in meningococcal disease. Science 301(5631):373–375

Johansson L, Rytkonen A, Wan H, Bergman P, Plant L, Agerberth B, Hokfelt T, Jonsson AB (2005) Human-like immune responses in CD46 transgenic mice. J Immunol 175(1):433–440

Jones SA, Gibson T, Maltby RC, Chowdhury FZ, Stewart V, Cohen PS, Conway T (2011) Anaerobic respiration of *Escherichia coli* in the mouse intestine. Infect Immun 79(10):4218–4226

Kamdar K, Khakpour S, Chen J, Leone V, Brulc J, Mangatu T, Antonopoulos DA, Chang EB, Kahn SA, Kirschner BS, Young G, DePaolo RW (2016) Genetic and metabolic signals during acute enteric bacterial infection alter the microbiota and drive progression to chronic inflammatory disease. Cell Host Microbe 19(1):21–31

Karita M, Kouchiyama T, Okita K, Nakazawa T (1991) New small animal model for human gastric *Helicobacter pylori* infection: success in both nude and euthymic mice. Am J Gastroenterol 86(11):1596–1603

Karita M, Tsuda M, Okita K, Sugiyama T, Nakazawa T (1995) Evaluation of a new bismuth-free triple therapy in nude mice and humans. Eur J Gastroenterol Hepatol 1:S31–S34

Kim DH, Kim SW, Song YJ, Oh TY, Han SU, Kim YB, Joo HJ, Cho YK, Kim DY, Cho SW, Kim MW, Kim JH, Hahm KB (2003) Long-term evaluation of mice model infected with *Helicobacter pylori*: focus on gastric pathology including gastric cancer. Aliment Pharmacol Ther 18(Suppl 1):14–23

Klose KE (2000) The suckling mouse model of cholera. Trends Microbiol 8(4):189–191

Koedel U, Paul R, Winkler F, Kastenbauer S, Huang PL, Pfister HW (2001) Lack of endothelial nitric oxide synthase aggravates murine pneumococcal meningitis. J Neuropathol Exp Neurol 60(11):1041–1050

Koedel U, Scheld WM, Pfister HW (2002) Pathogenesis and pathophysiology of pneumococcal meningitis. Lancet Infect Dis 2(12):721–736

Kotloff KL, Winickoff JP, Ivanoff B, Clemens JD, Swerdlow DL, Sansonetti PJ, Adak GK, Levine MM (1999) Global burden of *Shigella* infections: implications for vaccine development and implementation of control strategies. Bull World Health Organ 77(8):651–666

Kraus MD, Amatya B, Kimula Y (1999) Histopathology of typhoid enteritis: morphologic and immunophenotypic findings. Mod Pathol 12(10):949–955

Kumar A, Haery C, Paladugu B, Kumar A, Symeoneides S, Taiberg L, Osman J, Trenholme G, Opal SM, Goldfarb R, Parrillo JE (2006) The duration of hypotension before the initiation of antibiotic treatment is a critical determinant of survival in a murine model of *Escherichia coli* septic shock: association with serum lactate and inflammatory cytokine levels. J Infect Dis 193(2):251–258

Kuzushit N, Rogers AB, Monti NA, Whary MT, Park MJ, Aswad BI, Shirin H, Koff A, Eguchi H, Moss SF (2005) p27kip1 deficiency confers susceptibility to gastric carcinogenesis in *Helicobacter pylori*-infected mice. Gastroenterology 129(5):1544–1556

Lane MC, Alteri CJ, Smith SN, Mobley HL (2007) Expression of flagella is coincident with uropathogenic *Escherichia coli* ascension to the upper urinary tract. Proc Natl Acad Sci U S A 104(42):16669–16674

Lebeis SL, Sherman MA, Kalman D (2008) Protective and destructive innate immune responses to enteropathogenic *Escherichia coli* and related A/E pathogens. Future Microbiol 3(3):315–328

Leib SL, Tauber MG (1999) Pathogenesis of bacterial meningitis. Infect Dis Clin N Am 13(3):527–548

Lopez-Medina E, Fan D, Coughlin LA, Ho EX, Lamont IL, Reimmann C, Hooper LV, Koh AY (2015) *Candida albicans* inhibits *Pseudomonas aeruginosa* virulence through suppression of pyochelin and pyoverdine biosynthesis. PLoS Pathog 11(8):e1005129

Maaser C, Housley MP, Iimura M, Smith JR, Vallance BA, Finlay BB, Schreiber JR, Varki NM, Kagnoff MF, Eckmann L (2004) Clearance of *Citrobacter rodentium* requires B cells but not secretory immunoglobulin A (IgA) or IgM antibodies. Infect Immun 72(6):3315–3324

Maestrone G, Sadek S, Kubacki E, Mitrovic M (1975) *Sphaerophorus necrophorus*: laboratory model for the evaluation of chemotherapeutic agents in mice. Cornell Vet 65(2):187–204

Marchetti M, Arico B, Burroni D, Figura N, Rappuoli R, Ghiara P (1995) Development of a mouse model of *Helicobacter pylori* infection that mimics human disease. Science 267(5204):1655–1658

Mastropaolo MD, Evans NP, Byrnes MK, Stevens AM, Robertson JL, Melville SB (2005) Synergy in polymicrobial infections in a mouse model of type 2 diabetes. Infect Immun 73(9):6055–6063

Mathan MM, Mathan VI (1991) Morphology of rectal mucosa of patients with shigellosis. Rev Infect Dis 13(Suppl 4):S314–S318

Mathur R, Oh H, Zhang D, Park SG, Seo J, Koblansky A, Hayden MS, Ghosh S (2012) A mouse model of *Salmonella typhi* infection. Cell 151(3):590–602

Maurer KJ, Ihrig MM, Rogers AB, Ng V, Bouchard G, Leonard MR, Carey MC, Fox JG (2005) Identification of cholelithogenic enterohepatic *Helicobacter* species and their role in murine cholesterol gallstone formation. Gastroenterology 128(4):1023–1033

McCardell BA, Madden JM, Stanfield JT (1986a) Effect of iron concentration on toxin production in *Campylobacter jejuni* and *Campylobacter coli*. Can J Microbiol 32(5):395–401

McCardell BA, Madden JM, Stanfield JT (1986b) A mouse model for the measurement of virulence of species of *Campylobacter*. J Infect Dis 153(1):177

McHugh DR, Cotton CU, Moss FJ, Vitko M, Valerio DM, Kelley TJ, Hao S, Jafri A, Drumm ML, Boron WF, Stern RC, McBennett K, Hodges CA (2018) Linaclotide improves gastrointestinal transit in cystic fibrosis mice by inhibiting sodium/hydrogen exchanger 3. Am J Physiol Gastrointest Liver Physiol 315(5):G868–G878

Mead PS, Slutsker L, Dietz V, McCaig LF, Bresee JS, Shapiro C, Griffin PM, Tauxe RV (1999) Food-related illness and death in the United States. Emerg Infect Dis 5(5):607–625

Medina E, North RJ (1998) Resistance ranking of some common inbred mouse strains to *Mycobacterium tuberculosis* and relationship to major histocompatibility complex haplotype and Nramp1 genotype. Immunology 93(2):270–274

Mohammadi M, Redline R, Nedrud J, Czinn S (1996) Role of the host in pathogenesis of *Helicobacter*-associated gastritis: *H. felis* infection of inbred and congenic mouse strains. Infect Immun 64(1):238–245

Monack DM, Mueller A, Falkow S (2004) Persistent bacterial infections: the interface of the pathogen and the host immune system. Nat Rev Microbiol 2(9):747–765

Mook-Kanamori B, Geldhoff M, Troost D, van der Poll T, van de Beek D (2012) Characterization of a pneumococcal meningitis mouse model. BMC Infect Dis 12(1):71

Morens DM, Taubenberger JK, Fauci AS (2008) Predominant role of bacterial pneumonia as a cause of death in pandemic influenza: implications for pandemic influenza preparedness. J Infect Dis 198(7):962–970

Nagaoka K, Yanagihara K, Harada Y, Yamada K, Migiyama Y, Morinaga Y, Izumikawa K, Kakeya H, Nakashima M, Nishimura M, Kohno S (2013) Establishment of a new murine model of liver abscess induced by *Fusobacterium necrophorum* injected into the caudal vein. J Med Microbiol 62(Pt 11):1755–1759

Neely AN, Law EJ, Holder IA (1986) Increased susceptibility to lethal *Candida* infections in burned mice preinfected with *Pseudomonas aeruginosa* or pretreated with proteolytic enzymes. Infect Immun 52(1):200–204

Newman JV, Zabel BA, Jha SS, Schauer DB (1999) *Citrobacter rodentium* espB is necessary for signal transduction and for infection of laboratory mice. Infect Immun 67(11):6019–6025

Nomura A, Stemmermann GN, Chyou PH, Perez-Perez GI, Blaser MJ (1994) *Helicobacter pylori* infection and the risk for duodenal and gastric ulceration. Ann Intern Med 120(12):977–981

Nygren E, Li BL, Holmgren J, Attridge SR (2009) Establishment of an adult mouse model for direct evaluation of the efficacy of vaccines against *Vibrio cholerae*. Infect Immun 77(8):3475–3484

O'Brien AD (1982) Innate resistance of mice to *Salmonella typhi* infection. Infect Immun 38(3):948–952

Peters BM, Noverr MC (2013) *Candida albicans-Staphylococcus aureus* polymicrobial peritonitis modulates host innate immunity. Infect Immun 81(6):2178–2189

Phalipon A, Sansonetti PJ (2007) Shigella's ways of manipulating the host intestinal innate and adaptive immune system: a tool box for survival? Immunol Cell Biol 85(2):119–129

Pineyro P, Zhou X, Orfe LH, Friel PJ, Lahmers K, Call DR (2010) Development of two animal models to study the function of *Vibrio parahaemolyticus* type III secretion systems. Infect Immun 78(11):4551–4559

Rabelo-Goncalves EM, Nishimura NF, Zeitune JM (2005) Development of a BALB/c mouse model of *Helicobacter pylori* infection with fresh and frozen bacteria. Biol Res 38(1):101–109

Reese TA, Bi K, Kambal A, Filali-Mouhim A, Beura LK, Burger MC, Pulendran B, Sekaly RP, Jameson SC, Masopust D, Haining WN, Virgin HW (2016) Sequential infection with common pathogens promotes human-like immune gene expression and altered vaccine response. Cell Host Microbe 19(5):713–719

Ritchie JM, Rui H, Bronson RT, Waldor MK (2010) Back to the future: studying cholera pathogenesis using infant rabbits. MBio 1(1):e00047

Rivera MF, Lee J-Y, Aneja M, Goswami V, Liu L, Velsko IM, Chukkapalli SS, Bhattacharyya I, Chen H, Lucas AR, Kesavalu LN (2013) Polymicrobial infection with major periodontal pathogens induced periodontal disease and aortic atherosclerosis in hyperlipidemic ApoE null mice. PLoS One 8(2):e57178

Rosen DA, Pinkner JS, Jones JM, Walker JN, Clegg S, Hultgren SJ (2008a) Utilization of an intracellular bacterial community pathway in *Klebsiella pneumoniae* urinary tract infection and the effects of FimK on type 1 pilus expression. Infect Immun 76(7):3337–3345

Rosen DA, Pinkner JS, Walker JN, Elam JS, Jones JM, Hultgren SJ (2008b) Molecular variations in *Klebsiella pneumoniae* and *Escherichia coli* FimH affect function and pathogenesis in the urinary tract. Infect Immun 76(7):3346–3356

Saito M, Iwawaki T, Taya C, Yonekawa H, Noda M, Inui Y, Mekada E, Kimata Y, Tsuru A, Kohno K (2001) Diphtheria toxin receptor-mediated conditional and targeted cell ablation in transgenic mice. Nat Biotechnol 19(8):746–750

Sakagami T, Dixon M, O'Rourke J, Howlett R, Alderuccio F, Vella J, Shimoyama T, Lee A (1996) Atrophic gastric changes in both *Helicobacter felis* and *Helicobacter pylori* infected mice are host dependent and separate from antral gastritis. Gut 39(5):639–648

Savkovic SD, Villanueva J, Turner JR, Matkowskyj KA, Hecht G (2005) Mouse model of enteropathogenic *Escherichia coli* infection. Infect Immun 73(2):1161–1170

Sawai N, Kita M, Kodama T, Tanahashi T, Yamaoka Y, Tagawa Y, Iwakura Y, Imanishi J (1999) Role of gamma interferon in *Helicobacter pylori*-induced gastric inflammatory responses in a mouse model. Infect Immun 67(1):279–285

Sawasvirojwong S, Srimanote P, Chatsudthipong V, Muanprasat C (2013) An adult mouse model of *Vibrio cholerae*-induced diarrhea for studying pathogenesis and potential therapy of cholera. PLoS Negl Trop Dis 7(6):e2293

Schaeffer AJ, Schwan WR, Hultgren SJ, Duncan JL (1987) Relationship of type 1 pilus expression in *Escherichia coli* to ascending urinary tract infections in mice. Infect Immun 55(2):373–380

Schierle CF, De la Garza M, Mustoe TA, Galiano RD (2009) Staphylococcal biofilms impair wound healing by delaying reepithelialization in a murine cutaneous wound model. Wound Repair Regen 17(3):354–359

Schluter D, Chahoud S, Lassmann H, Schumann A, Hof H, Deckert-Schluter M (1996) Intracerebral targets and immunomodulation of murine *Listeria monocytogenes* meningoencephalitis. J Neuropathol Exp Neurol 55(1):14–24

Semaniakou A, Croll RP, Chappe V (2018) Animal models in the pathophysiology of Cystic fibrosis. Front Pharmacol 9:1475

Sereny B (1957) Experimental keratoconjunctivitis shigellosis. Acta Microbiol Acad Sci Hung 4(4):367–376

Shah VS, Meyerholz DK, Tang XX, Reznikov L, Abou Alaiwa M, Ernst SE, Karp PH, Wohlford-Lenane CL, Heilmann KP, Leidinger MR, Allen PD, Zabner J, McCray PB Jr, Ostedgaard LS, Stoltz DA, Randak CO, Welsh MJ (2016) Airway acidification initiates host defense abnormalities in cystic fibrosis mice. Science 351(6272):503–507

Shapiro MA, Donovan KD, Gage JW (2000) Comparative efficacy of clinafloxacin in a pneumococcal meningitis mouse model. J Antimicrob Chemother 45(4):489–492

Sharma D, Yagnik B, Baksi R, Desai N, Padh H, Desai P (2016) Shigellosis murine model established by intraperitoneal and intranasal route of administration: a comparative comprehension overview. Microbes Infect 19(1):47–54

Siegrist CA (2001) Neonatal and early life vaccinology. Vaccine 19(25-26):3331–3346

Simmons CP, Clare S, Ghaem-Maghami M, Uren TK, Rankin J, Huett A, Goldin R, Lewis DJ, MacDonald TT, Strugnell RA, Frankel G, Dougan G (2003) Central role for B lymphocytes and CD4+ T cells in immunity to infection by the attaching and effacing pathogen *Citrobacter rodentium*. Infect Immun 71(9):5077–5086

Singer M, Sansonetti PJ (2004) IL-8 is a key chemokine regulating neutrophil recruitment in a new mouse model of Shigella-induced colitis. J Immunol 173(6):4197–4206

Song J, Willinger T, Rongvaux A, Eynon EE, Stevens S, Manz MG, Flavell RA, Galan JE (2010) A mouse model for the human pathogen *Salmonella typhi*. Cell Host Microbe 8(4):369–376

Spano S, Gao X, Hannemann S, Lara-Tejero M, Galan JE (2016) A bacterial pathogen targets a host Rab-family GTPase defense pathway with a GAP. Cell Host Microbe 19(2):216–226

Sprinz H, Gangarosa EJ, Williams M, Hornick RB, Woodward TE (1966) Histopathology of the upper small intestines in typhoid fever. Biopsy study of experimental disease in man. Am J Dig Dis 11(8):615–624

Stahl M, Ries J, Vermeulen J, Yang H, Sham HP, Crowley SM, Badayeva Y, Turvey SE, Gaynor EC, Li X, Vallance BA (2014) A novel mouse model of *Campylobacter jejuni* gastroenteritis reveals key pro-inflammatory and tissue protective roles for toll-like receptor signaling during infection. PLoS Pathog 10(7):e1004264

Stanfield JT, McCardell BA, Madden JM (1987) *Campylobacter* diarrhea in an adult mouse model. Microb Pathog 3(3):155–165

Stecher B, Hapfelmeier S, Muller C, Kremer M, Stallmach T, Hardt WD (2004) Flagella and chemotaxis are required for efficient induction of *Salmonella enterica* serovar Typhimurium colitis in streptomycin-pretreated mice. Infect Immun 72(7):4138–4150

Sun X, Wang H, Zhang Y, Chen K, Davis B, Feng H (2011) Mouse relapse model of *Clostridium difficile* infection. Infect Immun 79(7):2856–2864

Ta A, Thakur BK, Dutta P, Sinha R, Koley H, Das S (2017) Double-stranded RNA induces cathelicidin expression in the intestinal epithelial cells through phosphatidylinositol 3-kinase-protein kinase Czeta-Sp1 pathway and ameliorates shigellosis in mice. Cell Signal 35:140–153

Tan TQ, Smith CW, Hawkins EP, Mason EO Jr, Kaplan SL (1995) Hematogenous bacterial meningitis in an intercellular adhesion molecule-1-deficient infant mouse model. J Infect Dis 171(2):342–349

Thompson LJ, Danon SJ, Wilson JE, O'Rourke JL, Salama NR, Falkow S, Mitchell H, Lee A (2004) Chronic *Helicobacter pylori* infection with Sydney strain 1 and a newly identified mouse-adapted strain (Sydney strain 2000) in C57BL/6 and BALB/c mice. Infect Immun 72(8):4668–4679

Tu S, Bhagat G, Cui G, Takaishi S, Kurt-Jones EA, Rickman B, Betz KS, Penz-Oesterreicher M, Bjorkdahl O, Fox JG, Wang TC (2008) Overexpression of interleukin-1beta induces gastric inflammation and cancer and mobilizes myeloid-derived suppressor cells in mice. Cancer Cell 14(5):408–419

Velsko IM, Chukkapalli SS, Rivera-Kweh MF, Zheng D, Aukhil I, Lucas AR, Larjava H, Kesavalu L (2015) Periodontal pathogens invade gingiva and aortic adventitia and elicit inflammasome activation in αvβ6 integrin-deficient mice. Infect Immun 83(12):4582–4593

Virna S, Deckert M, Lutjen S, Soltek S, Foulds KE, Shen H, Korner H, Sedgwick JD, Schluter D (2006) TNF is important for pathogen control and limits brain damage in murine cerebral listeriosis. J Immunol 177(6):3972–3982

Walker RI, Caldwell MB, Lee EC, Guerry P, Trust TJ, Ruiz-Palacios GM (1986) Pathophysiology of *Campylobacter* enteritis. Microbiol Rev 50(1):81–94

Wang TC, Dangler CA, Chen D, Goldenring JR, Koh T, Raychowdhury R, Coffey RJ, Ito S, Varro A, Dockray GJ, Fox JG (2000) Synergistic interaction between hypergastrinemia and *Helicobacter* infection in a mouse model of gastric cancer. Gastroenterology 118(1):36–47

Wang E, Ouellet N, Simard M, Fillion I, Bergeron Y, Beauchamp D, Bergeron MG (2001) Pulmonary and systemic host response to *Streptococcus pneumoniae* and *Klebsiella pneumoniae* bacteremia in normal and immunosuppressed mice. Infect Immun 69(9):5294–5304

Ward JM, Anver MR, Haines DC, Melhorn JM, Gorelick P, Yan L, Fox JG (1996) Inflammatory large bowel disease in immunodeficient mice naturally infected with *Helicobacter hepaticus*. Lab Anim Sci 46(1):15–20

Wassenaar TM, Blaser MJ (1999) Pathophysiology of *Campylobacter jejuni* infections of humans. Microbes Infect 1(12):1023–1033

Watters C, DeLeon K, Trivedi U, Griswold JA, Lyte M, Hampel KJ, Wargo MJ, Rumbaugh KP (2013) *Pseudomonas aeruginosa* biofilms perturb wound resolution and antibiotic tolerance in diabetic mice. Med Microbiol Immunol 202(2):131–141

Whitney CG (2016) Changing epidemiology of pneumococcal disease in the era of conjugate vaccines. Curr Epidemiol Rep 3(2):125–135

Wilkins TD, Smith LD (1974) Chemotherapy of an experimental *Fusobacterium* (*Sphaerophorus*) *necrophorum* infection in mice. Antimicrob Agents Chemother 5(6):658–662

Wotherspoon AC, Ortiz-Hidalgo C, Falzon MR, Isaacson PG (1991) *Helicobacter pylori*-associated gastritis and primary B-cell gastric lymphoma. Lancet 338(8776):1175–1176

Yamakawa T, Mitsuyama J, Hayashi K (2002) *In vitro* and *in vivo* antibacterial activity of T-3912, a novel non-fluorinated topical quinolone. J Antimicrob Chemother 49(3):455–465

Yamamoto Y, Klein TW, Newton CA, Widen R, Friedman H (1988) Growth of *Legionella pneumophila* in thioglycolate-elicited peritoneal macrophages from A/J mice. Infect Immun 56(2):370–375

Yang JY, Lee SN, Chang SY, Ko HJ, Ryu S, Kweon MN (2013) A mouse model of shigellosis by intraperitoneal infection. J Infect Dis 209(2):203–215

Yen ML, Cairns S, Camilli A (2017) A cocktail of three virulent bacteriophages prevents *Vibrio cholerae* infection in animal models. Nat Commun 8:14187

Yoshida S, Goto Y, Mizuguchi Y, Nomoto K, Skamene E (1991) Genetic control of natural resistance in mouse macrophages regulating intracellular *Legionella pneumophila* multiplication *in vitro*. Infect Immun 59(1):428–432

Zhang S, Moss SF (2012) Rodent models of *Helicobacter* infection, inflammation, and disease. Methods Mol Biol 921:89–98

Zhang Z, Jin L, Champion G, Seydel KB, Stanley SL Jr (2001) *Shigella* infection in a SCID mouse-human intestinal xenograft model: role for neutrophils in containing bacterial dissemination in human intestine. Infect Immun 69(5):3240–3247

Zhao G, Hochwalt PC, Usui ML, Underwood RA, Singh PK, James GA, Stewart PS, Fleckman P, Olerud JE (2010) Delayed wound healing in diabetic (db/db) mice with *Pseudomonas aeruginosa* biofilm challenge: a model for the study of chronic wounds. Wound Repair Regen 18(5):467–477

Zwijnenburg PJ, van der Poll T, Florquin S, van Deventer SJ, Roord JJ, van Furth AM (2001) Experimental pneumococcal meningitis in mice: a model of intranasal infection. J Infect Dis 183(7):1143–1146

Nonhuman Primate Model Use in Understanding Infectious Diseases

25

Poulomi Ghosh and Saprativ P. Das

Contents

Abstract

Due to high similarity in the immune system, humans have a closely related phylogenic relationship with nonhuman primates. The large family of nonhuman primate permits suitable assortment for diverse microbial infections. The primates, excluding humans have naturally related oral structure as human and can easily be implemented in dental studies. Nonhuman primate models are used in understanding pathogenicity involved in infectious diseases due to their short life cycle, rapid turnover in generation and genetic variation. The models of nonhuman primates are also applied in regulating disease pathogenesis and protection

P. Ghosh · S. P. Das (✉)
Department of Biotechnology, Brainware University, Kolkata, West Bengal, India
e-mail: spd.bt@brainwareuniversity.ac.in

B. Siddhardha et al. (eds.), *Model Organisms for Microbial Pathogenesis, Biofilm Formation and Antimicrobial Drug Discovery*,
https://doi.org/10.1007/978-981-15-1695-5_25

mechanism significance. Nonhuman primates play a pivotal role in producing vaccines for infectious disease and prevent the disease based on host–pathogen interactions. They help to understand viral pathogenesis and therapeutic safety of antiviral drugs. They are also actively engaged in the "3Rs" of animal research viz., replacement, reduction, and refinement. The nonhuman primates viz., Macaques due to their enhanced metabolic activity are mostly used to develop vaccines against infections like-HIV/AIDS, group-A *Streptococcus*, tuberculosis and influenza. The notable attributes of utility and the host pathogenicity interaction helps in the development of nonhuman primate models for further development for a great research area in biomedical research.

Keywords
Phylogenic relationship · Pathogenicity · Macaque · Replacement · Reduction · Refinement

25.1 Introduction

Owing to higher similarity in immune system, humans have a closely related phylogenic relationship with nonhuman primates. The large family of nonhuman primate permits suitable assortment for diverse microbial infections. The primates, excluding humans have naturally related oral structure as human and can easily be implemented in dental studies. Nonhuman primate models are used in understanding pathogenicity involved in infectious diseases due to their short life cycle, rapid turnover in generation, and genetic variation. Nonhuman primate models are also applied to determine the importance of disease pathogenesis and protection mechanism. Nonhuman primates play a pivotal role in producing vaccines for infectious disease and prevent the disease based on host–pathogen interactions (Boussaoud et al. 2005). They help in understanding viral pathogenesis and antiviral therapeutic safety. They are also actively engaged in the 3Rs of animal research viz., replacement, reduction, and refinement. The nonhuman primates, macaques; due to their enhanced metabolic activity are mostly used to develop vaccines against infectious diseases like AIDS, group-A *Streptococcus*, tuberculosis, and influenza (Ostrowski 1998).

25.2 Phylogenetic Tree

Microbial taxa within bacteria and archaea form discrete genealogical clustered groups that can be illustrated in phylogenetic trees (Ciccarelli et al. 2006). Evolutionary trees exhibit conditional phylogenetic connections of numerous diverging pedigrees linked by nodes. The organism whose nucleotide sequences have been analyzed is identified at the end of every division. Each node represents a separate occurrence, and the branches' lengths symbolize the amount of molecular fluctuations that have occurred between the two nodes (Hodge and Cope 2000). Often sequences are obtained from well classified microbes that have been grown in pure culture; however, this is not always the case. The sequences of small subunit ribosomal ribonucleic acid (SSU rRNA) have become predominantly important in both identifying microbes in nature and constructing phylogenetic trees to describe their evolutionary relationships (Podani 2019).

There are five stages in constructing a phylogenetic tree. The alignment of the nucleotide or amino acid sequence is the first important stage. Although manual inspection of alignment is important, it is usually done using an online resource (Ciccarelli et al. 2006). In addition to SSU rRNA gene sequences, protein-coding genes may be analyzed in which alignment of amino acids is performed. This is because the genetic code is degenerated (Mayr 2009). The sequence of amino acid may be conserved irrespective of the nucleotide sequence. Next, the orientation must be scrutinized for a phylogenetic sign to determine if it is appropriate in continuing with tree building. There are two extremes in this regard; the perfect alignment of the sequences at one end of the spectrum, with the absence of any matches whatsoever at the other (Mayr 2009). Phylogenetic analysis can only be performed on those sequences that fall in the middle, with a combination of arbitrary and coordinated points (Santamaria and Theron 2009). The third step is the toughest that involves the choosing of the appropriate usage of the tree building technique. The last two steps involve the computer-assisted applications, finally followed by physical inspection of the resultant tree to make sure it makes sense (Santamaria and Theron 2009).

The different approaches to building a phylogenetic tree can be divided into two broad categories, a distance-based (phenetic) approach and a character-based (cladistic) approach. Distance-based approaches are the most intuitive (Penny et al. 1992). Here the alterations amid the associated sequences are reckoned for each pair with subsequent summarization into an only sequence, which is roughly the percent difference between the two sequences (Woese 2002). A tree is then generated by serially linking pairs that are ever more distantly related (i.e., start with those with the least number of sequence differences and move to those with the most) (Townsend et al. 2012) This is called cluster analysis and should be carefully applied as it has the unattractive capability of generating trees even in the absence of evolutionary relationships. Neighbor joining is another distances-based method that uses a slightly different matrix that attempts to avoid this problem by modifying the distance between each pair of nodes based on the average divergence from all other nodes (Woese 2002).

Character-based methods for phylogenetic tree building are more complicated but generate more robust trees. These approaches commence with three assumptions viz., the pathway of evolution, inferring the ancestors at the each node and choosing the best tree according to a specific model of evolutionary change (Woese 2002). These methods include maximum closeness, which assumes that the least percentage of fluctuations happened amongst ancestors with existing (living) organisms. Another approach is called maximum likelihood (Arenas and Posada 2010).

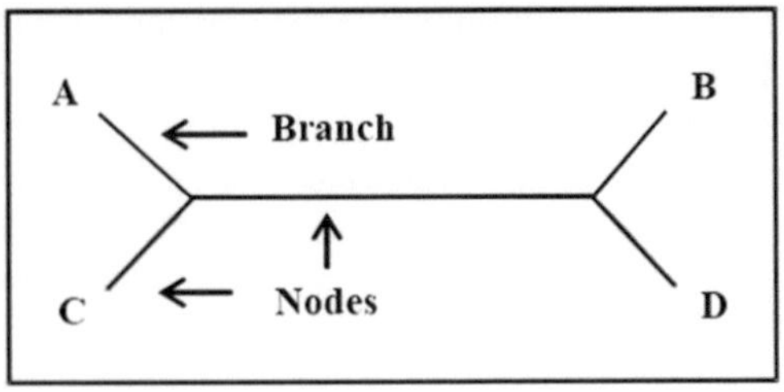

Fig. 25.1 Unrooted tree joining four taxonomic units

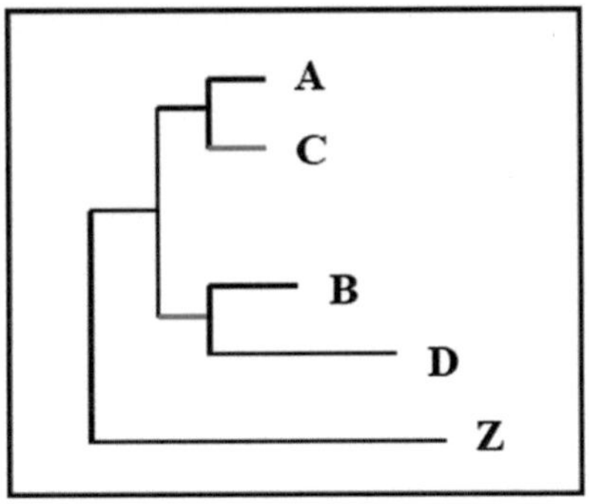

Fig. 25.2 Unrooted tree can be rooted by adding an outgroup, represented by "*Z*"

This necessitates a big data set as evaluation of probability (i.e., the likelihood) based on certain evolutionary and molecular information needed to build each possible tree. Importantly, a tree can be rooted or unrooted. An unrooted tree merely signifies phylogenetic interactions yet does not offer an evolutionary track (Fig. 25.1). An unrooted tree can be rooted by adding data from an outgroup—a species known to be very distantly related to all the species in the tree (Fig. 25.2). The root is determined by the point of the tree where the outgroup joins, providing a point of reference to identify the oldest node on the tree, which is the node closest to the group (Arenas and Posada 2010).

25.3 Pathogenicity

25.3.1 Pathogenicity and Infectious Disease

Relationships between two organisms can be very complex. A large living entity that maintenances the growth and survival of a smaller living body are called the host. A pathogen is an organism that causes disease and the pathogen's ability to cause infection is known as pathogenicity. Microbiota associated with the gut become pathogens when they are present in a location within the host (Casadevall and Pirofski 2014). Infectious disease results from the antagonism with microbial agents such as viruses, bacteria, fungi, and protozoa (Bernstein et al. 2018). Many microbiologists define the infectious disease process as the host–pathogen relationship. The host is known to have an infection while a pathogen is developing and proliferating within or on the host surface (Nathan 2015). The infection can fluctuate extensively with reference to location, rigorousness, and the organism involved. A contagious disease is described as any health change in which a part or whole of the host body is unable to carry its ordinary function owing to the existence of a pathogen or its yields, toxins.

The fundamental process of infections is essentially a competition for resources. The host represents an opportunity for the infective agent to obtain protection, nutrients, and energy to use for its own survival. Infectious agents must thereby develop mechanisms to access and exploit their hosts. Furthermore, to continue survival, pathogens must also conceive approaches to pass on to an improved environment once the immediate one declines in value. There must be occurrence of the pathogen's distribution to another point in one host or into another host (Bernstein et al. 2018). The communicable disease chain or the infection cycle characterizes these happenings in the arrangement of an interesting secret, where understanding the disease process in only revealed when all of the links of the chain are known.

A prior contact is necessary for a microorganism to cause disease in human beings. Pathogens come from a source that represents the site from which the pathogen is instantaneously transferred to the host either directly from the environment or incidentally through a transitional mediator. The source can be humans, animals or water, food, and soil (Nathan 2015). The period during which the source is communicable or is disseminating the pathogen defines the infectivity period. If the infection source can be eradicated or controlled, the infectious disease cycle itself will be intermittent and the pathogen transmission will be prohibited.

The natural environmental site in which the pathogens normally reside defines a reservoir. It is also the place from where the pathogen is attained by the source or where uninterrupted host infection can occur, implying a reservoir occasionally plays the role of a source. Reservoirs can be living or nonliving. The increasing impingements of humans on the environment and increased exposure to antibiotics and mutagens have played significant roles in the substantive changes in reservoirs over the last 100 years (Casadevall and Pirofski 2014).

Zoonosis is the transmission of infectious disease from animals to humans viz., transfer of rabies infection from rabid dog to humans. In addition, being bitten by arthropod vectors such as masavuitoes, ticks, mites, or biting flies, can lead to infections like equine encephalomyelitis, Lyme disease, fever, and plague. A microorganism's contact along with its survival inside the host is a prerequisite for a microbe to cause disease. To survive, the microorganism needs a suitable environment, proper source of nutrients, and protection from harmful elements. From the microorganism's point of view, it is simply expressing genetic information that allows it to survive. But from the host's point of view, the relationship in one that results in disease (Casadevall and Pirofski 2014).

All microorganisms have specific physical requirements for optimal growth and reproduction. Its natural environment, as well as its new host environment, must accommodate its genetic predisposition for temperature, pH, moisture, oxygen concentration, etc. Despite the wide diversity in metabolic strategies displayed by microorganisms, all microbial pathogens are predominantly chemoheterotrophic.

Furthermore, some are mesophiles, some are acidophiles and others are halophiles, depending on where they reside. In the absence of nutrients, they can "persist" with minimal metabolism, relocate, or die. When a microorganism is introduced to the human host as its new environment, it will adapt to meet its metabolic and physiological needs. In its need for nutrients, the microorganism competes with eukaryotic host cells. The successful pathogen will compete with the host cells and evade host immune defenses through the use of virulence factors (Bien et al. 2011; Casadevall and Pirofski 2014).

Survival of the microorganism in the host also depends primarily on its abilities to "find shelter," avoiding innate resistance factors and evading detection along with removal of immune response (Fig. 25.3). Some microorganisms survive and replicate inside host cells; in some cases, they reside in the cells that are supposed to destroy them (Fig. 25.3). Other organisms attach to and between host cells, make capsules and secrete exopolysaccharides from communal shelters called biofilms (Casadevall and Pirofski 2014). Some microorganisms produce enzymes that inactivate resistance factors. Still other organisms have evolved specialized protein secretion systems to selectively kill host cells. These physical and chemical characteristics reflect microbial virulence, facilitating microbial survival.

Many microorganisms use a strategy that poisons other cells, the usage of toxins. A toxin (poison) is an element that modifies the regular metabolism of host cells with harmful effects on the host. The entry of toxins inside the host blood causes toxemia. Toxigenicity is the pathogens' capability to yield toxins. Intoxications are ailments resulting from a definite toxin viz., botulin toxin, tetanus toxin produced by the pathogen. Intoxication disease does not require the presence of the actively growing pathogen (Casadevall and Pirofski 2014). Infectious diseases are typically not detected until distinct physical or physiological events alert the host of cellular battle. Infectious diseases regularly have characteristics signs and symptoms. The signs are objective changes that can be directly observed in the body such as fever or rash. On the contrary, symptoms are subjective modifications that a patient experiences viz., pain and loss of appetite. The term symptom is regularly castoff to

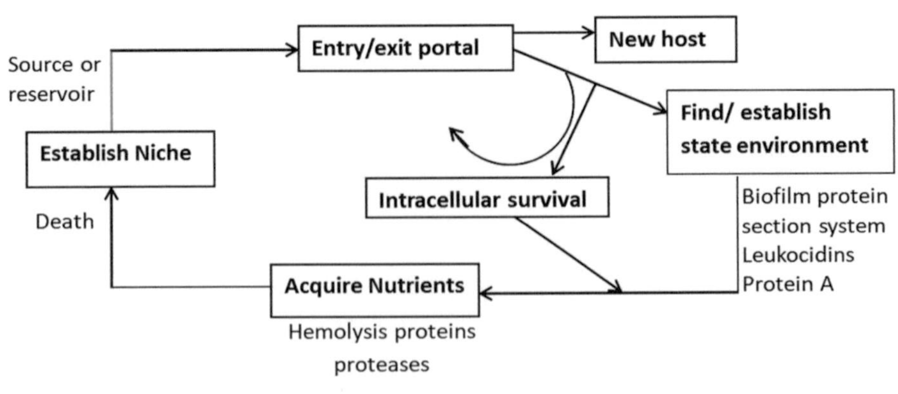

Fig. 25.3 The "infection" process

comprise the clinical sign. A disease condition is a combination of signs and symptoms that are specific for the disease. Frequent laboratory investigations are essential for analysis (Alberts et al. 2002).

Clinically, an infectious disease course usually has a typical outline and can be divided into numerous phases (Fig. 25.4). The incubation stage is the time between entry of the pathogen and the signs and symptoms' progress characterized by the spreading of the pathogen not to that appropriate level to source clinical manifestation. The period length varies with the disease. Next, the prodromal stage arises with the commencement of symptoms and signs that are not yet particularly sufficient to make a diagnosis. However, the patient often is transmittable. This is trailed by the illness period when the disease is utmost severe and shows typical signs and symptoms. The host immune response is typically activated at this stage. Finally, the signs and symptoms begin to go away during the decline period. The recovery stage is often stated as convalescence (Alberts et al. 2002).

25.3.2 Virulence and Virulence Factors

Pathogenicity is a general term that states to the prospect of an organism to cause disease. Virulence refers more specifically to the magnitude of pathogenicity caused by a particular microorganism. The extent to which the pathogen causes damage, including invasiveness and infectivity, defines its virulence. Thus, any structural or soluble product that increases pathogenicity is a virulence factor. For bacteria, protozoa, and fungi, virulence is directly interrelated with a pathogen's capability to persist well outside its host without the usage of a vector; it depends on survival of the host and will incline to be less virulent (Webb and Kahler 2008). When a cellular pathogen can endure for long phases outside its host, it can afford to vacate the host and wait for a new one to arise besides (Bien et al. 2011). This appears to stimulate improved virulence. The health of the host is not perilous, but broad proliferation within the host will upsurge the effectiveness of transmission viz., in case of tuberculosis and diphtheria, the microbes *Mycobacterium tuberculosis* and *Corynebacterium diphtheriae*, respectively, stay alive for a relatively long time from hours to days and the aspects that enable persistence are determined in

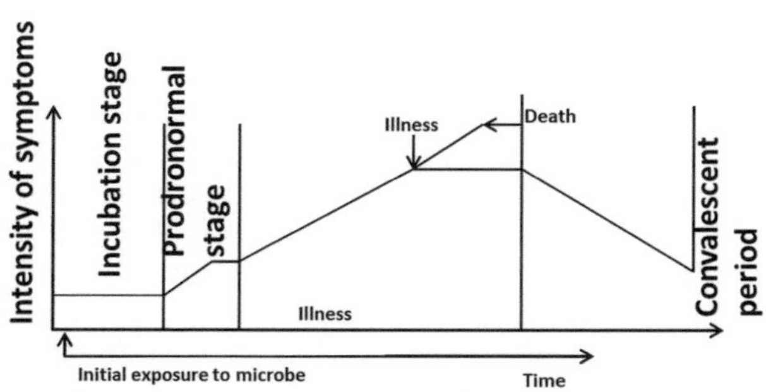

Fig. 25.4 Characteristic pattern and phases of an infectious disease

exclusive sequences of DNA that are voluntarily changed between bacteria through conjugation, transduction, or transformation (horizontal gene transfer).

Very large segments (10–200 kilobases) of bacterial chromosomal and plasmid DNA have been found to encode virulence factors. These DNA segments are called pathogenicity islands, as they appear to have been inserted into the existing DNA. Many bacteria (e.g., *Yersinia* sp., *Pseudomonas aeruginosa*, *Shigella flexneri*, and enteropathogenic *Escherichia coli*) carry at least one pathogenicity island (Alan 2008). Pathogenicity islands commonly augment bacterial virulence and are absent in nonpathogenic associates of the similar genus or species viz., the *Staphylococcus aureus* pathogenicity island encodes several superantigen genes, including the gene for the toxin that causes toxic shock syndrome. Pathogenicity islands can be documented by numerous common sequence features:

(a) The 3′ and 5′ ends of the islands comprise insertion like elements, suggesting their promiscuity as genetic elements. In fact, the horizontal transfer of pathogenicity islands is now well documented.
(b) The G + C nucleotide contains islands of pathogenicity differing considerably from that of the residual bacterial genome.
(c) Pathogenicity island DNA similarly displays numerous open reading frames, signifying further putative genes. Interestingly, pathogenicity islands are typically inserted into the host genome near tRNA genes. An exceptional instance of virulence genes passed on pathogenicity islands is seen in protein secretion systems (Casadevall and Pirofski 2014; Thrall and Burdon 2003). Virulence factors enable the microbes to carry out specific roles whose outcome on the host can cause damage to the tissue or the host cell. Virulence factors are upregulated by specific host triggers, enabling the microbe to persist well its place in the host (Deborah et al. 2005).

25.3.2.1 Adherence and Colonization Factors

The leading step in the infectious disease process is the entry along with attachment of the microbe to a vulnerable host. Entry into the host may be through respiratory system, skin, gastrointestinal system, urogenital system, or the eye conjunctiva. Some pathogens enter into the host by blood transfusions, sexual contact, and organ transplants, or by insect vectors. Bacteria, fungi, and protozoa require a portal of entry into the host to consume nutrients, viruses go into the host's cells to replicate and some bacteria enter the host cell but mostly endure in the tissue spaces nearby the cell. So, as for the pathogen, effective escape from the host is just as significant as its preliminary entry. Unless a successful escape occurs, the disease cycle will be interrupted and the pathogen will not be able to perpetuate (Deborah et al. 2005).

Subsequently being transferred to an appropriate host, the pathogen must be able to adhere to and inhabit host cells or tissues. In this perspective, colonization means the formation of a microbial replication site on or within a host. It does not affect in tissue damage or invasion. Colonization depends on the pathogen's ability to survive in the new host environment and to compete effectively with the host cells and host's normal microbiota for vital nutrients. Specialized structures permit microorganisms to contend for surface attachment sites, which is also essential for colonization (Casadevall and Pirofski 2014; Thrall and Burdon 2003). Pathogens adhere with a great specificity to a specific tissue. Adherence structures such as fimbriae, pili, capsule materials, and focused adhesion molecules on the occupying cell surface help bind to the host site. The microbial products and structural constituents add to infectivity, thus they are one type of virulence factor.

25.3.2.2 Invasion Factors

Pathogens can be described in terms of their infectivity and invasiveness. Infectivity defines the microbe's ability to establish a discrete focal point of infection whereas; invasiveness is the capability of the microbe to extent to adjoining or other tissues. For certain pathogens, a localized infection is adequate to cause a disease. However, most pathogens attack other tissues. Some bacterial and fungal pathogens enter into tissue for existence as well as for multiplication (Casadevall and Pirofski 2014; Thrall and Burdon 2003).

Pathogens can either vigorously or submissively enter the hosts' mucous membranes and epithelium. Dynamic permeation may be achieved through fabrication of lytic substances that alter the host tissue by attacking the extracellular matrix and intestinal lining; degrading carbohydrate–protein complex between cells or on the cell surface and disrupting the host cell surface (Webb and Kahler 2008). Passive mechanisms of penetration are not related to the pathogen itself like minor breaks, lesions, or ulcers in a mucous membrane that permit initial entry; burns or wounds abrasions' on the skin membrane; arthropod vectors that generate small wounds during feeding and tissue damage produced by other organisms (Alan 2008).

When below the mucous membrane, a pathogen may enter into the inner tissue and endure circulation throughout the host body by specific structures stimulating its distribution, thus, representing one type of virulence factor. Bacteria may also go into the lesser terminal lymphatic capillaries surrounding the epithelial cells. These capillaries unite and form large lymphatic vessels that ultimately channel into circulatory system. The bacteria upon reaching the circulatory system have access to all organs and systems of the host. The occurrence of living bacteria in the bloodstream is known as bacteremia or fungal toxins in the blood is termed septicemia (Deborah et al. 2005).

Invasiveness varies among pathogen. For example, *Clostridium tetani*, causing tetanus produces a variety of virulence factors (toxins and proteolytic enzymes) but is considered noninvasive because it does not spread from one tissue to another. *Bacillus anthracis* causing anthrax and *Yersinia pestis* inducing plague produce highly invasive virulence factors.

25.3.2.3 Exotoxins

The bacterial pathogen on propagating release soluble, heat labile proteins viz., exotoxins that usually get deactivated at 60–80 °C. Most of the exotoxins are formed by Gram-positive bacteria, although some Gram-negative bacteria too make exotoxins. Often exotoxins move from the infection site to other body tissues or target cells, exhibiting their effects (Deborah et al. 2005). Exotoxins are usually produced by specific bacteria having plasmids or prophage bearing toxin genes. Exotoxins are associated with the disease and named for the disease they generally produce (e.g., diphtheria toxin). Exotoxins exert their biological activity by specific mechanism and are grouped by their proteins' structure or their mechanism of action (Alberts et al. 2002). One kind is the AB toxin, which gets its name from the fact that it has two different toxin components, an "A" and a "B" both can be a sole polypeptide or different polypeptides. The B portion of the toxin fixes to a host cell receptor and is distinct from A portion, which move into the cell and has enzyme activity causing the toxicity. Thus, the B component regulates the type of cell and the effect caused by the toxin, whereas the A component employ the damaging effect. AB toxins act on cells by different mechanisms (Webb and Kahler 2008).

Exotoxin is the channel (pore) making toxins. They disrupt membrane veracity thereby, making the cell lyse and die. As like proteins, toxins are simply documented by the host immune system, which yields antitoxin antibodies that removes their respective toxin by identifying, binding, and inactivating them (Bien et al. 2011). Finally, exotoxins called superantigens act by stimulating as many as 30% of host T cells to overexpress and discharge enormous quantities of cytokines from further host immune cells in the absence of a specific antigen. The excessive concentration of cytokines causes multiple host organs to fail, providing pathogen the dissemination time (Webb and Kahler 2008).

25.3.2.4 Endotoxins

Gram-negative bacteria contain lipopolysaccharide (LPS) in the outer membrane that can be toxic to hosts. This LPS is known an endotoxin as it is bound to the bacterium and is unrestricted when the microorganism lyse. Some endotoxins are free during bacterial proliferation. The toxic component of the LPS in the lipid protein is called lipid A. It is not a single macromolecular assembly; rather it is an intricate collection of lipid residues. Lipid A is heat stable and toxic in monogram amounts but only weakly immunogenic (Bien et al. 2011).

Unlike the structural and functional diversity of exotoxins, the lipid A of various Gram-negative bacteria produces similar systemic effects regardless of the microbe from which it is derived. These include fever (endotoxin is pathogenic), shock, blood coagulation, weakness, diarrhea, inflammation, and fibrinolysis. The key biological consequence of lipid A is an indirect one, facilitated by host molecule and systems rather than by lipid A itself. For example, endotoxins can primarily activate a protein called the Hageman factor (blood clotting factor 11), which in turn stimulates and over excites up to four humoral systems, coagulation, complement, fibrinolytic, and kininogen systems, the result is unregulated blood clotting within capillaries (distributed intravascular coagulation) and multiorgan failure (Bien et al. 2011). Endotoxins also incidentally persuade a fever in the host by producing macrophages to discharge endogenous pyrogen that reset the hypothalamic thermostat. One significant endogenous pyrogen is the cytokine interleukin-1 (IL-1). Other cytokines freed by macrophages also produce fever. The net effect is often called septic shock and can also be prompted by certain Gram-positive bacteria and pathogenic fungi. All drugs, especially those given intravenously or intramuscularly, must be free of endotoxin for this reason (Alberts et al. 2002).

25.3.2.5 Mycotoxins

Mycotoxins are toxins produced by fungi viz., *Aspergillus flavus* and *A. parasiticus*, produce aflatoxins and *A. stachybotrys* produce satratoxins. These fungi generally pollute food crops and water-damaged buildings, respectively. An estimated 4.5 billion people in developing countries may be exposed chronically to aflatoxins through their diet. Exposure to aflatoxins is known to cause both chronic and acute liver disease and liver cancer (Alberts et al. 2002). Aflatoxins are extremely carcinogenic, mutagenic, and immunosuppressive. Approximately 18 different types of aflatoxins exist. Aflatoxins are categorized into two wide groups rendering to their chemical structure. The Stachybotrys trichothecene mycotoxins are potent inhibitors of DNA, RNA, and protein synthesis. They induce inflammation, disrupt surfactant phospholipids in the lungs leading to pathological changes in tissue. The fungus, *Claviceps purpurea*, also produce toxic substances. The products are generically referred to as ergots reflecting the name of the tuber-like structure of the fungi. The ergot is a fungal resting stage and is composed of a compact mass of hyphae (Alberts et al. 2002). The ergots from various *Claviceps* sp. produce alkaloids that have varying physiological effects on humans.

25.4 Macaque Classification

Under the kingdom animalia and class mammalia, the macaques are the oldest monkeys in the world. They belong to the genus *Macaca* and subfamily *Cercopithecinae* (Ostrowski 1998; Boussaoud et al. 2005; Groves 2005).

The 20 species of macaque are existing all over Asia, North Africa, and Gibraltar. They basically take seeds, flowers, leaves, tree bark, and certain macaque take crabs as their food (Zhen 2018). Their taxonomic analysis and different species are represented in Tables 25.1 and 25.2, respectively. In general, macaques are widely used in biomedical research (Groves 2005). In 1950, crab-eating macaques were used in studies for the development of polio vaccine. Through embryonic cell nuclear transfer, it became the first cloned primate and plays an important role in cloning (Ostrowski 1998; Groves 2005; Boussaoud et al. 2005).

Table 25.1 Taxonomic analysis of Macaque monkey

Suborder	*Haplorrhini*
Intraorder	Simiiformer
Superfamily	Cercopithecoidea
Family	Cercopithecidae
Subfamily	Cecopithecine
Genus	*Macaca*

Table 25.2 Different species of the Macaque monkey

1	Rhesus Macaque	*Macaca mulatta*
2	Japanese Macaque	*Macaca fuscata*
3	Crab eating Macaque	*Macaca fasicularis*
4	Barbary Macaque	*Macaca sylvanus*
5	Lion coiled Macaque	*Macaca silenus*
6	Southern pig Tailed	*Macaca nemestrina*
7	Celebs crested Macaque	*Macaca nigra*
8	Formosan rock Macaque	*Macaca cyclopis*
9	Stup tailed Macaque	*Macaca arctoides*
10	Assam Macaque	*Macaca assamensis*
11	Tibetian Macaque	*Macaca thibetana*
12	Tonkean Macaque	*Macaca tonkeana*
13	Northern pig tailed Macaque	*Macaca leonin*
14	Gorontalo Macaque	*Macaca nigrescens*
15	Siberut Macaque	*Macaca siberu*
16	Crested black Macaque	*Macaca nigra*
17	Toque Macaque	*Macaca sinica*
18	Moor Macaque	*Macaca maura*
19	Bonnet Macaque	*Macaca radiata*
20	Bokkoi	*Macaca pagensis*

25.5 Prospect of Using Nonhuman Primate Model in Pathogenesis

Infectious diseases are basically caused by virus, bacteria, and parasites that are mainly communicable. Analyzing data shows that 14.9 million deaths are responsible for the infectious disease (Ostrowski 1998). The five single agent killers in infectious disease which are responsible are HIV/AIDS, tuberculosis, malaria, leishmania, and dengue (Basaraba 2008). Basically nonhuman proteins have a long out brakes to understand due to their higher immune system comprising natural killer cell receptor, Toll-like receptor (TLR), carbohydrate requisite lectins, and adhesion between rodent and primates which is also used to understand the immunogenicity of virus-based vaccines. Primates are used in various purposes to understand the host–pathogen interactions, infection, and chronic disease as well as in science and medical research. By using the nonhuman primates, we can prevent or use the critical analysis and their effective vaccines to control the infection.

25.5.1 HIV/AIDS

Acquired Immunodeficiency Syndrome (AIDS) was first recognized in 1981 and discovered as an infection in humans in the year of 1983. The HIV virus has its own generic code. The HIV virus directly attack the host cell and by replication it engulfs the RNA genetic code and make their own viral DNA and can continuously copy itself, over and over when the human cell normally divides. It is highly infectious disease due to the number of infected people, normally above 3 million per year. HIV is primarily transmitted through sexual conduction as well as contracted by perinatal transmission from mother to child. It can be also spread by infected blood usually through contaminated syringes and blood transfusion.

Basically, AIDS is characterized by a constant exhaustion of the immune system of the body, decrease in the level of specific subsets of blood lymphocytes, CD4 cells. The intricate pathophysiology of HIV's effects on the immune system has been studied by numerous experts. HIV is mainly caused by two viruses:

a. HIV-1—The type 1 is more virulent transferred to human from the chimpanzee subspecies pan troglodytes.
b. HIV-2—It is geographically constrained to West Africa. It is thoroughly connected in sequence to SIV mac from macaques.

Though most of the nonhuman primates have several limitations as model system for the study of HIV/AIDS, the only susceptible ones in the understanding of HIV-1 are the Pigtail macaques and Chimpanzees. The nonhuman primate is applied to understanding infectious HIV because of their high immune system and when the primates are infected with HIV-1, the advancement to state of immunodeficiency. Chimpanzees played an important role toward determining HIV infection in nonhuman primates as whether HIV-1 does not affect or infect the Chimpanzees, but HIV virus produces infection or clinical sings inside the nonhuman primates. Current research is focused on Simian Immunodeficiency Virus (SIV) or SHIV using nonhuman primates for understanding bacterial pathogenesis. African nonhuman primates are used in SIV whereas Asian nonhuman primates are used to understand AIDS.

Mainly HIV virus enters into human body through CD4 and CCR5 or CD4 and CXCR4 receptors, which too show the effect of SIV infection in macaques. Research revealed that nonhuman primates allow the antibodies to diminish $CD8^+$ immune T cells and control to acid infection. Nonhuman primates have been used to develop antiviral drugs. Besides toxicological studies, antiviral investigation can be done through nonhuman primates. Two research areas are particularly involved in nonhuman primates' studies viz., the HIV evolved protein stimulating immunological antibody production and T cell development by stimulating vaccines. *Macaca mulatta* is highly effective to form the antibody to acute infection and regenerate the CD8+ve antibodies (Keen 2012).

25.5.2 Malaria

The species involved to cause malaria in human body are *Plasmodium vivax*, *P. malaria*, *P. falciparum*, and *P. ovale*. They mostly cause infection by attacking the parasites. Malaria is a most infectious disease that approximately affects 160 million people each year, and occurs mainly in children. The virus enters into the host body by biting of the female anopheles mosquito. Subsequently entering the host body, it directly invades the liver cells and then it is transmitted into blood cells and adhere to the blood vessels' walls causing organ damage. Damaging the organ does premature death of organ (Schmidt 1966; Larsen et al. 2009). Particularly, *P. vivax* and *P. ovale* are dormant in the liver for months or even years giving rise to very sudden outbreaks, nevertheless *P. falciparum* is characterized by coma, multiorgan failure, anemia, and other complications. In the development of malaria vaccine, nonhuman primate model involved is *Macaca mulatta*, used to develop the vaccine that includes the immune response and manifested through CD8 and CD4 T cells. Nowadays, priming with DNA encoding malaria proteins and then heightening the same DNA in a virus vector produces very strong T cell in macaques. It is the protein by which blood transmission and infection can be prevented (Schmidt 1966; Larsen et al. 2009).

P. falciparum gives rise to a greater infection due to their severity and mortality. A research in the pathological treatment shows difficulties, and still now it is not likely to advance a continuous in vitro culture system for the progress of the blood stages. Though *Macaca mulatta* were routinely used model system for the malaria vaccination studies; rhesus macaques model systems were found critical to understand the invasion pathway in 1970s due to human genomic sequence and blood group (Schmidt 1966; Larsen et al. 2009). Nonhuman primates presently play a significant part to control the *P. vivax* malaria. The Duffy binding protein (DBP) binds the red blood cell and creates in vitro assays inside the primate's body. It shows a huge response in clinical studies so, nonhuman models are future to the vaccination of malaria disease.

25.5.3 Tuberculosis

Tuberculosis, a highly infectious disease, caused the death of millions between 1700 and 1950 in the developed world (Basaraba, 2008). But presently TB is under control in many developing countries. It presents a major health challenge for commons. *M. tuberculosis* mainly develops in lungs and spreads throughout the body parts and evades immune system (Dharmadhikari and Nardell 2008). The main causes of the disease are alcohol abuse, dietary insufficiency, genetic factors, etc. the virus remains same or becomes more active for several years.

In the twentieth century, vaccine was developed against tuberculosis. The current approaches and treatments like, chemotherapy and vaccinations are used in tuberculosis (Wolf et al. 1988). Currently, WHO developed dots strategy (directly observed therapy, short course) or chemotherapeutic agents has led to vaccination of tuberculosis (Scanga and Flynn 2014). The literature survey shows that the clinical value of enzyme-linked immunosorbent assay (ELISA), lymphocyte stimulation test, and erythrocyte sedimentation rates are very effective in nonhuman primates (Schmidt 1966; Larsen et al. 2009). In accordance with the care and use of laboratory animals (NRC1996) primates were given the strain of *M. tuberculosis* by injection and studied under Biosafety at level 3 (Kaushal et al. 2012). The primates were monitored daily on the basis of behavior, appetite, and general health. Blood was collected every week and PCR was done for culture sample from abdomen protein derivative. Also, hematology lymphocyte proliferation assays were done. Finally after 7 months, the infection affected the rhesus monkey and in due course, some symptoms like necrosis, failure of fibrous connective tissue leading to death (Capuano et al. 2003; Flynn 2006; Mehra et al. 2013; Darrah et al. 2014).

25.5.4 Leishmania

As per the World Health Organization estimates, among 359 million people 12 million people are highly affected by "leishmania," a major infectious disease affecting poorest region of the world. Cutaneous leishmaniasis (CL) is caused by extremely pathogenic parasite. The parasite is highly contagious due to their chronicity, dormancy, and inclination to metastasize by causing skin abrasions with mucosal association. Mostly in humans, Leishmania breaks the epidemiological pattern and disintegrates the immunological system.

Most of the recognized circumstances of human CL are initiated by *L. major*. For investigational infection in macaques persuaded by parasite of the alike species sources a self-limiting rigorousness and provides the most ethical acceptance for trials of the vaccines (Amaral et al. 2002; Campos-Neto et al. 2001) (Table 25.3). The macaque (*Macaque mulatta*) was first infested with *L. amazonensis* developing more lesions gradually (Amaral et al. 2002) (Table 25.3). Experiments showed the live skin abrasions confined amastigotes with plasma cells, lymphocytes, and infiltrate macrophages. Affected macaques with *L. amazonensis* showed that CD4+/CD8+ T-cell helps to heal with the skin lesions (Parrot et al. 1927; Herrer et al. 1973; Kenney et al. 1999; Amaral et al. 2002; Campos-Neto et al. 2001).

25.5.5 Dengue

Dengue virus is mainly transmitted to humans by vector *Aedes* sp. mosquitoes. It causes a widespread array of illness comprising dengue fever and dengue hemorrhagic fever basically leading to life-threatening dengue shock syndrome (DSS) (Gubler 1998; Fan 2009). Among 100 countries, 50–100 million human beings are annually affected in tropical and subtropical regions (Gubler 1998; Fan 2009; Simmons et al. 2012).

Old world monkeys are the human representatives in understanding dengue. Old world monkeys are further related to humans due to the reproducible symptoms, genetic relationship along with immuno response histocompatibility complexes alleles (Gubler 1998; Miura 2013). Bonnet monkeys (*M. radiata*) are basically used to understand the pathogenesis and thereby employed for the development of antiviral drug and vaccines (Kyle et al. 2008; Asada et al. 2002). The method involved in studies of dengue use NHP models. The isolation of peripheral blood

Table 25.3 Case studies of human cutaneous leishmaniasis (CL)

Macaca species, number of monkeys involved	Leishmania vaccine	Vaccination practice	Challenges infection, route	Protection conferred by vaccines	References
Macaca fascicularis, 2	HKLV	Immunization of the monkeys were done with only one dose of *L. tropica* ama, SC	Animals were confronted at third week post vaccination through viable *L. tropica* ama, SC	Lack of protection	Parrot et al. (1927)
Macaca mulatta, 12	HKLV	Immunization of the animals with a single SC dose of killed *L. amazonensis* prom (0.25–1 mg) along with rhIL-12 (2 μm) and alum (0.125–0.5 mg)	Animals were challenged on week after vaccination with 10^7 metacyclic *L. amazonensis* prom, id	Complete	Kenney et al. (1999)
Macaca mulatta, 8	HKLV	Primary dosing of the monkeys along with booster (twice) dose at 1 month interval by inoculation of ALM (1 mg) along with BCG	Animals were challenged at 18th week post vaccination by 10^7 viable *L. major* prom, id	Lack of protection	Amaral et al. (2002)
Macaca mulatta, 8	LAV	Vaccination of the monkeys with only dosage of 10^8 attenuated *L. major*	Confrontation of animals at 18th week post vaccination with 10^7 viable *L. major* prom, id	Absence of protection	Amaral et al. (2002)
Macaca mulatta, 6	SUPV	Primates were vaccinated at every 1 month interval, with a mixture of LlmsTI1 (25 μg), TSA (25 μg), rhIL-12 (2 μg), and alum (200 μg). Booster dosage was given to the monkeys 1 month later along only with the antigens and alum (without IL-12)	Animals were challenged at 5th week after vaccination with 10^7 viable *L. major* prom, id	Complete	Campos-Neto et al. (2001)

Table 25.4 Trials on monkeys to understand dengue

Sl. no.	Method	Outcomes	References
1.	Analysis of viral replication in PBMC of *M. radiata* Strains were isolated from PMBC of bonnet monkey and infected with DENV. After 2 days of the infection viral titers were measured by plaque assay and supernatants were harvested	Analyses of phylogenetic relationship revealed that strains were highly infected and progeny viruses were replicated in primary cells	Gubler (1998)
2.	Infection of *M. radiata* with DENV-4 Accordingly with the regulation, *in M. radiata,* the strain of DENV-4. 9–48 were inoculated into the 3 monkeys. The monkeys with 6–7 kg in body weight after anesthetization with ketamine hydrochloride; blood samples were collected at 2, 3, 5, 7, 10, 7, 14, and 24 days of post inoculation Blood samples were centrifuged and separated from cell to serum	Using NSI antigen and ELISA examination, *M. radiata* showed some clinical symptoms like weight loss in plasma; quantitative RT-PCR exhibited the viremia of the RNA infection	Kyle et al. (2008)
3.	Profiles of leukocytes and platelets in blood. The platelets and number of leukocytes were measured by using automatic blood cell counting device	Platelets were slightly decreased for some and blood samples and cell properties were determined by using an automated hematology analysis	Parga-Lozano et al. (2009)

mononuclear cells derived from *M. radiata* is most commonly used to develop a NHP model for dengue (Table 25.4) (Parga-Lozano et al. 2009). Most replicative DENV 9–48 strains of bonnet monkeys is used by analyzing phylogenetic analysis studies based on plasma membrane also helps to determine the replication of the virus and development of the vaccine (Table 25.4) (Kyle et al. 2008; Asada et al. 2002; Parga-Lozano et al. 2009).

25.6 Conclusion

Nonhuman primate models not only play an important role for invasive methodologies but also help in understanding bacterial pathogenesis. Macaque group of animals are similar to humans due to their immune system and brain circuits. In some cases without that type of functional mapping, it may be impossible to understand the particular mechanism of pathogenesis. Nonhuman primates also constitute a significant reduction in toxicology. In pharmaceutical studies, beyond the early developmental stage, novel drugs are tested inside the primate body for efficacy and safety. WHO has taken up some major efforts to control the infectious disease by providing proper treatment. The welfare impacts of laboratory lies in housing animals to develop a condition in understanding pathogenesis by blood sampling followed by an immune response. For trials of potential vaccines against SIV and SHIV, macaques are similar to humans infected with HIV. Genetic studies show an improvement in rodent models to test the vaccine by identifying the malaria genome drug target. Though nonhuman primates have been employed to understand tuberculosis by programming patterns inside the body, rodents are mostly affected due to decrease in growth regulators. Some ethical arguments come into existence in use of nonhuman primates in large numbers for research and clinical trials.

Acknowledgments Poulomi Ghosh and Saprativ P. Das are thankful to Brainware University for providing the necessary funding for research. The contribution of Bijayata Patra toward the manuscript preparation is also thankfully accredited. The authors gratefully acknowledge Prof. Subrata Kumar Dey for providing the essential infrastructural facilities.

References

Alan S (2008) What is a virulence factor. Crit Care 12(6):197

Alberts B, Johnson A, Lewis J (2002) Introduction to pathogens: molecular biology of the cell, 4th edn. Garland Science Publishers, New York

Amaral VF, Teva A, Oliveira-Neto MP, Silva AJ, Pereira MS, Cupolillo E, Porrozzi R, Coutinho SG, Pirmez C, Beverley SM, Grimaldi G Jr (2002) Study of the safety, immunogenicity and efficacy of attenuated and killed *Leishmania* (*Leishmania*) major vaccines in a rhesus monkey (*Macaca mulatta*) model of the human disease. Mem Inst Oswaldo Cruz 97:1041–1048

Arenas M, Posada D (2010) The effect of recombination on the reconstruction of ancestral sequences. Genetics 184(4):1133–1139

Asada Y, Kawamoto Y, Shotake T, Terao K (2002) Molecular evolution of IgG subclass among nonhuman primates: implication of differences in antigenic determinants among Apes. Primates 43:343–349

Basaraba RJ (2008) Experimental tuberculosis: the role of comparative pathology in the discovery of improved tuberculosis treatment strategies. Tuberculosis 88:35–47

Bernstein H, Bernstein C, Michod RE (2018) Sex in microbial pathogens. Infect Genet Evol 57:8–25

Bien J, Sokolova O, Bozko P (2011) Characterization of virulence factors of *Staphylococcus aureus*: novel function of known virulence factors that are implicated in activation of airway epithelial proinflammatory response. J Pathog 2011:601905

Boussaoud D, Tanné-Gariépy J, Wannier T, Rouiller EM (2005) Callosal connections of dorsal versus ventral premotor areas in the macaque monkey: a multiple retrograde tracing study. BMC Neurosci 6:67

Campos-Neto A, Porrozzi R, Greeson K, Coler RN, Webb JR, Seiky YA, Reed SG, Grimaldi G Jr (2001) Protection against cutaneous leishmaniasis induced by recombinant antigens in murine and non-human primate models of the human disease. Infect Immun 69:4103–4108

Capuano SV, Croix DA, Pawar S, Zinovik A, Myers A, Lin PL, Bissel S, Fuhrman C, Klein E, Flynn JL (2003) Experimental *Mycobacterium tuberculosis* infection of cynomolgus macaques closely resembles the various manifestations of human *M. tuberculosis* infection. Infect Immun 71:5831–5844

Casadevall A, Pirofski LA (2014) Ditch the term pathogen. Nature 516(7530):165–166

Ciccarelli FD, Doerks T, Von MC, Creevey CJ, Snel B, Bork P (2006) Toward automatic reconstruction of a highly resolved tree of life. Science 311(5765):1283–1287

Darrah PA, Bolton DL, Lackner AA, Kaushal D, Aye PP, Mehra S, Blanchard JL, Didier PJ, Roy CJ, Rao SS, Hokey DA, Scanga CA, Sizemore DR, Sadoff JC, Roederer M, Seder RA (2014) Aerosol vaccination with AERAS-402 elicits robust cellular immune responses in the lungs of rhesus macaques but fails to protect against high-dose *Mycobacterium tuberculosis* challenge. J Immunol 193:1799–1811

Deborah TH, Elizabeth A, Shakhnovich EP, John JM (2005) Small-molecule inhibitor of *Vibrio cholerae* virulence and intestinal colonization. Science 310(5748):670–674

Dharmadhikari AS, Nardell EA (2008) What animal models teach humans about tuberculosis. Am J Respir Cell Mol Biol 39:503–508

Fan S (2009) Immunogenicity and protective efficacy of a live attenuated H5N1 vaccine in nonhuman primates. PLoS Pathog 5:e1000409

Flynn JL (2006) Lessons from experimental *Mycobacterium tuberculosis* infections. Microbes Infect 8:1179–1188

Groves CP (2005) Mammal species of the world: a taxonomic and geographic reference, 3rd edn. Johns Hopkins University Press, Baltimore, pp 161–165

Gubler DJ (1998) Dengue and dengue hemorrhagic fever. Clin Microbiol Rev 11:480–496

Herrer A, Christensen HA, Beumer RJ (1973) Reservoir hosts of cutaneous leishmaniasis among Panamanian forest mammals. Am J Trop Med Hyg 22:585–591

Hodge T, Cope M (2000) A myosin family tree. J Cell Sci 113(19):3353–3354

Kaushal D, Mehra S, Didier PJ, Lackner AA (2012) The non-human primate model of tuberculosis. J Med Primatol 41:191–201

Keen EC (2012) Paradigms of pathogenesis: targeting the mobile genetic elements of disease. Front Cell Infect Microbiol 2:161

Kenney RT, Sacks DL, Sypek JP, Vilela L, Gam AA, Evans-Davis K (1999) Protective immunity using recombinant human IL-12 and alum as adjuvants in a primate model of cutaneous leishmaniasis. J Immunol 163:4481–4488

Kyle JL, Balsitis SJ, Zhang L, Beatty PR, Harris E (2008) Antibodies play a greater role than immune cells in heterologous protection against secondary dengue virus infection in a mouse model. Virology 380:296–303

Larsen MH, Biermann K, Chen B, Hsu T, Sambandamurthy VK, Lackner AA, Aye PP, Didier P, Huang D, Shao L, Wei H, Letvin NL, Frothingham R, Haynes BF, Chen ZW, Jacobs WR Jr (2009) Efficacy and safety of live attenuated persistent and rapidly cleared *Mycobacterium tuberculosis* vaccine candidates in non-human primates. Vaccine 27:4709–4717

Mayr E (2009) Cladistic analysis or cladistic classification? J Zool Syst Evol Res 12:94–128

Mehra S, Alvarez X, Didier PJ, Doyle LA, Blanchard JL, Lackner AA, Kaushal D (2013) Granuloma correlates of protection against TB and mechanisms of immune modulation by *Mycobacterium tuberculosis*. J Infect Dis 207:1115–1127

Miura M (2013) Characterization of simian T-cell leukemia virus type 1 in naturally infected Japanese macaques as a model of HTLV-1 infection. Retrovirology 10:118

Nathan C (2015) From transient infection to chronic disease. Science 350(6257):161

Ostrowski SR (1998) B-virus from pet macaque monkeys: an emerging threat in the United States? Emerging infectious diseases. Emerg Infect Dis 4(1):117–121

Parga-Lozano C, Reguera R, Gomez-Prieto P, Arnaiz-Villena A (2009) Evolution of major histocompatibility complex G and C and natural killer receptors in primates. Hum Immunol 70:1035–1040

Parrot L, Donatien A, Lestoquard F (1927) Notes expérimentales sur le bouton d'Orient et sur la leishmaniose canine viscérale. Arch Inst Pasteur Alger 5:120–130

Penny D, Hendy MD, Steel MA (1992) Progress with methods for constructing evolutionary trees. Trends Ecol Evol 7(3):73–79

Podani J (2019) The coral of life. Evol Biol 46:123–144

Santamaria R, Theron R (2009) Tree evolution: visual analysis of phylogenetic trees. Bioinformatics 25(15):1970–1971

Scanga CA, Flynn JL (2014) Modeling tuberculosis in nonhuman primates. Cold Spring Harb Perspect Med 4:a018564

Schmidt L (1966) Studies on the antituberculous activity of ethambutol in monkeys. Ann N Y Acad Sci 135:747–758

Simmons CP, Farrar JJ, Nguyen V, Wills B (2012) Dengue. N Engl J Med 366:1423–1432

Thrall PH, Burdon JJ (2003) Evolution of virulence in a plant host-pathogen metapopulation. Science 299:1735–1737

Townsend JP, Su Z, Tekle Y (2012) Phylogenetic signal and noise: predicting the power of a data set to resolve phylogeny. Genetics 61(5):835–849

Webb SAR, Kahler CM (2008) Bench-to-bedside review: bacterial virulence and subversion of host defences. Crit Care 12:234

Woese C (2002) On the evolution of cells. Proc Natl Acad Sci U S A 99(13):8742–8747

Wolf RH, Gibson SV, Watson EA, Baskin GB (1988) Multidrug chemotherapy of tuberculosis in rhesus-monkeys. Lab Anim Sci 38:25–33

Zhen L (2018) Cloning of macaque monkeys by somatic cell nuclear transfer. Cell 172(4):881–887

Understanding Biofilm Dynamics: In Vitro and In Vivo Models

26

Rabbani Syed, Mohammed S. Alqahtani, Ayesha Mateen, Fawaz Alasmari, and Youssef Sari

Contents

R. Syed (✉)
Department of Pharmaceutics, College of Pharmacy, King Saud University, Riyadh, Saudi Arabia

Department of Clinical Laboratory Sciences, College of Applied Medical Sciences, King Saud University, Riyadh, Saudi Arabia
e-mail: rsyed@ksu.edu.sa

M. S. Alqahtani
Department of Pharmaceutics, College of Pharmacy, King Saud University, Riyadh, Saudi Arabia

A. Mateen
Department of Clinical Laboratory Sciences, College of Applied Medical Sciences, King Saud University, Riyadh, Saudi Arabia

F. Alasmari (✉)
Department of Pharmacology and Toxicology, College of Pharmacy, King Saud University, Riyadh, Saudi Arabia
e-mail: ffalasmari@ksu.edu.sa

Y. Sari
Department of Pharmacology and Experimental Therapeutics, College of Pharmacy and Pharmaceutical Sciences, University of Toledo, Toledo, OH, USA

B. Siddhardha et al. (eds.), *Model Organisms for Microbial Pathogenesis, Biofilm Formation and Antimicrobial Drug Discovery*,
https://doi.org/10.1007/978-981-15-1695-5_26

Abstract

Biofilms are bacterial cells, especially planktonic bacteria, structured as a community; they group, grow, and develop together by way of complex chemical and molecular communication. The biochemical communication in a biofilm is termed "quorum sensing," and constitutes the most vital component of their structural and functional stability. In addition to this constant chemical communication, they also exchange nutrients and sometimes genetic components, giving rise to daughter cells with altered genetic makeup. Bacterial biofilms can occur in different contexts, including dental plaque on teeth and pond scum in the environment. Bacterial biofilms are usually infectious and pose considerable challenges to treat. After sufficient growth and expansion, biofilms start discharging their "virulence" in either of two forms: (1) planktonic shower or (2) fragment detachment. The huge mass of bacterial cells in a biofilm is immensely stronger than the corresponding singular planktonic pattern, and their collective virulence presents a heavy challenge to the host's immune cells. Biofilm bacteria are usually resistant to most common antibiotics, and present a great threat of ruinous infection. Infectious biofilms can sometimes also demand surgical resection of the infected tissue or organ. In addition to the bacterial biofilms, fungi and viral particles integrate to form large communities, producing concomitantly challenging infections. Many species of fungi including *Aspergillus*, *Cryptococcus*, *Trichosporon*, and *Pneumocystis* are all capable of forming fungal biofilms and collectively exhibit resistance to antifungal therapies. However, biofilms are not always menacing to human life. Their beneficial effects and biotechnological applications are very active fields of research. Nonetheless, biofilms of bacterial, fungal, and viral origin all constitute major platforms for chronic infections and diseases. As a mechanistic approach, scientists believe that the destruction or inhibition of "quorum sensing" can enable disruption of bacterial biofilms. Research also supports that pyruvate is essential for the construction and maintenance of biofilms, and its depletion can degrade their strength. This chapter will explain in detail the molecular signaling, construction, and maintenance of biofilms, their potential threats and beneficial applications, and current research goals.

Keywords

Biofilms · Quorum sensing · Biofilm formation · Biofilm application · Biofilm dynamics

26.1 Introduction

Biofilms are formed by several different microorganisms when they adhere together to living or nonliving surfaces through the generation of an extracellular polymeric substance (Vu et al. 2009). One of the attributes that regulate the attachment of the microorganisms is growth surface (Chen et al. 2012; Eginton et al. 1995). Initially, a biofilm begins as a microcolony, which then attracts a variety of other microbes, thereby rendering the biofilm rather complex with multiple monomeric agents. The biofilm's constituent microbes usually follow a life cycle pattern consisting of attachment, growth, development, and maturation followed by detachment in order to be able to exert virulence and cause infection (Sandal et al. 2007). During their life cycle, biofilm cells exchange nutrients and communicate by exchanging genetic material, thereby producing daughter cells of varied genetic makeup (Lim et al. 2017). "Quorum sensing" is a well-known mode of communication that is prevalent in bacterial biofilms where different bacterial species/cells coordinate with each other (Popat et al. 2012).

Research into biofilms is extensive as they pose major threats to public health in various forms, especially those species which are found in hospitals (Li and Tian 2012). Biofilm-forming bacteria are infectious in nature and known to contribute to multiple types of infections (Donlan 2002; Jamal et al. 2015). Biofilms are also difficult to treat as they are a group of individual entities which possess unique cellular properties both individually and collectively (Lechevallier et al. 1988; Stoodley et al. 1994; Thomen et al. 2017). Furthermore, most bacterial biofilms are highly resistant to extant antibiotics (Balcázar et al. 2015). These challenges demand substantial and rapid research strategies by which to improve patient management.

26.2 History of Biofilms

The history of biofilm research began in the seventeenth century when a Dutch researcher, Antoni van Leeuwenhoek, observed "animalcule" on tooth surfaces using a simple microscope. Subsequent study of biofilms did not take off until the twentieth century, when researchers investigated their underlying components from composition to growth, development, and virulence (Donlan 2001a, b). Evaluation of trickling filters from wastewater treatment systems using scanning and transmission electric microscopy also confirmed that biofilms are composed of multiple organisms (De Carvalho and da Fonseca 2007). This study also revealed that a polysaccharide is formed by the enclosed cells and surrounding matrix. It is suggested that biofilms are highly resistant to disinfectants due to microbial slimes. Several other studies afterwards investigated the mechanisms of biofilm attachment to a surface, their character, co-existence, and detachment; some few studies have also demonstrated techniques to screen and characterize biofilms, ranging from standard microscopical techniques to the recent establishment of confocal laser scanning microscopy methods (Petrova and Sauer 2016).

26.3 Composition

Biofilms can be produced by microorganisms including bacteria, viruses, and fungi. Bacterial biofilms display intra- and inter-species variation from their planktonic counterparts. Biofilms can adhere to a variety of surfaces such as medical machines, tissues, and natural aquatic systems (Johnson et al. 1991). The extracellular polymeric substances (EPS) synthesized by biofilm microbes consist of protein, DNA, RNA, and polysaccharides each at $<1–2\%$, $<1\%$, $<1\%$, and 1–2%, respectively. Water constitutes the majority (97%) of biofilm structure, allowing nutrient diffusion throughout the matrix (Jamal et al. 2015). The high-water content of biofilms dilutes antimicrobial agents, neutralizing their effects and resulting in prolonged treatment (Stewart 2015). Biofilm microbes are also able to regulate their gene expression to maintain collective homeostasis under extreme conditions including temperature or nutritional changes, pH and osmolarity, and cell density (Loo et al. 2003).

26.4 Formation of Biofilms

There are generally four steps in the construction and maintenance of biofilms:

1. Attachment
2. Micro-colony formation
3. Three-dimensional structure formation and maturation
4. Detachment.

Figure 26.1 illustrates in detail the life cycle of biofilm: its formation, attachment, development, and dispersion.

26.4.1 Attachment

Biofilms usually form and develop in an aqueous medium, which is an ideal environment that enhances their attachment and growth. When a bacterium or any other microorganism comes upon a substratum within the medium, it slows down and makes a reversible connection with the substrate surface or with an already-adhered microbe (Carniello et al. 2018). Microbial structures such as flagella, pili, fimbriae, and extracellular proteins or polysaccharides may promote biofilm formation. The attachment of such structures to rough, hydrophilic, and coated surfaces has been established to provide better environment for their growth and survival (Carniello et al. 2018). To better understand biofilms, it is essential to further elucidate the contributions of other aspects such as the substratum material, conditioning films on the substratum, hydrodynamics of the aqueous medium, properties of the medium, and characteristics of the microbial cell surface (Hellal et al. 2016).

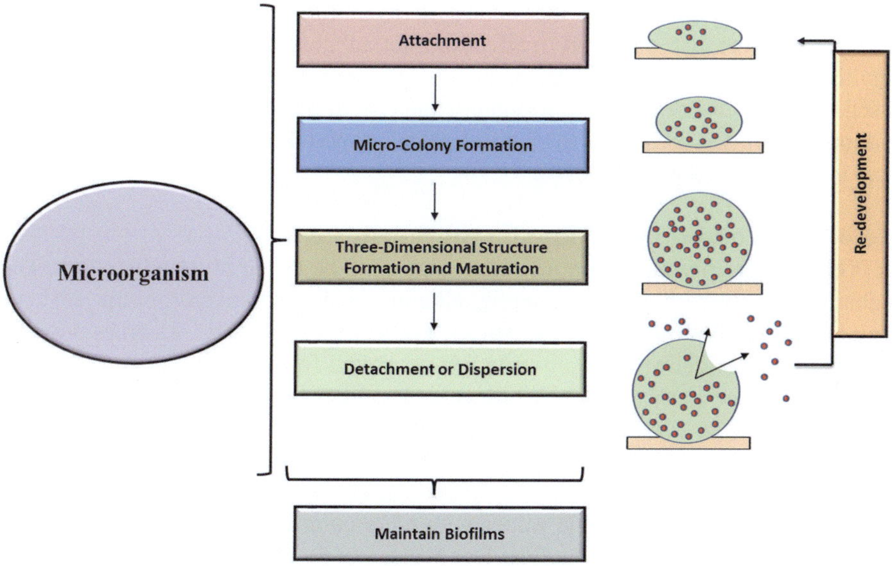

Fig. 26.1 A detailed life cycle of biofilm, its formation, attachment, development, and dispersion

26.4.1.1 Substrate

Several substrate attributes affect the attachment process. For example, it has been shown that microbial colonization increases with increasing surface roughness. This appears to be attributable to diminishment of shear forces due to the greater surface area provided by rough surfaces (Fang et al. 2017). We suggest here that physicochemical properties of the surface may affect attachment. Studies of biofilm attachment to hydrophobic surfaces have produced contradictory results; nonetheless, between the substratum and cell surface could help the cell to resist specific forces (Renner and Weibel 2011).

26.4.1.2 Hydrodynamics of the External and Internal Aqueous Media

It is highly essential to understand hydrodynamic characteristics of the medium, such as flow rates and shear forces, in order to precisely characterize biofilm formation, attachment, spread, and survival. This knowledge could be highly useful for addressing many fundamental questions and for predicting various effects on biofilms. Environmental and systemic conditions produce hydrodynamic forces which influence several biofilm properties such as cellular adherence to water on a surface, motile cell adherence to the surface itself, colonization, and motility (Shen et al. 2015).

Notably, biofilm growth is influenced by the internal hydrodynamics of its molecular transport system, which controls oxygen availability and delivery (Lagos et al. 2016). Therefore, mechanical stress might be reflected in critical effects within the hydrodynamic system of the biofilm. It has been observed that the hydrodynamic properties of microorganisms that contribute to biofilms vary (Yates 1986). These differences were also observed to affect biofilm adherence to and detachment from multiple types of surfaces (Kaplan et al. 2003). Computational fluid dynamics can be used to effectively simulate the formation and execution of hydrodynamic states for various biofilms, which helps to explore critical properties of signaling toward effective targeting (Lemos et al. 2016).

26.4.1.3 Conditioning Films on the Substratum

It is extremely important for a microbial biofilm to be aware of its substratum. The substratum may be highly wet, acidic, alkaline, hydrophobic/hydrophilic, or have any of a number of other characteristics (Maki et al. 2000). Thus, it comes down to the microbes to identify the substratum most suited for their growth and development. For instance, old bacterial biofilms are less likely to settle on less wet surfaces as compared with surfaces of high wettability (Sneed et al. 2014). Thus, surface hydration may be a vital method to control biofilm properties. It is important to note that correlations exist between water contact angle and the standardized harmonic mean or between contact angle values with other parameters such as bacterial density, thickness, and dry weight (Van Der Westen et al. 2018). In combination with the environment, other features of the substratum such as its nature, texture, and the presence of organic material influence various characteristics of the biofilm (Hellal et al. 2016). The features of the formation of biofilms from the solid–liquid interface colonization have been well investigated by existing literature (Auger et al. 2009). However, the solid–air interface colonization has been less extensively studied, which could also be important in terms of biofilms formation. Other attributes that impact biofilm attachment include the growth medium, hydrodynamics of the aqueous medium, and characteristics of the medium and the cell surface (Hellal et al. 2016). These factors, individually or together, may help the microbes irreversibly to attach to the substratum even if there are any active repulsive forces around the substratum. The substratum and its surrounding layers can be split into various levels, some of which are negligible. There is negligible flow velocity in the area that is adjacent to the substratum–liquid interface (Donlan 2002). This zone is referred to as the hydrodynamic boundary layer, and its thickness is dependent on linear velocity (Donlan 2002). Specifically, boundary layer will become thinner with higher velocity (Taherzadeh et al. 2012). Characteristics of the next layer out include considerable mixing and extreme turbulence, which may substantially affect cell–substratum interactions (Yang et al. 2017). Cells behave like particles in liquid, and therefore, velocity is one parameter that plays an important role in the rate of microbial settling and association with a submerged surface (Taherzadeh et al. 2012). Other factors that affect cellular attachment to the substratum include nutrient levels, pH, temperature, and ionic strength; these may also play significant roles (Renner and Weibel 2011). Additionally, cellular structures known as fimbriae also play a major part in biofilm attachment, helping bypass the electrostatic repulsion barrier that initially exists between the cell and the substratum (Garrett et al. 2008). Microbial genes and their expression are also being studied extensively for prospective roles in governing the process of biofilm attachment (Garrett et al. 2008).

26.4.1.4 Properties of the Cellular Surface in a Biofilm

The physicochemical properties of the cellular surface are critical for the growth and development of biofilms. The formation of a biofilm is feasible only when isolated or planktonic cells attach to either an inert or a living surface that possesses altering properties (Whitehead and Verran 2009).

Thus, the properties of both the cellular surface and the attachment platform dictate the survival potential of a biofilm (Veerachamy et al. 2014). The availability of a solid attachment surface is a prerequisite for the initial formation and microcolonization of biofilms (Crouzet et al. 2014). Cellular properties such as the presence of an S-layer and cell surface hydrophobicity are understood to be vital to the mechanism of biofilm formation (Garrett et al. 2008; Auger et al. 2009). A well-established cellular component that assists in the successful adhesion and colonization of bacteria is the exopolysaccharide produced abundantly by the specific species *Pseudomonas aeruginosa* (Ghafoor et al. 2011). More generally, surface area is an important factor; the growth and development of biofilms are improved with better oxygen exchange of a larger surface area. Negatively charged cell surfaces give specific strains of biofilms nonacidic character, which could also be linked to the presence of carboxylate and sulfate groups on the surface of the microbiota (Kostakioti et al. 2013).

26.4.1.5 Quorum Sensing: A Universal Mode of Microbial Communication

Every biofilm comprises two layers of cells: the topmost, motile, free-flowing cells and the bottom layer of cells attached to the surface with which they must communicate steadily and effectively (Veerachamy et al. 2014). Quorum sensing is a well-documented mechanism that enables communication between microorganisms, especially bacteria (Woodard and Saleh 2008). Quorum sensing also occurs in many other microorganisms, such as fungi and viruses; recent evidence indicates that viruses integrated together as a community communicate with a mechanism that could resemble quorum sensing (Callaway 2017). The properties of individual cells are shared by means of molecular signaling, which could drive complex processes ranging from nutrient and water exchange to genetic transfer, producing daughter cells with altered genetic makeup, thereby influencing microbial structure and biofilm differentiation (Remuzgo-Martínez et al. 2015). Discovery of this complex communication network has revealed that microorganisms organize into groups and form well-organized communities, which was once believed to be possible only in multicellular organisms (Palková 2004). These well-organized communities are sometimes referred to as "cities" of microorganisms and are controlled and regulated by the architecture of the biofilm. Quorum sensing holds great importance when the microorganisms are exposed to extreme conditions, allowing them to take collective action, benefit one another, and improve survival for each and all (Cvitkovitch et al. 2003). Additionally, quorum sensing is vital for ensuring that the bed of microorganisms reaches a significant cell density before it expresses virulence, thereby increasing its effect against host defense mechanisms (De Kievit and Iglewski 2000).

There are a number of examples in nature that demonstrate both the potential benefits and the hazards of collective microbial biofilms. One such beneficial biofilm is formed by bacteria *Vibrio fischeri*, whose bioluminescence is controlled by quorum sensing. These bacteria live in symbiotic relationships with their marine host (Dobretsov et al. 2013) and produce bioluminescence for host activities such as attracting prey, avoiding predators, and finding a mate. Thus, the bacterial biofilm benefits the host, while also taking advantage of living in the nutrient-rich environment it provides (Toyofuku et al. 2016). Another such instance is the "milky sea," an intensive, uniform, and sustained glow on the surface of the sea produced by massive numbers of the marine bacterium *V. harveyi*, which lives in association with some colonies of microalgae (Grasland et al. 2003). Whether beneficial or pathogenic, quorum sensing is what enables the effective social communication between individual entities that is necessary for a biofilm to grow and develop.

26.4.2 Microcolony Formation

Microorganisms in the marine water bed colonize both biotic and abiotic surfaces. It has been documented that the formation of biofilms on marine surfaces offers numerous benefits to these organisms, thereby supporting critical ecological and geochemical functions (Jones and Bennett 2017). Various aquatic organisms may participate in the formation of biofilms and colonization; of these, bacteria are the most diverse and significant in terms of their composition, dynamics, and functions (Jones and Bennett 2017). We suggest here that bacteria exhibit distinct mechanisms in their colonization and functional properties. The most significant advantage in the context of a marine environment is their access to resources; microcolonization is highly dependent on high-energy-carbohydrate availability. Colonization helps promote the synthesis of extracellular enzymes, maintenance of enzyme structural integrity, and physiological homeostasis (Hall and Mah 2017). As colonization promotes the hydrolytic activities of secreted extracellular enzymes, it increases microbial opportunities to utilize nutrients and organic matter (Decho and Gutierrez 2017). Colonization also acts as shielding for the biofilm matrix, promoting its protective effects against antibiotics and other toxins. Colonization may support the transition of bacterial cells to a competent state, in which the transfer of extracellular DNA may be effective, and the secretion and expression of biopolymer-targeted extracellular enzymes are triggered (Seper et al. 2011).

26.4.3 Three-Dimensional Structure and Maturation of Biofilms

The structure of a biofilm depends strongly upon its adaptation to the environment and the factors that determine cellular decisions to remain intact (Toyofuku et al. 2016). Each stage of the biofilm life cycle is highly regulated and is affected by various environmental factors. Biofilm structure is especially impacted by the attachment surface, properties of individual microorganisms, hydrodynamic conditions, and the availability of nutrients. Environmental attributes are linked to gene expression and regulation through second messengers such as cAMP and c-di-GMP. Biofilms are characterized by a heterogeneous interior, which enables the precise observation of gene expression profiles specific to a cell's location (Finelli et al. 2003). As biofilms are multicellular, cellular communication is also a major factor in shaping the 3D structure of biofilm, enabling the development of a social network with division of labor. Sociomicrobiology better explains the concept of social conditioning in biofilms. Within the biofilm structure, the cells with low volume/structure values have easier access to nutrients than those having high volume/surface ratios. Slow growth has also been documented with odd-numbered n-alkanes, due to a low odd-numbered chain of carbon atoms in fatty acids.

Biofilm maturation is a process that involves reversible and irreversible stages with numerous conserved and/or species-specific factors. EPS, as also mentioned in the previous section, promotes the establishment of a structured architecture in the maturation stage. Development of antibiotic resistance is mediated by increased EPS and other factors such as oxygen, which is involved in the dysregulation of metabolism and growth of cell-associated biofilms. Individual cellular contact with the surface also triggers gene expression changes, promoting sessility in the formation of the extracellular matrix. It is also established that the maturation of a biofilm with a complex communication system leads to increase in eDNA (Gloag et al. 2013). Maturation is followed by dispersal stage, when planktonic microbes leave the biofilm. It is also noteworthy that individual cells are prone to dispersal due to shear stress and other environmental changes as they mature.

26.4.4 Detachment

Detachment is the process of dispersal of the aggregated microbes in a biofilm. Detachment events can occur for a number of reasons, including nutrient deprivation, increased virulence, and active maturation (Hunt et al. 2004). The detached microbes offer increased virulence against the human immune system (Le et al. 2018). The mechanisms of biofilm dispersal can be grouped into two categories: active and passive. When the microbe initiates the process of detachment, it is termed active. On the other hand, when detachment is induced by external factors such as fluid shear, human intervention, predator grazing, or abrasion, it is termed passive (Horn et al. 2003). Studies have additionally demonstrated three distinct models of biofilm dispersal: erosion, sloughing, and seeding (Lee and Yoon 2017). If the biofilm continuously releases single cells or clusters of cells throughout its formation and development, this is termed erosion. In contrast, if the biofilm suddenly sheds a portion of its structure, especially at later stages, this is referred to as sloughing. Seeding dispersal denotes the release of live cells from the centers of microcolonies.

The processes of biofilm growth and detachment are distinct, and detachment evidently involves the dispersion of non-aggregated cells inside the colony of microbes in a biofilm. Indeed, microbial biofilms utilize different pathways for detachment due to simple physical shearing at different time points depending on the microenvironment, whether the cause is cell lysis, flow of fluid, enzymatic dissolution, or something else. The complex mechanism of detachment is being extensively researched so as to provide methods of targeting and destruction. Detachment also depends on the biofilm structure and the rate and extent of its accumulation. Several integrative approaches involving experimental and computational studies are highly in demand to explore the factors contributing to and regulating biofilm detachment (Toyofuku et al. 2016).

The process of dispersal utilizes active enzymes, including glycosidases, proteases, and deoxyribonucleases, that degrade endogenous matrix components. For instance, mucoid strains of the human pathogen *Pseudomonas aeruginosa* produce alginate (a biofilm polysaccharide) and alginate lyase (an enzyme that degrades alginate). Exogenous addition of such a component to *Pseudomonas aeruginosa* biofilms in culture promoted antibiotic activities due to enhancing detachment. Sodium nitroprusside also shows enhanced detachment of *Pseudomonas aeruginosa* biofilm colonies in vitro, and several other small molecules have been demonstrated efficacious in inducing detachment, proving to have clinical applications (Worthington et al. 2012).

26.5 Biofilms and Public Health Risk

The increase of biofilm-associated infections over time is associated with serious consequences (Galie et al. 2018). Often, the impact of biofilms might be underestimated. Notably, persisters in biofilms are involved in the development of recurrent infections, and they can resist antibiotics (Michiels et al. 2016). A number of public health risks are posed by biofilms (Miquel et al. 2016), some of which are detailed as follows.

26.5.1 Food Industry

The food industry constitutes a major biohazard zone in terms of biofilms contributing to serious economic and health issues. The development of biofilms along food manufacturing surfaces may result in serious economic losses (Galie et al. 2018). Food spoilage induced by microbial species can also result in serious hygienic problems and financial losses (Galie et al. 2018). Therefore, the impact of biofilms on food safety is a concern for human health and well-being. Bacterial biofilms have been commonly identified in ready-to-eat food products (RTE) and raw food, including seafood (Choi et al. 2015). Biofilms on seafood are generally seen in crabs, Pacific oysters, and shrimp. Microbial biofilms in seafood can be composed of numerous viruses, bacteria, and parasites that can exist on food contact surfaces and in water, where they can remain attached for a longer period and develop antibiotic resistance (Galie et al. 2018). Microbial biofilms are also common in dairy perishables including yogurt, butter, and cheese, and in semi-perishables including casein and milk powder. Lack of proper cleaning and sanitization can constitute a major cause of biofilms in milk processing units, leading to both health and economic loss (Licker et al. 2017). The major bacterial genera affecting the dairy industry are *Enterobacter*, *Micrococcus*, *Listeria*, *Streptococcus*, *Bacillus*, and *Pseudomonas*. Biofilms have also been observed on the surfaces of equipment, where they can reduce transfer of heat, increase the rate of corrosion, and increase resistance to friction (Lindsay and Geornaras 1996). These factors can affect equipment functionality and safety.

26.5.2 Biofilms in Infectious Diseases Field

Secure water must be accessible to all as it is the most important necessity for human life; the contamination of water with biofilms results in serious health problems. Aquatic biofilms are a potential cause for water-borne diseases in both developing and developed countries (Del Pozo 2018). In fact, many health issues such as diarrhea and death relate to drinking water, especially in infants in developing countries. Biofilms are known to be responsible for infections in animals and humans (Balqadi et al. 2018), and are being extensively researched in major fields including food safety, animal welfare and health, and disease control (Nadell et al. 2008). Infections in humans mainly include middle-ear infections, urinary tract infections (UTIs), catheter infections, dental plaque, gingivitis, cystic fibrosis, endocarditis, and infections in heart valves and joint prostheses (Mahami and Adu-Gyamfi 2011). To date, 60% of UTIs have been connected to protozoan, bacterial, and fungal biofilms; of these, *Helicobacter pylori* is predominant, affecting 20% of infected persons. UTIs are dangerous if not properly treated, as they can give rise to other complications such as bacteremia, acute pyelonephritis, vaginosis, renal infections, bladder cancer, and death. Inflammation and tissue damage are also caused by many strains of single species that can be found in polymicrobial communities (He et al. 2016).

26.5.3 Biofilm in Medical Devices

Biofilms have been established as major causative agents of diseases ranging from allergies to systemic infections (Fastenberg et al. 2016). This is not surprising, as around 99% of microorganisms on the planet live in biofilm communities (Davey and O'Toole 2000). The bacterial species that target medical devices include *Staphylococcus aureus*, *Enterococcus faecalis*, *Staphylococcus epidermidis*, *Escherischia coli*, *Streptococcus viridans*, *Klebsiella pneumoniae*, *Proteus mirabilis*, and *Pseudomonas aeruginosa*. Urinary catheters are most prone to biofilm infections, which may be composed of multiple species of bacteria that have been isolated from and documented on indwelling medical devices (Murugan et al. 2016). In cardiovascular devices (heart valves or catheters), *Staphylococcus epidermidis and Staphylococcus aureus* are the most commonly observed, causing a significant proportion of infection cases (Chambers 2005). Overall, biofilms are a serious global threat for public health, potentially compounded by factors such as poor sanitary conditions and low economic resources, which elevate the chance of infection in developing countries (Licker et al. 2017). The import of biofilms to public health demands the development of effective control measures for the treatment and prevention of biofilm formation.

26.6 Applications of Biofilms

While biofilm-forming microbes can be pathogenic and feature chronic infections, biofilms also have practical implementations in biotechnological applications. Due to their inherent properties of long-term activity, self-immobilization, and high resistance to reactants, biofilms have been suggested as a great industrial resource and workhorse for continuous processing. Biofilm reactors are used in commercial treatment of wastewater and off-gas (Mitra et al. 2014). Biofilms are also beneficial in numerous other sectors including synthetic chemistry, bioenergy, biologics, and the food industry. Incorporating an N_2-fixing rhizobial strain in fungal–rhizobial biofilms (FRB) has potential applications in biofertilizers and the biocontrol of plants (Seneviratne et al. 2008). Another topic of recent research interest is the application of biofilms to energy and drug discovery studies, as well as environmental and toxicological research (Bjarnsholt et al. 2013). Intertidal and marine biofilms have also been utilized in various marine and intertidal applications (Dang and Lovell 2016). In particular, marine biofilms have been used as electrodes in microbial fuel cells. Biofilms have also been studied extensively for the synthesis of various biotechnology compounds such as surfactants and ethanol (Rasamiravaka et al. 2015; Berlanga and Guerrero 2016; Rosche et al. 2009). Thus, understanding biofilms can allow the design of systems that can facilitate improved yields and productivity, as well as improving on the successful applications achieved so far.

26.7 Enabling Future Therapeutics Against Resistant Biofilms

Microbial resistance mechanisms have been described as the means by which an antimicrobial agent is prevented from interacting with its target (Garnett and Matthews 2012). Various terms have conventionally been used to describe biofilm resistance mechanisms, including "recalcitrance," "tolerance," and "resistance." All of these terms encompass the decreased susceptibility of biofilm cells to antibiotic agents; biofilm resistance to antibiotics is significant, around 1000-fold greater than for its planktonic counterparts. This resistance can be explained by various mechanisms, but enzymes and extracellular polymers could be the major determinants (Molobela et al. 2010). Other factors involved in the development of microbial defenses include lower nutrient content and the low penetration rate of antibiotics (Singh et al. 2017). All told, biofilm-specific antibiotic resistance is multifactorial and resistance mechanisms vary depending on bacterial strain, the specific antimicrobial compound, the developmental stage and age of the biofilm, and growth conditions. Several studies also highlight that resistance could be a factor of genes and mutations that accumulate within cells and the community. It is hypothesized that these genetic mutations can be transferred between species, enabling stronger defenses against microbial killing (Modi et al. 2017). In particular, specific mutations for increased efflux pump expression and chromosomal beta-lactamase are conventional resistance mechanisms that have been found to contribute to biofilm resistance and survival (Soto 2013).

Increasing resistance in biofilms promotes the persistence of chronic biofilm-associated infections despite antibiotic therapy. These persistent cells are usually more resistant to antibiotics, which also play an important role in supporting the reestablishment of the biofilm community after therapy (Vlamakis et al. 2008). In addition, mutant biofilms sometimes offer impaired resistance with varied action mechanisms. For instance, endogenous oxidative stress in *P. aeruginosa* biofilm is a major cause for promoting mutagenesis by mutagenic repair of DNA double strand breaks (Elsen et al. 2013). The biofilm matrix could also play a vital role in antibiotic resistance mechanisms (Stewart 2002). Furthermore, cells in dormant parts of a biofilm may not be much affected by antibiotics, as has been observed with β-lactams, ciprofloxacin, tetracycline, and tobramycin (Hall and Mah 2017). In addition to being tolerant against antibiotics, bacterial biofilms causing chronic infections also resist phagocytosis and various other components of the host's defense system (Stewart 2002). However, while in vitro studies have provided evidence regarding antibiotic resistance and chronic infections in microbial biofilms (Sedlacek and Walker 2007), it is suggested that laboratory medium and conditioning factors are far simpler than the in vivo conditions pathogenic biofilms inhabit, and may not accurately reflect the nutrient and pharmacokinetic fluctuations encountered by biofilms in vivo. Thus, in vitro cultures may not reflect the behavioral phenotypic shift through which microbial cells take on a lifestyle with sessile properties (Rollet et al. 2009).

To target biofilm infections that are refractory to current therapy, it is essential to disable biofilm resistance. A topic of serious interest is addressing biofilms that infect wounds or medical devices, which are antibiotic resistant and difficult to treat. Similarly, "persister" cells present a challenge, being highly virulent, antibiotic resistant, and metabolically dormant; they can resist antibiotic therapy and sometimes repopulate biofilms (Lewis 2012). One approach for resistant biofilms is to pursue early aggressive antibiotic chronic supportive treatment or prophylaxis, such as using quorum sensing inhibitors to increase susceptibility to antibiotics. Genetic mutations could also be a pharmacological target for the development of novel chemotherapeutic compounds (Kostakioti et al. 2013), and antimicrobial photodynamic therapy could be an effective modality against resistant microbes (Cieplik et al. 2014). Photoactivation has been effective against resistant biofilms, although bacterial pigments have shown resistance mechanisms (Rao et al. 2017) and may protect biofilms by acting as antioxidants. A linkage between virulence and pigment production has been demonstrated in some bacterial species, and also correlates with natural light (Mohammadi et al. 2012). Phages are another potential mechanism for acting against bacterial biofilms, and are sufficiently economical as to be explored for large-scale use in food preparation (Oliver et al. 2005). However, while all of these measures are proving to be effective against microbial biofilms in vitro, there is at present insufficient robustness and clinical evidence for them to be widely adopted.

26.8 Conclusion

A biofilm structure is not always made of one type of bacteria, and therefore these bacteria are usually multilingual, with the ability to transfer and receive signals within and between different types of bacteria. Extensive scientific research is underway to effectively address the antibiotic resistance, virulence, and infectious properties of these biofilms both in vitro and in vivo (through animal models). The potential for several different combinations of drugs and antibiotics to effectively destroy biofilms of various species is being tested. These include combinations of anticancer, antidepressant, and anti-inflammatory drugs such as sodium diclofenac, aspirin, and ibuprofen, which have shown efficacy against *Staphylococcus aureus*, *Escherichia coli*, and *Candida albicans*, respectively.

Acknowledgments The authors extend their appreciation to the International Scientific Partnership Program-146 (ISPP-146) at deanship of scientific research—King Saud University for the support.

Conflict of Interest The authors declare no conflict of interest.

References

Auger S, Ramarao N, Faille C, Fouet A, Aymerich S, Gohar M (2009) Biofilm formation and cell surface properties among pathogenic and nonpathogenic strains of the *Bacillus cereus* group. Appl Environ Microbiol 75(20):6616–6618

Balcázar JL, Subirats J, Borrego CM (2015) The role of biofilms as environmental reservoirs of antibiotic resistance. Front Microbiol 6:1216

Balqadi AA, Salama AJ, Satheesh S (2018) Microfouling development on artificial substrates deployed in the Central Red Sea. Oceanologia 60(2):219–231

Berlanga M, Guerrero R (2016) Living together in biofilms: the microbial cell factory and its biotechnological implications. Microb Cell Factories 15(1):165

Bjarnsholt T, Ciofu O, Molin S, Givskov M, Høiby N (2013) Applying insights from biofilm biology to drug development—can a new approach be developed? Nat Rev Drug Discov 12(10):791

Callaway E (2017) Do you speak virus? Phages caught sending chemical messages. Nature. https://doi.org/10.1038/nature.2017.21313

Carniello V, Peterson BW, van der Mei HC, Busscher HJ (2018) Physico-chemistry from initial bacterial adhesion to surface-programmed biofilm growth. Adv Colloid Interface Sci 261:1–14

Chambers ST (2005) Diagnosis and management of staphylococcal infections of pacemakers and cardiac defibrillators. Intern Med J 35:S63–S71

Chen S, Cheng X, Zhang X, Sun D (2012) Influence of surface modification of polyethylene biocarriers on biofilm properties and wastewater treatment efficiency in moving-bed biofilm reactors. Water Sci Technol 65(6):1021–1026

Choi NY, Bae YM, Lee SY (2015) Cell surface properties and biofilm formation of pathogenic bacteria. Food Sci Biotechnol 24(6):2257–2264

Cieplik F, Tabenski L, Buchalla W, Maisch T (2014) Antimicrobial photodynamic therapy for inactivation of biofilms formed by oral key pathogens. Front Microbiol 5:405

Crouzet M, Le Senechal C, Brözel VS, Costaglioli P, Barthe C, Bonneu M, Vilain S (2014) Exploring early steps in biofilm formation: set-up of an experimental system for molecular studies. BMC Microbiol 14(1):253

Cvitkovitch DG, Li YH, Ellen RP (2003) Quorum sensing and biofilm formation in *Streptococcal* infections. J Clin Invest 112(11):1626–1632

Dang H, Lovell CR (2016) Microbial surface colonization and biofilm development in marine environments. Microbiol Mol Biol Rev 80(1):91–138

Davey ME, O'Toole GA (2000) Microbial biofilms: from ecology to molecular genetics. Microbiol Mol Biol Rev 64(4):847–867

De Carvalho CC, da Fonseca MMR (2007) Assessment of three-dimensional biofilm structure using an optical microscope. BioTechniques 42(5):616–620

De Kievit TR, Iglewski BH (2000) Bacterial quorum sensing in pathogenic relationships. Infect Immun 68(9):4839–4849

Decho AW, Gutierrez T (2017) Microbial extracellular polymeric substances (EPSs) in ocean systems. Front Microbiol 8:922

Del Pozo JL (2018) Biofilm-related disease. Expert Rev Anti-Infect Ther 16(1):51–65

Dobretsov S, Abed RM, Teplitski M (2013) Mini-review: inhibition of biofouling by marine microorganisms. Biofouling 29(4):423–441

Donlan RM (2001a) Biofilms and device-associated infections. Emerg Infect Dis 7(2):277

Donlan RM (2001b) Biofilm formation: a clinically relevant microbiological process. Clin Infect Dis 33(8):1387–1392

Donlan RM (2002) Biofilms: microbial life on surfaces. Emerg Infect Dis 8(9):881

Eginton PJ, Gibson H, Holah J, Handley PS, Gilbert P (1995) The influence of substratum properties on the attachment of bacterial cells. Colloids Surf B Biointerfaces 5(3–4):153–159

Elsen S, Collin-Faure V, Gidrol X, Lemercier C (2013) The opportunistic pathogen *Pseudomonas aeruginosa* activates the DNA double-strand break signaling and repair pathway in infected cells. Cell Mol Life Sci 70(22):4385–4397

Fang H, Chen Y, Huang L, He G (2017) Analysis of biofilm bacterial communities under different shear stresses using size-fractionated sediment. Sci Rep 7(1):1299

Fastenberg JH, Hsueh WD, Mustafa A, Akbar NA, Abuzeid WM (2016) Biofilms in chronic rhinosinusitis: pathophysiology and therapeutic strategies. W J Otorhinolaryngology Head Neck Surg 2(4):219–229

Finelli A, Gallant CV, Jarvi K, Burrows LL (2003) Use of in-biofilm expression technology to identify genes involved in *Pseudomonas aeruginosa* biofilm development. J Bacteriol 185(9):2700–2710

Galie S, García-Gutiérrez C, Miguélez EM, Villar CJ, Lombó F (2018) Biofilms in the food industry: health aspects and control methods. Front Microbiol 9:898

Garnett JA, Matthews S (2012) Interactions in bacterial biofilm development: a structural perspective. Curr Protein Pept Sci 13(8):739–755

Garrett TR, Bhakoo M, Zhang Z (2008) Bacterial adhesion and biofilms on surfaces. Prog Nat Sci 18(9):1049–1056

Ghafoor A, Hay D, Rehm BH (2011) Role of exopolysaccharides in *Pseudomonas aeruginosa* biofilm formation and architecture. Appl Environ Microbiol 77(15):5238–5246

Gloag ES, Turnbull L, Huang A, Vallotton P, Wang H, Nolan LM, Monahan LG (2013) Self-organization of bacterial biofilms is facilitated by extracellular DNA. PNAS 110(28):11541–11546

Grasland B, Mitalane J, Briandet R, Quemener E, Meylheuc T, Linossier I, Haras D (2003) Bacterial biofilm in seawater: cell surface properties of early-attached marine bacteria. Biofouling 19(5):307–313

Hall CW, Mah TF (2017) Molecular mechanisms of biofilm-based antibiotic resistance and tolerance in pathogenic bacteria. FEMS Microbiol Rev 41(3):276–301
He S, Li X, Li R, Fang L, Sun L, Wang Y, Wu M (2016) Annexin A2 modulates ROS and impacts inflammatory response via IL-17 signaling in polymicrobial sepsis mice. PLoS Pathog 12(7):e1005743
Hellal J, Michel C, Barsotti V, Laperche V, Garrido F, Joulian C (2016) Representative sampling of natural biofilms: influence of substratum type on the bacterial and fungal communities structure. Springerplus 5(1):822
Horn H, Reiff H, Morgenroth E (2003) Simulation of growth and detachment in biofilm systems under defined hydrodynamic conditions. Biotechnol Bioeng 81(5):607–617
Hunt SM, Werner EM, Huang B, Hamilton MA, Stewart PS (2004) Hypothesis for the role of nutrient starvation in biofilm detachment. Appl Environ Microbiol 70(12):7418–7425
Jamal M, Tasneem U, Hussain T, Andleeb S (2015) Bacterial biofilm: its composition, formation and role in human infections. Res Rev 4:1–14
Johnson CR, Sutton DC, Olson RR, Giddins R (1991) Settlement of crown-of-thorns starfish: role of bacteria on surfaces of coralline algae and a hypothesis for deepwater recruitment. Mar Ecol Prog Ser 71:143–162
Jones AA, Bennett PC (2017) Mineral ecology: surface specific colonization and geochemical drivers of biofilm accumulation, composition, and phylogeny. Front Microbiol 8:491
Kaplan JB, Meyenhofer MF, Fine DH (2003) Biofilm growth and detachment of *Actinobacillus actinomycetemcomitans*. J Bacteriol 185(4):1399–1404
Kostakioti M, Hadjifrangiskou M, Hultgren SJ (2013) Bacterial biofilms: development, dispersal, and therapeutic strategies in the dawn of the postantibiotic era. Cold Spring Harb Perspect Med 3(4):a010306
Lagos ME, White CR, Marshall DJ (2016) Biofilm history and oxygen availability interact to affect habitat selection in a marine invertebrate. Biofouling 32(6):645–655
Le KY, Park MD, Otto M (2018) Immune evasion mechanisms of *Staphylococcus epidermidis* biofilm infection. Front Microbiol 9:359
LeChevallier MW, Cawthon CD, Lee RG (1988) Inactivation of biofilm bacteria. Appl Environ Microbiol 54(10):2492–2499
Lee K, Yoon SS (2017) *Pseudomonas aeruginosa* biofilm, a programmed bacterial life for fitness. J Microbiol Biotechnol 27(6):1053–1064
Lemos M, Wang S, Ali A, Simões M, Wilson DI (2016) A fluid dynamic gauging device for measuring biofilm thickness on cylindrical surfaces. Biochem Eng J 106:48–60
Lewis K (2012) Persister cells: molecular mechanisms related to antibiotic tolerance. In: Antibiotic resistance. Springer, Berlin, Heidelberg, pp 121–133
Li YH, Tian X (2012) Quorum sensing and bacterial social interactions in biofilms. Sensors 12(3):2519–2538
Licker M, Moldovan R, Hogea E, Muntean D, Horhat F, Baditoiu L, Zambori C (2017) Microbial biofilm in human health—an updated theoretical and practical insight. Rom J Lab Med 25(1):9–26
Lim JH, Song SH, Park HS, Lee JR, Lee SM (2017) Spontaneous detachment of *Streptococcus mutans* biofilm by synergistic effect between zwitterion and sugar alcohol. Sci Rep 7(1):8107
Lindsay D, Geornaras I (1996) Biofilms associated with poultry processing equipment. Microbios 86(347):105–116
Loo CY, Mitrakul K, Voss IB, Hughes CV, Ganeshkumar N (2003) Involvement of the ADC operon and manganese homeostasis in *Streptococcus gordonii* biofilm formation. J Bacteriol 185(9):2887–2900
Mahami T, Adu-Gyamfi A (2011) Biofilm-associated infections: public health implications. Int Res J Microbiol 2(10):375–381
Maki JS, Ding L, Stokes J, Kavouras JH, Rittschof D (2000) Substratum/bacterial interactions and larval attachment: films and exopolysaccharides of *Halomonas marina* (ATCC 25374) and their effect on barnacle cyprid larvae, *Balanus amphitrite* Darwin. Biofouling 16(2–4):159–170

Michiels JE, Van den Bergh B, Verstraeten N, Michiels J (2016) Molecular mechanisms and clinical implications of bacterial persistence. Drug Resist Updat 29:76–89

Miquel S, Lagrafeuille R, Souweine B, Forestier C (2016) Anti-biofilm activity as a health issue. Front Microbiol 7:592

Mitra S, Sana B, Mukherjee J (2014) Ecological roles and biotechnological applications of marine and intertidal microbial biofilms. In: Productive biofilms. Springer, Cham, pp 163–205

Modi BP, Teves ME, Pearson LN, Parikh HI, Haymond-Thornburg H, Tucker JL, Strauss JF III (2017) Mutations in fetal genes involved in innate immunity and host defense against microbes increase risk of preterm premature rupture of membranes (PPROM). Mol Genet Genomic Med 5(6):720–729

Mohammadi M, Burbank L, Roper MC (2012) Biological role of pigment production for the bacterial phytopathogen *Pantoea stewartii* subsp. stewartii. Appl Environ Microbiol 78(19):6859–6865

Molobela IP, Cloete TE, Beukes M (2010) Protease and amylase enzymes for biofilm removal and degradation of extracellular polymeric substances (EPS) produced by *Pseudomonas fluorescens* bacteria. Afr J Microbiol Res 4(14):1515–1524

Murugan K, Selvanayaki K, Al-Sohaibani S (2016) Urinary catheter indwelling clinical pathogen biofilm formation, exopolysaccharide characterization and their growth influencing parameters. Saudi J Biol Sci 23(1):150–159

Nadell CD, Xavier JB, Levin SA, Foster KR (2008) The evolution of quorum sensing in bacterial biofilms. PLoS Biol 6(1):e14

Oliver SP, Jayarao BM, Almeida RA (2005) Foodborne pathogens in milk and the dairy farm environment: food safety and public health implications. Foodborne Pathog Dis 2(2):115–129

Palková Z (2004) Multicellular microorganisms: laboratory versus nature. EMBO Rep 5(5):470–476

Petrova OE, Sauer K (2016) Escaping the biofilm in more than one way: desorption, detachment or dispersion. Curr Opin Microbiol 30:67–78

Popat R, Crusz SA, Messina M, Williams P, West SA, Diggle SP (2012) Quorum-sensing and cheating in bacterial biofilms. Proc Biol Sci 279(1748):4765–4771

Rao N, Prabhu M, Xiao M, Li WJ (2017) Fungal and bacterial pigments: secondary metabolites with wide applications. Front Microbiol 8:1113

Rasamiravaka T, Labtani Q, Duez P, El Jaziri M (2015) The formation of biofilms by *Pseudomonas aeruginosa*: a review of the natural and synthetic compounds interfering with control mechanisms. Biomed Res Int 2015:759348

Remuzgo-Martínez S, Lázaro-Díez M, Mayer C, Aranzamendi-Zaldumbide M, Padilla D, Calvo J, Ramos-Vivas J (2015) Biofilm formation and quorum-sensing-molecule production by clinical isolates of *Serratia liquefaciens*. Appl Environ Microbiol 81(10):3306–3315

Renner LD, Weibel DB (2011) Physicochemical regulation of biofilm formation. MRS Bull 36(5):347–355

Rollet C, Gal L, Guzzo J (2009) Biofilm-detached cells, a transition from a sessile to a planktonic phenotype: a comparative study of adhesion and physiological characteristics in *Pseudomonas aeruginosa*. FEMS Microbiol Lett 290(2):135–142

Rosche B, Li XZ, Hauer B, Schmid A, Buehler K (2009) Microbial biofilms: a concept for industrial catalysis? Trends Biotechnol 27(11):636–643

Sandal I, Hong W, Swords WE, Inzana TJ (2007) Characterization and comparison of biofilm development by pathogenic and commensal isolates of *Histophilus somni*. J Bacteriol 189(22):8179–8185

Sedlacek MJ, Walker C (2007) Antibiotic resistance in an in vitro subgingival biofilm model. Oral Microbiol Immunol 22(5):333–339

Seneviratne G, Zavahir JS, Bandara WMMS, Weerasekara MLMAW (2008) Fungal-bacterial biofilms: their development for novel biotechnological applications. World J Microbiol Biotechnol 24(6):739

Seper A, Fengler VH, Roier S, Wolinski H, Kohlwein SD, Bishop AL, Schild S (2011) Extracellular nucleases and extracellular DNA play important roles in *Vibrio cholerae* biofilm formation. Mol Microbiol 82(4):1015–1037
Shen Y, Monroy GL, Derlon N, Janjaroen D, Huang C, Morgenroth E, Nguyen TH (2015) Role of biofilm roughness and hydrodynamic conditions in *Legionella pneumophila* adhesion to and detachment from simulated drinking water biofilms. Environ Sci Technol 49(7):4274–4282
Singh S, Singh SK, Chowdhury I, Singh R (2017) Understanding the mechanism of bacterial biofilms resistance to antimicrobial agents. Open Microbiol J 11:53–62
Sneed JM, Sharp KH, Ritchie KB, Paul VJ (2014) The chemical cue tetrabromopyrrole from a biofilm bacterium induces settlement of multiple Caribbean corals. Proc Biol Sci 281(1786):20133086
Soto SM (2013) Role of efflux pumps in the antibiotic resistance of bacteria embedded in a biofilm. Virulence 4(3):223–229
Stewart PS (2002) Mechanisms of antibiotic resistance in bacterial biofilms. Int J Med Microbiol 292(2):107–113
Stewart PS (2015) Antimicrobial tolerance in biofilms. Microbiol Spectr 3(3)
Stoodley P, DeBeer D, Lewandowski Z (1994) Liquid flow in biofilm systems. Appl Environ Microbiol 60(8):2711–2716
Taherzadeh D, Picioreanu C, Horn H (2012) Mass transfer enhancement in moving biofilm structures. Biophys J 102(7):1483–1492
Thomen P, Robert J, Monmeyran A, Bitbol AF, Douarche C, Henry N (2017) Bacterial biofilm under flow: first a physical struggle to stay, then a matter of breathing. PloS One 12(4):e0175197
Toyofuku M, Inaba T, Kiyokawa T, Obana N, Yawata Y, Nomura N (2016) Environmental factors that shape biofilm formation. Biosci Biotechnol Biochem 80(1):7–12
Van der Westen R, Sjollema J, Molenaar R, Sharma PK, Van der Mei HC, Busscher HJ (2018) Floating and tether-coupled adhesion of bacteria to hydrophobic and hydrophilic surfaces. Langmuir 34(17):4937–4944
Veerachamy S, Yarlagadda T, Manivasagam G, Yarlagadda PK (2014) Bacterial adherence and biofilm formation on medical implants: a review. Proc Inst Mech Eng H 228(10):1083–1099
Vlamakis H, Aguilar C, Losick R, Kolter R (2008) Control of cell fate by the formation of an architecturally complex bacterial community. Genes Dev 22(7):945–953
Vu B, Chen M, Crawford R, Ivanova E (2009) Bacterial extracellular polysaccharides involved in biofilm formation. Molecules 14(7):2535–2554
Whitehead KA, Verran J (2009) The effect of substratum properties on the survival of attached microorganisms on inert surfaces. In: Marine and industrial biofouling. Springer, Berlin, Heidelberg, pp 13–33
Woodard B, Saleh MA (2008) Three-dimensional quantitative structure activity relationships of quorum-sensing and biofilm inhibitors in gram-negative bacteria. J Environ Sci Health B 43(4):281–287
Worthington RJ, Richards JJ, Melander C (2012) Small molecule control of bacterial biofilms. Org Biomol Chem 10(37):7457–7474
Yang J, Guo X, Ding D, Gao W, Huang D, Chen Y, Liang X (2017) Interactions between natural biofilm, substratum surface wettability, and mussel plantigrade settlement. Sci China Earth Sci 60(2):382–390
Yates GT (1986) How microorganisms move through water: the hydrodynamics of ciliary and flagellar propulsion reveal how microorganisms overcome the extreme effect of the viscosity of water. Am Sci 74(4):358–365

Model Organisms and Antimicrobial Drug Discovery

27

Jobina Rajkumari, Madhu Dyavaiah, Asad Syed, and Busi Siddhardha

Contents

Abstract

Numerous mammalian and non-mammalian hosts have been extensively employed to study the action of virulence determinants, mechanism of host defense, and pharmacokinetics of drugs. several synthetic, as well as natural antimicrobial compounds, have been identified using mammalian as well as non-mammalian infection models. In vivo screening approach using animal models allows concurrent assessment of drug efficacy and toxicity of the antimicrobial drugs. Mice model has been intensively used to explore the virulence attributes

J. Rajkumari · B. Siddhardha (✉)
Department of Microbiology, School of Life Sciences, Pondicherry University, Puducherry, India

M. Dyavaiah
Department of Biochemistry and Molecular Biology, School of Life Sciences, Pondicherry University, Puducherry, India

A. Syed
Department of Botany and Microbiology, College of Sciences, King Saud University, Riyadh, Saudi Arabia

B. Siddhardha et al. (eds.), *Model Organisms for Microbial Pathogenesis, Biofilm Formation and Antimicrobial Drug Discovery*,
https://doi.org/10.1007/978-981-15-1695-5_27

and understand the infection process. It also facilitates in-depth comparative studies which are highly relevant for clinical applications. However, the large-scale screening of antimicrobial agents using rodents suffers from high cost and ethical constrains. Non-mammalian host models such as the larvae of *Caenorhabditis elegans* and zebra fish allows non-invasive real-time imaging of the bacterial cells at various stages of the infection. Due to their strong functional and structural similarities with an immune response that of the mammals, insect models like *Drosophila melanogaster*, *Galleria mellonella*, and *Bombyx mori* have also been extensively used to study the virulence phenotypes and therapeutic efficacy of novel antimicrobial drugs. *Saccharomyces cerevisiae* is a popular invertebrate host model for screening and identification of antimicrobial agents due to its highly conserved and homologous gene function to that of human subjects.

Keywords
Rodents · *Caenorhabditis elegans* · *Drosophila melanogaster* · *Danio rerio* · *Saccharomyces cerevisiae*

27.1 Introduction

The ever-escalating cases of drug-resistant pathogens and the disappointingly slow pace in the launch of new antibiotics is a big concern to public health (Peterson and Pukkila-Worley 2018). Hence, the development of new classes of anti-infective agents becomes a necessity. Traditionally, the potential antimicrobial efficacy of an antibiotic was evaluated using in vitro assays against the target pathogen (Moy et al. 2009). This approach, though effective in identifying novel drugs, the rate of new antibiotic discovery is slow. Moreover, the bioactive compounds identified are often inefficient due to a lack of effectiveness in the in vivo host (Peterson and Pukkila-Worley 2018). With the recent advancement in the field of microbial genomics, genomics-assisted or target-based antimicrobial screening for anti-infectives is possible. However, since drug discovery usually takes place in the absence of the host, further in vivo clinical trials and toxicological studies of the promising bioactive agent is imperative (Mukhopadhyay and Peterson 2006). Moreover, in the absence of an ideal host model, it is difficult to predict the bioavailability and metabolism of a drug in the host system (Peterson and Pukkila-Worley 2018). Hence, the establishment of ideal infectious disease models becomes a necessity in antimicrobial drug discovery.

The use of animal models in clinical research is inevitable to understand the pathogenesis of human disease, thereby facilitating the assessment and development of new control measures (Lieschke and Currie 2007). Various animals including rodents, insects, zebra fish, and *Caenorhabditis elegans* have been predominantly used as a model system in the study of human diseases. These animal models allow direct assessment of drug efficacy with the added advantage to evaluate the toxicity of the compounds simultaneously. Moreover, unlike conventional in vitro screening technique, this approach also facilitates in understanding the action of the drugs and the response manifested in complex host/pathogen relationship (Moy et al. 2009). This chapter focuses on the role model organisms employed in the screening and evaluation of antimicrobial agents.

27.2 Mammalian Hosts—Rodents

Rodents have always been the most popular and established in vivo model system since they can be easily housed and bred. This mammalian model is also suitable for evaluating possible induction of immune response in the host, elicited by the pathogen. Murine models are extensively adopted in the pharmacokinetics study for drugs targeted at an infection site. Mice model has been intensively used to explore the virulence attributes and to understand the mechanisms of pathogenicity in cases of skin and intra-abdominal infection, pneumonia, and septicaemia. They have also been commonly incorporated to investigate the effect of antimicrobial therapy in chronic *Pseudomonas aeruginosa* pneumonia and urinary tract infections (Zhao et al. 2016).

27.2.1 Application of Rodents in the Screening of Antimicrobial Drugs

Rodents including rats, mice, hamsters, and guinea pigs are widely employed for the preclinical test and development of new antimicrobial agents. Some of the antimicrobial agents which were studied for their antimicrobial efficacy in the rodent model are presented in Table 27.1. The antimicrobial effect of Epinecidin-1 (Epi-1) against *Helicobacter pylori* infection was studied using mice model. On exposure to Epi-1, a considerable reduction in bacterial burden was achieved (Narayana et al. 2015). Similarly, Gupta et al. (2015) reported that verapamil enhances the efficacy of bedaquiline toward *Mycobacterium tuberculosis* using mice model. It was suggested that the enhanced bactericidal effect resulted due to the inhibition in efflux pump by verapamil, consequently resulting in accumulation of antimycobacterial drug in the bacterial cell (Gupta et al. 2015). The in vivo potency of cadazolid in the prevention of diarrhea and mortality caused by *Clostridium difficile* was demonstrated using mice and hamster models (Locher et al. 2014).

The combined effect of rifampin and fosfomycin (FS10) toward the biofilm formation of methicillin-resistant *Staphylococcus aureus* (MRSA) was evaluated using male albino guinea pigs (Mihailescu et al. 2014). The synergistic therapeutic effect of FS10 and tigecycline in the treatment of staphylococcal wound infection was also evaluated in a mice model. An enhanced wound healing process indicated by high epithelialization and collagen formation was observed on the application of FS10-soaked foam dressing supplemented with tigecycline (Simonetti et al. 2016). The

Table 27.1 Antimicrobial agents which were studied for their antimicrobial efficacy in rodent model

Drug/compounds	Target pathogen	Infection model	Reference
Epinecidin-1	*H. pylori*	Mice	Narayana et al. (2015)
Verapamil + bedaquiline	*M. tuberculosis*	Mice	Gupta et al. (2015)
Cadazolid	*C. difficile*	Hamster, mice	Locher et al. (2014)
Fosfomycin + rifampin	Methicillin-resistant *S. aureus*	Guinea pigs	Mihailescu et al. (2014)
RNAIII inhibiting peptide derivative, FS10 and tigecycline	Methicillin-resistant and Methicillin-sensitive *S. aureus*	Mice (wound infection model)	Simonetti et al. (2016)
Verdinexor	*Influenza virus*	Mice and ferrets (intranasally inoculated)	Perwitasari et al. (2016)
Melimine-coated titanium	*P. aeruginosa*, *S. aureus*	Mice and rat (subcutaneous infection models)	Chen et al. (2016)
Silver nanoparticles	*P. aeruginosa*	Murine (skin infection model)	Escarcega-Gonzalez et al. (2018)
Curcumin-encapsulated nanoparticles	Methicillin-resistant *S. aureus* and *P. aeruginosa*	Mice (wound model)	Krausz et al. (2015)
Photodynamic inactivation, methylene blue (MB)	*C. albicans*	Mice (infected via the lateral caudal vein)	Kato et al. (2013)
Ramoplanin and vancomycin	*C. difficile*	Hamster model	Freeman et al. (2005)

efficacy of the antiviral drug, verdinexor toward influenza virus infection was evaluated in mice and ferrets. In both the rodent model, verdinexor effectively reduced virus burden and inflammatory cytokine expression (Perwitasari et al. 2016). The antimicrobial activity of the antimicrobial peptide, melimine-coated titanium toward *Staphylococcus aureus* and *P. aeruginosa* was assessed in mice and rat subcutaneous infection models (Chen et al. 2016). The antagonistic effect of the silver nanoparticles (AgNPs) toward *P. aeruginosa* was evaluated in murine skin infection model (Escarcega-Gonzalez et al. 2018). Topical administration of curcumin nanoparticles significantly reduced the growth of MRSA and *P. aeruginosa.* Moreover, the nanoparticles also accelerated the wound healing process in a murine burn wound model (Krausz et al. 2015).

The pathogenicity of *Candida albicans* exposed to sublethal antimicrobial photodynamic inactivation (aPDI) with methylene blue as a photosensitizer was determined in mice systemic infection model (Kato et al. 2013). The antagonistic activity of vancomycin and a novel glycolipodepsipeptide, ramoplanin against *Clostridium difficile* infection was studied in the hamster model. Both the test compounds were found to rapidly suppress the symptoms in infected host (Freeman et al. 2005). Recent advancement in the new transgenic strategy has also facilitated the use of mice models in better understanding of the progression of human diseases (Lieschke and Currie 2007).

27.3 Non-mammalian Hosts

Numerous compounds have been identified using mammalian infection models including mice and rabbits. However, the large-scale screening of antimicrobial agents using rodents are costly and labor intensive (Mukhopadhyay and Peterson 2006). On the contrary, the non-mammalian hosts like *Caenorhabditis elegans*, *Drosophila melanogaster*, and *Danio rerio* extend ethical and economical advantages over the mammalian models. These genetically tractable multicellular organisms have been of great benefit in elucidating virulence factors of many pathogens as well as host defenses. It also allows screening of antimicrobial drugs and analyses of the dynamic transcriptomes of both host and microbe through the course of infection (Lee et al. 2018).

27.3.1 *Caenorhabditis elegans*

In the area of drug discovery, *C. elegans* has been extensively studied for decades. The adult nematode normally attains ~1 mm in length and a single self-fertilizing hermaphrodite worm can generate up to 300 progeny in a life cycle of 3 days (Ewbank and Zugasti 2011). The rapid generation, low cost, and easy maintenance allow maintenance of large numbers of these tiny free-living nematodes in laboratory conditions (Peterson and Pukkila-Worley 2018). An estimated 50% of *C. elegans* genes have orthologs in humans. Unlike higher model organisms such as rodents, there exist no ethical or logistical concerns in using *C. elegans* models to study host–pathogen relationship (Zhang et al. 2017).

27.3.1.1 *C. elegans* as an Infection Model for Host–Pathogen Interactions

C. elegans has been extensively employed in the study of microbial pathogenicity, host defense, and drug discovery (Ewbank and Zugasti 2011). Many of the virulence factors employed by the pathogens in establishing infection in humans were also found to be fatal in the nematode model. Numerous bacterial pathogens including *P. aeruginosa*, *Salmonella enterica*, *Enterococcus faecalis*, *S. aureus* and fungal pathogen like *Cryptococcus neoformans* have been reported for their ability to infect and kill *C. elegans*. The pathogenicity of each pathogen was assessed based on a feeding-based pathogenicity assay wherein the usual food source (*Escherichia coli*) was replaced with the test pathogen and survival rate of the nematodes were recorded. In addition, *C. elegans* were also found to mount pathogen-specific innate immune response identical to the mammals (Moy et al. 2006). *C. elegans* also serve as a convenient tool in the study of the virulence factors of *Candida orthopsilosis*, *C. parapsilosis*, and *C. metapsilosis*. These Candida species causes organ protrusion and hyphae formation, thereby resulting in the killing of the nematode (Souza et al. 2018). Hence, these translucent nematodes provide new insights into the mechanism of accumulation and dissemination of pathogen with the possibility of direct application in the management of human infections (Peterson and Pukkila-Worley 2018).

27.3.1.2 Application of *C. elegans* in the Screening of Antimicrobial Drugs

C. elegans represents a suitable tool for in vivo large-scale screening of antimicrobial compounds (Table 27.2). In addition, to ease in cultivation and possibility of automated handling, *C. elegans* allows assessment of both the antimicrobial efficacy of the test compound and the compound's toxicity simultaneously. Using *C. elegans* infection model, 16 compounds and 9 extracts with the ability to promote survival of nematode infected with *Enterococcus faecalis* were identified (Moy et al. 2006). The ability of synthetic molecules, meta-bromo-thiolactone (mBTL), in attenuating the pathogenicity of *P. aeruginosa* was assessed using *C. elegans* fast-kill assay. Apart from the attenuation in the production of pyocyanin and biofilm development, mBTL also protected the *C. elegans* from killing by *P. aeruginosa* (O'Loughlin et al. 2013).

The bactericidal effect of antimicrobial peptides, temporins and esculentin toward MDR-*P. aeruginosa* and their mode of action were investigated using the *C. elegans* model. The test peptides were highly effective in enhancing the survival rate of *Pseudomonas*-infected *C. elegans* (Uccelletti et al. 2010). *C. elegans–Acinetobacter baumannii* liquid killing assay was adopted to simultaneously screen insect-derived antimicrobial peptides (AMPs) for their anti-infective potential as well as toxicity toward the host. Cecropin AMPs and BR003-cecropin A were found to enhance the survival rate of the nematode host (Jayamani et al. 2015). Jayamani

Table 27.2 Antimicrobial agents which were studied for their antimicrobial efficacy in the *C. elegans* model

Drug/compounds	Target pathogen	Reference
Meta-bromothiolactone	*P. aeruginosa*	O'Loughlin et al. (2013)
Frog skin antimicrobial peptides	*P. aeruginosa*	Uccelletti et al. (2010)
Insect-derived cecropins	*A. baumannii*	Jayamani et al. (2015)
Diflunisal	*F. tularensis*	Jayamani et al. (2017)
Clove oil	*P. aeruginosa*, *A. hydrophila*	Husain et al. (2013)
Closantel	MRSA and vancomycin-resistant staphylococci	Rajamuthiah et al. (2014)
Tea polyphenol	*P. aeruginosa*	Yin et al. (2016)
Hamamelitannin	*S. aureus*	Brackman et al. (2016)

et al. (2017) documented the use of *C. elegans* to test the effect of the nonsteroidal anti-inflammatory drug (NSAID) diflunisal against *Francisella tularensis* (Jayamani et al. 2017). *P. aeruginosa* PAO1 pre-infected *C. elegans* exhibited an extended survival rate on exposure to 1.6% v/v of clove oil suggesting that clove oil down-regulates the virulence attributes of *P. aeruginosa* PAO1 (Husain et al. 2013). Based on a *C. elegans*-based liquid assay, the anti-helminthic drug, closantel exhibited anti-staphylococcal efficacy and reduced the mortality of the nematodes infected with methicillin-resistant and vancomycin-resistant *S. aureus* (Rajamuthiah et al. 2014). Tea polyphenol causes a considerable reduction in the *P. aeruginosa* pathogenicity and subsequent killing of *C. elegans* (Yin et al. 2016). In another *C. elegans*-based study, Hamamelitannin, a potent antagonist of the quorum sensing system significantly increased the susceptibility of *S. aureus* (Brackman et al. 2016).

27.3.2 Zebra Fish

Zebra fish has gained considerable interest in recent times in the identification and development of antimicrobial drugs. This tiny genetically tractable vertebrate model is easily manipulated, maintained, and involves less stringent ethical and practical restrictions (Saralahti and Ramet 2015). In addition, the high fecundity and low cost of zebra fish embryo make them an ideal system for fast and large-scale screening of anti-infective. The zebra fish embryos are transparent in the first several days post fertilization, followed by ~3 weeks where the larvae are translucent. Being optically transparent, zebra fish embryos enable visualization of bacterial load and the route of dispersal in real time, using a fluorescent stereomicroscope (Clatworthy et al. 2009). Zebra fish also possess well-developed adaptive and innate immune systems relatively similar to mammals, which allows in-depth comparative studies which are highly relevant for clinical applications (Meijer et al. 2014).

27.3.2.1 Zebra fish as an Infection Model for Host–Pathogen Interactions

Zebra fish (*Danio rerio*) has proven to a promising noninvasive infection model in deciphering the infection process and host responses evoked by streptococcal infections in humans (Mukhopadhyay and Peterson 2006). Adult zebra fish has been exploited in the identification of various virulence factors associated with the infections of *S. agalactiae*, *S. pyogenes*, and *S. pneumoniae*. *P. aeruginosa* is known to cause lethal infection in zebra fish embryos, thus facilitating the evaluation of *P. aeruginosa* pathogenesis with the added advantage of genetic tractability and host defense (Clatworthy et al. 2009). Zebra fish embryo enables detailed real-time analysis of the dynamics of host–pathogen interaction in cases of pneumococcal meningitis (Jim et al. 2016). Similarly, zebra fish embryos infected with DsRed-labelled *Salmonella typhimurium* allows non-invasive, fast, and real-time observation of the bacterial localization and fate of the pathogen in the model host, using a multi-dimensional digital imaging microscopy (van der Sar et al. 2003). The study of the *Mycobacterium marinum*–zebra fish infection model has generated valuable insights into the pathogenesis of human tuberculosis, with much medical relevance (Berg and Ramakrishnan 2012).

Zebra fish embryo serves an excellent host model to study the different virulence factors contributing to the different steps of infection of *Vibrio cholerae*. The colonization of human pathogen, *V. cholerae* in the model organism was found to occur in close proximity to the intestinal epithelium similarly to the colonization observed in mammals (Runft et al. 2014). Zebra fish larvae have also been also used in the study of *Aeromonas hydrophila* infections. The exceptionally transparent zebra fish larvae allowed non-invasive imaging of the bacterial cells and various steps of the infection process (Saraceni et al. 2016).

27.3.2.2 Application of Zebra fish in the Screening of Antimicrobial Drugs

Zebra fish embryos and early larvae serve as excellent infection models in identifying novel drugs at the preclinical phase of drug discovery (Clatworthy et al. 2009). The zebra fish-based screening approach allows concurrent analysis of bioavailability and toxicity of antimicrobial drugs (Meijer et al. 2014). Table 27.3 represents few antimicrobial agents which were studied for their antimicrobial efficacy using zebra fish model. The use of nanoparticles (NPs) for drug delivery is believed to possess enormous potential in controlling different diseases. The efficacy of Afzelin and Quercetin capped AgNPs in controlling the proliferation of *Salmonella typhi* was studied in zebra fish (Lotha et al. 2018b). In another report, capsaicinoid-coated AgNPs were found to inhibit both planktonic and biofilm mode of growth of *S. aureus* in the zebra fish (Lotha et al. 2018a). The antibacterial efficacy of gelatin

Table 27.3 Antimicrobial agents which were studied for their antimicrobial efficacy in the zebra fish model

Drugs/compounds	Target pathogen	Infection model	Reference
Quercetin- and Afzelin-capped silver nanoparticles	*S. enterica* serovar Typhi	Adult zebra fish	Lotha et al. (2018b)
Silver nanoparticles capped with capsaicinoids	*S. aureus*	Adult zebra fish	Lotha et al. (2018a)
Vancomycin-loaded gelatin nanospheres	*S. aureus*	Zebra fish larvae	Zhang et al. (2018)
Antimicrobial peptide, Bactenecin 5	*M. marinum*	Embryo	Price et al. (2019)
Antimicrobial peptide, Epinecidin-1	*V. vulnificus*	Adult zebra fish	Pan et al. (2011)
The bioactive compound from *Streptomyces* sp. ASK2	*K. pneumoniae*	Adult zebra fish	Cheepurupalli et al. (2017)
Epicatechin gallate	*S. aureus*	Embryo	Stevens et al. (2015)
Rifampicin, Pretomanid, Delamanid	*M. marinum*	Zebra fish larvae	Dalton et al. (2017)
Cell-free supernatant of *B. subtilis* subsp. *spizizenii* and *B. thuringiensis*	*Vibrio cholerae*, *S. aureus*	Embryo	Ravindran et al. (2016)

nanospheres was evaluate using *S. aureus*-infected zebra fish larvae and its ability in facilitating delivery of vancomycin was also studied. The nanospheres significantly enhance the survival of *S. aureus*-infected zebra fish larvae, suggesting considerable therapeutic effect of vancomycin against *S. aureus* infection (Zhang et al. 2018).

The anti-mycobactericidal activity of the antimicrobial peptide, Bactenecin 5 was demonstrated using the zebra fish model. A single dose of the peptide was able to inhibit the infection of zebra fish embryos with *M. marinum* (Price et al. 2019). The protective effect of another antimicrobial peptide, Epinecidin-1 against acute bacterial infection caused by *Vibrio vulnificus* was also studied in zebra fish. Epinecidin-1 was found to significantly decrease the mortality and modulated some immune-related genes in zebra fish without exerting any cytotoxicity (Pan et al. 2011). Bioactive molecule from *Streptomyces* sp. was assessed for drug-induced toxicity and efficacy in *Klebsiella pneumonia* infected adult zebra fish. An increase in the survival rate of infected zebra fish suggested the potential therapeutic effect of the bioactive compound in curtailing the bio-burden of *K. pneumonia* (Cheepurupalli et al. 2017). Zebra fish embryo model was also used to study the capacity of Epicatechin gallate (ECg) in altering the course of infection of *β*-lactam-resistant *S. aureus*. On pretreatment with ECg, the lethality of MRSA for zebra fish embryos was reduced considerably in the presence as well as in the absence of the β-lactam oxacillin (Stevens et al. 2015).

Zebra fish larvae infected with bioluminescent *M. marinum* was adopted for the rapid screening of rifampicin- and nitroimidazole-based anti-mycobacterial compounds (Dalton et al. 2017). Cell-free supernatant of *Bacillus subtilis* subsp. *spizizenii* and *Bacillus thuringiensis* exhibited profound antagonistic effect against *Vibrio cholerae* and *S. aureus*, respectively, in the zebra fish model (Ravindran et al. 2016). Hence, the use of zebra fish allows successful preclinical drug evaluation and toxicity analysis against numerous human pathogens.

27.3.3 Insect Models

For decades, conventional in vivo experiments have relied on the use of vertebrates including mice, ferrets, rats, and guinea pigs to assess the efficacy of antimicrobial agents against a range of pathogens. Although these animals facilitate direct application to humans, the cost and ethical constrains associated has led to the exploration of alternative model systems. Due to the strong functional and structural similarities between the immune response of insects to that of mammals, insects like *Drosophila melanogaster*, *Galleria mellonella*, and *Bombyx mori* are now being employed to study the changes in the virulence phenotypes and therapeutic efficacy of novel antimicrobial drugs (Kavanagh and Fallon 2010).

27.3.3.1 *Drosophila melanogaster*

Fruit fly (*D. melanogaster*) has been used as an in vivo model organism in numerous areas of research for nearly a century. Few of the important aspects of fruit fly which offer great advantages over mammals in antibacterial drug discovery include: (1)

high fecundity and inexpensive (Vodovar et al. 2004), (2) requires little effort in maintenance and raises almost no ethical issues (Ahsan et al. 2019), (3) a high degree of sequence similarity (up to 75%) with human genome, and (4) innate immune system is remarkably similar to that of humans (Needham et al. 2004).

27.3.3.1.1 Drosophila as an Infection Model for Host–Pathogen Interactions

D. melanogaster is considered as a pioneer in vivo model organism used in elucidating the molecular mechanisms of host–microbe interaction and immunity on instigation by infection (Ahsan et al. 2019). Fruit flies are also used to study the gene expression during the infection process and allows high-throughput screening of novel anti-infectives. *D. melanogaster* has been recognized as an infection model for a diverse range of bacterial pathogens including *B. subtilis*, *E. faecalis*, *P. aeruginosa*, *M. marinum*, *S. aureus*, *L. monocytogenes*, and *Streptococcus* sp. (Needham et al. 2004). The oral infection of Drosophila by *P. aeruginosa* results in biofilm development in the crop region of the host (Mulcahy et al. 2011). *S. aureus* causes lethal systemic infection to the fruit flies, thereby facilitating high-throughput screening of virulence factors (Needham et al. 2004). The screening of genetic variants of *P. aeruginosa* using Drosophila has also led to the identification of the pil–chp transduction system which plays a significant role in *P. aeruginosa* pathogenicity (Vodovar et al. 2004).

27.3.3.1.2 Application of *D. melanogaster* in the Screening of Antimicrobial Drugs

D. melanogaster is regarded as one of the suitable model organisms in drug discovery and widely used in the study of emerging human diseases (Ahsan et al. 2019). This non-mammalian model serves as a cost-effective system for the development and validation of low toxicity and high efficacy of numerous antimicrobial compounds (Table 27.4). *D. melanogaster* model was used to investigate the antifungal activity and the toxicity of ellagic acid (EA) against *C. albicans*. EA was found to exhibit a protective effect and reduce the bio-burden of *C. albicans* in *D. melanogaster* (Sampaio et al. 2018). The efficacy of novel antibacterial peptides and peptoids in the treatment of methicillin-resistant *S. aureus* (MRSA) infections were evaluated using *D. melanogaster*. The study resulted in the identification of two lantibiotics, NAI-107 and nisin which were able to rescue adult flies from the fatal

Table 27.4 Antimicrobial agents which were studied for their antimicrobial efficacy in the *D. melanogaster* model

Drugs/compound	Target pathogen	Reference
Ellagic acid	*C. albicans*	Sampaio et al. (2018)
Lantibiotic NAI-107	Methicillin-Resistant *S. aureus*	Thomsen et al. (2016)
Ulva reticulata extract	*S. aureus*	Nainu et al. (2018)
Hibiscus sabdariffa L. calyx extract	*S. aureus*	Ahsan et al. (2019)
Fucoidan coated ciprofloxacin loaded chitosan nanoparticles	*Salmonella paratyphi* A	Elbi et al. (2017)

infections of *S. aureus* without causing adverse effect to the model host (Thomsen et al. 2016). The anti-staphylococcal effect of *Ulva reticulata* extract *reticulata* and *Hibiscus sabdariffa* L. against *S. aureus* was studied using genetically tractable *D. melanogaster* (Nainu et al. 2018; Ahsan et al. 2019). On evaluation of in vivo antimicrobial efficacy and toxicity of Fucoidan coated ciprofloxacin loaded chitosan nanoparticles using *D. melanogaster*, the nanoparticles were found to show significant antibacterial activity and effectively dispersed *Salmonella paratyphi* A gallstone biofilm (Elbi et al. 2017).

27.3.3.2 Silkworm (*Bombyx mori*)

Silkworm is a typical representative of Lepidoptera insects which is widely used in drug discovery. This simple insect has a short generation time but an efficient immune system, with an innate immune response having a close resemblance to mammals. It possesses numerous genes that are homologous to human. Silkworm has diverse mutant strains and morphological mutations. The moderate body size of silkworm allows easy dissection to obtain organs and tissues such as silk gland, midgut, and hemolymph, for experimental purposes. Moreover, oral administration and intravenous injection experiments similarly to those of mammals are successfully conducted using silkworm models (Meng et al. 2017).

27.3.3.2.1 Silkworm as an Infection Model for Host–Pathogen Interactions

As an infection model, silkworm provides a screening platform for simultaneous evaluation of both the toxicity and the pharmacokinetics of the test compounds with less stringent ethical issues (Matsumoto et al. 2012). Studies have found that silkworm is highly sensitive to pathogenic fungal and bacterial strains of humans. *A. fumigatus* and *Candida* sp, including *C. albicans*, *C. glabrata*, *C. neoformans*, and *C. tropicalis* are known to cause death in silkworm within a few days of injection (Ishii et al. 2017). Intravenous injection of *S. aureus*, *P. aeruginosa*, or *V. cholerae* in the fifth instar larvae of silkworm resulted in the death of over 90% of the infected larvae in 2 days of treatment (Kaito et al. 2002). The use of silkworm infection model led to identification of numerous virulence genes (*cvfA*, *cvfB*, and *cvfC*) of *S. aureus*. Interestingly, *cvfA*-deficient mutants of *S. pyogenes* were also found to produce less exotoxin and showed attenuated virulence in both silkworm and mice model. This findings provided possibility for obtaining similar results upon clinical trial on human subjects (Kaito et al. 2005).

27.3.3.2.2 Application of Silkworm in the Screening of Antimicrobial Drugs

Silkworm serves as an inexpensive model system for determining the pharmacokinetic parameters of drug candidates at an early stage of drug development. Some of the antimicrobial agents which were studied for their antimicrobial efficacy in silkworm model is presented in Table 27.5. Paudel et al. (2018) used silkworm infection model to illustrate the correlation between the pharmacokinetic parameters and therapeutic activity of GPI0039, a novel nitrofuran dichloro-benzyl ester (Paudel

Table 27.5 Antimicrobial agents which were studied for their antimicrobial efficacy in silkworm model

Drug/compound	Target pathogen	Reference
Amphotericin B, fluconazole, ketoconazole, flucytosine	*C. neoformans*	Matsumoto et al. (2012)
Terbinafine, itraconazole	*A. vanbreuseghemii*	Ishii et al. (2017)
Vancomycin, linezolid, oxazolidinones	*S. aureus*	Barman et al. (2008)
Lysocin E	*S. aureus*	Hamamoto et al. (2015)
Nosokomycins	Methicillin-resistant *S. aureus*	Uchida et al. (2010)
Chloramphenicol, vancomycin, tetracycline, kanamycin	*S. aureus*	Hamamoto et al. (2004)
Antimicrobial peptides, lariatin A, calpinactam, lysocin E, propeptin	*M. smegmatis*	Yagi et al. (2017)
Staphylococcal phages, S25-3 and S13	*S. aureus*	Takemura-Uchiyama et al. (2013)

et al. 2018). Hamamoto et al. (2009) demonstrated that the lethal dose and metabolic pathway of the cytotoxic compound, umbelliferone, 4-methyl umbelliferone, and 7-ethoxycoumarine in silkworms were consistent with those in the mammalian mice model (Hamamoto et al. 2009). The therapeutic effects of antifungal drugs, i.e., flucytosine, amphotericin B, fluconazole, and ketoconazole were quantitatively validated using *C. neoformans*-infected silkworm as a model system (Matsumoto et al. 2012). Ishii et al. (2017) adopted silkworm dermatophyte infection model to validate the antifungal efficacy of itraconazole and terbinafine towards *A. vanbreuseghemii* (Ishii et al. 2017). Barman et al. (2008) used silkworm larvae to investigate the therapeutic effects of vancomycin, linezolid, and oxazolidinones against *S. aureus* (Barman et al. 2008). Hamamoto et al. (2015) used silkworm model to successfully determine the membrane damaging effect of lysocin E, against *S. aureus* (Hamamoto et al. 2015). The antagonistic effect of novel antibiotics, nosokomycin and ASP2397 toward methicillin-resistant *S. aureus* (MRSA) was screened rapidly using silkworm larvae (Uchida et al. 2010). The therapeutic effects of chloramphenicol, vancomycin, tetracycline, and vancomycin were evaluated quantitatively using silkworm infected with *S. aureus*, *S. maltophilia*, *C. albicans*, or *C. tropicalis* (Hamamoto et al. 2004). The antimycobacterial effect of four microbial peptides (lariatin A, calpinactam, lysocin E, and propeptin) against *M. smegmatis* were identified using silkworm infection assay (Yagi et al. 2017). A significant enhancement in the survival rate in the silkworm larvae infected with *S. aureus* was observed as a result of bacteriophage therapy (Takemura-Uchiyama et al. 2013).

27.3.4 *Saccharomyces cerevisiae*

Saccharomyces cerevisiae is a popular and widely used model organism due to its low generation time, ease of maintenance, and well-characterized genetic system.

The whole *S. cerevisiae* carries its genetic information in the form of ~6200 open reading frames (ORFs) which contain a multitude of selective markers, including markers for nutritional selection, drug susceptibility, and drug resistance. The remarkable conservation of gene function throughout evolution makes yeast an excellent model organism for drug discovery (Auerbach et al. 2005). Further, the molecular mechanisms regulating cellular processes that are affected during a bacterial infection such as DNA metabolism, cell cycle, apoptosis, cytoskeletal dynamics, and membrane traffic are relatively conserved from yeast to mammals (Valdivia 2004).

27.3.4.1 Application of *S. cerevisiae* in the Screening of Antimicrobial Drugs

S. cerevisiae has gained much popularity in recent times as a screening platform to identify novel bioactive compounds. Engineered yeast strains are used to assess the toxicity of drug candidates and to establish large-scale protein interaction maps for identification of novel drug candidates that modify or inhibit protein–protein interactions (Auerbach et al. 2005). Due to its tractable genetic systems, yeast offers rapid characterization of the host targets of bacterial virulence factors and in understanding host responses to the infections (Valdivia 2004).

S. cerevisiae was used as a model organism to elucidate the fungicidal action of thymol and eugenol. The antifungal compounds were found to involve in alteration of both cell wall and cell membrane of the yeast cells (Bennis et al. 2004). For *S. cerevisiae*, studies on genome-wide loss of function have been made possible by the creation of yeast deletion libraries created by the *Saccharomyces* Genome Deletion Project consortium (Auerbach et al. 2005). *S. cerevisiae* mutant strains were used to identify new genes affecting susceptibility to the antifungal agent, caspofungin. The antifungal agent was found to specifically interfere with cell wall formation and glucan synthesis of the target pathogen (Markovich et al. 2004). The molecular target and mechanism of antifungal action of rapamycin against opportunistic fungal pathogen, *Mucor circinelloides* was established using *S. cerevisiae* mutant strain (Bastidas et al. 2012).

References

Ahsan M, Gonsales AV, Sartini S, Wahyudin E, Nainu F (2019) *In vivo* anti-staphylococcal activity of roselle (*Hibiscus sabdariffa* L) calyx extract in *Drosophila* model of infection. J Herbmed Pharmacol 8(1):41–46

Auerbach D, Arnoldo A, Bogdan B, Fetchko M, Stagljar T (2005) Drug discovery using yeast as a model system: a functional genomic and proteomic view. Curr Proteomics 2(1):1–13

Barman TK, Arora P, Rao M, Bhadauriya T, Upadhyay DJ (2008) Utilization of *Bombyx mori* larvae as a surrogate animal model for evaluation of the anti-infective potential of oxazolidinones. J Infect Chemother 14(2):166–169

Bastidas RJ, Shertz CA, Lee SC, Heitman J, Cardenas ME (2012) Rapamycin exerts antifungal activity *in vitro* and *in vivo* against *Mucor circinelloides via* FKBP12-dependent inhibition of Tor. Eukaryot Cell 11(3):270–281

Bennis S, Chami F, Chami N, Bouchikhi T, Remmal A (2004) Surface alteration of *Saccharomyces cerevisiae* induced by thymol and eugenol. Lett Appl Microbiol 38(6):454–458

Berg RD, Ramakrishnan L (2012) Insights into tuberculosis from the zebrafish model. Trends Mol Med 18(12):689–690

Brackman G, Breyne K, De Rycke R, Vermote A, Van Nieuwerburgh F, Meyer E, Coenye T (2016) The quorum sensing inhibitor hamamelitannin increases antibiotic susceptibility of *Staphylococcus aureus* biofilms by affecting peptidoglycan biosynthesis and eDNA release. Sci Rep 6:20321

Cheepurupalli L, Raman T, Rathore SS, Ramakrishnan J (2017) Bioactive molecule from *Streptomyces* sp mitigates MDR *Klebsiella pneumoniae* in zebrafish infection model. Front Microbiol 8:614

Chen R, Willcox MD, Ho KK, Smyth D, Kumar N (2016) Antimicrobial peptide melimine coating for titanium and its *in vivo* antibacterial activity in rodent subcutaneous infection models. Biomaterials 85:142–151

Clatworthy AE, Lee JS, Leibman M, Kostun Z, Davidson AJ, Hung DT (2009) *Pseudomonas aeruginosa* infection of zebrafish involves both host and pathogen determinants. Infect Immun 77(4):1293–1303

Dalton JP, Uy B, Okuda KS, Hall CJ, Denny WA, Crosier PS, Wiles S (2017) Screening of anti-mycobacterial compounds in a naturally infected zebrafish larvae model. J Antimicrob Chemother 72(2):421–427

Elbi S, Raveendran N, Rajan VK, Baranwal G, Biswas R, Jayakumar R, Sathianarayanan S (2017) Fucoidan coated ciprofloxacin loaded chitosan nanoparticles for the treatment of intracellular and biofilm infections of *Salmonella*. Colloids Surf B Biointerfaces 160:40–47

Escarcega-Gonzalez CE, Garza-Cervantes JA, Vazquez-Rodriguez A, Montelongo-Peralta LZ, Trevino-Gonzalez MT, Diaz Barriga Castro E, Morones-Ramirez JR (2018) *In vivo* antimicrobial activity of silver nanoparticles produced via a green chemistry synthesis using *Acacia rigidula* as a reducing and capping agent. Int J Nanomedicine 13:2349–2363

Ewbank JJ, Zugasti O (2011) *C. elegans*: model host and tool for antimicrobial drug discovery. Dis Model Mech 4(3):300–304

Freeman J, Baines SD, Jabes D, Wilcox MH (2005) Comparison of the efficacy of ramoplanin and vancomycin in both *in vitro* and *in vivo* models of clindamycin-induced *Clostridium difficile* infection. J Antimicrob Chemother 56(4):717–725

Gupta S, Tyagi S, Bishai WR (2015) Verapamil increases the bactericidal activity of bedaquiline against *Mycobacterium tuberculosis* in a mouse model. Antimicrob Agents Chemother 59(1):673–676

Hamamoto H, Kurokawa K, Kaito C, Kamura K, Manitra Razanajatovo I, Kusuhara H, Sekimizu K (2004) Quantitative evaluation of the therapeutic effects of antibiotics using silkworms infected with human pathogenic microorganisms. Antimicrob Agents Chemother 48(3):774–779

Hamamoto H, Tonoike A, Narushima K, Horie R, Sekimizu K (2009) Silkworm as a model animal to evaluate drug candidate toxicity and metabolism. Comp Biochem Physiol C Toxicol Pharmacol 149(3):334–339

Hamamoto H, Urai M, Ishi K, Yasukawa J, Paudel A, Murai M, Sekimizu K (2015) Lysocin E is a new antibiotic that targets menaquinone in the bacterial membrane. Nat Chem Biol 11(2):127–133

Husain FM, Ahmad I, Asif M, Tahseen Q (2013) Influence of clove oil on certain quorum-sensing-regulated functions and biofilm of *Pseudomonas aeruginosa* and *Aeromonas hydrophila*. J Biosci 38(5):835–844

Ishii M, Matsumoto Y, Yamada T, Abe S, Sekimizu K (2017) An invertebrate infection model for evaluating anti-fungal agents against dermatophytosis. Sci Rep 7(1):12289

Jayamani E, Rajamuthiah R, Larkins-Ford J, Fuchs BB, Conery AL, Vilcinskas A, Mylonakis E (2015) Insect-derived cecropins display activity against *Acinetobacter baumannii* in a whole-animal high-throughput *Caenorhabditis elegans* model. Antimicrob Agents Chemother 59(3):1728–1737

Jayamani E, Tharmalingam N, Rajamuthiah R, Coleman JJ, Kim W, Okoli I, Mylonakis E (2017) Characterization of a *Francisella tularensis-Caenorhabditis elegans* pathosystem for the evaluation of therapeutic compounds. Antimicrob Agents Chemother 61(9)

Jim KK, Engelen-Lee J, van der Sar AM, Bitter W, Brouwer MC, van der Ende A, Vandenbroucke-Grauls CM (2016) Infection of zebrafish embryos with live fluorescent *Streptococcus pneumoniae* as a real-time pneumococcal meningitis model. J Neuroinflammation 13(1):188

Kaito C, Akimitsu N, Watanabe H, Sekimizu K (2002) Silkworm larvae as an animal model of bacterial infection pathogenic to humans. Microb Pathog 32(4):183–190

Kaito C, Kurokawa K, Matsumoto Y, Terao Y, Kawabata S, Hamada S, Sekimizu K (2005) Silkworm pathogenic bacteria infection model for identification of novel virulence genes. Mol Microbiol 56(4):934–944

Kato IT, Prates RA, Sabino CP, Fuchs BB, Tegos GP, Mylonakis E, Ribeiro MS (2013) Antimicrobial photodynamic inactivation inhibits *Candida albicans* virulence factors and reduces *in vivo* pathogenicity. Antimicrob Agents Chemother 57(1):445–451

Kavanagh K, Fallon JP (2010) *Galleria mellonella* larvae as models for studying fungal virulence. Fungal Biol Rev 24(1):79–83

Krausz AE, Adler BL, Cabral V, Navati M, Doerner J, Charafeddine RA, Friedman AJ (2015) Curcumin-encapsulated nanoparticles as innovative antimicrobial and wound healing agent. Nanomedicine 11(1):195–206

Lee YJ, Jang HJ, Chung IY, Cho YH (2018) *Drosophila melanogaster* as a polymicrobial infection model for *Pseudomonas aeruginosa* and *Staphylococcus aureus*. J Microbiol 56(8):534–541

Lieschke GJ, Currie PD (2007) Animal models of human disease: zebrafish swim into view. Nat Rev Genet 8(5):353–367

Locher HH, Seiler P, Chen X, Schroeder S, Pfaff P, Enderlin M, Keck W (2014) *In vitro* and *in vivo* antibacterial evaluation of cadazolid, a new antibiotic for treatment of *Clostridium difficile* infections. Antimicrob Agents Chemother 58(2):892–900

Lotha R, Shamprasad BR, Sundaramoorthy NS, Ganapathy R, Nagarajan S, Sivasubramanian A (2018a) Zero valent silver nanoparticles capped with capsaicinoids containing *Capsicum annuum* extract, exert potent anti-biofilm effect on food borne pathogen *Staphylococcus aureus* and curtail planktonic growth on a zebrafish infection model. Microb Pathog 124:291–300

Lotha R, Sundaramoorthy NS, Shamprasad BR, Nagarajan S, Sivasubramanian A (2018b) Plant nutraceuticals (Quercetrin and Afzelin) capped silver nanoparticles exert potent antibiofilm effect against food borne pathogen *Salmonella enterica* serovar Typhi and curtail planktonic growth in zebrafish infection model. Microb Pathog 120:109–118

Markovich S, Yekutiel A, Shalit I, Shadkchan Y, Osherov N (2004) Genomic approach to identification of mutations affecting caspofungin susceptibility in *Saccharomyces cerevisiae*. Antimicrob Agents Chemother 48(10):3871–3876

Matsumoto Y, Miyazaki S, Fukunaga DH, Shimizu K, Kawamoto S, Sekimizu K (2012) Quantitative evaluation of cryptococcal pathogenesis and antifungal drugs using a silkworm infection model with *Cryptococcus neoformans*. J Appl Microbiol 112(1):138–146

Meijer AH, van der Vaart M, Spaink HP (2014) Real-time imaging and genetic dissection of host-microbe interactions in zebrafish. Cell Microbiol 16(1):39–49

Meng X, Zhu F, Chen K (2017) Silkworm: a promising model organism in life science. J Insect Sci 17(5)

Mihailescu R, Furustrand Tafin U, Corvec S, Oliva A, Betrisey B, Borens O, Trampuz A (2014) High activity of Fosfomycin and Rifampin against methicillin-resistant *Staphylococcus aureus* biofilm *in vitro* and in an experimental foreign-body infection model. Antimicrob Agents Chemother 58(5):2547–2553

Moy TI, Ball AR, Anklesaria Z, Casadei G, Lewis K, Ausubel FM (2006) Identification of novel antimicrobials using a live-animal infection model. Proc Natl Acad Sci U S A 103(27):10414–10419

Moy TI, Conery AL, Larkins-Ford J, Wu G, Mazitschek R, Casadei G, Ausubel FM (2009) High-throughput screen for novel antimicrobials using a whole animal infection model. ACS Chem Biol 4(7):527–533

Mukhopadhyay A, Peterson RT (2006) Fishing for new antimicrobials. Curr Opin Chem Biol 10(4):327–333

Mulcahy H, Sibley CD, Surette MG, Lewenza S (2011) *Drosophila melanogaster* as an animal model for the study of *Pseudomonas aeruginosa* biofilm infections *in vivo*. PLoS Pathog 7(10):e1002299

Nainu F, Asri RM, Arsyad A, Manggau MA, Amir MN (2018) *In vivo* antibacterial activity of green algae *Ulva reticulata* against *Staphylococcus aureus* in *Drosophila* model of infection. Pharmacog J 10(5):993–997

Narayana JL, Huang HN, Wu CJ, Chen JY (2015) Epinecidin-1 antimicrobial activity: *In vitro* membrane lysis and *in vivo* efficacy against *Helicobacter pylori* infection in a mouse model. Biomaterials 61:41–51

Needham AJ, Kibart M, Crossley H, Ingham PW, Foster SJ (2004) *Drosophila melanogaster* as a model host for *Staphylococcus aureus* infection. Microbiology 150(Pt 7):2347–2355

O'Loughlin CT, Miller LC, Siryaporn A, Drescher K, Semmelhack MF, Bassler BL (2013) A quorum-sensing inhibitor blocks *Pseudomonas aeruginosa* virulence and biofilm formation. Proc Natl Acad Sci U S A 110(44):17981–17986

Pan CY, Wu JL, Hui CF, Lin CH, Chen JY (2011) Insights into the antibacterial and immunomodulatory functions of the antimicrobial peptide, epinecidin-1, against *Vibrio vulnificus* infection in zebrafish. Fish Shellfish Immunol 31(6):1019–1025

Paudel A, Panthee S, Urai M, Hamamoto H, Ohwada T, Sekimizu K (2018) Pharmacokinetic parameters explain the therapeutic activity of antimicrobial agents in a silkworm infection model. Sci Rep 8(1):1578

Perwitasari O, Johnson S, Yan X, Register E, Crabtree J, Gabbard J, Tripp RA (2016) Antiviral efficacy of verdinexor *in vivo* in two animal models of influenza A virus infection. PLoS One 11(11):e0167221

Peterson ND, Pukkila-Worley R (2018) *Caenorhabditis elegans* in high-throughput screens for anti-infective compounds. Curr Opin Immunol 54:59–65

Price RL, Bugeon L, Mostowy S, Makendi C, Wren BW, Williams HD, Willcocks SJ (2019) *In vitro* and *in vivo* properties of the bovine antimicrobial peptide, Bactenecin 5. PLoS One 14(1):e0210508

Rajamuthiah R, Fuchs BB, Jayamani E, Kim Y, Larkins-Ford J, Conery A, Mylonakis E (2014) Whole animal automated platform for drug discovery against multi-drug resistant *Staphylococcus aureus*. PLoS One 9(2):e89189

Ravindran C, Varatharajan GR, Rajasabapathy R, Sreepada RA (2016) Antibacterial activity of marine *Bacillus* substances against *V cholerae* and *S aureus* and *in vivo* evaluation using embryonic zebrafish test system. Indian J Pharm Sci 78:417–422

Runft DL, Mitchell KC, Abuaita BH, Allen JP, Bajer S, Ginsburg K, Withey JH (2014) Zebrafish as a natural host model for *Vibrio cholerae* colonization and transmission. Appl Environ Microbiol 80(5):1710–1717

Sampaio ADG, Gontijo AVL, Araujo HM, Koga-Ito CY (2018) *In vivo* efficacy of ellagic acid against *Candida albicans* in a *Drosophila melanogaster* infection model. Antimicrob Agents Chemother 62(12)

Saraceni PR, Romero A, Figueras A, Novoa B (2016) Establishment of infection models in zebrafish larvae (*Danio rerio)* to study the pathogenesis of *Aeromonas hydrophila*. Front Microbiol 7:1219

Saralahti A, Ramet M (2015) Zebrafish and streptococcal infections. Scand J Immunol 82(3):174–183

Simonetti O, Cirioni O, Cacciatore I, Baldassarre L, Orlando F, Pierpaoli E, Offidani A (2016) Efficacy of the quorum sensing inhibitor FS10 alone and in combination with tigecycline in an animal model of staphylococcal infected wound. PLoS One 11(6):e0151956

Souza ACR, Fuchs BB, Alves VS, Jayamani E, Colombo AL, Mylonakis E (2018) Pathogenesis of the *Candida parapsilosis* complex in the model host *Caenorhabditis elegans*. Genes (Basel) 9(8):401

Stevens CS, Rosado H, Harvey RJ, Taylor PW (2015) Epicatechin gallate, a naturally occurring polyphenol, alters the course of infection with beta-lactam-resistant *Staphylococcus aureus* in the zebrafish embryo. Front Microbiol 6:1043

Takemura-Uchiyama I, Uchiyama J, Kato S, Inoue T, Ujihara T, Ohara N, Matsuzaki S (2013) Evaluating efficacy of bacteriophage therapy against *Staphylococcus aureus* infections using a silkworm larval infection model. FEMS Microbiol Lett 347(1):52–60

Thomsen TT, Mojsoska B, Cruz JC, Donadio S, Jenssen H, Lobner-Olesen A, Rewitz K (2016) The lantibiotic NAI-107 efficiently rescues *Drosophila melanogaster* from infection with methicillin-resistant *Staphylococcus aureus* USA300. Antimicrob Agents Chemother 60(9):5427–5436

Uccelletti D, Zanni E, Marcellini L, Palleschi C, Barra D, Mangoni ML (2010) Anti-*Pseudomonas* activity of frog skin antimicrobial peptides in a *Caenorhabditis elegans* infection model: a plausible mode of action *in vitro* and *in vivo*. Antimicrob Agents Chemother 54(9):3853–3860

Uchida R, Iwatsuki M, Kim YP, Ohte S, Omura S, Tomoda H (2010) Nosokomycins, new antibiotics discovered in an *in vivo*-mimic infection model using silkworm larvae I: fermentation, isolation and biological properties. J Antibiot (Tokyo) 63(4):151–155

Valdivia RH (2004) Modeling the function of bacterial virulence factors in *Saccharomyces cerevisiae*. Eukaryot Cell 3(4):827–834

van der Sar AM, Musters RJ, van Eeden FJ, Appelmelk BJ, Vandenbroucke-Grauls CM, Bitter W (2003) Zebrafish embryos as a model host for the real time analysis of *Salmonella typhimurium* infections. Cell Microbiol 5(9):601–611

Vodovar N, Acosta C, Lemaitre B, Boccard F (2004) *Drosophila*: a polyvalent model to decipher host-pathogen interactions. Trends Microbiol 12(5):235–242

Yagi A, Uchida R, Hamamoto H, Sekimizu K, Kimura KI, Tomoda H (2017) Anti-*Mycobacterium* activity of microbial peptides in a silkworm infection model with *Mycobacterium smegmatis*. J Antibiot (Tokyo) 70(5):685–690

Yin H, Deng Y, Wang H, Liu W, Zhuang X, Chu W (2016) Corrigendum: tea polyphenols as an antivirulence compound disrupt quorum-sensing regulated pathogenicity of *Pseudomonas aeruginosa*. Sci Rep 6:17987

Zhang J, Holdorf AD, Walhout AJ (2017) *C. elegans* and its bacterial diet as a model for systems-level understanding of host-microbiota interactions. Curr Opin Biotechnol 46:74–80

Zhang X, Song J, Klymov A, Zhang Y, de Boer L, Jansen JA, Leeuwenburgh SC (2018) Monitoring local delivery of vancomycin from gelatin nanospheres in zebrafish larvae. Int J Nanomedicine 13:5377–5394

Zhao M, Lepak AJ, Andes DR (2016) Animal models in the pharmacokinetic/pharmacodynamic evaluation of antimicrobial agents. Bioorg Med Chem 24(24):6390–6400

Yeast: A Model Organism for Antimicrobial Drug Discovery

28

Priyanka Sharma

Contents

Abstract

The application and development of genomic technologies using *Saccharomyces cerevisiae* as a valuable model organism have provided novel and useful insights for the identification of mechanism of action of small molecules and drugs in vivo. In this review, we focus on the development of yeast-based functional genomic and proteomic approaches for advancing the effectiveness of using yeast as an excellent model organism in the drug discovery process.

Keywords

Yeast · *Saccharomyces cerevisiae* · Antimicrobial · Drug discovery

---

P. Sharma (✉)
Malaria Drug Discovery Laboratory, International Centre for Genetic Engineering and Biotechnology (ICGEB), New Delhi, India

B. Siddhardha et al. (eds.), *Model Organisms for Microbial Pathogenesis, Biofilm Formation and Antimicrobial Drug Discovery*,
https://doi.org/10.1007/978-981-15-1695-5_28

28.1 Introduction

For millennia, nature has been the best source of medicinal products with many useful bioactive products developed from microbial sources. Microbes have been an exceptional contributor to human well-being across the globe. Alexander Fleming, in 1928, discovered a compound that was produced by a mold that killed *Staphylococcus aureus* in a Petri dish. The mold was recognized as *Penicillium notatum* that produced the active agent, named penicillin. During World War II, penicillin was used as a powerful antibacterial substance (Bennett and Chung 2001). A large portion of genomes from microbes are dedicated to the production of these valuable natural products. A single microbe can make 30–50 natural product compounds (Katz and Baltz 2016). The number of biologically active secondary metabolites discovered from the fungal origin is ~8600, which represents 38% of all known microbial products. Discovered in 1896 by Gosio, the first crystalline fungal product was mycophenolic acid, which is a secondary metabolite, derived from *Penicillium glaucoma* (Berdy 2005). We have entered the "dark ages" in the past few decades in terms of novel antibiotic discovery. The reason is that the same compounds are frequently rediscovered mostly due to redundancy of the isolation samples in addition to isolation of strain and screening technologies (Baltz 2006; Wright 2015; Sharma et al. 2016). For novel drug development, it may take 10–15 years, with an estimated cost of 800 million USD approximately (DiMasi et al. 2003).

28.2 The "Budding Yeast"

Saccharomyces cerevisiae, also known as the "budding yeast" is a unicellular eukaryotic fungus with a short generation time (~90 min). It is one of the simplest forms of organism; it is inexpensive to maintain and it can grow in solid or liquid growth media. The cells of this organism are robust in size. The life cycle of *S. cerevisiae* alternates between haploid and diploid stages, following the life cycle of any sexual eukaryote (Fig. 28.1). The life cycle seems to be stable in both its diploid plus haploid forms in yeast (Duina et al. 2014). The haploid genome of this organism is small in size with low genome complexity, packaged into 16 well-characterized chromosomes (Menacho-Marquez and Murguia 2007). These simple features of yeast make it the first eukaryote to be sequenced (12.5 Mb) and its 6466 ORFs are usable readily (Hudson et al. 1997; Kowalczuk et al. 1999). It is identified as "generally recognized as safe" (GRAS) organism (Goffeau et al. 1996). These characteristics make yeast a perfect vehicle for drug discovery against antifungal infections. Its cell wall is made up of chitin, glucans, and mannoproteins. The cell wall does not allow molecules with Mr >700 to pass through it. There are also reports which confirmed that the molecules with considerably high Mr (up to 400,000) can pass through the yeast cell wall (De Nobel and Barnett 1991). There is a little amount of data available studying the effect of its cell wall on the bioavailability of tiny molecules. When there is sufficient amount of nutrients, yeast cultures are able to grow

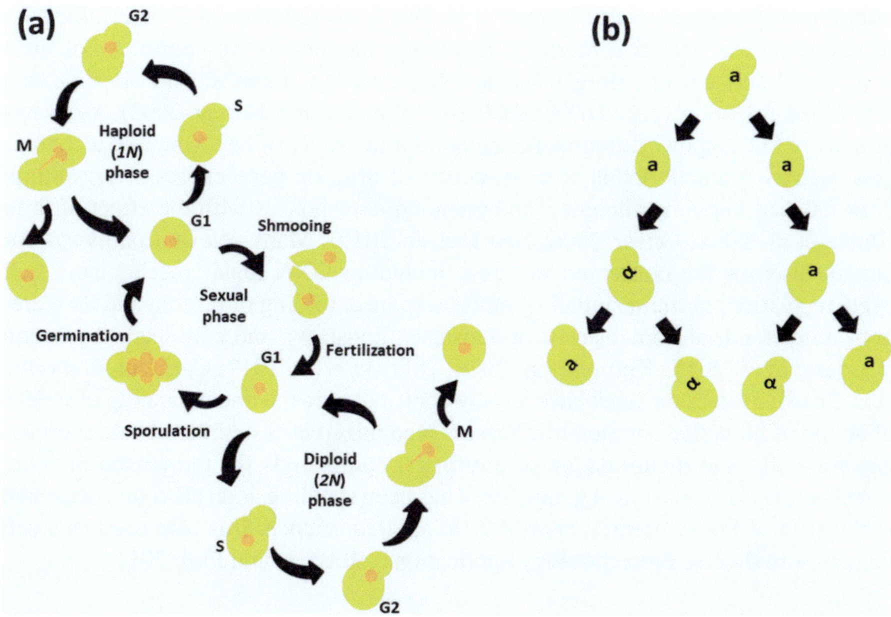

Fig. 28.1 Life cycle of *Saccharomyces cerevisiae*. (**a**) Alternation of generations and the vegetative growth. (**b**) Switching of mating type in homothallic (Ho+) haploids. Mother cells switch in the G1 phase of life cycle

in an exponential manner by asymmetric budding from mother to daughter cells. Mother cells preserve its damaged cellular material during cell division which accumulates over time. Thus, the mother cells take care of the health of their daughter cells. Ultimately, the mother cells die after 20–25 cell divisions and their cellular material is released into the environment (Steinkraus et al. 2008; Longo and Fabrizio 2012). When there is deficiency of nutrients, yeast stops dividing by entering into a stationary phase. Depending upon the yeast strain and the culturing conditions, the life span of this organism varies from few days up to several weeks in the postmitotic phase. This short life span is very suitable for drug discovery (Zimmermann et al. 2018). Penicillin is made by *P. chrysogenum* using a non-ribosomal peptide synthetase (NRPS). Awan et al. (2017) genetically engineered *S. cerevisiae* with the gene for NRPS and four fungal genes for the secretion of penicillin (Awan et al. 2017). The broth used for culturing the yeast contained the antibiotic which blocked the growth of *Streptococcus*.

Yeast has turned out to be a prominent testing ground for the research on human diseases and their pathways. Out of the ~6000 genes in yeast, about 90% have been already well represented and around 30% of its genome is preserved in humans (Stefanini et al. 2013). Almost half of the disease genes in human exhibit the orthologues of yeast and at least 31% of the yeast genome encoding proteins have human orthologue (Botstein and Fink 2011). Several drugs such as rapamycin, wortmannin, CsA, and FK506 targeting human proteins are also known to inhibit the orthologous protein in yeast (Sinclair et al. 1997). The genome of *S. cerevisiae* possesses numerous selective markers, including markers for susceptibility of drug (e.g., *cyhR* and *URA3*), drug resistance (e.g., *natMX*, *kanMX*, and *patMX*), and nutritional selection (e.g., *HIS3* and *URA3*) (Snyder and Kumar 2002). Yeast has turned out to be a powerful model organism for the study of new genomic technologies such as synthetic lethal screens, action of drug on gene expression profiling, drug-induced haploinsufficiency, and phenotypic responses with the effect of drug (Baetz et al. 2004; Oliver 2006; Smith et al. 2010). Many relevant pathways for human diseases are conserved in yeast, including DNA repair mechanisms, cell cycle regulation, nutrient signaling, stress response, folding of protein and its secretion, mitochondrial homeostasis, proteostasis, lipostasis, and cell death regulation (Lasserre et al. 2015; Knorre et al. 2016; Bilinski et al. 2017; Carmona-Gutierrez et al. 2018). Studies on yeast have largely contributed to the understanding of modes of action of bioactive compounds. Yeast is recognized as a significant model organism for study and identification of antifungal compounds for the reason of functional conservation of fungal genes. Yeast has been standing as an all-round organism in the field of basic research as well as in applied science. It is also used as a cell factory with diverse biotechnology applications (Botstein and Fink 2011).

28.3 Drug Screening Using Yeast Deletion Strain Collection

The gene deletion collection was formed by the *Saccharomyces* Genome Deletion Project consortium (http://sequencewww.stanford.edu/group/yeast_deletion_project/). This project is a community-based genomics attempt which accurately and scientifically removes all the 6000 genes in yeast from start-to-stop codon. Dual 20 base pair sequences are tagged to each deletion strain. Chemical selection in co-culture against these yeast mutants is performed with the help of this method as it serves as unique strain identifiers (Giaever et al. 2002). A second collection comprises of a group of heterozygous-diploid yeast strains removed in either essential or nonessential genes. This was done to unlock the gene sequence information in yeast for its functional understanding (Winzeler et al. 1999; Giaever et al. 1999). This opened the doors toward further understanding the "omic" platforms, including gene ontology standardization, gene expression analysis, metabolomics, proteomics, ChIP–Chip methods, and many more (Dolinski and Botstein 2005; Botstein and Fink 2011; Giaever and Nislow 2014). *S. cerevisiae* will always be a prime model organism in biology for the understanding of proteins and its pathways that can be transformed to eliminate disease (Auerbach et al. 2005; Dixon and Stockwell 2009).

Development of genomic approaches in yeast has been extensively applied for the screening of antifungals as well as for the analysis of their mechanism of action. The mutant strains with a deletion in yeast have provided maximum insight into the amalgamation of empirical screening and target-based approaches in the field of antimicrobial research. The abundance of strain-specific barcodes is used to measure relative abundance of each yeast strain and they are quantified by barcode sequencing or by microarray hybridization (Shoemaker et al. 1996; Pierce et al. 2007; Smith et al. 2009). These target-based assays allow the discovery of all genes essential for the survival of the yeast in any extreme condition, including chemical disturbance. The study of high-throughput phenotypic analyses of the mutants in yeast is enabled by the haploid and heterozygous yeast strain collections. This type of cell-based assays offers two benefits. Firstly, native conformation and then its union with the networking partners are favored by a protein target that is restricted in the cellular environment. Secondly, chemical compounds can pass through the cellular envelope of the yeast but are not exported or inactivated from the cell. Such information is time friendly and cost effective toward the path of drug development. Two examples are cited for the prospective applications of yeast haploid deletion mutant collection. In the study reported by Tucker and Fields (2004), sensitivity of the deletion strains in yeast was quantified to menadione, ibuprofen, mefloquine, and hydrogen peroxide. Six hundred and fifty-nine strains were found to be sensitive to at least one of the chemicals used (Tucker and Fields 2004). *YBR216C* was determined as a novel constituent of the response to oxidative stress from the result of large-scale two-hybrid screening and data correlation attained with hydrogen peroxide. In another approach, a team of researchers identified the genes which demonstrated transformed sensitivity to wortmannin, a clinical anti-inflammatory agent and an immunosuppressor. In this context, 4850 haploid deletion strains and 1175 strains responsible for heterozygous diploid deletion were first tested for resistance or hypersensitivity to wortmannin. Meanwhile, 1067 genes were revealed in the wortmannin responsive network. Complex connections that are novel were revealed between numerous biological modes including DNA replication and checkpoint response that are induced by DNA damage and the metabolic pathways of phosphatidylinositol (Zewail et al. 2003).

28.4 Drug-Induced Haploinsufficiency

In yeast, the result of gene dosage for drug target isolation was reported by Rine et al. (1983) for the first time (Rine et al. 1983). It was reported that the protein targets of tunicamycin and compactin were identified with the help of genomic library screening for genes which provide drug resistance when it is over-expressed. Yeast clones that possess several copies of a target gene exhibited less sensitivity to the compound compared to those clones that expressed the nontarget genes. This particular approach is used for the application of any compound targeting a yeast protein but some yeast genes are toxic when it is over-expressed. This demands the advancement of a process that allowed for genes recognition by sensitizing a yeast cell using a drug by decreasing the dosage of the target gene. The drug sensitivity screening presents an exceptional challenge to classical genetic approach. In yeast genetic screening, a mutated strain is relocated from rich media to media that contain drug with the help of replica plating. Genomic library is used for transformation of hypersensitive colonies for clonal identification exhibiting the phenotype by functional complementation of the mutated gene (Armour and Lum 2005). As a result of reduction in gene dosage of a drug target from double copies to single copy, heterozygous strains of yeast become more sensitive to inhibitory chemicals of growth with respect to the wild-type strains. This phenomenon is called drug-induced haploinsufficiency (Marton et al. 1998). A diploid cell becomes sensitive to drug when the number of copies of a drug target gene reduces from two to one (Giaever et al. 1999). Application of drug-induced haploinsufficiency profiling (HIP) confirmed to be extremely specific. Targets of antimicrobial agents that are well-characterized were recognized as the most sensitive heterozygous deletion strains from the pool of 6000 heterozygous deletion strains (Giaever et al. 2004; Lum et al. 2004; Balibar and Roemer 2016). HIP is advantageous as it helps to identify the compound that is inhibitory in nature and its candidate targets. The potency of this assay has been established in natural product extracts, novel compounds, and well-characterized compounds (Hillenmeyer et al. 2008; Lain et al. 2008; Oh et al. 2010). Screening strategy of HIP presents a genome-wide outlook of drug or chemical specificity since all the targets of the cells are presented for identification. In in vivo, single-gene products may not be targeted by the most specific drugs. Thus, mechanism of action is well understood by revealing the secondary and tertiary targets of diverse compounds via HIP. Multi-copy suppression profiling approach is widely used for the identification of drug–target interactions (Fig. 28.2) (Hoon et al. 2008; Magtanong et al. 2011).

28.5 Small-Molecule Three-Hybrid Approach

The three-hybrid system of yeast takes advantage of both genetic and molecular tools developed for protein–protein interactions detection (Fields and Song 1989). This approach can be useful for the query of the proteome of mammals for the purpose of small-molecule targets. Two small molecules (hybrid) comprising of a compound is useful for the reconstitution of a transcription factor which is functional for driving the expression of reporter gene (Fig. 28.3). Two types of interactions occur for transcriptional components dimerization. A well-known ligand–receptor interaction (e.g., dexamethasone–glucocorticoid receptor) is used as an anchor used for test compound screening for protein interaction. Interaction of the protein target with the test compound results in the reporter construct expression. In a study conducted by Becker et al. (2004), methotrexate was used as an anchor for the identification of both the novel and the recognized targets of cyclin-dependent kinase inhibitors from numerous libraries of human cDNA. Both the DNA-binding domain of LexA and the activation domain of GAL4 formed the center of a positive selection system with the help of the selectable auxotrophic marker HIS3 acting as reporter. Careful selection of yeast strain background and construction of anchor compound–protein interactions are essential determinants of assay specificity and sensitivity. The three-hybrid method is a significant tool for the future efforts of drug development (Becker et al. 2004).

28.6 Antimicrobial Target Space in Yeast

The prime focus on the discovery of potential antimicrobial targets is the yeast deletion strains. Using high-throughput functional genomics, robotic means are systematically used for the construction of double deletion mutants of the Yeast

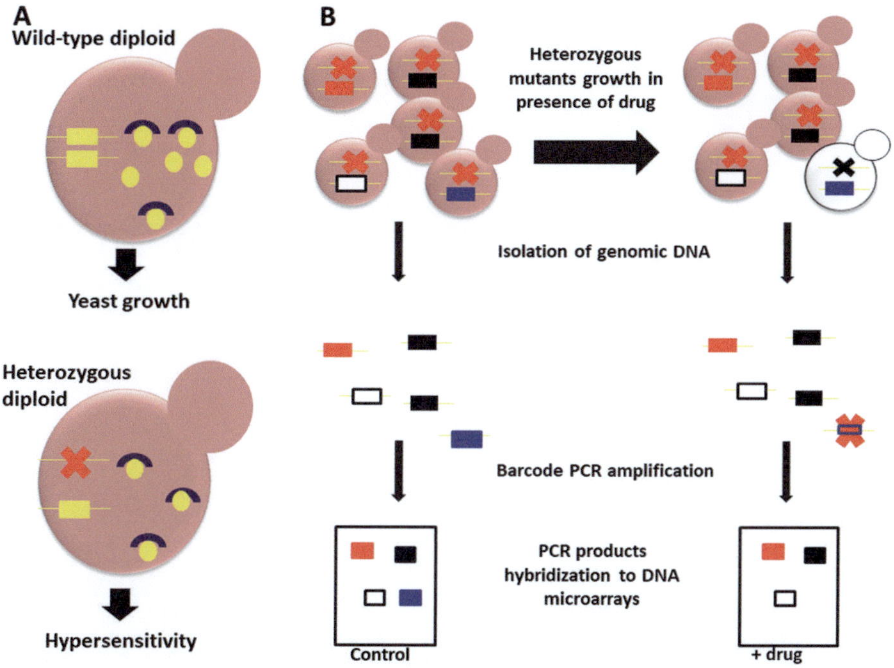

Fig. 28.2 (**a**) Drug-induced haploinsufficiency. Yeast cells that possess single copy of a drug target gene requires minimum drug (red) for inhibition of its target (blue) as compared to wild-type yeast that possess double copies. As a result, the heterozygous diploid will turn to be more sensitive when a concentration of sublethal drug is present. (**b**) Haploinsufficiency profiling. After addition of drug, hypersensitivity is displayed by the drug target gene mutants only (yeast with blue tag). After the competitive growth and the isolation of genomic DNA, PCR amplification is performed for the tagged regions. Relative abundance of heterozygotes prior to and after the drug treatment is estimated by PCR products hybridization on an array of oligonucleotide. Yeasts that possess one copy of the drug target gene (with blue tag) will not be able to grow when the drug is present and will only be detected in absence of the drug (control)

Deletion Collection (Baryshnikova et al. 2010; Kuzmin et al. 2014). This is the way in which the genome-wide genetic interactions are surveyed for the discovery of genes not necessary for cell growth but conditionally essential if there is inactivation of other genes or pathways. This enormous arrangement of synthetic lethal genetic interactions (>170,000) provide an idea for the consideration of combination agents displaying precise synergistic antimicrobial action as each of the agents inhibits exclusively one of the two genetically established synthetically lethal targets. It yields more targets, such as biofilm factors, virulence determinants, and nonessential genes. When one target act as an important factor of virulence or each agent targets an essential protein then there is an enormous implication to alleviate drug resistance. Synthetic lethal interactions in nonessential genes offer a new target set from where combination agents (cognate inhibitors) may be screened for the first time.

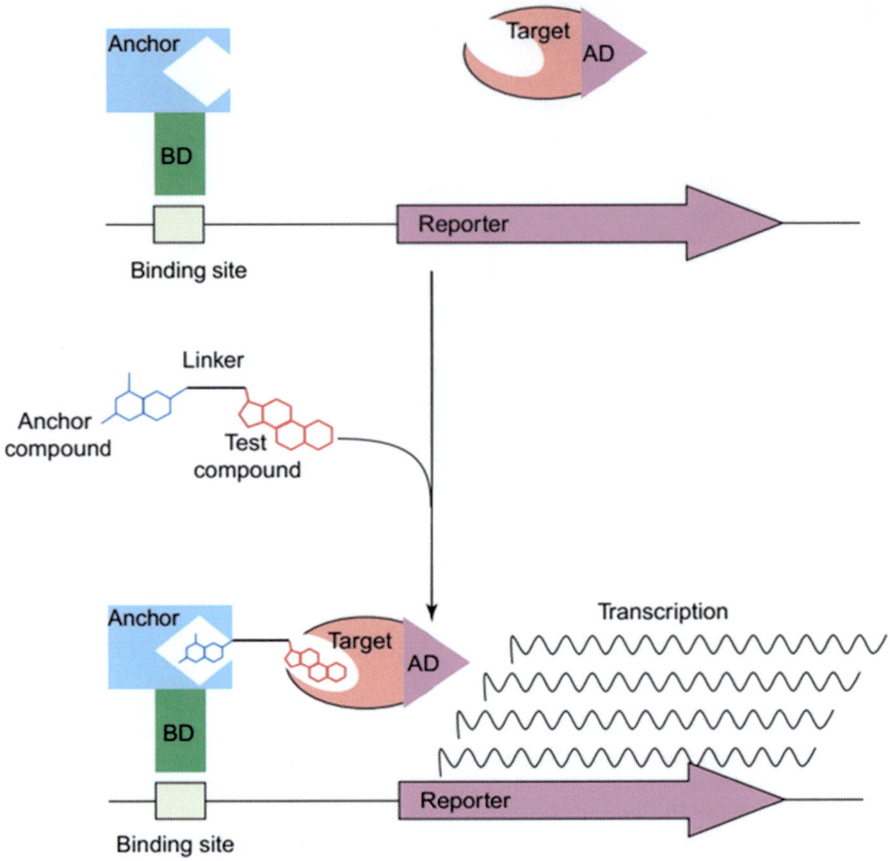

Fig. 28.3 Small-molecule three-hybrid system used to detect interactions of compound–protein. Whenever there is an interaction between an unknown target protein with a test compound, the activation of a reporter gene takes place. *BD* DNA-binding domain, *AD* transcriptional activation domain (reproduced from Armour and Lum 2005)

28.7 Yeast for High-Throughput Screening for Drug Discovery

High-throughput screening has revolutionized the process of drug discovery as this method aids in the screening of numerous compounds for biological activity (typically ≥100,000 compounds/day). High-throughput screening assays for drug discovery are performed by protein target-based and parasite cell-based in vitro screening procedures from corresponding protein targets to unicellular and multicellular organisms (Siqueira-Neto et al. 2012; Frearson et al. 2010; Gilbert 2014). As compared to target-directed platforms for eukaryotic cell-based assays for the purpose of high-throughput screening, yeast platforms are more advantageous as compared to the target-based assays and the usual cell-based assays in vitro (Table 28.1). Even though cell-based systems are significantly advantageous over the usual target-based methods in vitro, their use in the application of parasitic neglected tropical diseases is restricted by their tractability.

28.8 Future Prospects

Recent innovations like genomic technologies are integrated with drug discovery with hope for bigger success. But the effect of the revolution of genomics had a contradictory effect on antibacterial discovery. The reason being that on the basis of a superficial consideration and the simplicity for development of the biochemical assays in vitro, targets were initially chosen. Thousands of molecules were screened and hundreds of high-throughput screenings were performed but no single target-based program could produce a synthetic antibiotic of clinical use. The effort of successful

Table 28.1 Comparison of high-throughput screening strategies (Denny and Steel 2015)

Platform	Yeast-based HTS	Cellular HTS (parasites)	In vitro HTS
Advantages	• No requirement of purified target • Eukaryotic cellular context that allows to select early the drug-like compounds • Manipulation ease • Growth speed • Culture cost is low • Genetic manipulation is straightforward which allows construction of sophisticated, discriminatory platforms	• No requirement of validated target • Purified target not required • True cellular context that allows to select early the drug-like compounds • Naive unbiased approach that challenges all targets at once	• Chance of specific, simple, and sensitive assay system • SAR facilitation on the basis of target molecular recognition
Disadvantages	• Requirement of validated target • Chances of insensitivity as a result of thick cell wall, membrane barriers, and drug pumps • Reliability of substitution platform on the capability of a heterologous protein to be functional in an axenic system • Testing of target protein in a non-native cellular context • Expression in axenic context leads to target redundancy possibility	• Drug pumps and membrane barriers cause insensitivity • Costly, failure to culture significant life cycle stages • Reliability of assays on nontarget-specific and phenotypic output • Target knowledge not required for SAR • Posteriori approach for target deconvolution	• Requirement of validated target • Purified target required, limits the assay of tough purification/assay targets • Protein purification and reagents for assay makes it costly • Suitable substrates and reporters are required • Inhibitors not able to penetrate cell membrane are likely to be identified

antimicrobial discovery using yeast system requires integration of genomic technologies along with pathway-centric focus with the phenotypic screening and emphasis on natural product libraries. With the drug discovery innovations in yeast, the researchers of the bacterial community have developed a wide range of genomic resources in whole-cell context for the discovery of small molecules that are pathway specific, identification of targets, and the study of nontraditional targets (Barker et al. 2010; Farha and Brown 2015). The role of classical forward genetics studies in yeast is also of great significance wherein there is isolation of drug-resistant mutants and with the help of high-coverage next-generation sequencing method, mapping of mutations can be performed to their individual target which is an easy and unbiased technique. This technique is being used for identifying novel inhibitors of lysyl tRNA synthetase, LolCDE lipoprotein transporter, and DNA gyrase (Gomez et al. 2007; Bax et al. 2010; McLeod et al. 2015; Montgomery et al. 2015). By the use of reverse chemical genetic strategies, whole-cell screening can be multiplexed and more considerately modified for the identification of inhibitors of particular pathways and elimination of non-desirable targets or compounds. Also, the mapping of intricate lethal genetic interactions (synthetic in nature) between processes and pathways inside the cell is essential for the purpose of identification of novel antimicrobial targets that is possibly necessary, synergistic with the immune system or for virulence. This approach of system biology is also proving to be of extreme importance for the treatment of microbial infections by minimal disturbance to the microbiome.

References

Armour CD, Lum PY (2005) From drug to protein: using yeast genetics for high-throughput target discovery. Curr Opinion Chem Biol 9:20–24

Auerbach D, Arnoldo A, Bogdan B, Fetchko M, Stagljar I (2005) Drug discovery using yeast as a model system: a functional genomic and proteomic view. Curr Proteomics 2:1–13

Awan AR, Blount BA, Bell DJ, Shaw WM, Ho JCH et al (2017) Biosynthesis of the antibiotic nonribosomal peptide penicillin in baker's yeast. Nat Commun 8:15202

Baetz K, McHardy L, Gable K, Tarling T, Rebérioux D et al (2004) Yeast genome-wide drug-induced haploinsufficiency screen to determine drug mode of action. Proc Natl Acad Sci U S A 101:4525–4230

Balibar CJ, Roemer T (2016) Yeast: a microbe with macro-implications to antimicrobial drug discovery. Brief Funct Genomics 15:147–154

Baltz RH (2006) Marcel Faber Roundtable: is our antibiotic pipeline unproductive because of starvation, constipation or lack of inspiration? J Ind Microbiol Biotechnol 33:507–513

Barker CA, Farha MA, Brown ED (2010) Chemical genomic approaches to study model microbes. Chem Biol 17:624–632

Baryshnikova A, Costanzo M, Dixon S, Vizeacoumar FJ, Myers CL et al (2010) Synthetic genetic array (SGA) analysis in *Saccharomyces cerevisiae* and *Schizosaccharomyces pombe*. Methods Enzymol 470:145–179

Bax BD, Chan PF, Eggleston DS, Fosberry A, Gentry DR et al (2010) Type IIA topoisomerase inhibition by a new class of antibacterial agents. Nature 466:935–940

Becker F, Murthi K, Smith C, Come J, Costa-Roldán N et al (2004) A three-hybrid approach to scanning the proteome for targets of small molecule kinase inhibitors. Chem Biol 1:211–223

Bennett JW, Chung KT (2001) Alexander Fleming and the discovery of penicillin. Adv Appl Microbiol 49:163–184

Berdy J (2005) Bioactive microbial metabolites: a personal view. J Antibiot 58:1–26

Bilinski T, Bylak A, Zadrag-Tecza R (2017) The budding yeast *Saccharomyces cerevisiae* as a model organism: possible implications for gerontological studies. Biogerontology 18:631–640

Botstein D, Fink GR (2011) Yeast: an experimental organism for 21st century biology. Genetics 189:695–704

Carmona-Gutierrez D, Bauer MA, Zimmermann A, Aguilera A, Austriaco N et al (2018) Guidelines and recommendations on yeast cell death nomenclature. Microb Cell 5:4–31

De Nobel JG, Barnett JA (1991) Passage of molecules through yeast cell walls: a brief essay-review. Yeast 7:313–323

Denny PW, Steel PG (2015) Yeast as a potential vehicle for neglected tropical disease drug discovery. J Biomol Screen 20:56–63

DiMasi JA, Hansen RW, Grabowski HG (2003) The price of innovation: new estimates of drug development costs. J Health Econ 22:151–185

Dixon SJ, Stockwell BR (2009) Identifying druggable disease-modifying gene products. Curr Opin Chem Biol 13:549–555

Dolinski K, Botstein D (2005) Changing perspectives in yeast research nearly a decade after the genome sequence. Genome Res 15:1611–1619

Duina AA, Miller ME, Keeney JB (2014) Budding yeast for budding geneticists: a primer on the *Saccharomyces cerevisiae* model system. Genetics 197:33–48

Farha MA, Brown ED (2015) Unconventional screening approaches for antibiotic discovery. Ann N Y Acad Sci 1354:54–66

Fields S, Song O (1989) A novel genetic system to detect protein-protein interactions. Nature 340:245–246

Frearson JA, Brand S, McElroy SP et al (2010) N-myristoyltransferase inhibitors as new leads to treat sleeping sickness. Nature 464:728–732

Giaever G, Nislow C (2014) The yeast deletion collection: a decade of functional genomics. Genetics 197:451–465

Giaever G, Shoemaker DD, Jones TW, Liang H, Winzeler EA et al (1999) Genomic profiling of drug sensitivities via induced haploinsufficiency. Nat Genet 21:278–283

Giaever G, Chu AM, Ni L, Connelly C, Riles L et al (2002) Functional profiling of the *Saccharomyces cerevisiae* genome. Nature 418:387–391

Giaever G, Flaherty P, Kumm J et al (2004) Chemogenomic profiling: identifying the functional interactions of small molecules in yeast. Proc Natl Acad Sci U S A 101:793–798

Gilbert IH (2014) Target-based drug discovery for human African trypanosomiasis: selection of molecular target and chemical matter. Parasitology 141:28–36

Goffeau A, Barrell B, Bussey H, Davis RW, Dujon B et al (1996) Life with 6000 genes. Science 274:546–551

Gomez L, Hack MD, Wu J, Wiener JJ, Venkatesan H et al (2007) Novel pyrazole derivatives as potent inhibitors of type II topoisomerases. Part 1: synthesis and preliminary SAR analysis. Bioorg Med Chem Lett 17:2723–2727

Hillenmeyer ME, Fung E, Wildenhain J, Pierce SE, Hoon S et al (2008) The chemical genomic portrait of yeast: uncovering a phenotype for all genes. Science 320:362–365

Hoon S, Smith AM, Wallace IM et al (2008) An integrated platform of genomic assays reveals small-molecule bioactivities. Nat Chem Biol 4:498–506

Hudson JRJ, Dawson EP, Rushing KL, Jackson CH, Lockshon D et al (1997) The complete set of predicted genes from *Saccharomyces cerevisiae* in a readily usable form. Genome Res 7:1169–1173

Katz L, Baltz RH (2016) Natural product discovery: past, present and future. J Ind Microbiol Biotechnol 43:155–176

Knorre DA, Sokolov SS, Zyrina AN, Severin FF et al (2016) How do yeast sense mitochondrial dysfunction? Microb Cell 3:532–539

Kowalczuk M, Mackiewicz P, Gierlik A, Dudek MR, Cebrat S (1999) Total number of coding open reading frames in the yeast genome. Yeast 15:1031–1034

Kuzmin E, Sharifpoor S, Baryshnikova A, Costanzo M, Myers CL et al (2014) Synthetic genetic array analysis for global mapping of genetic networks in yeast. Methods Mol Biol 1205:143–168
Lain S, Hollick JJ, Campbell J, Staples OD, Higgins M et al (2008) Discovery, *in vivo* activity and mechanism of action of a small-molecule p53 activator. Cancer Cell 13:454–463
Lasserre JP, Dautant A, Aiyar RS, Kucharczyk R, Glatigny A et al (2015) Yeast as a system for modeling mitochondrial disease mechanisms and discovering therapies. Dis Model Mech 8:509–526
Longo VD, Fabrizio P (2012) Chronological aging in *Saccharomyces cerevisiae*. Subcell Biochem 57:101–121
Lum PY, Armour CD, Stepaniants SB, Cavet G, Wolf MK et al (2004) Discovering modes of action for therapeutic compounds using a genome-wide screen of yeast heterozygotes. Cell 116:121–137
Magtanong L, Ho CH, Barker SL, Jiao W, Baryshnikova A et al (2011) Dosage suppression genetic interaction networks enhance functional wiring diagrams of the cell. Nat Biotechnol 29:505–511
Marton MJ, DeRisi JL, Bennett HA, Iyer VR, Meyer MR et al (1998) Drug target validation and identification of secondary drug target effects using DNA microarrays. Nat Med 4:1293–1301
McLeod SM, Fleming PR, MacCormack K, McLaughlin RE, Whiteaker JD et al (2015) Small-molecule inhibitors of gram-negative lipoprotein trafficking discovered by phenotypic screening. J Bacteriol 197:1075–1082
Menacho-Marquez M, Murguia JR (2007) Yeast on drugs: *Saccharomyces cerevisiae* as a tool for anticancer drug research. Clin Transl Oncol 9:221–228
Montgomery JI, Smith JF, Tomaras AP, Zaniewski R, McPherson CJ et al (2015) Discovery and characterization of a novel class of pyrazolopyrimidinedione tRNA synthesis inhibitors. J Antibiot 68:361–367
Oh J, Fung E, Schlecht U et al (2010) Gene annotation and drug target discovery in *Candida albicans* with a tagged transposon mutant collection. PLoS Pathog 6:e1001140
Oliver SG (2006) From genomes to systems: the path with yeast. Philos Trans R Soc Lond Ser B Biol Sci 361:477–482
Pierce SE, Davis RW, Nislow C, Giaever G (2007) Genome-wide analysis of barcoded *Saccharomyces cerevisiae* gene-deletion mutants in pooled cultures. Nat Protoc 2:2958–2974
Rine J, Hansen W, Hardeman E, Davis RW (1983) Targeted selection of recombinant clones through gene dosage effects. Proc Natl Acad Sci U S A 80:6750–6754
Sharma P, Kalita MC, Thakur D (2016) Broad spectrum antimicrobial activity of forest-derived soil Actinomycete, *Nocardia* sp. PB-52. Front Microbiol 7:347
Shoemaker DD, Lashkari DA, Morris D, Mittmann M, Davis RW (1996) Quantitative phenotypic analysis of yeast deletion mutants using a highly parallel molecular bar-coding strategy. Nat Genet 14:450–456
Sinclair DA, Mills K, Guarente L (1997) Accelerated aging and nucleolar fragmentation in yeast *sgs1* mutants. Science 277:1313–1316
Siqueira-Neto JL, Moon S, Jang J et al (2012) An image-based high-content screening assay for compounds targeting intracellular *Leishmania donovani* amastigotes in human macrophages. PLoS Negl Trop Dis 6:1671
Smith AM, Heisler LE, Mellor J et al (2009) Quantitative phenotyping via deep barcode sequencing. Genome Res 19:1836–1842
Smith AM, Ammar R, Nislow C, Giaever G (2010) A survey of yeast genomic assays for drug and target discovery. Pharmacol Ther 127:156–164
Snyder M, Kumar A (2002) Yeast genomics: past, present and future promise. Funct Integr Genomics 2:135–137
Stefanini I, De Filippo C, Cavalieri D (2013) Yeast as a model in high-throughput screening of small-molecule libraries. In Trabocchi A (ed) Diversity-oriented synthesis. Wiley, Hoboken, pp 455–482
Steinkraus KA, Kaeberlein M, Kennedy BK (2008) Replicative aging in yeast: the means to the end. Annu Rev Cell Dev Biol 24:29–54

Tucker CL, Fields S (2004) Quantitative genome-wide analysis of yeast deletion strain sensitivities to oxidative and chemical stress. Comp Funct Genom 5:216–224

Winzeler EA, Shoemaker DD, Astromoff A et al (1999) Functional characterization of the *S. cerevisiae* genome by gene deletion and parallel analysis. Science 285:901–906

Wright GD (2015) Solving the antibiotic crisis. ACS Inf Dis 1:80–84

Zewail A, Xie MW, Xing Y, Lin L, Zhang PF (2003) Novel functions of the phosphatidylinositol metabolic pathway discovered by a chemical genomics screen with wortmannin. Proc Natl Acad Sci U S A 100:3345–3350

Zimmermann A, Hofer S, Pendl T, Kainz K, Madeo F et al (2018) Yeast as a tool to identify anti-aging compounds. FEMS Yeast Res 18:1–16

Caenorhabditis elegans: A Tool for Antimicrobial Drug Discovery

29

Kitlangki Suchiang and Ramatchandirane Mahesh

Contents

K. Suchiang (✉) · R. Mahesh
Department of Biochemistry and Molecular Biology, Pondicherry University, Puducherry, India

B. Siddhardha et al. (eds.), *Model Organisms for Microbial Pathogenesis, Biofilm Formation and Antimicrobial Drug Discovery*,
https://doi.org/10.1007/978-981-15-1695-5_29

Abstract

The increasing popularity of *C. elegans* model in the field of novel drug discovery and research is attributed to its transparent small size, simple anatomy, short life cycle, and rapid cultivation time. Worms' transparent body allows for an in vivo monitoring of cells or whole organism in real time where both host and pathogen interactions can be visualized during an entire course of the experiment. The ease of worms culturing techniques aided with its capacity to generate unlimited numbers of progeny in minimum time is ideal for automated high-throughput screening. Multiple endpoints that ranges from survival and death to a mere abnormal locomotive behavior of worms allow for accurate measurements and validations of a lead antimicrobial compound, and this can be translated to higher mammalian like a human. Interestingly, *C. elegans* by itself is a source of a repertoire of antimicrobial modulators/compounds and this has contributed to the identification of novel anti-infectives, host-directed immunomodulators, and effector molecules-like antimicrobial proteins (AMPs) and peptides. Screening for antimicrobial activity of either natural or synthetic origin can be conducted at the cellular and molecular levels involving different parameters from virulence factors to enzymes and pathways with ease and precision. Taken together, *C. elegans* as a model organism has numerous contributions in uncovering host candidates for drug discovery. Thus, precise correlations of the molecular mechanisms involved between the activities of a compound identified with that of a more complex mammalian infection can be utilized for translational purposes.

Keywords

C. elegans · Antimicrobial · Quorum sensing · Immune system · Drug discovery

29.1 Introduction

The introduction of antimicrobial agents for treatment of various serious life-threatening bacterial infections and ailments has dramatically reduced mortality and may be considered a landmark in the advancement of medical science and modern medicine. Antimicrobial agents kill or antagonize microorganism growth and are distinguished pharmacotherapeutic agents among other medicinal classes for their unique specific target on bacteria rather than their host components like human tissues, organelles, or biomolecules (Gould and Bal 2013). The first antimicrobial agent in the world was Salvarsan, considered as "magic bullet" to treat syphilis and was introduced by Ehrlich in 1911. Subsequently, in 1928, Fleming discovered penicillin and sulfonamides were developed by Domagk and others in 1935. It was reported that with the availability and usability of sulfanilamide for the subcutaneous treatment of acute meningococcal meningitis, a remarkable decline in death rate to approximately 10% was observed from the previous alarmingly 70–90% mortality rate of the pre-antibiotic era (Schwentker et al. 1937). However, the last decades of the twentieth century, with inappropriate use and the emergence of numbers of new antimicrobials with different targets into clinical use, the issue of antimicrobial resistance that develops in no time has surpassed our imagination and is a big matter of significant global health threat (Levy 1998). Thus, to tackle the emergence of new strains and to halt the spreading of an overwhelming and deadly antimicrobial resistance, there is an urgent need both at the level of regulation of antibiotics for their appropriate use only when needed, and for the faster and precision-based development and discovery of new agents.

The discovery of new novel compounds or active biomolecules requires hundred millions of dollars with minimum 10–20 years time (Dickson and Gagnon 2004). It is practically stages of the journey that usually begin with either rational planning of new prototype compounds using computational methods for prediction or an in vitro screening and, the testing of lead compounds in vivo in higher animal model (Brown and Wright 2016). These stages of the drug discovery require time and investment and often end up in failure due to poor biological activities and/or toxicity when tested in higher animals or in clinical trials. Another faster approach to curtail the time-consuming procedures, high cost, and scientific ethical constraints are to evaluate for an in vivo activity of library of compounds directly by using different available model organisms manually or in a high-throughput-based screening methods (Arvanitis et al. 2013).

29.2 *C. elegans*: A Model Organism

With an aim to further understand developmental genetics, Sydney Brenner first introduced *Caenorhabditis elegans* (*C. elegans*) as a model organism in the early 1960s. This contribution has awarded him and his colleagues with Nobel Prize in 2002 (Brenner 1974; Nigon and Félix 2017). *C. elegans* is a free-living bacteriovore that is nonparasitic in nature, with an adult hermaphrodite transparent body measuring about 1 mm in length and 50 μm in diameter. It thrives and lives in temperate soil environments, feeds on microbes that are digested with the help of a pharynx. The worm lacks respiratory or circulatory systems and its intestine is composed of epithelial cells in the form of a tube that passes through the entire length of the animal body (McGhee 2013). In the laboratory, it is cultured on a Nematode Growth Media (NGM) with nonpathogenic *E. coli* OP50 as a food source (Brenner 1974; Hall and Altun 2008) (Fig. 29.1).

As a multicellular organism, a mature adult *C. elegans* hermaphrodite has 959 cells that include 302 nerve cells of its nervous system, 213 hypodermis cells, and 34 cells are for intestines. Worms have an average life span of approximately 2–3 weeks and a reproduction time of 3.5 days. The worm's population is composed of

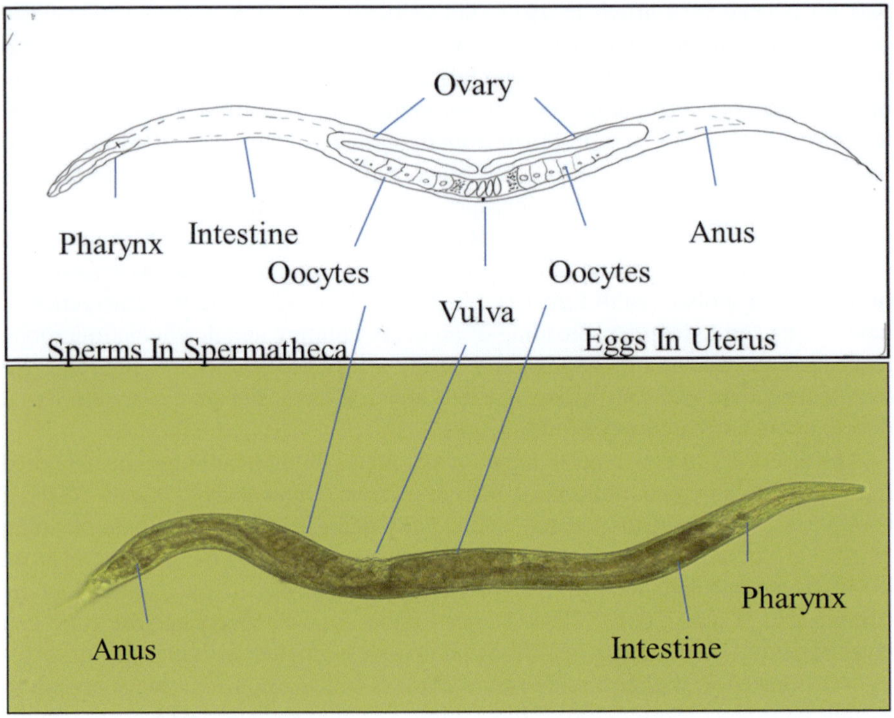

Fig. 29.1 Diagram and images depicting *C. elegans* body with different organs and tissues

two natural sexes: rare occurrence XO males and XX hermaphrodites which dominates the population. The hermaphrodites produce both oocytes and sperm that allows for its self-fertilization and its male character allows for the production of the limited number of sperm that can be used for self-fertilization internally. Male worms, on the other hand, can be produced for up to 50% from the mating between a hermaphrodite and a male worm and rarely (0.1%) during hermaphrodite reproduction from spontaneous X chromosome nondisjunction events (Sulston and Horvitz 1977; Sulston and Hodgkin 1988).

One unique and appealing feature of *C. elegans* is its defined tissues like intestines, muscles, and gonads amidst its simple body plan. Briefly, at the outer body surface lies the epidermis which encloses the ventral and dorsal nerve cords and bands of muscle which helps in control and coordinating movements of the organism. In addition, *C. elegans* has pseudocoelomic cavity that is formed out of six cells called coelomocytes (Grant and Sato 2006). Thus, specific morphological and cellular changes in any tissues such as the muscle (Herndon et al. 2002), gonad (Garigan et al. 2002; Golden et al. 2007), pharynx (Garigan et al. 2002), and hypodermis (Golden et al. 2007; Haithcock et al. 2005) can be used as parameter in any drug discovery study. Extensive and detailed descriptions of *C. elegans* body systems with a clear depiction of various cell types by electron micrographs are also available freely in WormAtlas (www.wormatlas.org).

29.2.1 *C. elegans'* Life Cycle

A unique and short life cycle of *C. elegans* proceeds with different larval stages that are temperature dependent. This life cycle which includes a reproductive life cycle (egg to egg-laying parent) is complete in around 2.5 days at 25 °C to about 5.5 days at 15 °C. The worm eggs which are laid at about 40 cells stage arises out of self-fertilization by the adult hermaphrodite. When the eggs are hatched, the animals have to grow through four larval stages designated as L1–L4 stages, and each of which has different sizes compared to adulthood. Adult hermaphrodite worms can self-fertilize where each worm can bring about 300 progeny each during its reproductive active life cycle. Thus, with a generation time of around 3.5 days at 20 °C, the laboratory strain of wild type (N2) *C. elegans* can have an average life span around 2–3 weeks. Alternatively, when integrating environmental conditions are not cordial as when population density is crowded, scarcity of food supply, and in extreme prevailing living temperature, an alternative life form called "dauer" or dauer arrest is adopted by the larval stage worms. This developmental decision to exist as dauer is late in the L1 larval stage of the worms and they have a characteristic phenotype of thin movable body with a specialized cuticle (Cassada and Russell 1975). During this state, worms cannot eat since their mouth have an internal plug that closes their oral orifice (Riddle et al. 1981) and their pharynxes do not pump and remain constricted (Vowels and Thomas 1992; Cassada and Russell 1975). Interestingly, dauers can remain viable and can roam around for months and they appear to be viable and non-aging. When prevailing environmental conditions are suitable as when they encounter a food source, dauer can reenter the L4 stage and live for about 15 more days (Klass 1977; Kenyon et al. 1993) (Fig. 29.2).

29.3 Relevance of the *C. elegans* Model for Antimicrobial Drug Discovery: Why?

Much of our current knowledge and in our future predictions within the field of biological and allied sciences rely heavily on scientific experimentations and cumulative data of in vivo and in vitro research models. Thus choosing an in vivo research model plays a crucial part in experimental success, its precision, and predictability. Preference of *C. elegans* for antimicrobial drug discovery has increased rapidly in recent years, and this is attributed to our recent knowledge of its similarity to higher multicellular organisms across evolutionary time point. This close similarity is seen not only in the architecture of its outer body cuticle/epithelial cells to human skin epidermis but at a deeper level in the form of its organelle functions and the presence of different homologs and orthologs of important signaling pathways. The ease of carrying out the latest genetic manipulations has further made *C. elegans* a superior model in the field of basic biology and in different utilities like new discovery and translational research. On a broader view, why C. *elegans* is relevant and appealing as a model organism in antimicrobial drug discovery is outlined in the description below.

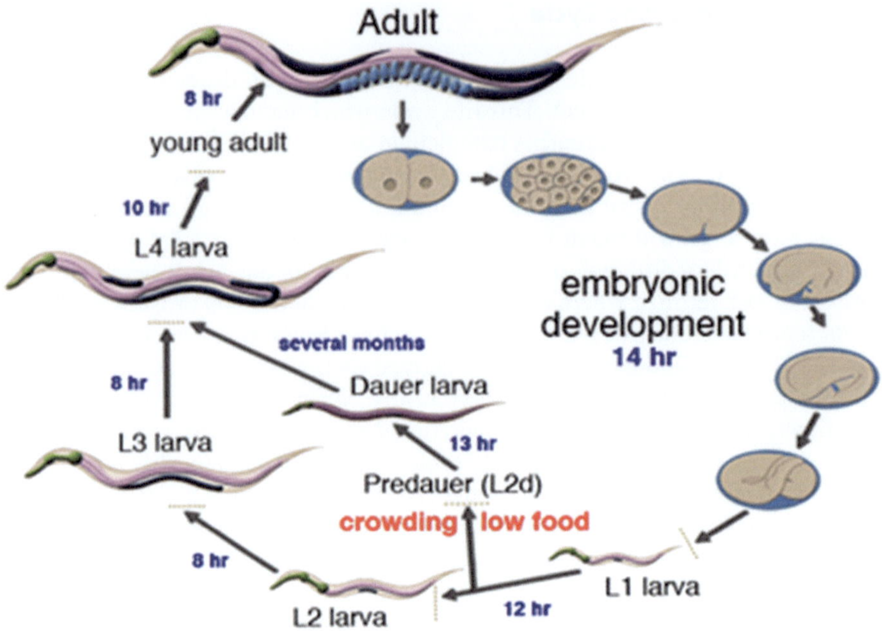

Fig. 29.2 Life cycle of *C. elegans* at 22 °C and its growth and development (Adopted from http://wormatlas.org)

29.3.1 General Experimental Considerations

Execution of a successful antimicrobial drug discovery research program requires patience, a prior commitment to scientific regulations, and ethical criteria aided with continuous source of financial support (Zak and O'Reilly 1993). Importantly, with no ethical constraints, *C. elegans* model counteracts the limitations of different available in vivo models both in terms of accessibility and versatility. In the laboratory setting, an unending number of animals or progeny can be reproduced from its short life and reproductive cycle. Thus, it is feasible for carrying out both small- and large-scale drug screens, and this can be completed in a short time span. Its maintenance involves culturing on low-cost culture of nematode growth medium (NGM) agar (both liquid and solid), and feed on nonpathogenic *E. coli* OP50 strain. For daily experimental needs, handling of *C. elegans* involves simple techniques and their small size which fits in 96- or 384-well plates is ideal for high-throughput experiments. For long term, it can be preserved or stored at −80 °C. In carrying out anti-infectives screens, simple, specific, and standardized infection protocols are available, which usually only require the exposure of the wild type (N2) or mutant strains of worms to different pathogens followed by monitoring their interactions. Thus, the effect of a tested compound after experimental manipulations like down-regulation/deletion of genes involved in bacterial virulence or pathogenicity can be studied and monitored real time and in whole animal system (Anyanful et al. 2009). Furthermore, by utilizing its complex organ and tissue system, accuracy and precision of drug targeting research can be reached and this can be correlated and translated to that of a higher mammalian system like humans. Therefore, in general, *C. elegans* represents a convenient whole animal model system that can be used in antimicrobial drug discovery (Powell and Ausubel 2008).

29.3.2 *C. elegans* as a Model Organism for Genetic Studies and Manipulation

Well known for having the first genome completely sequenced amongst multicellular organisms, *C. elegans* annotated genome has surprisingly many conserved genes and signaling pathways with that of human (www.wormbase.org) (*C. elegans* Sequencing Consortium 1998; Wilson et al. 1994). Its entire body cell lineage has been elucidated from a fertilized egg development till it reached adulthood and this has been correlated clearly with its anatomy. Furthermore, its anatomical ultrastructure including cell types and developmental biology has been described in a detailed manner by thin-section electron microscopy (Sulston et al. 1983; White et al. 1986).

On the forefront of gene expression studies and analysis, *C. elegans* is at the forefront of other species. The availability of its entire sequenced genome has simplified functional genomic approaches where manipulations like forward and reverse genetics can be carried out with ease. The availability and accessibility of WormBase, the information-rich central well-annotated genomic database of *C. elegans* serves as an interface for new discovery and predictability in drug discovery and related research (Harris et al. 2004). Forward genetics can be performed easily in *C. elegans* to discover which particular gene when mutated is responsible for a particular observed phenotype. On the contrary, reverse genetic screens can also be easily carried out, where the function of a normal gene is modulated or changed to establish gene function in terms of observing the resulting phenotypes comparatively (Pujol et al. 2001).

In terms of genetic analysis, the ease of generating mutations in *C. elegans* has attracted many researchers. The two major and common genetic screens employed are chemical mutagen (ethyl methane sulfatonate; EMS) screens which work very efficiently on the worm and, genome-wide RNA interference (RNAi) screens. However, mutations can also be generated with other means available like by ionizing radiation, by transposon hopping and others. These techniques allow for an in-depth dissection of nematode genetics that has translational applications in diverse areas of biology from aging, development, DNA damage and repair, signal transduction, drug discovery amongst other biological processes (Zipperlen et al. 2001; Hamilton et al. 2005; Jorgensen and Mango 2002).

Since the advent of RNAi, knocking down of *C. elegans* genes for identifying genes required for a variety of biological processes has been regularly reported by different researchers. Indeed, it was first characterized using *C. elegans* and Andrew Fire and Craig C. Mello shared the 2006 Nobel Prize in Physiology or Medicine for their work on RNAi using *C. elegans* (Fire et al. 1998). Notably, RNAi feeding protocols (either by injection, soaking, or feeding with dsRNA) can be carried out with ease without any sophisticated instruments, and systemic targeting of any gene of interest can be accurate. To further enhance the pace and development in *C. elegans* genetics research, libraries covering almost 94% of the 20,000 genes have been established and are available for researchers (Kamath and Ahringer 2003; Lamitina et al. 2006; Rual et al. 2004). The transparency of the worm body when the gene was tagged with reporter gene GFP (green fluorescent protein) enables real-time monitoring, where qualitative and quantitative measurements of expression can be carried out. Similarly, real-time monitoring for the effectiveness of tested drugs and/or pathogenicity of a particular bacteria can be studied. Transgenic *C. elegans* strains can be generated easily via microinjection of DNA and genome-wide of *C. elegans* gene expressions projects have been carried out and documented (Aballay et al. 2000; Labrousse et al. 2000).

Recently, to facilitate further our understanding of gene function and editing, unique technology of "clustered regularly interspaced short palindromic repeats (CRISPR-Cas9)" enzyme system has also been applied for genome engineering in *C. elegans* model system (Hsu et al. 2014; Chen et al. 2013). Thus, over 3000 phenotypically defined mutant strains are available for supply for researchers around the globe from the *Caenorhabditis* Genetics Centre (CGC), University of Michigan. Indeed, several important discoveries in biomedical sciences have been achieved by using *C. elegans*. This includes apoptosis (Yuan et al. 1993; Ellis et al. 1991), RNAi (Fire et al. 1998), and GFP applications (Chalfie et al. 1994). Recently, the development of the first array Comparative Genome Hybridization (aCGH) was established in *C. elegans*. This has speeded up the identification of novel induced deletion or insertion mutant strains for the *C. elegans* Knockout Consortium; besides it allows for the comparative detection of natural gene content variation between different strains of *C. elegans* (Maydan et al. 2007). Whole-Genome Sequencing (WGS) has also been reported in *C. elegans* for a generation of clone mutants and serves as a platform to different biological investigations from unraveling mutational signatures of carcinogens to deficiency of repair mechanism (Meier et al. 2014). With enormous knowledge base available on its genes and genomes, (Lai et al. 2000; Lehner et al. 2006a, b), worm model is undoubtedly a powerful model of disease modeling, and in antimicrobial drug discovery and its development (Markaki and Tavernarakis 2010; O'Reilly et al. 2014; Dengg and van Meel 2004).

29.3.3 *C. elegans* Homologous Genes and Concordant Pathways

The prediction and assessment of genes inferred from sequence analysis estimated that 60–80% of *C. elegans* genes are shared and conserved with human genes (Harris et al. 2004; Lai et al. 2000; Sonnhammer and Durbin 1997). Additionally, many genes that are linked to human diseases are also found in *C. elegans* (Kaletta and Hengartner 2006; Markaki and Tavernarakis 2010). For example, the different components of apoptosis, its modes of regulations, activation and execution necessary for *C. elegans* development and homeostasis are found to be very similar or conserved with that of humans (Vaux et al. 1992). Similarly, the significant and central roles of insulin/IGF-1 signaling (IIS) pathway in regulations of life span, metabolism, and growth are observed in *C. elegans* through numbers of ortholog and homolog genes. Indeed, the first defined life-extension IGF1/FOXO pathway was proposed through genetic studies on *C. elegans* first long-lived mutants (Klass 1977). These similarities are extended to the levels of pharmacological interventions as seen in the case of metformin which can modulate IIS components to improve health and life span in both human being and *C. elegans* (Onken and Driscoll 2010).

Different signaling molecules and pathways involved in *C. elegans* development and to encounter pathogens are also strikingly similar to those in metazoans and humans (Ewbank 2006; Irazoqui et al. 2010; Sullivan-Brown et al. 2016). Most interestingly, worms do utilize the same neurotransmitters and neuronal signaling pathways for the overall maintenance and functions of its nervous system as that of higher vertebrates (Kaletta and Hengartner 2006; Peterson et al. 2008). To move efficiently in different environments, to transit from one form of locomotion to another during its specific crawling and swimming gaits, biogenic amines like dopamine and serotonin are required to initiate and set the transition phase of its movement (Vidal-Gadea et al. 2011). Availability of worms genes closely associated with Alzheimer's disease (AD) etiology, its proteolytic pathways processing and regulations have been explored extensively to understand the different complex process of AD (Alexander et al. 2014; Lu et al. 2014). For these reasons, *C. elegans* has been employed regularly as whole animal to model various complex human diseases such as diabetes (Ogg et al. 1997), AD (Levitan et al. 1996), Parkinson's disease (Braungart et al. 2004), and even as infection model (Markaki and Tavernarakis 2010).

29.3.4 *C. elegans* as a Model Organism for Drug Toxicity Studies

To strengthen toxicity studies, different biological systems and model organisms are required to present strengths and limitations of the tested sample. It is ideal to have a combination of toxicologically validated tools and different mammalian model organisms, since studies in only one model can result in unspecificity and poor response when tested in humans (Olson et al. 2000; Nass and Hamza 2007; Hartung 2009; Knight et al. 2009; Tralau et al. 2012). Thus there is a need for evaluation of toxicity in more than one mammalian species to maintain cost and to increase productivity and predictivity (Olson et al. 2000). In comparison to different in vitro assays, toxicity assays in *C. elegans* scientific data can be generated easily from the different lethal and sublethal endpoints of an intact and metabolically active animal with different tissue and organ systems (Boyd et al. 2010; Corsi 2006). On the same line of usability, *C. elegans* model has served as an indicator or predictor of chemical toxicity, for different environmental pollutants, metal oxides, organic solvents, and toxins (Sochova et al. 2006). Williams and Dusenbery (1988), based on their early ranking and adult mortality in *C. elegans*, reported that the toxicity levels of different metal salts correlated well with rat and mouse oral $LD_{50.}$ Notably, in comparison, they also found that the LC_{50s} in *C. elegans* were higher to LD_{50s} of rodent subjects used (Williams and Dusenbery 1988). In addition, specific target screens using *C. elegans* successful evaluation and detection of specific types of toxins have been reported. Using transgenic reporter strain, Allard et al. (2013) reported that exposure to different specific aneugens, toxins that change the numbers of chromosomes per cell resulted in an increase in the number of male XO male progeny in *C. elegans* population (Allard et al. 2013). In terms of neuropharmacology and neurotoxicity studies, many of the neurons that express specific neurotransmitters have also been identified and evaluated (Rand and Nonet 1997). In term of practicality, the availability of established, enhanced, and easily quantifiable *C. elegans*-based toxicity assays in the form of tracking and analyze locomotive changes, AChE assay, behavioral assays allows for the identification of deadly neurotoxic compounds and to classify their levels of toxicity accordingly. Recently in an extensive study, toxicity studies were performed by exposing *C. elegans* to over 900 chemicals. These data were used for comparative studies with ToxCast data from rabbits, zebrafish, and rats (Boyd et al. 2016).

29.3.5 *C. elegans* as an Infection Model Organism for Different Human Pathogens

Reinforcement of pathogen in the host and the initiation of the infection process is the outcome from interactions of different levels between the host and the corresponding pathogens (Mason et al. 1995). The very basic principle in assaying and underpinning the use of *C. elegans* in anti-infective/anti-microbes drug discovery is that the same pathogen that causes infection in humans infects *C. elegans* too. Indeed, it is estimated that more than 40 human pathogenic strains reported so far can infect *C. elegans* too (Sifri et al. 2005). Reports on *C. elegans* infected with *Pseudomonas aeruginosa*, *Vibrio cholera, Salmonella enterica*, *Enterococcus faecalis*, and *Staphylococcus aureus* are amongst the well-studied strains (Table 29.1). Additionally, pathogens of clinical importance such as facultative *Cryptococcus neoformans*, and the human bacterial pathogens *Shigella spp.*, *Salmonella enterica* and *Listeria spp.* are reported to cause infections in *C. elegans* (Mylonakis et al. 2002; Burton et al. 2006; Kesika and Balamurugan 2012). In terms of elucidation of evolutionarily conserved pathways associated with microbial pathogenesis, *C. elegans* has been a pioneered model. Mechanisms of worm killing identified by different pathogens include extracellular skin infection, persistent intestinal colonization, biofilm formation, and toxin-mediated killing. (Smith et al. 2002; Dunbar et al. 2012). Generally, these Gram-positive and Gram-negative bacteria can infect worms through the pharynx and sometimes through vulva or epidermis. In several cases, it has been reported that certain pathogenic bacteria could accumulate extracellularly and accumulate on the surface of the body before their entry into the worm (Mahajan-Miklos et al. 1999; Sifri et al. 2003) (Table 29.1).

Another appeal of the *C. elegans* model in drug discovery is the availability of the established infection model. Establishment of a *C. elegans–Vibrio alginolyticus*, *C. elegans–P. aeruginosa* infection model to name a few have been reported. Indeed, as a host model *C. elegans* platforms are ideal to investigate on the different

Table 29.1 Examples of some common pathogens between humans and *C. elegans*

Pathogens	Human diseases	Mode of pathogenesis	Findings	Reference(s)
Pseudomonas aeruginosa	Broad spectrum of opportunistic infections including urinary tract, respiratory, dermatitis, bacteremia, gastrointestinal infections, etc.	Fast and slow killing mechanism depending on the culture medium osmolarity	Paralytic killing by diffusible toxins Phenazines, ROS generation, induces p38 MAPK pathway. Virulence is strain dependent	Cezairliyan et al. (2013); Mahajan-Miklos et al. (1999); Darby et al. (1999); Kim et al. (2002)
Listeria monocytogenes	Listeriosis (sepsis, meningitis, or encephalitis)	Extracellularly, accumulation in the worm intestine, decreases the number of eggs laid	The virulence factors gene *PrfA* and *DegU* are required for virulence in *C. elegans*	Thomsen et al. (2006)
Staphylococcus aureus	Skin infections, Sepsis, endocarditis, pneumonia, and toxic shock syndrome	Colonization of the nematode intestine. Worms death occurred after feeding on *S. aureus* for 48–72 h	*C. elegans esp-2* and *esp-8* mutants exhibited increased susceptibility to infection thus, p38 MAPK is involved	Sifri et al. (2003); Irazoqui et al. (2010)
Cronobacter sakazakii	Bacteraemia, meningitis, necrotizing enterocolitis	Worms die in several days as a consequence of an accumulation of bacteria in the host intestine	Induction of host immunity mediated antimicrobial genes (*lys-7, clec-60 and clec-87*)	Sivamaruthi et al. (2011)
Enterococcus faecalis	Endocarditis, Sepsis, urinary tract infections (UTIs), meningitis etc.	Colonize the nematode intestine with a 50% lethal time of around 4 days, after establishing a persistent infection in the intestines	Virulence factors, including cytolysin, gelatinase and the fsrABC regulatory system is required for full pathogenicity in *C. elegans*	Sifri et al. (2002); Garsin et al. (2001)
Burkholderia pseudomallei	Melioidosis (wide range of localized, pulmonary, blood and disseminated infections)	Limited intestinal lumen colonization	Kills by a mechanism involving a neuromuscular endotoxin that targets Ca^{2+} signaling mediated by L-type Ca^{2+} channels	O'Quinn et al. (2001)

(continued)

Table 29.1 (continued)

Pathogens	Human diseases	Mode of pathogenesis	Findings	Reference(s)
Vibrio cholerae	Watery diarrhea	Colonization of the intestinal tract	Hemolysin, protease (PrtV) is required for lethality. Cause developmental delay and intestinal vacuolation	Cinar et al. (2010); Sahu et al. (2012); Vaitkevicius et al. (2006)
Yersinia enterocolitica	Yersiniosis (fever, diarrhea, erythema nodosum, appendicitis, bacteremia)	Infection of *C. elegans* by biofilm formation on the worm head	*tcaA* gene that encodes an insecticidal toxin plays a great role in the nematocidal activity	Spanier et al. (2010)
Klebsiella pneumoniae	Bowel disease, Pneumoniae, Suppurative infections, bacteremia, septicemia, and urinary tract infections, etc.	Accumulation and propagation of the pathogen inside the worms' intestine	Downregulating of p38 MAPK, inhibiting the production of antimicrobial genes such as *nlp*, *lys*, and C-type lectins	Kamaladevi and Balamurugan (2015)

mechanisms involved in host resistance to *P. aeruginosa* and *S. aureus* infection and to identify new bacteriocins in combatting infections from both the strains (Kim et al. 2002; Kong et al. 2014a, b). Identification of *P. aeruginosa* virulence-related genes was also reported using genome-wide *C. elegans* infection model (Feinbaum et al. 2012). Furthermore, with combinations of powerful new optogenetic techniques and vision-based tracking systems, automatic and precise quantification of all behavioral changes, in-depth dissections of these infections have given rise to new useful scientific information (Husson et al. 2013; McDiarmid et al. 2018). For example, by tagging with a genetically encoded mini singlet oxygen generator and visualization by fluorescence microscopy, specific protein–protein interactions between host and pathogen could be observed and manipulated. Similarly, applications of high-resolution imaging by electron microscopy have been used to determine the detrimental consequences of protein destabilization resulted by enhanced ROS generation of invaded pathogens (Qi et al. 2012; Shu et al. 2011).

29.3.6 *C. elegans* Innate Immunity as a Target for Drug Discovery

Some compounds can have modulatory activity for the host immune system and this, in turn, can give protection to the host. Thus, studies on the responsiveness of the host immune response to a particular bioactive compound or immunomodulators represent a new scope in the search for novel anti-infectives. The nematode *C.*

elegans shaped by its interactions with environmental microbes is under constant exposure to pathogens attack and responds to them constantly. Thus, it has evolved with inducible innate immune response machinery that is ideal for studying molecular mechanisms involving microbe-specific responses. Numerous studies have pinpointed on the needs of inducible immune defenses in *C. elegans* and these pathways are tightly regulated to ensure host survival (Cox and Walter 1996; Schröder and Kaufman 2005; Mori et al. 1996, Ewbank 2007).

Innate immunity or native immunity initiates nonspecific defense mechanisms thereby providing the host with early defense against microbes. Amidst the absence of cell-mediated adaptive immunity and specialized immune cells, *C. elegans* are effective in responses to different pathogen invasions with its simple and ancestral innate immunity. Its components share strikingly many similarities with higher mammalian immune systems (Alper et al. 2008). *C. elegans* complex innate immune approach for disease resistance is well established including avoidance behaviors, antimicrobial peptides (AMPs), proteins and physical barriers was reported and well studied (Pradel et al. 2007; Pujol et al. 2001; Gravato-Nobre et al. 2011). At the molecular level, responses to bacterial infections with different conserved signal transduction and important regulatory pathways which include the p38 Mitogen-Activated Protein Kinase (p38MAPK) pathway, programmed cell death pathway; the Transforming Growth Factor-β (TGF-β) pathway and the DAF-2 insulin/IGF-I like signaling pathway (Kim et al. 2002; Mallo et al. 2002; Ewbank 2006; Pukkila-Worley et al. 2012; Visvikis et al. 2014). For example, studies showed that *C. elegans* with mutations in *esp-2* and *esp-8* genes are prone to *S. aureus* infection, and this indicates that the conserved p38 MAP kinase signaling pathway has a role to play against *S. aureus*, as previously shown for *P. aeruginosa* (Sifri et al. 2003).

29.3.7 *C. elegans* as a Model Organism for High-Throughput Drug Screening (HTS)

A large population of *C. elegans* is usually cultured on solid agar NGM plates or in bulk liquid. With its rapid development time, the capability to generate synchronized age of wild-type, mutant or transgenic worms, a size that can be dispensed in 96-, 384-, and 1536-well plate formats, *C. elegans* has the requirements of high-throughput screening methods (HTS) ideal for in vivo model drug screens. The first initiative by Lehner and his colleagues that began in 2006 has led to the possibility for the development of a liquid HTS in *C. elegans* using 96-well plates. Moy and colleagues performed the first antimicrobial compounds liquid-based screens on *C. elegans–Enterococcus faecalis* infection model (Lehner et al. 2006a, b; Moy et al. 2006, 2009a, b). Since then, automated worm transfer method by the application of Complex Object Parametric Analyzer and Sorter (COPAS™ BIOSORT, Union Biometrica) was reported in a screening of thousands of compounds as antimicrobial (Kwok et al. 2006). Furthermore, as part of tackling the multidrug-resistant strain of *P. aeruginosa–C. elegans* HTS model have been applied to screen bioactivity of 1300 extracts. This has led to the identification of 36 active fractions and 4 of which showed significant anti-psuedomonas activity both in vitro and in vivo when tested (Zhou et al. 2011). Recently, a modified version of reusable HTS 96 well plate with gold microelectrodes integrated on the bottom has been reported for anti-helmintic drug screening. This has promising applications in resistance study, whereby locomotion and motility endpoint of *C. elegans* can be monitored in real time (Smout et al. 2010).

For enhancing worm transfer and sorting thousands of animals in short time, different methods have also come up including BioTek MicroFlo volumetric dispenser and Fluorescence-Activated Cell Sorter (FACS) (Fernandez et al. 2010; Leung et al. 2013). Furthermore, an alternative WormScan method is available for measuring mortality, morphology, and movement of worms in an automated mode using microscopy (Mathew et al. 2012). Stroustrup et al. (2013) represented with an improved version called "Lifespan Machine" to validate custom images of thousands of worms scanned under a particular treatment for creating libraries of life span data (Stroustrup et al. 2013). Recently, microfluidic devices have revolutionized *C. elegans* HTS where isolation, compound treatments, and different manipulations can be carried out side by side with microscopy real-time monitoring (Ben-Yakar et al. 2009; Clausell-Tormos et al. 2008; Shi et al. 2010). The feasibility of *C. elegans* HTS platform in the discovery and identification of therapeutics against *Burkholderia pseudomallei* and *A. baumannii* has been reported (Lakshmanan et al. 2014; Jayamani et al. 2015). Considering the clinical significances of these strains, this discovery can have translational benefits and significant social impact in the form of reducing mortality and long-term health complications.

29.4 Drug Discovery and Experimental Approaches Using C. *elegans*

Compounds and drugs can interact with different components of an organism. As a model organism, *C. elegans* allows for the detection of those interactions at an organism level with multiple endpoints (e.g., behavior and locomotion, reproduction, life span, etc). Thus, *C. elegans* can serve the roles exhibited by both in vitro and in vivo model systems used for drug screens and toxicology research. However, it must be reiterated that experimental strategy should be clear as it plays a central and crucial part in studies and researches on the discovery of antimicrobial compounds involving *C. elegans* as a whole animal model organism. Several preliminary issues must be addressed concerning the exposure of the animals to any compound. These include whether the exposure is to be brief for a short period or chronic and whether the exposure fits better in a liquid medium or on agar plates or both. For general toxicity studies in *C. elegans*, where the biological assay is aimed to monitor the overall growth or development of the worms, then a chronic exposure is most frequently used. If, however for physiological, behavioral, or specific biological studies like changes in pharyngeal pumping rate or paralysis, induction or cessation of egg-laying, or loss of chemotaxis response, then a brief (minute to hours exposure) to a drug may be more appropriate. Importantly, the age of the animals to be tested is also important, because different developmental stages often have different drug sensitivities and, in some cases, different responses. Thus, for protocols involving brief drug exposure and studies involving life span analysis, it is important to use synchronous populations. Subsequently, comparative studies can also be done to determine stage-specific effects directly.

Another strategy in drug discovery is to investigate the action of a compound/drug and to characterize particular aspects of *C. elegans* in phenotypic analysis of both wild-type and mutant animals. This strategy also uses hit compounds as screening or selective agents to isolate new drug-resistant or hypersensitive mutants and, thus to identify genes with altered drug responses. The last strategy involves comprehensive and in-depth use of *C. elegans* (either wild-type N2, mutants or both) to investigate the mechanism of action of uncharacterized or poorly characterized compounds.

At the molecular level, the mechanism of infection of different pathogens can be studied with specificity and accuracy with the application of different standardized *C. elegans* infection models as mentioned before (Aballay and Ausubel 2002). For example, *C. elegans*–*E. faecalis* infection assay has been reported in antimicrobial HTS, where numbers of compounds and extracts have been identified based on their ability to increased worm's survival and resistance to infection (Moy et al. 2009a, b). Additionally, specific and sensitive types of assays are available for nematode killing, for example, *P. aeruginosa*, the so-called "fast killing" assay where the killing of worms within an hour is carried out on high-osmolarity medium and the "slow-killing" assay where pathogenicity or killing of worms in several days are carried out on low-nutrient media (Mahajan-Miklos et al. 1999). This allows for an accurate study and hence the specific elucidation of the target of a screened compound.

29.4.1 *C. elegans*-Based Assays in Drug Screens

The typical workflow of the manual (small-scale) and high-throughput (large-scale) antimicrobial screening platform involving *C. elegans* is similar to as shown in Fig. 29.1. Among different assays available in the worm model, the notable survival assays have been useful assay that has been used regularly prior to in-depth identification of novel genetic factors, molecules, or signaling pathways involved (Hae-Eun et al. 2017). As part of its application, exact environmental conditions for different survival assays (life span, abiotic, and pathogen resistance assay) should be evaluated first depending on the requirements of a typical survival assay to be employed (Amrit et al. 2014; Mahajan-Miklos et al. 1999; Keith et al. 2014). Subsequently, for survival analysis, two widely used curves are simple survival curves/mortality rate with Kaplan–Meier survival plots to illustrate the percentage of animals alive at different time scale (Kaplan and Meier 1958). Additionally, log-rank test and Fisher's exact test and other statistical methods are also employed for analyzing survival curves (Fisher 1990; Mantel 1966) (Fig. 29.3).

The methods employed in a particular *C. elegans* survival assay do not depend only on the pathogen under study and care should be taken regarding the aim of the study needed to be conducted. For example, *P. aeruginosa* where two mechanisms have been used by the pathogen in establishing slow and fast killing. If the study is aimed at anti-quorum sensing/virulence of a particular antimicrobial agent, then slow killing assays should be carried out. Else, survival analysis for the fast killing assay can be carried out to investigate the effect of a particular antimicrobial agent on the secretion of lethal toxins secreted by the pathogen like phenazines. Another recently developed assay for PA14 pathogenicity is the assay where worms are cultured in a nutrient deficient condition and the rate of killings of the worms can be monitored within 48 h (Kirienko et al. 2014). A simple HTS chemical screening for effects on *C. elegans* life span (Lucanic et al. 2018) and specifically to identify

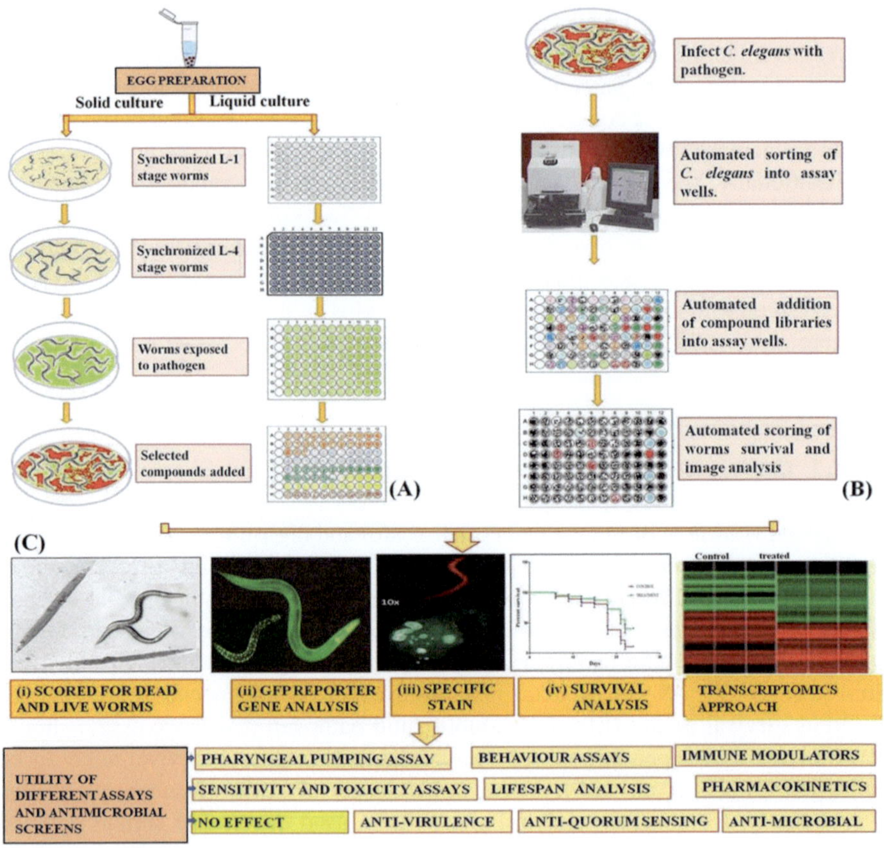

Fig. 29.3 Overview schematic diagram illustrating experimental approaches in antimicrobial drug screens employing *C. elegans* as in vivo whole model organism; (**a**) Manual and small-scale screening using both liquid and solid culture; (**b**) Large-scale automated high-throughput screening (HTS) using liquid culture; (**c**) Different assays employed with their utility and endpoints

small molecules that are active against Methicillin-resistant *Staphylococcus aureus* (MRSA) was also developed in liquid infection assays (Rajamuthiah et al. 2014).

Different techniques and assays are also available for measuring *C. elegans* responses and viability after treatment with drug candidates such as pharyngeal pumping assay, larval assays, or different biochemical assays (Anderson et al. 2004; James and Davey 2007; Rand and Johnson 1995). Staining techniques such as propidium iodide, Sytox to measure cellular viability (Hunt et al. 2012; Roth et al. 1997); nile red and BODIPY-labeled fatty acids stain for lipid studies (Ashrafi et al. 2003; Mak et al. 2006) have also been routinely used with great specificity and sensitivity. Antibody-based histochemical stains with various detection systems have also been applied in *C. elegans* (Duerr 2006).

To monitor the effect of a tested compound at the genetic level, visualization of a specific gene and its expression profiling can be carried out by green fluorescent protein (GFP) and β-GAL (LacZ) fusion (Chalfie et al. 1994; Fisher 1990). Therefore, GFP tagged transgenic strains simplifies the process of monitoring biological activities of tested compounds right from expression pattern analysis and protein localization during cellular and physiological responses (Chalfie et al. 1994; Hobert 2002; Haithcock et al. 2005) (Fig. 29.3). Another recent approach that can provide in-depth knowledge into the activating functions of different genes within different tissues is by determining the spatiotemporal pattern of gene expression. Unique transcriptional expression patterns for the effect of screen compounds can be evaluated in *C.elegans* model on a large scale by the application of microarrays or "serial analysis of gene expression techniques" (SAGE) (Ruzanov and Riddle 2010; Velculescu et al. 1995). *C. elegans* toolkit is exponentially growing in past years and powerful studies like the optogenetics field which is favorable to the transparent body and known nervous system of the worms are ideal for different physiological studies required in the process of drug discovery.

29.5 Contributions of *C. elegans* Model System on Antimicrobial Drug Discovery

29.5.1 *C. elegans* in the Understanding of Bacterial Virulence Factors

The outcome of antagonistic encounters between pathogenic microbes and their hosts is the predicament of what kind of virulence factors that microbes employed and how efficient the host is in response to these different factors. "Virulence, a term often used interchangeably with pathogenicity, refers to the degree of pathology caused by the organism. The extent of the virulence is usually correlated with the ability of the pathogen to multiply within the host and may be affected by other factors (ie, conditional)." Pathogenic microorganism's capability to invade, evade, or cause diseases in the host is by means of virulence factors they produced either in the form of adherence factors, invasion factors, capsules, secretion systems, and other toxins (Clatworthy et al. 2007). Thus studies using different virulence biomarkers or factors such as pseudolysin (LasB) encoded by the *lasB* gene (Morihara et al. 1965), the presence of pili, flagella and the ability to initiate biofilm formation, etc. (Gellatly and Hancock 2013). For example, *P. aeruginosa* is avirulent, unless coordinated expressions of its different virulence factors are mediated (Mesaros et al. 2007).

Recently, the significant roles of virulence factors in pathogenesis process were highlighted and it is considered as an emerging second alternative for finding novel anti-infectives (Barczak and Hung 2009; Clatworthy et al. 2007). Inhibition of virulence factors by any compound/agent would quash bacteria's capability to take advantage of the host and its immune system. Examples of antivirulence compounds include an isoquinoline alkaloid, Virstatin that inhibits cholera toxin (CT) and virulence transcriptional activator ToxT protein; lethal factor inhibitors from *Bacillus anthracis* and type three secretion system (T3SS) chemical inhibitors such as caminoside B and salicylidene acylhydrazides (Zambelloni et al. 2017) was also successful in disarming the pathogens.

C. elegans allows for the identification of compounds that can reduce or alter the degree of virulence of different pathogens apart from the studies that measure killing or growth of the pathogen, for example, *P. aeruginosa,* one of the so-called "ESKAPE' pathogens, highlighted by the Infectious Diseases Society of America (ISDA) as being the greatest threat to human health" (Boucher et al. 2009). A molecular mechanism of its virulence that is depending on the medium and strains was elucidated using a *P. aeruginosa–C. elegans* infection model (Mahajan-Miklos et al. 1999). Pinpointing and identification of virulence-related genes of *P. aeruginosa* PA14 and *Staphylococcus aureus* through genome-wide studies on *C. elegans* infection model was also reported (Feinbaum et al. 2012; Begun et al. 2005). Similarly, identification of well-known virulence factors such as lasR and phoPQ, in *P. aeruginosa* and *S. enterica* (Aballay et al. 2000, Tan et al. 1999) and quorum sensing protein Cyl and FsrB, was reported and validated with the use of *C. elegans* (Garsin et al. 2001).

Recently, the involvement of bacterial quorum sensing (QS) has attracted significant research interest in tackling antimicrobial resistance (Hastings and Nealson 1977). "QS refers to a series of signaling pathways used by microbes to communicate with each other and alter their properties depending on the density of their population at a certain time. It enhances production and sensing of small extracellular molecules known as autoinducers (AIs)," that is released in proportion to cell density (Papenfort and Bassler 2016). The *C. elegans* model has elucidated the link between QS and virulence in many Gram-negative bacteria including *Chromobacterium violaceum* (Swem et al. 2009), *E. coli* O157:H7 (Lee et al. 2011), *Yersinia pseudotuberculosis* (Atkinson et al. 2011), *Burkholderia cenocepacia* (Deng et al. 2012), or *Burkholderia pseudomallei* (Song et al. 2005). Using *C. elegans* Atkinson et al. (2012) reported that the inhibition of the initiation of type III secretion system by quorum sensing enhanced the initiation of biofilm development on the surface of the worms by *Yersinia* (Atkinson et al. 2012). Similarly, using this model it was elucidated that acyl homoserine lactones, an autoinducer of the *P. aeruginosa* quorum sensing can have a crucial part in the activity of its virulence factors and biofilm-associated genes (Williams et al. 2007).

29.5.2 *C. elegans* in Attenuation of Virulence Factors

The virulence dynamic and intricate regulatory network that determines virulence is controlled by a number of transcription factors (such as sigma factors RpoS and RpoN, and quorum sensing regulators) (Balasubramanian et al. 2013). Additionally, the roles of sigma factor, σ54, or RpoN in regulations of quorum sensing, biofilm formation, and production of different virulence factors were also reported (Heurlier et al. 2003; Shao et al. 2018; Potvin et al. 2008; Caiazza and O'Toole 2004; Cai et al. 2015). RpoN was also linked to *P. aeruginosa* resistance to several antibiotics (Viducic et al. 2017). Lloyd et al. (2019) reported that using a *P. aeruginosa–C. elegans* infection model the blocking of RpoN by cis-acting peptide inhibits virulence of *P. aeruginosa* cultures of cystic fibrosis patients, thereby increasing antibiotic sensitivity observed with retarded movement and less degree of virulence (Lloyd et al. 2019). Disarming and attenuation of *P. aeruginosa* virulence factor LasB by leveraging a *C. elegans* infection model and LasB knockout strain was also visualized and reported (Zhu et al. 2015). Similarly, identification and modulation of a virulence factor Nif3-family protein YqfO03 from *P. syringae* wild-type strain MB03 against *C. elegans* reduced brood size, inhibition of L1 growth, and abnormal locomotion was also reported (Manan et al. 2018).

29.5.3 *C. elegans* in the Discovery of Natural Compounds with Anti-Quorum Sensing/Virulence Factors Activity

Compounds derived from plants and natural sources with QS inhibitory activity have been used as medicine in traditional medicine since ancient times. A different portion of plants are rich sources of compounds and exudates that can be a candidate that can retaliate or interfere with QS. Plant secondary metabolites in the form of phenolics, flavonoids, saponins and etc. have a broad spectrum of biological applications which can benefit the host (Choo et al. 2006). In a recent proof of evidence for the feasibility of *C. elegans* as an in vivo model to screen natural products for anti-Qs/virulence factor, the anti-biofilm property of several natural products have been identified employing *S. aureus–C. elegans* model of host–pathogen interactions. "Hamamelitannin, the active component isolated from American witch hazel" can inhibit different aspects of quorum sensing mechanism (Lee et al. 2011, 2014). Using *C. elegans* and mouse as infection models, this compound and its analogs are reported to be effective in increasing sensitivity of *S. aureus* biofilms to broad-spectrum antibiotic treatment (Brackman et al. 2016) (Table 29.2). Sarabhai et al. (2013) reported that derivatives from *Terminalia chebula* Retz. components, ellagic acid can counteract *P. aeruginosa* PAO1 virulence through repression of many of its quorum sensing genes (Sarabhai et al. 2013).

29.5.4 Immune-Stimulatory Compounds as Antimicrobial

Immunostimulants or immunostimulators help in improvement of the body's natural resistance to infections by inducing nonspecific activation of the immune system components (Kumar et al. 2011). In interesting findings recently, a small molecule designated as RPW-24 was reported to have enhanced *C. elegans* survival upon *P. aeruginosa* infection. This is dictated in a dose-dependent manner and this is associated with upregulation of pathway and genes of the immune response pathway (p38 MAPK pathway and the transcription factor ATF-7) (Pukkila-Worley et al. 2012; Shivers et al. 2010). Similar reports include modulation or stimulation of the *C. elegans* immune response by Harmane, an alkaloid (Jakobsen et al. 2013) and UE-12 extract from *Orthosiphon stamineus* leaves that is mediated primarily by the p38 MAPK and secondary on the DAF-2 signaling pathways (Kong et al. 2014a, b). Kandasamy et al. (2012) reported that immune response-associated gene *lys-7* is activated by a plant extract and this helps in giving protection to the worms from the invading bacterial species (Kandasamy et al. 2012). Similarly, based on the versatility and ability of colistin to modulate p38/PMK-1 pathway and transcription factors like FOXO/DAF-16 and SKN-1 during whole animal chemical screen in *C. elegans*, it is expected that the same drug can have the same capability in activating immunity of higher organisms when consumed (Cai et al. 2015).

29.6 *C. elegans*: Hub of Antimicrobial Peptides and Proteins

Antimicrobial peptides (AMPs) are short evolutionary conserved low molecular weight molecules (below 5 kDa) that range in size from 10 to 150 amino acids. Most commonly, AMPs antimicrobial activity is seen directly in the form of attacking the surface of the bacterial membrane and/or by acting internally inside the cytoplasm of bacteria through their ability to translocate across its target cell membrane (Hancock and Sahl 2006). Most significantly, host AMPs' and proteins' roles in reducing or preventing infections is seen in *C. elegans* model system as varieties of AMPs and proteins are reported to be produced as part of normal and stimulate activation of the defense system. Their expression can be either constitutive as part of the innate immune system of the worm or are inducible at the time of infection or upon encounter with different pathogens (Kato et al. 2002; Ren et al. 2017; Alegado and Tan 2008). Tarr (2012) reported that the ability of bacterial strains to form a colony in the intestine of *C. elegans* is an indirect measurement of how high or low is the expression of AMPs, such as Abf-2 and Spp-1 present (Tarr 2012). Based on sequence similarity, *C. elegans* genome has various candidate AMPs and most of them are transcriptionally regulated upon infection. Therefore, it is of suggested view that to solve the problem of prevailing antibacterial drug resistance and in the development of new strategies and therapeutics, studies on available AMPs of *C. elegans* and its interactions with different innate components of the worms can offer a clue of the mechanistic process involved and in the possibility of discovering new AMPs in the near future.

Table 29.2 *C. elegans* as tool for investigation of natural compounds against virulence factors and quorum sensing

Compound name	Infection model-pathogens	Virulence factor(s)/QS involved and findings	Reference(s)
Chlorogenic acid	*C. elegans–P. aeruginosa* PAO1	Attenuation of virulence factors by regulating quorum sensing. Downregulate many *las and pqs genes*, protease, elastase, rhamnolipid, and reduces pyocyanin production	Wang et al. (2019)
Betulin and Betulinic acid	*C. elegans–P. aeruginosa* PAO1	Even at sublethal concentration. There was a reduction in the levels of virulence factors. Biofilm formation reduced, an inhibitor for QS receptors like LasR and RhlR	Rajkumari et al. (2018a, b)
Fukugiside	*C. elegans–S. pyogenes*	Expressions of different genes that encode virulence factor. It has versatile roles in inhibiting biofilm formation either by making the cell surface less hydrophobic in nature and can interfere with existing biofilm stability	Nandu et al. (2018)
Hamamelitannin	*C. elegans–S. aureus*	Block QS through the TraP, reduces the quorum sensing regulator RNAIII system, increase *S. aureus* biofilm susceptibility toward a broad-spectrum antibiotic, vancomycin	Brackman et al. (2016); Kiran et al. (2008)
Baicalin	*C. elegans–P. aeruginosa* PAO1	"Dose-dependent inhibitory effects on virulence phenotypes (LasA protease, LasB elastase, pyocyanin, rhamnolipid, motilities, and exotoxin A) regulated by QS in *P. aeruginosa*"	Luo et al. (2017)
Berberine	*C. elegans–S. typhimurium; C. elegans–P. aeruginosa* PA01	Significant reduction in adhesion, invasion, and infection in worm through possible modulation and interaction with the quorum sensing-mediated receptors	Aswathanarayana and vittal (2018)
Trans-stilbene	*C. elegans–S. aureus*	Attenuates virulence in the nematode *C. elegans* by its antagonize activity of hemolysis, reduces biofilm formation, repression of the *hla* gene and *ica* operon	Lee et al. (2014)

(continued)

Table 29.2 (continued)

Compound name	Infection model-pathogens	Virulence factor(s)/QS involved and findings	Reference(s)
Curcumin	*C. elegans–S. aureus* *C. elegans–P. aeruginosa* PAO1	A broad spectrum of positive effects. It can ameliorate PAO1 virulence by interfering and reducing levels of different virulence factors, entire quorum sensing signaling, can repress biofilms genes	Kong et al. (2014a, b); Rudrappa and Bais (2008)
Cinnamic acid	*C. elegans–P. aeruginosa* PAO1	Reduced worms mortality, a competitive inhibitor for las receptors. Thus different Las ligands cannot bind to their natural receptors to amplify the quorum sensing-mediated response	Rajkumari et al. (2018a, b)
Ellagic acid derivatives	*C. elegans–P. aeruginosa* PAO1	Downregulated quorum sensing and their receptors	Sarabhai et al. (2013)
Eugenol	*C. elegans–Escherichia coli* O157:H7	Reduced virulence factors "(violacein, elastase, pyocyanin, biofilm formation)," increased worms survival. inhibits quorum sensing at even minimal concentrations	Kim et al. (2016); Zhou et al. (2013)

29.6.1 Neuropeptide-Like Proteins

The *C. elegans* genomes encode at least 32 neuropeptide-like proteins (NLPs) with different unique bioactive motifs or subgroups and are reported to be present and expressed predominantly in its neurons and other endocrine active tissues (Nathoo et al. 2001). Numerous reports have shown the significant involvement of NLPs in preventing fungal or bacterial intrusion in worms through enhanced expressions of various *nlp* genes. For example, increased expression levels of peptides encoded by *nlp-29, 31,* and *33* were observed to be effective in fighting against bacterial and fungal infections (Couillault et al. 2004). Similarly, antimicrobial functions of the peptides encoded by *nlp-24, 25, 27, 28*, and *30* were also reported (Couillault et al. 2004). Additionally, apart from their presumed roles as antimicrobial agents in the worms, there were region-specific expressions of different NLPs with *nlp-29* reported to be abundant in the hypodermis, *nlp-31* most commonly in the developing embryos and hypoderm (Nathoo et al. 2001), whereas *nlp-33* is reported to be expressed specifically in the hypoderm (Couillault et al. 2004). The presence of some NLPs like *nlp-24* and *27* genes in neurons indicates that besides their role as anti-infectives they may have an additional role to serve as neuropeptides (Nathoo et al. 2001).

29.6.2 Caenacins

Caenacins or *Caenorhabditis* bacteriocin is very similar to NLP with mature peptides usually observed with the presence of glycine, aromatic amino acids, and unique motifs of the cleavage site. They were first reported in *C. elegans* with overexpression of genes in the "*cnc-2* cluster" (*cnc-1*, *cnc-2*, *cnc-3*, *cnc-4*, *cnc5*, and *cnc-11*) increases resistance to fungus *D. coniospora* (Zugasti and Ewbank 2009). Although phylogenetically and structurally and related to the NLPs, these *cnc genes* are regulated in a very distinct way (Zugasti et al. 2009). For example, unlike NLPs, osmotic stress has only little effect on the expression of *cnc-11* and no effect on the other members of the *cnc-2* cluster. Out of the 11 caenacins expressed by *C. elegans*, caenacin 2 (*cnc-2*) and the genes of the "cnc-2 cluster" have been the most investigated and was more strongly induced upon infection than upon wounding and is exclusively expressed in the epidermis. Furthermore, Zugasti et al. (2009) reported that the induced expression observed was pathogen specific as seen only upon infection with *D. coniospora* but not with *S. marescens* and *Pseudomonas aeruginosa*. The important role of transforming growth factor β ortholog DBL-1 in the regulation of expression of *cnc-2* gene is also reported (Zugasti et al. 2009).

29.6.3 Antibacterial Factor (ABF) Peptides

The *C. elegans* genome encodes highly conserved six homologs antibacterial factor (ABF-1 to ABF-6) (Zhang et al. 2000; Kato and Komatsu 1996). All ABF peptides belong to a cysteine-stabilized α-helix and β-sheet (CSαβ) superfamily, share conserved cysteine residues, and a secretory signal sequence at the amino end. The biosynthesis and expressions of ABFs are considered to be a part as requirements of *C. elegans* innate immunity to infections. Kato et al. (2002) showed that recombinant ABF-2 has antimicrobial activity (membrane splitting) to a broad spectrum of bacteria in vitro, and this is more pronounced especially in Gram-positive bacteria (Kato et al. 2002; Zhang et al. 2000). Interestingly, as part of its roles in normal digestion and regular protection, ABFs are constitutively expressed even in the lack of immune challenge. These constitutive expressions are dictated in a tissue-specific manner wherein pharyngeal has specifically high ABF-1 and ABF-2 expressions in comparison to the intestine where expressions of both ABF-1 and ABF-3 are observed (Kato et al. 2002; Alper et al. 2007). On the contrary, there were reports of pathogen-specific upregulation of different ABF peptides depending on the type of pathogen that causes the infection. For example, as part of its defense response to thwart or reduce *Salmonella typhimurium* in the intestinal lumen of the worms, the expression of ABF-2 is reported to be significantly upregulated (Alegado and Tan 2008). On the other hand, upregulation of both ABF-1 and ABF-2 is seen in the case of *Cryptococcus neoformans* and a little expression of ABF-3 is seen after *Staphylococcus aureus* infection in worms (Alper et al. 2007). With numerous evidence pinpointing that ABF peptides have a role in giving protection to *C. elegans* from various group of pathogens, it is of utmost scientific goals to decipher the involvement of different molecules or signaling pathways in the process.

29.6.4 Caenopores

Based on their similarity to functions and structures of amoebapores, Caenopores are pore-forming polypeptides saposin-like proteins (spp) found in *C. elegans*. In terms of structure similarity, the caenopores belong to the SAPLIP (saposin-like protein), such as the amoebapores first characterized in *Entamoeba histolytica* (Banyai and Patthy 1998; Leippe 1999) with diverse protein family, unique super-secondary structure and their ability to interact with phospholipid membranes (Bruhn 2005). It is a huge family, with 28 different genes coding for at least 33 distinct proteins. In *C. elegans*, 23 members are reported and these have antibacterial activity by their ability to form pore on the surface of the bacteria (Banyai and Patthy 1998; Roeder et al. 2010). Caenopores are observed most commonly in the *C. elegans* intestine in the intestinal lumen since their activity is modulated at an acidic environment, and this is ideal for giving protection to worms from any unwanted pathogens that reach the digestive tract (Alper et al. 2010; Roeder et al. 2010). In this protection role, caenopores unitedly work together with other AMPs and proteins and their level of expressions is dictated by the types of pathogen they encountered (Alegado and Tan 2008; Wong et al. 2007; Evans et al. 2008; Roeder et al. 2010).

29.6.5 Lysozymes

Lysozymes are known for defencing roles and in digestion in many organisms including *C. elegans* (Jolle's 1996). The diversity of lysozymes found in *C. elegans* is enormous, with different levels of sequence divergence within a single species. They contain up to 16 different lysozyme genes that produce two very distinct lysozyme types. These include the six invertebrate-type (*ilys-1* to *ilys-6)* and another ten of the genes that is characterized by distinct sequence and similarity to various protist taxa (*lys-1* to *lys-10*) (Schulenburg and Boehnisch 2008). These genes which can be inducible or constitutive in nature are expressed in a tissue-specific manner. Their activations and repressions are known during feeding, starvation, or bacterial infection (Mallo et al. 2002; Alper et al. 2008). For example, ILYS-3 protein was found to be expressed constitutively in the pharynx as daily requirements for grinding of food bacteria, coelomocytes for protective against invading bacterial pathogens, and dynamically in the intestine for digestion. Lysozyme genes like *lys-1* are reported to be expressed more when worms are infected with *Serratia marcescens* (Mallo et al. 2002). Gravato-Nobre et al. (2011) reported that when worms are facing with any Gram-positive pathogens, there were increased activation of *ilys-3* gene in the intestine and this activation is dependent upon the ERK-MAPK pathway and the crucial involvement of MPK-1 activity in the pharynx rather than in the intestine. Other reports suggest that some types of lysozymes can protect *C. elegans* against both Gram-positive and Gram-negative bacterial pathogens (Mallo et al. 2002; O'Rourke et al. 2006), and different lysozyme genes can be regulated differentially by different microbes (Dierking et al. 2016). The involvement of *C. elegans* immune signaling pathways such as DAF-16 (FOXO) and TGF-beta pathway in the regulations of several lysozyme genes was also reported (Murphy et al. 2003) whereby inactivation of these pathways as seen in mutant worm strains predisposed them to *M. nematophilum* and *P. aeruginosa* infections (Mochii et al. 1999; Alper et al. 2007). Overall, in *C. elegans,* lysozyme genes contribute to both defense roles as inherent bactericidal effectors in a microbe-specific form in *C. elegans.*

29.6.6 Lectins

Biological recognition phenomena at subcellular or surface level play a crucial role in the mediation of pathogen attachment to their specific target. These can have biological consequences in the form of easy identification and economical self-defense to the host against its invading pathogens. Lectins help in performing the roles of identification and are "carbohydrate-binding proteins, macromolecules that are highly specific for sugar moieties of other molecules." The *C. elegans* genome contains 283 gene family members of the C-Type Lectin-Like Domain genes (CTLD) (*clec genes*) and is the seventh most abundant gene family in *C. elegans.* CTLD have a unique region known as carbohydrate recognition domain (CRD and can exist as either soluble and/or membrane-bound proteins. They exhibit multiple patterns in their modes of the regulation when encountered with pathogens (O'Rourke et al. 2006; Wong et al. 2007; Troemel et al. 2006). The protective roles of lectins are diverse, with some lectins known to be involved in identification of a particular pathogen (via binding to specific patterns of surface on microbial sugars), others for the process of presenting microbes for destruction by an immune cell or opsonization, whereas others seem to protect or mask host epitopes from binding to specific microbial effectors (Schulenburg et al. 2008; Ideo et al. 2009). Proteomics study on *C. elegans* infected with *Aeromonas hydrophila* also confirmed the involvement of C-type lectins in the recognition and protection of *C. elegans* (Bogaerts et al. 2010). Thus, drugs that can target mammalian C-type lectin receptors can modulate cell-specific and host–pathogen interactions (Lepenies et al. 2013; Maglinao et al. 2014).

29.7 *C. elegans* Autophagy Roles as Antimicrobial

Recently, studies have shown that *C. elegans* has progressed substantially through changes in innate immunity or alternatively through the different evolving signaling mechanism. Another form of response that can be triggered by pathogen invasion is the process of autophagy or self-eating where infected or damaged cellular components is engulfed in autophagosomes, fuse with the lysosome, and degraded for recycling. Accumulating pieces of evidence on *C. elegans* has demonstrated a significant role of autophagy in protective roles against different infections. For example, effective autophagy process in *C. elegans* is reported to be important for worms' resistance from *Salmonella enterica* serovar Typhimurium infection (Jia et al. 2009; Curt et al. 2014), *N. parisii*, and *S. aureus* (Visvikis et al. 2014), whereas induction of autophagy through RNAi against let-363, the worm ortholog of TOR (Target of Rapamycin) led to a dramatic decrease in *N. parisii* cells load (Bakowski et al. 2014). Zou et al. (2014) reported that autophagy participates in host defense against a pathogenic bacterium *P. aeruginosa* and the induction of autophagy through a conserved extracellular signal-regulated kinase (ERK) protect *C. elegans* against necrosis during infection (Zou et al. 2014). Similarly, mitophagy (the specific degradation or removal of mitochondria by autophagy) has been linked with increased resistance to the killing of *P. aeruginosa* in worms (Kirienko et al. 2015).

29.8 Other Immune Effectors

Irrespective of numerous studies on the ill effects of ROS, there were reports that suggest enhanced production of ROS can positively contribute to immune responses and defense system of an organism. In *C. elegans* this can be mediated through single enzyme dual oxygenase (DUOX, called Ce-Duox1/BLI-3), to generate ROS during infection. The enhanced activation of BLI-3 during pathogen infection has been reported during bacterial and fungal pathogens, and this excess ROS generated can serve as a protective force of both intestinal and epidermal immunity (McCallum and Garsin 2016; Chavez et al. 2009). Other factors that are reported to have a regulating role in the immune response of the worms during infections include the involvement of endoplasmic stress and unfolding protein response (UPR) (Singh and Aballay 2017).

29.9 Conclusion

Emphasizing on the rise of new and deadly antimicrobial-resistant bacterial strains, an in-depth understanding of the whole process involved is crucial for new targeted drug discovery. With numerous clear pieces of evidence of the commonness of different human pathogens that are pathogenic to C. *elegans*, the employability of *C. elegans* model system to understand the whole process of pathogenesis is encouraging. At the forefront of development strategies for anti-infective therapeutics, by utilizing the integrated network of *C. elegans'* innate immune response to infections, a great deal of new knowledge is waiting to be tapped in the area of the host–pathogen interface. On the practical point of view, with the ease of its culturing techniques, short life cycle, and genetically rich resources, *C. elegans* is undoubtedly an attractive model for high-throughput screening thereby effectively reducing the cost and time involved in the discovery of novel antimicrobial compounds. Additionally, with the availability of established handling techniques, *C. elegans* may connect and fill in the limitation gaps of different in vitro and mammalian toxicity assays at a cheaper cost and predictability. At the molecular and cellular levels, with the availability of numerous homologs and ortholog genes of human and higher vertebrates, *C. elegans* model represents itself as a realistic model for elucidation and direct targeting approaches. Lastly, in the development of alternative anti-infectives which can only enhancing the resistance of the host to infections, further researches on *C. elegans* APPs is needed. Thus, a better understanding of its innate immune system and its modulational susceptibility can be utilized to uncover hosts of candidates for new drug discovery and targeted therapeutics in the near future.

References

Aballay A, Ausubel FM (2002) *Caenorhabditis elegans* as a host for the study of host–pathogen interactions. Curr Opin Microbiol 5:97–101

Aballay A, Yorgey P, Ausubel FM (2000) *Salmonella typhimurium* proliferates and establishes a persistent infection in the intestine of *Caenorhabditis elegans*. Curr Biol 10:1539–1542

Alegado RA, Tan MW (2008) Resistance to antimicrobial peptides contributes to persistence of *Salmonella typhimurium* in the *C. elegans* intestine. Cell Microbiol 10:1259–1273

Alexander AG, Marfil V, Li C (2014) Use of *Caenorhabditis elegans* as a model to study Alzheimer's disease and other neurodegenerative diseases. Front Genet 5:279

Allard P, Kleinstreuer NC, Knudsen TB, Colaiacovo MP (2013) A *C. elegans* screening platform for the rapid assessment of chemical disruption of germline function. Environ Health Perspect 121:717–724

Alper S, McBride SJ, Lackford B, Freedman JH, Schwartz DA (2007) Specificity and complexity of the *Caenorhabditis elegans* innate immune response. Mol Cell Biol 27:5544–5553

Alper S, Laws R, Lackford B, Boyd WA, Dunlap P, Freedman JH, Schwartz DA (2008) Identification of innate immunity genes and pathways using a comparative genomics approach. Proc Natl Acad Sci U S A 105:7016–7021

Alper S, McElwee MK, Apfeld J, Lackford B, Freedman JH, Schwartz DA (2010) The *Caenorhabditis elegans* germ line regulates distinct signaling pathways to control ifespan and innate immunity. J Biol Chem 285:1822–1828

Amrit FR, Ratnappan R, Keith SA, Ghazi A (2014) The *C. elegans* lifespan assay toolkit. Methods 68:465–475

Anderson GL, Cole RD, Williams PL (2004) Assessing behavioral toxicity with *Caenorhabditis elegans*. Environ Toxicol Chem 23:1235–1240

Anyanful A, Easley KA, Benian GM, Kalman D (2009) Conditioning protects *C. elegans* from lethal effects of enteropathogenic E. coli by activating genes that regulate lifespan and innate immunity. Cell Host Microbe 5:450–462

Arvanitis M, Glavis-Bloom J, Mylonakis E (2013) *C. elegans* for anti-infective discovery. Curr Opin Pharmacol 13:769–774

Ashrafi K, Chang FY, Watts JL, Fraser AG, Kamath RS, Ahringer J, Ruvkun G (2003) Genome-wide RNAi analysis of *Caenorhabditis elegans* fat regulatory genes. Nature 421:268–272

Aswathanarayana JB, Vittal RR (2018) Inhibition of biofilm formation and quorum sensing mediated phenotypes by berberine in *Pseudomonas aeruginosa* and *Salmonella typhimurium*. RSC Adv 8:36133–36141

Atkinson S, Goldstone RJ, Joshua GWP, Chang CY, Patrick HL, Cámara M, Wren BW, William P (2011) Biofilm development on *Caenorhabditis elegans* by *Yersinia* is facilitated by quorum sensing-dependent repression of type III secretion. PLoS Pathog 7(1):e1001250

Bakowski MA, Desjardins CA, Smelkinson MG, Dunbar TL, Lopez-Moyado IF, Rifkin SA, Cuomo CA, Troemel ER, Schneider DS (2014) Ubiquitin-mediated response to microsporidia and virus infection in C. elegans. PLoS Pathog 10(6):e1004200

Balasubramanian D, Schneper L, Kumari H, Mathee K (2013) A dynamic and intricate regulatory network determines *Pseudomonas aeruginosa* virulence. Nucleic Acids Res 41:1–20

Banyai L, Patthy L (1998) Amoebapore homologs of *Caenorhabditis elegans*. Biochim Biophys Acta 1429:259–264

Barczak AK, Hung DT (2009) Productive steps toward an antimicrobial targeting virulence. Curr Opin Microbiol 12:490–496

Begun J, Sifri CD, Goldman S, Calderwood SB, Ausubel FM (2005) *Staphylococcus aureus* virulence factors identified by using a high-throughput *Caenorhabditis elegans*-killing model. Infect Immun 73:872–877

Ben-Yakar A, Chronis N, Lu H (2009) Microfluidics for the analysis of behavior, nerve regeneration, and neural cell biology in *C. elegans*. Curr Opin Neurobiol 19:561–567

Bogaerts A, Beets I, Schoofs L, Verleyen P (2010) Antimicrobial peptides in *Caenorhabditis elegans*. Invert Surviv J 7:45–52

Boucher HW, Talbot GH, Bradley JS, Edwards JE, Gilbert D, Rice LB, Scheld M, Spellberg B, Bartlett J (2009) Bad bugs, no drugs: no ESKAPE! An update from the Infectious Diseases Society of America. Clin Infect Dis 48(1):1–12

Boyd WA, McBride SJ, Rice JR, Snyder DW, Freedman JH (2010) A high throughput method for assessing chemical toxicity using a *Caenorhabditis elegans* reproduction assay. Toxicol Appl Pharmacol 245:153–159

Boyd WA, Smith MV, Co CA, Pirone JR, Rice JR, Shockley KR, Freedman JH (2016) Developmental effects of the ToxCast™ phase I and phase II chemicals in *Caenorhabditis elegans* and corresponding responses in Zebrafish, rats, and rabbits. Environ Health Perspect 124:586–593

Brackman G, Breyne K, De Rycke R, Vermote A, Van Nieuwerburgh F, Meyer E et al (2016) The quorum sensing inhibitor hamamelitannin increases antibiotic susceptibility of *Staphylococcus aureus* biofilms by affecting peptidoglycan biosynthesis and eDNA release. Sci Rep 6:20321

Braungart E, Gerlach M, Riederer P, Baumeister R, Hoener MC (2004) *Caenorhabditis elegans* MPP+ model of Parkinson's disease for high-throughput drug screenings. Neurodegener Dis 1:175–183

Brenner S (1974) Genetics of *Caenorhabditis elegans*. Genetics 77:71–94

Brown ED, Wright GD (2016) Antibacterial drug discovery in the resistance era. Nature 529:336–343

Bruhn H (2005) A short guided tour through functional and structural features of saposin-like proteins. Biochem J 389:249–257

Burton EA, Pendergast AM, Aballay A (2006) The *Caenorhabditis elegans* ABL-1 tyrosine kinase is required for Shigella flexneri pathogenesis. Appl Environ Microbiol 72:5043–5051

C. elegans Sequencing Consortium (1998) Genome sequence of the nematode *C. elegans*: a platform for investigating biology. Science 282:2012–2018

Cai Z, Liu Y, Chen Y, Yam JK, Chew SC, Chua SL, Wang K, Givskov M, Yang L (2015) RpoN regulates virulence factors of *Pseudomonas aeruginosa* via modulating the PqsR quorum sensing regulator. Int J Mol Sci 16:28311–28319

Caiazza NC, O'Toole GA (2004) SadB is required for the transition from reversible to irreversible attachment during biofilm formation by Pseudomonas aeruginosa PA14. J Bacteriol 186:4476–4485

Cassada RC, Russell RL (1975) The dauer larva, a post-embryonic developmental variant of the nematode *Caenorhabditis elegans*. Dev Biol 46:326–342

Cezairliyan B, Vinayavekhin N, Grenfell-Lee D, Yuen GJ, Saghatelian A, Ausubel FM (2013) Identification of *Pseudomonas aeruginosa* phenazines that kill *Caenorhabditis elegans*. PLoS Pathog 9(1):e1003101

Chalfie M, Tu Y, Euskirchen G, Ward WW, Prasher DC (1994) Green fluorescent protein as a marker for gene expression. Science 263:802–805

Chavez V, Mohri-Shiomi A, Garsin DA (2009) Ce-Duox1/BLI-3 generates reactive oxygen species as a protective innate immune mechanism in *Caenorhabditis elegans*. Infect Immun 77:4983–4989

Chen C, Fenk LA, de Bono M (2013) Efficient genome editing in *Caenorhabditis elegans* by CRISPR-targeted homologous recombination. Nucleic Acids Res 41:e193

Choo JH, Rukayadi Y, Hwang JK (2006) Inhibition of bacterial quorum sensing by vanilla extract. Lett Appl Microbiol 42:637–641

Cinar HN, Kothary M, Datta AR, Tall BD, Sprando R, Bilecen K, Yildiz F, McCardellet B (2010) *Vibrio cholerae* hemolysin is required for lethality, developmental delay, and intestinal vacuolation in Caenorhabditis elegans. PLoS One 5:e11558. https://doi.org/10.1371/journal.pone.0011558

Clatworthy AE, Pierson E, Hung DT (2007) Targeting virulence: a new paradigm for antimicrobial therapy. Nat Chem Biol 3:541–548

Clausell-Tormos J, Lieber D, Baret JC, El-Harrak A, Miller OJ, Frenz L, Blouwolff J, Humphry KJ, Koster S, Duan H, Holtze C, Weitz DA, Griffiths AD, Merten CA (2008) Droplet-based microfluidic platforms for the encapsulation and screening of mammalian cells and multicellular organisms. Chem Biol 15:427–437

Corsi AK (2006) A biochemist's guide to *Caenorhabditis elegans*. Anal Biochem 359:1–17

Couillault C, Pujol N, Reboul J, Sabatier L, Guichou J-F, Kohara Y, Ewbank JJ (2004) TLR-independent control of innate immunity in *Caenorhabditis elegans* by the TIR domain adaptor protein TIR-1, an ortholog of human SARM. Nat Immunol 5:88–494

Cox JS, Walter P (1996) A novel mechanism for regulating activity of a transcription factor that controls the unfolded protein response. Cell 87:391–404

Curt A, Zhang J, Minnerly J, Jia K (2014) Intestinal autophagy activity is essential for host defense against Salmonella typhimurium infection in *Caenorhabditis elegans*. Dev Comp Immunol 45:214–218

Darby C, Cosma CL, Thomas JH, Manoil C (1999) Lethal paralysis of Caenorhabditis elegans by *Pseudomonas aeruginosa*. Proc Natl Acad Sci U S A 96:15202–15207

Deng Y, Schmid N, Wang C, Wang J, Pessi G, Wu D, Lee J, Aguilar C, Ahrens CH, Chang C, Song H, Eberl L, Zhang LH (2012) Cis-2-dodecenoic acid receptor RpfR links quorum-sensing signal perception with regulation of virulence through cyclic dimeric guanosine monophosphate turnover. Proc Natl Acad Sci U S A 109:15479–15484

Dengg M, van Meel JCA (2004) *Caenorhabditis elegans* as model system for rapid toxicity assessment of pharmaceutical compounds. J Pharmacol Toxicol Methods 50:209–214

Dickson M, Gagnon JP (2004) The cost of new drug discovery and development. Discov Med 4:172–179

Dierking K, Yang W, Schulenburg H (2016) Antimicrobial effectors in the nematode *Caenorhabditis elegans*: an outgroup to the Arthropoda. Philos Trans R Soc Lond B Biol Sci 371:20150299

Duerr JS (2006) Immunohistochemistry. In: WormBook (ed) The *C. elegans* research community, WormBook. doi: https://doi.org/10.1895/wormbook.1.105.1., http://www.wormbook.org

Dunbar TL, Yan Z, Balla KM, Smelkinson MG, Troemel ER (2012) *C. elegans* detects pathogen induced translational inhibition to activate immune signaling. Cell Host Microbe 11:375–386

Ellis RE, Yuan JY, Horvitz HR (1991) Mechanisms and functions of cell-death. Annu Rev Cell Biol 7:663–698

Evans EA, Kawli T, Tan MW (2008) *Pseudomonas aeruginosa* suppresses host immunity by activating the DAF-2 insulin-like signaling pathway in *Caenorhabditis elegans*. PLoS Pathog 4:e1000175

Ewbank JJ (2006) Signaling in the immune response. WormBook:1–12

Ewbank JJ (2007) The genetics of pathogen avoidance in *Caenorhabditis elegans*. Mol Microbiol 66:563–570

Feinbaum RL, Urbach JM, Liberati NT, Djonovic S, Adonizio A, Carvunis AR, Ausubel FM (2012) Genome-wide identification of *Pseudomonas aeruginosa* virulence-related genes using a Caenorhabditis elegans infection model. PLoS Pathog 8(7):e1002813

Fire A, Xu SQ, Montgomery MK, Kostas SA, Driver SE, Mello CC (1998) Potent and specific genetic interference by double-stranded RNA in *Caenorhabditis elegans*. Nature 391:806–811
Fisher RA (1990) In: Bennett JH (ed) Statistical methods, experimental design, and scientific inference. Oxford University Press, Oxford
Garigan D, Hsu A-L, Fraser AG, Kamath RS, Ahringer J, Kenyon C (2002) Genetic analysis of tissue aging in *Caenorhabditis elegans*: a role for heat-shock factor and bacterial proliferation. Genetics 161:1101–1112
Garsin DA, Sifri CD, Mylonakis E, Qin X, Singh KV, Murray BE, Calderwood SB, Ausubel FM (2001) A simple model host for identifying gram-positive virulence factors. Proc Natl Acad Sci U S A 98:10892–10897
Gellatly SL, Hancock RE (2013) *Pseudomonas aeruginosa*: new insights into pathogenesis and host defenses. Pathog Dis 67:159–173
Golden TR, Beckman KB, Lee AHJ, Dudek N, Hubbard A, Samper E, Melov S (2007) Dramatic age-related changes in nuclear and genome copy number in the nematode *Caenorhabditis elegans*. Aging Cell 6:179–188
Gould IM, Bal AM (2013) New antibiotic agents in the pipeline and how they can help overcome microbial resistance. Virulence 4:185–191
Grant BD, Sato M (2006) Intracellular trafficking. In: WormBook (ed) The *C. elegans* research community, WormBook, doi: https://doi.org/10.1895/wormbook.1.77.1
Gravato-Nobre MJ, Stroud D, O'Rourke D, Darby C, Hodgkin J (2011) Glycosylation genes expressed in seam cells determine complex surface properties and bacterial adhesion to the cuticle of *Caenorhabditis elegans*. Genetics 187:141–155
Hae-Eun HP, Yoonji J, Seung-Jae VL (2017) Survival assays using *Caenorhabditis elegans*. Mol Cells 40:90–99
Haithcock E, Dayani Y, Neufeld E, Zahand AJ, Feinstein N, Mattout A, Gruenbaum Y, Liu J (2005) Age-related changes of nuclear architecture in *Caenorhabditis elegans*. Proc Natl Acad Sci U S A 102:16690–16695
Hall DH, Altun Z (2008) *C. elegans* Atlas. Cold Spring Harbor Laboratory Press, Cold Spring Harbor, p 340
Hamilton B, Dong Y, Shindo M, Liu W, Odell I, Ruvkun G, Lee SS (2005) A systematic RNAi screen for longevity genes in *C. elegans*. Genes Dev 19:1544–1555
Hancock REW, Sahl HG (2006) Antimicrobial and host-defense peptides as new anti-infective therapeutic strategies. Nat Biotechnol 24:1551–1557
Harris TW, Chen NS, Cunningham F, Tello-Ruiz M, Antoshechkin I, Bastiani C, Bieri T, Blasiar D, Bradnam K, Chan J, Chen CK, Chen WJ, Davis P, Kenny E, Kishore R, Lawson D, Lee R, Muller HM, Nakamura C, Ozersky P, Petcherski A, Rogers A, Sabo A, Schwarz EM, Van Auken K, Wang Q, Durbin R, Spieth J, Sternberg PW, Stein LD (2004) WormBase: a multi-species resource for nematode biology and genomics. Nucleic Acids Res 32:D411–D417
Hartung T (2009) Toxicology for the twenty-first century. Nature 460:208–212
Hastings JW, Nealson KH (1977) Bacterial bioluminescence. Annu Rev Microbiol 31:549–595
Herndon LA, Schmeissner PJ, Dudaronek JM, Brown PA, Listner KM, Sakano Y, Paupard MC, Hall DH, Driscoll M (2002) Stochastic and genetic factors influence tissue-specific decline in ageing *C. elegans*. Nature 419:808–814
Heurlier K, Denervaud V, Pessi G, Reimmann C, Haas D (2003) Negative control of quorum sensing by RpoN (sigma54) in *Pseudomonas aeruginosa* PAO1. J Bacterial 185:2227–2235
Hobert O (2002) PCR fusion-based approach to create reporter gene constructs for expression analysis in transgenic *C. elegans*. Biotechniques 32:728–730
Hsu PD, Lander ES, Zhang F (2014) Development and applications of CRISPR-Cas9 for genome engineering. Cell 157:1262–1278
Hunt PR, Olejnik N, Sprando RL (2012) Toxicity ranking of heavy metals with screening method using adult *Caenorhabditis elegans* and propidium iodide replicates toxicity ranking in rat. Food Chem Toxicol 50:3280–3290
Husson SJ, Gottschalk A, Leifer AM (2013) Optogenetic manipulation of neural activity in C. elegans: from synapse to circuits and behaviour. Biol Cell 105(6):235–250

Ideo H, Fukushima K, Gengyo-Ando K, Mitani S, Dejima K, Nomura K, Yamashita K (2009) A *Caenorhabditis elegans* glycolipid-binding galectin functions in host defense against bacterial infection. J Biol Chem 284:26493–26501

Irazoqui JE, Urbach JM, Ausubel FM (2010) Evolution of host innate defence: insights from *Caenorhabditis elegans* and primitive invertebrates. Nat Rev Immunol 10:47–58

Jakobsen H, Bojer MS, Marinus MG, Xu T, Struve C, Krogfelt KA, Løbner-Olesen A (2013) The alkaloid compound harmane increases the lifespan of Caenorhabditis elegans during bacterial infection, by modulating the nematode's innate immune response. PLoS One 8:e60519

James CE, Davey MW (2007) A rapid colorimetric assay for the quantitation of the viability of free-living larvae of nematodes in vitro. Parasitol Res 101:975–980

Jayamani E, Rajamuthiah R, Larkins-Ford J, Fuchs BB, Conery AL, Vilcinskas A, Ausubel FM, Mylonakis E (2015) Insect-derived cecropins display activity against *Acinetobacter baumannii* in a whole-animal high-throughput *Caenorhabditis elegans* model. Antimicrob Agents Chemother 59:1728–1737

Jia K, Thomas C, Akbar M, Sun Q, Adams-Huet B, Gilpin C, Levine B (2009) Autophagy genes protect against *Salmonella typhimurium* infection and mediate insulin signaling-regulated pathogen resistance. Proc Natl Acad Sci U S A 106:14564–14569

Jorgensen EM, Mango SE (2002) The art and design of genetic screens: *caenorhabditis elegans*. Nat Rev Genet 3:356–369

Kaletta T, Hengartner MO (2006) Finding function in novel targets: *C. elegans* as a model organism. Nat Rev Drug Discov 5:387–398

Kamaladevi A, Balamurugan K (2015) Role of PMK-1/p38 MAPK defense in *Caenorhabditis elegans* against *Klebsiella pneumoniae* infection during host-pathogen interaction. Pathog Dis 75:1–9

Kamath RS, Ahringer J (2003) Genome-wide RNAi screening in *Caenorhabditis elegans*. Methods 30:313–321

Kandasamy S, Khan W, Evans F, Critchley AT, Prithiviraj B (2012) Tasco(R): a product of *Ascophyllum nodosum* enhances immune response of *Caenorhabditis elegans* against *Pseudomonas aeruginosa* infection. Mar Drugs 10:84–105

Kaplan EL, Meier P (1958) Nonparametric estimation from incomplete observations. J Am Stat Assoc 53:457–481

Kato Y, Komatsu S (1996) ASABF, a novel cysteine-rich antibacterial peptide isolated from the nematode *Ascaris suum*. Purification, primary structure and molecular cloning of cDNA. J Biol Chem 271:30493–30498

Kato Y, Aizawa T, Hoshino H, Kawano K, Nitta K, Zhang H (2002) Abf-1 and abf-2, ASABF-type antimicrobial peptide genes in *Caenorhabditis elegans*. Biochem J 361:221–230

Keith SA, Amrit FR, Ratnappan R, Ghazi A (2014) The *C. elegans* healthspan and stress-resistance assay toolkit. Methods 68:476–486

Kenyon C, Chang J, Gensch E, Rudner A, Tabtiang R (1993) A *C. elegans* mutant that lives twice as long as wild type. Nature 366:461–464

Kesika P, Balamurugan K (2012) Studies on *Shigella boydii* infection in *Caenorhabditis elegans* and bioinformatics analysis of immune regulatory protein interactions. Biochim Biophys Acta 1824:1449–1456

Kim DH, Feinbaum R, Alloing G, Emerson FE, Garsin DA, Inoue H, Tanaka-Hino M, Hisamoto N, Matsumoto K, Tan MW, Ausubel FM (2002) A conserved p38 MAP kinase pathway in *Caenorhabditis elegans* innate immunity. Science 297:623–626

Kim YG, Lee JH, Gwon G, Kim SI, Park JG, Lee J (2016) Essential oils and eugenols inhibit biofilm formation and the virulence of *Escherichia coli* O157:H7. Sci Rep 6:36377

Kiran MD, Adikesavan NV, Cirioni O, Giacometti A, Silvestri C, Scalise G, Ghiselli R, Saba V, Orlando F, Shoham M, Balaban N (2008) Discovery of a quorum-sensing inhibitor of drug-resistant staphylococcal infections by structure-based virtual screening. Mol Pharmacol 73:1578–1586

Kirienko NV, Cezairliyan BO, Ausubel FM, Powell JR (2014) *Pseudomonas aeruginosa* PA14 pathogenesis in *Caenorhabditis elegans*. Methods Mol Biol 1149:653–669

Kirienko NV, Ausubel FM, Ruvkun G (2015) Mitophagy confers resistance to siderophore mediated killing by *Pseudomonas aeruginosa*. Proc Natl Acad Sci U S A 112:1821–1826

Klass MR (1977) Aging in the nematode *Caenorhabditis elegans*: major biological and environmental factors influencing life span. Mech Ageing Dev 6:413–429

Knight AW, Little S, Houck K, Dix D, Judson R, Richard A, McCarroll N, Akerman G, Yang C, Birrell L, Walmsley RM (2009) Evaluation of highthroughput genotoxicity assays used in profiling the US EPA ToxCast chemicals. Regul Toxicol Pharmacol 55:188–199

Kong C, Yehye WA, Abd Rahman N, Tan MW, Nathan S (2014a) Discovery of potential anti-infectives against *Staphylococcus aureus* using a *Caenorhabditis elegans* infection model. BMC Complement Altern Med 14(4)

Kong C, Tan MW, Nathan S (2014b) *Orthosiphon stamineus* protects Caenorhabditis elegans against Staphylococcus aureus infection through immunomodulation. Biol Open 3:644–655

Kumar S, Gupta P, Sharma S, Kumar D (2011) A review on immunostimulatory plants. J Chin Integr Med 9:117–128

Kwok TC, Ricker N, Fraser R, Chan AW, Burns A, Stanley EF, McCourt P, Cutler SR, Roy PJ (2006) A small-molecule screen in *C. elegans* yields a new calcium channel antagonist. Nature 441:91–95

Labrousse A, Chauvet S, Couillault C, Kurz CL, Ewbank JJ (2000) *Caenorhabditis elegans* is a model host for *Salmonella typhimurium*. Curr Biol 10:1543–1545

Lai CH, Chou CY, Ch'ang LY, Liu CS, Lin W (2000) Identification of novel human genes evolutionarily conserved in *Caenorhabditis elegans* by comparative proteomics. Genome Res 10:703–713

Lakshmanan U, Yap A, Fulwood J, Yichun L, Hoon SS, Lim J, Ting A, Sem XH, Kreisberg JF, Tan P, Tan G, Flotow H (2014) Establishment of a novel whole animal HTS technology platform for melioidosis drug discovery. Comb Chem High Throughput Screen 17:790–803

Lamitina T, Huang CG, Strange K (2006) Genome-wide RNAi screening identifies protein damage as a regulator of osmoprotective gene expression. Proc Natl Acad Sci U S A 103:12173–12178

Lee KM, Lim J, Nam S, Yoon MY, Kwon YK, Jung BY, Park Y, Park S, Yoon SS (2011) Inhibitory effects of broccoli extract on *Escherichia coli* O157:H7 quorum sensing and *in vivo* virulence: effects of broccoli extract on bacterial virulence. FEMS Microbiol Lett 321:67–74

Lee K, Lee JH, Ryu SY, Cho MH, Lee J (2014) Stilbenes reduce *Staphylococcus aureus* hemolysis, biofilm formation and virulence. Foodborne Pathog Dis 11:710–717

Lehner B, Crombie C, Tischler J, Fortunato A, Fraser AG (2006a) Systematic mapping of genetic interactions in *Caenorhabditis elegans* identifies common modifiers of diverse signaling pathways. Nat Genet 38:896–903

Lehner B, Tischler J, Fraser AG (2006b) RNAi screens in *Caenorhabditis elegans* in a 96-well liquid format and their application to the systematic identification of genetic interactions. Nat Protoc 1:1617–1620

Leippe M (1999) Antimicrobial and cytolytic polypeptides of amoeboid protozoa-effector molecules of primitive phagocytes. Dev Comp Immunol 23:267–279

Lepenies B, Lee J, Sonkaria S (2013) Targeting C-type lectin receptors with multivalent carbohydrate ligands. Adv Drug Deliv Rev 65:1271–1281

Leung CK, Wang Y, Malany S, Deonarine A, Nguyen K, Vasile S, Choe KP (2013) An ultra high-throughput, whole-animal screen for small molecule modulators of a specific genetic pathway in *Caenorhabditis elegans*. PLoS One 8:e62166

Levitan D, Doyle TG, Brousseau D, Lee MK, Thinakaran G, Slunt HH, Sisodia SS, Greenwald I (1996) Assessment of normal and mutant human presenilin function in *Caenorhabditis elegans*. Proc Natl Acad Sci U S A 93:14940–14944

Levy SB (1998) The challenge of antibiotic resistance. Sci Am 278:32–39

Lloyd MG, Vossler JL, Nomura CT, Moffat JF (2019) RpoN reduces virulence of *Pseudomonas aeruginosa* isolated from cystic fibrosis patients and increases antibiotic sensitivity in a laboratory strain. Sci Rep 9(1):6677

Lu T, Aron L, Zullo J, Pan Y, Kim H, Chen Y, Yang TH, Kim HM, Drake D, Liu XS, Bennett DA, Colaiacovo MP, Yankner BA (2014) REST and stress resistance in ageing and Alzheimer's disease. Nature 507:448–454

Lucanic M, Garrett T, Gill MS, Lithgow GJ (2018) A simple method for high throughput chemical screening in *Caenorhabditis Elegans*. J Vis Exp 133:e56892

Luo J, Dong B, Wang K, Cai S, Liu T, Cheng X, Lei D, Chen Y, Li Y, Kong J, Chen Y (2017) Baicalin inhibits biofilm formation, attenuates the quorum sensing-controlled virulence and enhances *Pseudomonas aeruginosa* clearance in a mouse peritoneal implant infection model. PLoS One 12(4):e0176883. https://doi.org/10.1371/journal.pone.0176883

Maglinao M, Eriksson M, Schlegel MK, Zimmermann S, Johannssen T, Götze S, Seeberger PH, Lepenies B (2014) A platform to screen for C-type lectin receptor-binding carbohydrates and their potential for cell-specific targeting and immune modulation. J Control Release 175:36–42

Mahajan-Miklos S, Tan MW, Rahme LG, Ausubel FM (1999) Molecular mechanisms of bacterial virulence elucidated using a *Pseudomonas aeruginosa-Caenorhabditis elegans* pathogenesis model. Cell 96:47–56

Mak HY, Nelson LS, Basson M, Johnson CD, Ruvkun G (2006) Polygenic control of *Caenorhabditis elegans* fat storage. Nat Genet 38:363–368

Mallo GV, Kurz CL, Couillault C, Pujol N, Granjeaud S, Kohara Y, Ewbank JJ (2002) Inducible antibacterial defense system in *C. elegans*. Curr Biol 12:1209–1214

Manan A, Bazai ZA, Fan J, Yu H, Li L (2018) The Nif3-family protein YqfO03 from *Pseudomonas syringae* MB03 has multiple nematicidal activities against *Caenorhabditis elegans* and *Meloidogyne incognita*. Int J Mol Sci 19:3915

Mantel N (1966) Evaluation of survival data and two new rank order statistics arising in its consideration. Cancer Chemother Rep 50:163–170

Markaki M, Tavernarakis N (2010) Modeling human diseases in *Caenorhabditis elegans*. Biotechnol J 5:1261–1276

Mason CM, Kolls JK, Nelson S (1995) Pathogenesis and host defense in pulmonary infections. Curr Opin Pulm Med 1:163–170

Mathew MD, Mathew ND, Ebert PR (2012) WormScan: a technique for high throughput phenotypic analysis of *Caenorhabditis elegans*. PLoS One 7:e33483

Maydan JS, Flibotte S, Edgley ML, Lau J, Selzer RR, Richmond TA, Pofahl NJ, Thomas JH, Moerman DG (2007) Efficient high-resolution deletion discovery in *Caenorhabditis elegans* by array comparative genomic hybridization. Genome Res 17:337–347

McCallum KC, Garsin DA (2016) The role of reactive oxygen species in modulating the *Caenorhabditis elegans* immune response. PLoS Pathog 12(11):e1005923

McDiarmid TA, Au V, Loewen AD, Liang J, Mizumoto K, Moerman DG, Rankin CH (2018) CRISPR-Cas9 human gene replacement and phenomic characterization in to understand the functional conservation of human genes and decipher variants of uncertain significance. Dis. Model Mech 11(12):dmm036517

McGhee JD (2013) The *Caenorhabditis elegans* intestine. Wiley Interdiscip Rev Dev Biol 2:347–367

Meier B, Cooke SL, Weiss J, Bailly AP, Alexandrov LB, Marshall J, Raine K, Maddison M, Anderson E, Stratton MR, Gartner A, Campbell PJ (2014) *C. elegans* whole-genome sequencing reveals mutational signatures related to carcinogens and DNA repair deficiency. Genome Res 24:1624–1636

Mesaros N, Nordmann P, Plesiat P, Roussel-Delvallez M, Van Eldere J, Glupczynski J, VanLaethem Y, Jacobs F, Lebecque P, Malfroot A, Tulkens PM, Van Bambek F (2007) *Pseudomonas aeruginosa*: resistance and therapeutic options at the turnof the new millennium. Clin Microbiol Infect 13:560–578

Mochii M, Yoshida S, Morita K, Kohara Y, Ueno N (1999) Identification of transforming growth factor-beta- regulated genes in *Caenorhabditis elegans* by differential hybridization of arrayed cDNAs. Proc Natl Acad Sci U S A 96:15020–15025

Mori K, Kawahara T, Yoshida H, Yanagi H, Yura T (1996) Signalling from endoplasmic reticulum to nucleus: transcription factor with a basic-leucine zipper motif is required for the unfolded protein response pathway. Genes Cells 1:803–817

Morihara K, Tsuzuki H, Oka T, Inoue H, Ebata M (1965) *Pseudomonas aeruginosa* elastase solation, crystallization, and preliminary characterization. J Biol Chem 240:3295–3304

Moy TI, Ball AR, Anklesaria Z, Casadei G, Lewis K, Ausubel FM (2006) Identification of novel antimicrobials using a live-animal infection model. Proc Natl Acad Sci U S A:10414–10419

Moy TI, Conery AL, Larkins-Ford J, Wu G, Mazitschek R, Casadei G, Lewis K, Carpenter AE, Ausubel FM (2009a) High-throughput screen for novel antimicrobials using a whole animal infection model. ACS Chem Biol 4:527–533

Moy TI, Ball AR, Anklesaria Z, Casadei G, Lewis K, Ausubel FM (2009b) Identification of novel antimicrobials using a live-animal infection model. Proc Natl Acad Sci U S A 103:10414–10419

Murphy CT, McCarroll SA, Bargmann CI, Fraser A, Kamath RS, Ahringer J, Li H, Kenyon C (2003) Genes that act downstream of DAF-16 to influence the lifespan of *Caenorhabditis elegans*. Nature 424:277–283

Mylonakis E, Ausubel FM, Perfect JR, Heitman J, Calderwood SB (2002) Killing of *Caenorhabditis elegans* by *Cryptococcus neoformans* as a model of yeast pathogenesis. Proc Natl Acad Sci U S A 99:15675–15680

Nandu TG, Subramenium GA, Shiburaj S, Viszwapriya D, Iyer PM, Balamurugan K, Rameshkumar KB, Karutha Pandian S (2018) Fukugiside, a biflavonoid from *Garcinia travancorica* inhibits biofilm formation of *Streptococcus pyogenes* and its associated virulence factors. J Med Microbiol 67:1391–1401

Nass R, Hamza I (2007) The nematode *C. elegans* as an animal model to explore toxicology *in vivo*: solid and axenic growth culture conditions and compound exposure parameters. Curr Protoc Toxicol. Chapter 1. https://doi.org/10.1002/0471140856.tx0109s31

Nathoo AN, Moeller RA, Westlund BA, Hart AC (2001) Identification of neuropeptide-like protein gene families in *Caenorhabditis elegans* and other species. Proc Natl Acad Sci U S A 98:14000–14005

Nigon VM, Félix MA (2017) History of research on *C. elegans* and other free-living nematodes as model organisms. In: Sternberg PW (eds) The *C. elegans* research community, WormBook, pp 1–84

O'Quinn AL, Wiegand EM, Jeddeloh JA (2001) *Burkholderia pseudomallei* kills the nematode *Caenorhabditis elegans* using an endotoxin-mediated paralysis. Cell Microbiol 3:381–393

O'Reilly LP, Luke CJ, Perlmutter DH, Silverman GA, Pak SC (2014) *C. elegans* in high-throughput drug discovery. Adv Drug Deliv Rev 69–70:247–253

O'Rourke D, Baban D, Demidova M, Mott R, Hodgkin J (2006) Genomic clusters, putative pathogen recognition molecules, and antimicrobial genes are induced by infection of *C. elegans* with *M nematophilum*. Genome Res 16:1005–1016

Ogg S, Paradis S, Gottlieb S, Patterson GI, Lee L, Tissenbaum HA, Ruvkun G (1997) The fork head transcription factor DAF-16 transduces insulin-like metabolic and longevity signals in *C. elegans*. Nature 389:994–999

Olson H, Betton G, Robinson D, Thomas K, Monro A, Kolaja G, Lilly P, Sanders J, Sipes G, Bracken W, Dorato M, Van Deun K, Smith P, Berger B, Heller A (2000) Concordance of the toxicity of pharmaceuticals in humans and in animals. Regul Toxicol Pharmacol 32:56–67

Onken B, Driscoll M (2010) Metformin induces a dietary restriction-like state and the oxidative stress response to extend *C. elegans* healthspan via AMPK, LKB1, and SKN-1. PLoS One 5:e8758.10.1371

Papenfort K, Bassler BL (2016) Quorum sensing signal-response systems in Gram-negative bacteria. Nat Rev Microbiol 14:576–588

Peterson RT, Nass R, Boyd WA, Freedman JH, Dong K, Narahashi T (2008) Use of non-mammalian alternative models for neurotoxicological study. Neurotoxicology 29:546–555

Potvin E, Sanschagrin F, Levesque RC (2008) Sigma factors in *Pseudomonas aeruginosa*. FEMS Microbiol Rev 32:38–55

Powell JR, Ausubel FM (2008) Models of *Caenorhabditis elegans* infection by bacterial and fungal pathogens. Methods Mol Biol 415:403–427

Pradel E, Zhang Y, Pujol N, Matsuyama T, Bargmann CI, Ewbank JJ (2007) Detection and avoidance of a natural product from the pathogenic bacterium Serratia marcescens by *Caenorhabditis elegans*. Proc Natl Acad Sci U S A 104:2295–2300

Pujol N, Link EM, Liu LX, Kurz CL, Alloing G, Tan MW, Ray KP, Solari R, Johnson CD, Ewbank JJ (2001) A reverse genetic analysis of components of the toll signaling pathway in *Caenorhabditis elegans*. Curr Biol 11:809–821

Pukkila-Worley R, Feinbaum R, Kirienko NV, Larkins-Ford J, Conery AL, Ausubel FM (2012) Stimulation of host immune defenses by a small molecule protects *C. elegans* from bacterial infection. PLoS Genet 8:e1002733

Qi YB, Garren EJ, Shu X, Tsien RY, Jin Y (2012) Photo-inducible cell ablation in *Caenorhabditis elegans* using the genetically encoded singlet oxygen generating protein miniSOG. Proc Natl Acad Sci U S A 109:7499–7504

Rajamuthiah R, Fuchs BB, Jayamani E, Kim Y, Larkins-Ford J, Conery A, Ausubel FM, Mylonakis E (2014) Whole animal automated platform for drug discovery against multi-drug resistant *Staphylococcus aureus*. PLoS One 9:e89189

Rajkumari J, Borkotoky S, Muralib A, Suchiang K, Mohanty SK, Busi S (2018a) Attenuation of quorum sensing controlled virulence factors and biofilm formation in *Pseudomonas aeruginosa* by pentacyclic triterpenes, betulin and betulinic acid. Microb Pathog 118:48–60

Rajkumari J, Borkotoky S, Murali A, Suchiang K, Mohanty SK, Busi S (2018b) Cinnamic acid attenuates quorum sensing associated virulence factors and biofilm formation in *Pseudomonas aeruginosa* PAO1. Biotechnol Lett 40:1087–1100

Rand JB, Johnson CD (1995) Chapter 8 genetic pharmacology: interactions between drugs and gene products in *Caenorhabditis elegans*. Methods Cell Biol 48:187–204

Rand BR, Nonet ML (1997) Neurotransmitter assignment for specific neurons. In: Riddle DL, Blumenthal T, Meyer BJ, Priess JR (eds) *C elegans* II. Cold Spring Harbor Laboratory Press, Plainview, pp 1049–1052

Ren M, Zhao L, Lv X, Wang D (2017) Antimicrobial proteins in the response to graphene oxide in *Caenorhabditis elegans*. Nanotoxicology 11:578–590

Riddle DL, Swanson MM, Albert PS (1981) Interacting genes in nematode dauer larva formation. Nature 290:668–671

Roeder T, Stanisak M, Gelhaus C, Bruchhaus I, Grotzinger J, Leippe M (2010) Caenopores are antimicrobial peptides in the nematode *Caenorhabditis elegans* instrumental in nutrition and immunity. Dev Comp Immunol 34:203–209

Roth BL, Poot M, Yue ST, Millard PJ (1997) Bacterial viability and antibiotic susceptibility testing with SYTOX green nucleic acid stain. Appl Environ Microbiol 63:2421–2431

Rual JF, Ceron J, Koreth J, Hao T, Nicot AS, Hirozane-Kishikawa T, Vandenhaute J, Orkin SH, Hill DE, van den Heuvel S, Vidal M (2004) Toward improving *Caenorhabditis elegans* phenome mapping with an ORFeome-based RNAi library. Genome Res 14:2162–2168

Rudrappa T, Bais HP (2008) Curcumin, a known phenolic from Curcuma longa, attenuates the virulence of *Pseudomonas aeruginosa* PAO1 in whole plant and animal pathogenicity models. J Agric Food Chem 56:1955–1962

Ruzanov P, Riddle DL (2010) Deep SAGE analysis of the *Caenorhabditis elegans* transcriptome. Nucleic Acids Res 38:3252–3262

Sahu SN, Lewis J, Patel I, Bozdag S, Lee JH, LeClerc JE, Cinar HN, Schulenburg H (2012) Genomic analysis of immune response against *Vibrio cholerae* Hemolysin in *Caenorhabditis elegans*. PLoS One 7(5):e38200. https://doi.org/10.1371/journal.pone.0038200

Sarabhai S, Sharma P, Capalash N (2013) Ellagic acid derivatives from *Terminalia chebula* Retz. Downregulate the expression of quorum sensing genes to attenuate *Pseudomonas aeruginosa* PAO1 virulence. PLoS One 8(1):e53441. https://doi.org/10.1371/journal.pone.0053441

Schröder M, Kaufman RJ (2005) The mammalian unfolded protein response. Annu Rev Biochem 74:739–789

Schulenburg H, Boehnisch C (2008) Diversification and adaptive sequence evolution of Caenorhabditis lysozymes (Nematoda: Rhabditidae). BMC Evol Biol 8(1):114

Schulenburg H, Hoeppner MP, Weiner J, Bornberg-Bauer E (2008) Specificity of the innate immune system and diversity of C-type lectin domain (CTLD) proteins in the nematode *Caenorhabditis elegans*. Immunobiology 213:237–250

Schwentker FF, Gelman S, Long PH (1937) The treatment of meningococcic meningitis. JAMA 108:1407–1408

Shao X, Zhang X, Zhang Y, Zhu M, Yang P, Yuan J, Xie Y, Zhou T, Wang W, Chen S, Liang H, Deng X (2018) RpoN-dependent direct regulation of quorum sensing and the type VI secretion system in *Pseudomonas aeruginosa* PAO1. J Bacteriol 200:e00205–e0018J

Shi W, Wen H, Lu Y, Shi Y, Lin B, Qin J (2010) Droplet microfluidics for characterizing the neurotoxin-induced responses in individual *Caenorhabditis elegans*. Lab Chip 10:2855–2863

Shivers RP, Pagano DJ, Kooistra T, Richardson CE, Reddy KC, Whitney JK, Kamanzi O, Matsumoto K, Hisamoto N, Kim DH (2010) Phosphorylation of the conserved transcription factor ATF-7 by PMK-1 p38 MAPK regulates innate immunity in *Caenorhabditis elegans*. PLoS Genet 6(4):e1000892

Shu X, Lev-Ram V, Deerinck TJ, Qi Y, Ramko EB, Davidson MW, Jin Y, Ellisman MH, Tsien RY (2011) A genetically encoded tag for correlated light and electron microscopy of intact cells, tissues, and organisms. PLoS Biol 9(4):e1001041

Sifri CD, Begun J, Ausubel FM, Calderwood SB (2003) *Caenorhabditis elegans* as a model host for *Staphylococcus aureus* pathogenesis. Infect Immun 71:2208–2217

Sifri CD, Begun J, Ausubel FM (2005) The worm has turned microbial virulence modeled in *Caenorhabditis elegans*. Trends Microbiol 13:119–127

Singh J, Aballay A (2017) Endoplasmic reticulum stress caused by lipoprotein accumulation suppresses immunity against bacterial pathogens and contributes to immunosenescence. MBio 8:e00778–e00717

Sivamaruthi BS, Ganguli A, Kumar M, Bhaviya S, Pandian SK, Balamurugan K (2011) *Caenorhabditis elegans* as a model for studying Cronobacter sakazakii ATCC BAA-894 pathogenesis. J Basic Microbiol 51:540–549

Smith MP, Laws TR, Atkins TP, Oyston PCF, dePomerai DI, Titball W (2002) A liquid based method for the assessment of bacterial pathogenicity using the nematode *Caenorhabditis elegans*. FEMS Microbiol Lett 210:181–185

Smout MJ, Kotze AC, McCarthy JS, Loukas A (2010) A novel high throughput assay for anthelmintic drug screening and resistance diagnosis by real-time monitoring of parasite motility. PLoS Negl Trop Dis 4:e885

Sochova I, Hofman J, Holoubek I (2006) Using nematodes in soil ecotoxicology. Environ Int 32:374–383

Song Y, Xie C, Ong YM, Gan YH, Chua KL (2005) The BpsIR quorum-sensing system of *Burkholderia pseudomallei*. J Bacteriol 187:785–790

Sonnhammer EL, Durbin R (1997) Analysis of protein domain families in *Caenorhabditis elegans*. Genomics 46:200–216

Spanier B, Starke M, Higel F, Scherer S, Fuchs TM (2010) *Yersinia enterocolitica* infection and *tcaA*-dependent killing of *Caenorhabditis elegans*. Appl Environ Microbiol 76:6277–6285

Stroustrup N, Ulmschneider BE, Nash ZM, Lopez-Moyado IF, Apfeld J, Fontana W (2013) The Caenorhabditis elegans lifespan machine. Nat Methods 10:665–670

Sullivan-Brown JL, Tandon P, Bird KE, Dickinson DJ, Tintori SC, Heppert JK, Meserve JH, Trogden KP, Orlowski SK, Conlon FL, Goldstein B (2016) Identifying regulators of morphogenesis common to vertebrate neural tube closure and *Caenorhabditis elegans* gastrulation. Genetics 202:123–139

Sulston J, Hodgkin J (1988) Methods. In: Wood WB (ed) The nematode *C. elegans*. Cold Spring Harbor Laboratory Press, Cold Spring Harbor, pp 587–606

Sulston JE, Horvitz HR (1977) Post-embryonic cell lineages of the nematode *Caenorhabditis elegans*. Dev Biol 56:110–156

Sulston JE, Schierenberg E, White JG, Thomson JN (1983) The embryonic cell lineage of the nematode *Caenorhabditis elegans*. Dev Biol 100:64–119

Swem LR, Swem DL, O'Loughlin CT, Gatmaitan R, Zhao B, Ulrich SM, Bassler BL (2009) A quorum sensing antagonist targets both membrane-bound and cytoplasmic receptors and controls bacterial pathogenicity. Mol Cell 35:143–153

Tan MW, Mahajan-Miklos S, Ausubel FM (1999) Killing of *Caenorhabditis elegans* by *Pseudomonas aeruginosa* used to model mammalian bacterial pathogenesis. Proc Natl Acad Sci U S A 96:715–720

Tarr DE (2012) Distribution and characteristics of ABFs, cecropins, nemapores, and lysozymes in nematodes. Dev Comp Immunol 36:502–520

Thomsen LE, Slutz SS, Tan MW, Ingmer H (2006) *Caenorhabditis elegans* is a model host for *Listeria monocytogenes*. Appl Environ Microbiol 72:1700–1701

Tralau T, Riebeling C, Pirow R, Oelgeschlager M, Seiler A, Liebsch M, Luch A (2012) Wind of change challenges toxicological regulators. Environ Health Perspect 120:1489–1494

Troemel ER, Chu SW, Reinke V, Lee SS, Ausubel FM, Kim DH (2006) p38 MAPK regulates expression of immune response genes and contributes to longevity in *C. elegans*. PLoS Genet 2(11):e183. https://doi.org/10.1371/journal.pgen.0020183

Vaitkevicius K, Lindmark B, Ou G, Song T, Toma C, Iwanaga M, Zhu J, Andersson A, Hammarström ML, Tuck S, Wai SN (2006) A *Vibrio cholerae* protease needed for killing of *Caenorhabditis elegans* has a role in protection from natural predator grazing. Proc Natl Acad Sci U S A 103:9280–9285

Vaux DL, Weissman IL, Kim SK (1992) Prevention of programmed cell death in *Caenorhabditis elegans* by human bcl-2. Science 258:1955–1957

Velculescu VE, Zhang L, Vogelstein B, Kinzler KW (1995) Serial analysis of gene expression. Science 270:484–487

Vidal-Gadea A, Topper S, Young L, Crisp A, Kressin L, Elbel E, Maples T, Brauner M, Erbguth K, Axelrod A, Gottschalk A, Siegel D, Pierce-Shimomura JT (2011) *Caenorhabditis elegans* selects distinct crawling and swimming gaits via dopamine and serotonin. Proc Natl Acad Sci U S A 108:17504–17509

Viducic D, Murakami K, Amoh T, Ono T, Miyake Y (2017) RpoN promotes *Pseudomonas aeruginosa* survival in the presence of tobramycin. Front Microbiol 8:839

Visvikis O, Ihuegbu N, Labed SA, Labed SA, Luhachack LG, Alves AF, Wollenberg AC, Stuart LM, Stormo GD, Irazoqui JE (2014) Innate host defense requires TFEB-mediated transcription of cytoprotective and antimicrobial genes. Immunity 40:896–909

Vowels JJ, Thomas JH (1992) Genetic analysis of chemosensory control of dauer formation in *Caenorhabditis elegans*. Genetics 130:105–123

Wang H, Chu W, Ye C, Gaeta B, Tao H, Wang M, Qiu Z (2019) Chlorogenic acid attenuates virulence factors and pathogenicity of *Pseudomonas aeruginosa* by regulating quorum sensing. Appl Microbiol Biotechnol 103:903–915

White JG, Southgate E, Thomson JN, Brenner S (1986) The structure of the nervous-system of the nematode *Caenorhabditis-Elegans*. Philos Trans R Soc Lond B Biol Sci 314:1–340

Williams PL, Dusenbery DB (1988) Using the nematode Caenorhabditis elegans to predict mammalian acute lethality to metallic salts. Toxicol Ind Health 4:469–478

Williams P, Winzer K, Chan WC, Camara M (2007) Look who's talking communication and quorum sensing in the bacterial world. Philos Trans R Soc Lond B Biol Sci 362:1119–1134

Wilson R, Ainscough R, Anderson K, Baynes C, Berks M, Bonfield J, Burton J, Connell M, Copsey T, Cooper J (1994) 2.2 Mb of contiguous nucleotide sequence from chromosome III of *C. elegans*. Nature 3:32–38

Wong D, Bazopoulou D, Pujol N, Tavernarakis N, Ewbank JJ (2007) Genome-wide investigation reveals pathogen-specific and shared signatures in the response of *Caenorhabditis elegans* to infection. Genome Biol 8:R194, 2007

Yuan JY, Shaham S, Ledoux S, Ellis HM, Horvitz HR (1993) The *C- elegans* cell death gene *ced-3* encodes a protein similar to mammalian interleukin-1-beta- converting enzyme. Cell 75:641–652

Zak O, O'Reilly T (1993) Animal infection models and ethics—the perfect infection model. J Antimicrob Chemother 31:193–205

Zambelloni R, Connolly JPR, Uribe AH, Burgess K, Marquez R, Roe AJ (2017) Novel compounds targeting the enterohemorrhagic *Escherichia coli* type three secretion system reveal insights into mechanisms of secretion inhibition. Mol Microbiol 105:606–619

Zhang H, Yoshida S, Aizawa T, Murakami R, Suzuki M, Koganezawa N (2000) *In vitro* antimicrobial properties of recombinant ASABF, an antimicrobial peptide isolated from the 51 nematode *Ascaris suum*. Antimicrob Agents Chemother 44:2701–2705

Zhou YM, Shao L, Li JA, Han LZ, Cai WJ, Zhu CB, Chen DJ (2011) An efficient and novel screening model for assessing the bioactivity of extracts against multidrug-resistant *Pseudomonas aeruginosa* using *Caenorhabditis elegans*. Biosci Biotechnol Biochem 75:1746–1751

Zhou L, Zheng H, Tang Y, Yu W, Gonget Q (2013) Eugenol inhibits quorum sensing at sub-inhibitory concentrations. Biotechnol Lett 35:631

Zhu J, Cai X, Harris TL, Gooyit M, Wood M, Lardy M, Janda KD (2015) Disarming Pseudomonas aeruginosa virulence factor LasB by leveraging a *Caenorhabditis elegans* infection model. Chem Biol 22:483–491

Zipperlen P, Fraser AG, Kamath RS, Martinez-Campos M, Ahringer J (2001) Roles for 147 embryonic lethal genes on *C.elegans* chromosome I identified by RNA interference and ideo microscopy. EMBO J 20:3984–3992

Zou CG, Ma YC, Dai LL, Zhang KQ (2014) Autophagy protects C. elegans against necrosis during Pseudomonas aeruginosa infection. Proc Natl Acad Sci U S A 111(34):12480–12485

Zugasti O, Ewbank JJ (2009) Neuroimmune regulation of antimicrobial peptide expression by a noncanonical TGF-beta signaling pathway in *Caenorhabditis elegans* epidermis. Nat Immunol 10:249–256

Zebrafish Model System in Antimicrobial Drug Discovery

30

Rajesh Pamanji

Contents

Abstract

Zebrafish (*Danio rerio*) has already been used in disease modelling and drug discovery. The development of zebrafish as an infectious disease model for studying human diseases has paved the path in understanding pathogenesis and host–pathogen interactions. Many pathogenic zebrafish models were generated using different species of bacteria, parasites, fungus, and virus which are pathogenic to humans and studied their pathogenesis, but very few people tested the antimicrobial drugs on the diseased models. In the present chapter we have been trying to discuss common diseases of zebrafish along with infectious diseases modelled and the importance of zebrafish in antimicrobial drug discovery, how relevant its immune system with that of humans is in defense against pathogens and drug metabolism. The limitations like how antibiotics are harmful to early

R. Pamanji (✉)
Molecular Reproduction, Development and Genetics, Indian Institute of Science, Bangalore, India
e-mail: prajesh@iisc.ac.in

B. Siddhardha et al. (eds.), *Model Organisms for Microbial Pathogenesis, Biofilm Formation and Antimicrobial Drug Discovery*,
https://doi.org/10.1007/978-981-15-1695-5_30

stages of development and relevance of using the larval and adult zebrafish were also discussed.

Keywords
Zebrafish · Antimicrobial · Drug discovery · Infectious diseases

30.1 Introduction

Humankind is confronted with many new infectious diseases on a global scale with the environmental changes, and despite of increased number of diseases and antimicrobial resistance, the discovery of new antibiotics is declining. Advancement in modern genomics and chemical synthesis has led to the discovery and search of a novel class of antibiotic drugs. However, there are many issues in identifying and approving the novel antibiotics, majority includes compounds identified by in vitro and in vivo failure at later stages of drug screening against the bacteria. Therefore, a major setback in antimicrobial drug discovery is the lack of ideal model for drug screening (Takaki et al. 2013). The present chapter describes the generation of zebrafish infectious disease models in high-throughput and natural aquatic environment with different types of pathogens and, drug screening in vitro and in vivo platforms to identify potent antimicrobial drugs (Fig. 30.1).

30.2 Relevance of Zebrafish

Aquatic environment is the home of many infectious pathogens. Though they are native to aquatic environment, they are not infectious until environmental conditions favor their growth, i.e., they are like opportunistic. So, analyses of aquatic microbes in the zebrafish will shed light on how fish acts as an environmental reservoir for human infections (Rowe et al. 2014).

Zebrafish is a vertebrate teleost fish mostly researched model organism for the past four decades because of virtues like small size, easy breeding, laying an awful number of embryos, development within 72 hpf (hours post fertilized) and mainly translucent during development (Bradford et al. 2017). A more comprehensive elucidation of the resemblances and variances between human and zebrafish has to be validated for better use of zebrafish in drug development. Advancement in technologies like high-throughput screening has made drug testing on zebrafish simple, but still it is a question of how relevant is the output from zebrafish screens to human (MacRae and Peterson 2015), though it consists of 25 chromosomes with a genome size of around 1.5 (10 billion base pairs which is half of the size of the human genome but it shares 71% protein coding genes and 82% disease-associated genes (Howe et al. 2013). It has been used as a model organism in many fields like

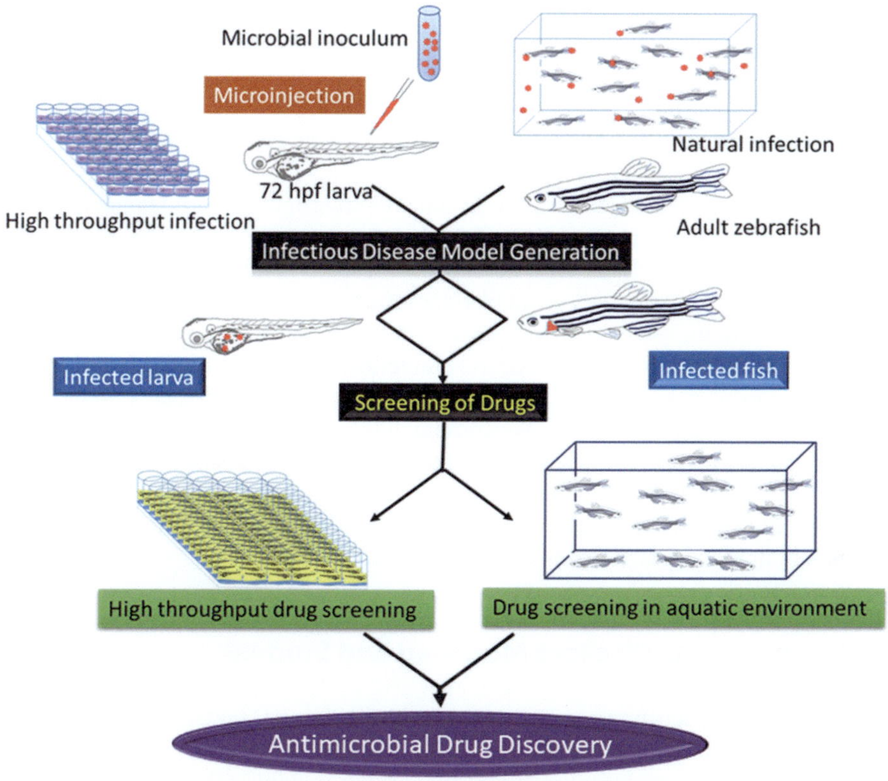

Fig. 30.1 Graphical representation of zebrafish infectious disease model generation and antimicrobial drug discovery

toxicology, developmental biology, cancer genetics, disease modelling, and drug discovery (Meyers 2018).

The main criterion to study any pathological condition of an organism is its immune system. The zebrafish has both innate and adaptive immune responses, but during the early embryonic development, it relies upon innate immunity for its defense against microbes. In zebrafish innate immune response elements like neutrophils, macrophages, NK cells, dendritic cells, eosinophils, basophils, and neutrophils have been identified and characterized, most of which mirrors mammalian system in morphology and function. Moreover, the zebrafish has highly conserved innate immune signal transduction pathways (Goody et al. 2014). The drug metabolism is also an important task to be considered in testing the antibiotic drugs in an organism. Compared to zebrafish embryo or larval biotransformation ability, adult zebrafish has some similarities and differences in relations to metabolites generated in humans (Saad et al. 2017; de Souza et al. 2018). Generation of the zebrafish infectious disease model will provide insights into the disease progression and host–pathogen interactions. Though this model has been used for the last 15 years for the analysis of different pathogens (Neely 2017) still there is a lack of literature on many disease modelling along with the antimicrobial drug discovery.

30.3 Common Diseases of Zebrafish in Research Facilities

The factors which promote infection in aquatic environments are stagnant water, high densities, and excess decaying of feed or fecal matter (Matthews 2004). The common diseases associated with zebrafish in research facilities include bacterial (mycobacteriosis, gill disease, swim bladder infections, etc.), Protozoan (velvet disease, neon tetra disease, white Spot disease, etc.), and metazoan parasitic diseases (capillariasis, renal myxosporidiosis). Noninfectious and idiopathic diseases include nephrocalcinosis, pericardial effusion and dilated cardiomyopathy, hepatic megalocytosis and many other diseases (Westerfield 2007).

Few species of bacterial pathogens like *Aeromonas hydrophila*, *Edwardsiella tarda*, *Mycobacterium marinum*, *Streptococcus agalactiae*, *Streptococcus iniae*, *Vibrio alginolyticus*, *Vibrio cholera*, and *Vibrio vulnificus* are capable of causing diseases both in fish and human (Rowe et al. 2014).

30.4 Infectious Diseases Modelled and Studied

The infectious bacterium or any microbial disease can be modelled at the duct of cuvier, caudal vein, notochord, or muscle of zebrafish by using the simple microinjection technique in embryo/larval and by syringe in adult fishes as described by Neely (2017). Microbial species like bacterial, fungal, viral, and protozoan modelled in zebrafish till date are tabulated and presented in Tables 30.1, 30.2, 30.3, and 30.4, respectively. From the existing literature it was clear that zebrafish infection models were used for study of various factors like study of in vivo assessment of drug efficacy, vertebrate innate immune system in host–pathogen interactions, assessing the immune cell response and cell signalling, phage therapy against infection, testing the toxicity of compounds, transcript profiling, neuroimmune response, predicting the metabolism of xenobiotics, imaging of the innate immune response, behavioral response, high-throughput translational medicine screen and antibiotics toxicity but not antimicrobial drug discovery.

30.5 In Vivo Antimicrobial Drugs Testing

A simple antimicrobial drug screening platform requires maintenance of infected larvae/adults and single application of drugs in well plates or in aquarium tanks by static or semi-static methods and quantifying the bacterial burden and disease recovery (Fig. 30.1). As the lead compounds from pharma industries cost more, it is necessary to reduce the amount of the drug to be used for testing. High-throughput screening of drugs can be performed by placing zebrafish embryos/larvae in 96 or

Table 30.1 Bacterial species modelled in Zebrafish

S. No.	Organism	References
1	*Aeromonas hydrophila*	Saraceni et al. (2016); Cantas et al. (2012)
2	*Edwardsiella tarda*	Lee et al. (2015); van Soest et al. (2011); Pressley et al. (2005)
3	*Klebsiella pneumoniae*	Marcoleta et al. (2018); Cheepurupalli et al. (2017)
4	*Listeria monocytogenes*	Ding et al. (2019); Luukinen et al. (2018a, b); Shan et al. (2015); Corbo et al. (2014); Levraud et al. (2009)
5	*Mycobacterium abscessus*	Bernut et al. (2016); Bernut et al. (2015); Bernut et al. (2014)
6	*Mycobacterium leprae*	Madigan et al. (2017a); Madigan et al. (2017b)
7	*Mycobacterium marinum*	Luukinen et al. (2018a, b); Chen et al. (2018); Takaki et al. (2013); Swaim et al. (2006); Prouty et al. (2003)
8	*Pseudomonas aeruginosa*	Cafora et al. (2019); Rocker et al. (2015); Phennicie et al. (2010); Clatworthy et al. (2009)
9	*Salmonella typhimurium*	Varas et al. (2017a, b); Tyrkalska et al. (2016)
10	*Shigella flexneri*	Boucontet et al. (2018); Willis et al. (2018); Duggan and Mostowy (2018); Mostowy et al. (2013)
11	*Staphylococcus aureus*	Zhang et al. (2019); Pollitt et al. (2018); Connolly et al. (2017); Li and Hu (2012); Prajsnar et al. (2008, 2012)
12	*Streptococcus agalactiae*	Saralahti and Ramet (2015); Kim et al. (2015); Patterson et al. (2012); Neely et al. (2002)
13	*Streptococcus pneumoniae*	Jim et al. (2016); Rounioja et al. (2012)
14	*Streptococcus iniae*	Vincent et al. (2017); Harvie and Huttenlocher (2015); Harvie et al. 2013)
15	*Vibrio alginolyticus*	Liu et al. (2019); Yang et al. (2018)
16	*Vibrio cholerae*	Logan et al. (2018); Nag et al. (2018); Mitchell et al. (2017); Runft et al. (2014)
17	*Vibrio vulnificus*	Pan et al. (2011)

Table 30.2 Fungal species modelled in zebrafish

S. No.	Organism	Reference
1	*Aspergillus fumigatus*	Jones et al. (2019); Wiemann et al. (2017); Knox et al. (2014)
2	*Candida albicans*	Chen et al. (2015), Chen et al. (2013); Chao et al. (2010)
3	*Candida auris*	Nett (2019); Rossato and Colombo (2018); Johnson et al. (2018)
4	*Cryptococcus neoformans*	Fu et al. (2018); Davis et al. (2016); Bojarczuk et al. (2016); Sabiiti et al. (2012)
5	*Talaromyces marneffei*	Ellett et al. (2018); Rosowski et al. (2018);
6	*Pneumocystis jiroveci*	Pinho et al. (2013); Chiu et al. (2008)

Table 30.3 Virus species modelled in zebrafish

S. No.	Organism	Reference
1	Influenza A	Goody et al. (2017); Sullivan et al. (2017); Gabor et al. 2014
2	Herpes simplex type 1	Burnham et al. (2016); Zou et al. (2014); Yakoub et al. (2014); Antoine et al. (2014); Burgos et al. (2008)
3	Hepatitis C	Ding et al. (2015), (2011); Zhao et al. (2013)
4	Chikungunya	Passoni et al. (2017); Palha et al. (2013)

Table 30.4 Protozoan species modelled in zebrafish

Sl. No.	Organism	Reference
1	*Pleistophora hyphessobryconis*	Sanders et al. (2010)
2	*Toxoplasma gondii*	Sanders et al. (2015)

384 well plates. But when it comes to study uptake of the drugs, whole larva lysis is not suitable because some of the drugs may adhere to skin even after washing leading to false–positive results of drug uptake. To overcome this issue, Ordas et al. (2015) has developed a microneedle sampling from the yolk followed by mass spectrometry analysis. In adult zebrafish injecting a particular dose of antibiotics directly into the body by syringe injection or through oral administration (Sridevi et al. 2014) is highly suggestible to overcome the diseased condition like in humans.

30.6 Repurposing Antibiotic Drug Screening on Zebrafish

Though much research was not done on antimicrobial drug discovery using zebrafish, researchers are trying to screen the existing antibiotics on zebrafish infectious models. One such highly studied model is tuberculosis. Zebrafish is highly prone to tuberculosis caused by *Mycobacterium marinum* in the aquatic environment. Because of many limitations in drug discovery and multidrug resistance, people are interested in zebrafish to study pathophysiology of tuberculosis and drug screening. Antibiotics like rifampicin, moxifloxacin, and 15 preclinical GSK compounds (Ordas et al. 2015), compound like PBTZ169 benzothiazinone derivative (Makarov et al. 2014) and a variety of nitroimidazole-based next-generation anti-mycobacterial drugs were tested against *M. marinum* infectious zebrafish model (Dalton et al. 2017). Zebrafish HCV (hepatitis C virus) disease model was created by injecting HCV sub-replicons and tested ribavirin and oxymatrine drugs for their efficiency in reducing the replication of virus (Ding et al. 2011).

30.7 Disadvantages

The major difference of zebrafish with that of humans is it's ectothermic nature. Moreover, many organs are not present in zebrafish, like breast, lungs, prostate, and hair. Zebrafish are highly sensitive to the surrounding environment which was greatly influenced by the factors like temperature, light, nutrition, water quality, and population density; these must be highly controlled in order to accurately interpret data obtained (Ali et al. 2011).

One of the major setbacks of testing antibiotics during development of zebrafish is their teratogenic effects and oxidative stress inducing nature apart from acting against the microorganisms. Zebrafish embryos exposed to antibiotics like tetracycline (Zhang et al. 2015), oxytetracycline and amoxicillin (Oliveira et al. 2013), and Cefazolin Sodium (Chen et al. 2017) displayed developmental toxic effects like hatching delay, shorter body length, increased yolk sac area, uninflated swim bladder, oxidative stress-induced apoptosis, and cardio and neural toxicities. Zhao et al. (2011) observed that aminoglycoside antibiotics like gentamycin, neomycin, and streptomycin are inducing severe ototoxicity and nephrotoxicity during embryonic development in zebrafish.

30.8 Future Directions

Zebrafish has already addressed many infectious disease modelling, but still there is a lack of studies on many pathogens which are infectious to humans and other organisms. Researchers have overlooked the field of antimicrobial drug discovery using animal models, particularly of zebrafish, though it gained a lot of importance in disease modelling. As zebrafish embryo/larval condition is too sensitive to most of the antibiotics, it is better to test on adult zebrafish infectious models.

Acknowledgments The author is thankful to Director, Indian Institute of Science, Bangalore, India for providing the facilities for ongoing research work and to Science and Engineering Research Board (SERB), Govt. of India, for providing financial assistance in the form of National postdoctoral fellowship PDF/2017/002688.

References

Ali S, Champagne DL, Spaink HP, Richardson MK (2011) Zebrafish embryos and larvae: a new generation of disease models and drug screens. Birth Defects Res C 93:115–133

Antoine TE, Jones KS, Dale RM, Shukla D, Tiwari V (2014) Zebrafish: modeling for herpes simplex virus infections. Zebrafish 11(1):17–25

Bernut A, Moigne VL, Lesne T, Lutfalla G, Herrmann JL, Kremer L (2014) In vivo assessment of drug efficacy against *Mycobacterium abscessus* using the embryonic zebrafish test system. Antimicrob Agents Chemother 58(7):4054–4063

Bernut A, Dupont C, Sahuquet A, Herrmann JL, Lutfalla G, Kremer L (2015) Deciphering and imaging pathogenesis and cording of *Mycobacterium abscessus* in zebrafish embryos. J Vis Exp 103:e53130

Bernut A, Nguyen-Chi M, Halloum I, Herrmann JL, Lutfalla G, Kremer L (2016) *Mycobacterium abscessus*-induced granuloma formation is strictly dependent on TNF signaling and neutrophil trafficking. PLoS Pathog 12(11):e1005986

Bojarczuk A, Miller KA, Hotham R, Lewis A, Ogryzko NV, Kamuyango AA, Frost H, Gibson RH, Stillman E, May RC, Renshaw SA, Johnston SA (2016) *Cryptococcus neoformans* intracellular proliferation and capsule size determines early macrophage control of infection. Sci Rep 6:21489

Boucontet L, Passoni G, Thiry V, Maggi L, Herbomel P, Levraud JP, Colucci-Guyon E (2018) A model of superinfection of virus-infected zebrafish larvae: increased susceptibility to bacteria associated with neutrophil death. Front Immunol 9:1084

Bradford YM, Toro S, Ramachandran S, Ruzicka L, Howe DG, Eagle A, Kalita P, Martin R, Taylor Moxon SA, Schaper K, Westerfield M (2017) Zebrafish models of human disease: gaining insight into human disease at ZFIN. ILAR J 58(1):4–16

Burgos JS, Ripoll-Gomez J, Alfaro JM, Sastre I, Valdivieso F (2008) Zebrafish as a new model for herpes simplex virus type 1 infection. Zebrafish 5(4):323–333

Burnham LA, Jaishankar D, Thompson JM, Jones KS, Shukla D, Tiwari V (2016) Liposome-mediated herpes simplex virus uptake is glycoprotein-d receptor-independent but requires Heparan sulfate. Front Microbiol 7:973

Cafora M, Deflorian G, Forti F, Ferrari L, Binelli G, Briani F, Ghisotti D, Pistocchi A (2019) Phage therapy against *Pseudomonas aeruginosa* infections in a cystic fibrosis zebrafish model. Sci Rep 9(1):1527

Cantas L, Midtlyng PJ, Sorum H (2012) Impact of antibiotic treatments on the expression of the R plasmid tra genes and on the host innate immune activity during pRAS1 bearing *Aeromonas hydrophila* infection in zebrafish (*Danio rerio*). BMC Microbiol 112:37

Chao CC, Hsu PC, Jen CF, Chen IH, Wang CH, Chan HC, Tsai PW, Tung KC, Wang CH, Lan CY, Chuang YJ (2010) Zebrafish as a model host for *Candida albicans* infection. Infect Immun 78(6):2512–2521

Cheepurupalli L, Raman T, Rathore SS, Ramakrishnan J (2017) Bioactive molecule from *Streptomyces* sp. Mitigates MDR *Klebsiella pneumoniae* in zebrafish infection model. Front Microbiol 8:614

Chen YY, Chao CC, Liu FC, Hsu PC, Chen HF, Peng SC, Chuang YJ, Lan CY, Hsieh WP, Wong DS (2013) Dynamic transcript profiling of *Candida albicans* infection in zebrafish: a pathogen-host interaction study. PLoS One 8(9):e72483

Chen YZ, Yang YL, Chu WL, You MS, Lo HJ (2015) Zebrafish egg infection model for studying *candida albicans* adhesion factors. PLoS One 10(11):e0143048

Chen B, Gao ZQ, Liu Y, Zheng YM, Han Y, Zhang JP, Hu CQ (2017) Embryo and developmental toxicity of cefazolin sodium impurities in zebrafish. Front Pharmacol 8:403

Chen Z, Shao XY, Wang C, Hua MH, Wang CN, Wang X, Wang QJ, Yao JY, Fan YH, Qin YW (2018) *Mycobacterium marinum* infection in zebrafish and microglia imitates the early stage of tuberculous meningitis. J Mol Neurosci 64(2):321–330

Chiu LL, Cunningham LL, Raible DW, Rubel EW, Ou HC (2008) Using the zebrafish lateral line to screen for ototoxicity. J Assoc Res Otolaryngol 9(2):178–190

Clatworthy AE, Lee JS, Leibman M, Kostun Z, Davidson AJ, Hung DT (2009) *Pseudomonas aeruginosa* infection of zebrafish involves both host and pathogen determinants. Infect Immun 77(4):1293–1303

Connolly J, Boldock E, Prince LR, Renshaw SA, Whyte MK, Foster SJ (2017) Identification of *Staphylococcus aureus* factors required for pathogenicity and growth in human blood. Infect Immun 85(11):e00337–e00317

Corbo CP, Gaylets CE, Molesan AJ, Rivera WA (2014) *Listeria monocytogenes* infects the zebrafish central nervous system; A useful model to analyze the neuroimmune response. Microsc Microanal 20(Suppl 3):1324–1325

Dalton JP, Uy B, Okuda KS, Hall CJ, Denny WA, Crosier PS, Swift S, Wiles S (2017) Screening of anti-mycobacterial compounds in a naturally infected zebrafish larvae model. J Antimicrob Chemother 72(2):421–427

Davis JM, Huang M, Botts MR, Hull CM, Huttenlocher A (2016) A zebrafish model of cryptococcal infection reveals roles for macrophages, endothelial cells, and neutrophils in the establishment and control of sustained fungemia. Infect Immun 84(10):3047–3062

de Souza AC, Sardela VF, de Sousa VP, Pereira HMG (2018) Zebrafish (*Danio rerio*): a valuable tool for predicting the metabolism of xenobiotics in humans? Comp Biochem Physiol C Toxicol Pharmacol 212:34–46

Ding CB, Zhang JP, Zhao Y, Peng ZG, Song DQ, Jiang JD (2011) Zebrafish as a potential model organism for drug test against hepatitis C virus. PLoS One 6(8):e22921

Ding CB, Zhao Y, Zhang JP, Peng ZG, Song DQ, Jiang JD (2015) A zebrafish model for subgenomic hepatitis C virus replication. Int J Mol Med 35(3):791–797

Ding C, Liu Q, Li J, Ma J, Wang S, Dong Q, Xu D, Qiu J, Wang X (2019) Attenuated *Listeria monocytogenes* protecting zebrafish (*Danio rerio*) against Vibrio species challenge. Microb Pathog 132:38–44

Duggan GM, Mostowy S (2018) Use of zebrafish to study Shigella infection. Dis Model Mech 11(2):dmm032151

Ellett F, Pazhakh V, Pase L, Benard EL, Weerasinghe H, Azabdaftari D, Alasmari S, Andrianopoulos A, Lieschke GJ (2018) Macrophages protect *Talaromyces marneffei* conidia from myeloperoxidase-dependent neutrophil fungicidal activity during infection establishment in vivo. PLoS Pathog 14(6):e1007063

Fu MS, Coelho C, De Leon-Rodriguez CM, Rossi DCP, Camacho E, Jung EH, Kulkarni M, Casadevall A (2018) *Cryptococcus neoformans* urease affects the outcome of intracellular pathogenesis by modulating phagolysosomal pH. PLoS Pathog 14(6):e1007144

Gabor KA, Goody MF, Mowel WK, Breitbach ME, Gratacap RL, Witten PE, Kim CH (2014) Influenza A virus infection in zebrafish recapitulates mammalian infection and sensitivity to anti-influenza drug treatment. Dis Model Mech 7(11):1227–1237

Goody MF, Sullivan C, Kim CH (2014) Studying the immune response to human viral infections using zebrafish. Dev Comp Immunol 46(1):84–95

Goody M, Jurczyszak D, Kim C, Henry C (2017) Influenza A virus infection damages zebrafish skeletal muscle and exacerbates disease in zebrafish modeling Duchenne muscular dystrophy. PLoS Curr 9: ecurrents.md.8a7e35c50fa2b48156799d3c39788175

Harvie EA, Huttenlocher A (2015) Non-invasive Imaging of the Innate Immune Response in a Zebrafish Larval Model of Streptococcus iniae Infection. J Vis Exp 98:e52788

Harvie EA, Green JM, Neely MN, Huttenlocher A (2013) Innate immune response to Streptococcus iniae infection in zebrafish larvae. Infect Immun 81(1):110–121

Howe K, Clark MD, Torroja CF, Torrance J, Berthelot C, Muffato M, Collins JE, Humphray S et al (2013) The zebrafish reference genome sequence and its relationship to the human genome. Nature 496(7446):498–503

Jim KK, Engelen-Lee J, van der Sar AM, Bitter W, Brouwer MC, van der Ende, Veening JW, van de Beek D, Vandenbroucke-Grauls CM (2016) Infection of zebrafish embryos with live fluorescent Streptococcus pneumoniae as a real-time Pneumococcal meningitis model. J Neuroinflammation 13(1):188

Johnson CJ, Davis JM, Huttenlocher A, Kernien JF, Nett JE (2018) Emerging fungal pathogen Candida auris evades neutrophil attack. mBio 9(4):e01403–e01418

Jones CN, Ellett F, Robertson AL, Forrest KM, Judice K, Balkovec JM, Springer M, Markmann JF, Vyas JM, Warren HS, Irimia D (2019) Bifunctional small molecules enhance neutrophil activities against *Aspergillus fumigatus in vivo* and *in vitro*. Front Immunol 10:644

Kim BJ, Hancock BM, Del Cid N, Bermudez A, Traver D, Doran KS (2015) *Streptococcus agalactiae* infection in zebrafish larvae. Microb Pathog 79:57–60

Knox BP, Deng Q, Rood M, Eickhoff JC, Keller NP, Huttenlocher A (2014) Distinct innate immune phagocyte responses to *Aspergillus fumigatus* conidia and hyphae in zebrafish larvae. Eukaryot Cell 13(10):1266–1277

Lee SB, Choe Y, Chon TS, Kang HY (2015) Analysis of zebrafish (Danio rerio) behavior in response to bacterial infection using a self-organizing map. BMC Vet Res 11:269

Levraud JP, Disson O, Kissa K, Bonne I, Cossart P, Herbomel P, Lecuit M (2009) Real-time observation of *Listeria monocytogenes*-phagocyte interactions in living zebrafish larvae. Infect Immun 77(9):3651–3660

Li YJ, Hu B (2012) Establishment of multi-site infection model in zebrafish larvae for studying *Staphylococcus aureus* infectious disease. J Genet Genomics 39(9):521–534

Liu SR, Peng XX, Li H (2019) Metabolic mechanism of ceftazidime resistance in *Vibrio alginolyticus*. Infect Drug Resist 12:417–429

Logan SL, Thomas J, Yan J, Baker RP, Shields DS, Xavier JB, Hammer BK, Parthasarathy R (2018) The *Vibrio cholerae* type VI secretion system can modulate host intestinal mechanics to displace gut bacterial symbionts. Proc Natl Acad Sci USA 115(16):E3779–E3787

Luukinen H, Hammarén MM, Vanha-Aho LM, Parikka M (2018a) Modeling tuberculosis in *Mycobacterium marinum* infected adult zebrafish. J Vis Exp 140:e58299

Luukinen H, Hammarén MM, Vanha-Aho LM, Svorjova A, Kantanen L, Järvinen S, Luukinen BV, Dufour E, Rämet M, Hytönen VP, Parikka M (2018b) Priming of innate antimycobacterial immunity by heat-killed *Listeria monocytogenes* induces sterilizing response in the adult zebrafish tuberculosis model. Dis Model Mech 11(1):dmm031658

MacRae CA, Peterson RT (2015) Zebrafish as tools for drug discovery. Nat Rev Drug Discov 14(10):721–731

Madigan CA, Cambier CJ, Kelly-Scumpia KM, Scumpia PO, Cheng TY, Zailaa J, Bloom BR, Moody DB, Smale ST, Sagasti A, Modlin RL, Ramakrishnan L (2017a) A macrophage response to *Mycobacterium leprae* phenolic glycolipid initiates nerve damage in leprosy. Cell 170(5):973–985

Madigan CA, Cameron J, Ramakrishnan L (2017b) A Zebrafish Model of *Mycobacterium leprae* granulomatous infection. J Infect Dis 216(6):776–779

Makarov V, Lechartier B, Zhang M, Neres J, van der Sar AM, Raadsen SA, Hartkoorn RC, Ryabova OB, Vocat A, Decosterd LA, Widmer N, Buclin T, Bitter W, Andries K, Pojer F, Dyson PJ, Cole ST (2014) Towards a new combination therapy for tuberculosis with next generation benzothiazinones. EMBO Mol Med 6:372–383

Marcoleta AE, Varas MA, Ortiz-Severín J, Vásquez L, Berríos-Pastén C, Sabag AV, Chávez FP, Allende ML, Santiviago CA, Monasterio O, Lagos R (2018) Evaluating different virulence traits of *Klebsiella pneumoniae* using *Dictyostelium discoideum* and zebrafish larvae as host models. Front Cell Infect Microbiol 8:30

Matthews JL (2004) Common diseases of laboratory zebrafish. Methods Cell Biol 77:617–643

Meyers JR (2018) Zebrafish: development of a vertebrate model organism. Curr Protocols Essential Lab Tech 16(1):e19

Mitchell KC, Breen P, Britton S, Neely MN, Withey JH (2017) Quantifying *Vibrio cholerae* enterotoxicity in a zebrafish infection model. Appl Environ Microbiol 83(16):e52788

Mostowy S, Boucontet L, Mazon Moya MJ, Sirianni A, Boudinot P, Hollinshead M, Cossart P, Herbomel P, Levraud JP, Colucci-Guyon E (2013) The zebrafish as a new model for the in vivo study of *Shigella flexneri* interaction with phagocytes and bacterial autophagy. PLoS Pathog 9(9):e1003588

Nag D, Mitchell K, Breen P, Withey JH (2018) Quantifying *Vibrio cholerae* colonization and diarrhea in the adult zebrafish model. J Vis Exp 137:e57767

Neely MN (2017) The Zebrafish as a Model for Human Bacterial Infections. Methods Mol Biol 1535:245–266

Neely MN, Pfeifer JD, Caparon M (2002) *Streptococcus*-zebrafish model of bacterial pathogenesis. Infect Immun 70(7):3904–3914

Nett JE (2019) *Candida auris*: an emerging pathogen "incognito"? PLoS Pathog 15(4):e1007638

Oliveira R, McDonough S, Ladewig JC, Soares AM, Nogueira AJ, Domingues I (2013) Effects of oxytetracycline and amoxicillin on development and biomarkers activities of zebrafish (*Danio rerio*). Environ Toxicol Pharmacol 36(3):903–912

Ordas A, Raterink RJ, Cunningham F, Jansen HJ, Wiweger MI, Jong-Raadsen S, Bos S, Bates RH, Barros D, Meijer AH, Vreeken RJ, Ballell-Pages L, Dirks RP, Hankemeier T, Spaink HP (2015) Testing tuberculosis drug efficacy in a zebrafish high-throughput translational medicine screen. Antimicrob Agents Chemother 59(2):753–762

Palha N, Guivel-Benhassine F, Briolat V, Lutfalla G, Sourisseau M, Ellett F, Wang CH, Lieschke GJ, Herbomel P, Schwartz O, Levraud JP (2013) Real-time whole-body visualization of Chikungunya virus infection and host interferon response in zebrafish. PLoS Pathog 9(9):e1003619

Pan CY, Wu JL, Hui CF, Lin CH, Chen JY (2011) Insights into the antibacterial and immunomodulatory functions of the antimicrobial peptide, epinecidin-1, against *Vibrio vulnificus* infection in zebrafish. Fish Shellfish Immunol 31(6):1019–1025

Passoni G, Langevin C, Palha N, Mounce BC, Briolat V, Affaticati P, De Job E, Joly JS, Vignuzzi M, Saleh MC, Herbomel P, Boudinot P, Levraud JP (2017) Imaging of viral neuroinvasion in the zebrafish reveals that Sindbis and chikungunya viruses favour different entry routes. Dis Model Mech 10(7):847–857

Patterson H, Saralahti A, Parikka M, Dramsi S, Trieu-Cuot P, Poyart C, Rounioja S, Ramet M (2012) Adult zebrafish model of bacterial meningitis in *Streptococcus agalactiae* infection. Dev Comp Immunol 38(3):447–455

Phennicie RT, Sullivan MJ, Singer JT, Yoder JA, Kim CH (2010) Specific resistance to *Pseudomonas aeruginosa* infection in zebrafish is mediated by the cystic fibrosis transmembrane conductance regulator. Infect Immun 78(11):4542–4550

Pinho BR, Santos MM, Fonseca-Silva A, Valentão P, Andrade PB, Oliveira JM (2013) How mitochondrial dysfunction affects zebrafish development and cardiovascular function: an *in vivo* model for testing mitochondria-targeted drugs. Br J Pharmacol 169(5):1072–1090

Pollitt EJG, Szkuta PT, Burns N, Foster SJ (2018) *Staphylococcus aureus* infection dynamics. PLoS Pathog 14(6):e1007112

Prajsnar TK, Cunliffe VT, Foster SJ, Renshaw SA (2008) A novel vertebrate model of *Staphylococcus aureus* infection reveals phagocyte-dependent resistance of zebrafish to non-host specialized pathogens. Cell Microbiol 10(11):2312–2325

Prajsnar TK, Hamilton R, Garcia-Lara J, McVicker G, Williams A, Boots M, Foster SJ, Renshaw SA (2012) A privileged intraphagocyte niche is responsible for disseminated infection of *Staphylococcus aureus* in a zebrafish model. Cell Microbiol 14(10):1600–1619

Pressley ME, Phelan PE, Witten PE, Mellon MT, Kim CH (2005) Pathogenesis and inflammatory response to *Edwardsiella tarda* infection in the zebrafish. Dev Comp Immunol 29(6):501–513

Prouty MG, Correa NE, Barker LP, Jagadeeswaran P, Klose KE (2003) Zebrafish-*Mycobacterium marinum* model for mycobacterial pathogenesis. FEMS Microbiol Lett 225(2):177–182

Rocker AJ, Weiss AR, Lam JS, Van Raay TJ, Khursigara CM (2015) Visualizing and quantifying *Pseudomonas aeruginosa* infection in the hindbrain ventricle of zebrafish using confocal laser scanning microscopy. J Microbiol Methods 117:85–94

Rosowski EE, Knox Benjamin P, Archambault LS, Huttenlocher A, Keller NP, Wheeler RT, Davis JM (2018) The zebrafish as a model host for invasive fungal infections. J Fungi 4(4):E136

Rossato L, Colombo AL (2018) *Candida auris*: what have we learned about its mechanisms of pathogenicity? Front Microbiol 9:3081

Rounioja S, Saralahti A, Rantala L, Parikka M, Henriques-Normark B, Silvennoinen O, Rämet M (2012) Defense of zebrafish embryos against *Streptococcus pneumoniae* infection is dependent on the phagocytic activity of leukocytes. Dev Comp Immunol 36(2):342–348

Rowe HM, Withey JH, Neely MN (2014) Zebrafish as a model for zoonotic aquatic pathogens. Dev Comp Immunol 46(1):96–107

Runft DL, Mitchell KC, Abuaita BH, Allen JP, Bajer S, Ginsburg K, Neely MN, Withey JH (2014) Zebrafish as a natural host model for *Vibrio cholerae* colonization and transmission. Appl Environ Microbiol 80(5):1710–1717

Saad M, Matheeussen A, Bijttebier S, Verbueken E, Pype C, Casteleyn C, Van Ginneken C, Apers S, Maes L, Cos P, Van Cruchten S (2017) *In vitro* CYP-mediated drug metabolism in the zebrafish (embryo) using human reference compounds. Toxicol In Vitro 42:329–336

Sabiiti W, May RC, Pursall ER (2012) Experimental models of cryptococcosis. Int J Microbiol 2012:626745

Sanders JL, Lawrence C, Nichols DK, Brubaker JF, Peterson TS, Murray KN, Kent ML (2010) *Pleistophora hyphessobryconis* (Microsporidia) infecting zebrafish *Danio rerio* in research facilities. Dis Aquat Organ 91(1):47–56

Sanders JL, Zhou Y, Moulton HM, Moulton ZX, McLeod R, Dubey JP, Weiss LM, Kent ML (2015) The zebrafish, *Danio rerio*, as a model for *Toxoplasma gondii*: an initial description of infection in fish. J Fish Dis 38(7):675–679

Saraceni PR, Romero A, Figueras A, Novoa B (2016) Establishment of infection models in zebrafish larvae (*Danio rerio*) to study the pathogenesis of *Aeromonas hydrophila*. Front Microbiol 7:1219

Saralahti A, Ramet M (2015) Zebrafish and Streptococcal infections. Scand J Immunol 82(3): 174–183

Shan Y, Fang C, Cheng C, Wang Y, Peng J, Fang W (2015) Immersion infection of germ-free zebrafish with *Listeria monocytogenes* induces transient expression of innate immune response genes. Front Microbiol 6:373

Sridevi JP, Anantaraju HS, Kulkarni P, Yogeeswari P, Sriram D (2014) Optimization and validation of *Mycobacterium marinum*-induced adult zebrafish model for evaluation of oral antituberculosis drugs. Int J Mycobacteriol 3(4):259–267

Sullivan C, Jurcyzszak D, Goody MF, Gabor KA, Longfellow JR, Millard PJ, Kim CH (2017) Using zebrafish models of human Influenza A Virus infections to screen antiviral drugs and characterize host immune cell responses. J Vis Exp 119:e55235

Swaim LE, Connolly LE, Volkman HE, Humbert O, Born DE, Ramakrishnan L (2006) *Mycobacterium marinum* infection of adult zebrafish causes caseating granulomatous tuberculosis and is moderated by adaptive immunity. Infect Immun 74(11):6108–6117

Takaki K, Davis JM, Winglee K, Ramakrishnan L (2013) Evaluation of the pathogenesis and treatment of *Mycobacterium marinum* infection in zebrafish. Nat Protoc 8(6):1114–1124

Tyrkalska SD, Candel S, Angosto D, Gómez-Abellan V, Martín-Sanchez F, García-Moreno D, Zapata-Perez R, Sanchez-Ferrer A, Sepulcre MP, Pelegrin P, Mulero V (2016) Neutrophils mediate *Salmonella Typhimurium* clearance through the GBP4 inflammasome-dependent production of prostaglandins. Nat Commun 7:120e77

van Soest JJ, Stockhammer OW, Ordas A, Bloemberg GV, Spaink HP, Meijer AH (2011) Comparison of static immersion and intravenous injection systems for exposure of zebrafish embryos to the natural pathogen *Edwardsiella tarda*. BMC Immunol 12:58

Varas M, Farina A, Díaz-Pascual F, Ortíz-Severín J, Marcoleta AE, Allende ML, Santiviago CA, Chávez FP (2017a) Live-cell imaging of *Salmonella typhimurium* interaction with zebrafish larvae after injection and immersion delivery methods. J Microbiol Methods 135:20–25

Varas M, Ortíz-Severín J, Marcoleta AE, Díaz-Pascual F, Allende ML, Santiviago CA, Chávez FP (2017b) *Salmonella typhimurium* induces cloacitis-like symptomsin zebrafish larvae. Microb Pathog 107:317–320

Vincent WJB, Harvie EA, Sauer JD, Huttenlocher A (2017) Neutrophil derived LTB4 induces macrophage aggregation in response to encapsulated *Streptococcus iniae* infection. PLoS One 12(6):e0179574

Westerfield M (2007) The Zebrafish Book; A guide for the laboratory use of zebrafish (Danio rerio), 5th edn. University of Oregon Press, Eugene

Wiemann P, Perevitsky A, Lim FY, Shadkchan Y, Knox BP, Landero Figueora JA, Choera T, Niu M, Steinberger AJ, Wüthrich M, Idol RA, Klein BS, Dinauer MC, Huttenlocher A, Osherov N, Keller NP (2017) *Aspergillus fumigatus* copper export machinery and reactive oxygen intermediate defense counter host copper-mediated oxidative antimicrobial offense. Cell Rep 19(5):1008–1021

Willis AR, Torraca V, Gomes MC, Shelley J, Mazon-Moya M, Filloux A, Lo Celso C, Mostowy S (2018) *Shigella*-induced emergency granulopoiesis protects zebrafish larvae from secondary infection. MBio 9(3):e00933–e00918

Yakoub AM, Rawal N, Maus E, Baldwin J, Shukla D, Tiwari V (2014) Comprehensive analysis of herpes simplex virus 1 (HSV-1) entry mediated by zebrafish 3-O-sulfotransferase isoforms: implications for the development of a zebrafish model of HSV-1 infection. J Virol 88(21): 12915–12922

Yang MJ, Cheng ZX, Jiang M, Zeng ZH, Peng B, Peng XX, Li H (2018) Boosted TCA cycle enhances survival of zebrafish to *Vibrio alginolyticus* infection. Virulence 9(1):634–644

Zhang Q, Cheng J, Xin Q (2015) Effects of tetracycline on developmental toxicity and molecular responses in zebrafish (*Danio rerio*) embryos. Ecotoxicology 24(4):707–719

Zhang X, de Boer L, Stockhammer OW, Grijpma DW, Spaink HP, Zaat SAJ (2019) A zebrafish embryo model for in vivo visualization and intravital analysis of biomaterial-associated *Staphylococcus aureus* infection. J Vis Exp 143:e58523

Zhao Z, Tong JW, Zhang JP, You XF, Jiang JD, Hu CQ (2011) Zebrafish model for the study on drug ototoxicity of aminoglycoside antibiotics. Acta Pharmaceutica Sinica 46(8):928–935

Zhao Y, Qin W, Zhang JP, Hu ZY, Tong JW, Ding CB, Peng ZG, Zhao LX, Song DQ, Jiang JD (2013) HCV IRES-mediated core expression in zebrafish. PLoS One 8(3):e56985

Zou M, De Koninck P, Neve RL, Friedrich RW (2014) Fast gene transfer into the adult zebrafish brain by herpes simplex virus 1 (HSV-1) and electroporation: methods and optogenetic applications. Front Neural Circuits 8:41

Murine Models for Development of Anti-Infective Therapeutics

31

Angela E. Peter, Pola Sudhakar, B. V. Sandeep, B. Ganga Rao, and V. Lakshmi Kalpana

Contents

A. E. Peter · P. Sudhakar (✉) · B. V. Sandeep · V. L. Kalpana
Department of Biotechnology, College of Science and Technology, Andhra University, Visakhapatnam, India

B. G. Rao
College of Pharmaceutical Sciences, Andhra University, Visakhapatnam, India

B. Siddhardha et al. (eds.), *Model Organisms for Microbial Pathogenesis, Biofilm Formation and Antimicrobial Drug Discovery*,
https://doi.org/10.1007/978-981-15-1695-5_31

Abstract
The key to the medicine chest of the society is in the active discovery of new anti-infective therapeutics. The process of development of novel anti-infective therapeutics involves in vitro studies on the anti-infective agent and its properties along with in silico study of its interaction with binding sites. There is however lacunae, as in vitro and in silico studies alone are insufficient to validate the use of an anti-infective agent in human subjects. In vivo studies are imperative to analyze the mechanism of action and the other possible interactions that can take place when the drug enters a mammalian system. The use of murine models (mice and rats) as model organisms to study an infectious disease and possible drug molecules that can be effective as a prophylactic or as a treatment for cure has aided in the development of anti-infective agents. Murine models have increased understanding of the pathogenesis of the disease and have facilitated a better understanding of the mechanism of protective action exerted by the drug molecules on the host organism. This chapter discusses briefly the mouse models and rat models that are in use. It also elaborates on specific bacterial, fungal, viral, protozoal, and helminth diseases modeled in mouse and rat models for testing the efficacy of a particular drug molecule under study.

Keywords
Mouse models · Rat models · Mammalian system model organisms · Antibacterial agents · Antifungal agents · Antiviral agents · Antiprotozoal agents · Antihelminthic agents

Abbreviations

5-FU	5-fluorouracil
AIDS	Acquired immunodeficiency syndrome
AMB	amphotericin B
AWI	Adult worms in intestine
BDCRB	2-bromo-5,6-dichloro-(1-β-D-ribofuranosyl) benzimidazole
BHI	Brain Heart Infusion
BHIB	Brain Heart Infusion Broth
CFU	Colony forming units
CIP	ciprofloxacin
DENV	Dengue virus
EDTA	Ethylenediaminetetraacetic acid
ED_{50}	50% Effective dose
FFU	focus-forming units
FK/MFG	micafungin
FL	fluconazole
HCMV	Human cytomegalovirus
HIV	Human immunodeficiency virus
HRBC	Human red blood cells

HSV	Herpes simplex virus
ICZ/ITZ	itraconazole
ID_{50}	50% Infective dose
iRBC	Infected red blood cells
i.v.	intravenous
L3	Third larval stage, which is infective
LD_{50}	50%-Lethal dose
LD_{90}	90% Lethal dose
LPG	Larvae per gram of muscle tissue
MIC	Minimum inhibitory concentration
MID	Mouse infectious dose, sub-lethal
MIU	Million International Units
M-H	Mueller-Hinton
MLD	Mouse lethal dose
MRS	De Man, Rogosa and Sharpe
NB	Nutrient broth
NIK	Nikkomycin Z
NO-np	Nitric oxide-releasing nanoparticle system
Np	Control nanoparticles
NS1	nonstructural influenza viral protein
p24	viral capsid core protein
PBS	Phosphate buffered saline
Per os	oral administration
PFU	Plague forming units
p.i.	post infection
PSC	posaconazole
rpm	rotations per minute
RPMI	Roswell park memorial institute
SAM	ampicillin-sulbactum
SDB	Sabouraud dextrose broth
TRB	terbinafine
TSB	Tryptic soy broth
TZP	piperacillin-tazobactum
VAN	vancomycin
VDJ recombination	The random recombination occurs in the primary lymphoid organs, rearranging the variable (V), joining (J) and in some cases the diversity (D) gene segments
VRC	voriconazole

31.1 Introduction

The invasion and growth of a microorganism in an animal or a human with any resulting host response is termed as an infection. Infection arises when organisms like bacteria, viruses, or other microbes enter the body and begin to multiply. Upon invasion, the immune system in the body starts multiple mechanisms in response to the infection in order to protect the body and to eliminate the foreign invader. Symptoms like fever, rashes, headache, and malaise typically observed during an infection are essentially an outcome of the different activities of the immune system to rid the body of the agent that is causing the infection. Bacteria, viruses, fungi, protozoans, and helminths are the different infectious agents that can enter the body, multiply, and cause damage to the cells hence leading to disease. An anti-infective agent may be described as a substance that prevents the spread of infections by killing infectious organisms or agents. In other words, it is a substance of natural origin or that is chemically synthesized that is used on humans or animals to arrest infection by inhibiting the activity or destroying the harmful infectious agents or microorganisms. The Oxford dictionary defines the word therapeutics as the branch of medicine concerned with the treatment of diseases. Anti-infective therapeutics is the branch of science that deals with the discovery and development of compounds that are used to target disease-causing infectious agents. Most of the anti-infective agents discovered are of natural origin. These are either secondary metabolites synthesized by plants or are toxins produced by microorganisms like bacteria, fungi, and viruses or are toxins produced by marine sources like algae, molluscs, snails, and other fishes. Sometimes, anti-infective agents are accidentally discovered. The current trend, however, is to scientifically evaluate or screen the different possible sources of anti-infective agents. The pharmacological activity of terrestrial plants is assessed. Screening is either done randomly or those plants that are known to be poisonous owing to the production of toxic compounds are evaluated. The different organisms from the marine environment ranging from the microscopic ones like actinomycetes to the macroscopic algae like *Ulva, Gracilaria, Enteromorpha*, and *Sargassum,* Gastropods like molluscs, cones, and starfish and members of the fish family like puffer fish, toadfish, stargazers, and stonefish, that are known to be poisonous have been explored for the toxins that they produce and the pharmacological activity of the respective toxins are evaluated. The rise and increase in drug-resistant pathogens has only served to further the need and cause in the search for novel anti-infective therapeutics. As a result, all possible sources, terrestrial and marine organisms producing toxins and those that do not are evaluated scientifically in the hope of discovering potent anti-infective therapeutics that specifically target the infective agent and cause minimal and/or no harmful side effects to humans.

The ability to provide clinicians with advanced and sufficient knowledge in order to be able to predict the pathology of the disease and to facilitate selection of the appropriate treatment procedure is one of the most essential goals of biomedical research. Scientific evaluation of different natural sources of bioactive compounds is the first step in the development of anti-infective therapeutics. The activity of the bioactive compound is established and quantified by different in vitro procedures. Computer modeling procedures are also used to simulate the disease condition, in particular to simulate the interactions of the drug molecules with the possible receptors on the cell surfaces of cells, known as in silico studies. Though in vitro and in silico studies are very useful and are also indispensable, they only provide a window to the actual mechanisms and reactions taking place in the body. The groundwork or the primary work which is carried out in vitro and in silico is very productive and is indicative of therapeutic value, but it is limited in the sense that the data is only one-dimensional. This data is not sufficient to qualify the use of the bioactive compounds in either human or animal subjects. In other words, this information obtained from in vitro and in silico studies does not take into consideration the complexity of living organisms and the possibility of interactions that could take place because of the whole-organism systems.

Further extensive research is needed in suitable living models that can mimic the exact disease condition for better understanding of how the bioactive compound is working and to rule out the possibility that the compound is toxic to body cells. This is where in vivo studies come into the picture. In vivo studies are experimental procedures carried out in a living organism like a laboratory organism. This part of the analysis is the most crucial and critical step in the research procedure as the dose of the bioactive compound, the safety of usage, the actual mechanism of action, and targeting of the infectious agent are all established during this step of the study. The use of animal models to test relationships and mechanisms under controlled experimental conditions is of critical advantage as it can be extrapolated to predict clinical outcomes in humans with the same disease conditions. While several animals can be possibly used for in vivo studies as disease models, only a few are considered as feasible. The size of the animal, life span, and period of growth are some of the parameters that are the decisive factors for the selection of an appropriate model. These parameters account for constraints like space management, time factor for the course of study, and the manageability of model animals. Some of the commonly used animal models are mice or rats, hamsters and guinea pigs. Among these animal models, mouse or rat models, collectively referred to as murine models are very frequently used. Their small size, short life span, reliable breeding, and the relative genetic similarity to humans make them the first choice as a suitable animal model to mimic disease conditions. Another feature is that there is a similarity in many physiological characteristics, consequently presenting similar organ system biology for studying diseases of the immune system, endocrine system, cardiovascular system, and other biological systems. A significant point of advantage is that they are amenable to genetic manipulation. Transgenic mice models can be established easily and they facilitate the study of the interactions of specific molecules like drugs or proteins with cell surface receptors and the mechanism of host susceptibility and disease symptoms. Humanized mouse models created by transgenesis take the study of human response to infectious agents to the next level, unraveling the mysteries of pathogenic mechanisms of infectious disease.

Murine models of human disease are highly valuable and indispensable tools in biomedical research. It is imperative, however, that there is appropriate validation and careful consideration of the compatibility of the data with the particular disease condition under study. The extent to which the mouse model exhibits the phenotype that mimics the clinical characteristics of the human disease is to be established for validation of that particular mouse model. The development of powerful tools has facilitated the comparative study at the genome level and has enabled the comprehensive genomic manipulation possible in order to produce different murine models to mimic human disease pathology.

This chapter elaborates the different murine models available to test the efficacy of anti-infective therapeutic agents that are in the process of development. First, different mouse and rat models that are commonly used are listed out. The next part of the chapter discusses some selected studies where mice or rats were specifically employed as model organisms for different bacterial, fungal, viral, protozoal, and helminth diseases to assess the potency of the respective anti-infective agents. The information presented in this chapter is not exhaustive, but is a representation of some of the latest studies carried out by several researchers.

31.2 Murine Models

Some of the commonly available murine models that are routinely used as in vivo animal models for studying the efficacy of anti-infective agents in the process of development of novel anti-infective therapeutics are listed below.

31.2.1 A 129 Mouse

This is a group of transgenic mice commonly represented as "IFNα/βR−/− mice." The coat color of these mice is agouti. This group of transgenic mice is produced by knockout technology carried out on embryonic stem cells by electroporation. The gene which was targeted is the ubiquitous membrane receptor; the interferon α/β receptor (type I interferon). Since the genotype of these mice is homozygous, they are more susceptible to viral infections (The Jakson Laboratory).

31.2.2 AG 129 Mouse

Also referred to as "INFα/β/γR−/− mice," this group of transgenic mice is produced by double knockout technology carried out on embryonic stem cells by electroporation. As the genotype is homozygous, the strain shows increased susceptibility to viral infections. Their coat color is agouti (Marshall Bioresources).

31.2.3 BALB/c Mouse

Also known as the "Bagg albino," it was developed by H. J. Bagg from the stock of a home pet dealer in Ohio in 1913. It is an inbred albino mouse strain. In addition to the advantage of easy breeding with this variety, there is also a minimum weight difference between the males and females of this variety. Being a great multipurpose model, it is especially useful for studying infectious disease models (Charles River Laboratories).

31.2.4 B6.129S-Cybb^{tm1Din}/J Mouse

It is a highly specific mouse model, also known as "gp91 phox-." The affected hemizygous male mice cannot produce phagocyte superoxide. Consequently, the mice have an increased susceptibility to infection with *Aspergillus fumigates* and *Staphylococcus aureus*. It also causes an altered inflammatory response in thioglycollate peritonitis. These special features make this mouse a suitable model to study and explore drugs for the treatment of chronic granulomatous disease and to analyze the role played by phagocyte-derived oxidants in the inflammatory process (The Jackson Laboratory).

31.2.5 C57BL/6 Mouse

A widely used inbred strain, the C57BL/6 is commonly referred to as "Black 6" or "C57 black 6" and is abbreviated as "B6." This was the first inbred strain to have its genome sequenced. In addition to being employed for various researches like cardiovascular research, developmental biology research, obesity and diabetes research, and neurobiology research, it is also used for studying infectious diseases, in particular, Tuberculosis and *Salmonella* infections (Charles River Laboratories).

31.2.6 CD-1 ICR Mouse

ICR stands for Institute of Cancer Research, in Philadelphia where the stock for this general multipurpose model animal was initiated. CD stands for Caesarean Derived. This is a mouse of Swiss origin that is used extensively in pharmaceutical and oncological research as a general purpose stock. It is used to test the safety and efficacy of drugs and in particular anti-infective agents. It is easy to handle and is relatively docile. The gene for coat color is "c," hence it is an albino. The different body parameters like body weight, complete blood profile, and biochemical parameters like sodium level, potassium level, phosphate level, glucose level, bilirubin level, creatinine level, cholesterol level, triglyceride level, and hemoglobin count have been well established at different growth stages for both males and females (Charles River Laboratories).

31.2.7 CDF 1 Mouse

This is a hybrid strain, also referred to as the "CD2F1" mouse. It is obtained by a cross between two inbred strains, the female being BALB/c and the male being DBA/2. The coat color of this resulting CDF 1 mouse is brown. The growth of the male in terms of mean weight (in grams) is distinctly more than that of the females in this strain. In addition to being used for monoclonal antibody production and transplantation research, it is also used for testing the safety and efficacy of drug candidates (Charles River Laboratories).

31.2.8 DBA-1 Mouse

It is an inbred strain that is the oldest of all inbred mice strains. The coat color is dilute brown, non-agouti. Dr. C. C. Little developed this model in 1909 at the

Jackson Laboratory. It is used for pharmacological testing and evaluation and as animal models for inflammatory studies (The Jackson Laboratory).

31.2.9 DBA/2 Mouse

This mouse strain was developed in 1909 by C. C. Little from stock segregating for coat color. The coat color is dilute brown (Charles River Laboratories).

31.2.10 OF 1 Mouse

OF stands for Oncins France. This mouse strain was developed in 1935 from a line of mice which were productive and which had vigor in Carworth Farms. It is an outbred albino strain. This strain is characterized by its rapid growth rate and its good breeding. It is used for pharmacology, toxicology, and physiology studies (Charles River Laboratories).

31.2.11 SCID Mouse

This mouse model has severe combined immunodeficiency, as denoted by the name SCID, which affects both B and T lymphocytes. This model has a genetic autosomal recessive mutation, designated as $Prkdc^{scid}$. They are homozygous for this allele, which is mapped to the centromeric end of chromosome number 16. An impaired VDJ rearrangement leads to their lack of functional lymphocytes. The SCID mouse model serves as an excellent in vivo model for the study of infectious diseases. It is incredibly versatile and hence is exceedingly valuable for biomedical research (Charles River Laboratories).

31.2.12 Swiss Model

The Swiss mouse model, named so because it originated from the "Centre Anticancéreux de Romand," located in Lausanne, Switzerland, is one of the most commonly used animal models for biomedical research and efficacy studies. It is an albino outbred mouse. The different body parameters like body weight, complete blood profile, and biochemical parameters like sodium level, potassium level, phosphate level, glucose level, bilirubin level, creatinine level, cholesterol level, triglyceride level, and hemoglobin count have been well established at different growth stages for both males and females (Charles River Laboratories).

31.2.13 Swiss Webster Mouse

This is an albino mouse whose stock was produced using selective inbreeding carried out by Dr. Leslie Webster. The foundation animals for this stock were Swiss mice which were maintained at Rockefeller Institute after obtaining them from Switzerland in 1926. The growth rate in case of males is much more than that of females. These animals are suitable animal models for general purposes and in particular for safety and efficacy testing of different drug molecules (Taconic Biosciences).

31.2.14 Cotton Rat

The cotton rat, *Sigmondon hispidus* which was developed in 1996 by the National Institutes of Health, Bethesda, Maryland and Virion Systems Inc. is a New World rodent. The coat color varies from grey to black to brown. It is a very important murine model as it is frequently used in the study of infectious diseases. Other murine models are not so susceptible to human pathogens and allow their replication only at low levels or need rodent-adapted pathogenic strains. The unique susceptibility of cotton rat to human pathogens and its ability to simulate a significant number of parameters of human disease makes it a preferred animal model for the study of therapeutic drug candidates for infectious diseases. It is an ideal animal model for research leading to the development of vaccines and for the study of gene therapy based on adenoviral vectors. The growth parameters of both male and female cotton rats are well established (Envigo).

31.2.15 Fischer Rat

It is an albino rat, commonly represented as the "CDF" rat. It was produced by the mating of #344 rats obtained by purchase from a local breeder, Fischer. This colony of rats was originated by M. R. Curtis in 1920 at the Columbia University Institute for Cancer Research. It is a general multipurpose animal model which is used for surgical modeling, oncology studies, nutrition studies, and to test the safety and efficacy of drug molecules. The growth rate of males and females is almost similar for the first 6 weeks, after which the males show a marked increase in growth when compared to females. The different hematological and biochemical parameters of both males and females are well established (Charles River Laboratories).

31.2.16 Sprague-Dawley Rat

It is an outbred, albino rat with its origin from the Central Institute for Laboratory Animals in Germany. It was created by R. W. Dawley in 1925 from an albino female (probably Wistar) and a hooded male hybrid whose origin is not known.

The resulting progeny was then crossed with the female's progeny for seven generations. This rat has an elongated head and the tail is longer than its body. It is employed virtually in all the disciplines of biomedical research and especially in pharmacology and toxicology studies. It is an excellent general multipurpose model and can be used efficiently testing the safety and efficacy of drug molecules. It is docile, easy to handle, and has a fast growth rate. The growth of males is distinctly more than that of the females. The clinical chemistry, hematology, specific organ weights, and urinalysis in both males and females have been standardized (Taconic Biosciences).

31.2.17 Wistar Rat

The Wistar rat is a very widely used murine model as a general multipurpose animal model and as a surgical model. It is also used extensively as a specific model for several studies including infectious disease research and safety and efficacy testing of drug molecules under study. It is an outbred albino rat with the coat color white and is a good breeder. Its origin is from the Central Institute for Laboratory Animals in Germany. It is easy to handle and has good longevity making it an ideal model for infectious disease research. The various hematological and biochemical parameters in both males and females are established. The growth of both males and females has been quantified, and the growth of males is significantly more that of females (Charles River Laboratories).

31.2.18 Wistar Furth Rats

This albino rat model was developed in 1945 by J. Furth from a commercial Wistar stock that he obtained. In creating this model, he endeavored to produce a rat model with high susceptibility/incidence of leukemia. This model is most commonly used for cancer research studies like carcinogenesis and leukemia. It is also a suitable macrothrombocytopenia model and is ideal for grand mal seizure studies. This rat model is also resistant to chronic renal disease. The growth rate is similar for males and females only for the first 4 weeks after which the growth of males exceeds that of females (Envigo).

All the murine animal models discussed above are available for purchase from different standard establishments like CSIR-Indian Institute of Integrative Medicine, Jammu; National Institute of Biologicals, Noida; and CSIR-Centre for Cellular and Molecular Biology, Secunderabad in India. Jackson Laboratories, Maine; Charles River Laboratories, Massachusetts; and Taconic Biosciences, USA are some of the important suppliers in the USA. Janvier Labs, France; Envigo, UK; and JMSR (Japan Mouse/Rat Strain Resources Database, Japan are some other standard labs supplying murine models in other regions of the world.

31.3 Murine Models for Antibacterial Therapeutics

Bacterial infections constitute a major segment of infectious diseases plaguing humans. Infections caused by *Staphylococcus, Klebsiella, Pseudomonas, Enterococci, Escherichia,* and *Helicobacter* are some of the most frequently occurring persistent bacterial infections. Delayed or inefficient treatment of these bacterial infections causes a prolonged inflammatory state that can trigger cancers due to the production of carcinogenic bacterial metabolites or due to the chronic inflammatory state of the body. The emergence of multidrug resistant bacteria has served to increase the burden on the existing antibiotics. As a result, the development of novel antibacterial therapeutics is of utmost importance. Different murine models are used to simulate specific bacterial disease conditions to test the efficacy of candidate drug molecules. Clinically isolated bacteria are cultured and a suitable "challenge" dose is administered to the murine model. The test drug molecules are administered in varying doses and animals are sacrificed after the period of study to assess bacterial load in different tissues by histopathological studies. The efficacy is then determined by comparing efficiency with that of existing drug molecules. Some studies on murine models for the development of antibacterial therapeutics are discussed in brief below.

Silva et al. (2016) engineered a synthetic peptide, derived from a marine tunicate antimicrobial peptide, which they named clavanin-MO. The in vivo anti-infective activity of this engineered peptide was assessed in 6-week-old female C57BL/6 mice models. The bacterial strains *E. coli* ATCC 8739, *E. coli* KPC-positive ID No. 1812446, *S. aureus* ATCC29213, and *S. aureus* (MRSA) ATCC 33591 were used for this study. All four were clinical isolates. These clinical isolates were plated on BHI agar from frozen stock and grown for 24 h. After this incubation period, three colonies were isolated from each plate and transferred to 1 ml of BHIB. The broth cultures were incubated for 12–16 h overnight at an optimum temperature of 37 °C with shaking. Aggressive model of infection was established in the mouse models by infecting 200 μl of one of the above bacterial culture suspended in PBS via intraperitoneal injection. For Gram-negative bacteria, effective challenge dose which was found to be ~2×10^7 CFU/mouse in PBS was used and for Gram-positive bacteria, ~2×10^9 CFU/mouse in PBS was employed. The synthetic peptide clavanin-MO was found to exert a protective effect on the mouse models infected with the lethal strains of *E. coli* and *S. aureus.* The peptide was observed to not only kill the lethal bacteria but to also resolve the infection because of its immunomodulatory property.

S. aureus is frequently found to be the causative agent behind bacterial abscess. Han et al. (2009) evaluated the therapeutic use of nitric oxide releasing nanoparticles in a mouse model of abscess infection caused by *S. aureus.* Female BALB/c mice (6–8 weeks old) were clean shaved on their flanks after providing anesthesia (100 mg/kg ketamine and 10 mg/kg xylazine). The skin was then disinfected with iodine. A clinical isolate of *S. aureus*, MRSA 6498 was collected from a patient's wounds from the Montefiore Medical Centre, the Bronx, NY and stored in BHI broth containing 40% glycerol at −80 °C until use. Before infection, it was cultured on TSB at 37 °C in a shaker, with rotations at 150 rpm. The optical density at 600 nm was ascertained to monitor the bacterial growth. A 100 μl suspension with 10^7 MRSA in PBS was subcutaneously injected into each flank of the murine models (two abscesses per mouse). The following day, 5 mg/ml of nitric oxide-releasing nanoparticle system (NO-np) or control nanoparticles (np) dissolved in PBS was topically applied or injected into the abscesses resulting from challenge dose of MRSA injection. Mice that were treated by topical application were treated with a second topical application of NO-np or np the following day. Mice that were treated by topical application only were treated with a second topical application of NO-np or just np the next day. Both intradermal and topical use of NO-np was found to reduce the area of abscess and the bacterial load, along with improving the skin architecture of the region. The use of NO-np also caused a reduction of angiogenesis thus preventing the dissemination of bacteria from the abscess to the surrounding tissue regions.

You et al. (2009) tested the efficacy of a novel aminoglycoside antibiotic vertilmicin in CD-1 ICR mice. Intraperitoneal infection with 0.5 ml of bacterial suspension in 5% mucin consists of 100 times the medium lethal dose of infection. Two Gram-negative and two Gram-positive organisms were used to cause systemic bacterial infection in the animal models. The challenge dose (CFU/mouse) varied for the bacteria, as represented in Table 31.1. The relatively potent broad-spectrum antibacterial activity was established by studying the resulting MIC and ED_{50} values. The efficacy of vertilmicin in comparison with that of gentamicin to treat ascending urinary tract infection caused by *E. coli* 9612 was also studied in female CD-1 ICR mice. This was done by first restricting the water consumption of mice for a day (24 h). Following this, the mice were subjected to transurethral injection of 0.05 ml of bacterial suspension in 5% mucin, the challenge dose being 2.2×10^9 CFU/mouse into the bladder. This was performed using a round-point needle under the influence of pentobarbital anesthesia. While both exhibited high dose-dependent efficacy, the activity of vertilmicin was higher than gentamicin at a similar concentration.

Table 31.1 Mouse systemic infection models

Sl. no.	Bacteria	Challenge dose
1	*E. coli* ATCC 25922	1.0×10^5
2	*E. coli* ATCC 9612	3.7×10^4
3	*E. coli* ATCC 1515	2.4×10^6
4	*K. Pneumoniae* ATCC 935	3.0×10^5
5	*K. Pneumoniae* ATCC 967	2.6×10^5
6	*S. aureus* ATCC 29213[a]	1.5×10^5
7	*S. aureus* ATCC 9344[b]	2.5×10^5
8	*S. aureus* ATCC 15[a]	2.5×10^5
9	*E. faecalis* ATCC 29212	2.8×10^7
10	*E. faecalis* ATCC HH22	1.8×10^8

[a]*MSSA* Methicillin-sensitive *S. aureus*
[b]*MRSA* Methicillin-resistant *S. aureus*

Sarkar et al. (2003) studied the antibacterial activity of Dobutamine hydrochloride in male Swiss mice models against *S. typhimurium* NCTC 74. A 50 MLD challenge where the LD_{50} corresponds to 0.95×10^9 CFU/mouse suspended in 0.5 ml of NB was given to the animal models by repeated passage to cause infection. Dobutamine was found to significantly reduce the viable bacterial count in the liver, spleen, and heart blood of the infected mouse models 2 h and 18 h post administration of bacterial challenge when compared to the control (saline-treated) mouse models.

The activity of a new cephalosporin, RJW-54428 was evaluated by Griffith et al. (2003) in comparison to existing antibiotic vancomycin, penicillin G, cefotaxime, and ampicillin in different mouse models. Male Swiss mice were employed as an animal model for studying sepsis caused by *S. aureus* strains; pneumonia caused by *S. pneumoniea* and neutropic condition of the thigh caused by *S. aureus*. BALB/c mouse model was used to study polynephritis caused about by *E. faecalis.* To attain sepsis, *S. aureus* Smith ATCC 13709 and methicillin-resistant *S. aureus* (MRSA) 076 were used. Both the strains of *S. aureus* were grown overnight in brain heart infusion broth at an optimum temperature of 37 °C. After washing well with PBS, the cells were adjusted to 10^8 CFU/ml by correlating the absorbance at 600 nm to predetermined plate counts. The resulting inoculum was mixed with an equal proportion of sterile 14% hog gastric mucin. This can be stored in an ice bath for up to 1 h. The mice were intraperitoneally infected with 0.5 ml of the bacterial suspension 100 times the 100% lethal dose; 1.4×10^7 CFU/mouse. Todd-Hewitt broth was used for growing *S. pneumoniae* ATCC 6301. The bacteria were allowed to grow for 22 h post inoculation in T-H broth at 35 °C. After the 22 h growing period, the inoculums for infection was prepared to 6×10^6 CFU/ml by correlating the absorbance at 600 nm to predetermined plate counts. 0.05 ml of this inoculum was then instilled intranasally to cause pneumonia. The challenge dose for this was 3.0×10^5 CFU/mouse. BHIB was used for growing *E. faecalis* ATCC 23241 aerobically overnight at an optimum temperature of 35 °C. The next day, the broth was centrifuged in order to harvest the bacterial cells. The inoculum was adjusted to 10^8 CFU/ml by checking the absorbance at 600 nm as opposed to the predetermined plate counts. Polynephritis model was obtained by injection of 0.1 ml of prepared inoculum of *E. faecalis* into the lateral tail vein of male BALB/c mice. The challenge dose was 1.2×10^8 CFU/mouse. Male Swiss mice were injected with 0.1 ml of inoculum of MRSA strain COL and VISA strain HIP-5836 in each thigh (four thighs per group per time point) intramuscularly (on day 5) after being made nephrotic by injecting 150 mg per kg body weight (on day 1 and 4) of cyclophosphamide intraperitoneally. Both the bacteria were grown overnight in BHIB at 35°C. The next morning, they were subcultured into fresh BHIB and incubated at 35 °C for 4 h. Before infection of the animal models, the inoculum was adjusted to a final concentration of ~5.0×10^6 CFU/ml by correlating the absorbance at 600 nm to that of predetermined plate counts. The new cephalosporin RJW-54428 was found to be effective in all the different infective animal models, with activity on par of more than that of the standard reference antibiotics used, vancomycin, penicillin G, cefotaxime, and ampicillin.

One of the underlying causes of endocarditis is nosocomial bacteremia. The three leading causative agents for infective endocarditis are streptococci, staphylococci, and enterococci. *E. faecalis* GC6181, *E. faecalis* GC6191, *E. faecalis* GC6207, and MRSA 8-4 were used to induce endocarditis in male Wistar rats (200–250 g). The efficacy of a new glycylcycline antibiotic, GAR-936 was comparatively studied with vancomycin for the treatment of bacterial-induced endocarditis by Murphy et al. (2000). The mating of *E. faecalis* GC6181 (vancomycin-susceptible) and a vancomycin-resistant clinical isolate of *E. faecalis* by a modified filter method with colony selection carried out on vancomycin agar and verification on Riboprint led to the construction of *E. faecalis* GC6191. *E. faecalis* GC6207 was obtained from Cleveland VA Hospital. A clinical isolate of MRSA from the hospital was used. The bacteria were cultured and maintained in M-H II broth. A sealed polyethylene cannula (PE10) was first inserted through the right carotid artery into the left ventricle and sutured on in rat models. Forty-eight hours post cannula implantation, a 5-h bacterial culture was diluted in sterile saline to get a challenge dose of 10^5–10^6 CFU/ml and 1 ml of this bacterial suspension was intravenously injected. 24/36 h post injection of bacterial challenge dose, antibacterial agents were administered subcutaneously every 12 h for 3 days. The dose for GAR-936 ranged 0.5–80 mg/kg/day and for vancomycin was ranging 40–240 mg/kg/day. The therapeutic potential of GAR-936 in in vivo rat models for the treatment of bacterial endocarditis was established by measuring bacterial titers in the aseptically removed, homogenized, and serially diluted heart samples of the treated animals.

A few more examples of murine models of bacterial infections are represented in Table 31.2.

31.4 Murine Models for Antifungal Therapeutics

Antifungal infections represent a class of infections to humans that occur with persistence but whose seriousness is often underrated. Some of the common disease-causing fungi are *Candida* sps, *Aspergillus* sps, *Mucor* sps, and *Rhizopus* sps. Fungal diseases are often overlooked and neglected. The characterization of novel antifungal agents to combat infection is also not as much under focus as in the case of bacterial and viral infections. As a result, the therapeutics available to treat fungal infections is not satisfactory demanding newer strategies and novel molecules to curb the occurrence and spread of fungal infections. Different fungal diseases have been simulated in murine models to evaluate the efficacy of existing drugs in comparison with newer more potent test molecules. Clinical isolates of fungi are cultured and the conidia or fungal cells are injected into the murine models at a dose sufficient to cause disease. The procedures for establishing various fungal models of disease in mouse or rat models are discussed below.

Calvo et al. (2011) studied the comparative potency of different antifungal drugs like terbinafine (TRB) and itraconazole (ITZ), which are traditionally used in the treatment of the chronic fungal infection of the skin, Chromoblastomycosis when compared to newer drugs like posaconazole (PSC) and voriconazole (VRC) in mice

Table 31.2 Murine models for the development of antibacterial therapeutics

S. no.	Disease	Murine model	Bacteria	Route of infection	Challenge dose	Antibacterial agent	Treatment	Result	Reference
1	Bacteremia	Albino Wistar rats; 8–9 weeks old (140–180 g)	MRSA USA300	Intravenous (tail vein)	2.5 ± 0.5 × 10^{10} CFU/ml	P128, a 27 kDa bacteriophage-derived ectolysin	200 µl of P128 (10, 2.5, 0.5 mg/kg) 5 min/2 h after bacterial challenge (intravenous infusion via cannula)/200 µl of 2.5, 0.5, 0.25 mg/kg administered intravenously 2 h post infection/VAN, 120 mg/kg administered subcutaneously on the day of infection	P128 is stable and highly effective in treating MRSA induced bacteremia	Channabasappa et al. (2017)
2	Septic arthritis	CD1 mice; 10/5 weeks old	*S. aureus* Newman strain and bioluminescent *S. aureus* Xen34, Xen8.1, Xen29	Intravenous (lateral tail vein)	Newman strain—2E+06 CFU/mouse; bioluminescent strain—1-2E+07 CFU/mouse	4C-staph vaccine formulation (adjuvant protein combination vaccine)	Active/passive immunization	Reduced bacterial burden in joints, reduced inflammation, protection against bacterial-mediated cellular toxicity	Corrado et al. (2016)

(continued)

Table 31.2 (continued)

S. no.	Disease	Murine model	Bacteria	Route of infection	Challenge dose	Antibacterial agent	Treatment	Result	Reference
3	Pneumonia	Immunocompetent C57Bl/6NCrlBR male mice; 8–10 weeks old	*P. aeruginosa* MDR-RP73 (clinically isolated)	Intratracheal instillation	Acute infection—1×10^7 CFU; chronic infection—1×10^6 CFU	POL7001, macrocycle antibiotic with novel and potent activity against *P. aeruginosa*; CIP	Intratracheal (with cannula)1 dose/day for acute infection/ endotracheal (with microsprayer) every 2 days for chronic infection/ subcutaneous parenteral administration	Reduced bacterial load in lung, progressive body weight recovery, lower inflammatory marker levels	Cigana et al. (2016)
4	Sepsis, Lung infection, Skin infection	BALB/c female mice, 8 week old (20 g)	Sepsis/Lung Infection—*P. aeruginosa* PAO1; Skin Infection—*P. aeruginosa* P1242	Sepsis—intraperitoneal administration; Lung infection—intravenous and intratracheal administration; Skin infection—bacterial suspension deposited on abraded skin	Sepsis—500 µl of 1.5×10^3 CFU/ mouse; Lung infection—50 µl of $1–3 \times 10^3$ CFU/ mouse (i.v. models), 50 µl of 1×10^6 CFU/ mouse (i.t. models); Skin infection—20 µl of 5×10^3 CFU/ mouse	SET-M33 L (pegylated form)	Sepsis—5 mg/kg SET-M33 L administered intravenously 24 and 72 h post infection; Lung infection—5 mg/kg SET-M33L administered intravenously 1 and 6 h post infection (i.v. model), 5 mg/kg SET-M33 L administered intratracheally once; Skin infection—50 µl of M33 based-lotion (10 mg/ml) applied topically at infection site daily	Increased survival rate in sepsis and lung infection models and complete healing in skin infection model	Brunetti et al. (2016)

5	Enterococcus infection for establishing pharmacodynamics of VAN, SAM, TZP	Udea:ICR (CD-2) mice, 6 weeks old (23–27 g), female	*E. faecalis* ATCC 29212, *E. faecium* ATCC 1934	Intramuscular (in each thigh); 0.1 ml	~5 $\log_{10}$ CFU/g	VAN, SAM, TZP	VAN—23.4–600 mg/kg/day; SAM—18.75–2400 mg/kg/day; TZP—18.75–4800 mg/kg/day; 2 h post infection	Mouse model for optimized thigh infection established	Rodriguez et al. (2015)
6	Peritonitis model	CD-1 female mice (18–21 g)	*P. aeruginosa* 13 s strain	Intraperitoneal injection; 0.5 ml	10^6–10^7 CFU/ml	R2 pyocin isolated from *P. aeruginosa* PAO1	0.1 ml of 3×10^{11} pyocin administered once intraperitoneally/intravenously (dorsal tail vein) 1 h post infection	Potential immunogenic property of R-pyocins established	Scholl and Martin (2008)
7	Bacterial sepsis	Swiss Webster mice (25–30 g)	*P. aeruginosa* PAO1	Intravenous; 0.1 ml	3–4 × 10^7 CFU	WLBU2—De novo–derived cationic antimicrobial peptide	1, 1.5, 3, 4 mg/kg WLBU2 ~1 h after bacterial challenge	Protective effect established	Deslouches et al. (2007)
8	Foreign body invasion	Wistar rats (outbred), Fischer rats (inbred), an inbred cross	*S. epidermidis* 10b	Catheter segments prior incubated in bacterial suspension for 2 h at 0 °C inserted in the subcutis region	6 $\log_{10}$ CFU	Teicoplanin, teicoplanin + rifampicin	Teicoplanin—50 mg/kg, teicoplanin + rifampicin (25 mg/kg); 5 h post implantation to 16 h prior to explanation for totally 7 doses	Foreign body infection is preventable and treatable. Rifampicin and teicoplanin combination more potent than only teicoplanin	Wijngaerden et al. (1999)

models. For the establishment of the disease, a clinical isolate of the fungi *F. pedrosoi* (FMR 6630) was used. Two-week-old cultures of *F. pedrosoi* on PDA were suspended in sterile saline. In order to remove clumps of cells or hyphae, this suspension was filtered through sterile gauze and based on hemocytometer counts, and the suspension so obtained was adjusted to attain the required inoculum for the challenge dose of infection. CD-1 athymic male mice served as the animal model for the experiment. Chronic infection was established by subcutaneous infection with *F. pedrosoi* (FMR 6630) at a challenge dose of 3×10^6 CFU over the left thigh. For the infection to develop sufficiently the treatment with antifungal agents only began 3 weeks post administration of the challenge dose and it lasted for 4 months. Calipers were used to measure the skin lesions caused due to infection. Treatment with PSC at a dose of 20 mg/kg was found to be the most effective antifungal agent, with significantly smaller lesions post treatment. Treatment with TRB (250 mg/kg) and ITZ (50 mg/kg) also was effective with a reduction in lesion size when compared to the control group. The inflammatory response was higher in animals treated with VRC when compared to the other anti-infectives, TRB, ITZ, and PSC.

Hata et al. (2011) evaluated the antifungal activity of a broad-spectrum antifungal agent with a novel mechanism of action (inhibition of fungal glycosylphosphatidylinositol biosynthesis), E1210 in murine models of oropharyngeal and disseminated candidiasis, pulmonary aspergillosis and disseminated fusariosis. The six fungal strains used were *Candida albicans* IFM49971, *C. albicans* IFM49738 and *C. tropicalis* E83037, *Aspergillus flavus* IFM50915, *Aspergillus fumigatus* IFM51126, and *Fusarium solani* IFM50956. Administration of E1210 orally was found to exert a dose-dependent efficacy on the infections caused by the fungi in the animal models under study. The details of the animal models, a challenge dose of fungi for causing infection, and the antifungal therapeutic details are mentioned in Table 31.3.

The efficacy of three antifungal agents, posaconazole (PSC), voriconazole (VRC), and amphotericin B (AMB) was tested in a murine model of systemic infection of *Cryptococcus gatti,* which is known to cause life-threatening diseases of the central nervous system and the pulmonary system in humans by Calvo et al. (2010). Male OF 1 mice models with a mean weight of 30 g were infected with three clinical isolates of *C. gatti*, FMR 8394, FMR 8396, and FMR 8410. One day before inducing infection, the mice were immunosuppressed by a single intraperitoneal dose of cyclophosphamide (200 mg/kg body weight) plus 5-fluorouracil (150 mg/kg body weight) intravenously. All mice were subcutaneously injected with ceftazidime (5 mg/day) from days 1–7 after the introduction of fungal infection to prevent bacterial infections. Mice were injected with 0.2 ml of a challenge dose of conidial suspension (2×10^5 CFU) in sterile saline solution into the lateral tail vein. AMB (1.5 mg/kg body weight) was administered once/day intraperitoneally, PSC (10, 20, or 40 mg/kg body weight) once/day was administered orally and VRC (10, 40, or 60 mg/kg body weight) was administered orally once/day. The treatments lasted for 10 days, beginning 1 day after administration of the challenge dose of fungal suspension. Among the three drugs, AMB exhibited the highest efficacy, with increased survival rate and reduced tissue burden in the organs tested.

Table 31.3 Murine models for antifungal therapeutic study performed by Hata et al. (2011)

S. no.	Murine model	Immunosuppression	Antibacterial therapy	Fugal Sps. administered	Challenge dose	Antifungal therapy
1	Oropharyngeal candidiasis model—female ICR mice (5 weeks old, ~25 g	Cortisone acetate (4 mg, subcutaneously; prior to and 3 days post infection)	Tetracycline hydrochloride (1 mg/ml, via drinking water from the day of immunosuppression)	*C. albicans* IFM49971	4×10^5 CFU/mouse	E1210 (twice/day) or fluconazole (once/day) administered orally, for 3 days
2	Disseminated candidiasis model—female ICR mice (5 weeks old, ~25 g	5-FU (200 mg/kg, subcutaneously; 6 days prior to infection)	Ciprofloxacin (0.1 mg/ml, via drinking water; 2–3 days prior to infection-5–7 days post infection)	*C. albicans* IFM49971, *C. albicans* IFM49738, *C. tropicalis* E83037	3×10^5 CFU/mouse	E1210/ voriconazole (2–3 times/day) or fluconazole (once/day) administered orally or Caspofungin/ liposomal amphotericin B (once/day) intravenously, for 3 days
3	Pulmonary aspergillosis model—female DBA/2 N mice (8 weeks old, ~18 g	5-FU (200 mg/kg, subcutaneously; 5–6 days prior to infection)	Ciprofloxacin (0.1 mg/ml, via drinking water; 3–4 days prior to infection-7 days post infection)	*Aspergillus flavus* IFM50915, *Aspergillus fumigatus* IFM51126	*A. flavus*—3×10^4 conidia/mouse *A. fumigatus*—6×10^4 conidia/mouse	E1210/ voriconazole (2 times/day) administered orally or Caspofungin/ liposomal amphotericin B (once/day) intraperitoneally, for 4/7 days
4	Disseminated fusariosis model—female DBA/2 N mice (8 weeks old, ~18 g	5-FU (200 mg/kg, subcutaneously; 5–6 days prior to infection)	Ciprofloxacin (0.1 mg/ml, via drinking water; 3 days prior to infection-7 days post infection)	*Fusarium solani* IFM50956	5×10^3 cells/mouse	E1210/voriconazole (3 times/day) administered orally, for 5 days

A persistent source of disseminated infections in high-risk patients is caused by fungal biofilms. They are also recalcitrant to antifungal therapy. The potency of anidulafungin on mature *Candida albicans* biofilms was studied in rat models in comparison with the activity of fluconazole by Kucharíková et al. (2010). Female Sprague-Dawley rats (200 g) were treated with 1 mg/l dexamethasone via drinking water to render them immunocompromised. One centimeter pieces (20 pieces per test concentration) of serum-coated polyurethane catheters incubated in serum overnight were suspended in the challenge dose of *C. albicans* SC5314 strain, 5×10^4 cells/ml at 37 °C for 90 min. The polyurethane catheter pieces were then washed and subcutaneously implanted on the lower back of the immunosuppressed rats. Antifungal therapy was initiated only after 48 h, giving time for the biofilms to mature. Anidulafungin (10 mg/kg) and fluconazole (125 mg/kg) were intraperitoneally administered daily, for 7 days. Post this treatment regime, the catheters were retrieved, subjected to washing and sonication, and biofilm formation was quantified by calibrating the cell numbers by CFU counting. Treatment with anidulafungin resulted in lower cell numbers when compared to those treated with fluconazole.

The efficacy of three antifungal agents, micafungin (MFG), amphotericin B (AMB), and voriconazole (VRC) were evaluated alone, and in double and triple combinations by Rodríguez et al. (2009) in a murine model of disseminated infection brought about by the opportunistic fungal pathogen, *Scedodporium prolificans*. This fungus is an emerging pathogen that causes infection in both immunocompetent individuals and resistant life-threatening infections in immunocompromised individuals. A clinical isolate, *S. prolificans* FMR 6719, was obtained and used to cause infection in the male OF1 mice. The fungal culture was suspended in sterile saline and subjected to filtration using a sterile gauze facilitating the removal of hyphae and clumps of cells if any. The suspension so obtained contains ≥95% of viable conidia. Using a hemocytometer, the inoculum was adjusted to the desired concentration as required. Culture on potato dextrose agar can be used to confirm the hemocytometercount. OF1 mouse models (weight approx. 30 g) used were treated with cyclophosphamide 1 day before infection in order to render them immunocompromised. The appropriate challenge dose for causing acute infection was determined by experimenting with three doses, 1×10^4 CFU/ml, 7.5×10^4 CFU/ml, and 1×10^5 CFU/ml. A dose of 7.5×10^4 CFU/ml was found to cause substantial infection and was used as a challenge dose. 0.2 ml of 7.5×10^4 CFU/ml of conidial suspension was introduced via injection into the lateral tail vein. One day after administration of the challenge dose, treatment started and continued for 10 days post introduction of infection. The single doses were composed of MFG at a dose of 10 mg/kg body weight which was administered intraperitoneally once, everyday; AMB at a dose of 1.5 mg/kg body weight which was given intraperitoneally, daily once and VRC at a dose of 60 mg/kg body weight which was given once, daily by oral gavage method. MFG plus AMB, MFG plus VRC, and AMB plus VRC were the double combination doses given similar to the monotherapies. The triple combination involved all the three antifungal test agents given together, MFG plus AMB and VRC, the dose and route of administration similar to that of the monotherapy. It was observed that the survival rate improved and the fungal load to the brain and kidneys reduced significantly in the combined treatment method with a combination of the two antifungal drugs: micafungin combined with voriconazole and micafungin combined with amphotericin B. These double combinations of the drugs exhibited more significant protective effects than the triple combination of the drugs. The survival rate was also prolonged with the monotherapies, but at lower levels than the double combination therapies.

Zygomycosis is a fatal yet rare skin disease caused by filamentous fungi. Immunocompromised individuals are generally the ones who succumb to this disease. Dannaoui et al. (2002) tested the efficacy of three antifungal agents, amphotericin B, itraconazole, and terbinafine on mouse models as suitable treatment options for zygomycosis. Outbred female CD-1 mice that were 5–7 weeks old (weight 20–22 kg) were used. Three fungal strains were used to cause infection, *Rhizopus microspores* var. *rhizopodi formis* (AZN 1185) that was isolated from an invasive infection, *Absidia corymbifera* (AZN 4095 [CBS 271.65]) that was from an unknown origin and *Apophysomyces elegans* (AZN 1829 [CBS 658.93 and ATCC 90757]) which was isolated from a patient suffering with osteomyelitis. All three isolates were stored at −80 °C in 10% glycerol as conidial suspensions till required. Before infection, the isolates of *R. microspores* and *A. corymbifera* were cultured on Sabouraud dextrose agar supplemented with 0.02% chloramphenicol at 35 °C for 7 days. The surface of the agar was subsequently washed with sterile saline containing 0.05% Tween 80 for harvesting of the spores of the fungi. Sporulation of *A. elegans* was attained by culturing the mycelium in sterile distilled water that was supplemented with 0.1% yeast extract at 37 °C for 10 days. The surface of the agar slants was then washed with 1 ml of sterile 0.9% saline containing 0.05% Tween 80 in order to prepare the inoculum. A nylon filter (11 μm pore size) was used for filtering the spore suspensions. Once the fungal suspension of the different infective agents was prepared, a hemocytometer was used for counting. The above preparation can be stored only for a maximum period of 24 h at 4 °C. Viability can be determined prior to infection and post storage by plating dilutions prepared in saline with 0.05% Tween 80. After incubating the plates at 35 °C, CFU is counted at 24 h. On the day of infection, saline is used to adjust the spore suspension to the desired concentration. 0.1 ml of the colloidal spore suspension is injected into a lateral tail vein of CD-1 mice models with the dose being the LD_{90} level. For *R. microspores*, LD_{90} is 7×10^6 CFU/mouse, for *A. corymbifera*, LD_{90} is 7.5×10^5 CFU/mouse and for *A. elegans*, LD_{90} is 3×10^3 CFU/mouse. Treatment with antifungal drugs began 2 h after inducing infection and continued for 10 days. Amphotericin B desoxycholate was administered in 5% glucose intraperitoneally. Itraconazole and terbinafine hydrochloride was given by gavage after diluting in sterile water. All three fungal isolates were found to cause lethal and accurate infection. A high rate of survival of 90–100% was observed in all infected animals treated with amphotericin B. Though terbinafine was absorbed as a drug, no beneficial effect was observed in mice infected with *R. microspores* and *A. corymbifera*. Itraconazole was found to reduce the mortality rate significantly in those mice infected with *A. corymbifera* and *A.*

Table 31.4 Murine models for the development of antifungal therapeutics

S. no.	Disease	Murine model	Fungi	Route of infection	Challenge dose	Antifungal agent	Treatment	Result	Reference
1	Acute invasive fungal rhinosinusitis	Sprague-Dawley female rats, 6–8 weeks old (230 ± 10 g)	*A. fumigatus* AF9732	Intranasal inoculation	100 μl of conidial suspension. Doses—5 × 10^7 conidia/ml, 1 × 10^7conidia/ml, 1 × 10^6 conidia/ml	–	–	Acute invasive fungal Rhinosinusitis established for all 3 doses experimented	Yan et al. (2014)
2	Disseminated candidiasis	CD1 male mice, 8 weeks old (30–34 g)	*C. albicans* CaCi-2 (clinical isolate)	200 μl intravenous inoculation via tail vein	1 × 10^6 CFU/ml	Genetic compromise of Hsp90 + FL	Genetic compromise of Hsp90–200 μl of 1 × 10^7 cells of *tetO-HSP90/hsp90Δ* FL—2 mg/kg, administered intraperitoneally 1 h post infection, once a day for 4 days	Genetic compromise of *C. albicans* Hsp90 expression augments the therapeutic efficacy of FL	Cowen et al. (2008)
3	Candidiasis	Mice	*C. albicans*	Intravenous inoculation via lateral tail vein	2 × 10^5 yeast of *C. albicans*	RsAFP2, a plant defensin	50 μl injected intravenously 1 h before/after challenge dose administration, once every 24 h for 4 days. Dose-7/14 mg/kg body weight	Prophylactic activity of RsAFP2 in mouse model established	Tavares et al. (2008)
4	Invasive pulmonary aspergillosis	Sprague-Dawley male rats (120–175 g)	*Aspergillus fumigatus* AF293	Intranasal delivery of 300 μl inoculum via a sterile micropipette	1 × 10^7 conidia	AMB and EDTA alone and in combination with amphotericin B lipid complex (ABLC)	Treatment initiated 18 h after infection and continued for 7 days AMB—1 mg, given intravenously; ABLC—5 mg/kg given intravenously; EDTA—30 mg/kg administered intraperitoneally; EDTA (30 mg/kg; intraperitoneal injection) + ABLC (5 mg/kg; intravenous injection)	Reduced mortality, increased survival, lower *A. fumigatus* load in lungs and decreased severity of lesions in lungs in rats treated with EDTA+ABLC than AMB/ABLC/EDTA alone	Hachem et al. (2006)

5	Chronic granulomatous disease	CGD mice	*Aspergillus fumigatus*	Intratracheal inoculation of 25 μl of conidial suspension via cannula	1.25×10^4 CFU/mouse	AMB/micafungin/AMB + micafungin	Treatment initiated on the day of challenge and continued for 4 days. AMB—1 mg/kg body weight, administered intraperitoneally; micafungin—10 mg/kg administered intravenously	Combination of AMB and micafungin resulted in prolonged survival post infection than treatment with AMB or micafungin alone	Dennis et al. (2006)
6	Systemic aspergillosis	CD-1 mice, 7 week old females (mean weight—24.9/26.4 g)	*Aspergillus fumigatus* 10AF	Intravenous inoculation of conidia	Initial study—8×10^6 conidia Combination study—6.7×10^6 conidia	Micafungin (FK), alone and in combination with amphotericin B, itraconazole, Nikkomycin Z	Therapy—1 day post infection for 12 days FK—1/3/10 mg/kg body weight/dose; 2/day administered subcutaneously Combination therapy—FK + AMB (0.8 mg/kg; 1/day administered intravenously); FK + ICZ (100 mg/kg; 1/day given orally); FK + NIK (200 mg/kg; 1/day administered subcutaneously)	Prolonged survival and reduced CFU in the brain and kidney on treatment with FK alone. Increased survival on treatment of FK + AMB/FK + ICZ. Treatment with FK + NIK exhibited greater efficacy than treatment with FK/NIK alone	Luque et al. (2003)
7	Oral and vulvovaginal candidiasis	Sprague-Dawley female rats, 6 weeks old (200 g)	*C. albicans* 4711E	Oral model-cotton swab rolled over all regions of mouth twice; Vulvovaginal model—Intravaginal inoculation	Oral model—5×10^8 cells; Vulvovaginal model—10^7 cells	GW471552/GW471558 (two new azasordarins)	0.5 ml administered subcutaneously Dose—1, 5, 10 mg/kg body weight 4 days post infection Oral model—every 8 h for 7 days; Vulvovaginal model—every 4 h for 3 days	Reduction/eradication of fungal load at the infected area	Martinez et al. (2001)

(continued)

Table 31.4 (continued)

S. no.	Disease	Murine model	Fungi	Route of infection	Challenge dose	Antifungal agent	Treatment	Result	Reference
8	Pulmonary mucormycosis	ICR male mice (~20 g)	*Rhizopus oryzae*	50 µl of conidial suspension inoculated into the nares of the mouse. Mouse held upright for 2–3 min for proper aspiration of fungi	4×10^7 conidia/ml	AMB, FL, trovafloxacin, ciprofloxacin	Treatment initiated 24 h post infection and continued for 10 days Dose: AMB—1 mg/kg body weight/day; FL—80 mg/kg body eight, 2/day; trovafloxacin—80 mg/kg body weight, 2/day; ciprofloxacin—80 mg/kg body weight, 2/day	Aggressive mucormycosis could be efficiently controlled by treatment with FL in combination with trovafloxacin or ciprofloxacin	Sugar and Liu (2000)

elegans, but was inefficient and showed no protective activity in mice infected with *R. microspores*.

Some more examples of murine models of fungal infections are represented in Table 31.4.

31.5 Murine Models for Anti-Viral Therapeutics

Viral infections represent a wide range of human diseases, some that are trivial and can be easily managed medically to others that are severe and lethal, which if not treated immediately can lead to death. *Influenza* virus, HIV, Ebola, Pox virus are some viruses responsible for viral infections of varying intensities in humans. The search for novel antiviral therapeutics is an ongoing process as a consequence of the emergence of new pathogenic viruses. Murine models that mimic the disease state caused by these pathogenic viruses are of great importance. The antiviral activity of different test drug molecules is studied in mouse/rat models that are administered with a sufficient dose of virus particles to cause viral infection. Some murine models of viral infections to assess the efficacy of antiviral drug molecules are discussed below.

Perwitasari et al. (2016) studied the efficacy of a novel nuclear export inhibitor and antiviral drug verdinexor in vivo in BALB/c mouse models infected with different influenza A virus strains. Parental and mouse-adapted influenza A/California/04/09, A/Philippines/2/82/X-79, and A/WSN/33 were the influenza-causing viral strains used for the study. Six to eight weeks old female mice were used for the experimental procedures described. Mice were infected by intranasal inoculation of 0.1 ml of A/California/04/09 or A/Philippines/2/82/X-79 at a dose of 10×MID_{50}. Mice were also infected with 10×LD_{50} of mouse-adapted influenza A/California/04/09 intranasally to study the survival rate. Verdinexor reduces pulmonary virus burden, limits virus shedding, reduces pulmonary pro-inflammatory cytokine expression, and modulates leukocyte infiltration into the bronchoalveolar spaces, thus reducing the burden of the viral infection on the animal.

Titova et al. (2015) made use of immunocompetent ICR mice and immunodeficient SCID mice as animal models to test the efficacy of antiviral drug candidates NIOCH-14 and ST-246 against the viral disease, smallpox. They comparatively studied the sensitivity and susceptibility to infection of ICR mice and SCID mice to smallpox-causing virus VARV strain IND-3a. Following intranasal challenge, more significant difference in the accumulation of virus was observed in the ICR mouse. ICR and SCID mice were intranasally challenged with the VARV strain at a dose of 3.7 $\log_{10}$ PFU (10 ID_{50}) and 4.5 $\log_{10}$ PFU (10 ID_{50}), respectively. Both the drugs NIOCH-14 and ST-246 caused a reduction in the production of VARV in the lungs of the infected ICR mice p.i. as opposed to the control. Treatment of infected SCID mice with both the drugs was also found to result in a reduced titre of virus in mouse lungs 4 days p.i. when compared to the control.

A novel antifungal drug, Favipiravir (T-705) that directly inhibits only viral RNA polymerase but has no such inhibitory action on cellular DNA or RNA polymerases were tested in Wistar-Furth rat models infected with Rift Valley Fever virus strain ZH501 by Caroline et al. (2014). The virus was propagated on VeroE6 cells. Female Wistar-Furth rat models (8–10 weeks old) were infected with RVFV ZH501 (LD_{50} ~2 PFU) by subcutaneous or aerosol exposure. Aerosol exposure was carried out in a whole-body aerosol chamber for 10 min to small particle aerosols created by a 3-jet Collision nebulizer controlled by the AeroMP aerosol exposure control system. 0.4% carboxymethylcellulose and 10% sucrose were used for making dilutions of T-705 to the concentrations 20, 50, and 100 mg/kg. Two hundred microliters of the antifungal suspension was orally administered twice daily. The first dose was administered within 1 h of viral exposure and continued for the next 14 days, twice each day. Ninety-two percent of the virus-infected rats that received T-705 treatment survived lethal infection with no significant weight loss or fever.

A dengue fever viremia model was studied in AG 129 mice deficient for the interferon –α/β and –γ receptors (7–9 weeks old) by Schul et al. (2007). Dengue fever is a disease found to occur very frequently in the tropical and subtropical regions, where it is spread via mosquitoes. It often becomes fatal due to poor treatment measures or lack of proper immediate medical care. C6/36 mosquito cells (ATCC CRL-1660) were grown at 28 °C in RPMI 1640 medium plus 5% fetal bovine serum to produce DENV. Four unadapted tissue culture produced strains of DENV strains were used for the assay: Hawaii (DENV1, ATCC VR-71), New Guinea C (NGC, DENV2, ATCC VR-1255), H241 (DENV4, ATCC VR-1490), and TSV01 (DENV2) passaged in cell cultures 10–15 times. The mice were infected intraperitoneally with 0.4 ml of a viral suspension of 5×10^6 PFU/AG 129 mouse. Among the four, TSV01 exhibited highest viremia, so it was selected for further study. A dose range from 1×10^5 to 1×10^8 PFU/mouse was studied to determine the optimum dose for infection. To attain dose levels higher than 1×10^7, cell culture supernatants were subjected to centrifugation at 30,000 g for 1 h and resuspended in RPMI 1640 medium. Higher dose levels of 1×10^6, 1×10^7, and 1×10^8 resulted in increasing levels of virus in the blood. The presence of the characteristic trait of dengue infection in humans, NS1 protein in the blood samples of DENV-infected mice indicated that the animals model the viral infection aptly. The efficacy of antiviral drugs, ribavirin (~40 μmol/l), 7-deaza-2′-C-methyl-adenosine (15 μmol/l), N-nonyl-deoxynojirimycin (75 mg/kg) and 6-O-butanoylcastanospermine (75 mg/kg) was then tested in a standard infection model; mice were infected with 2×10^6 PFU/mouse of TSV01 strain of DENV. Reduced viremia even after delayed treatment in a dose-dependent manner resulted in a reduction of pro-inflammatory cytokine levels and splenomegaly in infected animals.

Dandri et al. (2005) demonstrated the use of mouse models for antiviral therapy to evaluate the efficacy of antiviral drugs against hepatitis B virus. uPA transgenic mice and RAG-2 knockout mice served as mouse models. Primary hepatocytes from tree shrew *Tupaia belangeri* were isolated by collagenase perfusion method and transplanted immediately. If storage is necessary, they can be stored by cryopreservation in liquid nitrogen for up to several months. Once the cells are needed, they are first thawed by placing the vials with frozen cells in a water bath at 37 °C for a short period and then put back on ice. RPMI culture media is added to the thawed cells and the viability is determined by trypan blue exclusion. Fifty microliters of PBS containing ~5 × 10^5 viable tupaia hepatocytes is then transplanted via intrasplenic injection into $uPA^{+/-}$ /$RAG\text{-}2^{-/-}$mice. For HBV infection, HBV positive serum (1 × 10^8 HBV-DNA genome equivalents/ml) was used. This was obtained from a HBsAg-positive chronic carrier (genotype D). Woolly monkey serum, which contained 1 × 10^9 WM-HBV DNA genome equivalents/ml was another source used to cause viral infection. Both these serum samples were stored at −70 °C until use. The mice were then subjected to infection by injecting 20 μl of HBV or WM-HBV-positive serum intraperitoneally. Among the infected mouse models, the efficacy of antiviral agent adefovir dipivoxil was tested in the engrafted mice that showed stable viral titers of ≥5 × 10^5 HBV or WM-HBV genome equivalents. Following transplantation with tupaia hepatocytes, mouse liver repopulation ranging 30–80% was observed. The presence of HBsAg was established by ELISA in both mouse models receiving HBV and WM-HBV within 6 weeks of challenge. Adefovir dipivoxil exhibited significant antiviral activity and inhibited the synthesis of viral DNA in the repopulated uPA/RAG-2 mice livers.

Kern et al. (2004) used the SCID mouse model for the evaluation of three benzimidazole nucleosides (BDCRB, GW275175X, and GW257406X) as suitable antiviral therapeutics against human cytomegalovirus (CMV) infection. Two different mouse models were prepared by different human fetal tissues at different regions of the SCID mouse. The human fetal tissues were obtained from a standard facility (Advanced Biosciences Resources, Alameda, California). In the first model, human fetal retinal tissue was implanted in the anterior chamber of the eye of the SCID mouse. In the second model, human fetal thymus and liver tissues were implanted under the kidney capsule of the SCID mouse. For the first experimental infection, male SCID mice that were 4–8 weeks old were used. They were given anesthesia by injecting ketamine (100 mg/kg body weight) and xylazine (15 mg/kg) intraperitoneally and proparacaine-HCL (0.5%) by topically instilling in the eye. Then, the infusion containing mechanically dissociated human fetal retinal tissue was inserted into the nasal sclera and into the anterior chamber by using a 27-by ½-in-gauge winged infusion needle. Approximately 5 μl of tissue was injected at the temporal side of the anterior chamber and the needle was removed. The mice were undisturbed for 6–9 weeks to allow the implants to grow. Following this period, the mice were administered with 10 μl of 2000–7500 PFU of HCMV (the amount depending on the experimental parameters) via sterile injection into the anterior chamber containing the implants, under the influence of anesthesia. For the other model, 4–6 week old male SCID mice were used. Following administration of anesthesia, an 18-gauge trocar was used for the implantation of fragments of human fetal thymus/liver under the capsule of one of the kidneys. The implant was then allowed to grow for 12–14 weeks. After this period, the grafts were inoculated with 2000–9000 PFU of HCMV (depending on the experiment). The Toledo strain of HCMV was used for both the experimental models. Since benzimidazole nucleosides had the efficacy to treat both the kinds of animal models in the experimentation procedure, it can be concluded that this class of compounds can be effective against the different HCMV infections occurring in the immunocompromised individuals/hosts.

A few more murine models developed to test the efficacy of antiviral agents are listed in Table 31.5.

31.6 Murine Models for Antiprotozoal Therapeutics

The burden caused by protozoal diseases on human health is substantial, affecting more than 500 million people in the world. Malaria, trypanosomiasis, leishmaniasis, and intestinal infections like giardiasis are some of the most commonly occurring protozoal infections that can be lethal if not treated properly. Many of the drugs used frequently for the treatment of protozoal infections are conventional that have been in use for several decades. Poor safety, drug resistance, and low efficacy are some of the limiting factors of these drugs. The great need for newer, more potent antiprotozoal agents, and the increased understanding of protozoans accelerated by the advances in the field of science and technology has propelled many studies for the discovery, synthesis, and development of many novel antiprotozoal drug candidates. Some studies for assessing the efficacy of diverse antiprotozoal agents in murine models are represented below.

Marango et al. (2017) tested the antiprotozoal efficacy of the extracts of *Tephrosia vogelii* in BALB/c mice in comparison to standard drugs, pentostam and amphotericin B. Protozoal infection was induced by the *Leishmnia major* strain IDUB/KE/83 = NLB-144, that was isolated originally from a female *Phlebotmus dubosqi* in 1983 from Kenya. Promastigotes were subjected to incubation at 25 °C to facilitate growth to stationary phase for the generation of infective metacyclic forms, obtained by the 6th day of culture. A hemocytometer with a Nikon optiphot optical microscope at 40× magnification was used to count the pomastigotes in the medium. Eight-week-old female BALB/c mice (20 ± 2 g) were injected with 1×10^6 stationary phase *L. major* promastigotes in 50 μl PBS into the left hind footpad. Treatment was initiated at the beginning of the 5th week post infection and continued for 28 days. One group of mice received *T. vogelii* orally and another group was administered with the extract intraperitoneally. The dose of the extract for both groups was 0.2 mg/ml (200 μg/ml). Standard drugs pentostam and amphotericin B were administered to two other groups, each at a dose of 100 μg/ml. Mice that were treated with *T. vogelii* intraperitoneally exhibited greater reduction in lesion size and significant control of parasite burden when compared to those that received the plant extract orally. However, higher activity was exhibited by the standard drugs used, pentostam and amphotericin B.

Pegoraro et al. (2017) developed two compounds, SC81458 and SC83288 with optimized pharmacological and antiparasite properties from an amicarbalide. The efficacy of these two compounds was studied in a mouse model of malarial infection. Humanized NSG mice (9–11 weeks) were subjected to infection with *Plasmodium falciparum* wild-type strain 3D7 or the multidrug-resistant strain W2. In addition to this model, the rodent malaria model caused by the parasites *Plasmodium vinckei* and *Plasmodium berghei* were also used to study the efficacy of the two novel compounds developed. The blood cultures of the parasites were

Table 31.5 Murine models for the development of antiviral therapeutics

S. no.	Disease	Murine model	Virus	Route of infection	Challenge dose	Antiviral agent	Treatment	Result	Reference
1	Ebola virus infection	IFNAR$^{-/-}$ C57BL/6 mice and INFNAR$^{-/-}$ 129/Sv mice (mice lacking type I IFN-α/β receptor (INFR$^{-/-}$)	Wild-type Zaire EBOV Mayinga 1976 strain	Intranasal inoculation	1000 FFU of EBOV	T-705 (favipiravir, a pyrazinecarboxamide derivative)	Treatment administered twice daily per os by using a stomach probe IFNAR$^{-/-}$ C57BL/6 mice—300 mg/(kg × d) initiated 6 days p.i./ 8 days p.i.; INFNAR$^{-/-}$ 129/Sv mice—30 mg/(kg × d) from days 2–9/ 300 mg/(kg × d) from days 2–9/ days 4–11/ days 6–13	Administration 6 days p.i. resulted in rapid virus clearance, reduction of biochemical parameters of disease severity and prevented lethal outcome in 100% of treated mice models	Oestereich et al. (2014)
2	Influenza infection	BALB/c female mice, 6–8 weeks old	A/Puerto Rico/8/1934 (H1N1) and A/Udorn/307/72 (H3N2)	Intranasal instillation	50 μl ~ 350 PFU	LL-37, the human cathelicidin	Treatment began 1 day prior to infection and lasted for 7 days post infection. LL-37 introduced via nebulization	Lower concentration of pro-inflammatory cytokines than untreated animals	Barlow et al. (2011)
3	Respiratory syncytial viral infection	BALB/c female mice, 8–12 weeks old	Respiratory syncytial virus RSV A2 strain	Intranasal administration	100 μl of 2×10^6 PFU	TMC353121, a novel small molecule RSV inhibitor	Intravenous administration at doses of 0.5–10 mg/kg	Reduced viral replication and inhibition of lung inflammation	Olszewska et al. (2011)

(continued)

Table 31.5 (continued)

S. no.	Disease	Murine model	Virus	Route of infection	Challenge dose	Antiviral agent	Treatment	Result	Reference
4	Filovirus infection (Hemorrhagic fevers)	C57BL/6 mice, 8–10 weeks old/ BALB/c mice, 8–12 weeks old	*Zaire ebolavirus* (ZEBOV) and *Marburgvirus* (MARV-Ravn)	ZEBOV infection—Intraperitoneal injection of mouse-adapted EBOV in C57BL/6 mice MARV-Ravn infection—Intraperitoneal injection of mouse adapted MARV-Ravn in BALB/c mice	1000 PFU/ mouse for both viruses	FGI-103, a low molecular weight inhibitor	10 mg/kg, 24 h post infection as a single dose administered intraperitoneally	Reduced viremia and viral burden in tissues	Warren et al. (2010)
5	Monkeypox viral infection	C57BL/6 mice	Monkeypox virus—MPXV-ZAI-79	Intranasal infection to seed the upper respiratory tract	5 µl/nare, 5000PFU	CMX001, ST-246	Treatment initiated on the day of infection (4 h post infection) via oral gavage CMX01–10 mg/kg on the day of infection followed by 2.5 mg/kg every alternate day after first dose until 14th day ST-246—100 ma/kg, daily for 10 days	Negligible weight loss in treated animals. Improved survival rate in CMX01 treated animals when compared to ST-246 treated ones	Stabenow et al. (2010)

6	Herpes simplex viral infection	BALB/c female mice, 4 weeks old	HSV-1 SC16	Suspension placed on a shaved region of the skin at the base of the right ear 0.5 cm lateral to the ventral mid-line	10 µl of 5×10^4 PFU	BAY 57–1293 (a new class of potent inhibitors, targeting virus helicase-primase complex) and famciclovir	Treatment initiated 1 day post infection. BAY 57–1293—15 mg/kg, 1/day administered for 4 days intraperitoneally/orally; famciclovir—140–200 mg/kg/day administered for 7 days via drinking water	Virus titres in skin drastically reduced post treatment	Biswas et al. (2007)
7	Myocarditis and Epicarditis	BALB/c mice, 6 weeks old	CVB3, a cardiotropic strain of woodruff H3 variant derived from murine heart-passaged infectious cDNA copies of the wild-type viral genome	Intraperitoneal injection, 1 h post administration of mIF	100 µl of 30,000 PFU	Murine interferon-β and -α_2 (mIF)	Intraperitoneal injection of mIF, followed by subcutaneous injection every other day 2.5–10 MIU/kg of mIF-β and mIF-α_2)	Reduced weight loss, increased survival rate, elimination of cardiac viral load, injury protection of monocytes	Wang et al. (2007)
8	HIV infection	Sprague-Dawley rats (hCD4/hCCR5transgenic rats—expressing HIV-1 receptor complex on CD4 T cells and macrophages)	Clinically isolated HIV-$1_{YU\text{-}2}$	Intravenous injection to the tail vein through a plastic catheter on the third day of antiviral therapy	I: Efavirenz-treated group—6.7–10^6 TZM-BL IU; 5000 ng of p24 per rat; II: Enfuvirtide-treated group—6.7–10^5 TZM-BL IU; 500 ng of p24 per rat	Efavirenz (nonnucleoside reverse transcriptase inhibitor) and Enfuvirtide (peptide fusion inhibitor)	Efavirenz—25 mg/kg/day, 1/day by oral gavage; Enfuvirtide—4 mg/kg/day, 1/day or 2/day by subcutaneous injection Administration continued 4 days post challenge	Prophylactic treatment resulted in significant reduction of cDNA load in spleen 4 days post HIV-1 challenge	Goffinet et al. (2007)

maintained under controlled conditions of CO_2, O_2, N_2, and humidity: 3%, 5%, 92%, and 95%, respectively. Following mouse humanization, infection was induced in mice by injecting 0.5 ml of *P. falciparum*-infected HRBC at a parasitemia of 0.3%. Seven days post infection all mice exhibited a patent parasitemia. Mice were then injected with SC81458 or SC83288 intraperitoneally, at varying doses of 2.5, 5.0, or 10 mg/kg body weight, once a day for the next 4 days. For the *P. vinckei* model of infection, 20 mg/kg body weight of SC83288 or 30 mg/kg body weight of SC81458 was intraperitoneally administered once a day for 4 days. A dose of 30 mg/kg body weight of SC83288 and SC81458 was administered for 4 days following *P berghei* infection. In case of *P. falciparum* infection, SC81458 and SC83288 were found to cause a significant reduction in the parasitemia burden of over 90% within the first 24 h of treatment. Recrudescence of infection was observed in the case of the mice infected with *P. falciparum* 3D7 and treated with SC81458 at a dose of 2.5 mg/kg. In case of *P. vinckei* infection, treatment with SC83288 cured all the mice and no recrudescence was observed. Treatment with SC81458 reduced parasite load, but infection persisted and recrudescent parasitemia resulted in death of all the treated mice following the last administration of the compound. In case of the *P. berghei,* both SC81458 and SC83288 showed absolutely no effect in the treatment of disease.

Babesiosis is a disease that results in the infection of the red blood cells caused by a protozoal parasite of the Genus *Babesia* which is transmitted through the bite of ticks. The inhibitory effect of 19 antiprotozoal drugs and antibiotics on *Babesia microti* infection was studied in female BALB/c mice (18 ± 2 g) by Yao et al. (2015). *B. Microti* strain ATCC°,RPRA-99 was used for the study. Infection was brought about by intraperitoneal inoculation of 0.2 ml blood containing 1×10^7 parasitized erythrocytes. Each drug was administered at doses of 25, 50, and 100 mg/kg/day for 4 days starting on the day of infection, 4 h post inoculation. Azithromycin, atovaquone, primaquine, and robenidine hydrochloride exhibited significant suppression of parasitemia among all the 19 drugs used in the study. Treatment with atovaquone and azithromycin showed high inhibitory rates but microscopy studies revealed that the blood of the recovered mice was still infectious. Best results were observed in the mice treated with robenidine hydrochloride, with a 100% inhibitory rate at a dosage of 100 mg/kg and the blood from the recovered mice not causing parasitemia in subpassage experiments.

Sharma and Shukla (2014) studied the protective effect of sulphadoxine pyrimethamine and chloroquine in *Plasmodium berghei*-infected pregnant BALB/c mice. *P. berghei* NK65 strain was used for the study and was maintained in mice by serial passage of red blood cells infected with *P. berghei*. Six to eight weeks old pregnant female BALB/c mice (20–22 g) were inoculated with 1×10^6 iRBC intraperitoneally. Two hours post malarial infection mice were treated with a single dose of sulphadoxine pyrimethamine orally (26 mg/kg body weight) or with chloroquine orally (6.25 mg/kg body weight) for 4 days. Hundred percent survival rate, normal delivery, and a significantly lower percentage of parasitemia were observed in the mice treated with sulphadoxine pyrimethamine and chloroquine. The placentae of the infected mice treated with these antimalarial agents also exhibited no morphological or cellular alterations. Antimalarial treatment with sulphadoxine pyrimethamine and chloroquine was also found to produce significantly reduced malondialdehyde levels, a lower measure of lipid peroxidation, and reduced number of apoptotic cells in the placentae of infected mice.

Giardiasis is an intestinal infection caused by the protist *Giardia intestinalis* (syn, *G. lamblia* and *G. duodenalis*). Shukla et al. (2013) comparatively studied the efficacy of the treatment of *Giardia* infection with antiprotozoal drugs alone and in combination with the probiotic *Lactobacillus casei* in BALB/c mice. The trophozoites of *G. intestinalis* Portland strain I were used for infecting BALB/c mice, 5–6 weeks old (18–20 g). The trophozites were grown in TYI-S-33 medium. Prior to sterilization with a 0.22 μm Seitz filter, the pH of the medium was adjusted to 6.9. Actively growing trophozoites, 48–72 h old were chilled in ice for 15 min, centrifuged for 15 min, and suspended in PBS. A single dose of *Gardia* trophozoites were administered to mice orally, with the challenge dose being 5×10^6/0.1 ml. Treatment was initiated 1 day post challenge dose administration. Treatment regimes were as follows: a single dose of tinidazole (500 μg/mouse) administered orally, nitazoxanide (80 μg/mouse) was administered orally twice a day for three consecutive days, metronidazole (250 μg/mouse) was administered thrice a day for three consecutive days, a single dose of albendazole (130 μg/mouse) was administered orally for five consecutive days, oral administration of probiotic *L. casei* (1×10^9 CFU/ml) till the end of the experiment, treatment with albendazole (130 μg/mouse) for 5 days and *L. casei* till the end of the experiment and treatment with albendazole (65 μg/mouse) for 5 days and *L. casei* till the end of the experiment. *L. casei* MTCC 1423 at a concentration of 1×10^9/0.1 ml was used as a probiotic. MRS agar was used for *L. casei* culture. For inoculation, 18 h culture was cold centrifuged for 15 min at 8000 rpm and resuspended in PBS. Among the four antiprotozoal agents used, albendazole was the most potent, reducing both the severity and the duration of giardiasis. Administration of the probiotic along with albendazole facilitated restoration of gut morphology thus reducing infection and enhancing recovery.

The infection with the protozoan parasite *Cryptosporidium* significantly contributes to diarrheal disease. In immunocompromised individuals like AIDS patients, this infection can lead to chronic diarrhea which can spread to extraintestinal locations and can be life-threatening. Downey et al. (2008) evaluated the efficacy of pyrvinium pamoate against *Cryptosporidium parvum* infection in neonatal mice. *C. parvum* Iowa isolate oocysts was obtained by experimental infection of a female Holstein calf. Using continuous flow centrifugation, the oocytes were extracted from the feces, purified by cesium chloride-grade centrifugation, and stored in PBS at 4 °C. 10^5 oocysts were orally inoculated into 3-day-old BALB/c mice. Treatment was initiated 3 days post infection. Pyrvinium at doses of 5 mg/kg/day or 12.5 mg/kg/day was administered orally for four or six consecutive days. Paromomycin at a dose of 100 mg/kg/day administered for four or six consecutive days served as the positive control. The number of oocytes in fecal smears and the intensity of trophozoite infection in the ileocecal intestinal regions of treated and untreated mice were compared. A high reduction in the intensity of infection (>90%) in mice treated with pyrvinium was observed in addition to significant reduction in tissue pathology.

A few more murine models developed to study the efficacy of antiprotozoal agents are listed in Table 31.6.

31.7 Murine Models for Antihelminthic Therapeutics

Parasitic, infectious worms are known as helminths. In the developing regions of the world, they are one of the most common causative agents of infectious diseases. The global burden of diseases caused by helminths is substantially high. The disease burden as a consequence of helminths is substantially high globally. Some of the helminths are frequently found to cause infections in humans. Many of the helminth infections are known to have direct and indirect effects on other comorbidities like AIDS and malaria. The transmission of helminths is also seamless and incredibly effortless; it penetrates through the skin and infected eggs are ingested or infected hosts are consumed. This increases the susceptibility range of humans to helminth infections. While children are more susceptible, adults with poor hygiene and immunocompromised individuals are prone to infection. Parasitic worms are known to manipulate the immune system of the host which escalates their importance as pathogens and subsequently makes their eradication and control a daunting task. Expanding large-scale treatment programs in developing countries increases the drug pressure on parasite populations causing the risk of drug resistance. These factors necessitate the development of novel antihelmintic agents to adroitly bring about a reduction in the global burden of infectious diseases caused by helminths. Some murine models developed for determining of the potentiality of antihelminthic agents are described below.

The in vivo antihelminthic efficiency of ethanol extract of the seeds of *Curcubita pepo* was assessed in CBA mice models by Grzbek et al. (2016). CBA mice were infected with a suspension of 100 *H. bakeri* stage three larvae cultured, isolated previously and adjusted to the desired concentration in 0.2 ml distilled water. Twenty-one days post infection, the patent infection period initiates. Animals were then administered via gavage with 2, 4, 6, and 8 g/kg body weight of ethanol extract of *C. pepo*. A decrease in fecal egg counts and adult worm burdens indicated the antihelminthic activity of the extract of *C. pepo*.

The nematicidal efficacy of Cry5B protein from *Bacillus thuringiensis* in 6-weeks-old female Swiss Webster mice (~25 g) was assessed by Hu et al. (2010). One of the best models of chronic intestinal nematode models for rodents, mice infected with *Heligmosomoides bakeri* was used in this study. The larvae of *H. bakeri* were drawn into a pipette tip and transferred into glass test tubes using a blunt-ended syringe for storing until gavage after counting the number of larvae under a microscope. Mice were subjected to *H. bakeri* infection per os with a 100 ± 10 suspension of *H. bakeri* larvae in 0.1 ml of distilled water. Treatment began 15 days post infection. Two Cry5B producing strains of *B. thuringiensis* were used, HD1 and 4Q7. For the HD1 strain, a concentration of 2.5 mg/ml was used and for the 4Q7 strain, a concentration of 2.25 mg/mg was used. Tribendimidine was used as standard to comparatively study the nematicidal activity of the Cry proteins. A single

Table 31.6 Murine models for the development of antiprotozoal therapeutics

S. no.	Disease	Murine model	Protozoan	Route of infection	Challenge dose	Antiprotozoal agent	Treatment	Result	Reference
1	Trypanosomiasis—acute stage STIB900 mouse model	NMRI mice (20–25 g)	*Trypanosoma bruceirhodesiense* strain STIB900	Intraperitoneal injection	5×10^3 STIB900 bloodstream forms	28DAP010, a novel diamidine	Treatment initiated 3 days p.i. Dose—10, 20, 25, 40, 100 mg/kg for 1/5/10 days administered intraperitoneally/per os	High efficacy against African trypanosomes	Wenzler et al. (2014)
2	African Trypanosomiasis	NMRI female mice (20–25 g)	*Trypanosoma brucei gambiense* ITMAP141267, STIB930, 130R and 45Rstrains		10^5 bloodstream forms	DB829, a novel aza-diamidine	Drug administration initiated 3 days p.i. and continued for 4 days	Significant antitrypanosomal activity exhibited	Wenzler et al. (2013)
3	Trypanosoma mouse CNS model	NMRI female mice (20–25 g)	*Trypanosoma brucei brucei*GVR35	Intraperitoneal inoculation	2×10^4 bloodstream forms	Fexinidazole, a 5-nitroimidazole and its principle metabolites sulfoxide and sulfone	Treatment initiated 21–25 days p.i. intraperitoneally or per os Dose: Fexinidazole—20, 25, 50, 100 mg/kg for 5 days 1 or 2/day Fexinidazole sulfoxide and sulfone—50, 100 mg/kg for 5 days 1 or 2/day	Effective second stage curing of trypanosomiasis at the dosage of 200 mg/kg/day for 5 days	Kaiser et al. (2011)
4	Babesiosis	ICR mice 8–12 weeks old (~30 g)	*Babesia rodhaini* Australian strain	Intravenous (tail vein) or intraperitoneal injection	1×10^6 iRBC	Clindamycin phosphate, oxytetracycline, ganazeq	Clindamycin phosphate—50, 100, 200 mg/kg body weight administered through gastric tube, oxytetracycline—100 mg/kg body weight administered subcutaneously, ganazeq—25 mg/kg/day for 3 days	A combination of two of the drugs more potent than single drug alone	Wijaya et al. (2000)

(continued)

Table 31.6 (continued)

S. no.	Disease	Murine model	Protozoan	Route of infection	Challenge dose	Antiprotozoal agent	Treatment	Result	Reference
5	Leishmaniasis	BALB/c mice, 8 weeks old (18–22 g)	*Leishmania amazonensis* strain IFLA/BR/67/PH8 and MHOM/GF/84/CAY-H-142	Subcutaneously in right rear footpad	200 µl containing 1×10^6 amastigotes	Quinoline alkaloids isolated from stem bark of *Galipea longiflora* (2-phenylquinoline and 2-*n*-pentlyquinoline), Glucantime as standard	Treatment initiated 1 day p.i. and continued for 14 days Glucantime—56 mg/kg/day, quinoline alkaloids—100 mg/kg/day	Activity of alkaloids as potent as standard against H-142 strain, but less potent than standard against PH8 strain	Fournet et al. (1994)
6	Cryptosporidiosis	Sprague-Dawley rats (200–250 g) immunosuppressed by administering 25 mg hydrocortisone acetate subcutaneously 2/week	*C. parvum*	Inoculation by oral gavage	10^5 oocysts per animal	Sinefungin produced by *Streptomyces incarnatus*	Dose—0.01–10 mg/kg of body weight/day administered orally	Dose dependent suppression of oocyte shedding, oocyte disappearance from ileal sections. Prevention of infection if administered prior to/on the day of challenge	Brasseur et al. (1993)

dose of Cry5B protein from either of the bacterial strain when administered was found to cause ~98% reduction in egg production by the parasite and an ~70% reduction in the worm burden when administered per os at ~700 nmoles/kg. The activity of standard drug tribendimidine at a dose of 2.2 μmoles/kg was found to be similar to that of Cry5B protein at a dose of 700 nmoles/kg. This suggests the potential activity of Cry5B as a suitable antihelminth agent for humans.

The parasitic nematode *Angiostrongylus cantonensis* is one of the most common causes of eosinophilic meningitis and eosinophilic meningoencephalitis. The mature adults live in pulmonary arteries of rats. The efficacy of albendazole in conjunction with the extract of *Artemisia capillaries*, traditionally used in Chinese Herbal Medicine was evaluated by Lai (2006) in 5-week-old female BALB/c mice (20–25 g). The infective larvae of *A. cantonensis* was collected from wild giant African snails and propagated in the lab by cycling through rats and *Biomphalaria glabrata* snails. For retrieval, the shells of snails were crushed, tissues homogenized, digested in pepsin–HCl, and incubated at 37 °C in a water bath for 2 h with agitation. The sample was then centrifuged at 1400 g for 10 min to remove host cellular debris. The larval sediment was collected by serial washing in double distilled water and counted under a microscope. Treatment was initiated 2, 4, 6, 8, 10, 12, and 14 post infection for 7 days continuously. Albendazole at a dose of 10 mg/kg/day was administered alone or in combination with *A. capillaries* extract at a dose of 100 mg/kg/day. Worm recovery, a histopathological score of the fourth ventricle, tissue-type plasminogen activator, urokinase-type plasminogen activator, matrix metalloproteinase-9, cerebrospinal fluid total protein, leukocyte counts and pro-inflammatory cytokines were assessed to be evaluated as therapeutic parameter indicators. Albendazole used in conjunction with *A. capillaries* extract was found to reduce all the above parameters.

Deori and Yadav (2016) evaluated the antihelmintic activity of the crude methanol extract (1.95% *w*/w) of the stem bark of *Oroxylum indicum* on juvenile and adult stages of *Hymenolepis diminuta* in albino Wistar rats (200 g). *H. diminuta* was maintained in the laboratory in alternating hosts, Wistar arts and flour beetles. Four weeks after beetles are fed with gravid proglottids of *H. diminuta*, they were dissected to collect cysticercoids which were used to infect rats for antihelmintic study. Rats were inoculated orally with four cysticercoids each. For testing the activity against the juvenile stage of *H. diminuta,* treatment was given on days 3–7 post infection and to test the efficacy against adult stage of the parasite, treatment was given on days 21–25 post infection with cysticercoids. The animals were administered with 250, 500, and 1000 mg/kg body weight of *O. indicum* extract or with a single dose 10 mg/kg body weight of the reference standard praziquantel. The eggs per gram count in faces and the worm count was assessed to estimate the activity of the standard or plant extract. The extract was more potent against juveniles when compared to adults. At a dose of 1000 mg/kg body weight, the plant extract exhibited substantial reduction in worm count and eggs per gram faces.

The efficacy of albendazole to treat meningitis caused by *Angiostrongylus cantonensis* was studied in BALB/c mice infected with the third larval stage of this nematode by Lan et al. (2004). The infective larval stage was collected from *Achatina fulica* (giant African snails) in southern Taiwan. Three-week-old BALB/c mice were orally inoculated with 60 larvae/animal. Treatment was initiated 5 days post infection. Mice were administered with albendazole at a dose of 10 mg/kg via a stomach tube for 7 or 14 consequent days. Pronounced larvicidal activity and reduced gelatinolytic activity was observed in the mice treated with albendazole.

Trichuris trichiura is one of the most ubiquitous gastrointestinal helminths of man that causes trichuriasis or whipworm infection. A mouse model of trichuriasis was produced by Else et al. (1988). Six to eight-week-old male C57/BL/10/ScSc/Ola (B10), B10.BR/Ola, DBA/2/Ola, and BALB/c/Ola mice were infected with 400 *T. muris* eggs. Methyridine is an antihelmintic drug that causes expulsion of all worms within hours of administration. The infected mice were subcutaneously administered with methyridine at a dose of 500 mg/kg body weight. The use of different strains of mice facilitated the study of the predisposition of mice to trichuriasis infection.

Some murine models for helminth infections that were developed are represented in Table 31.7. A few murine models developed to study the potency of drug candidates as antihelmintic agents are listed in Table 31.8.

31.8 Conclusion

The use of murine models to simulate the pathogenesis of an infectious disease in pursuit of the discovery of an appropriate anti-infectious agent in a suitable host system has been in vogue for the past few decades. The ability to uniquely manipulate their genome to render them resistant or susceptible to a particular pathogen further increased the versatility of the use of murine models, making them the powerhouse for biomedical research. Mouse and rat models with specific bacterial sepsis and inflammation, fungal disease conditions like candidiasis, aspergillosis, viral

Table 31.7 Murine models for helminth infections (Murthy et al. 2011)

S. no.	Disease	Animal model	Helminth	Infective agent/ route	Outcome
1	Filarial infection	Cotton rat	*Litomosoides carinii*	Larvae (stage3)/ subcutaneous	Discovery of diethylcarbamazine
2	Nonhuman filariasis	CD1 and BALB/c mice	*Brugiapahangi*	–	–
3	Human filariasis	BALB/c mice	*B. malayi*	Larvae (stage 3) or adult worms/ intraperitoneal	–
4	Ascariasis	Mouse	*Ascaris suis*	Eggs/per os	–
5	Taeniasis	Mouse	*Hymenolepis nana*	Eggs/per os	–
6	Schistosomiasis	Mouse	*Schistosoma mansoni, S. japonicum*	Cercariae/skin	–

Table 31.8 Murine models for the development of antihelminth therapeutics

S. no.	Disease	Murine model	Helminth	Route of infection	Challenge dose	Antihelmintic agent	Treatment	Result	Reference
1	Roundworm infection	Albino Wistar mice (12.50–27.80 g)	*Heligmosomoides polygyrus*	Via esophageal intubation	250 L3 stage of *H. polygyrus*	*Securidaca longepedunculata* crude extract in 70% aqueous methanol and pyrantel embonate as standard	Treatment administered from 18–21 days p.i. *S. longepedunculata*—500, 1000, 2000 mg/kg body weight Pyrantel embonate—100 mg/kg body weight	While *S. longepedunculate* exhibited in vitro larvicidal activity, the in vivo result in terms of number of worm eggs per gram in fecal matter and total worm burden was not significant. Standard exhibited high efficiency in vivo	Adeile et al. (2013)
2	Taeniasis	Wistar rats (100–120 g)	*Hymenolepis diminuta*	Oral infection	4 cysticercoids	5.71% crude methanol extract of *Clerodendrum colebrookianum* and praziquantel as standard	Dose: Praziquantel—5 mg/kg, administered per os Plant extract—200, 400, 800 mg/kg body weight Treatment for larval stage—on days 2–6 p.i., treatment for immature stage—on days 8–12 p.i., treatment of adult stage—on days 21–25 p.i.	Dose dependent efficacy against larvae, immature and adult stages of *H. dimunuta*. Standard exhibited higher activity than plant extract	Yadav and Nil (2012)
3	Trichinosis	NIH female mice (24 g)	*Trichinella spiralis* ISS3 strain	Oral inoculation	25, 50, 200 infective larvae (stage 1)/mouse	*Lactobacillus casei* ATCC7469 strain	10^9 CFU administered intraperitoneally once a week for 6 weeks, prior to larval infection	At the lower doses of infection (25 and 50 *T. spiralis* infective larvae), AWI reduction upto 100% was observed on days 4 and 10 p.i. and LPG at day 30 p.i. reduced by 100%. At higher dose of infection (200 *T. spiralis* infective larvae), AWI reduced to 78.6% and 76.7%, 4 and 10 days p.i., resp. and 80.9% reduction of LPG	Gómez et al. (2011)

(continued)

Table 31.8 (continued)

S. no.	Disease	Murine model	Helminth	Route of infection	Challenge dose	Antihelmintic agent	Treatment	Result	Reference
4	Tapeworm infection	Wistar rats (100–120 g)	*Hymenolepis diminuta*	Oral inoculation	5 cysticercoids	7.56% crude methanol extract of ripe fruit extract of *Solanum myriacanthum* and praziquantel as standard	200, 400, 800 mg/kg body weight of *S. myriacanthum* or 5 mg/kg of praziquantel (per os) administered on days 2–4/8–10/21–23 p.i.	Dose-dependent worm count and eggs per gram observed on treatment with *S. myriacanthum* extract.	Yadav and Tangpu (2012)
5	Trichinosis	Sprague-Dawley labial rats (175–200 g)	*Trichinella spiralis*	Oral inoculation	3000 larvae (stage 1) of *T. spiralis*	Crude methanol extract (90%) of fruits of *Balanites aegyptiaca* and albendazole as standard	*B. aegyptiaca* extract—1000 mg/kg body weight administered orally, albendazole—10 mg/kg body weight. Treatment initiated 24 h/15 days/30 days p.i. and continued for 5 successive days	Marked reduction in migration and encysted larval rate in the muscular tissue in *B. aegyptiaca*-extract treated mice when compared to treatment with albendazole. Treatment with albendazole resulted in significant eradication of adult worms; *B. aegyptiaca* extract was less effective against adult worms	Shalaby et al. (2010)
6	Trichostrongylid parasite infection	Wistar rats (112–182 g)	*Nippostrongylus brasiliensis*	Subcutaneous inoculation in the cervical region	200 viable L3 of *N. brasiliensis*	Crude methanol extract of *Xylopia aethiopica*	Treatment given for 3 days, from the 5th day p.i. Dose—0.8, 1.0, 1.2, 1.4, 1.7 and 2.0 g/kg of *X. Aethiopica* extract	Significant deparasitization rates observed in the rats treated with the plant extract	Suleiman et al. (2005)

diseases like Ebola and HIV, protozoal infections like giardiasis, trypanosomiasis and helminth infections like trichuriasis and taeniasis have all been created and studied for the development of specific antibiotics. A better understanding of the intricacies of the pathogenesis of a disease, the immune response and the various immune cells and immune cell-mediated compounds produced and their mode of action can be obtained by using murine models of disease. The use of murine models of disease also facilitates the progressive study of the mechanism of action of anti-infective agents, the cells and the immune responses and inflammatory processes on which they exert their influence in the host throughout the course of the treatment. All this make murine models an indispensable tool in the study for the development of novel anti-infective therapeutics. However, there is a possibility that a murine model developed might not predict the human response to a particular drug molecule under study. Care should be taken to study the literature well on murine models established prior the study. Statistical analysis should be applied to facilitate a more accurate conclusion of study. A systematic and progressive analysis of all possible parameters throughout the course of the disease model can minimize the scope for errors and unreliable results, thereby increasing the scientific validity of the study.

References

Adeile RC, Fakae BB, Isuzu IU (2013) Antihelmintic activity of *Securidaca longepedunculata* (Family: Polygalaceae) root extract in mice, *in vitro* and *in vivo*. Asian Pac J Trop Med 6:841–846. https://doi.org/10.1016/S1995-7645(13)60150-9

Barlow PG, Svoboda P, Mackellar A et al (2011) Antiviral activity and increased host defense against influenza infection elicited by the human cathelicidin LL-37. PLoS One 6:e25333. https://doi.org/10.1371/journal.pone.0025333

Biswas S, Jennens L, Field HJ (2007) The helicase primase inhibitor, BAY 57-1293 shows potent therapeutic antiviral activity superior to famciclovir in BALB/c mice infected with herpes simplex virus. Antivir Res 75:30–35. https://doi.org/10.1016/j.antiviral.2006.11.006

Brasseur P, Lemeteil D, Ballet JJ (1993) Curative and preventive anticryptosporidium properties of sinefungin in an immunosuppressed adult rat model. Antimicrob Agents Chemother 37:889–892. https://doi.org/10.1128/aac.37.4.889

Brunetti J, Falciani C, Roscia G et al (2016) *In vitro* and *in vivo* efficacy, toxicity, bio-distribution and resistance selection of a novel antibacterial drug candidate. Sci Rep 6:26077. https://doi.org/10.1038/srep26077

Calvo E, Pastor FJ, Mayayo E et al (2011) Antifungal therapy in an athymic murine model of chromoblastomycosis by *Fonsecaea pedrosoi*. Antimicrob Agents Chemother 55:3709–3713. https://doi.org/10.1128/AAC.01662-10

Calvo E, Pastor FJ, Rodríguez MM et al (2010) Antifungal therapy in a murine model of disseminated infection by *Cryptococcus gattii*. Antimicrob Agents Chemother 54:4047–4077. https://doi.org/10.1128/AAC.00172-10

Caroline AL, Powell DS, Bethel LM et al (2014) Broad spectrum antiviral activity of favipiravir (T-705): protection form highly lethal inhalational Rift Valley Fever. PLoS Negl Trop Dis 8:e2790. https://doi.org/10.1371/journal.pntd.0002790

Channabasappa S, Durgaiah M, Chkkamadaiah R et al (2017) Efficacy of novel antistaphylococcal ectolysinon P128 in a rat model of methicillin-resistant *Staphylococcus aureus* bacteremia. Antimicrob Agents Chemother 62:e01385–e01317. https://doi.org/10.1128/AAC.01358-17

Charles River Laboratories. https://www.criver.com/products-services/find-model
Cigana C, Bernardini F, Facchini M et al (2016) Efficacy of the novel antibiotic POL7001 in preclinical models of *Pseudomonas aeruginosa* pneumonia. Antimicrob Agents Chemother 60:4991–5000. https://doi.org/10.1128/AAC.00390-16
Corrado A, Donato P, Maccari S et al (2016) *Staphylococcus aureus*-dependent septic arthritis in murine knee joints: local immune response and beneficial effects of vaccination. Sci Rep 6:38043. https://doi.org/10.1038/srep38043
Cowen LE, Singh SD, Köhler JR et al (2008) Harnessing Hsp90 function as a powerful, broadly effective therapeutic strategy for fungal infectious disease. Proc Natl Acad Sci U S A 106:2818–2823. https://doi.org/10.1073/pnas.0813394106
Dandri M, Burda MR, Zuckerman DM et al (2005) Chronic infection with hepatitis B viruses and antiviral drug evaluation in uPA mice after liver repopulation with tupaia hepatocytes. J Hepatol 42:84–60. https://doi.org/10.1016/j.jhep.2004.09.021
Dannaoui E, Mouton JW, Meis JFGM et al (2002) Efficacy of antifungal therapy in a nonneutropenic murine model of zygomycosis. Antimicrob Agents Chemother 46:1953–1959. https://doi.org/10.1128/AAC.46.6.1953-1959.2002
Dennis CG, Greco WR, Brun Y (2006) Effect of amphotericin B and micafungin combination on survival, histopathology and fungal burden in experimental aspergillosis in the p47$^{phox-/-}$ mouse model of chronic granulomatus disease. Antimicrob Agents Chemother 50:422–427. https://doi.org/10.1128/AAC.50.2.422-427.2006
Deori K, Yadav AK (2016) Antihelmintic effects of *Oroxylum indicum* stem bark extract on juvenile and adult stages of *Hymenolepis diminuta* (Cestoda), an *in vitro* and *in vivo* study. Parasitol Res 115:1275–1285. https://doi.org/10.1007/s00436-015-4864-6
Deslouches B, Gonzalez IA, DeAlmeida D et al (2007) *De novo*-derived cationic antimicrobial peptide activity in a murine model of *Pseudomonas aeruginosa* bacteraemia. J Antimicrob Chemother 60:669–672. https://doi.org/10.1093/jac/dkm/253
Downey AS, Chong CR, Graczyk TK, Sullivan DJ (2008) Efficacy of pyrvinium pamoate against *Cryptosporidium parvum* infection *in vitro* and in a neonatal mouse model. Antimicrob Agents Chemother 52:3106–3112. https://doi.org/10.1128/AAC.00207-08
Else KJ, Wakelin D, Raoch TIA (1988) Host predisposition to trichuriasis: the mouse *T. muris* model. Parasitology 98:275–282. https://doi.org/10.1017/S0031182000062193
Envigo. https://www.envigo.com/products-services/research-models-services/models/. Accessed 30 May 2019
Fournet A, Barrios AA, Muñoz V et al (1994) Antiprotozoal activity of quinoline alkaloids isolated from *Galipea longiflora,* a Bolivian plant used as a treatment for cutaneous leishmaniasis. Phytother Res 8:174–178. https://doi.org/10.1002/ptr2650080312
Goffinet C, Allespach I, Keppler OT (2007) HIV-susceptible transgenic rats allow rapid preclinical testing of antiviral compounds targeting virus entry or reverse transcription. Proc Natl Acad Sci 104:1015–1020. https://doi.org/10.1073/pnas.0607414104
Gómez FM, Castro BEF, Garfias CRB (2011) The intraperitoneal inoculation of *Lactobacillus casei* in mice induces total protection against *Trichinella spiralis* infection at low challenge doses. Parasitol Res 109:1607–1617. https://doi.org/10.1007/s00436-011-2432-2
Griffith DC, Harford L, Williams R et al (2003) *In vivo* antibacterial activity of RWJ-54428 a new cephalosporin with activity against gram-positive bacteria. Antimicrob Agents Chemother 47:43–47. https://doi.org/10.1128/AAC.47.1.43–47.2003
Grzbek M, Koch WK, Strachecka A et al (2016) Evaluation of antihelminthtic activity and composition of pumpkin seeds (*Cucurbita pepo* L.) seed extracts- *in vitro* and *in vivo* studies. Int J Mol Sci 17:1456. https://doi.org/10.33990/ijms17091456
Hachem R, Bahna P, Hanna H et al (2006) EDTA as an adjunct antifungal agent for invasive pulmonary aspergillosis in a rodent model. Antimicrob Agents Chemother 50:1823–1827. https://doi.org/10.1128/AAC.50.5.1823-1827.2006
Han G, Martinez LR, Mihu HR et al (2009) Nitric oxide releasing nanoparticles are therapeutic for *Staphylococcus aureus* abscess in a murine model of infection. PLoS One 4:e7804–e7813. https://doi.org/10.1371/journal.pone.0007804

Hata K, Horii T, Miyazaki M et al (2011) Efficacy of oral E1210, a new broad-spectrum antifungal with a novel mechanism of action, in murine models of candidiasis, aspergillosis and fusariosis. Antimicrob Agents Chemother 55:4543–4551. https://doi.org/10.1128/AAC.00366-11

Hu Y, Georghiou SB, Kelleher AJ, Aroian RV (2010) *Bacillus thuringiensis* Cry5 protein is highly efficacious as a single-dose therapy against an intestinal roundworm infection in mice. PLoS Negl Trop Dis 43:e614. https://doi.org/10.1371/journal.pntd.0000614

Kaiser M, Bray MA, Cal M et al (2011) Antitrypanosomal activity of fexinidazole, a new oral nitroimidazole drug candidate for the treatment of sleeping sickness. Antimicrob Agents Chemother 55:5602–5608. https://doi.org/10.1128/AAC.00246-11

Kern ER, Hartline CB, Rynbak RJ et al (2004) Activities of benzimidazole-D and L-ribonucleosides in animal models of cytomegalovirus infections. Antimicrob Agents Chemother 48:1749–1755. https://doi.org/10.1128/AAC.48.5.1749–1755.2004

Kucharíková S, Tournu H, Holtappels M et al (2010) *In vivo* efficacy of anidulafungin against mature *Candida albicans* biofilms in a novel rat model of catheter-associated Candadiasis. Antimicrob Agents Chemother 54:4474–4475. https://doi.org/10.1128/AAC.00697-10

Lan KP, Wang CJ, Lai SC et al (2004) The efficacy of therapy with albendazole in mice with parasitic meningitis caused by *Angiostrongylus cantonensis*. Parasitol Res 93:311–317. https://doi.org/10.1007/s00436-004-1105-09

Lai SC (2006) Chinese herbal medicine Yin-Chen-extract as an adjunct to antihelmentic albendazole used against *Angiostrongylus cantonensis*-induced eosinophilic meningitis or meningoencephalitis. Am J Trop Med Hyg 75:556–562. https://doi.org/10.4269/ajtmh.2006.75.556

Luque JC, Clemons KV, Stevens DA (2003) Efficacy of micafungin alone or in combination against systemic murine aspergillosis. Antimicrob Agents Chemother 47:1452–1455. https://doi.org/10.1128/AAC.47.4.1452-1455.2003

Marango SN, Wandabwa CK, Makwali JA et al (2017) Experimental therapeutic assays of *Tephrosia vogelii* against *Leishmania major* infection in murine model: *in vitro* and *in vivo*. BMC Res Notes 10:698. https://doi.org/10.1186/s13104-017-3022-x

Marshall Bioresources. https://www.marshallbio.com/mice

Martinez A, Ferrer S, Santos I et al (2001) Antifungal activities of two new azasordarins, GW471552 and GW471558, in experimental models of oral and vulvovaginal candidiasis in immunosuppresed rats. Antimicrob Agents Chemother 45:3304–3309. https://doi.org/10.1128/AAC.45.12.3304-3309

Murphy TM, Deitz JM, Petersen PJ et al (2000) Theraeutic efficacy of GAR-936, a novel glycylcycline, in a rat model of experimental endocarditis. Antimicrob Agents Chemother 4:3022–3027. https://doi.org/10.1128/aac.44.11.3022-3027.2000

Murthy PK, Joseph SK, Murthy PSR (2011) Plant products in the treatment and control of filariasis and other helminth infections ad assay systems for antifilarial/antihelminthic activity. Planta Med 77:647–661. https://doi.org/10.1055/s-0030-1250452

Oestereich L, Lüdtke A, Wurr S et al (2014) Successful treatment of Ebola virus infection with T-705 (favipiravir) in a small animal model. Antivir Res 105:17–21. https://doi.org/10.1016/j.antiviral.2014.02.014

Olszewska W, Ispas G, Schnoeller C et al (2011) Antiviral and lung protective activity of a novel respiratory syncytial virus fusion inhibitor in a mouse model. Eur Respir J 38:401–408. https://doi.org/10.1183/09031936.00005610

Pegoraro S, Duffey M, Otto TD et al (2017) SCSC83288 is a clinical development candidate for the treatment of severe malaria. Nat Commun 8:14193. https://doi.org/10.1083/ncomms14193

Perwitasari O, Johnson S, Yan X et al (2016) Antiviral efficacy of verdinexor *in vivo* in two animal models of influenza A virus infection. PLoS One 11:e0.167221. https://doi.org/10.1371/journal.pone.0167221

Rodriguez CA, Agudelo M, Gonzalez JM et al (2015) An optimized mouse thigh infection model for enterococci and its impact on antimicrobial pharmacodynamics. Antimicrob Agents Chemother 59:233–238. https://doi.org/10.1128/AAC.02352-13

Rodríguez MM, Calvo E, Serena C et al (2009) Effects of double and triple combinations of antifungal drugs in a murine model of disseminated infection by *Scedosporium prolificans*. Antimicrob Agents Chemother 53:2153–2155. https://doi.org/10.1128/AAC.01477-08

Sarkar A, Kumar KA, Dutta NK et al (2003) Evaluation of *in vitro* and *in vivo* antibacterial activity of dobutamine hydrochloride. Indian J Med Microbiol 21:172–178

Scholl D, Martin DW (2008) Antibacterial activity of R-type pyocins towards *Pseudomonas aeruginosa* in murine peritonitis model. Antimicrob Agents Chemother 52:167–1652. https://doi.org/10.1128/AAC.01479-07

Schul W, Liu W, Xu HY et al (2007) A dengue fever viremia model in mice shows reduction in viral replication and suppression of the inflammatory response after treatment with antiviral drugs. J Infect Dis 195:665–674. https://doi.org/10.1086/511310

Shalaby MA, Moghazy FM, Shalaby HA, Nasr SM (2010) Effect of methanolic extract of *Balanites aegyptiaca* fruits on enteral and parenteral stages of *Trichinella spiralis* in rats. Parasitol Res 107:17–25. https://doi.org/10.1007/s00436-010-1827-9

Sharma L, Shukla G (2014) Treatment of pregnant BALB/c mice with sulphadoxine pyrimethamine or chloroquine abrogates *Plasmodium berghei*-induced placental pathology. Parasitol Int 63:49–56. https://doi.org/10.1016/j.parint.2013.08.016

Shukla G, Kaur H, Sharma L (2013) Comparative therapeutic effect of probiotic *Lactobacillus casei* alone and in conjunction with antiprotozoal drugs in giardiasis. Parasitol Res 112:2143–2149. https://doi.org/10.1007/s00436-0133394-3

Silva ON, Núñez CF, Haney EF et al (2016) An anti-infective synthetic peptide with dual antimicrobial and immunomodulatory properties. Sci Rep 6:35465. https://doi.org/10.1038/srep35465

Stabenow J, Buller RM, Schriewer J et al (2010) A mouse model of lethal infection for evaluating prophylactics and therapeutics against Monkeypox virus. J Virol 84:3909–3920. https://doi.org/10.1128/JVI.0201-09

Sugar AM, Liu XP (2000) Combination antifungal therapy in treatment of murine pulmonary mucormycosis: roles of quinolones and azoles. Antimicrob Agents Chemother 44:2004–2006. https://doi.org/10.1128/AAC.44.7.2004-2006.2000

Suleiman MM, Mammam M, Aliu YO, Ajanusi JO (2005) Antihelmintic activity of the crude methanol extract of *Xylopia aethiopica* against *Nippostrongylus brasiliensis* in rats. Veterinarski Arhiv 75:487–495

Taconic Biosciences. https://www.taconic.com/find-your-model/ Accessed 30 May 2019

Tavares PM, Thevissen K, Cammue BPA et al (2008) *In vitro* activity of the antifungal plant defensin RsAFP2 against *Candida* isolates and its *in vivo* efficacy in prophylactic murine models of candidiasis. Antimicrob Agents Chemother 52:4522–4525. https://doi.org/10.1128/AAC.00448-08

Titova KA, Sergeev AA, Zamedyanskaya AS et al (2015) Using ICR and SCID mice as animal models for small pox to assess antiviral drug efficacy. J Gen Virol 96:2832–2843. https://doi.org/10.1099/vir.0.000216

The Jackson Laboratory. https://www.jax.org/jax-mice-and-services/find-and-order-jax-mice

Wang YX, Cunha V, Vincelette J et al (2007) Antiviral and myocyte protective effects of murine interferon $-\beta$ and α_2 in coxsackievirus B-3 induced myocarditis and epicarditis in BALB/c mice. Am J Physiol Heart Circ Physiol 293:H69–H76. https://doi.org/10.1152/ajpheart.00154.2007

Warren TK, Warfield KL, Wells J et al (2010) Antiviral activity of a small-molecule inhibitor of filovirus infection. Antimicrob Agents Chemother 54(5):2152–2159. https://doi.org/10.1128/AAC.01315-09

Wenzler T, Yang S, Patrick DA et al (2014) *In vitro* and *in vivo* evaluation of 28DAP010, a novel diamidine for treatment of second-stage African sleeping sickness. Antimicrob Agents Chemother 58:4452–4463. https://doi.org/10.1128/AAC.02309-13

Wenzler T, Yang S, Braissant O et al (2013) Pharmacokinetics, *Trypanosoma brucei gambiense* efficacy and time of drug action of DB829, a preclinical candidate for treatment of second-stage human African trypanosomiasis. Antimicrob Agents Chemother 57:5330–5343. https://doi.org/10.1128/AAC.00398-13

Wijaya A, Wulansari R, Ano H et al (2000) Therapeutic effect of clidamycin and tetracycline on *Babesia rodhaini* infectionin mouse model. Parasitology 62:835–839. https://doi.org/10.1292/jvms.62.835

Wijngaerden EV, Peetermans WE, Vandersmissen J et al (1999) Foreign body infection: a new rat model for prophylaxis and treatment. J Antimicrob Chemother 44:669–674. https://doi.org/10.1093/jac/44.5669

Yadav AK, Nil T (2012) *In vivo* antihelmintic activity of *Clerodendrum colebrookianum* Wapl., a traditionally used taenicidal plant in Northeast India. Parasitol Res 111:1841–1846. https://doi.org/10.1007/s00436-012-2908-8

Yadav AK, Tangpu V (2012) Antihelmintic activity of ripe fruit extract of *Solanum myriacanthum* Dunal (Solanaceae) against experimentally induced *Hymenolepis diminuta* (Cestoda) infections in rats. Parasitol Res 110:1047–1053. https://doi.org/10.1007/s00436-011-2596-9

Yan Y, Zhao Z, Wan H et al (2014) A novel fungus concentration-dependent rat model for acute invasive fungal rhinosinusitis: an experimental study. BMC Infect Dis 14:3856. https://doi.org/10.1186/s12879-014-0713-y

Yao JM, Zhang HB, Liu CS et al (2015) Inhibitory effects of 19 antiprotozoal drugs and antibiotics on *Babesia microti* infection in BALB/c mice. J Infect Dev Ctries 9:1004–1010. https://doi.org/10.3855/jidc.5500

You XF, Li CR, Yang XY et al (2009) *In vivo* antibacterial activity of Vertilmicin, a new aminoglycoside antibiotic. Antimicrob Agents Chemother 53:4525–4528. https://doi.org/10.1128/AAC.00223-09

Nonhuman Primate Models for Antimicrobial Drug Discovery

32

Anushka Khasnobish and Hidetoshi Morita

Contents

Abstract

Animal models are an indispensable part of biomedical research and drug discovery. Relating the research findings using small animals (like murine models) to the human is difficult given the difference between in vivo physiological and kinetic conditions. Thus, nonhuman primates (NHPs) are better candidates for biomedical research in understanding pathogenesis of various diseases and this would in turn expedite the formulation of drug discovery strategies. For including the NHPs in a research, proper care must be taken for selection, housing, maintenance, and handling of the models. A debate concerning the usage of the NHP models in biomedical research is rising in the world research community and is supported by ethical measures. This chapter summarizes the current trend in use of NHP models in antimicrobial drug discovery, the knowledge to be

A. Khasnobish · H. Morita (✉)
Laboratory of Animal Applied Microbiology, Graduate School of Environmental and Life Science, Okayama University, Okayama, Japan
e-mail: hidetoshi-morita@okayama-u.ac.jp

B. Siddhardha et al. (eds.), *Model Organisms for Microbial Pathogenesis, Biofilm Formation and Antimicrobial Drug Discovery*,
https://doi.org/10.1007/978-981-15-1695-5_32

acquired for proper handling of the NHPs, and also explores briefly the ethical aspects of such usage.

Keywords
Nonhuman primate · Antimicrobial drug discovery strategy · Ethics in animal handling

32.1 Introduction

Biomedical research can be defined as the area of research governing biological processes and disease pathogenesis and experts like doctors (including veterinarians), scientists, computer engineers, statisticians, and technicians are involved at various stages of the research. This type of research has various stages of development—basic research, applied research, in vitro research, ex vivo research, and in vivo research (California Biomedical Research Association Factsheet—What is Biomedical Research? 2018). Preclinical trials and clinical trials fall under the category of *in vivo research.* In the medical and biomedical field of research, preclinical trials is the stage at which the findings from basic, applied, in vitro, and ex vivo research activities are accumulated to prepare a research design that can be first tested on animal models (hence in vivo) and if the results are favorable to the hypothesis, the research design is then extended toward clinical studies. In case of drug discovery and development, the preclinical trials can assess different aspects of the drug such as effects of drug on the body of the organism (pharmacodynamics), countereffect of the body functions on the drug (pharmacokinetics), or toxicology test. It is required by ethical committees across the globe that all the preclinical trials must conduct good laboratory practices (GLP) test.

The animal models chosen for every research study are species specific, which implies that not all type of animals can be used for any research study. One key step in the animal model selection for research design and methodology of preclinical trials is the inclusion of the 3 *R*s—*R*eplacement, *R*eduction, and *R*efinement. These is the widely accepted principles in various animal handling laws, guidelines, ethical doctrines and was first proposed by William Russell and Rex Burch (Franco et al. 2018; Russell and Burch 1959). These 3Rs can be briefly described as follows:

- Replacement stands for avoiding use of animals in research studies by using alternative techniques such as in vitro models or use of lower species animals only when no other alternatives are available.
- Refinement is associated with animal handling and housing, and also with optimizing experimental protocols to minimize suffering of the animals.
- Reduction is seen as budgeting branch of 3R approach where the study design is modulated in a way that required information is retrieved with inclusion of

minimal number of animals in the experiments (Animal Welfare and the Three Rs: Replacement, Refinement and Reduction (Leaflet) 2016).

In the biomedical research, researchers are often recommended to focus on the "replacement" segment of the 3R approach. But the decision at the preclinical stage of biomedical research is usually not whether to choose an animal model or not; instead the question at this stage is which species of animal is appropriate for providing conclusive results from the study (Wood and Hart 2007). An informed decision to identify the best animal model for the specific study not just follows through the replacement strategy but also reduces the need for inclusion of more than a fixed number of animal experiments and thus refines the final outcome by providing a detailed outlook from a set number of animals while abiding by animal welfare practices (Wood and Hart 2007). Therefore, it is clear that proper practice of the inclusion of the 3Rs in strategizing preclinical studies leads to conservation of animal life as the preclinical trials involve animal models like rodents and NHPs.

32.2 What Are Nonhuman Primate Models?

A researcher arrives on a decision to work with model organism in biomedical research after reviewing basic biology literature, availability of funds, staffs and machinery, and ethical committee rulings. However, despite possessing higher number of technical, ethical, and financial challenges, researchers often advocate for the use of NHPs in the disease biology advancement.

Let us try to understand what are the NHP models and why they are considered as important element of biomedical research. According to 9 CFR 1.1 in US Law, NHPs are defined as "any non-human member of the highest order of mammals including prosimians, monkeys, and apes." Thus, among the animal models that are used in the preclinical trials, NHP models are the closest relatives of humans in the evolutionary tree. As a part of biomedical research, NHP model species including the genera *Macaca, Pan (Homo), Papio, Aotus, Callithric, Saimiri, Chlorocebus, Saquinus, and Cercocebus* have been developed and are used as animal models (Anderson 2008). Broadly classified, the NHPs in the biomedical research belong to New World species like marmosets (such as *Callithrix jaccus*) or Old World species like crab-eating macaque or rhesus macaque or baboons (though baboons are very rare in the research scene) (Chatfield and Morton 2018).

32.3 NHP Models and Biomedical Research

With recent stringent legislations regarding the animal research, there has been a drastic drop in animal use, especially NHPs in research environment in the European Union (Chatfield and Morton 2018; EU 2010). However, this situation has led to the increased overseas collaboration with laboratories and research center that have infrastructure and legislation that allows animal use in research. While the use of

NHPs have dropped in the EU, the rest of the world has seen an exponential hike in the percentage. According to the data released by the United States Department of Agriculture, until September 2017, 9.57% of total animals housed in animal research facilities were NHPs. The animal research facilities included in this report were involved in regulated activities that included activities that did not involve pain and did not administer pain-relieving drugs; activities that involving pain and administered pain-relieving drugs; and activities involving pain or distress and for which administering pain-relieving drugs would adversely affect results. Though it is difficult to get an exact number of animal research studies worldwide, but it is somewhere close to exceeding 100,000 experiments each year that involves NHPs at some point of the study protocol (Chatfield and Morton 2018). The argument in favor of inclusion of NHPs in the research is the genetic similarity between human and NHPs which makes them the best candidates for preclinical drug testing and for exploring disease pathogenesis as well as healthy functioning of the neurological system, among other sections of biomedical research fields (Chatfield and Morton 2018; Wong 2014). Now that we have seen the rising percentage of NHP models in the research community, let us explore infectious disease research which utilize NHP models and understand why NHPs are important for this research theme.

32.3.1 Infectious Diseases and NHPs

According to WHO, malady caused by pathogenic microorganisms like bacteria, fungi, parasites, or viruses are defined as infectious diseases and these diseases can be transmitted directly or indirectly from one person to another. Due to the transmissibility feature of these diseases, a large number of populations gets affected upon the onset of these diseases and this impacts the public health as well as overall GDP of the country. According to WHO Global Health Estimates 2016, 14.8% of the total deaths were contributed to infectious diseases like tuberculosis, STDs including Human Immunodeficiency Virus/Acquired Immunodeficiency syndrome (HIV/AIDS), diarrhoeal diseases, measles, whooping cough, meningitis, encephalitis, hepatitis, malaria, and dengue. Among these, tuberculosis, HIV/AIDS, and malaria are amongst the highest scorers in the death toll chart. However, the percentage of total death by tuberculosis, HIV/AIDS, and malaria has decreased from 3.2%, 2.8%, 1.5% in 2006 to 2.3%, 1.8%, 0.8% in 2016. Amongst the various reasons that has led to this decrease in death toll due to these diseases, dynamic and robust research studies to understand the disease pathogenicity and immune response of the host, which led to optimizing health care, is one of the probable reasons. This can be understood if we seek the ongoing animal research in the disease-associated fields of research.

32.3.1.1 HIV/AIDS

Despite the drop in the percentage of total deaths caused by the HIV/AIDS globally, according to WHO global fact sheets, as of 2017, 940,000 people died and 36.9 million people are living with HIV/AIDS. In HIV/AIDS research the contribution and

Table 32.1 Glossary of HIV/AIDS

Terms to remember	
Human immunodeficiency virus	Human immunodeficiency virus or HIV is a class of retroviruses (viruses whose genetic material is Ribonucleic acid, RNA). HIV is responsible for causing acquired immunodeficiency syndrome (AIDS) by attacking $CD4^+$ T cells which are primary soldiers of the body's acquired immune system. HIV-1 is the globally distributed virulent strain in this category which is believed to be first transmitted from Chimpanzee and HIV-2 is closely related to the simian immunodeficiency virus isolated from macaques and mangabeys (Weatherall et al. 2006).
Simian immunodeficiency virus	Simian immunodeficiency virus or SIV belongs to the same class of retroviruses as HIV and causes AIDs-like manifestation in nonhuman primates.
Simian human immunodeficiency virus	Simian human immunodeficiency virus or SHIV is a chimeric virus which was constructed by combining genetic material from SIV and HIV-1. Due to the combinatorial features, SHIV has broad spectrum of hosts and acts as a pathogen to all the NHPs.
Antiretroviral therapy	According to WHO, treatment module that consists of at least three antiretroviral (ARV) drugs in order to stop progression of infection caused by HIV is defined as antiretroviral therapy or ART. Zidovudine (azidothymidine or AZT) was the first antiretroviral drug that was tested and was approved by United States Food and Drug Administration (FDA) for AIDS treatment (Broder 2010).

usage of NHP models can be broadly categorized into two groups—in understanding pathogenesis and virology of HIV and in drug therapy and vaccine development for treatment of AIDS. The following segment is a summarized description of these two categories in relation to NHP models and HIV/AIDS research. Also, Table 32.1 defines terminology associated with HIV/AIDS research that would enable you to understand the context of the discussion.

In the nineteenth and twentieth centuries, animal research was considered to be fundamental in proving the etiology of the infectious diseases (Madeley 2008; Veazey and Lackner 2017). In 1981, there were series of reports of immunodeficiency-associated diseases characterized by opportunistic pathogens and was seen as endemic, "insidious" plague across the globe and the subsequent years of research recognized these as first reports of infections caused by HIV (Siegal et al. 1981; Masur et al. 1981; Gottlieb et al. 1981; Veazey and Lackner 2017). As these reports shared the timeline with discovery of first human retroviruses in T cells of a leukemia patient (Levy et al. 1984), the researchers hypothesized the existence of a retroviral agent attacking the T cells causing AIDS and this was followed by both in vitro research as well as animal model research which eventually paved the path for isolating putative retroviral agents causing AIDS. With animal experiments, especially studies in NHPs (Table 32.2), the origin of pandemic strain of HIV-1 (group M) from simian immunodeficiency virus (SIV), infecting chimpanzees and group O from SIV infecting gorillas has been established (Bailes et al. 2003; Keele et al. 2006; D'Arc et al. 2015; Veazey and Lackner 2017). Distinct SIVs with NHP hosts

Table 32.2 List of studies conducted for HIV/AIDS research which involved NHP model usage

NHP models used	Comments	References
Chimpanzees	Used for understanding HIV pathogenesis and virology	Lubeck et al. (1997); Boyer et al. (1997); Berman et al. (1996), (1990b); Fultz et al. (1992); Girard et al. (1991); Emini et al. (1992); Girard et al. (1995); Bruck et al. (1994); Boyer et al. (1996); Estaquier et al. (1994); Koopman et al. (1999); Schuitemaker et al. (1993); Zarling et al. (1990); Keele et al. (2009); Novembre et al. (1997), (2001); O'Neil et al. (2000); Shedlock et al. (2009); Greenwood et al. (2015); Gilden et al. (1986); Fultz et al. (1991); Etienne et al. (2011); Rudicell et al. (2010); SAXINGER et al. (1987); Berman et al. (1990a); Veazey and Lackner (2017)
Rhesus macaques		Farell (2018); Letvin et al. (1985); Reimann et al. (1996); Daniel et al. (1985); Sutjipto et al. (1990); Veazey et al. (1998); Smit-McBride et al. (1998); Letvin and King (1990a); Estaquier et al. (1994); Gougeon et al. (1997); Xiao et al. (2015); Shiver et al. (2002a, b); Barouch et al. (2000); Ourmanov et al. (2000); Horton et al. (2002); Casimiro et al. (2005); Liang et al. (2005); Letvin and King (1990b); Nishimura et al. (2004); Miller et al. (1989); Kanki et al. (1985); Veazey and Lackner (2017)
Chimpanzees	Used for vaccine development	Berman et al. (1990b); Veazey and Lackner (2017)
Rhesus macaques		Reimann et al. (1996); Hanke et al. (1999), (2005); Schmitz et al. (1999); Amara et al. (2001); Shingai et al. (2013); Barouch et al. (2013); Veazey and Lackner (2017); Weatherall et al. (2006)
Rhesus macaques (Indian)	Used for antiretroviral therapy (including cART) research	Horiike et al. (2012); Leone et al. (2010); zur Megede et al. (2008); Deere et al. (2010); Hofman et al. (2004); North et al. (2005), (2010); Pandrea et al. (2011); Del Prete and Lifson (2013)
Rhesus macaques (Chinese)		(Jasny et al. 2012; Vagenas et al. 2010; Del Prete and Lifson 2013)
Pig-tailed macaque		Shen et al. (2003), (2007); Ambrose et al. (2007); Boltz et al. (2012); Kearney et al. (2011); Shao et al. (2009); Hatziioannou et al. (2009); Thippeshappa et al.(2011); Del Prete and Lifson (2013)
Crab-eating macaque		Benlhassan-Chahour et al. (2003); Bourry et al. (2010); Moreau et al. (2012); Del Prete and Lifson (2013)

were isolated and identified from chimpanzees, sooty mangabeys, gorillas, and some other macaque species (Farell 2018). However, most of these SIVs were found to be nonpathogenic to their host species and this made SIV_{mac}, isolated from rhesus macaques, stand out from the crowd for being able to induce retroviral infection with AIDS-like manifestation (Farell 2018). This explains the importance of Old World monkeys, especially *Macaca mulata,* in the HIV/AIDS research.

A group of researchers at Harvard and New England Regional Primate Research Centre, who were responsible for isolating the SIV from captive rhesus macaques, reported that the antibodies isolated from AIDS patients were able to recognize the envelope protein of SIV and this laid out the basic outline of primate origin of HIV (Letvin et al. 1985; Kanki et al. 1985; Daniel et al. 1985). Envelope (Env) glycoproteins of HIV are the main elements that interact with receptors on T cells and then enters them. In 1986 scientists reported that CD4 transgene, which encoded for CD4 receptors on the surface of immune cells in humans, was responsible for disposing the T cells to HIV-1 infection (Maddon et al. 1986; Dalgleish et al. 1984; Farell 2018). This was followed by studies reporting the antigenic differences of Env glycoproteins of SIV and HIV-1 and this was the setup that eventually led to the development of chimeric simian–human immunodeficiency viruses (SHIV) which were, in contrast to its predecessor, able to invade the primate immune system, decrease $CD4^+$ T cells, eventually causing death of the infected animals (Reimann et al. 1996; Joag et al. 1996; Farell 2018). These studies are often presented as cruelties of animal research by animal welfare organizations. Being a chimeric virus, SHIV is made up of genetic materials from HIV-1 and SIV and is thus a culmination of genes that encodes Env glycoproteins and other regulatory factors for successful infection and maintenance of the HIV-1 in the host, and genes that encode the backbone of SIV (Farell 2018). Thus, successful construction of SHIV, a virus which was able to infect NHPs and present AIDS manifestation, prepared the base for research studies that led to the development of vaccines and drugs targeting HIV-1 Env by using SHIV-infected monkeys (Farell 2018). From HIV research involving macaques, it is now known that SIV infection in the macaques is aggressive in the gut-associated lymphocytes as opposed to the blood lymphocytes (Weatherall et al. 2006). This can be explained from the fact that CCR5, a co-receptor for infection and maintenance of HIV-1 in the host, is expressed on the activated $CD4^+$ T lymphocytes in the mucosal linings of gut and vagina as these areas are in continuous contact with microbial cells and are rare to find on the $CD4^+$ T lymphocytes in the blood of healthy human or nonhuman primate hosts (Poles et al. 2001; Bleul et al. 1997; Lackner and Veazey 2007). This finding led to ethical acceptance of biopsy of gut of HIV infected patients which has in turn changed the perspective in the understanding of disease progression (Weatherall et al. 2006). Numerous studies in NHPs for understanding HIV-1 and SIV, revealing homology ratio between HIV-1, SIV, and HIV-2, has assisted in developing SHIV hybrids expressing HIV-1 gp120 to test vaccines developed against viral envelope proteins (Weatherall et al. 2006).

A culmination of these events led to the development of prevention and treatment modules for HIV and has also paved the path for development of new vaccines for HIV (Veazey and Lackner 2017). The NHPs have been used scarcely in the development of antiretroviral therapy (except toxicology studies) (Weatherall et al. 2006). Research on SIV–macaque model for effects on drug tenofovir exhibited protection of macaque's offspring from SIV infection and this drug was further developed for human use (Van Rompay et al. 1999; Weatherall et al. 2006). The main research attempts in HIV/AIDS field which used the NHP models extensively is for development of herd immunity and the best way to achieve this is by developing an efficient and affordable vaccine. In the HIV/AIDS vaccine research there are broadly two different approaches which involve NHP models—to check the effect of live attenuated SIV against superinfection with homologous virulent virus strain; and developing challenge viruses by incorporating information of anti-envelope glycoprotein vaccines of the SHIV viruses (Weatherall et al. 2006). For testing the effects and efficiency of these vaccines, it is essential to assess deliberate viral challenge in a biological system and researchers agree that SIV–macaque model is the best candidate for this job (Weatherall et al. 2006) and is an indispensable step before time- and cost-consuming clinical trials. In scientific research and biomedical research in particular, an experiment that fails also leads to some important findings that propagates the future research aspects. For example, a cytomegalovirus-vectored SIV vaccine tested on NHP could not immunize the macaque against SIV infection but shortly after vaccination, 50% of the animals controlled plasma viremia followed by viral clearance (Hansen et al. 2013a; Veazey and Lackner 2017). The same group of researchers also revealed that this viral clearance in the NHPs was associated with lesser known MHC-II-linked immune responses instead of MHC-1-linked responses that are usually seen in case of microbial infections and vaccination studies (Hansen et al. 2013b; Veazey and Lackner 2017). Thus, despite failure as an experiment to develop vaccine for HIV infection, this study provided a perspective into the acquired immune response as a result of vaccine administration (Veazey and Lackner 2017). Another group of researchers reported that upon administration of an adenoviral-vectored vaccine in macaques, 50% of the study animals were protected against repetitive rectal challenges a CCR5 tropic SHIV (Stephenson et al. 2016; Barouch et al. 2015; Veazey and Lackner 2017).

Recent potential strategy in developing cure for HIV and other viral infection is a combination of antimicrobial drug therapy with vaccination or other strategies. For example, researchers have demonstrated control of viremia in SIV–macaques system upon administration of standard ART in combination with a monoclonal antibody (anti-$\alpha4\beta7$) (Byrareddy et al. 2016; Veazey and Lackner 2017). In contrast to this, one of the control groups which were administered with standalone ART exhibited sustained viremia progressing to AIDS (Veazey and Lackner 2017). These examples project the importance of NHP in HIV/AIDS research and biomedical research community is hopeful of arriving at a possible cure and/or a preventive vaccine against HIV infection thanks to SIV- and HIV-infected NHP models.

Another antimicrobial drug-based strategy that has transformed HIV infection from progressive, fatal infection to a treatable chronic condition is called highly active antiretroviral therapy (HAART) or combination antiretroviral therapy or cART (Del Prete and Lifson 2013), which uses three or more antiretroviral drugs for treatment. It has been noted that HAART is responsible for reduction of AIDS mortality rates by 50% in the last decade (Roy et al. 2015). In the HIV/AID landscape, HAART has reduced vertical transmission of HIV, decreased co-occurrence of opportunistic infections and has halved the incident rates of HIV-associated dementia (Hayashi et al. 2006; Roy et al. 2015). For the cART studies, NHP models like rhesus macaques, pig-tailed macaques, and cynomolgus macaques have been used to generate pathogenic model (Del Prete and Lifson 2013) and then studying them upon cART administration. Since NHP models, like SIVmac239-infected-Indian rhesus macaques, demonstrated reproducible, authentic pathogenesis identical to the infection progression of HIV infection in humans (Del Prete and Lifson 2013), they are considered to be the best choice for conducting cART-associated research. Pig-tailed macaques inoculated with mixture of immunosuppressive SIV/DeltaB670 viral quasispecies and CD4-independent, neurovirulent, macrophage-tropic infectious molecular clone SIV/17E-Fr is an alternative to SIV infection model and features fast progression of infection, increased viral load, precipitous declines in $CD4^+$ T lymphocytes and encephalitis within 90 days of infection if left untreated (Zink et al. 1999; Dinoso et al. 2009; Del Prete and Lifson 2013). HIV-associated neuroinfection in this latter model makes it a promising NHP candidate for assessing CNS-targeted ART (Zink et al. 1999, 2010; Dinoso et al. 2009; Del Prete and Lifson 2013). As drugs developed against HIV exhibit lower or no activity against SIV targets, it is critical to determine the antiretroviral drug potency against the virus used to inoculate NHP models prior drug testing and this can be done by quantifying plasma-adjusted inhibitory concentration of 95% (Del Prete and Lifson 2013). The pharmacological effect usually varies between primates, that is between different species of NHP models as well as between humans and NHP models and this can be addressed by determining effective dosing regimens, using in vivo models, that would maintain above threshold inhibitory concentrations; this step is essential to achieve intended drug efficacy (Del Prete and Lifson 2013). NHP models can help in estimating strategies that would maximize drug penetration to key sites of viral replication in vivo to achieve maximal viral suppression by cART drugs (Cohen 2011; Del Prete and Lifson 2013). Prior to establishing cART treatment regimens with three or more drugs together, in vivo confirmation of the antiretroviral activity of individual drugs by monotherapy is conducted and is an essential step to understand the probability and mechanism of post-inoculation drug resistance development (Del Prete and Lifson 2013). Initially effective in preventing SIV infection in NHP models, Tenofovir, a type of nucleoside reverse transcriptase inhibitor (NRTI) drug, is now the prime component of cART trials in NHP (Tsai et al. 1995, 1998; Del Prete and Lifson 2013). In monotherapy studies on SIV-infected monkeys, NRTIs like emtricitabine and lamivudine lowered plasma viremia (Van Rompay et al. 2002; Del Prete and Lifson 2013). To overcome the limitation of using only NRTI since non-nucleoside reverse transcriptase inhibitors (NNRT) was not effective against SIV or SIV-infected NHP, chimeric RT-SHIVs were constructed by inserting NNRTI-sensitive HIV-1 RT(reverse transcriptase) gene in SIVmac239 or SIVmne (related species of SIVmac) backbone (Balzarini et al. 1995, 1997; Ambrose et al. 2004; Uberla et al. 1995; Del Prete and Lifson 2013). In ART monotherapy, NNRTs like Efavirenz led to virologic suppression in RT-SHIVmac239 infected rhesus macaques and RT-SHIVmne infected pig-tailed macaques (Del Prete and Lifson 2013).

Compared to basic three drug regimens, cART treatment strategies that included at least four ART drugs were found to suppress viral loads in HIV-infected humans as well as the NHP models (Del Prete and Lifson 2013). For antiretroviral drug discovery research involving NHP models, the investigators should give special attention to drug:drug interactions leading to drug toxicity in NHP models when drugs are used in specific combinations (Del Prete and Lifson 2013). For example, in a regimen that involved didanosine and stavudine to treat SIV infection in macaques led to diabetes after more than 6 months of dosing and the researchers attribute this to didanosine toxicity (Vaccari et al. 2012; Dunham et al. 2013; Del Prete and Lifson 2013). To optimize the NHP model studies in HIV/AIDS research, molecular and biochemical virological assays like quantitative PCR/RT-PCR methods are indispensable and essential to demonstrate the efficiency of the cART regimen in reducing viremia and checking infection (Cline et al. 2005; Hansen et al. 2011; Del Prete and Lifson 2013). Utilizing real-time RT-PCR assays, ultrasensitive assays have been developed to measure plasma vRNA (virological RNA) within larger volumes of source specimen (Queen et al. 2011; Deere et al. 2010; Del Prete and Lifson 2013). From the accounts given in this chapter, it can be understood that in advancing HIV/AIDS research for containing as well as preventing the infection and its gradual chronic progression, strategies like novel therapeutic vaccine design and antiretroviral drug discovery by utilizing advanced basic HIV pathogenicity studies is essential and for these to be viable for clinical tests, preliminary tests for efficiency of the regimen on NHP models is necessary.

32.3.2 Tuberculosis

According to WHO Global Health Estimates 2016, despite a 0.9% decrease in global death rates from 2000 to 2016, tuberculosis (TB) continues to be the leading cause of death due to infectious diseases across the globe. It is reported that factors like poor quality of life, the increasing rate of incidence of multidrug resistant (MDR) TB strains, and the susceptibility to incurring TB in those who are suffering from retroviral infections like HIV, in combination with each other has increased TB-related threats in China, India, Eastern Europe, and sub-Saharan Africa (Weatherall et al. 2006). TB is a communicable disease and is caused by a bacterium, *Mycobacterium tuberculosis*. Transmission of *M. tuberculosis* occurs when an infected person coughs, sneezes, shouts, or sings and in the process releases *M. tuberculosis* containing airborne particles called droplet nuclei which then remains suspended in the air for several hours and is inhaled by a person (Centers for Disease Control and Prevention 2013). Upon inhalation, the droplet nuclei (1–5 μ in size) passes through respiratory tract and reaches alveoli in the lungs (Centers for Disease Control and Prevention 2013). What must be noted here is that mere exposure to the bacterium does not always lead to active TB and the outcome of the exposure (apart from active TB) varies from clearance of the pathogen by the immune system of the host without any infection to asymptomatic, latent TB (Scanga and Flynn 2014). Persons with latent TB do not have the TB disease; however, *M. tuberculosis* in the latent state can reactivate and cause active infection in the body (Centers for Disease Control and Prevention 2013; Scanga and Flynn 2014; Barry et al. 2009; Lin and Flynn 2010). In the case of latent TB, the host immune system encapsulates the extracellular tubercle bacilli leading to the formation of granuloma, which is the pathologic hallmark of TB (Scanga and Flynn 2014; Centers for Disease Control and Prevention 2013). With this brief background of TB pathogenesis, let us now move onto the treatment and cure regimens of TB.

One of the major issues reported about TB infection's antimicrobial regimen, which takes approximately 6–8 months for completion, is the rising incidence of multidrug resistance in *M. tuberculosis* (Scanga and Flynn 2014). With the evolving bacterial genetics, the TB vaccine, *Mycobacterium bovis* Bacillus Calmette-Gue'rin (BCG) is facing limited efficacy (Scanga and Flynn 2014; Andersen and Doherty 2005) in providing complete protection against the infection. This calls for new drug discovery and development. For the preclinical stages of the drug development research, a model system that can reiterate human TB infection with the following features is indispensable: (1) susceptible against human TB-associated *m. tuberculosis* strains; (2) exhibit the human TB infection-associated complex histopathology and disease presentations; (3) reproduce the entire range of potential human TB infection outcomes; (4) exhibit latent TB with identical human latent TB characteristics, such as triggers to reactivate to infective TB; (5) reflect genetic diversity and immunological similarity of humans; (6) similar to humans with respect to metabolism and physiology for ensuring translational antimicrobial drug development (Scanga and Flynn 2014). Years of rigorous TB research has concluded that NHP model is the sole research model that possesses all of the above features (Flynn et al. 2017; Scanga and Flynn 2014) and this explains their use over decades in TB research for understanding bacterial pathogenesis, multidrug resistance phenomenon, and for efficient drug and vaccine development.

All species of NHPs are potentially susceptible to incur *M. tuberculosis* infection with Old World species being the most susceptible and New World monkeys being the least susceptible (Scanga and Flynn 2014; Montali et al. 2001; RC 1984). The first report of NHP use for TB infection research was provided by Leon Schmidt who inoculated macaques intratracheally with 10^2–10^3 colony-forming units of *M. bacterium* and recorded active TB infection in all the animals under study (Schmidt et al. 1959; Schmidt 1956, 1966; Scanga and Flynn 2014). This is historically an important study in the world of NHP model usage in TB research, for this study established macaques as preferable model for *M. tuberculosis*-associated antimicrobial drugs (Scanga and Flynn 2014). There is a wide range of application of NHP models in current day TB research such as understanding HIV pathogenesis, investigate latent TB features and reactivation triggers in vivo, coinfection system of HIV–TB, and development of new age vaccines and antimicrobial drugs.

Here we briefly summarize anti-tubercle drug development with NHP models. Apart from vaccination and upgrading public health and hygiene conditions, chemotherapy is a major treatment regimen for TB infection (Weatherall et al. 2006; Dye and Floyd 2006). In the beginning of TB treatment regimen, effective treatment was ensured by administration of more than two antimycobacterial drugs for a duration of 6–12 months; however, difficulty in sustenance of strict public health care over such a long period reduced the effectivity of this regimen, especially in low-income and developing countries (Weatherall et al. 2006). An immediate result of this poor management of health care is prevalence of partial treatment or ineffective and extensive use of single antimycobacterial agents, which in turn has led to catastrophic increase in TB infection due to multidrug resistant *M. tuberculosis* (Weatherall et al. 2006). This is where the currently effective DOTS strategy (Directly observed Therapy, Short Course) came into existence which involved rapid diagnosis of *M. tuberculosis* by screening sputum, uninterrupted supply of antimycobacterial drugs worth 6–8 months of treatment regime and efficient health condition monitoring (Scanga and Flynn 2014; Weatherall et al. 2006). Even though the DOTS strategy is a step ahead, the duration of regime is still long enough to affect its effectivity by increasing chances of MDR upsurge in the areas which are most affected by TB infection in the population. There are various factors that favor drug testing in NHPs at preclinical trial stage—(1) Chemotherapeutic drugs display pharmacokinetics in NHP models which are similar to the ones observed in humans and this enables simple dose calculations for the treatment regime (McLeay et al. 2012; Lin et al. 2012; Scanga and Flynn 2014); (2) It is reasonable to test disparate drug efficacies on tubercle bacilli residing in distinct granulomas presented in *M. tuberculosis*-infected cynomolgus macaques (Lin et al. 2009; Scanga and Flynn 2014); (3) NHPs make efficacy testing of the single-drug chemotherapy possible (Scanga and Flynn 2014). Economic obstacles in housing and handling NHPs are often the concern of researchers as well as ethics committees; however, advanced biomedical imaging technologies enable periodic assessment of TB infection—before, during, and after—chemotherapeutic regimen in the same animal models, thus cutting the costs by reducing the sample size in animal research (Lin et al. 2013; Scanga and Flynn 2014). Another advantage of advanced imaging technologies is the analysis of individual lesions from the NHPs during drug development research, which further provides statistical reliability with limited animals (Lin et al. 2013; Scanga and Flynn 2014).

Marmosets, which can be infected with *M. tuberculosis* and exhibit all the advantageous features of TB pathogenesis similar to larger NHPs are being considered as new age NHP model for antimycobacterial drug development associated research (Via et al. 2013; Scanga and Flynn 2014). Marmosets and other NHPs used in TB research are advantageous for drug development against latent TB; however, since presentation of latent TB is often unpredictable in the natural life cycle of NHP model post infectious dosage, genetically modified or immunologically primed NHP models can be used for this purpose. For example, reactivation of latent TB can be initiated in the TB monkey models with TNF-α antibody (Zhan et al. 2017; Lin et al. 2012; Lin and Flynn 2010). In a drug development study, metronidazole was found to reduce *M. tuberculosis* load in the reactivated latency model (Gumbo et al. 2015) but did not accelerate the efficiency of rifampin or isoniazid in active TB models (Gumbo et al. 2015; Lin et al. 2012). By incorporating cumulative approach that integrates advances in immunology, molecular biology, genomics, metabolomics, and mathematical modeling (Marino et al. 2011; Scanga and Flynn 2014), to prepare animal models in general and NHP infection models in particular would cut down the expenses in housing and handling NHPs in TB research and would also help in better analysis of the outcome from limited sample size in the sector of drug discovery for latent TB and MDR TB-associated infections.

With the above information in mind, another important contribution of NHP models in infectious diseases research can be understood. People with HIV/AIDS are highly susceptible to TB infection and the severity of the disease in such individuals is very high. Monkeys are the preferred model for studying TB/HIV coinfection and researchers report that this relationship would also advance the understanding of latent TB pathogenesis and process of reactivation (Mehra et al. 2011; Pena and Ho 2015). It was noted that rhesus macaques and cynomolgus macaques with latent TB infection, induced by high-dose BCG (Mehra et al. 2011; Shen et al. 2001, 2002) or low-dose Erdman strain (Diedrich et al. 2010), respectively, presented reactivation when coinfected with SIV (Pena and Ho 2015). Studies with these models reported that early T lymphocytic depletion in the host, instead of injected viral load, is the reactivation trigger for latent TB in cynomolgus macaques (Diedrich et al. 2010; Pena and Ho 2015) and this explains restoration of *M. tuberculosis* specific cell mediated immune responses when SIV-BCG coinfected macaques were treated with antiretroviral agents (Mehra et al. 2011; Pena and Ho 2015). Newborn rhesus macaques have been established as a model for studying pediatric TB/HIV coinfection and disease manifestation (Cepeda et al. 2013; Pena and Ho 2015).

32.3.3 Malaria

According to WHO Global Health Estimates, Malaria was amongst the top 20 leading causes of death across the globe in 2000s. With decades of biomedical research to decode pathogenesis of malarial parasites, antimalarial drug development and improved public health regimen and awareness have worked in consortium to reduce the percentage of global death by 0.7% and removed malaria from the top 20 list in 2016 WHO Global Health Estimates report. Malaria is an infectious disease caused by parasites of genus *Plasmodium* which are known for their complex life cycle alternating between vertebrate and invertebrate hosts (Joyner et al. 2015). This mosquito-borne infection is transmitted to susceptible mammals, including humans and other primates and the parasite goes through growth and developmental stages in the patient's liver before entering the bloodstream and initiating clinical manifestations associated with malaria (Joyner et al. 2015). The dormant forms of malarial parasite, called hypnozoites, in the vertebrate host's liver was first reported in rhesus macaques (experimentally infected with *Plasmodium cynomolgi),* followed by reports from chimpanzees infected with *P. vivax* (Krotoski et al. 1982b; Krotoski et al. 1982a; Joyner et al. 2015). One of the main challenges in reducing malarial infections is preventing relapse of malaria resulting from reactivation of dormant parasites in the host's liver and since human clinical studies have various limitations like diet and therapeutic variables, NHPs are best candidate for studying liver stage forms (LSFs) of malarial parasite and the reactivation strategy of the parasite (Joyner et al. 2015). Table 32.3 lists out *Plasmodium* species and strains that are potential candidates for studying relapse of the infection.

Table 32.3 *Plasmodium* species and strains that are potential candidates for studying relapse of the infection (Joyner et al. 2015)

Plasmodium species/strains	First isolated from	Principal NHP host	Primary cycle (days)	Time between relapses
P. vivax (Brazil VII)	Human	*Aotus* sp. and *Saimiri* sp.	7–8	1–2 months
P. vivax (Chesson)				1 month
P. vivax (Salvador I)				2–4 months
P. vivax (North Korea)				6–12 months
P. cynomolgi (B, M, Berok, Ceylon)	Crab-eating macaque and toque macaque	*Macaca mulatta*	8–10	1–2 months
P. simiovale	Toque macaque		>12	Varied: 2–4 weeks to more than 2 years
P. fieldi	Southern pig-tailed macaque and crab-eating macaque		>12	Varied: 2–4 weeks to more than 1 year

Researchers report that the increasing use of malarial NHP models in China and Southeast Asia could have long-term effect in monitoring drug safety studies (Saravanan et al. 2015). To minimize these effects, many pharmaceutical companies utilize the qPCR assays to diagnose malarial infection, and chloroquine and primaquine have been established as effective treatment drugs for treating NHPs suffering from malaria (Strait et al. 2012). However, with decades of investigations using NHP models in malarial research, research communities are trying to combine other technologies like systems biology, genomics, and machine learning with the knowledge and data available from NHP models to reduce the exploitation of monkeys in the biomedical research associated with malaria drug discovery and development.

32.3.4 Other Viral Infections

Biomedical research advancements using genome sequencing, gene modification strategies as well as animal models including NHPs is assisting basic as well as applied biology and this is in turn preparing us for generating prevention as well as cure strategies against emerging infectious diseases like virus-inflicted diseases as shown in Fig. 32.1 (Veazey and Lackner 2017). For example, there is an unanimity in the scientists involved in vaccine development for rapidly emerging and expanding infectious diseases, like infection induced by Ebola and Zika virus, that use of NHP for accelerating development of antiviral vaccines against these infectious agents is indispensable (Veazey and Lackner 2017; Marzi et al. 2015; Abbink et al. 2016).

First human epidemic by Zika virus (ZIKV) was recognized in 2007 (Estes et al. 2018; Baud et al. 2017). According to CDC, Zika virus can be horizontally transmitted through the bite of an infected *Aedes* species mosquito, through blood

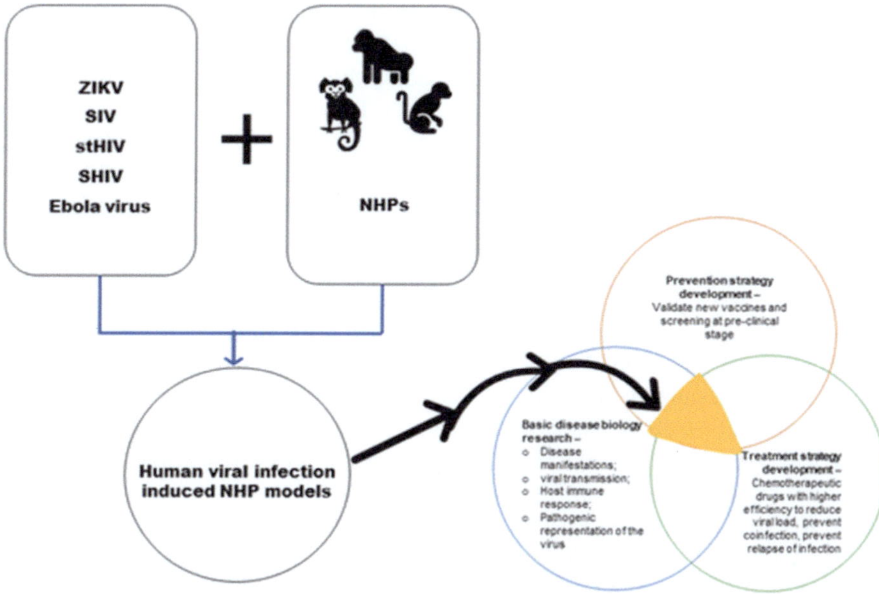

Fig. 32.1 Contribution of Nonhuman Primate models in human viral infectious diseases. Nonhuman primate (NHP) models which recapitulate human viral infection manifestations are selected and respective viral agent is used to experimentally infect the NHPs, thus forming human viral infection induced NHP models. These models play a central role in the biomedical research of human viral infections such as for understanding disease manifestations and host immune responses, for developing better prevention strategy in the form of new age vaccines and for developing efficient treatment strategy by designing and testing potential antiviral drug in NHPs. *SHIV* simian HIV, *SIV* simian immunodeficiency virus, *stHIV* simian-tropic HIV, *ZIKV* zika virus

transfusion, through laboratory and public health care facilities and through sex and vertically transmitted from mother to child and the spread of the infection is more common in the tropical zone as the weather condition is favorable for both mosquito breeding and viral growth (Estes et al. 2018). For understanding the pathogenesis of the causal agent, NHPs like cynomolgus macaques, rhesus macaques, and pigtail macaques are employed in the ZIKV infection investigation. ZIKV NHP models, prepared by subcutaneous injection of ZIKV, exhibits peak in plasma viremia between 2 and 6 days after infection and plasma clearance taken place within 10 days post infection (Estes et al. 2018; Dudley et al. 2016; Li et al. 2016; Osuna et al. 2016; Nguyen et al. 2017). As NHPs recapitulate human ZIKV infection, they are important for understanding the various complications arising from ZIKV infection. One such complication is the ability of ZIKV to cross placenta and affect the fetus (Estes et al. 2018). It was noted that viremia clearance took as long as 55 days in pregnant monkeys and this was similar to the human pregnancy case reports (Estes et al. 2018; Dudley et al. 2016; Driggers et al. 2016). ZIKV was detected in placenta and fetal brain and liver in studies involving pregnant pigtail macaques infected with ZIKV and NHP infection models of ZIKV also revealed the ZIKV RNA persistence in host tissues long time after viremia clearance (Estes et al. 2018; Koide et al. 2016; Hirsch et al. 2017; Aid et al. 2017). Such NHP-associated studies of ZIKV infection will pave way for understanding the vertical transmission of ZIKV and this would work as a primer for vaccine development for protection of pregnant women from ZIKV infection (Estes et al. 2018). Apart from vaccine development, chemotherapeutic drugs with anti-ZIKV characteristics have been identified; however, only two out of these (namely, 25-hydrocholesterol (25HC) and Galidesivir (BCX4430)) succeeded to preclinical studies that involved testing in NHPs (Osuna and Whitney 2017). 25HC, a product of cholesterol oxidation by cholesterol-25-hydroxylase, is a signaling molecule of innate immune system and has been characterized with antiviral activity (Osuna and Whitney 2017; Blanc et al. 2013).

Upon successful protection of model mice (BALB/c) against ZIKV infection upon 25HC pretreatment (Li et al. 2017), two rhesus macaques, administered with 25HC, were subjected to low dose of a Venezuelan ZIKV isolate 24 hours later. These RMs were treated with 25HC for one week and ZIKV detection test on blood and urine samples was performed daily. Opposed to control animals exhibiting ZIKV RNA load, blood and urine samples of the RMs undergoing chemotherapy showed viremia clearance. The second drug, BCZ4430, an adenosine nucleoside analogue that inhibits viral replication, has been previously reported as effective against ZIKV in mice and other small animal models, and against Ebola infections in both small animal models as well as in NHP. BCX4430 is undergoing preclinical trials for ZIKV treatment in NHPs (Osuna and Whitney 2017). Inclusion of NHPs in preclinical assessment for drug discovery against ZIKV treatment is expected to generate effective chemotherapy agents, especially to protect pregnant women and her fetus.

32.4 Conclusion

From the accounts given in this chapter, it can be said that the studies involving NHPs as a part of their research are often concentrated in only few fields like the ones noted above. Often times, the public and ethical committees raise the question as to why this trend of NHPs in biomedical research is seen. The question often directs the conversation in a way that says the researchers of the fields which uses NHPs are not considering their replacement strongly enough. However, biomedical research has worked over years for raising the health standards and quality therapeutics raising the life expectancy around the world and is not driven by any trend or fashion. In the report published in 2006, it was noted that even though cancer research is one of the dynamic field of biomedical research, NHPs are not usually considered for this stem of biology as over the years molecular biology, cell biology, biochemistry, and bioinformatics have developed to the stage where cancer mechanisms and drug discovery is conclusive with inclusion of in vitro and rodent models in the study (Weatherall et al. 2006). This establishes the fact that researchers are making a conscience and informed decision while choosing animal models, especially NHPs.

References

Abbink P, Larocca RA, De La Barrera RA, Bricault CA, Moseley ET, Boyd M, Kirilova M, Li Z, Ng'ang'a D, Nanayakkara O, Nityanandam R, Mercado NB, Borducchi EN, Agarwal A, Brinkman AL, Cabral C, Chandrashekar A, Giglio PB, Jetton D, Jimenez J, Lee BC, Mojta S, Molloy K, Shetty M, Neubauer GH, Stephenson KE, Peron JPS, Zanotto PMA, Misamore J, Finneyfrock B, Lewis MG, Alter G, Modjarrad K, Jarman RG, Eckels KH, Michael NL, Thomas SJ, Barouch DH (2016) Protective efficacy of multiple vaccine platforms against Zika virus challenge in rhesus monkeys. Science 353(6304):1129–1132

Aid M, Abbink P, Larocca RA, Boyd M, Nityanandam R, Nanayakkara O, Martinot AJ, Moseley ET, Blass E, Borducchi EN, Chandrashekar A, Brinkman AL, Molloy K, Jetton D, Tartaglia LJ, Liu J, Best K, Perelson AS, De La Barrera RA, Lewis MG, Barouch DH (2017) Zika virus persistence in the central nervous system and lymph nodes of rhesus monkeys. Cell 169(4):610–620.e614

Amara RR, Villinger F, Altman JD, Lydy SL, O'Neil SP, Staprans SI, Montefiori DC, Xu Y, Herndon JG, Wyatt LS, Candido MA, Kozyr NL, Earl PL, Smith JM, Ma H-L, Grimm BD, Hulsey ML, Miller J, McClure HM, McNicholl JM, Moss B, Robinson HL (2001) Control of a mucosal challenge and prevention of AIDS by a multiprotein DNA/MVA vaccine. Science 292(5514):69–74

Ambrose Z, Boltz V, Palmer S, Coffin JM, Hughes SH, Kewalramani VN (2004) In vitro characterization of a simian immunodeficiency virus-human immunodeficiency virus (HIV) chimera expressing HIV type 1 reverse transcriptase to study antiviral resistance in pigtail macaques. J Virol 78(24):13553–13561

Ambrose Z, Palmer S, Boltz VF, Kearney M, Larsen K, Polacino P, Flanary L, Oswald K, Piatak M Jr, Smedley J, Shao W, Bischofberger N, Maldarelli F, Kimata JT, Mellors JW, Hu SL, Coffin JM, Lifson JD, KewalRamani VN (2007) Suppression of viremia and evolution of human immunodeficiency virus type 1 drug resistance in a macaque model for antiretroviral therapy. J Virol 81(22):12145–12155

Andersen P, Doherty TM (2005) The success and failure of BCG—implications for a novel tuberculosis vaccine. Nat Rev Microbiol 3(8):656–662

Anderson DM (2008) The nonhuman primate as a model for biomedical research. In: Conn PM (ed) Sourcebook of models for biomedical research. Humana Press, Totowa, pp 251–258

Animal Welfare and the Three Rs: Replacement, Refinement and Reduction (Leaflet) (2016) Understanding animal research. http://www.understandinganimalresearch.org.uk/files/7914/1041/1800/05-The-Three-Rs-for-web.pdf. Accessed 25 May 2019

Bailes E, Gao F, Bibollet-Ruche F, Courgnaud V, Peeters M, Marx P, Hahn B, Sharp P (2003) Hybrid origin of SIV in chimpanzees. Science 300(5626):1713

Balzarini J, Weeger M, Camarasa MJ, De Clercq E, Uberla K (1995) Sensitivity/resistance profile of a simian immunodeficiency virus containing the reverse transcriptase gene of human immunodeficiency virus type 1 (HIV-1) toward the HIV-1-specific non-nucleoside reverse transcriptase inhibitors. Biochem Biophys Res Commun 211(3):850–856

Balzarini J, De Clercq E, Uberla K (1997) SIV/HIV-1 hybrid virus expressing the reverse transcriptase gene of HIV-1 remains sensitive to HIV-1-specific reverse transcriptase inhibitors after passage in rhesus macaques. J Acquir Immune Defic Syndr Hum Retrovirol 15(1):1–4

Barouch DH, Santra S, Schmitz JE, Kuroda MJ, Fu T-M, Wagner W, Bilska M, Craiu A, Zheng XX, Krivulka GR, Beaudry K, Lifton MA, Nickerson CE, Trigona WL, Punt K, Freed DC, Guan L, Dubey S, Casimiro D, Simon A, Davies M-E, Chastain M, Strom TB, Gelman RS, Montefiori DC, Lewis MG, Emini EA, Shiver JW, Letvin NL (2000) Control of viremia and prevention of clinical AIDS in rhesus monkeys by cytokine-augmented DNA vaccination. Science 290(5491):486–492

Barouch DH, Whitney JB, Moldt B, Klein F, Oliveira TY, Liu J, Stephenson KE, Chang H-W, Shekhar K, Gupta S, Nkolola JP, Seaman MS, Smith KM, Borducchi EN, Cabral C, Smith JY, Blackmore S, Sanisetty S, Perry JR, Beck M, Lewis MG, Rinaldi W, Chakraborty AK,

Poignard P, Nussenzweig MC, Burton DR (2013) Therapeutic efficacy of potent neutralizing HIV-1-specific monoclonal antibodies in SHIV-infected rhesus monkeys. Nature 503:224
Barouch DH, Alter G, Broge T, Linde C, Ackerman ME, Brown EP, Borducchi EN, Smith KM, Nkolola JP, Liu J, Shields J, Parenteau L, Whitney JB, Abbink P, Ng'ang'a DM, Seaman MS, Lavine CL, Perry JR, Li W, Colantonio AD, Lewis MG, Chen B, Wenschuh H, Reimer U, Piatak M, Lifson JD, Handley SA, Virgin HW, Koutsoukos M, Lorin C, Voss G, Weijtens M, Pau MG, Schuitemaker H (2015) Protective efficacy of adenovirus/protein vaccines against SIV challenges in rhesus monkeys. Science 349(6245):320–324
Barry CE 3rd, Boshoff HI, Dartois V, Dick T, Ehrt S, Flynn J, Schnappinger D, Wilkinson RJ, Young D (2009) The spectrum of latent tuberculosis: rethinking the biology and intervention strategies. Nat Rev Microbiol 7(12):845–855
Baud D, Gubler DJ, Schaub B, Lanteri MC, Musso D (2017) An update on Zika virus infection. Lancet 390(10107):2099–2109
Benlhassan-Chahour K, Penit C, Dioszeghy V, Vasseur F, Janvier G, Riviere Y, Dereuddre-Bosquet N, Dormont D, Le Grand R, Vaslin B (2003) Kinetics of lymphocyte proliferation during primary immune response in macaques infected with pathogenic simian immunodeficiency virus SIVmac251: preliminary report of the effect of early antiviral therapy. J Virol 77(23):12479–12493
Berman PW, Gregory TJ, Riddle L, Nakamura GR, Champe MA, Porter JP, Wurm FM, Hershberg RD, Cobb EK, Eichberg JW (1990a) Protection of chimpanzees from infection by HIV-1 after vaccination with recombinant glycoprotein gp120 but not gp160. Nature 345(6276):622–625
Berman PW, Gregory TJ, Riddle L, Nakamura GR, Champe MA, Porter JP, Wurm FM, Hershberg RD, Cobb EK, Eichberg JW (1990b) Protection of chimpanzees from infection by HIV-1 after vaccination with recombinant glycoprotein gp120 but not gp160. Nature 345(6276):622–625
Berman PW, Murthy KK, Wrin T, Vennari JC, Cobb EK, Eastman DJ, Champe M, Nakamura GR, Davison D, Powell MF, Bussiere J, Francis DP, Matthews T, Gregory TJ, Obijeski JF (1996) Protection of MN-rgp120-immunized chimpanzees from heterologous infection with a primary isolate of human immunodeficiency virus type 1. J Infect Dis 173(1):52–59
Blanc M, Hsieh WY, Robertson KA, Kropp KA, Forster T, Shui G, Lacaze P, Watterson S, Griffiths SJ, Spann NJ, Meljon A, Talbot S, Krishnan K, Covey DF, Wenk MR, Craigon M, Ruzsics Z, Haas J, Angulo A, Griffiths WJ, Glass CK, Wang Y, Ghazal P (2013) The transcription factor STAT-1 couples macrophage synthesis of 25-hydroxycholesterol to the interferon antiviral response. Immunity 38(1):106–118
Bleul CC, Wu L, Hoxie JA, Springer TA, Mackay CR (1997) The HIV coreceptors CXCR4 and CCR5 are differentially expressed and regulated on human T lymphocytes. Proc Natl Acad Sci 94(5):1925–1930
Boltz VF, Ambrose Z, Kearney MF, Shao W, Kewalramani VN, Maldarelli F, Mellors JW, Coffin JM (2012) Ultrasensitive allele-specific PCR reveals rare preexisting drug-resistant variants and a large replicating virus population in macaques infected with a simian immunodeficiency virus containing human immunodeficiency virus reverse transcriptase. J Virol 86(23):12525–12530
Bourry O, Mannioui A, Sellier P, Roucairol C, Durand-Gasselin L, Dereuddre-Bosquet N, Benech H, Roques P, Le Grand R (2010) Effect of a short-term HAART on SIV load in macaque tissues is dependent on time of initiation and antiviral diffusion. Retrovirology 7:78
Boyer JD, Wang B, Ugen KE, Agadjanyan M, Javadian A, Frost P, Dang K, Carrano RA, Ciccarelli R, Coney L, Williams WV, Weiner DB (1996) In vivo protective anti-HIV immune responses in non-human primates through DNA immunization. J Med Primatol 25(3):242–250
Boyer JD, Ugen KE, Wang B, Agadjanyan M, Gilbert L, Bagarazzi ML, Chattergoon M, Frost P, Javadian A, Williams WV, Refaeli Y, Ciccarelli RB, McCallus D, coney L, Weiner DB (1997) Protection of chimpanzees from high-dose heterologous HIV-1 challenge by DNA vaccination. Nat Med 3(5):526–532
Broder S (2010) The development of antiretroviral therapy and its impact on the HIV-1/AIDS pandemic. Antivir Res 85(1):1–18

Bruck C, Thiriart C, Fabry L, Francotte M, Pala P, Van Opstal O, Culp J, Rosenberg M, De Wilde M, Heidt P, Heeney J (1994) HIV-1 envelope-elicited neutralizing antibody titres correlate with protection and virus load in chimpanzees. Vaccine 12(12):1141–1148

Byrareddy SN, Arthos J, Cicala C, Villinger F, Ortiz KT, Little D, Sidell N, Kane MA, Yu J, Jones JW, Santangelo PJ, Zurla C, McKinnon LR, Arnold KB, Woody CE, Walter L, Roos C, Noll A, Van Ryk D, Jelicic K, Cimbro R, Gumber S, Reid MD, Adsay V, Amancha PK, Mayne AE, Parslow TG, Fauci AS, Ansari AA (2016) Sustained virologic control in SIV+ macaques after antiretroviral and α4β7 antibody therapy. Science 354(6309):197–202

California Biomedical Research Association Factsheet—What is Biomedical Research? (2018) California Biomedical Research Association. https://ca-biomed.org/wp-content/uploads/2018/02/FS-WhatBiomedical.pdf. Accessed 24 May 2019

Casimiro DR, Wang F, Schleif WA, Liang X, Zhang Z-Q, Tobery TW, Davies M-E, McDermott AB, O'Connor DH, Fridman A, Bagchi A, Tussey LG, Bett AJ, Finnefrock AC, Fu T-m, Tang A, Wilson KA, Chen M, Perry HC, Heidecker GJ, Freed DC, Carella A, Punt KS, Sykes KJ, Huang L, Ausensi VI, Bachinsky M, Sadasivan-Nair U, Watkins DI, Emini EA, Shiver JW (2005) Attenuation of simian immunodeficiency virus SIVmac239 infection by prophylactic immunization with DNA and recombinant adenoviral vaccine vectors expressing gag. J Virol 79(24):15547–15555

Centers for Disease Control and Prevention (2013) Core curriculum on tuberculosis: what the clinicians should know. Centers for Disease Control and Prevention, Atlanta. http://www.cdc.gov/tb/education/corecurr/pdf/corecurr_all.pdf. Accessed 1 Sept 2016

Cepeda M, Salas M, Folwarczny J, Leandro AC, Hodara VL, de la Garza MA, Dick EJ Jr, Owston M, Armitige LY, Gauduin MC (2013) Establishment of a neonatal rhesus macaque model to study Mycobacterium tuberculosis infection. Tuberculosis (Edinb) 93(Suppl):S51–S59

Chatfield K, Morton D (2018) The use of non-human primates in research. In: Schroeder D, Cook J, Hirsch F, Fenet S, Muthuswamy V (eds) Ethics dumping: case studies from north-south research collaborations. Springer International Publishing, Cham, pp 81–90

Cline AN, Bess JW, Piatak M Jr, Lifson JD (2005) Highly sensitive SIV plasma viral load assay: practical considerations, realistic performance expectations, and application to reverse engineering of vaccines for AIDS. J Med Primatol 34(5–6):303–312

Cohen J (2011) HIV/AIDS research. Tissue says blood is misleading, confusing HIV cure efforts. Science 334(6063):1614

D'Arc M, Ayouba A, Esteban A, Learn G, Boue V, Liegeois F, Etienne L, Tagg N, Leendertz F, Boesch C, Madinda N, Robbins M, Gray M, Cournil A, Ooms M, Letko M, Simon V, Sharp P, Hahn B, Delaporte E, Ngole EM, Peeters M (2015) Origin of the HIV-1 group O epidemic in western lowland gorillas. Proc Natl Acad Sci U S A 112:E1343–E1352

Dalgleish AG, Beverley PCL, Clapham PR, Crawford DH, Greaves MF, Weiss RA (1984) The CD4 (T4) antigen is an essential component of the receptor for the AIDS retrovirus. Nature 312(5996):763–767

Daniel M, Letvin N, King N, Kannagi M, Sehgal P, Hunt R, Kanki P, Essex M, Desrosiers R (1985) Isolation of T-cell tropic HTLV-III-like retrovirus from macaques. Science 228(4704):1201–1204

Deere JD, Higgins J, Cannavo E, Villalobos A, Adamson L, Fromentin E, Schinazi RF, Luciw PA, North TW (2010) Viral decay kinetics in the highly active antiretroviral therapy-treated rhesus macaque model of AIDS. PLoS One 5(7):e11640

Del Prete GQ, Lifson JD (2013) Considerations in the development of nonhuman primate models of combination antiretroviral therapy for studies of AIDS virus suppression, residual virus, and curative strategies. Curr Opin HIV AIDS 8(4):262–272

Diedrich CR, Mattila JT, Klein E, Janssen C, Phuah J, Sturgeon TJ, Montelaro RC, Lin PL, Flynn JL (2010) Reactivation of latent tuberculosis in cynomolgus macaques infected with SIV is associated with early peripheral T cell depletion and not virus load. PLoS One 5(3):e9611

Dinoso JB, Rabi SA, Blankson JN, Gama L, Mankowski JL, Siliciano RF, Zink MC, Clements JE (2009) A simian immunodeficiency virus-infected macaque model to study viral reservoirs that persist during highly active antiretroviral therapy. J Virol 83(18):9247–9257

Driggers RW, Ho C-Y, Korhonen EM, Kuivanen S, Jääskeläinen AJ, Smura T, Rosenberg A, Hill DA, DeBiasi RL, Vezina G, Timofeev J, Rodriguez FJ, Levanov L, Razak J, Iyengar P, Hennenfent A, Kennedy R, Lanciotti R, du Plessis A, Vapalahti O (2016) Zika virus infection with prolonged maternal viremia and fetal brain abnormalities. N Engl J Med 374(22):2142–2151

Dudley DM, Aliota MT, Mohr EL, Weiler AM, Lehrer-Brey G, Weisgrau KL, Mohns MS, Breitbach ME, Rasheed MN, Newman CM, Gellerup DD, Moncla LH, Post J, Schultz-Darken N, Schotzko ML, Hayes JM, Eudailey JA, Moody MA, Permar SR, O'Connor SL, Rakasz EG, Simmons HA, Capuano S, Golos TG, Osorio JE, Friedrich TC, O'Connor DH (2016) A rhesus macaque model of Asian-lineage Zika virus infection. Nat Commun 7:12204

Dunham RM, Gordon SN, Vaccari M, Piatak M, Huang Y, Deeks SG, Lifson J, Franchini G, McCune JM (2013) Preclinical evaluation of HIV eradication strategies in the simian immunodeficiency virus-infected rhesus macaque: a pilot study testing inhibition of indoleamine 2,3-dioxygenase. AIDS Res Hum Retrovir 29(2):207–214

Dye C, Floyd K (2006) Tuberculosis. In: Jamison DT, Breman JG, Measham AR et al (eds) Disease control priorities in developing countries, 2nd edn. Oxford University Press, New York

Emini EA, Schleif WA, Nunberg JH, Conley AJ, Eda Y, Tokiyoshi S, Putney SD, Matsushrta S, Cobb KE, Jett CM, Eichberg JW, Murthy KK (1992) Prevention of HIV-1 infection in chimpanzees by gpl20 V3 domain-specific monoclonal antibody. Nature 355(6362):728–730

Estaquier J, Idziorek T, De Bels F, Barre-Sinoussi F, Hurtrel B, Aubertin AM, Venet A, Mehtali M, Muchmore E, Michel P (1994) Programmed cell death and AIDS: significance of T-cell apoptosis in pathogenic and nonpathogenic primate lentiviral infections. Proc Natl Acad Sci U S A 91(20):9431–9435

Estes JD, Wong SW, Brenchley JM (2018) Nonhuman primate models of human viral infections. Nat Rev Immunol 18(6):390–404

Etienne L, Nerrienet E, LeBreton M, Bibila GT, Foupouapouognigni Y, Rousset D, Nana A, Djoko CF, Tamoufe U, Aghokeng AF, Mpoudi-Ngole E, Delaporte E, Peeters M, Wolfe ND, Ayouba A (2011) Characterization of a new simian immunodeficiency virus strain in a naturally infected Pan troglodytes troglodytes chimpanzee with AIDS related symptoms. Retrovirology 8(4)

EU (2010) Directive 2010/63/EU of the European Parliament and of the Council of 22 September 2010 on the protection of animals used for scientific purposes (text with EEA relevance). OJ L 276/33

Farell A (2018) Monkey insights. Nat Res. https://www.nature.com/magazine-assets/d42859-018-00007-7/d42859-018-00007-7.pdf. Accessed 27 May 2019

Flynn JL, Tsenova L, Izzo A, Kaplan G (2017) Experimental animal models of tuberculosis. In: Kaufmann SHE, Rubin E, Britton WJ, Helden PV (eds) Handbook of tuberculosis. Wiley, Weinheim. https://doi.org/10.1002/9783527611614.ch32

Franco NH, Sandoe P, Olsson IAS (2018) Researchers' attitudes to the 3Rs-An upturned hierarchy? PLoS One 13(8):e0200895

Fultz PN, Siegel RL, Brodie A, Mawle AC, Stricker RB, Swenson RB, Anderson DC, McClure HM (1991) Prolonged CD4+ lymphocytopenia and thrombocytopenia in a chimpanzee persistently infected with human immunodeficiency virus type 1. J Infect Dis 163(3):441–447

Fultz PN, Nara P, Barre-Sinoussi F, Chaput A, Greenberg ML, Muchmore E, Kieny M-P, Girard M (1992) Vaccine protection of chimpanzees against challenge with HIV-1-infected peripheral blood mononuclear cells. Science 256(5064):1687–1690

Gilden RV, Arthur LO, Robey WG, Kelliher JC, Graham CE, Fischinger PJ (1986) HTLV-III antibody in a breeding chimpanzee not experimentally exposed to the virus. Lancet 1(8482):678–679

Girard M, Kieny M-P, Pinter A, Barre-Sinoussi F, Nara P, Kolbe H, Kusumi K, Chaput A, Reinhart T, Muchmore E, Ronco J, Kaczorek M, Gomard E, Gluckman J-C, Fultz PN (1991) Immunization of chimpanzees confers protection aganist challenge with human immunodeficiency virus. Proc Natl Acad Sci U S A 88(2):542–546

Girard M, Meignier B, Barré-Sinoussi F, Kieny MP, Matthews T, Muchmore E, Nara PL, Wei Q, Rimsky L, Weinhold K (1995) Vaccine-induced protection of chimpanzees against infection by a heterologous human immunodeficiency virus type 1. J Virol 69(10):6239–6248

Gottlieb M, Schroff R, Schanker H, Weisman J, Fan P, Wolf R, Saxon A (1981) Pneumocystis carinii pneumonia and mucosal candidiasis in previously healthy homosexual men: evidence of a new acquired cellular immunodeficiency. N Engl J Med 305(24):1425–1431

Gougeon ML, Lecoeur H, Boudet F, Ledru E, Marzabal S, Boullier S, Roué R, Nagata S, Heeney J (1997) Lack of chronic immune activation in HIV-infected chimpanzees correlates with the resistance of T cells to Fas/Apo-1 (CD95)-induced apoptosis and preservation of a T helper 1 phenotype. J Immunol 158(6):2964–2976

Greenwood EJD, Schmidt F, Kondova I, Niphuis H, Hodara VL, Clissold L, McLay K, Guerra B, Redrobe S, Giavedoni LD, Lanford RE, Murthy KK, Rouet F, Heeney JL (2015) Simian immunodeficiency virus infection of chimpanzees (Pan troglodytes) shares features of both pathogenic and non-pathogenic lentiviral infections. PLoS Pathog 11(9):e1005146

Gumbo T, Lenaerts AJ, Hanna D, Romero K, Nuermberger E (2015) Nonclinical models for anti-tuberculosis drug development: a landscape analysis. J Infect Dis 211(Suppl 3):S83–S95

Hanke T, Samuel RV, Blanchard TJ, Neumann VC, Allen TM, Boyson JE, Sharpe SA, Cook N, Smith GL, Watkins DI, Cranage MP, McMichael AJ (1999) Effective induction of simian immunodeficiency virus-specific cytotoxic T lymphocytes in macaques by using a multiepitope gene and DNA prime-modified vaccinia virus Ankara boost vaccination regimen. J Virol 73(9):7524–7532

Hanke T, McMichael AJ, Dennis MJ, Sharpe SA, Powell LAJ, McLoughlin L, Crome SJ (2005) Biodistribution and persistence of an MVA-vectored candidate HIV vaccine in SIV-infected rhesus macaques and SCID mice. Vaccine 23(12):1507–1514

Hansen SG, Ford JC, Lewis MS, Ventura AB, Hughes CM, Coyne-Johnson L, Whizin N, Oswald K, Shoemaker R, Swanson T, Legasse AW, Chiuchiolo MJ, Parks CL, Axthelm MK, Nelson JA, Jarvis MA, Piatak M, Lifson JD, Picker LJ (2011) Profound early control of highly pathogenic SIV by an effector memory T-cell vaccine. Nature 473:523

Hansen SG, Jr MP, Ventura AB, Hughes CM, Gilbride RM, Ford JC, Oswald K, Shoemaker R, Li Y, Lewis MS, Gilliam AN, Xu G, Whizin N, Burwitz BJ, Planer SL, Turner JM, Legasse AW, Axthelm MK, Nelson JA, Früh K, Sacha JB, Estes JD, Keele BF, Edlefsen PT, Lifson JD, Picker LJ (2013a) Immune clearance of highly pathogenic SIV infection. Nature 502(7469):100–104

Hansen SG, Sacha JB, Hughes CM, Ford JC, Burwitz BJ, Scholz I, Gilbride RM, Lewis MS, Gilliam AN, Ventura AB, Malouli D, Xu G, Richards R, Whizin N, Reed JS, Hammond KB, Fischer M, Turner JM, Legasse AW, Axthelm MK, Edlefsen PT, Nelson JA, Lifson JD, Früh K, Picker LJ (2013b) Cytomegalovirus vectors violate CD8+ T cell epitope recognition paradigms. Science 340(6135):1237874

Hatziioannou T, Ambrose Z, Chung NP, Piatak M Jr, Yuan F, Trubey CM, Coalter V, Kiser R, Schneider D, Smedley J, Pung R, Gathuka M, Estes JD, Veazey RS, KewalRamani VN, Lifson JD, Bieniasz PD (2009) A macaque model of HIV-1 infection. Proc Natl Acad Sci U S A 106(11):4425–4429

Hayashi K, Pu H, Andras IE, Eum SY, Yamauchi A, Hennig B, Toborek M (2006) HIV-TAT protein upregulates expression of multidrug resistance protein 1 in the blood-brain barrier. J Cereb Blood Flow Metab 26(8):1052–1065

Hirsch AJ, Smith JL, Haese NN, Broeckel RM, Parkins CJ, Kreklywich C, DeFilippis VR, Denton M, Smith PP, Messer WB, Colgin LMA, Ducore RM, Grigsby PL, Hennebold JD, Swanson T, Legasse AW, Axthelm MK, MacAllister R, Wiley CA, Nelson JA, Streblow DN (2017) Zika virus infection of rhesus macaques leads to viral persistence in multiple tissues. PLoS Pathog 13(3):e1006219

Hofman MJ, Higgins J, Matthews TB, Pedersen NC, Tan C, Schinazi RF, North TW (2004) Efavirenz therapy in rhesus macaques infected with a chimera of simian immunodeficiency virus containing reverse transcriptase from human immunodeficiency virus type 1. Antimicrob Agents Chemother 48(9):3483–3490

Horiike M, Iwami S, Kodama M, Sato A, Watanabe Y, Yasui M, Ishida Y, Kobayashi T, Miura T, Igarashi T (2012) Lymph nodes harbor viral reservoirs that cause rebound of plasma viremia in SIV-infected macaques upon cessation of combined antiretroviral therapy. Virology 423(2):107–118

Horton H, Vogel TU, Carter DK, Vielhuber K, Fuller DH, Shipley T, Fuller JT, Kunstman KJ, Sutter G, Montefiori DC, Erfle V, Desrosiers RC, Wilson N, Picker LJ, Wolinsky SM, Wang C, Allison DB, Watkins DI (2002) Immunization of rhesus macaques with a DNA prime/modified vaccinia virus Ankara boost regimen induces broad simian immunodeficiency virus (SIV)-specific T-cell responses and reduces initial viral replication but does not prevent disease progression following challenge with pathogenic SIVmac239. J Virol 76(14):7187–7202

Jasny E, Geer S, Frank I, Vagenas P, Aravantinou M, Salazar AM, Lifson JD, Piatak M Jr, Gettie A, Blanchard JL, Robbiani M (2012) Characterization of peripheral and mucosal immune responses in rhesus macaques on long-term tenofovir and emtricitabine combination antiretroviral therapy. J Acquir Immune Defic Syndr 61(4):425–435

Joag SV, Li Z, Foresman L, Stephens EB, Zhao LJ, Adany I, Pinson DM, McClure HM, Narayan O (1996) Chimeric simian/human immunodeficiency virus that causes progressive loss of CD4+ T cells and AIDS in pig-tailed macaques. J Virol 70(5):3189–3197

Joyner C, Barnwell JW, Galinski MR (2015) No more monkeying around: primate malaria model systems are key to understanding Plasmodium vivax liver-stage biology, hypnozoites, and relapses. Front Microbiol 6:145–145

Kanki P, McLane M, King N, Letvin N, Hunt R, Sehgal P, Daniel M, Desrosiers R, Essex M (1985) Serologic identification and characterization of a macaque T-lymphotropic retrovirus closely related to HTLV-III. Science 228(4704):1199–1201

Kearney M, Spindler J, Shao W, Maldarelli F, Palmer S, Hu SL, Lifson JD, KewalRamani VN, Mellors JW, Coffin JM, Ambrose Z (2011) Genetic diversity of simian immunodeficiency virus encoding HIV-1 reverse transcriptase persists in macaques despite antiretroviral therapy. J Virol 85(2):1067–1076

Keele B, Heuverswyn FV, Li Y, Bailes E, Takehisa J, Santiago M, Bibollet-Ruche F, Chen Y, Wain L, Liegeois F, Loul S, Ngole E, Bienvenue Y, EDelaporte BJ, Sharp P, Shaw G, Peeters M, Hahn B (2006) Chimpanzee reservoirs of pandemic and nonpandemic HIV-1. Science 313(5786):523–526

Keele BF, Jones JH, Terio KA, Estes JD, Rudicell RS, Wilson ML, Li Y, Learn GH, Beasley TM, Schumacher-Stankey J, Wroblewski E, Mosser A, Raphael J, Kamenya S, Lonsdorf EV, Travis DA, Mlengeya T, Kinsel MJ, Else JG, Silvestri G, Goodall J, Sharp PM, Shaw GM, Pusey AE, Hahn BH (2009) Increased mortality and AIDS-like immunopathology in wild chimpanzees infected with SIVcpz. Nature 460:515

Koide F, Goebel S, Snyder B, Walters KB, Gast A, Hagelin K, Kalkeri R, Rayner J (2016) Development of a Zika virus infection model in cynomolgus macaques. Front Microbiol 7:2028–2028

Koopman G, Haaksma AGM, Velden JT, Hack CE, Heeney JL (1999) The relative resistance of HIV type 1-infected chimpanzees to AIDS correlates with the maintenance of follicular architecture and the absence of infiltration by CD8+ cytotoxic T lymphocytes. AIDS Res Hum Retroviruses 15(4):365–373

Krotoski WA, Collins WE, Bray RS, Garnham PC, Cogswell FB, Gwadz RW, Killick-Kendrick R, Wolf R, Sinden R, Koontz LC, Stanfill PS (1982a) Demonstration of hypnozoites in sporozoite-transmitted Plasmodium vivax infection. Am J Trop Med Hyg 31(6):1291–1293

Krotoski WA, Garnham PC, Bray RS, Krotoski DM, Killick-Kendrick R, Draper CC, Targett GA, Guy MW (1982b) Observations on early and late post-sporozoite tissue stages in primate malaria. I. Discovery of a new latent form of Plasmodium cynomolgi (the hypnozoite), and failure to detect hepatic forms within the first 24 hours after infection. Am J Trop Med Hyg 31(1):24–35

Lackner AA, Veazey RS (2007) Current concepts in AIDS pathogenesis: insights from the SIV/macaque model. Annu Rev Med 58(1):461–476

Leone A, Rohankhedkar M, Okoye A, Legasse A, Axthelm MK, Villinger F, Piatak M Jr, Lifson JD, Assouline B, Morre M, Picker LJ, Sodora DL (2010) Increased CD4+ T cell levels during IL-7 administration of antiretroviral therapy-treated simian immunodeficiency virus-positive macaques are not dependent on strong proliferative responses. J Immunol 185(3):1650–1659

Letvin NL, King NW (1990a) Immunologic and pathologic manifestations of the infection of rhesus monkeys with simian immunodeficiency virus of macaques. J Acquir Immune Defic Syndr 3(11):1023–1040

Letvin NL, King NW (1990b) Immunologic and pathologic manifestations of the infection of rhesus monkeys with simian immunodeficiency virus of macaques. J Acquir Immune Defic Syndr 3(11):1023–1040

Letvin N, Daniel M, Sehgal P, Desrosiers R, Hunt R, Waldron L, MacKey J, Schmidt D, Chalifoux L, King N (1985) Induction of AIDS-like disease in macaque monkeys with T-cell tropic retrovirus STLV-III. Science 230(4721):71–73

Levy JA et al (1984) Isolation of lymphocytopathic retroviruses from San Francisco patients with AIDS. Science 225:840–842

Li XF, Dong HL, Huang XY, Qiu YF, Wang HJ, Deng YQ, Zhang NN, Ye Q, Zhao H, Liu ZY, Fan H, An XP, Sun SH, Gao B, Fa YZ, Tong YG, Zhang FC, Gao GF, Cao WC, Shi PY, Qin CF (2016) Characterization of a 2016 clinical isolate of Zika virus in non-human primates. EBioMedicine 12:170–177

Li C, Deng YQ, Wang S, Ma F, Aliyari R, Huang XY, Zhang NN, Watanabe M, Dong HL, Liu P, Li XF, Ye Q, Tian M, Hong S, Fan J, Zhao H, Li L, Vishlaghi N, Buth JE, Au C, Liu Y, Lu N, Du P, Qin FX, Zhang B, Gong D, Dai X, Sun R, Novitch BG, Xu Z, Qin CF, Cheng G (2017) 25-Hydroxycholesterol protects host against Zika virus infection and its associated microcephaly in a mouse model. Immunity 46(3):446–456

Liang X, Casimiro DR, Schleif WA, Wang F, Davies M-E, Zhang Z-Q, Fu T-M, Finnefrock AC, Handt L, Citron MP, Heidecker G, Tang A, Chen M, Wilson KA, Gabryelski L, McElhaugh M, Carella A, Moyer C, Huang L, Vitelli S, Patel D, Lin J, Emini EA, Shiver JW (2005) Vectored Gag and Env but not Tat show efficacy against simian-human immunodeficiency virus 89.6P challenge in Mamu-A*01-negative rhesus monkeys. J Virol 79(19):12321–12331

Lin PL, Flynn JL (2010) Understanding latent tuberculosis: a moving target. J Immunol 185(1):15–22

Lin PL, Rodgers M, Smith L, Bigbee M, Myers A, Bigbee C, Chiosea I, Capuano SV, Fuhrman C, Klein E, Flynn JL (2009) Quantitative comparison of active and latent tuberculosis in the cynomolgus macaque model. Infect Immun 77(10):4631–4642

Lin PL, Dartois V, Johnston PJ, Janssen C, Via L, Goodwin MB, Klein E, Barry CE 3rd, Flynn JL (2012) Metronidazole prevents reactivation of latent Mycobacterium tuberculosis infection in macaques. Proc Natl Acad Sci U S A 109(35):14188–14193

Lin PL, Coleman T, Carney JP, Lopresti BJ, Tomko J, Fillmore D, Dartois V, Scanga C, Frye LJ, Janssen C, Klein E, Barry CE 3rd, Flynn JL (2013) Radiologic responses in cynomolgus macaques for assessing tuberculosis chemotherapy regimens. Antimicrob Agents Chemother 57(9):4237–4244

Lubeck MD, Natuk R, Myagkikh M, Kalyan N, Aldrich K, Sinangil F, Alipanah S, Murthy SCS, Chanda PK, Nigida SM, Markham PD, Zolla-pazner S, Steimer K, Wade M, Reitz MS, Arthur LO, Mizutani S, Davis A, Hung PP, Gallo RC, Eichberg J, Robert-Guroff M (1997) Long-term protection of chimpanzees against high-dose HIV-1 challenge induced by immunization. Nat Med 3(6):651–658

Maddon PJ, Dalgleish AG, McDougal JS, Clapham PR, Weiss RA, Axel R (1986) The T4 gene encodes the AIDS virus receptor and is expressed in the immune system and the brain. Cell 47(3):333–348

Madeley C (2008) Is it the cause?—Robert Koch and viruses in the 21st century. J Clin Virol 43(1):9–12

Marino S, Linderman JJ, Kirschner DE (2011) A multifaceted approach to modeling the immune response in tuberculosis. Wiley Interdiscip Rev Syst Biol Med 3(4):479–489

Marzi A, Robertson SJ, Haddock E, Feldmann F, Hanley PW, Scott DP, Strong JE, Kobinger G, Best SM, Feldmann H (2015) VSV-EBOV rapidly protects macaques against infection with the 2014/15 Ebola virus outbreak strain. Science 349(6249):739–742

Masur H, Michelis M, Greene J, Onorato I, Stouwe R, Holzman R, Wormser G, Brettman L, Lange M, Murray H, Cunningham-Rundles S (1981) An outbreak of communityacquired

Pneumocystis carinii pneumonia: initial manifestation of cellular immune dysfunction. N Engl J Med 305(24):1431–1438

McLeay SC, Morrish GA, Kirkpatrick CM, Green B (2012) The relationship between drug clearance and body size: systematic review and meta-analysis of the literature published from 2000 to 2007. Clin Pharmacokinet 51(5):319–330

Mehra S, Golden NA, Dutta NK, Midkiff CC, Alvarez X, Doyle LA, Asher M, Russell-Lodrigue K, Monjure C, Roy CJ, Blanchard JL, Didier PJ, Veazey RS, Lackner AA, Kaushal D (2011) Reactivation of latent tuberculosis in rhesus macaques by coinfection with simian immunodeficiency virus. J Med Primatol 40(4):233–243

Miller CJ, Alexander NJ, Sutjipto S, Lackner AA, Gettie A, Hendrickx AG, Lowenstine LJ, Jennings M, Marx PA (1989) Genital mucosal transmission of simian immunodeficiency virus: animal model for heterosexual transmission of human immunodeficiency virus. J Virol 63(10):4277–4284

Montali RJ, Mikota SK, Cheng LI (2001) Mycobacterium tuberculosis in zoo and wildlife species. Rev Sci Tech 20(1):291–303

Moreau M, Le Tortorec A, Deleage C, Brown C, Denis H, Satie AP, Bourry O, Deureuddre-Bosquet N, Roques P, Le Grand R, Dejucq-Rainsford N (2012) Impact of short-term HAART initiated during the chronic stage or shortly post-exposure on SIV infection of male genital organs. PLoS One 7(5):e37348

Nguyen SM, Antony KM, Dudley DM, Kohn S, Simmons HA, Wolfe B, Salamat MS, Teixeira LBC, Wiepz GJ, Thoong TH, Aliota MT, Weiler AM, Barry GL, Weisgrau KL, Vosler LJ, Mohns MS, Breitbach ME, Stewart LM, Rasheed MN, Newman CM, Graham ME, Wieben OE, Turski PA, Johnson KM, Post J, Hayes JM, Schultz-Darken N, Schotzko ML, Eudailey JA, Permar SR, Rakasz EG, Mohr EL, Capuano S III, Tarantal AF, Osorio JE, O'Connor SL, Friedrich TC, O'Connor DH, Golos TG (2017) Highly efficient maternal-fetal Zika virus transmission in pregnant rhesus macaques. PLoS Pathog 13(5):e1006378

Nishimura Y, Igarashi T, Donau OK, Buckler-White A, Buckler C, Lafont BAP, Goeken RM, Goldstein S, Hirsch VM, Martin MA (2004) Highly pathogenic SHIVs and SIVs target different CD4+T cell subsets in rhesus monkeys, explaining their divergent clinical courses. Proc Natl Acad Sci U S A 101(33):12324–12329

North TW, Van Rompay KK, Higgins J, Matthews TB, Wadford DA, Pedersen NC, Schinazi RF (2005) Suppression of virus load by highly active antiretroviral therapy in rhesus macaques infected with a recombinant simian immunodeficiency virus containing reverse transcriptase from human immunodeficiency virus type 1. J Virol 79(12):7349–7354

North TW, Higgins J, Deere JD, Hayes TL, Villalobos A, Adamson L, Shacklett BL, Schinazi RF, Luciw PA (2010) Viral sanctuaries during highly active antiretroviral therapy in a nonhuman primate model for AIDS. J Virol 84(6):2913–2922

Novembre FJ, Saucier M, Anderson DC, Klumpp SA, O'Neil SP, Brown CR, Hart CE, Guenthner PC, Swenson RB, McClure HM (1997) Development of AIDS in a chimpanzee infected with human immunodeficiency virus type 1. J Virol 71(5):4086–4091

Novembre FJ, de Rosayro J, Nidtha S, O'Neil SP, Gibson TR, Evans-Strickfaden T, Hart CE, McClure HM (2001) Rapid CD4+T-cell loss induced by human immunodeficiency virus type 1NC in uninfected and previously infected chimpanzees. J Virol 75(3):1533–1539

O'Neil SP, Novembre FJ, Hill AB, Suwyn C, Hart CE, Evans-Strickfaden T, Anderson DC, deRosayro J, Herndon JG, Saucier M, McClure HM (2000) Progressive infection in a subset of HIV-1—positive chimpanzees. J Infect Dis 182(4):1051–1062

Osuna CE, Whitney JB (2017) Nonhuman primate models of Zika virus infection, immunity, and therapeutic development. J Infect Dis 216(Suppl_10):S928–S934

Osuna CE, Lim SY, Deleage C, Griffin BD, Stein D, Schroeder LT, Omange RW, Best K, Luo M, Hraber PT, Andersen-Elyard H, Ojeda EF, Huang S, Vanlandingham DL, Higgs S, Perelson AS, Estes JD, Safronetz D, Lewis MG, Whitney JB (2016) Zika viral dynamics and shedding in rhesus and cynomolgus macaques. Nat Med 22(12):1448–1455

Ourmanov I, Brown CR, Moss B, Carroll M, Wyatt L, Pletneva L, Goldstein S, Venzon D, Hirsch VM (2000) Comparative efficacy of recombinant modified vaccinia virus Ankara expressing

simian immunodeficiency virus (SIV) Gag-Pol and/or Env in macaques challenged with pathogenic SIV. J Virol 74(6):2740–2751
Pandrea I, Gaufin T, Gautam R, Kristoff J, Mandell D, Montefiori D, Keele BF, Ribeiro RM, Veazey RS, Apetrei C (2011) Functional cure of SIVagm infection in rhesus macaques results in complete recovery of CD4+ T cells and is reverted by CD8+ cell depletion. PLoS Pathog 7(8):e1002170
Pena JC, Ho WZ (2015) Monkey models of tuberculosis: lessons learned. Infect Immun 83(3):852–862
Poles MA, Elliott J, Taing P, Anton PA, Chen ISY (2001) A preponderance of CCR5+CXCR4+ mononuclear cells enhances gastrointestinal mucosal susceptibility to human immunodeficiency virus type 1 infection. J Virol 75(18):8390–8399
Queen SE, Mears BM, Kelly KM, Dorsey JL, Liao Z, Dinoso JB, Gama L, Adams RJ, Zink MC, Clements JE, Kent SJ, Mankowski JL (2011) Replication-competent simian immunodeficiency virus (SIV) Gag escape mutations archived in latent reservoirs during antiretroviral treatment of SIV-infected macaques. J Virol 85(17):9167–9175
RC G (1984) Diseases in nonhuman primates. In: Kubica GP, Wayne LG (eds) The mycobacteria: a sourcebook. Dekker, New York, pp 903–921
Reimann KA, Li JT, Veazey R, Halloran M, Park IW, Karlsson GB, Sodroski J, Letvin NL (1996) A chimeric simian/human immunodeficiency virus expressing a primary patient human immunodeficiency virus type 1 isolate env causes an AIDS-like disease after in vivo passage in rhesus monkeys. J Virol 70(10):6922–6928
Roy U, Barber P, Tse-Dinh Y-C, Batrakova EV, Mondal D, Nair M (2015) Role of MRP transporters in regulating antimicrobial drug inefficacy and oxidative stress-induced pathogenesis during HIV-1 and TB infections. Front Microbiol 6:948
Rudicell RS, Holland Jones J, Wroblewski EE, Learn GH, Li Y, Robertson JD, Greengrass E, Grossmann F, Kamenya S, Pintea L, Mjungu DC, Lonsdorf EV, Mosser A, Lehman C, Collins DA, Keele BF, Goodall J, Hahn BH, Pusey AE, Wilson ML (2010) Impact of simian immunodeficiency virus infection on chimpanzee population dynamics. PLoS Pathog 6(9):e1001116
Russell WMS, Burch RL (1959) The principles of humane experimental technique. Methuen & Co Ltd, London
Saravanan C, Sasseville VG, Mansfield KG (2015) Chapter 10—nonhuman primate diseases of relevance in drug development and their impact on the interpretation of study findings. In: Bluemel J, Korte S, Schenck E, Weinbauer GF (eds) The nonhuman primate in nonclinical drug development and safety assessment. Academic Press, San Diego, pp 187–213
Saxinger C, Alter HJ, Eichberg JW, Fauci AS, Robey WG, Gallo RC (1987) Stages in the progression of HIV infection in chimpanzees. AIDS Res Hum Retrovir 3(4):375–385
Scanga CA, Flynn JL (2014) Modeling tuberculosis in nonhuman primates. Cold Spring Harb Perspect Med 4(12):a018564
Schmidt LH (1956) Some observations on the utility of simian pulmonary tuberculosis in defining the therapeutic potentialities of isoniazid. Am Rev Tuberc Pulm 74(2 Part 2):138–153. discussion 153–159
Schmidt LH (1966) Studies on the antituberculous activity of ethambutol in monkeys. Ann N Y Acad Sci 135(2):747–758
Schmidt L, Grover A, Hoffmann R (1959) The comparative therapeutic activities of thiocarbanidin and para-aminosalicylic acid administered alone and in combination with isoniazid. Trans 18th VA-Armed Forces Conf Chemother Tuberculosis 312–317
Schmitz JE, Kuroda MJ, Santra S, Sasseville VG, Simon MA, Lifton MA, Racz P, Tenner-Racz K, Dalesandro M, Scallon BJ, Ghrayeb J, Forman MA, Montefiori DC, Rieber EP, Letvin NL, Reimann KA (1999) Control of viremia in simian immunodeficiency virus infection by CD8+ lymphocytes. Science 283(5403):857–860
Schuitemaker H, Meyaard L, Kootstra NA, Dubbes R, Otto SA, Tersmette M, Heeney JL, Miedema F (1993) Lack of T cell dysfunction and programmed cell death in human immunodeficiency virus type 1-infected chimpanzees correlates with absence of monocytotropic variants. J Infect Dis 168(5):1140–1147

Shao W, Kearney M, Maldarelli F, Mellors JW, Stephens RM, Lifson JD, KewalRamani VN, Ambrose Z, Coffin JM, Palmer SE (2009) RT-SHIV subpopulation dynamics in infected macaques during anti-HIV therapy. Retrovirology 6(1):101

Shedlock DJ, Silvestri G, Weiner DB (2009) Monkeying around with HIV vaccines: using rhesus macaques to define 'gatekeepers' for clinical trials. Nat Rev Immunol 9(10):717–728

Shen Y, Shen L, Sehgal P, Zhou D, Simon M, Miller M, Enimi EA, Henckler B, Chalifoux L, Sehgal N, Gastron M, Letvin NL, Chen ZW (2001) Antiretroviral agents restore Mycobacterium-specific T-cell immune responses and facilitate controlling a fatal tuberculosis-like disease in Macaques coinfected with simian immunodeficiency virus and Mycobacterium bovis BCG. J Virol 75(18):8690–8696

Shen Y, Zhou D, Chalifoux L, Shen L, Simon M, Zeng X, Lai X, Li Y, Sehgal P, Letvin NL, Chen ZW (2002) Induction of an AIDS virus-related tuberculosis-like disease in macaques: a model of simian immunodeficiency virus- mycobacterium coinfection. Infect Immun 70(2):869–877

Shen A, Zink MC, Mankowski JL, Chadwick K, Margolick JB, Carruth LM, Li M, Clements JE, Siliciano RF (2003) Resting CD4+ T lymphocytes but not thymocytes provide a latent viral reservoir in a simian immunodeficiency virus-Macaca nemestrina model of human immunodeficiency virus type 1-infected patients on highly active antiretroviral therapy. J Virol 77(8):4938–4949

Shen A, Yang HC, Zhou Y, Chase AJ, Boyer JD, Zhang H, Margolick JB, Zink MC, Clements JE, Siliciano RF (2007) Novel pathway for induction of latent virus from resting CD4(+) T cells in the simian immunodeficiency virus/macaque model of human immunodeficiency virus type 1 latency. J Virol 81(4):1660–1670

Shingai M, Nishimura Y, Klein F, Mouquet H, Donau OK, Plishka R, Buckler-White A, Seaman M, Piatak M, Lifson JD, Dimitrov D, Nussenzweig MC, Martin MA (2013) Antibody-mediated immunotherapy of macaques chronically infected with SHIV suppresses viraemia. Nature 503:277

Shiver JW, Fu T-M, Chen L, Casimiro DR, Davies M-E, Evans RK, Zhang Z-Q, Simon AJ, Trigona WL, Dubey SA, Huang L, Harris VA, Long RS, Liang X, Handt L, Schleif WA, Zhu L, Freed DC, Persaud NV, Guan L, Punt KS, Tang A, Chen M, Wilson KA, Collins KB, Heidecker GJ, Fernandez VR, Perry HC, Joyce JG, Grimm KM, Cook JC, Keller PM, Kresock DS, Mach H, Troutman RD, Isopi LA, Williams DM, Xu Z, Bohannon KE, Volkin DB, Montefiori DC, Miura A, Krivulka GR, Lifton MA, Kuroda MJ, Schmitz JE, Letvin NL, Caulfield MJ, Bett AJ, Youil R, Kaslow DC, Emini EA (2002a) Replication-incompetent adenoviral vaccine vector elicits effective anti-immunodeficiency-virus immunity. Nature 415(6869):331–335

Shiver JW, Fu TM, Chen L, Casimiro DR, Davies ME, Evans RK, Zhang ZQ, Simon AJ, Trigona WL, Dubey SA, Huang L, Harris VA, Long RS, Liang X, Handt L, Schleif WA, Zhu L, Freed DC, Persaud NV, Guan L, Punt KS, Tang A, Chen M, Wilson KA, Collins KB, Heidecker GJ, Fernandez VR, Perry HC, Joyce JG, Grimm KM, Cook JC, Keller PM, Kresock DS, Mach H, Troutman RD, Isopi LA, Williams DM, Xu Z, Bohannon KE, Volkin DB, Montefiori DC, Miura A, Krivulka GR, Lifton MA, Kuroda MJ, Schmitz JE, Letvin NL, Caulfield MJ, Bett AJ, Youil R, Kaslow DC, Emini EA (2002b) Replication-incompetent adenoviral vaccine vector elicits effective anti-immunodeficiency-virus immunity. Nature 415(6869):331–335

Siegal F, Lopez C, Hammer G, Brown A, Kornfeld S, Gold J, Hassett J, Hirschman S, Cunningham-Rundles C, Adelsberg B (1981) Severe acquired immunodeficiency in male homosexuals, manifested by chronic perianal ulcerative herpes simplex lesions. N Engl J Med 305(24):1439–1444

Smit-McBride Z, Mattapallil JJ, McChesney M, Ferrick D, Dandekar S (1998) Gastrointestinal T lymphocytes retain high potential for cytokine responses but have severe CD4+ T-cell depletion at all stages of simian immunodeficiency virus infection compared to peripheral lymphocytes. J Virol 72(8):6646–6656

Stephenson KE, D'Couto HT, Barouch DH (2016) New concepts in HIV-1 vaccine. Development 41:39–46

Strait K, Else JG, Eberhard ML (2012) Chapter 4—parasitic diseases of nonhuman primates. In: Abee CR, Mansfield K, Tardif S, Morris T (eds) Nonhuman primates in biomedical research, 2nd edn. Academic Press, Boston, pp 197–297

Sutjipto S, Pedersen NC, Miller CJ, Gardner MB, Hanson CV, Gettie A, Jennings M, Higgins J, Marx PA (1990) Inactivated simian immunodeficiency virus vaccine failed to protect rhesus macaques from intravenous or genital mucosal infection but delayed disease in intravenously exposed animals. J Virol 64(5):2290–2297

Thippeshappa R, Polacino P, Yu Kimata MT, Siwak EB, Anderson D, Wang W, Sherwood L, Arora R, Wen M, Zhou P, Hu S-L, Kimata JT (2011) Vif substitution enables persistent infection of pig-tailed macaques by human immunodeficiency virus type 1. J Virol 85(8):3767–3779

Tsai C-C, Follis KE, Sabo A, Beck TW, Grant RF, Bischofberger N, Benveniste RE, Black R (1995) Prevention of SIV infection in macaques by (R)-9-(2-phosphonylmethoxypropyl)adenine. Science 270(5239):1197–1199

Tsai CC, Emau P, Follis KE, Beck TW, Benveniste RE, Bischofberger N, Lifson JD, Morton WR (1998) Effectiveness of postinoculation (R)-9-(2-phosphonylmethoxypropyl) adenine treatment for prevention of persistent simian immunodeficiency virus SIVmne infection depends critically on timing of initiation and duration of treatment. J Virol 72(5):4265–4273

Uberla K, Stahl-Hennig C, Bottiger D, Matz-Rensing K, Kaup FJ, Li J, Haseltine WA, Fleckenstein B, Hunsmann G, Oberg B et al (1995) Animal model for the therapy of acquired immunodeficiency syndrome with reverse transcriptase inhibitors. Proc Natl Acad Sci U S A 92(18):8210–8214

Vaccari M, Boasso A, Fenizia C, Fuchs D, Hryniewicz A, Morgan T, Weiss D, Doster MN, Heraud JM, Shearer GM, Franchini G (2012) Fatal pancreatitis in simian immunodeficiency virus SIV(mac251)-infected macaques treated with 2′,3′-dideoxyinosine and stavudine following cytotoxic-T-lymphocyte-associated antigen 4 and indoleamine 2,3-dioxygenase blockade. J Virol 86(1):108–113

Vagenas P, Aravantinou M, Williams VG, Jasny E, Piatak M Jr, Lifson JD, Salazar AM, Blanchard JL, Gettie A, Robbiani M (2010) A tonsillar PolyICLC/AT-2 SIV therapeutic vaccine maintains low viremia following antiretroviral therapy cessation. PLoS One 5(9):e12891

Van Rompay KK, Dailey PJ, Tarara RP, Canfield DR, Aguirre NL, Cherrington JM, Lamy PD, Bischofberger N, Pedersen NC, Marthas ML (1999) Early short-term 9-[2-(R)-(phosphonomethoxy)propyl]adenine treatment favorably alters the subsequent disease course in simian immunodeficiency virus-infected newborn Rhesus macaques. J Virol 73(4):2947–2955

Van Rompay KK, Matthews TB, Higgins J, Canfield DR, Tarara RP, Wainberg MA, Schinazi RF, Pedersen NC, North TW (2002) Virulence and reduced fitness of simian immunodeficiency virus with the M184V mutation in reverse transcriptase. J Virol 76(12):6083–6092

Veazey RS, Lackner AA (2017) Nonhuman primate models and understanding the pathogenesis of HIV infection and AIDS. ILAR J 58(2):160–171

Veazey RS, DeMaria M, Chalifoux LV, Shvetz DE, Pauley DR, Knight HL, Rosenzweig M, Johnson RP, Desrosiers RC, Lackner AA (1998) Gastrointestinal tract as a major site of CD4+ T cell depletion and viral replication in SIV infection. Science 280(5362):427–431

Via LE, Weiner DM, Schimel D, Lin PL, Dayao E, Tankersley SL, Cai Y, Coleman MT, Tomko J, Paripati P, Orandle M, Kastenmayer RJ, Tartakovsky M, Rosenthal A, Portevin D, Eum SY, Lahouar S, Gagneux S, Young DB, Flynn JL, Barry CE 3rd (2013) Differential virulence and disease progression following Mycobacterium tuberculosis complex infection of the common marmoset (Callithrix jacchus). Infect Immun 81(8):2909–2919

Weatherall D, Goodfellow P, Harris J, Hinde R, Johnson DL, Morris R, Ross N, Skehel J, Tickell C (2006) The use of non-human primates in research: a working group report chaired by Sir David Weatherall FRS FmedSci. London

Wong K (2014) The 1 percent difference. Sci Am 311:100. https://doi.org/10.1038/scientificamerican0914-100

Wood MW, Hart LA (2007) Selecting appropriate animal models and strains: making the best use of research, information and outreach. AATEX 14(Special Issue):303–306

Xiao Q, Li J, Yu Q, Bao R, Liu J, Liu H, Zhou L, Xian Q, Wang Y, Cheng-Mayer C, Wenzhe H, Zhuang K (2015) Distinct compartmentalization in the CNS of SHIVKU-1-infected Chinese rhesus macaque is associated with severe neuropathology. J Acquir Immune Defic Syndr 70(5):e168–e171

Zarling JM, Ledbetter JA, Sias J, Fultz P, Eichberg J, Gjerset G, Moran PA (1990) HIV-infected humans, but not chimpanzees, have circulating cytotoxic T lymphocytes that lyse uninfected CD4+ cells. J Immunol 144(8):2992–2998

Zhan L, Tang J, Sun M, Qin C (2017) Animal models for tuberculosis in translational and precision medicine. Front Microbiol 8:717–717

Zink MC, Suryanarayana K, Mankowski JL, Shen A, Piatak M, Spelman JP, Carter DL, Adams RJ, Lifson JD, Clements JE (1999) High viral load in the cerebrospinal fluid and brain correlates with severity of simian immunodeficiency virus encephalitis. J Virol 73(12):10480–10488

Zink MC, Brice AK, Kelly KM, Queen SE, Gama L, Li M, Adams RJ, Bartizal C, Varrone J, Rabi SA, Graham DR, Tarwater PM, Mankowski JL, Clements JE (2010) Simian immunodeficiency virus-infected macaques treated with highly active antiretroviral therapy have reduced central nervous system viral replication and inflammation but persistence of viral DNA. J Infect Dis 202(1):161–170

zur Megede J, Sanders-Beer B, Silvera P, Golightly D, Bowlsbey A, Hebblewaite D, Sites D, Nieves-Duran L, Srivastava R, Otten GR, Rabussay D, Zhang L, Ulmer JB, Barnett SW, Donnelly JJ (2008) A therapeutic SIV DNA vaccine elicits T-cell immune responses, but no sustained control of viremia in SIVmac239-infected rhesus macaques. AIDS Res Hum Retroviruses 24(8):1103–1116